Introduction to Paleobiology and the Fossil Record

Companion website

This book includes a companion website at:

www.blackwellpublishing.com/paleobiology

The website includes:

- An ongoing database of additional Practicals prepared by the authors
- Figures from the text for downloading
- Useful links for each chapter
- Updates from the authors

Introduction to Paleobiology and the Fossil Record

Michael J. Benton
University of Bristol, UK

David A. T. Harper
University of Copenhagen, Denmark

A John Wiley & Sons, Ltd., Publication

Blackwell Publishing was acquired by John Wiley & Sons in February 2007. Blackwell's publishing program has been merged with Wiley's global Scientific, Technical and Medical business to form Wiley-Blackwell.

Registered office: John Wiley & Sons Ltd, The Atrium, Southern Gate, Chichester, West Sussex, PO19 8SQ, UK

Editorial offices: 9600 Garsington Road, Oxford, OX4 2DQ, UK
The Atrium, Southern Gate, Chichester, West Sussex, PO19 8SQ, UK
111 River Street, Hoboken, NJ 07030-5774, USA

For details of our global editorial offices, for customer services and for information about how to apply for permission to reuse the copyright material in this book please see our website at www.wiley.com/wiley-blackwell

Library of Congress Cataloguing-in-Publication Data

Benton, M. J. (Michael J.)
Introduction to paleobiology and the fossil record / Michael J Benton, David A.T. Harper.
p. cm.
Includes bibliographical references and index.
ISBN 978-1-4051-8646-9 (hardback : alk. paper) – ISBN 978-1-4051-4157-4 (pbk. : alk. paper)
1. Evolutionary paleobiology. 2. Paleobiology. 3. Paleontology. I. Harper, D. A. T. II. Title.
QE721.2.E85B46 2008
560–dc22

2008015534

A catalogue record for this book is available from the British Library.

Set in 11 on 12 pt Sabon by SNP Best-set Typesetter Ltd, Hong Kong

Printed in Singapore by Markono Print Media Pte Ltd

4 2009

Contents

A companion resources website for this book is available at http://www.blackwellpublishing.com/paleobiology

Full contents

A companion resources website for this book is available at http://www.blackwellpublishing.com/paleobiology

Preface

The history of life is documented by fossils through the past 3.5 billion years. We need this long-term perspective for three reasons: ancient life and environments can inform us about how the world might change in the future; extinct plants and animals make up 99% of all species that ever lived, and so we need to know about them to understand the true scope of the tree of life; and extinct organisms did amazing things that no living plant or animal can do, and we need to explore their capabilities to assess the limits of form and function.

Every week, astonishing new fossil finds are announced – a 1 ton rat, a miniature species of human, the world's largest sea scorpion, a dinosaur with feathers. You read about these in the newspapers, but where do these stray findings fit into the greater scheme of things? Studying fossils can reveal the most astonishing organisms, many of them more remarkable than the wildest dreams (or nightmares) of a science fiction writer. Indeed, paleontology reveals a seemingly endless catalog of alternative universes, landscapes and seascapes that look superficially familiar, but which contain plants that do not look quite right, animals that are very different from anything now living.

The last 40 years have seen an explosion of paleontological research, where fossil evidence is used to study larger questions, such as rates of evolution, mass extinctions, high-precision dating of sedimentary sequences, the paleobiology of dinosaurs and Cambrian arthropods, the structure of Carboniferous coal-swamp plant communities, ancient molecules, the search for oil and gas, the origin of humans, and many more. Paleontologists have benefited enormously from the growing interdisciplinary nature of their science, with major contributions from geologists, chemists, evolutionary biologists, physiologists and even geophysicists and astronomers. Many areas of study have also been helped by an increasingly quantitative approach.

There are many paleontology texts that describe the major fossil groups or give a guided tour of the history of life. Here we hope to give students a flavor of the excitement of modern paleontology. We try to present all aspects of paleontology, not just invertebrate fossils or dinosaurs, but fossil plants, trace fossils, macroevolution, paleobiogeography, biostratigraphy, mass extinctions, biodiversity through time and microfossils. Where possible, we show how paleontologists tackle controversial questions, and highlight what is known, and what is not known. This shows the activity and dynamism of modern paleobiological research. Many of these items are included in boxed features, some of them added at the last minute, to show new work in a number of categories, indicated by icons (see below for explanation).

The book is intended for first- and second-year geologists and biologists who are taking courses in paleontology or paleobiology. It should also be a clear introduction to the science for keen amateurs and others interested in current scientific evidence about the origin of life, the history of life, mass extinctions, human evolution and related topics.

ACKNOWLEDGMENTS

We thank the following for reading chapters of the book, and providing feedback and comments that gave us much pause for thought, and led to many valuable revisions: Jan Audun Rasmussen

(Copenhagen), Mike Bassett (Cardiff), Joseph Botting (London), Simon Braddy (Bristol), Pat Brenchley (formerly Liverpool), Derek Briggs (Yale), David Bruton (Oslo), Graham Budd (Uppsala), Nick Butterfield (Cambridge), Sandra Carlson (Davis), David Catling (Bristol), Margaret Collinson (London), John Cope (Cardiff), Gilles Cuny (Copenhagen), Kristi Curry Rogers (Minnesota), Phil Donoghue (Bristol), Karen Dybkjær (Copenhagen), Howard Falcon-Lang (Bristol), Mike Foote (Chicago), Liz Harper (Cambridge), John Hutchinson (London), Paul Kenrick (London), Andy Knoll (Harvard), Bruce Liebermann (Kansas), Maria Liljeroth (Copenhagen), David Loydell (Portsmouth), Duncan McIlroy (St John's), Paddy Orr (Dublin), Alan Owen (Glasgow), Kevin Padian (Berkeley), Kevin Peterson (Dartmouth), Emily Rayfield (Bristol), Ken Rose (New York), Marcello Ruta (Bristol), Martin Sander (Bonn), Andrew Smith (London), Paul Taylor (London), Richard Twitchett (Plymouth), Charlie Wellman (Sheffield), Paul Wignall (Leeds), Rachel Wood (Edinburgh), Graham Young (Winnipeg) and Jeremy Young (London).

We are grateful to Ian Francis and Delia Sanderson together with Stephanie Schnur and Rosie Hayden for steering this book to completion, and to Jane Andrew for copy editing and to Mirjana Misina for guiding the editorial process. Last, but not least, we thank our wives, Mary and Maureen, for their help and forbearance.

Mike Benton
David Harper
February 2008

TYPES OF BOXES

Throughout the text you will find special topic boxes. There are five types of boxes, each with a distinguishing icon:

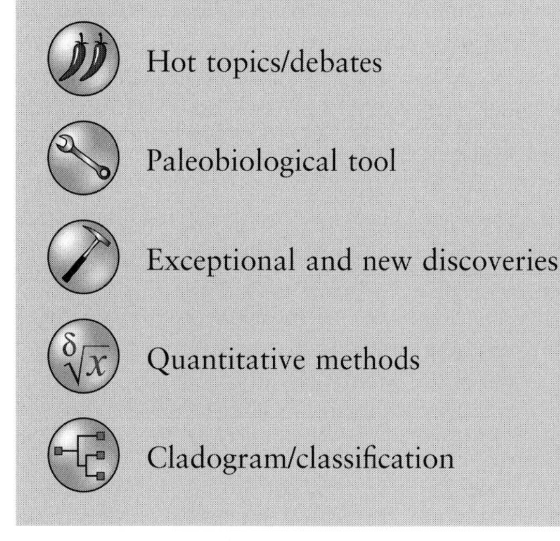

Chapter 1

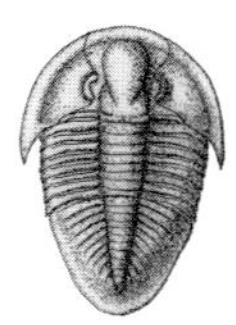

Paleontology as a science

Key points

- The key value of paleontology has been to show us the history of life through deep time – without fossils this would be largely hidden from us.
- Paleontology has strong relevance today in understanding our origins, other distant worlds, climate and biodiversity change, the shape and tempo of evolution, and dating rocks.
- Paleontology is a part of the natural sciences, and a key aim is to reconstruct ancient life.
- Reconstructions of ancient life have been rejected as pure speculation by some, but careful consideration shows that they too are testable hypotheses and can be as scientific as any other attempt to understand the world.
- Science consists of testing hypotheses, not in general by limiting itself to absolute certainties like mathematics.
- Classical and medieval views about fossils were often magical and mystical.
- Observations in the 16th and 17th centuries showed that fossils were the remains of ancient plants and animals.
- By 1800, many scientists accepted the idea of extinction.
- By 1830, most geologists accepted that the Earth was very old.
- By 1840, the major divisions of deep time, the stratigraphic record, had been established by the use of fossils.
- By 1840, it was seen that fossils showed direction in the history of life, and by 1860 this had been explained by evolution.
- Research in paleontology has many facets, including finding new fossils and using quantitative methods to answer questions about paleobiology, paleogeography, macroevolution, the tree of life and deep time.

All science is either physics or stamp collecting.

Sir Ernest Rutherford (1871–1937), *Nobel prize-winner*

Scientists argue about what is science and what is not. Ernest Rutherford famously had a very low opinion of anything that was not mathematics or physics, and so he regarded all of biology and geology (including paleontology) as "stamp collecting", the mere recording of details and stories. But is this true?

Most criticism in paleontology is aimed at the reconstruction of ancient plants and animals. Surely no one will ever know what color dinosaurs were, what noises they made? How could a paleontologist work out how many eggs *Tyrannosaurus* laid, how long it took for the young to grow to adult size, the differences between males and females? How could anyone work out how an ancient animal hunted, how strong its bite force was, or even what kinds of prey it ate? Surely it is all speculation because we can never go back in time and see what was happening?

These are questions about paleobiology and, surprisingly, a great deal can be inferred from fossils. **Fossils**, the remains of any ancient organism, may look like random pieces of rock in the shape of bones, leaves or shells, but they can yield up their secrets to the properly trained scientist. **Paleontology**, the study of the life of the past, is like a crime scene investigation – there are clues here and there, and the **paleontologist** can use these to understand something about an ancient plant or animal, or a whole **fauna** or **flora**, the animals or plants that lived together in one place at one time.

In this chapter we will explore the methods of paleontology, starting with the debate about how dinosaurs are portrayed in films, and then look more widely at the other kinds of inferences that may be made from fossils. But first, just what is paleontology for? Why should anyone care about it?

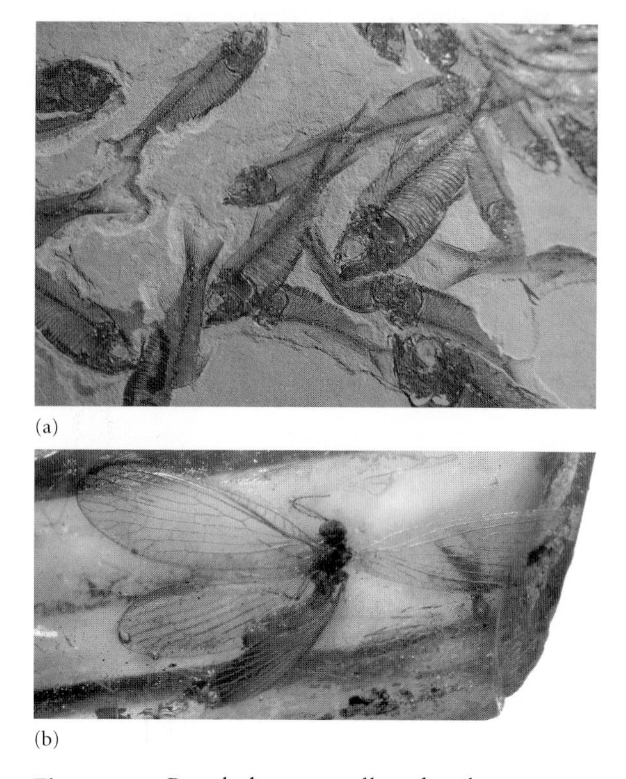

(a)

(b)

Figure 1.1 People love to collect fossils. Many professional paleontologists got into the field because of the buzz of finding something beautiful that came from a plant or animal that died millions of years ago. Fossils such as these tiny fishes from the Eocene of Wyoming (a), may amaze us by their abundance, or like the lacewing fly in amber (b), by the exquisite detail of their preservation. (Courtesy of Sten Lennart Jakobsen.)

PALEONTOLOGY IN THE MODERN WORLD

What is the use of paleontology? A few decades ago, the main purpose was to date rocks. Many paleontology textbooks justified the subject in terms of utility and its contribution to industry. Others simply said that fossils are beautiful and people love to look at them and collect them (Fig. 1.1). But there is more than that. We identify six reasons why people should care about paleontology:

1 *Origins*. People want to know where life came from, where humans came from, where the Earth and universe came from. These have been questions in philosophy, religion and science for thousands of years and paleontologists have a key role (see pp. 117–20). Despite the spectacular progress of paleontology, earth sciences and astronomy over the last two centuries, many people with fundamentalist religious beliefs deny all natural explanations of origins – these debates are clearly seen as hugely important.

2 *Curiosity about different worlds*. Science fiction and fantasy novels allow us to think about worlds that are different from what we see around us. Another way is to study paleontology – there were plants and animals in the past that were quite unlike

any modern organism (see Chapters 9–12). Just imagine land animals 10 times the size of elephants, a world with higher oxygen levels than today and dragonflies the size of seagulls, a world with only microbes, or a time when two or three different species of humans lived in Africa!

3 *Climate and biodiversity change*. Thinking people, and now even politicians, are concerned about climate change and the future of life on Earth. Much can be learned by studying the modern world, but key evidence about likely future changes over hundreds or thousands of years comes from studies of what has happened in the past (see Chapter 20). For example, 250 million years ago, the Earth went through a phase of substantial global warming, a drop in oxygen levels and acid rain, and 95% of species died out (see pp. 170–4); might this be relevant to current debates about the future?

4 *The shape of evolution*. The **tree of life** is a powerful and all-embracing concept (see pp. 128–35) – the idea that all species living and extinct are related to each other and their relationships may be represented by a great branching tree that links us all back to a single species somewhere deep in the Precambrian (see Chapter 8). Biologists want to know how many species there are on the Earth today, how life became so diverse, and the nature and rates of diversifications and extinctions (see pp. 169–80, 534–41). It is impossible to understand these great patterns of evolution from studies of living organisms alone.

5 *Extinction*. Fossils show us that extinction is a normal phenomenon: no species lasts forever. Without the fossil record, we might imagine that extinctions have been caused mainly by human interactions.

6 *Dating rocks*. **Biostratigraphy**, the use of fossils in dating rocks (see pp. 23–41), is a powerful tool for understanding deep time, and it is widely used in scientific studies, as well as by commercial geologists who seek oil and mineral deposits. Radiometric dating provides precise dates in millions of years for rock samples, but this technological approach only works with certain kinds of rocks. Fossils are very much at the core of modern stratigraphy, both for economic and industrial applications and as the basis of our understanding of Earth's history at local and global scales.

PALEONTOLOGY AS A SCIENCE

What is science?

Imagine you are traveling by plane and your neighbor sees you are reading an article about the life of the ice ages in a recent issue of *National Geographic*. She asks you how anyone can know about those mammoths and sabertooths, and how they could make those color paintings; surely they are just pieces of art, and not science at all? How would you answer?

Science is supposed to be about reality, about hard facts, calculations and proof. It is obvious that you can not take a time machine back 20,000 years and see the mammoths and sabertooths for yourself; so how can we ever claim that there is a scientific method in paleontological reconstruction?

There are two ways to answer this; the first is obvious, but a bit of a detour, and the second gets to the core of the question. So, to justify those colorful paintings of extinct mammals, your first answer could be: "Well, we dig up all these amazing skeletons and other fossils that you see in museums around the world – surely it would be pretty sterile just to stop and not try to answer questions about the animal itself – how big was it, what were its nearest living relatives, when did it live?" From the earliest days, people have always asked questions about where we come from, about origins. They have also asked about the stars, about how babies are made, about what lies at the end of the rainbow. So, the first answer is to say that we are driven by our insatiable curiosity and our sense of wonder to try to find out about the world, even if we do not always have the best tools for the job.

The second answer is to consider the nature of **science**. Is science only about certainty, about proving things? In mathematics, and many areas of physics, this might be true. You can seek to measure the distance to the moon, to calculate the value of pi, or to derive a set of equations that explain the moon's influence on the Earth's tides. Generation by generation, these measurements and proofs are tested and improved. But this approach does not work for most of the natural sciences. Here,

(a) (b)

Figure 1.2 Important figures in the history of science: (a) Sir Francis Bacon (1561–1626), who established the methods of induction in science; and (b) Karl Popper (1902–1994), who explained that scientists adopt the hypothetico-deductive method.

there have been two main approaches: induction and deduction.

Sir Francis Bacon (1561–1626), a famous English lawyer, politician and scientist (Fig. 1.2a), established the methods of **induction** in science. He argued that it was only through the patient accumulation of accurate observations of natural phenomena that the explanation would emerge. The enquirer might hope to see common patterns among the observations, and these common patterns would point to an explanation, or law of nature. Bacon famously met his death perhaps as a result of his restless curiosity about everything; he was traveling in the winter of 1626, and was experimenting with the use of snow and ice to preserve meat. He bought a chicken, and got out of his coach to gather snow, which he stuffed inside the bird; he contracted pneumonia and died soon after. The chicken, on the other hand, was fresh to eat a week later, so proving his case.

The other approach to understanding the natural world is a form of **deduction**, where a series of observations point to an inevitable outcome. This is a part of classical logic dating back to Aristotle (384–322 BCE) and other ancient Greek philosophers. The standard logical form goes like this:

All men are mortal.
Socrates is a man.
Therefore Socrates is mortal.

Deduction is the core approach in mathematics and in detective work of course. How does it work in science?

Karl Popper (1902–1994) explained the way science works as the **hypothetico-deductive method**. Popper (Fig. 1.2b) argued that in most of the natural sciences, proof is impossible. What scientists do is to set up **hypotheses**, statements about what may or may not be the case. An example of a hypothesis might be "*Smilodon*, the sabertoothed cat, was exclusively a meat eater". This can never be proved absolutely, but it could be refuted and therefore rejected. So what most natural scientists do is called **hypothesis testing**; they seek to **refute**, or disprove, hypotheses rather than to prove them. Paleontologists have made many observations about *Smilodon* that tend to confirm, or **corroborate**, the hypothesis: it had long sharp teeth, bones have been found with bite marks made by those teeth, fossilized *Smilodon* turds contain bones of other mammals, and so on. But it would take just one discovery of a *Smilodon* skeleton with leaves in its stomach area, or in its excrement, to disprove the hypothesis that this animal fed exclusively on meat.

Science is of course much more complex than this. Scientists are human, and they are subject to all kinds of influences and prejudices, just like anyone else. Scientists follow trends, they are slow to accept new ideas; they may prefer one interpretation over another because of some political or sociological belief. Thomas Kuhn (1922–1996) argued that science shuttles between so-called times of normal science and times of scientific revolution. Scientific revolutions, or **paradigm shifts**, are when a whole new idea invades an area of science. At first people may be reluctant to accept the idea, and they fight against it. Then some supporters speak up and support it, and then everyone does. This is summarized in the old truism – when faced with a new idea most people at first reject it, then they begin to accept it, and then they say they knew it all along.

A good example of a paradigm shift in paleontology was triggered by the paper by Luis Alvarez and colleagues (1980) in which they presented the hypothesis that the Earth had been hit by a meteorite 65 million years ago, and this impact caused the extinction of the dinosaurs and other groups. It took 10 years or more for the idea to become widely accepted as the evidence built up (see pp. 174–7). As another example, current attempts by religious fundamentalists to force their view of "intelligent design" into science will likely fail because they do not test evidence rigorously, and paradigm shifts only happen when the weight of evidence for the new theory overwhelms the evidence for the previous view (see p. 120).

So science is curiosity about how the world works. It would be foolish to exclude any area of knowledge from science, or to say that one area of science is "more scientific" than another. There is mathematics and there is natural science. The key point is that there can be no proof in natural science, only hypothesis testing. But where do the hypotheses come from? Surely they are entirely speculative?

Speculation, hypotheses and testing

There are facts and speculations. "The fossil is 6 inches long" is a fact; "it is a leaf of an ancient fern" is a speculation. But perhaps the word "speculation" is the problem, because it sounds as if the paleontologist simply sits back with a glass of brandy and a cigar and lets his mind wander idly. But speculation is constrained within the hypothetico-deductive framework.

This brings us to the issue of **hypotheses** and where they come from. Surely there are unknown millions of hypotheses that could be presented about, say, the trilobites? Here are a few: "trilobites were made of cheese", "trilobites ate early humans", "trilobites still survive in Alabama", "trilobites came from the moon". These are not useful hypotheses, however, and would never be set down on paper. Some can be refuted without further consideration – humans and trilobites did not live at the same time, and no one in Alabama has ever seen a living trilobite. Admittedly, one discovery could refute both these hypotheses. Trilobites were almost certainly not made from cheese as their fossils show cuticles and other tissues and structures seen in living crabs and insects. "Trilobites came from the moon" is probably an untestable (as well as wild) hypothesis.

So, hypotheses are narrowed down quickly to those that fit the framework of current observations and that may be tested. A useful hypothesis about trilobites might be: "trilobites walked by making leg movements like modern millipedes". This can be tested by studying ancient tracks made by trilobites, by examining the arrangement of their legs in fossils, and by studies of how their modern relatives walk. So, **hypotheses should be sensible and testable**. This still sounds like speculation, however. Are other natural sciences the same?

Of course they are. The natural sciences operate by means of hypothesis testing. Which geologist can put his finger on the atomic structure of a diamond, the core–mantle boundary or a magma chamber? Can we prove with 100% certainty that mammoths walked through Manhattan and London, that ice sheets once covered most of Canada and northern Europe, or that there was a meteorite impact on the Earth 65 million years ago? Likewise, can a chemist show us an electron, can an astronomer confirm the composition of stars that have been studied by spectroscopy, can a physicist show us a quantum of energy, and can a biochemist show us the double helix structure of DNA?

So, the word "speculation" can mislead; perhaps "**informed deduction**" would be a better way of describing what most scientists do. Reconstructing the bodily appearance and behavior of an extinct animal is identical to any other normal activity in science, such as reconstructing the atmosphere of Saturn. The sequence of observations and conjectures that stand between the bones of *Brachiosaurus* lying in the ground and its reconstructed moving image in a movie is identical to the sequence of observations and conjectures that lie between biochemical and crystallographic observations on chromosomes and the creation of the model of the structure of DNA. Both hypotheses (the image of *Brachiosaurus* or the double helix) may be wrong, but in both cases the models reflect the best fit to the facts. The critic has to provide evidence to refute the hypothesis, and present a replacement hypothesis that fits the data better. Refutation and skepticism are the gatekeepers of science – ludicrous hypotheses are quickly weeded out, and the remaining hypotheses have survived criticism (so far).

Fact and fantasy – where to draw the line?

As in any science, there are levels of certainty in paleontology. The fossil skeletons show the shape and size of a dinosaur, the rocks show where and when it lived, and associated fossils show other plants and animals of the time. These can be termed **facts**. Should a paleontologist go further? It is possible to think about a sequence of procedures a paleontologist uses to go from bones in the ground to a walking, moving reconstruction of an ancient organism. And this sequence roughly matches a sequence of decreasing certainty, in three steps.

The first step is to **reconstruct** the skeleton, to put it back together. Most paleontologists would accept that this is a valid thing to do, and that there is very little guesswork in identifying the bones and putting them together in a realistic pose. The next step is to reconstruct the muscles. This might seem highly speculative, but then all living vertebrates – frogs, lizards, crocodiles, birds and mammals – have pretty much the same sorts of muscles, so it is likely dinosaurs did too. Also, muscles leave scars on the bones that show where they attached. So, the muscles go on to the skeleton – either on a model, with muscles made from modeling clay, or virtually, within a computer – and these provide the body shape. Other soft tissues, such as the heart, liver, eyeballs, tongue and so on are rarely preserved (though surprisingly such tissues are sometimes exceptionally preserved; see pp. 60–5), but again their size and positions are predictable from modern relatives. Even the skin is not entirely guesswork: some mummified dinosaur specimens show the patterns of scales set in the skin.

The second step is to work out the basic biology of the ancient beast. The teeth hint at what the animal ate, and the jaw shape shows how it fed. The limb bones show how the dinosaurs moved. You can manipulate the joints and calculate the movements, stresses and strains of the limbs. With care, it is possible to work out the pattern of locomotion in great detail. All the images of walking, running, swimming and flying shown in documentaries such as *Walking with Dinosaurs* (see Box 1.2) are generally based on careful calculation and modeling, and comparison with living animals. The movements of the jaws and limbs have to obey the laws of physics (gravity, lever mechanics, and so on). So these broad-scale indications of paleobiology and biomechanics are defensible and realistic.

The third level of certainty includes the colors and patterns, the breeding habits, the noises. However, even these, although entirely unsupported by fossil data, are not fantasy. Paleontologists, like any people with common sense, base their speculations here on comparisons with living animals. What color was *Diplodocus*? It was a huge plant eater. Modern large plant eaters like elephants and rhinos have thick, gray, wrinkly skin. So we give *Diplodocus* thick, gray, wrinkly skin. There's no evidence for the color in the fossils, but it makes biological sense. What about breeding habits? There are many examples of dinosaur nests with eggs, so paleontologists know how many eggs were laid and how they were arranged for some species. Some suggested that the parents cared for their young, while others said this was nonsense. But the modern relatives of dinosaurs – birds and crocodilians – show different levels of parental care. Then, in 1993, a specimen of the flesh-eating dinosaur *Oviraptor* was found in Mongolia sitting over a nest of *Oviraptor* eggs – perhaps this was a chance association, but it seems most likely that it really was a parent brooding its eggs (Box 1.1).

Box 1.1 Egg thief or good mother?

How dramatically some hypotheses can change! Back in the 1920s, when the first American Museum of Natural History (AMNH) expedition went to Mongolia, some of the most spectacular finds were nests containing dinosaur eggs. The nests were scooped in the sand, and each contained 20 or 30 sausage-shaped eggs, arranged in rough circles, and pointing in to the middle. Around the nests were skeletons of the plant-eating ceratopsian dinosaur *Protoceratops* (see p. 457) and a skinny, nearly 2-meter long, flesh-eating dinosaur. This flesh eater had a long neck, a narrow skull and jaws with no teeth, and strong arms with long bony fingers. Henry Fairfield Osborn (1857–1935), the famed paleontologist and autocratic director of the AMNH, named this theropod *Oviraptor*, which means "egg thief". A diorama was constructed at the AMNH, and photographs and dioramas of the scene were seen in books and magazines worldwide: *Oviraptor* was the mean egg thief who menaced innocent little *Protoceratops* as she tried to protect her nests and babies.

Then, in 1993, the AMNH sent another expedition to Mongolia, and the whole story turned on its head. More nests were found, and the researchers collected some eggs. Amazingly, they also found a whole skeleton of an *Oviraptor* apparently sitting on top of a nest (Fig. 1.3). It was crouching down, and had its arms extended in a broad circle, as if covering or protecting the whole nest. The researchers X-rayed the eggs back in the lab, and found one contained an unhatched embryo. They painstakingly dissected the eggshell and sediment away to expose the tiny incomplete bones inside the egg – a *Protoceratops* baby? No! The embryo belonged to *Oviraptor*, and the adult over the nest was either incubating the eggs or, more likely, protecting them from the sandstorm that buried her and her nest.

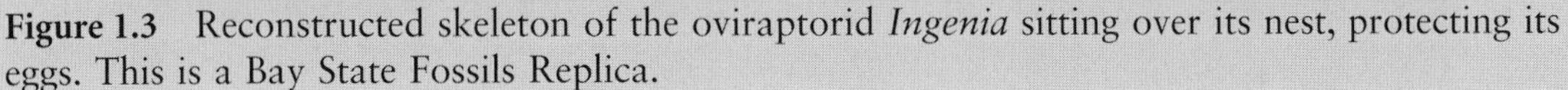

Figure 1.3 Reconstructed skeleton of the oviraptorid *Ingenia* sitting over its nest, protecting its eggs. This is a Bay State Fossils Replica.

As strong confirmation, an independent team of Canadian and Chinese scientists found another *Oviraptor* on her nest just across the border in northern China.

Read more about these discoveries in Norell et al. (1994, 1995) and Dong and Currie (1996), and at http://www.blackwellpublishing.com/paleobiology/.

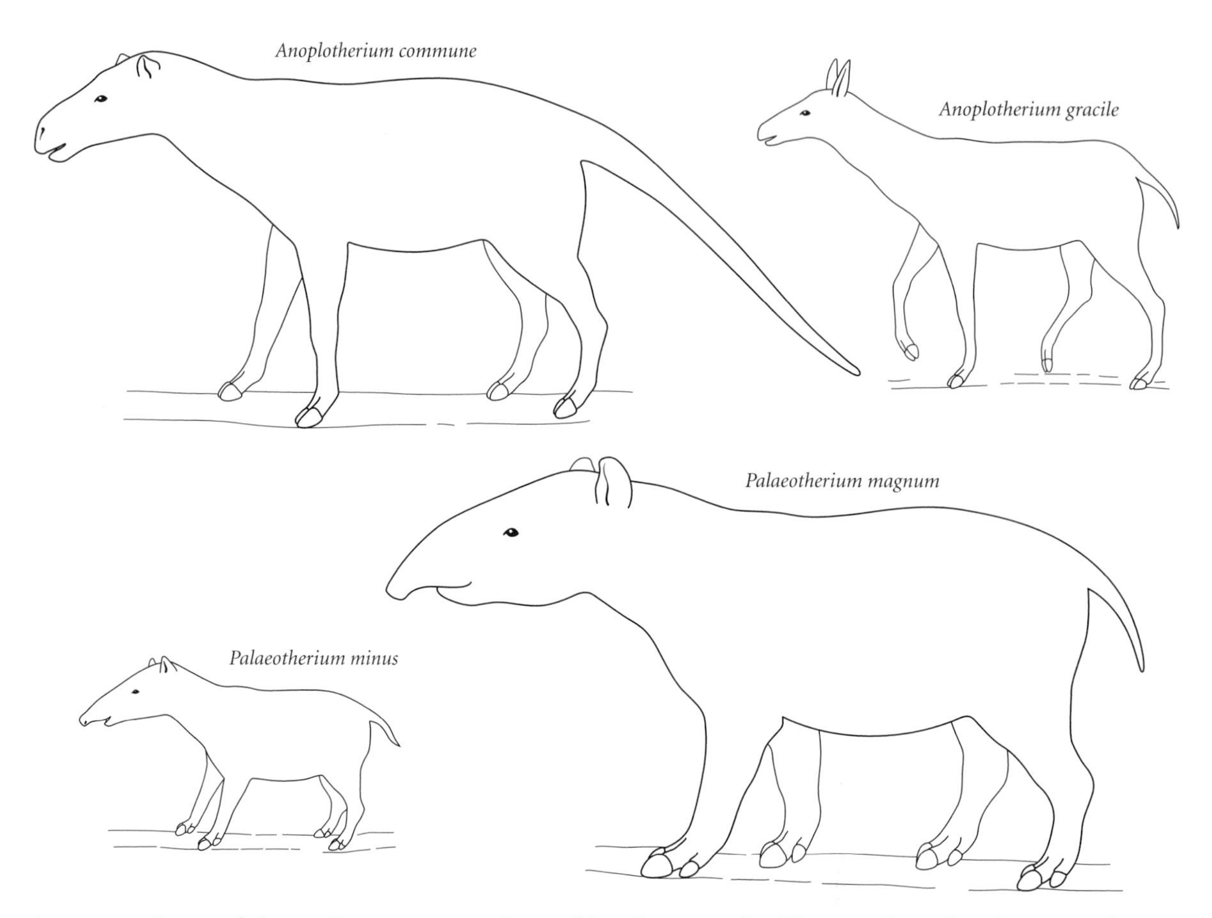

Figure 1.4 Some of the earliest reconstructions of fossil mammals. These outline sketches were drawn by C. L. Laurillard in the 1820s and 1830s, under the direction of Georges Cuvier. The image shows two species each of *Anoplotherium* and *Palaeotherium*, based on specimens Cuvier had reconstructed from the Tertiary deposits of the Paris Basin. (Modified from Cuvier 1834–1836.)

So, when you see a walking, grunting dinosaur, or a leggy trilobite, trotting across your TV screen, or featured in magazine artwork, is it just fantasy and guesswork? Perhaps you can now tell your traveling companion that it is a reasonable interpretation, probably based on a great deal of background work. The body shape is probably reasonably correct, the movements of jaws and limbs are as realistic as they can be, and the colors, noises and behaviors may have more evidence behind them than you would imagine at first.

Paleontology and the history of images

Debates about science and testing in paleontology have had a long history. This can be seen in the history of images of ancient life: at first, paleontologists just drew the fossils as they saw them. Then they tried to show what the perfect fossil looked like, repairing cracks and damage to fossil shells, or showing a skeleton in a natural pose. For many in the 1820s, this was enough; anything more would not be scientific.

However, some paleontologists dared to show the life of the past as they thought it looked. After all, this is surely one of the aims of paleontology? And if paleontologists do not direct the artistic renditions, who will? The first line drawings of reconstructed extinct animals and plants appeared in the 1820s (Fig. 1.4). By 1850, some paleontologists were working with artists to produce life-like paintings of scenes of the past, and even three-dimensional models for museums. The growth of museums, and improvements in printing processes, meant that by 1900 it was commonplace to see color paintings of scenes from ancient times, rendered by skilful artists and supervised by reputable paleontologists. Moving dinosaurs, of course, have had a long history in Hollywood movies through the 20th century, but paleontologists waited until the technology allowed more realistic computer-generated renditions in the 1990s, first in *Jurassic Park* (1993), and then in *Walking with Dinosaurs* (1999), and now in hundreds of films and documentaries each year (Box 1.2). Despite the complaints from some paleontologists about the mixing of fact and speculation in films and TV documentaries, their own museums often use the same technologies in their displays!

The slow evolution of reconstructions of ancient life over the centuries reflects the growth of paleontology as a discipline. How did the first scientists understand fossils?

STEPS TO UNDERSTANDING

Earliest fossil finds

Fossils are very common in certain kinds of rocks, and they are often attractive and beautiful objects. It is probable that people picked up fossils long ago, and perhaps even wondered why shells of sea creatures are now found high in the mountains, or how a perfectly preserved fish specimen came to lie buried deep within layers of rock. Prehistoric peoples picked up fossils and used them as ornaments, presumably with little understanding of their meaning.

Some early speculations about fossils by the classical authors seem now very sensible to modern observers. Early Greeks such as Xenophanes (576–480 BCE) and Herodotus (484–426 BCE) recognized that some fossils were marine organisms, and that these

Box 1.2 Bringing the sabertooths to life

Everyone's image of dinosaurs and ancient life changed in 1993. Steven Spielberg's film *Jurassic Park* was the first to use the new techniques of computer-generated imagery (CGI) to produce realistic animations. Older dinosaur films had used clay models or lizards with cardboard crests stuck on their backs. These looked pretty terrible and could never be taken seriously by paleontologists. Up to 1993, dinosaurs had been reconstructed seriously only as two-dimensional paintings and three-dimensional museum models. CGI made those superlative color images move.

Following the huge success of *Jurassic Park*, Tim Haines at the BBC in London decided to try to use the new CGI techniques to produce a documentary series about dinosaurs. Year by year, desktop computers were becoming more powerful, and the CGI software was becoming more sophisticated. What had once cost millions of dollars now cost only thousands. This resulted in the series *Walking with Dinosaurs*, first shown in 1999 and 2000.

Following the success of that series, Haines and the team moved into production of the follow-up, *Walking with Beasts*, shown first in 2001. There were six programs, each with six or seven key beasts. Each of these animals was studied in depth by consultant paleontologists and artists, and a carefully measured clay model (maquette) was made. This was the basis for the animation. The maquette was laser scanned, and turned into a virtual "stick model" that could be moved in the computer to simulate running, walking, jumping and other actions.

While the models were being developed, BBC film crews went round the world to film the background scenery. Places were chosen that had the right topography, climatic feel and plants. Where ancient mammals splashed through water, or grabbed a branch, the action (splashing, movement of the branch) had to be filmed. Then the animated beasts were married with the scenery in the studios of Framestore, the CGI company. This is hard to do, because shadowing and reflections had to be added, so the animals interacted with the backgrounds. If they run through a forest, they have to disappear behind trees and bushes, and their muscles have to move beneath their skin (Fig. 1.5); all this can be semiautomated through the CGI software.

Continued

Figure 1.5 The sabertooth *Smilodon* as seen in *Walking with Beasts* (2001). The animals were reconstructed from excellent skeletons preserved at Rancho La Brea in Los Angeles, and the hair and behavior were based on studies of the fossils and comparisons with modern large cats. (Courtesy of Tim Haines, image © BBC 2001.)

CGI effects are commonplace now in films, advertizing and educational applications. From a start in about 1990, the industry now employs thousands of people, and many of them work full-time on making paleontological reconstructions for the leading TV companies and museums.

Find out more about CGI at http://www.blackwellpublishing.com/paleobiology/.

provided evidence for earlier positions of the oceans. Other classical and medieval authors, however, had a different view.

Fossils as magical stones

In Roman and medieval times, fossils were often interpreted as mystical or magical objects. Fossil sharks' teeth were known as *glossopetrae* ("tongue stones"), in reference to their supposed resemblance to tongues, and many people believed they were the petrified tongues of snakes. This interpretation led to the belief that the glossopetrae could be used as protection against snakebites and other poisons. The teeth were worn as amulets to ward off danger, and they were even dipped into drinks in order to neutralize any poison that might have been placed there.

Most fossils were recognized as *looking like* the remains of plants or animals, but they were said to have been produced by a "plastic force" (*vis plastica*) that operated within the Earth. Numerous authors in the 16th and 17th centuries wrote books presenting this interpretation. For example, the Englishman Robert Plot (1640–1696) argued that ammonites (see pp. 344–51) were formed "by two salts shooting different ways, which by thwarting one another make a helical figure". These interpretations seem ridiculous now, but there was a serious problem in explaining how such specimens came to lie far from the sea, why they were often different from living animals,

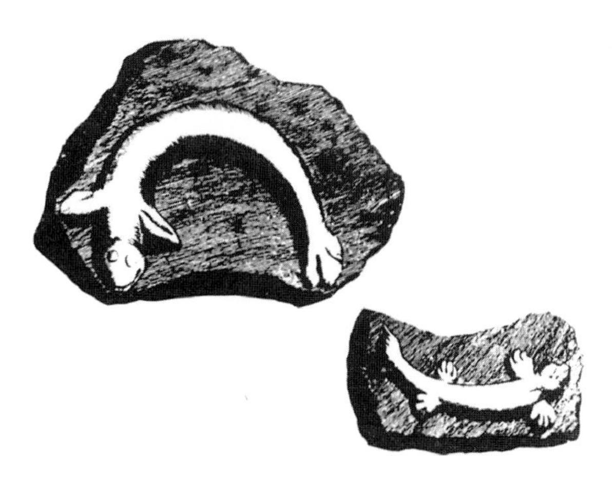

Figure 1.6 Lying stones: two of the remarkable "fossils" described by Professor Beringer of Wurzburg in 1726: he believed these specimens represented real animals of ancient times that had crystallized into the rocks by the action of sunlight.

and why they were made of unusual minerals.

The idea of plastic forces had been largely overthrown by the 1720s, but some extraordinary events in Wurzburg in Germany at that time must have dealt the final blow. Johann Beringer (1667–1740), a professor at the university, began to describe and illustrate "fossil" specimens brought to him by collectors from the surrounding area. But it turned out that the collectors had been paid by an academic rival to manufacture "fossils" by carving the soft limestone into the outlines of shells, flowers, butterflies and birds (Fig. 1.6). There was even a slab with a pair of mating frogs, and others with astrologic symbols and Hebrew letters. Beringer resisted evidence that the specimens were forgeries, and wrote as much in his book, the *Lithographiae Wirceburgensis* (1726), but realized the awful truth soon after publication.

Fossils as fossils

The debate about plastic forces was terminated abruptly by the debacle of Beringer's figured stones, but it had really been resolved rather earlier. Leonardo da Vinci (1452–1519), a brilliant scientist and inventor (as well as a great artist), used his observations of modern plants and animals, and of modern rivers and seas, to explain the fossil sea shells found high in the Italian mountains. He interpreted them as the remains of ancient shells, and he argued that the sea had once covered these areas.

Figure 1.7 Nicolaus Steno's (1667) classic demonstration that fossils represent the remains of ancient animals. He showed the head of a dissected shark together with two fossil teeth, previously called glossopetrae, or tongue stones. The fossils are exactly like the modern shark's teeth.

Later, Nicolaus Steno (or Niels Stensen) (1638–1686) demonstrated the true nature of glossopetrae simply by dissecting the head of a huge modern shark, and showing that its teeth were identical to the fossils (Fig. 1.7). Robert Hooke (1625–1703), a contemporary of Steno's, also gave detailed descriptions of fossils, using a crude microscope to compare the cellular structure of modern and fossil wood, and the crystalline layers in the shell of a modern and a fossil mollusk. This simple descriptive work showed that magical explanations of fossils were without foundation.

The idea of extinction

Robert Hooke was one of the first to hint at the idea of **extinction**, a subject that was hotly debated during the 18th century. The debate fizzed quietly until the 1750s and 1760s when accounts of fossil mastodon remains from North America began to appear. Explorers sent large teeth and bones back to Paris and London for study by the anatomic experts of the day (normal practice at the time, because the serious pursuit of science as a profession had not yet begun in North America). William Hunter noted in 1768 that the "American incognitum" was quite different from modern elephants and from mammoths, and was clearly an extinct animal, and a meat-eating one at that. "And if this animal was indeed carnivorous, which I believe cannot be doubted, though we may as philosophers regret it," he wrote, "as men we cannot but thank Heaven that its whole generation is probably extinct."

The reality of extinction was demonstrated by the great French natural scientist Georges Cuvier (1769–1832). He showed that the mammoth from Siberia and the mastodon from North America were unique species, and different from the modern African and Indian elephants (Fig. 1.8). Cuvier extended his studies to the rich Eocene mammal deposits of the Paris Basin, describing skeletons of horse-like animals (see Fig. 1.4), an opossum, carnivores, birds and reptiles, all of which differed markedly from living forms. He also wrote accounts of Mesozoic crocodilians, pterosaurs and the giant mosasaur of Maastricht.

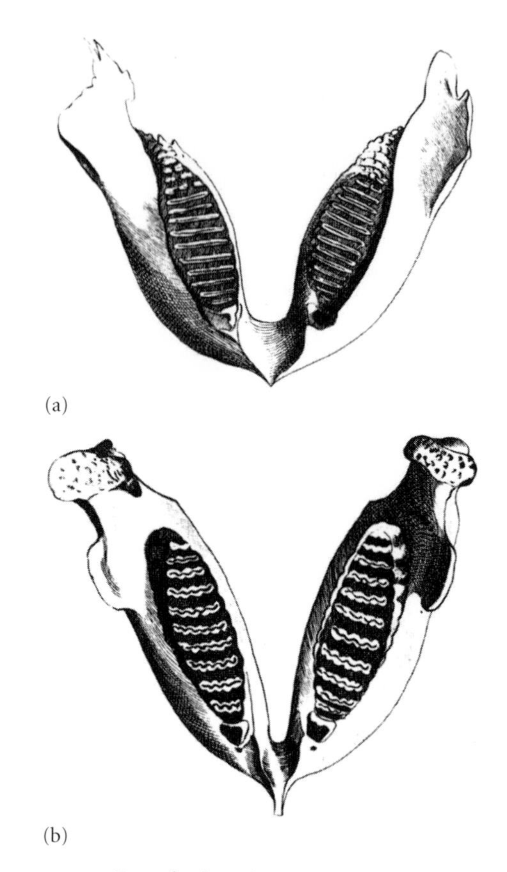

Figure 1.8 Proof of extinction: Cuvier's comparison of (a) the lower jaw of a mammoth and (b) a modern Indian elephant. (Courtesy of Eric Buffetaut.)

Cuvier is sometimes called the father of **comparative anatomy**; he realized that all organisms share common structures. For example, he showed that elephants, whether living or fossil, all share certain anatomic features. His public demonstrations became famous: he claimed to be able to identify and reconstruct an animal from just one tooth or bone, and he was usually successful. After 1800, Cuvier had established the reality of extinction.

The vastness of geological time

Many paleontologists realized that the sedimentary rocks and their contained fossils documented the history of long spans of time. Until the late 18th century, scientists accepted calculations from the Bible that the Earth was only 6000–8000 years old. This view was challenged, and most thinkers accepted an unknown, but vast, age for the Earth by the 1830s (see p. 23).

The geological periods and eras were named through the 1820s and 1830s, and geologists realized they could use fossils to recognize all major sedimentary rock units, and that these rock units ran in a predictable sequence everywhere in the world. These were the key steps in the foundations of **stratigraphy**, an understanding of geologic time (see p. 24).

FOSSILS AND EVOLUTION

Progressionism and evolution

Knowledge of the fossil record in the 1820s and 1830s was patchy, and paleontologists debated whether there was a **progression** from simple organisms in the most ancient rocks to more complex forms later. The leading British geologist, Charles Lyell (1797–1875), was an antiprogressionist. He believed that the fossil record showed no evidence of long-term, one-way change, but rather cycles of change. He would not have been surprised to find evidence of human fossils in the Silurian, or for dinosaurs to come back at some time in the future if the conditions were right.

Progressionism was linked to the idea of **evolution**. The first serious considerations of evolution took place in 18th century France, in the work of naturalists such as the Comte de Buffon (1707–1788) and Jean-Baptiste Lamarck (1744–1829). Lamarck explained the phenomenon of progressionism by a large-scale evolutionary model termed the "Great Chain of Being" or the Scala naturae. He believed that all organisms, plants and animals, living and extinct, were linked in time by a unidirectional ladder leading from simplest at the bottom to most complex at the top, indeed, running from rocks to angels. Lamarck argued that the Scala was more of a moving escalator than a ladder; that in time present-day apes would rise to become humans, and that present-day humans were destined to move up to the level of angels.

Darwinian evolution

Charles Darwin (1809–1882) developed the theory of evolution by **natural selection** in the 1830s by abandoning the usual belief that species were fixed and unchanging. Darwin realized that individuals within species showed considerable variation, and that there was not a fixed central "type" that represented the essence of each species. He also emphasized the idea of evolution by common descent, namely that all species today had evolved from other species in the past. The problem he had to resolve was to explain how the variation within species could be harnessed to produce evolutionary change.

Darwin found the solution in a book published in 1798 by Thomas Malthus (1766–1834), who demonstrated that human populations tend to increase more rapidly than the supplies of food. Hence, only the stronger can survive. Darwin realized that such a principle applied to all animals, that the surviving individuals would be those that were best fitted to obtain food and to produce healthy young, and that their particular **adaptations** would be inherited. This was Darwin's theory of evolution by natural selection, the core of modern evolutionary thought.

The theory was published 21 years after Darwin first formulated the idea, in his book *On the Origin of Species* (1859). The delay was a result of Darwin's fear of offending established opinion, and of his desire to bolster his remarkable insight with so many supporting facts that no one could deny it. Indeed, most scientists accepted the idea of evolution by common descent in 1859, or soon after, but very few accepted (or understood) natural selection. It was only after the beginning of modern genetics early in the 20th century, and its amalgamation with "natural history" (systematics, ecology, paleontology) in the 1930s and 1940s, in a movement termed the "Modern synthesis", that Darwinian evolution by natural selection became fully established.

PALEONTOLOGY TODAY

Dinosaurs and fossil humans

Much of 19th century paleontology was dominated by remarkable new discoveries. Collectors fanned out all over the world, and knowledge of ancient life on Earth increased enormously. The public was keenly interested then, as now, in spectacular new discoveries of dinosaurs. The first isolated dinosaur bones were described from England and Germany in the 1820s and 1830s, and tentative reconstructions were made (Fig. 1.9). However, it was only with the discovery of complete skeletons in Europe and North America in the 1870s that a true picture of these astonishing beasts could be presented. The first specimen of *Archaeopteryx*, the oldest bird, came to light in 1861: here was a true "missing link", predicted by Darwin only 2 years before.

Darwin hoped that paleontology would provide key evidence for evolution; he expected that, as more finds were made, the fossils would line up in long sequences showing the precise pattern of common descent. *Archaeopteryx* was a spectacular

Figure 1.9 The first dinosaur craze in England in the 1850s was fueled by new discoveries and dramatic new reconstructions of the ancient inhabitants of that country. This picture, inspired by Sir Richard Owen, is based on his view that dinosaurs were almost mammal-like. (Courtesy of Eric Buffetaut.)

start. Rich finds of fossil mammals in the North American Tertiary were further evidence. Othniel Marsh (1831–1899) and Edward Cope (1840–1897), arch-rivals in the search for new dinosaurs, also found vast numbers of mammals, including numerous horse skeletons, leading from the small four-toed *Hyracotherium* of 50 million years ago to modern, large, one-toed forms. Their work laid the basis for one of the classic examples of a long-term evolutionary trend (see pp. 541–3).

Human fossils began to come to light around this time: incomplete remains of Neandertal man in 1856, and fossils of *Homo erectus* in 1895. The revolution in our understanding of human evolution began in 1924, with the announcement of the first specimen of the "southern ape" *Australopithecus* from Africa, an early human ancestor (see pp. 473–5).

Evidence of earliest life

At the other end of the evolutionary scale, paleontologists have made extraordinary progress in understanding the earliest stages in the evolution of life. Cambrian fossils had been known since the 1830s, but the spectacular discovery of the Burgess Shale in Canada in 1909 showed the extraordinary diversity of soft-bodied animals that had otherwise been unknown (see p. 249). Similar but slightly older faunas from Sirius Passett in north Greenland and Chengjiang in south China have confirmed that the Cambrian was truly a remarkable time in the history of life.

Even older fossils from the Precambrian had been avidly sought for years, but the breakthroughs only happened around 1950. In 1947, the first soft-bodied Ediacaran fossils were found in Australia, and have since been identified in many parts of the world. Older, simpler, forms of life were recognized after 1960 by the use of advanced microscopic techniques, and some aspects of the first 3000 million years of the history of life are now understood (see Chapter 8).

Macroevolution

Collecting fossils is still a key aspect of modern paleontology, and remarkable new discoveries are announced all the time. In addition, paleontologists have made dramatic contributions to our understanding of large-scale evolution, **macroevolution**, a field that includes studies of rates of evolution, the nature of speciation, the timing and extent of mass extinctions, the diversification of life, and other topics that involve long time scales (see Chapters 6 and 7).

Studies of macroevolution demand excellent knowledge of time scales and excellent knowledge of the fossil species (see pp. 70–7). These two key aspects of the **fossil record**, our knowledge of ancient life, are rarely perfect: in any study area, the fossils may not be dated more accurately than to the nearest 10,000 or 100,000 years. Further, our knowledge of the fossil species may be uncertain because the fossils are not complete. Paleontologists would love to determine whether we know 1%, 50% or 90% of the species of fossil plants and animals; the eminent American paleontologist Arthur J. Boucot considered, based on his wide experience, that 15% was a reasonable figure. Even that is a generalization of course – knowledge probably varies group by group: some are probably much better known than others.

All fields of paleontological research, but especially studies of macroevolution, require quantitative approaches. It is not enough to look at one or two examples, and leap to a conclusion, or to try to guess how some fossil species changed through time. There are many quantitative approaches in analyzing paleontological data (see Hammer and Harper (2006) for a good cross-section of these). At the very least, all paleontologists must learn simple **statistics** so they can describe a sample of fossils in a reasonable way (Box 1.3) and start to test, statistically, some simple hypotheses.

Paleontological research

Most paleontological research today is done by paid **professionals** in scientific institutions, such as universities and museums, equipped with powerful computers, scanning electron microscopes, geochemical analytic equipment, and well-stocked libraries, and, ideally, staffed by lab technicians, photographers and artists. However, important work is done by **amateurs**, enthusiasts who are not paid to work as paleontologists, but frequently discover new sites and specimens, and many of whom develop expertise in a chosen group of fossils.

A classic example of a paleontological research project shows how a mixture of luck and hard work is crucial, as well as the

Box 1.3 Paleobiostatistics

Modern paleobiology relies on quantitative approaches. With the wide availability of microcomputers, a large battery of statistical and graphic techniques is now available (Hammer & Harper 2006). Two simple examples demonstrate some of the techniques widely used in taxonomic studies, firstly to summarize and communicate precise data, and secondly to test hypotheses.

The smooth terebratulide brachiopod *Dielasma* is common in dolomites and limestones associated with Permian reef deposits in the north of England. Do the samples approximate to living populations, and do they all belong to one or several species? Two measurements (Fig. 1.10a) were made on specimens from a single site, and these were plotted as a frequency polygon (Fig. 1.10a) to show the population structure. This plot can test the hypothesis that there is in fact only one species and that the specimens approximate to a typical single population. If there are two species, there should be two separate, but similar, peaks that illustrate the growth cycles of the two species.

Continued

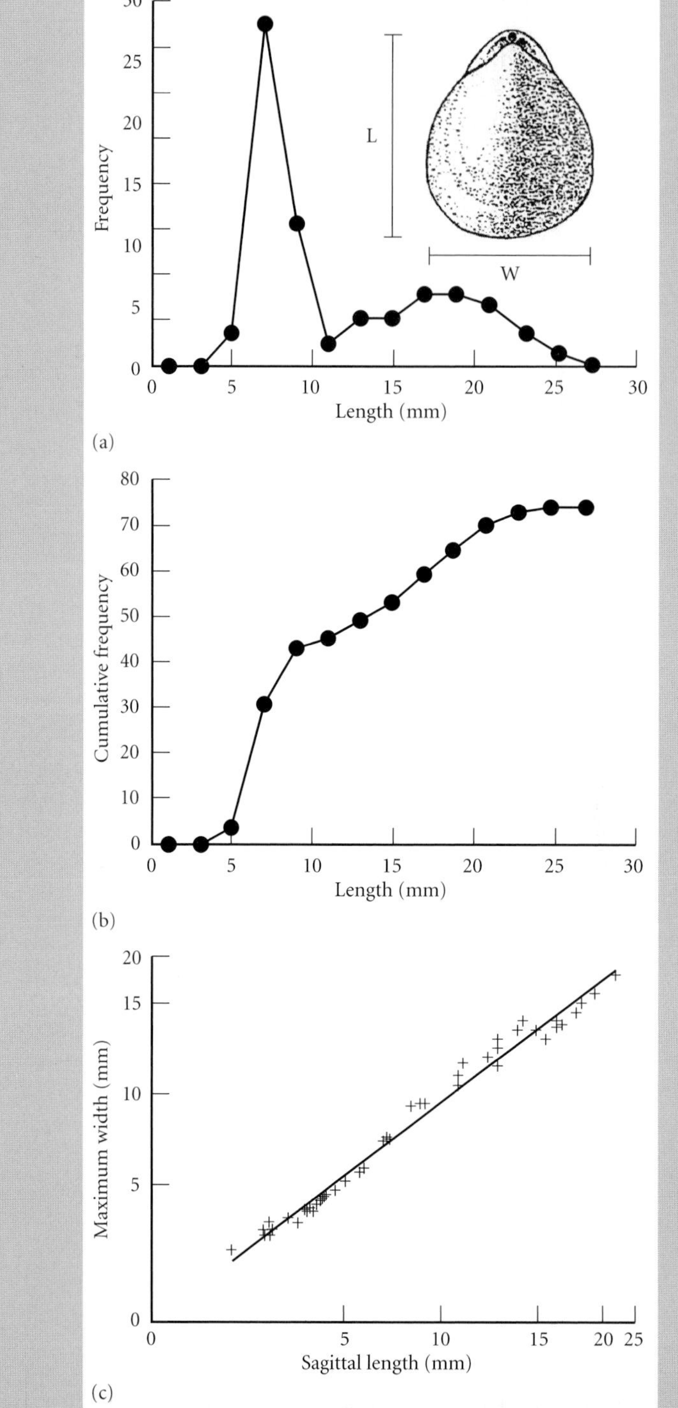

Figure 1.10 Statistical study of the Permian brachiopod *Dielasma*. Two measurements, sagittal length (L) and maximum width (W) were made on all specimens. The size–frequency distributions (a, b) indicate an enormous number of small shells, and far fewer large ones, thus suggesting high juvenile mortality. When the two shape measurements are compared (c), the plot shows a straight line ($y = 0.819x + 0.262$); on a previous logarithmic plot, the slope (α) did not differ significantly from unity, so an isometric relationship is assumed, and the raw data have been replotted.

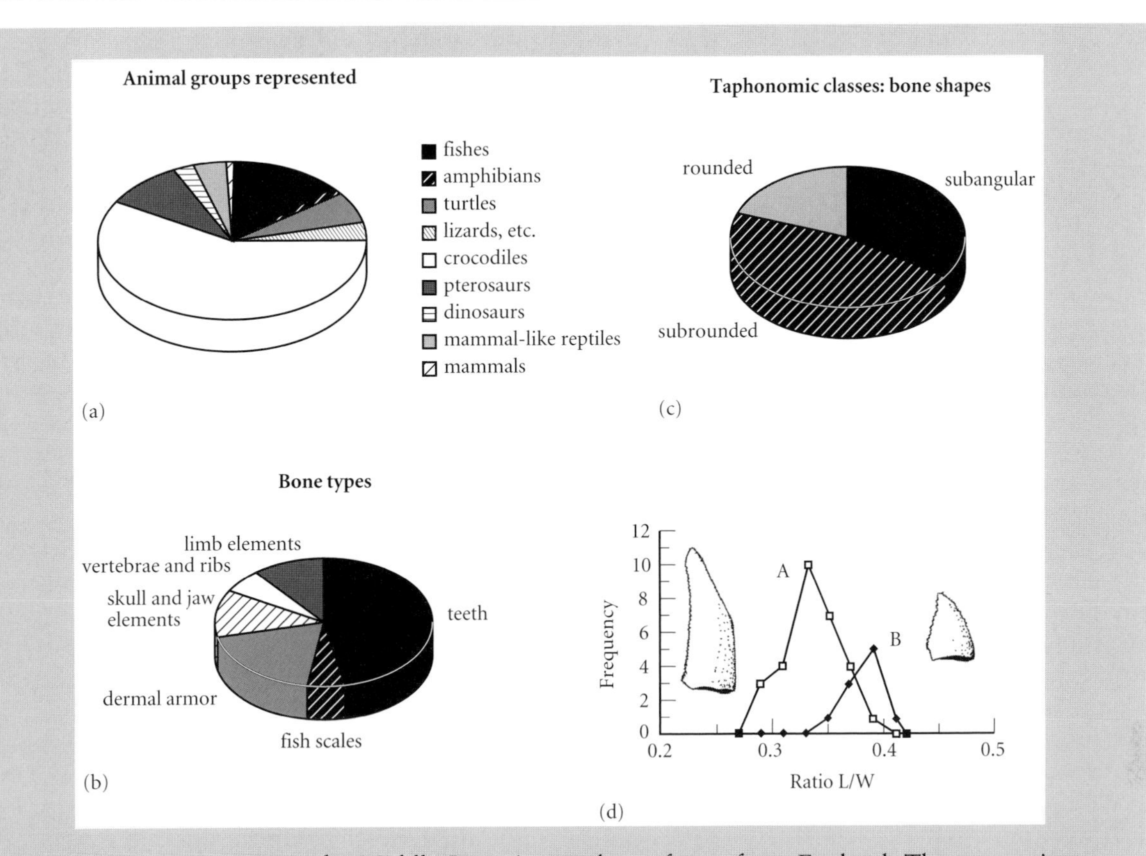

Figure 1.11 Composition of a Middle Jurassic vertebrate fauna from England. The proportions of the major groups of vertebrates in the fauna are shown as a pie chart (a). The sample can be divided into categories also of bone types (b) and taphonomic classes (c), which reflect the amount of transport. Dimensions of theropod dinosaur teeth show two frequency polygons (d) that are statistically significantly different (*t*-test), and hence indicate two separate forms.

The graph suggests that there is in fact a single species, but that the population has an imbalance (is skewed) towards smaller size classes, and hence that there was a high rate of juvenile mortality. This is confirmed when the frequency of occurrence of size classes is summed to produce a cumulative frequency polygon (Fig. 1.10b). It is possible to test ways in which this population diverges from a normal distribution (i.e. a symmetric "bell" curve with a single peak corresponding to the mean, and a width indicated by the standard deviation about the mean).

It is also interesting to consider growth patterns of *Dielasma*: did the shell grow in a uniform fashion, or did it grow more rapidly in one dimension than the other? The hypothesis is that the shell grew uniformly in all directions, and when the two measurements are compared on logarithmic scales (Fig. 1.10c), the slope of the line equals one. Thus, both features grew at the same rate.

In a second study, a collection of thousands of **microvertebrates** (teeth, scales and small bones) was made by sieving sediment from a Middle Jurassic locality in England. A random sample of 500 of these specimens was taken, and the teeth and bones were sorted into taxonomic groups: the results are shown as a pie chart (Fig. 1.11a). It is also possible to sort these 500 specimens into other kinds of categories, such as types of bones and teeth or taphonomic classes (Fig. 1.11b, c). A further analysis was made of the relatively abundant theropod (carnivorous dinosaur) teeth, to test whether they represented a single population of young and old animals, or whether they came from several species. Tooth lengths and widths were measured, and frequency polygons (Fig. 1.11d) show that there are two populations within the sample, probably representing two species.

cooperation of many people. The spectacular Burgess Shale fauna (Gould 1989; Briggs et al. 1994) was found by the geologist Charles Walcott in 1909. The discovery was partly by chance: the story is told of how Walcott and his wife were riding through the Canadian Rockies, and her horse supposedly stumbled on a slab of shale bearing beautifully preserved examples of *Marrella splendens*, the "lace crab". During five subsequent field seasons, Walcott collected over 60,000 specimens, now housed in the National Museum of Natural History, Washington, DC. The extensive researches of Walcott, together with those of many workers since, have documented a previously unknown assemblage of remarkable soft-bodied animals. The success of the work depended on new technology in the form of high-resolution microscopes, scanning electron microscopes, X-ray photography and computers to enable three-dimensional reconstructions of flattened fossils. In addition, the work was only possible because of the input of thousands of hours of time in skilled preparation of the delicate fossils, and in the production of detailed drawings and descriptions. In total, a variety of government and private funding sources must have contributed hundreds of thousands of dollars to the continuing work of collecting, describing and interpreting the extraordinary Burgess Shale animals.

The Burgess Shale is a dramatic and unusual example. Most paleontological research is more mundane: researchers and students may spend endless hours splitting slabs, excavating trenches and picking over sediment from deep-sea cores under the microscope in order to recover the fossils of interest. Laboratory preparation may also be tedious and long-winded. Successful researchers in paleontology, as in any other discipline, need endless patience and stamina.

Modern paleontological expeditions go all over the world, and require careful negotiation, planning and fund-raising. A typical expedition might cost anything from US$20,000 to $100,000, and field paleontologists have to spend a great deal of time planning how to raise that funding from government science programs, private agencies such as the National Geographic Society and the Jurassic Foundation, or from alumni and other sponsors. A typical high-profile example has been

Box 1.4 Giant dinosaurs from Madagascar

How do you go about finding a new fossil species, and then telling the world about it? As an example, we choose a recent dinosaur discovery from the Late Cretaceous of Madagascar, and tell the story step by step. Isolated dinosaur fossils had been collected by British and French expeditions in the 1880s, but a major collecting effort was needed to see what was really there. Since 1993, a team, led by David Krause of SUNY-Stony Brook, has traveled to Madagascar for nine field seasons with funding from the US National Science Foundation and the National Geographic Society. Their work has brought to light some remarkable new finds of birds, mammals, crocodiles and dinosaurs from the Upper Cretaceous.

One of the major discoveries on the 1998 expedition was a nearly complete skeleton of a titanosaurian sauropod. These giant plant-eating dinosaurs were known particularly from South America and India, though they have a global distribution, and isolated bones had been reported from Madagascar in 1896. The new fossil was found on a hillside in rocks of the Maevarano Formation, dated at about 70 million years old, in the Mahajanga Basin. The landscape is rough and exposed, and the bones were excavated under a burning sun. The first hint of discovery was a series of articulated tail vertebrae, but as the team reported, "The more we dug into the hillside, the more bones we found". Almost every bone in the skeleton was preserved, from the tip of the nose, to the tip of the tail. The bones were excavated and carefully wrapped in plaster jackets for transport back to the United States.

Back in the laboratory, the bones were cleaned up and laid out (Fig. 1.12). Kristi Curry Rogers worked on the giant bones for her PhD dissertation that she completed at SUNY-Stony Brook in 2001. Kristi, and her colleague Cathy Forster, named the new sauropod *Rapetosaurus krausei* in

(a)

(b)

Figure 1.12 Finding the most complete titanosaur, *Rapetosaurus*, in Madagascar: (a) Kristi Curry Rogers (front right) with colleagues excavating the giant skeleton; (b) after preparation in the lab, the whole skeleton can be laid out – this is a juvenile sauropod, so not as large as some of its relatives. (Courtesy of Kristi Curry Rogers.)

2001. It turned out to be different from titanosaurians already named from other parts of the world, and the specimen was unique in being nearly complete and in preserving the skull, which was described in detail by Curry Rogers and Forster in 2004. Its name refers to "rapeto", a legendary giant in Madagascan folklore. To date, *Rapetosaurus krausei* is the most complete and best-preserved titanosaur ever discovered.

Kristi Curry Rogers is now Curator and Head of Vertebrate Paleontology at the Science Museum of Minnesota, where she continues her work on the anatomy and relationships of sauropod dinosaurs, and on dinosaur bone histology. Read more about her at http://www.blackwellpublishing.com/paleobiology/. You can find out more about *Rapetosaurus* in Curry Rogers and Forster (2001, 2004) and at http://www.blackwellpublishing.com/paleobiology/.

Continued

a long-running program of study of dinosaurs and other fossil groups from the Cretaceous of Madagascar (Box 1.4).

Field expeditions attract wide attention, but most paleontological research is done in the laboratory. Paleontologists may be motivated to study fossils for all kinds of reasons, and their techniques are as broad as in any science. Paleontologists work with chemists to understand how fossils are preserved and to use fossils to interpret ancient climates and atmospheres. Paleontologists work with engineers and physicists to understand how ancient animals moved, and with biologists to understand how ancient organisms lived and how they are related to each other. Paleontologists work with mathematicians to understand all kinds of aspects of evolution and events, and the biomechanics and distribution of ancient organisms. Paleontologists, of course, work with geologists to understand the sequence and dating of the rocks, and ancient environments and climates.

But it seems that, despite centuries of study, paleobiologists have so much to learn. We don't have a complete tree of life; we don't know how fast diversifications can happen and why some groups exploded onto the scene and became successful and others did not; we don't know the rules of extinction and mass extinction; we don't know how life arose from non-living matter; we don't know why so many animal groups acquired skeletons 500 million years ago; we don't know why life moved on to land 450 million years ago; we don't know exactly what dinosaurs did; we don't know what the common ancestor of chimps and humans looked like and why the human lineage split off and evolved so fast to dominate the world. These are exciting times indeed for new generations to be entering this dynamic field of study!

Review questions

1 What kinds of evidence might you look for to determine the speed and mode of locomotion of an ancient beetle? Assume you have fossils of the whole body, including limbs, of the beetle and its fossilized tracks.

2 Which of these statements is in the form of a scientific hypothesis that may be tested and could be rejected, and which are non-scientific statements? Note, scientific hypotheses need not always be correct; equally, non-scientific statements might well be correct, but cannot be tested:
   - The plant *Lepidodendron* is known only from the Carboniferous Period.
   - The sabertoothed cat *Smilodon* ate plant leaves.
   - *Tyrannosaurus rex* was huge.
   - There were two species of *Archaeopteryx*, one larger than the other.
   - Evolution did not happen.
   - Birds and dinosaurs are close relatives that share a common ancestor.

3 Do you think scientists should be cautious and be sure they can never be contradicted, or should they make statements they believe to be correct, but that can be rejected on the basis of new evidence?

4 Does paleontology advance by the discovery of new fossils, or by the proposal and testing of new ideas about evolution and ancient environments?

5 Should governments invest tax dollars in paleontological research?

Further reading

Briggs, D.E.G. & Crowther, P.R. 2001. *Palaeobiology II*. Blackwell, Oxford.

Bryson, B. 2003. *A Short History of Nearly Everything*. Broadway Books, New York.

Buffetaut, E. 1987. *A Short History of Vertebrate Palaeontology*. Croom Helm, London.

Cowen, R. 2004. *The History of Life*, 4th edn. Blackwell, Oxford.

Curry Rogers, K. & Forster, C.A. 2001. The last of the dinosaur titans: a new sauropod from Madagascar. *Nature* **412**, 530–4.

Curry Rogers, K. & Forster, C.A. 2004. The skull of *Rapetosaurus krausei* (Sauropoda: Titanosauria) from the Late Cretaceous of Madagascar. *Journal of Vertebrate Paleontology* **24**, 121–44.

Dong Z.-M. & Currie, P.J. 1996. On the discovery of an oviraptorid skeleton on a nest of eggs at Bayan Mandahu, Inner Mongolia, People's Republic of China. *Canadian Journal of Earth Sciences* **33**, 631–6.

Foote, M. & Miller, A.I. 2006. *Principles of Paleontology*. W.H. Freeman, San Francisco.

Fortey, R. 1999. *Life: A Natural History of the First Four Billion Years of Life on Earth*. Vintage Books, New York.

Hammer, O. & Harper, D.A.T. 2005. *Paleontological Data Analysis*. Blackwell, Oxford.

Kemp, T.S. 1999. *Fossils and Evolution*. Oxford University Press, Oxford.

Mayr, E. 1991. *One Long Argument; Charles Darwin and the Genesis of Modern Evolutionary Thought*. Harvard University Press, Cambridge, MA.

Palmer, D. 2004. *Fossil Revolution: The Finds that Changed Our View of the Past*. Harper Collins, London.

Rudwick, M.J.S. 1976. *The Meaning of Fossils: Episodes in the History of Paleontology*. University of Chicago Press, Chicago.

Rudwick, M.J.S. 1992. *Scenes from Deep Time: Early Pictorial Representations of the Prehistoric World*. University of Chicago Press, Chicago.

References

Alvarez, L.W., Alvarez, W., Asaro, F. & Michel, H.V. 1980. Extraterrestrial causes for the Cretaceous-Tertiary extinction. *Science* **208**, 1095–108.

Beringer, J.A.B. 1726. *Lithographiae wirceburgensis, ducentis lapidum figuatorum, a potiori insectiformium, prodigiosis imaginibus exornatae specimen primum, quod in dissertatione inaugurali physico-historica, cum annexis corollariis medicis*. Fuggart, Wurzburg, 116 pp.

Briggs, D.E.G., Erwin, D.H. & Collier, F.J. 1994. *The Fossils of the Burgess Shale*. Smithsonian Institution Press, Washington.

Curry Rogers, K. & Forster, C.A. 2001. The last of the dinosaur titans: a new sauropod from Madagascar. *Nature* **412**, 530–4.

Curry Rogers, K. & Forster, C.A. 2004. The skull of *Rapetosaurus krausei* (Sauropoda: Titanosauria) from the Late Cretaceous of Madagascar. *Journal of Vertebrate Paleontology* **24**, 121–44.

Darwin, C.R. 1859. *On the Origin of Species by Means of Natural Selection, or the Preservation of Favoured Races in the Struggle for Life*. John Murray, London, 502 pp

Gould, S.J. 1989. *Wonderful Life. The Burgess Shale and the Nature of History*. Norton, New York.

Hammer, O. & Harper, D.A.T. 2005. *Paleontological Data Analysis*. Blackwell, Oxford.

Hunter, W. 1768. Observations on the bones commonly supposed to be elephant's bones, which have been found near the river Ohio, in America. *Philosophical Transactions of the Royal Society* **58**, 34–45.

Norell, M.A., Clark, J.M., Chiappe, L.M. & Dashzeveg, D. 1995. A nesting dinosaur. *Nature* **378**, 774–6.

Norell, M.A., Clark, J.M., Dashzeveg, D. et al. 1994. A theropod dinosaur embryo and the affinities of the Flaming Cliffs Dinosaur eggs. *Science* **266**, 779–82.

Chapter 2

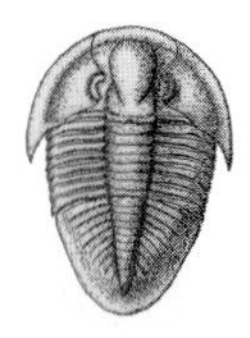

Fossils in time and space

Key points

- Scientists began to study the order and sequence of geological events during the Renaissance when artists rediscovered perspective.
- Lithostratigraphy is the establishment of rock units, forming the basis for virtually all geological studies; lithostratigraphic units are displayed on maps and measured sections.
- Biostratigraphy, using zone fossils, forms the basis for correlation and it can now be investigated using a range of quantitative techniques.
- Chronostratigraphy, global standard stratigraphy, is the division of geological time into workable intervals with reference to type sections in the field.
- Cyclostratigraphy and sequence stratigraphy can provide more refined frameworks that can also help understand biological change.
- Geochronometry is based on absolute time, measured in years before present by a range of modern, quantitative techniques.
- Paleobiogeography provides basic data to suggest and test plate tectonic and terrane models.
- Changes in geography allowed faunas and floras to migrate, and major groups to radiate and go extinct.
- The rhythmic joining and break up of continents through time has been associated with climate and diversity change.
- Fossils from mountain belts are significant in constraining the age and origin of tectonic events; fossil data have also provided estimates for finite strain and thermal maturation.

The Earth is immensely old, and the distribution of continents and oceans has changed radically over time. Early paleontologists did not know these things, and so they tried to pack the whole of the history of life into a relatively short span of time, vizualizing trilobites or dinosaurs inhabiting a world that was much as it is today.

Life on Earth, however, has been evolving for up to 4 billion years, and there has been a complex story of fossil groups coming and going, and continents moving from place to place. How do we develop geographic and temporal frameworks that are accurate and reliable enough to chart the distributions of fossil organisms through time and space? Fortunately, paleogeographers and stratigraphers are now equipped with a range of high-tech methods, virtually all computer-based, that provide a greater consensus for models describing the distributions of the continents, oceans and their biotas throughout geological time.

Fossils also store information on the **finite strain** and **thermal maturation** of rocks located in the planet's mountain belts, allowing the tectonic history of these ranges to be reconstructed; thermal maturation information is important in identifying the levels of thermal maturity of rocks and the gas and oil windows in hydrocarbon exploration. In some cases fossil shells also contain isotopes and other geochemical information that can identify changes in global climate (see p. 111).

FRAMEWORKS

> *Six distinct aspects of Tuscany we therefore recognize, two when it was fluid, two when level and dry, two when it was broken; and as I prove this fact concerning Tuscany by inference from many places examined by me, so do I affirm it with reference to the entire earth, from the descriptions of different places contributed by different writers.*
>
> Nicolaus Steno (1669) *The Prodromus of Nicolaus Steno's Dissertation Concerning a Solid Body Enclosed by Process of Nature Within a Solid*

Before the distributions of fossils in time and space can be described, analyzed and interpreted, fossil animals and plants must be described in their stratigraphic context. A rock stratigraphy is the essential framework that geologists and particularly paleontologists use to accurately locate fossil collections in both temporal and spatial frameworks. It seems, not surprisingly, that like a fine bottle of Italian wine, this can be traced back to the sunny, pastel landscapes of Tuscany and the Renaissance.

Leonardo's legacy

The origin of modern stratigraphy can be traced back to Leonardo da Vinci and his drawings. Pioneer work by the Danish polymath Nicolaus Steno (Niels Stensen) in northern Italy, during the late 17th century (see p. 11), established the simple fact that older rocks are overlain by younger rocks if the sequence has not been inverted (Fig. 2.1a). His law of **superposition of strata** is fundamental to all stratigraphic studies. In addition, Steno established in experiments that sediments are deposited horizontally and rock units can be traced laterally, often for considerable distances; remarkably simple concepts to us now, but earth shattering at the time. But what has this got to do with da Vinci?

Leonardo da Vinci (1452–1519) is famous for many things, and his contributions to science are refreshingly modern when we look back at them. In his art, da Vinci essentially rediscovered geological perspective, some 200 years before Steno, during the Renaissance (Rosenberg 2001). In his drawing of the hills of Tuscany, da Vinci portrayed a clear sequence of laterally-continuous, horizontal strata displaying the concept of superposition. Moreover, about a century after Steno, Giovanni Arduino recognized, again using superposition, three basically different rocks suites in the Italian part of the Alpine belt. A crystalline basement of older rocks, deformed during the Late Paleozoic Variscan orogeny, was overlain unconformably by mainly Mesozoic limestones deformed later during the Alpine orogeny; these in turn were overlain unconformably by poorly consolidated clastic rocks, mainly conglomerates. These three units constituted his primary, secondary and tertiary systems; the last term has been retained and formalized for the period of geological time

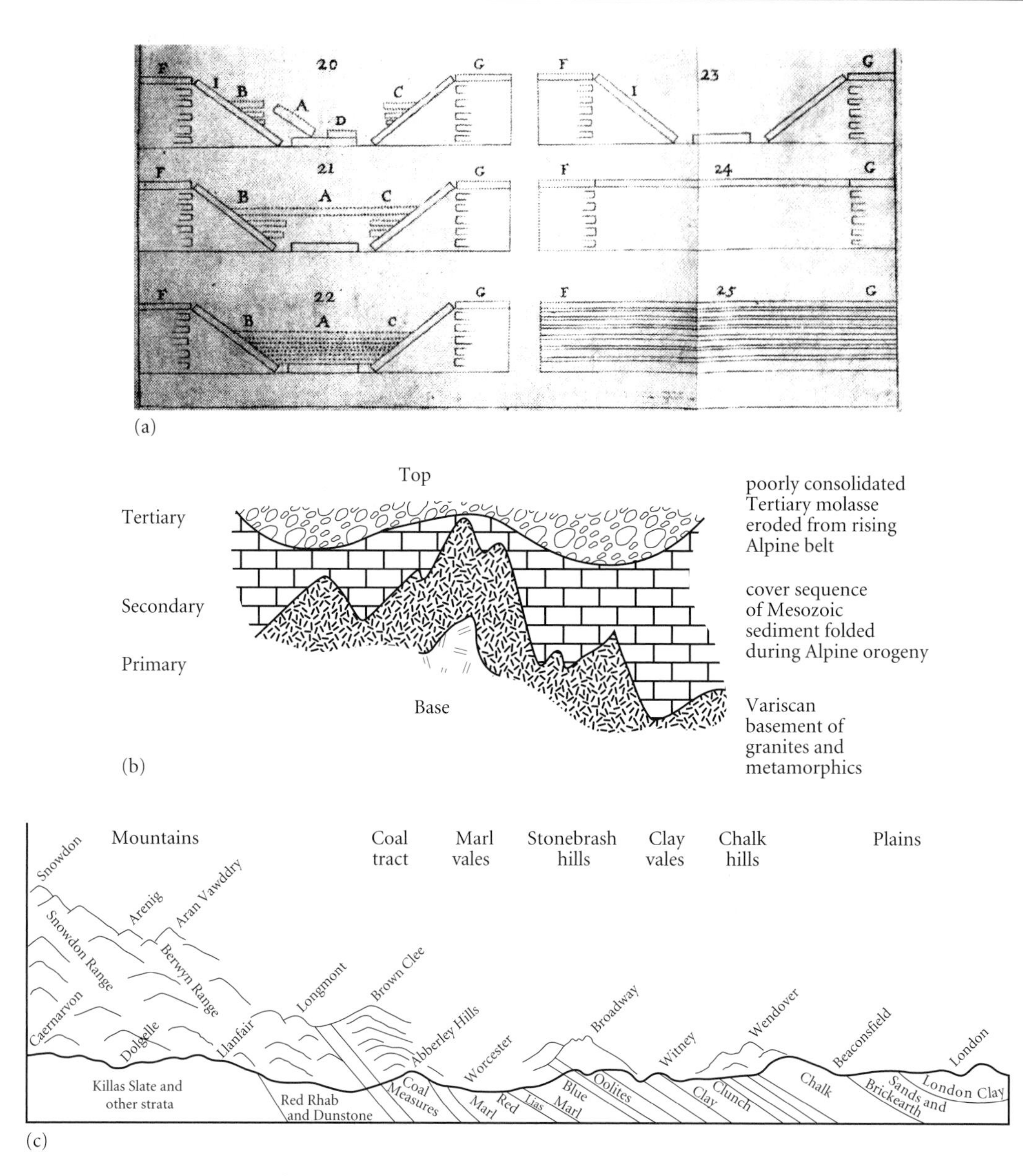

Figure 2.1 (a) Steno's series of diagrams illustrating the deposition of strata, their erosion and subsequent collapse (25, 24 and 23) followed by deposition of further successions (22, 21 and 20). These diagrams demonstrate not only superposition but also the concept of unconformity. (b) Giovanni Arduino's primary, secondary and tertiary systems, first described from the Apennines of northern Italy in 1760. These divisions were built on the basis of Steno's Law of Superposition of Strata. (c) Idealized sketch of William Smith's geological traverse from London to Wales; this traverse formed the template for the first geological map of England and Wales. Data assembled during this horse-back survey were instrumental in the formulation of the Law of Correlation by Fossils. (a, from Steno 1669; c, based on Sheppard, T. 1917. *Proc. Yorks. Geol. Soc.* **19**.)

succeeding the Cretaceous (Fig. 2.1b). These three divisions were used widely to describe rock successions elsewhere in Europe showing the same patterns, but these three systems were not necessarily the time correlatives of the type succession in the Apennines.

There is now a range of different types of stratigraphies based on, for example, lithology (lithostratigraphy), fossils (biostratigraphy), tectonic units, such as thrust sheets (tectonostratigraphy), magnetic polarity (magnetostratigraphy), chemical compositions (chemostratigraphy), discontinuities (allostratigraphy), seismic data (seismic stratigraphy) and depositional trends (cyclo- and sequence stratigraphies). The first two have most application in paleontological studies, although sequence and cyclostratigraphic frameworks are now providing greater insights into the climatic and environmental settings of fossil assemblages. Here, however, we concentrate on lithostratigraphy (rock framework), biostratigraphy (ranges of fossils) and chronostratigraphy (time dimension).

ON THE GROUND: LITHOSTRATIGRAPHY

All aspects of stratigraphy start from the rocks themselves. Their order and succession, or lithostratigraphy, are the building blocks for any study of biological and geological change through time. Basic stratigraphic data are first assembled and mapped through the definition of a lithostratigraphic scheme at a local and regional level. Lithostratigraphic units are recognized on the basis of rock type. The **formation**, a rock unit that can be mapped and recognized across country, irrespective of thickness, is the basic lithostratigraphic category. A formation may comprise one or several related lithologies, different from units above and below, and usually given a local geographic term. A **member** is a more local lithologic development, usually part of a formation, whereas a succession of contiguous formations, with some common characteristics is often defined as a **group**; groups themselves may comprise a **supergroup**. All stratigraphic units must be defined at a reference or **type section** in a specified area. Unfortunately, the entire thickness of many lithostratigraphic units is rarely exposed; instead of defining the whole formation, the bases of units are defined routinely in basal **stratotype** sections at a type locality and the entire succession is then pieced together later. These sections, like yardsticks or the holotypes of fossils (see p. 118), act as the definitive section for the respective stratigraphic units. These are defined within a rock succession at a specific horizon, where there is a lithologic boundary between the two units; the precise boundary is marked on a stratigraphic log. Since the base of the succeeding unit defines the top of the underlying unit, only basal stratotypes need ever be defined.

A stratigraphy, illustrated on a map and in measured sections, is required to monitor biological and geological changes through time and thus underpins the whole basis of Earth history. It is a simple but effective procedure. Successions of rock are often divided by gaps or **unconformities**. These surfaces separate an older part of the succession that may have been folded and uplifted before the younger part was deposited. Commonly there is a marked difference between the attitudes of the older and younger parts of the succession; but sometimes both parts appear conformable and only after investigation of their fossil content, is it clear that the surface represents a large gap in time.

Early geologists thought the Earth was very young, but the Scottish scientist James Hutton (1726–1797) noted the great cyclic process of mountain uplift, followed by erosion, sediment transport by rivers, deposition in the sea, and then uplift again, and argued that such processes had been going on all through Earth's history. He wrote in his *Theory of the Earth* (1795) that his understanding of geological time gave "no vestige of a beginning, – no prospect of an end". An example of Hutton's evidence is the spectacular unconformity at Siccar Point, Berwickshire, southern Scotland, where near-horizontal Old Red Sandstone (Devonian) strata overlie steeply-dipping Silurian greywackes. Beneath the unconformity, Hutton recognized the "ruins of an earlier world", establishing the immensity of geological time. This paved the way for our present concept of the Earth as a dynamic and changing system, a forerunner to the current Gaia hypothesis, which describes the Earth as a living organism in equilibrium with its biosphere. Although the Earth is not actually a living organism, this concept now forms the basis for Earth system science.

USE OF FOSSILS: DISCOVERY OF BIOSTRATIGRAPHY

Our understanding of the role of fossils in stratigraphy can be traced back to the work of William Smith in Britain and Georges Cuvier and Alexandre Brongniart in France. William Smith (1769–1839), in the course of his work as a canal engineer in England, realized that different rocks units were characterized by distinctive groups or **assemblages** of fossils. In a traverse from Wales to London, Smith encountered successively younger groups of rocks, and he documented the change from the trilobite-dominated assemblages of the Lower Paleozoic of Wales through Upper Paleozoic sequences with corals and thick Mesozoic successions with ammonites; finally he reached the molluskan faunas of the Tertiary strata of the London Basin (Fig. 2.1c). In France, a little later, the noted anatomist Georges Cuvier (see p. 12) together with Alexandre Brongniart (1770–1849), a leading mollusk expert of the time, ordered and correlated Tertiary strata in the Paris Basin using series of mainly terrestrial vertebrate faunas, occurring in sequences separated by supposed biological catastrophes.

These early studies set the scene for biostratigraphic correlation. In very broad terms, the marine Paleozoic is dominated by brachiopods, trilobites and graptolites, whereas the Mesozoic assemblages have ammonites, belemnites, marine reptiles and dinosaurs as important components, and the Cenozoic is dominated by mammals and molluskan groups, such as the bivalves and the gastropods. This concept was later expanded by John Phillips (1800–1874), who formally defined the three great eras, Paleozoic ("ancient life"), Mesozoic ("middle life") and Cenozoic ("recent life"), based on their contrasting fossils, each apparently separated by an extinction event. Many more precise biotic changes can, however, be tracked at the species and subspecies levels through morphological changes along phylogenetic lineages. Very accurate correlation is now possible using a wide variety of fossil organisms (see below).

Biostratigraphy: the means of correlation

Biostratigraphy is the establishment of fossil-based successions and their use in stratigraphic **correlation**. Measurements of the stratigraphic ranges of fossils, or assemblages of fossils, form the basis for the definition of **biozones**, the main operational units of a biostratigraphy. But the use of such zone fossils is not without problems. Critics have argued that there can be difficulties with the identifications of some organisms flagged as zone fossils; and, moreover, it may be impossible to determine the entire global range of a fossil or a fossil assemblage, so long as fossils can be reworked into younger strata by erosion and redeposition, but this is relatively rare. Nonetheless, to date, the use of fossils in biostratigraphy is still the best and usually the most accurate routine means of correlating and establishing the relative ages of strata. In order to correlate strata, fossils are normally organized into assemblage or range zones.

There are several types of **range zone** (Fig. 2.2); some are used more often than others. The concept of the range zone is based on the work of Albert Oppel (1831–1865). Oppel characterized successive lithologic units by unique associations of species; his zones were based on the consistent and exclusive occurrence of mainly ammonite species through Jurassic sections across Europe, where he recognized 33 zones in comparison with the 60 or so known today. His zonal scheme could be meshed with Alcide d'Orbigny's (1802–1857) stage classification of the system, based on local sections with geographic terms, further developed by Friedrich Quenstedt (1809–1889). Although William Smith had recognized the significance of fossils almost 50 years previously, Oppel established a modern and rigorous methodology that now underpins much of modern biostratigraphy.

The known range of a zone fossil (Box 2.1) is the time between its first and last appearances in a specific rock section, or **first appearance datum** (FAD) and **last appearance datum** (LAD). Clearly, it is unlikely that the entire global vertical range of the zone fossil is represented in any one section; nevertheless it is, in most cases, a workable approximation. This range, measured against the lithostratigraphy, is termed a **biozone**. It is the basic biostratigraphic unit, analogous to the lithostratigraphic formation. It too can be defined with reference to precise occurrences in the rock, and is defined again on the basis of a stratotype or basal stratotype section in a type area. Once biozones have been established, quantitative techniques may be used to understand the relationships between rock thickness and time, and to make links from locality to locality (Box 2.2).

This is all very well, of course, but the fossil record is rarely complete; only a small percentage of potential fossils are ever preserved. Stratigraphic ranges can also be influenced by the **Signor–Lipps effect** (Signor & Lipps 1982),

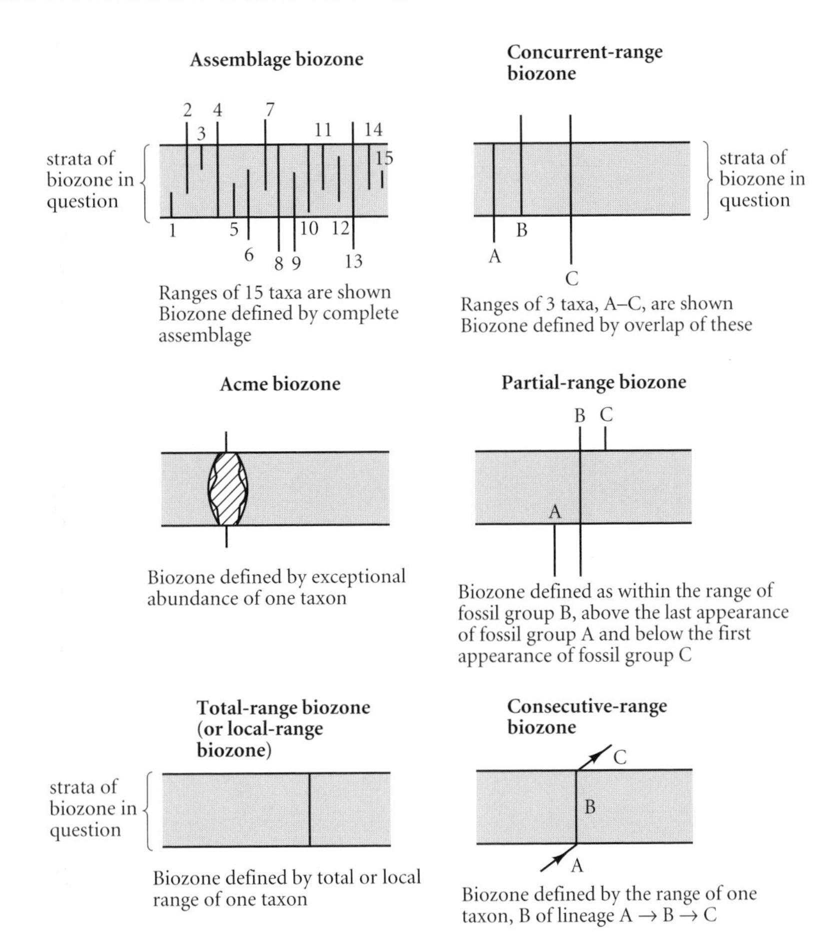

Figure 2.2 The main types of biozone, the operational units of a biostratigraphy. (Based on Holland 1986.)

the observation that stratigraphic ranges are always shorter than the true range of a species, i.e. you never find the last fossil of a species. So, incomplete sampling means that the disappearances of taxa may be "smeared" back in time from the actual point of disappearance. The Signor–Lipps effect is particularly relevant to mass extinctions, when this backsmearing can make relatively sudden extinction events appear gradual. This can be corrected to some extent by the use of statistical techniques to establish confidence intervals that are modeled on known sampling quality (see p. 165).

Many different animal and plant groups are used in biostratigraphic correlation (Fig. 2.5). Graptolites and ammonites are the best known and most reliable zone macrofossils with their respective biozones as short as 1 myr and 25 kyr, respectively. The most unusual zone fossils are perhaps those of pigs, which have been used to subdivide time zones in the Quaternary rocks of East Africa where hominid remains occur. Microfossil groups such as conodonts, dinoflagellates, foraminiferans and plant spores are now widely used (see pp. 209–32, 493–7), particularly in petroleum exploration. Microfossils approach the ideal zone fossils since they are usually common in small samples, such as drill cores and chippings, of many sedimentary lithologies and many groups are widespread and rapidly evolving. The only drawback is that some techniques used to extract them from rocks and sediments are specialized, involving acid digestion and thin sections.

Dividing up geological time: chronostratigraphy

Geological time was divided up by the efforts of British, French and German geologists between 1790 and 1840 (Table 2.1). The divisions were made first for practical reasons – one of the first systems to be named was the

Box 2.1 Zone fossils

The recognition and use of zone fossils is fundamental to biostratigraphic correlation. Fossil groups that were (i) rapidly evolving, (ii) widespread across different facies and biogeographic provinces, (iii) relatively common, and (iv) easy to identify make the ideal zone fossils. In the Early Paleozoic macrofauna, graptolites (see p. 412) are the closest to being ideal zone fossils, whereas during the Mesozoic, the ammonites (see p. 334) are most useful. The use of efficient zone fossils ensures that relatively short intervals of geological time can be correlated, often with a precision of a few hundred thousand years, over long distances through different facies belts around the world. In practice there are no ideal zone fossils. Most long-range correlations involve use of intermediate faunas with mixed facies.

For example, in Ordovician and Silurian rocks, deep-water facies are correlated by means of the rapidly-evolving and widespread graptolites; these fossils are rare in shallow-water shelf deposits where trilobites and brachiopods are much more common. Nevertheless, facies with both graptolite and shelly faunas may interdigitate in deep-shelf and slope sequences, allowing correlation through these mixed facies from deep to shallow water. Parallels can be drawn with the neritic ammonites and benthic bivalves and gastropods of the Mesozoic seas. Microfossils are widely used for correlation in hydrocarbon exploration; the amount of rock available in drill cores or cuttings is usually limited and a range of fossil microorganisms including foraminiferans and radiolarians together with dinoflagellates, spores and pollen form the basis for the correlation schemes used by petroleum companies.

On a simple plot of space against time (Fig. 2.3), an ideal zone fossil, such as an ammonite or graptolite, will represent a thin horizontal band reflecting a brief time duration but a widespread spatial distribution. In reality very few fossils approach the properties of an ideal zone fossil. The distribution of most is controlled to some degree by **facies**, the rocks that represent a particular life environment. A more typical **facies fossil**, such as a typical bivalve or gastropod, is not tightly constrained by time but appears to occur in a particular facies belt (Fig. 2.3).

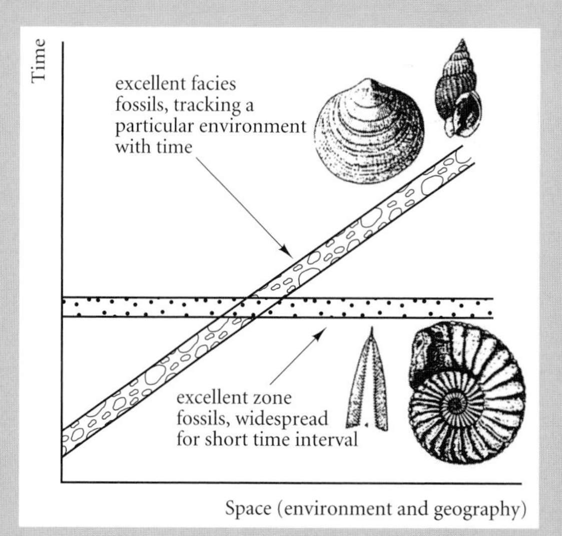

Figure 2.3 Behavior of ideal zone and facies fossils through a hypothetical global stratigraphic section.

Box 2.2 Quantitative biostratigraphy

Quantitative stratigraphy can be traced back to work by Charles Lyell (1797–1875), who plotted what we would now call decay curves (analogous to the decay curves for radiogeneic isotopes) for the molluskan and mammalian faunas of the Tertiary basins of northwest Europe. He wanted to look at "evolution in reverse", tracking back in time from the present day to see how proportions of living taxa changed the farther back you went into the rock record. He found that the proportions of modern to extinct forms declined the farther back in time he went, and he used this to define divisions in the Tertiary system. For example his Older Pliocene included only 10% of mammals and 50% of mollusks living today, whereas in the Newer Pliocene the respective figures are 90% and 80%. These ratios were used as a method of correlating Tertiary strata quantitatively. In recent years, driven by hydrocarbon and mineral exploration, a range of quantitative, computer-based techniques has become available (Hammer & Harper 2005). Three – graphic correlation, seriation and ranking and scaling – are outlined here.

A rigorous, numerically-based mode of correlation was developed by Alan Shaw when he was working in the petroleum industry during the 1950s. Because hydrocarbon reservoir and source rocks occur within stratigraphic successions, it is essential that the rocks in oil and gas fields are accurately correlated; geologists can then locate key horizons on the basis of biostratigraphy (Fig. 2.4a). **Graphic correlation** by Shaw's method requires fossil range data from two or more measured sections. Data of the first and last occurrences of fossil species are plotted against a measured stratigraphic section; this is repeated for a second section. Usually only the more common taxa are plotted. A bivariate scattergram is then drawn with section 1 along the x-axis and section 2 along the y-axis. The first and last occurrences are then plotted as x–y coordinates – for example the x coordinate represents the first appearance of species a along section 1 and the y coordinate its first appearance in section 2. A regression line is fitted to all the first (FAD) and last (LAD) appearance coordinates; this line of stratigraphic correlation can be used for interpolation, permitting the accurate correspondence of all levels in the two sections. A composite standard section can be constructed and refined by correlating it against additional actual sections.

Biostratigraphers also use techniques established by archeologists in the late 1800s. **Seriation** is an ordering technique designed to analyze gradients. Usually the gradients are temporal but biogeographic and environmental data have been investigated by seriation. Biostratigraphers tend to enter the ranges of organisms on range charts as sequential FADs. In simple terms seriation shuffles the original data matrix until the stratigraphically higher taxa are on the left hand side of the matrix and the stratigraphically lower taxa are on the right; any stratigraphic gradients in the data are then clearly visible (Fig. 2.4b) and can be interpreted.

Ranking and scaling (RASC) is a method of arranging a series of biostratigraphic events in order, and of estimating the stratigraphic distance between such events. The technique requires only first and last appearances measured in meters in a stratigraphic section, perhaps an exposure or oil well. Events are first ranked or ordered based on the majority of relative occurrences and then the distances between such events are calculated (Fig. 2.4c).

A dataset of Early Ordovician trilobite ranges is available at: http://www.blackwellpublishing.com/paleobiology. These data may be analyzed and manipulated using ranking and scaling, seriation and unitary associations; confidence intervals may also be calculated (see also Hammer & Harper 2005).

Continued

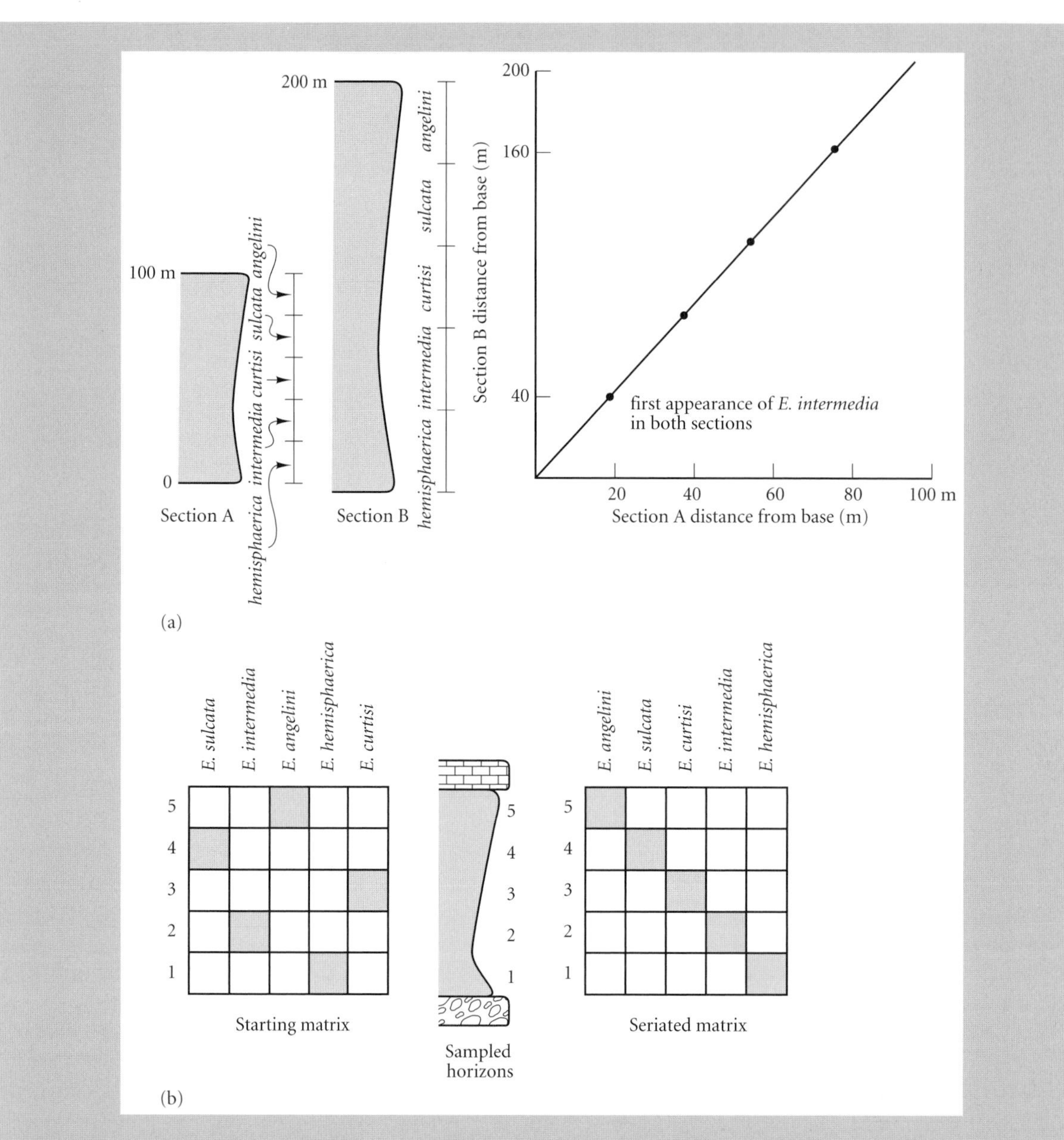

Figure 2.4 (a) Hypothetical and minimalist graphic correlation based on the stratigraphic distribution of the five apparent chronospecies of the Silurian brachiopod *Eocoelia*, in ascending order: *E. hemisphaerica*, *E. intermedia*, *E. curtisi*, *E. sulcata* and *E. angelini*; the first four range through the middle and upper Llandovery whereas the last is characteristic of the lower Wenlock. The ranges of these species are given from two artificial sections with the first appearances of each species plotted on both sections as x and y coordinates. The straight line fitted to the points allows a precise correlation between each part of the two sections. In this simple example all the points fit on a straight line; in practice a regression must be fitted to the scatter of data points. (b) Seriation of biostratigraphic data. The five *Eocoelia* species were collected from five horizons in a stratigraphic section; the data were collected and plotted randomly as a range chart. Seriation seeks to establish any structure, usually gradients, within the matrix by maximizing entries in the leading diagonal. The seriated matrix reveals the stratigraphic succession of *Eocoelia* species that is widely used for the correlation of Lower Silurian strata. Most seriations are based on much larger and more complex data matrices where any non-random structure, if present, is initially far from obvious.

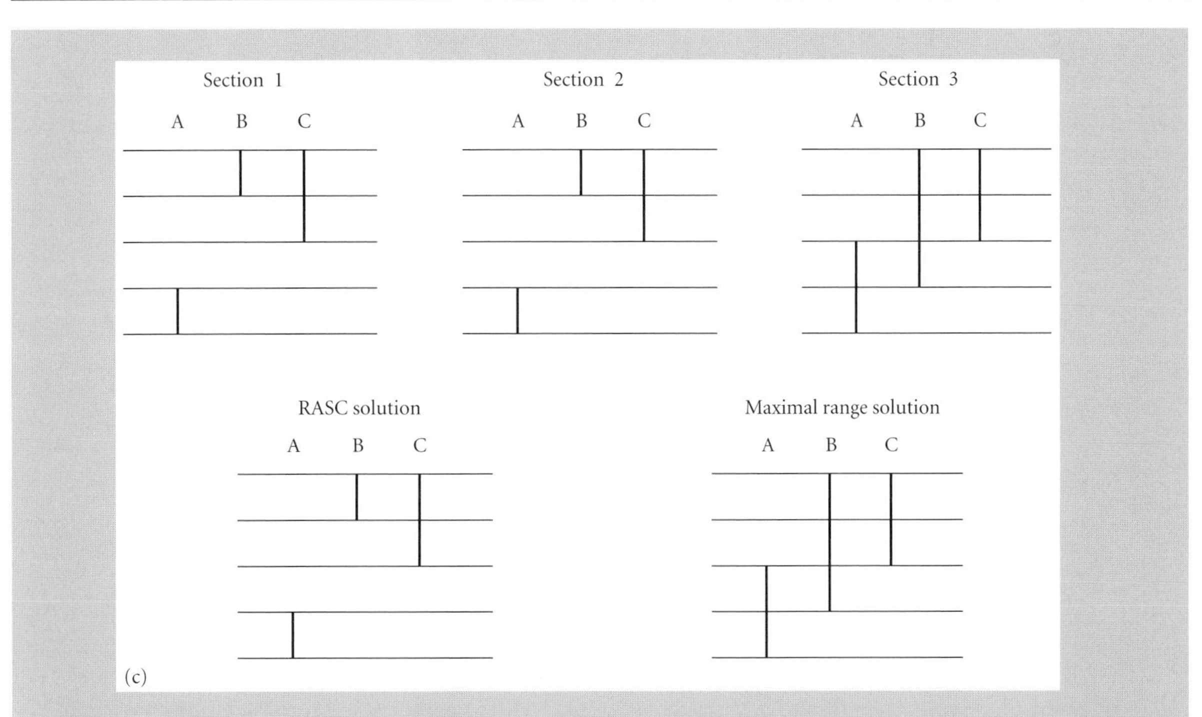

Figure 2.4 (***Continued***) (c) The RASC method predicts the solution most likely to occur in the next section based on previous data. Three sections (1–3) are presented and, based on a majority vote, the RASC solution is constructed; since the first two sections are similar they win over the third slightly different section. This is different to the maximum range solution that may be constructed by other methods. (c, based on Hammer & Harper 2005.)

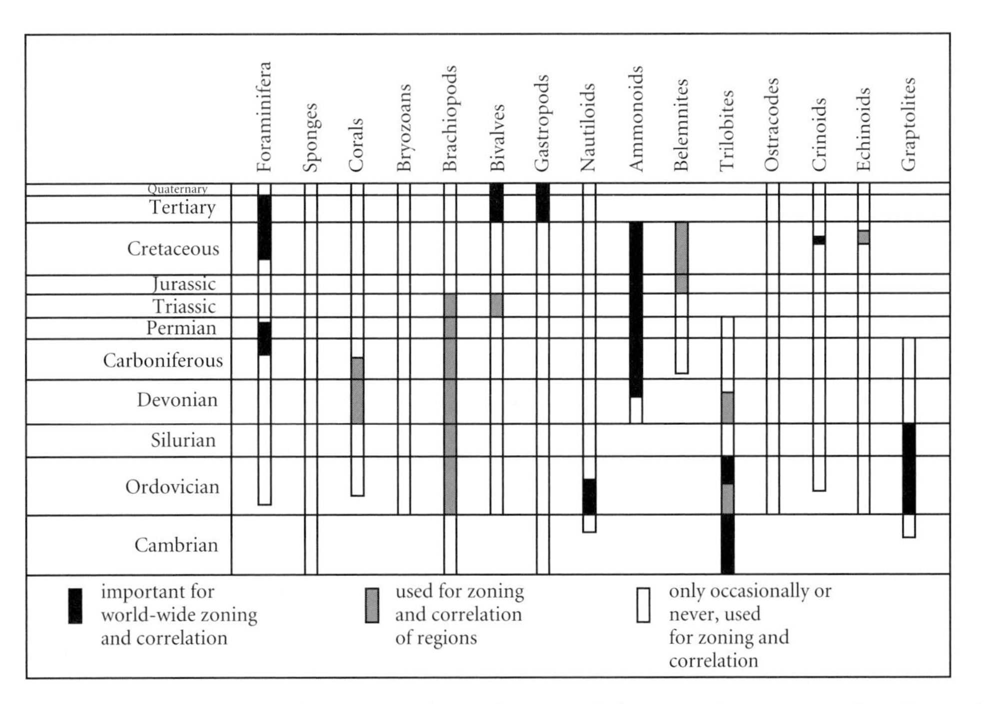

Figure 2.5 Approximate stratigraphic ranges through time of the main biostratigraphically useful invertebrate fossils groups. (Replotted from various sources.)

Table 2.1 Founding of the geological systems: systems, founders and the original type areas. In addition the Mississippian and Pensylvanian that equate with the Lower and Upper Carboniferous, respectively, were founded by Alexander Winchell (in 1870) and Henry Shaler Winchell (in 1891) based on rocks exposed in the Mississippi Valley and state of Pensylvania. The Paleogene and Neogene broadly correspond to the Lower and Upper Tertiary.

System	Founder, date	Original type area
Cambrian	Sedgwick, 1835	North Wales
Ordovician	Lapworth, 1879	Central Wales
Silurian	Murchison, 1835	South Wales and Welsh borders
Devonian	Murchison and Sedgwick, 1840	South England
Carboniferous	Coneybeare and Phillips, 1822	North England
Permian	Murchison, 1841	Western Russia
Triassic	Von Alberti, 1834	Germany
Jurassic	Von Humboldt, 1795	Switzerland
Cretaceous	D'Halloy, 1822	France
Tertiary	Arduino, 1760	Italy
Quaternary	Desnoyers, 1829	France

Carboniferous ("coal-bearing"), a unit of rock that early industrialists were keen to identify! In a mad rush in the 1830s, Roderick Murchison (1792–1871) and Adam Sedgwick (1785–1873) collaborated, and tussled, over the Lower Paleozoic. Sedgwick named the Cambrian and Murchison named the Silurian, based on sections in Wales. Each claimed the middle ground for his system, so what Murchison called the "Lower Silurian", Sedgwick called "Upper Cambrian". This territorial claim was resolved later by Charles Lapworth (1842–1920) who agreed with neither of them, and named the contentious rock successions the Ordovician in 1879. Ironically the Ordovician is one of the longest and most lithologically diverse of the geological systems but it was only formally accepted by the international community in 1960.

A problem with many of the original definitions of the geological systems was that they were separated from each other by unconformities. For the early workers, unconformities provided a convenient break between systems and, more importantly, it satisfied their view that the major divisions of Earth's history should be divided by global, catastrophic events. Unfortunately, many of these unconformities turned out to be only regional breaks that occurred in Europe, but not elsewhere. The bases of most systems then were represented by stratigraphic gaps, and gaps provide a poor basis for the global correlation of systemic boundaries.

All the system boundaries have been or are currently being reinvestigated by working groups of the International Union of the Geological Sciences (IUGS). The potential of each base for international correlation must be maximized. Thus the traditional bases of these systems must be placed within intervals of continuous sedimentation, with diverse and abundant faunas and floras in geographically and politically accessible areas that can be conserved and protected; ideally the sections should have escaped metamorphism and tectonism (Fig. 2.6). You can read more about the work of the IUGS at http://www.blackwellpublishing.com/paleobiology.

Chronostratigraphy or global standard stratigraphy is one of the most fundamental of all stratigraphic concepts. Everyday intervals of time, such as seconds, minutes and hours, are based on a universal time signal from an atomic clock. Units of geological time, such as the epoch and period, are much longer and of uneven lengths. The only standards available for the definition of these intervals are the rock successions themselves. Thus the rocks of the type section in the type area for the Silurian System act as an international standard for the Silurian Period, the time during which that system was deposited. The base of a chronostratigraphic interval is defined in a unique **stratotype section**, in a type area using the concept of a "**golden spike**" or marker point (Holland 1986). All the usual criteria for a workable stratotype

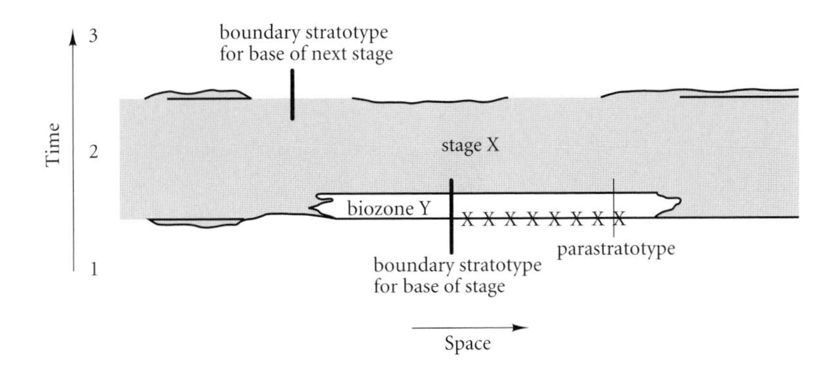

Figure 2.6 Key concepts in the definition of stratotypes and parastratotypes applicable to all stratigraphic units. The base of stage X is defined at an appropriate and suitable type section, coincident with the base of biozone Y, which can be used to correlate the base of the stage. The type section is usually conserved and further collecting across the boundary interval is restricted to the parastratotype section. The base of the stage is indicated as XXX. (Based on Temple, J.T. 1988. *J Geol. Soc. Lond.* **145**.)

<table>
<tr><th colspan="3">Chronostratigraphy</th><th>Lithostratigraphy</th><th>Biostratigraphy</th></tr>
<tr><th>Series or epoch</th><th>Stage or age</th><th>Chronozone</th><th>Formations and member</th><th>Graptolite biozones</th></tr>
<tr><td rowspan="9">Wenlock</td><td rowspan="3">Homerian</td><td rowspan="2">Gleedon</td><td>Much Wenlock Limestone Formation</td><td>ludensis</td></tr>
<tr><td>Farley Member of Coalbrookdale Formation</td><td>nassa</td></tr>
<tr><td>Whitwell</td><td rowspan="5">Coalbrookdale Formation</td><td>lundgreni</td></tr>
<tr><td colspan="2" rowspan="6">Sheinwoodian</td><td>ellesae</td></tr>
<tr><td>linnarssoni</td></tr>
<tr><td>rigidus</td></tr>
<tr><td>riccartonensis</td></tr>
<tr><td rowspan="2">Buildwas Formation</td><td>murchisoni</td></tr>
<tr><td>centrifugus</td></tr>
</table>

Figure 2.7 Stratigraphic case study: description and definition of the litho-, bio- and chronostratigraphy of the stratotype section of the Wenlock Series, along Wenlock Edge in Shropshire, UK. This is the internationally accepted standard for the Wenlock Epoch, the third time division of the Silurian Period.

section must, of course, be satisfied. The golden spike, which represents a point in the rock section and an instant in geological time, is then driven into the section, at least in theory. In practice the spike is usually adjusted to coincide with the first appearance (FAD) of a distinctive, recognizable fossil within a well-documented lineage. The ranges of all fossils occurring across the boundary are documented in detail as aids to correlating within the section and with sections elsewhere. Establishing stratotypes and golden spikes requires international agreement, and that can sometimes be hard to achieve (Box 2.3)! This horizon will then be the **global standard section and point** (GSSP) for this stratotype.

The Wenlock Epoch (time) was one of the first intervals of geological time to be defined with reference to a stratotype section for the Wenlock Series (rock) (Fig. 2.7). A lithostratigraphy was first established in the historic type area by definition of formations and members. On the basis of detailed collecting through the stratigraphy a succession of biozones was then defined, based on the ranges of characteristic graptolite faunas. Finally, a succession of stages was established together with two chronozones. This remains the international yardstick for Wenlock time. When discussing geological time, we generally use the adjectives early, mid and late, but when dealing with rock the use of lower, middle and upper is more appropriate.

Sequence stratigraphy: using transgressions and regressions

North American oil geologists developed a whole new system in the 1960s called **sequence stratigraphy**, an approach that emphasizes the importance of unconformities. In the early 1960s Larry Sloss recognized that the Phanerozoic rocks of the old North American continent could be split into six main cycles separated by unconformities (Fig. 2.9). These were large-scale cycles describing the major changes in sea level across an entire continent and through over 500 myr of Earth history. More minor sequences could be recognized within these major cycles. The fact that sedimentary rocks can be described as packets of strata, presumably deposited during **transgressive** events (when the sea floods the land), divided by periods of non-deposition during **regressions** (when the sea withdraws from the land), forms the basis for sequence stratigraphy.

Box 2.3 The Ordovician: a system on the move

The Ordovician System was born out of controversy, with Charles Lapworth taking the disputed overlapping strata between Sedgwick's Cambrian System and Murchison's Silurian System (see above). Despite the best efforts of British specialists (e.g. Fortey et al. 1995), they and many other international experts have pointed out that – although the classic British series and stages have wide global usage – they were based largely on endemic shelly faunas with only local and regional distributions, some units are bounded by **disconformities** (minor gaps in deposition, where the rocks below and above are oriented similarly, in contrast to the larger chunks of time represented by unconformities), and some have significant overlaps with adjacent series. Moreover many of the key sections are located in poorly exposed sections. In order to assemble a consolidated chronostratigraphy that would work internationally, definitions in new sections were necessary.

First, it was decided in the 1980s by the International Subcommission on Ordovician Stratigraphy, a group of highly-qualified experts drawn from all over the world, that basal stratotypes for chronostratigraphic units should be correlated by means of conodonts and graptolites, the most effective of all Paleozoic zone fossils. Second, there should only be three series, defined as lower, middle and upper; and third, new sections must be sought to define a new set of global stages: the first was ratified in 1987, and the last in 2007. This has not been without rancour. Colleagues from around the world have clashed noisily at meetings defending their "own" sections, and sometimes national pride and access to further research funding have influenced voting. Nevertheless, a consensus is emerging and all the new stages are defined and in place, based on diverse sections such as a road section and river bank in South China (Hirnantian) and the coast of western Newfoundland (Tremadocian). Some older names such as Hirnantian and Tremadocian have been retained with slightly different definitions, whilst some are new, such as Floian and Sandbian, both based on stratotype localities in Sweden. This new structure is already providing a much more accurate time framework to describe, analyze and model Ordovician Earth systems (Fig. 2.8).

More information of the work of the Subcommission and on the Ordovician System and its biotas is available at http://www.blackwellpublishing.com/paleobiology and on the home page of the related International Geological Correlation Program project 503 "Ordovician palaeogeography and palaeoclimate" linked at http://www.blackwellpublishing.com/paleobiology/.

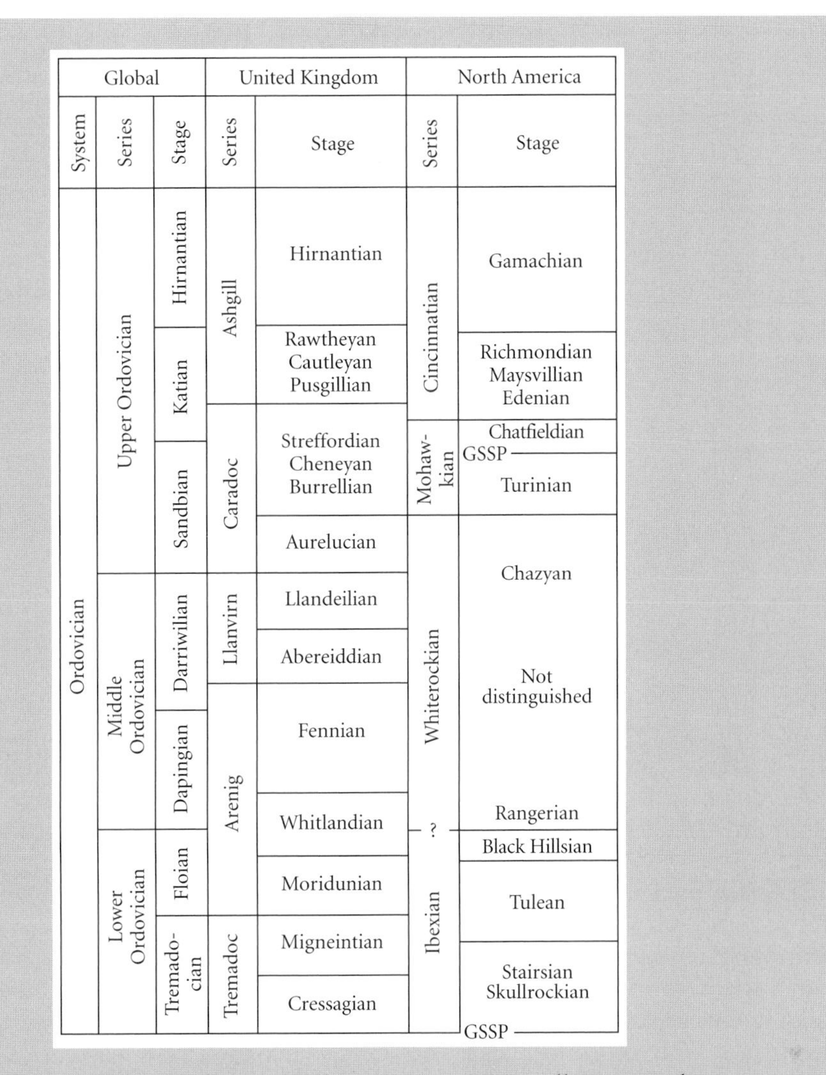

Figure 2.8 Current status of the development of a new, internationally accepted chronostratigraphy for the Ordovician System. New global series and stages are correlated with the comparable chronostratigraphic divisions used in North American and the United Kingdom and Ireland. GSSP, global standard section and point.

The dividing lines between transgressive and regressive system tracts are marked by various types and degrees of unconformities that may be recognized on seismic profiles. Whereas most major sequence boundaries are probably due to global **eustatic** changes in sea level associated with climatic change or fluctuations in seafloor spreading processes, sequences can also be generated by more local tectonic controls. Research teams in the Exxon Corporation expanded the concept of sequence stratigraphy to build global sea-level curves for the entire Phanerozoic during the 1980s and 1990s. The description of successions defined within unconformity-bounded sequences has proved valuable in hydrocarbon exploration, where sequence boundaries can be recognized at depth using seismic geophysics.

Sequence stratigraphers have developed their own specialist terminology (Fig. 2.10). A **sequence** is a unit of similar strata bounded by unconformities. Sequences are laid down in three-dimensional assemblages of lithofacies linked by common depositional processes that can be divided into individual **systems tracts**. The architecture of sequences is controlled by changes in sea level, whether eustat-

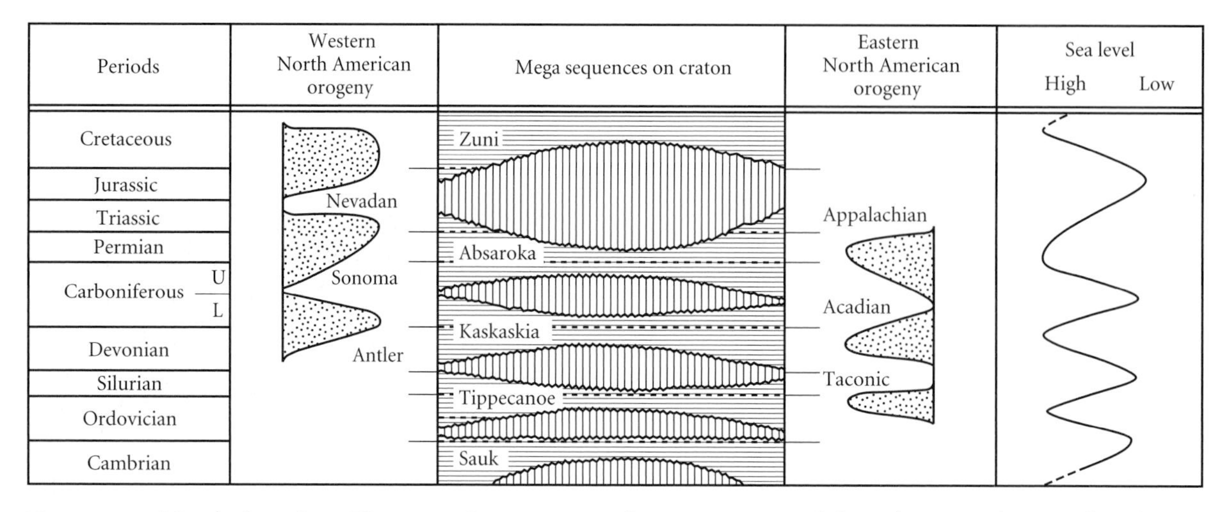

Figure 2.9 North American Phanerozoic sequences: the recognition of these large packages of rock or what are termed "megasequences" formed the basis for the modern discipline of sequence stratigraphy, established by the Exxon Corporation. (Based on various sources.)

ically or tectonically driven, or perhaps a mixture of both, and the room available for sediment, termed **accommodation space**. Normal regressions, driven by increased sediment supply, and forced regressions, driven by base level fall, will both generate falls in sea level, where **base level** is the level above which deposition is temporary and prone to erosion. Transgressions are prompted by base level rise, when this of course exceeds sedimentation rates. There are also six main types of surface: subaerial unconformity, basal surface of forced regression, regressive surface of marine erosion, maximum regressive surface, maximum flooding surface and ravinement surface; the first three are associated with base level fall and the last three with base level rise. Finally there is a variety of systems tracts (Fig. 2.10): lowstand, transgressive, highstand, falling stage and regressive systems tracts. Changes in sea level seem to have had major effects on the planet's marine biotas through time and sequence stratigraphy provides a framework to describe these effects (Box 2.4). For example, shell concentrations may be associated with stratigraphic condensation at maximum flooding surfaces, i.e. the deepest-water facies where deposition is very slow or they may lie near the top of highstand system tracts. Firmgrounds (see p. 522) and their biotas, that include usually burrowers and encrusters, favor major flooding surfaces. Moreover, diversity increases are often associated with marine transgressions as more shallow-water habitats are created when continents are flooded. On the other hand, marked regressive events have been associated with major extinctions through habitat loss. Nevertheless it has been suggested by some authors that such diversity changes are artificial. Transgressive units are generally more widespread across continental areas, so increasing the chance to collect fossils; the converse may be true for regressive events. But sampling biases alone cannot account for apparent changes in biodiversity through time; processes related to sea-level change and the formation and destruction of marine habitats have also provided controls on the origination and extinction of marine taxa (Peters 2005).

Cyclostratigraphy: finding the rhythm

Quaternary geologists have accepted for some time that recent climate change follows repeated cycles of astronomical change. These short-term patterns are called Milankovitch cycles, named after the Serbian mathematician Milutin Milankovitch (1879–1958). Such cycles are controlled by the additive effects of the Earth's movements through space (Fig. 2.12a) and can directly affect global sedimentation patterns. Three main types of movement occur: **eccentricity** (variation in the shape of the Earth's orbit from nearly circular

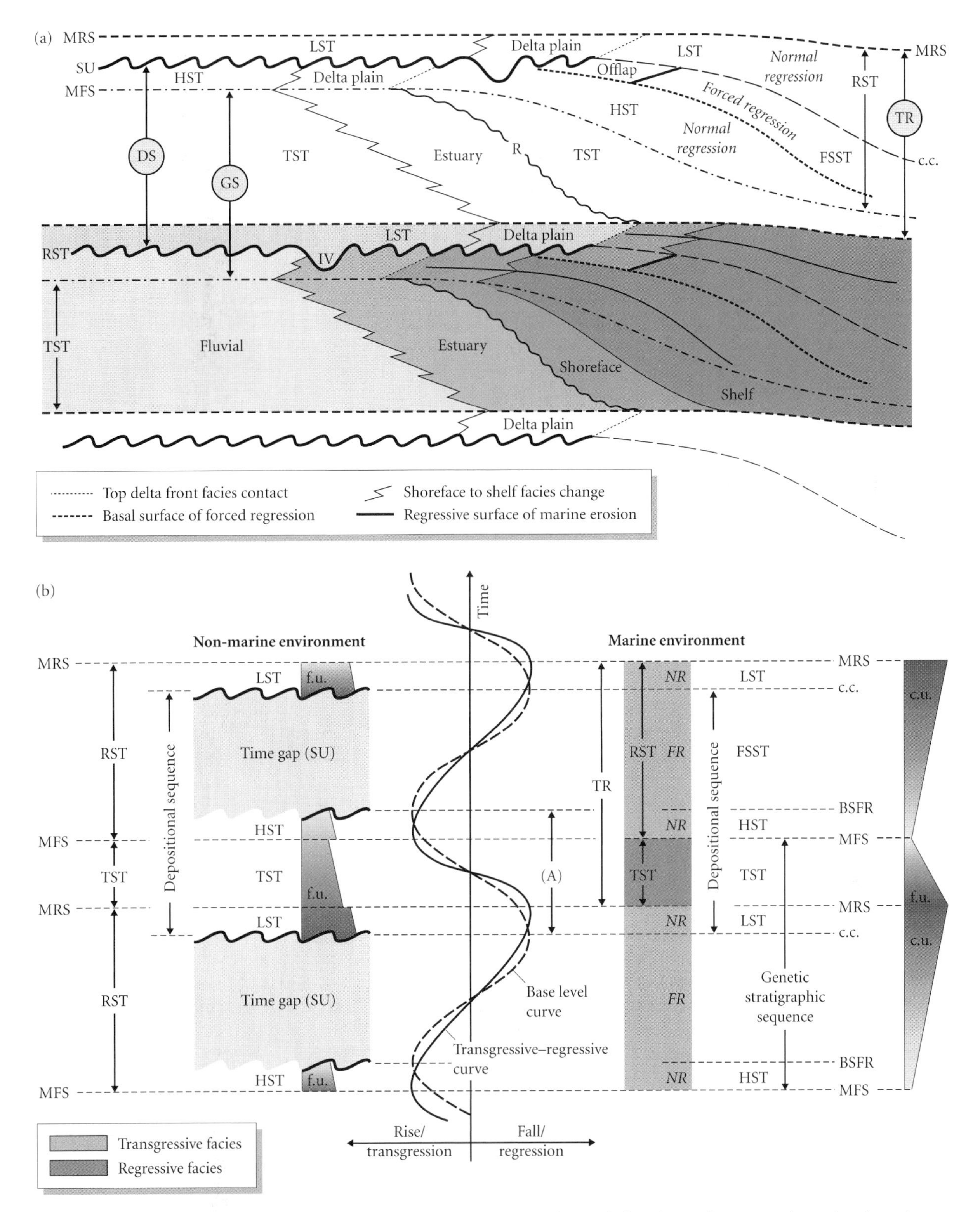

Figure 2.10 Sequences, system tracts and stratigraphic surfaces defined in relation to base level and transgression–regression curves: (a) stratal architecture across a non-marine to marine transect is related to (b) sequence stratigraphies in the non-marine and marine parts of the transect. (A), positive accommodation (base level rise); BSFR, basal surface of forced regression; c.c., correlative conformity; c.u., coarsening upward; DS, depositional sequence; FR, forced regression; FSST, falling stage systems tract; f.u., fining upward; GS, genetic stratigraphic sequence; HST, highstand systems tract; IV, incised valley; LST, lowstand systems tract; MFS, maximum flooding surface; MRS, maximum regressive surface; NR, normal regression; R, ravinement surface; RST, regressive systems tract; SU, subaerial unconformity; TR, transgressive–regressive sequence; TST, transgressive systems tract. (Based on Catuneanu, O. 2002. *J. African Earth Sci.* 35.)

to elliptical; 100 kyr cycle), **obliquity** (wobble of the Earth's axis; 41 kyr cycle) and **precession** (change in direction of the Earth's axis relative to the sun; 23 kyr cycle). Throughout the stratigraphic record there are many successions of rhythmically alternating lithologies, for example limestones and marls (calcareous shales), that may have been controlled by Milankovitch processes. Apart from their obvious value for correlation, such rhythms probably also effected changes in community compositions and structures together with the extinction and origination of taxa.

Some of the most extensive and remarkable decimeter-scale rhythms, probably controlled by precession cycles, have been detected in the Upper Cretaceous chalk facies, where individual couplets can be tracked from southern England to the Caucasus, a distance of some 3000 km. A cyclostratigraphic framework can be related to well-established ammonite, inoceramid bivalve and foraminiferan biozones together with carbon isotope excursions, providing a high-resolution and composite stratigraphy (Fig. 2.12b). The dark marly sediments may have been deposited during precession minima at eccentricity maxima during intervals of cool, wet climates (Gale et al. 1999).

Geological time scale: a common language

If we are to understand global events and rates of global processes, geologists must talk the same language when we correlate and date rocks (Box 2.5). Rapid developments in stratigraphy during the last few years (Gradstein & Ogg 2004) have prompted publication of GTS2004, an updated geological time scale (Gradstein et al. 2004). Over 50 of the 90 Phanerozoic boundaries are now properly defined in stratotype sections (GSSPs) and the new scale uses a spectrum of new stratigraphic methods, such as orbital tuning, together with more advanced radiometric dating techniques and new statistical tools (Fig. 2.13). Although traditional stratigraphic methods form the basis of the geological column and our understanding of the order of key biological events, the prospect of precisely defined radiometric dates makes it possible to determine the rates of many types of biological process. Not all the recommendations have met with universal approval, and they are only recommendations. For example, GTS2004 removed the Tertiary and Quaternary epochs from the chronostratigraphic column without the approval of the IUGS; but these terms are widely used and deeply embedded in the literature and are thus unlikely to disappear

Box 2.4 Sequences and fossils

There are eight brachiopod-dominated biofacies recognized across an onshore–offshore gradient in the Upper Ordovician rocks of Kentucky (Holland & Patzkowsky 2004). These assemblages were not discrete but rather formed part of a depth-related gradient, and the relative abundance of species varied through time. The development of these faunas across this part of the Appalachian Basin can be charted within sequence-stratigraphic frameworks. Figure 2.11 is a plot of the DCA (detrended correspondence analysis) axis 1 against the litho- and sequence stratigraphy of one of the key sections, the Frankfort composite section. The DCA axis is a proxy for taxa that were grouped together in the shallowest-water environments. Thus within the highstand system tracts, values for this axis are lower than those for the transgressive and system tracts and at the maximum flooding surface, where deeper-water taxa dominate. The upsection faunal changes show that the distribution of taxa was controlled by ecological factors dependent on sediment supply and sea-level changes, which in turn built the sequence stratigraphic architecture. Marked fluctuations in the faunas occurred during net regressive and transgressive events, emphasizing the depth-dependence of these assemblages.

The data used in this study are available at http://www.blackwellpublishing.com/paleobiology/.

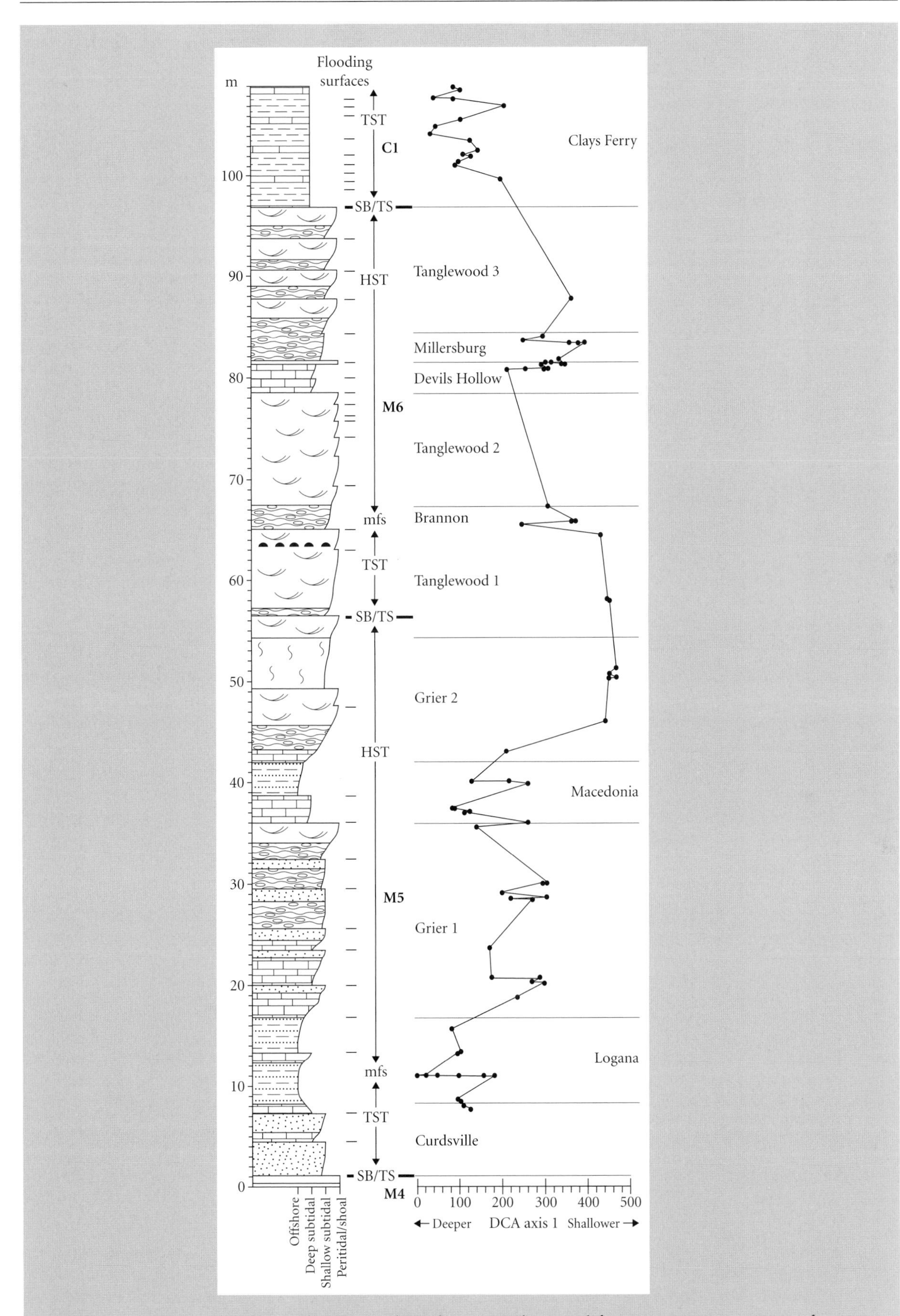

Figure 2.11 DCA axis 1 sample scores plotted against the Frankfort composite section. mfs, maximum flooding surface; HST, highstand systems tract; SB/TS, combined sequence boundary and transgressive surface; TST, transgressive systems tract. (From Holland & Patzkowsky 2004.)

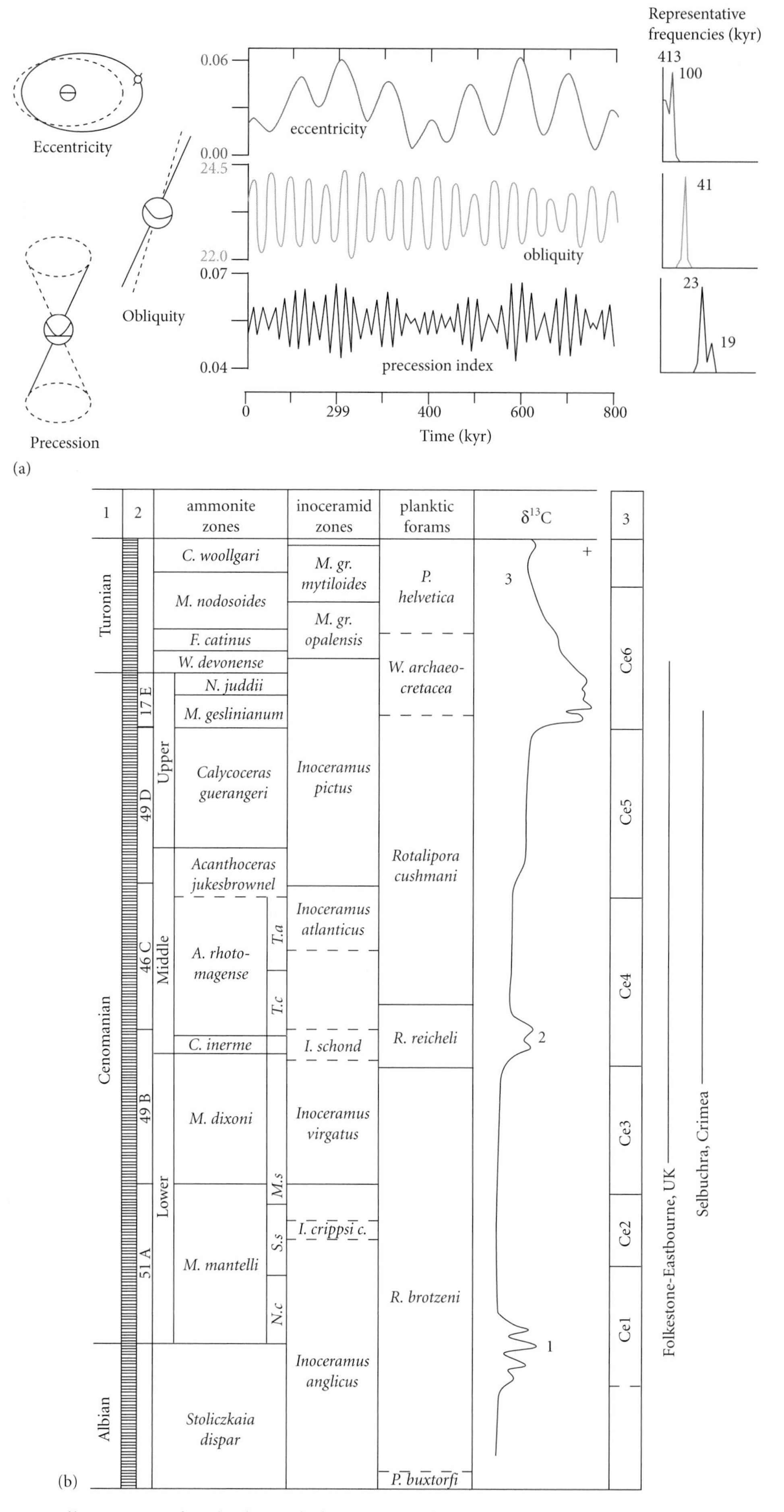

Figure 2.12 (a) Illustration of Milankovitch frequencies showing the relationships between eccentricity, obliquity and precession cycles. (b) Outline stratigraphy of Cenomanian Stage Upper Cretaceous chalk facies. Column 1, stages; column 2, cyclostratigraphy; column 3, sequences. (From Gale et al. 1999.)

from our stratigraphic charts in the near future.

PALEOBIOGEOGRAPHY

No man is an island, entire of itself;
every man is a piece of the continent,
a part of the main.
If a clod be washed away by the sea,
Europe is the less,
as well as if a promontory were,
as well as if a manor of thy friend's or of
thine own were.

John Dunne (1624) *Meditation*

All living organisms have a defined geographic range; the ranges may be large or small, and controlled by a variety of factors including climate and latitude. By the middle of the 1800s both Charles Darwin (1809–1882) and Alfred Russel Wallace (1823–1913) had recognized the reality of **biogeographic provinces** in their respective studies on the Galápagos islands and in the East Indies. The Earth today can be divided into six main provinces (Nearctic, Palearctic, Neotropical, Ethiopian, Oriental and Australasian) based on the perceptive work of Philip Sclater and Alfred Russel Wallace in the later 1800s.

Discrete biogeographic units are, however, defined by faunal and floral barriers. Provinces are characterized by their **endemic** (that is, local or regional) species that have restricted ranges in contrast to the more widespread **cosmopolitan** (worldwide) species. Continental configurations and positions have changed through time, as have faunal and floral provinces. Nevertheless, paleontological data were instrumental in demonstrating the drift of the wandering continents; the fit of the outlines of Africa, South America, India, Antarctica and Australia (Fig. 2.14) was clearly not a coincidence, nor was the matching of rocks and fossils among these continents. In the

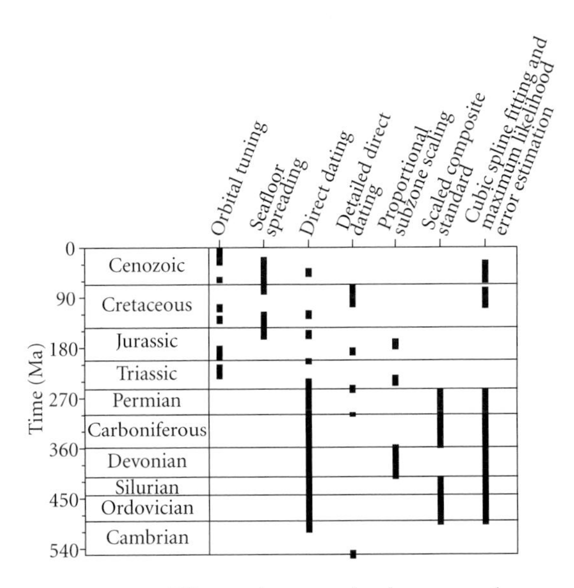

Figure 2.13 The various methods currently available to construct the geologic time scale 2004 (GTS2004).

Box 2.5 The Chronos initiative

There are a number of different geological time scales, developed by different groups of authors for different intervals of geological time, and many different ways to analyze time series data of this type. The Chronos (Greek for time) project is a web-based initiative that seeks to centralize all the various time scales and analytic tools through one web portal. This is a chronometric rather than chronostratigraphic system and thus deals with radiometric age rather than the relative order of events. Thus software is available to create your own geological time scale and to compare data from existing published sources. These facilities, together with the opportunities to build your own range charts and effect high-resolution correlation of strata, open many exciting opportunities. Real advances are now possible in dating the precise timing and rates of biological processes such as extinction and recovery rates together with the accurate timing of the origins of higher taxa and the velocity of morphological change along evolving lineages.

The site can be accessed through http://www.blackwellpublishing.com/paleobiology/.

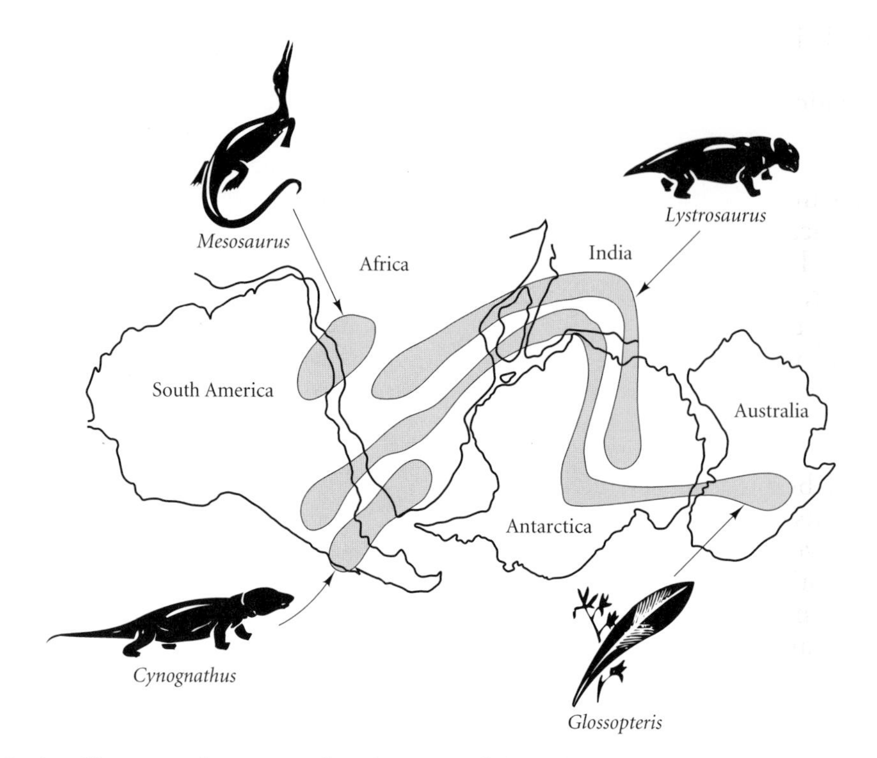

Figure 2.14 Carboniferous and Permian distributions of the *Glossopteris* flora and the *Mesosaurus* fauna and the fit of Gondwana. The tight fit of Gondwana and the correspondence of fossil faunas and floras across the southern continents suggested to Wegener and others that South America, Africa, India, Antarctica and Australia had drifted apart since the Permo-Triassic. (Based on Smith, P. 1990. *Geoscience Canada* **15**.)

early 1900s, the German scientist Alfred Wegener (1880–1930) suggested that the continents moved across the Earth's surface on a liquid core, suggesting that continents could in fact drift (although not through the oceans as he thought), some 50 years before the documentation of seafloor spreading and the plate tectonic revolution confirmed his theory (Wegener 1915); such data continue to be accumulated today as an integral part of paleogeographic analysis (Fortey & Cocks 2003). Our understanding of plate movements has been greatly advanced by a number of computerized paleogeographic systems; some, such as the Paleomap Project, even taking the Earth far into the future as well as deep into the past (linked at http://www.blackwellpublishing.com/paleobiology).

Faunal and floral barriers

Barriers of various types have partitioned biogeographic provinces through time. The first large-bodied organisms of the Late Neoproterozoic Ediacara faunas may have already developed their own provinces. George Gaylord Simpson (1902–1984) distinguished three types of passages: **corridors** were open at all times, **filters** allowed restricted access, whereas **sweepstake routes** opened only occasionally. In continental settings the barriers may be mountain ranges, inland seas or even rain forests. Marine faunas may be separated by wide expanses of deep ocean, swift ocean currents or land. In general terms the endemicity of most marine faunas decreases with depth; the more cosmopolitan faunas are located in deep-shelf and slope environments. But in the deeper basins, populated by specialized taxa, faunas are again endemic.

Faunal or floral provinces may be fragmented relatively rapidly if a barrier arises and the biotic responses may be quite sudden. For example, rifting and basin formation can split and isolate into fragments many existing terrestrial and fringing shelf provinces, whereas the same effects in the sea may be caused by the formation of an isthmus.

In some situations, the development of a barrier for some organisms may provide a corridor for others. The emergence of the Isthmus of Panama 3 Ma connected North and South America, but at the same time it separated the Atlantic and Pacific oceans. Before this event, South America had been isolated from North America for most of the past 70 myr, and was dominated by diverse, specialized, mammalian faunas consisting of unique marsupials, edentates, ungulates and rodents. However, the Isthmus of Panama provided a land bridge or corridor between the two continents and many terrestrial and freshwater taxa were free to move north and south across the isthmus (Fig. 2.15). The **great American biotic interchange** (GABI) allowed the North American fauna to invade the south and destabilize many of the continent's distinctive mammalian populations (Webb 1991). South American mammals were equally successful in the north and some such as the armadillo, opossum and porcupine still survive in North America.

The emergence of the isthmus also caused changes in the marine faunas of the Caribbean. Surprisingly, not many species became extinct, and there was a diversification of mollusks (Jackson et al. 1993). The emergence of the terrestrial land bridge and marine barrier may have initiated the upwelling of nutrients in the Caribbean area, and this in turn led to an increase in species diversity. Valentine (1973) had already drawn attention to a range of plate tectonic settings, including the spreading ridges, island arcs, subduction and fault zones, and the ways they can affect biological distributions. Thus tectonic features such as spreading ridges, transform faults and subduction zones create barriers for marine faunas whereas mid-plate island volcanoes can generate a series of stepping stones assisting the migration of animals and plants across great expanses of ocean. But there may be a more important relationship between tectonics and provinciality. There is a striking correlation between provinciality and continental fragmentation through time. Intervals when continents were many and dispersed apparently were times of increased provinciality, such as the Ordovician and the Cretaceous.

Island biogeography: alone and isolated?

Modern oceans are littered with islands. Most are transitory volcanic chains, developed above moving hotspots or at mid-oceanic ridges that will probably be subducted; some, however, are pieces of continental crust broken off adjacent continents. These lighter bits of crust are usually later imprisoned in mountain chains and can hold important paleontological data. The biogeography of modern islands is complex and it is hard to apply models based on modern islands to ancient examples (Box 2.6).

But islands and archipelagos play a number of biological roles. Most islands are isolated from the mainland, and they are important powerhouses of speciation (see p. 119). Some island chains play an important part in migrations, acting as stepping stones, where species and their larvae may move, sometimes over many hundreds or thousands of years, from one mainland to another. The vertebrate paleontologist Malcolm McKenna introduced some interesting analogies with ancient shipping. Moving island complexes that can allow the cross-latitude transfer of evolving animals and plants may have acted as "Noah's arks", just as Noah's biblical ship eventually beached on the summit of Mount Ararat with breeding pairs of all manner of contemporary life. The transit of India from Gondwana to Asia, together with its even-toed artiodactyls and odd-toed perissodactyls, is a possible example. In the longer term these complexes may function as "Viking funeral ships" (originally bound, of course, for Valhalla with decorated dead warriors) transporting exotic fossil assemblages to new locations. The occurrence of a Gondwanan Cambrian trilobite fauna in the Meguma Terrane of the Appalachians and an Ordovician trilobite fauna in Florida from the same high-latitude province, both now welded onto the North American continent, are remarkable examples.

Island **biotas** (faunas and floras) are often diverse, with many endemic species and commonly with evidence that these species came originally from one or more source continents. It is fascinating to study such modern islands and some, such as the Galápagos, or Aldabra, have become important sites for biologists to watch "evolution in action". It is much harder for paleontologists to

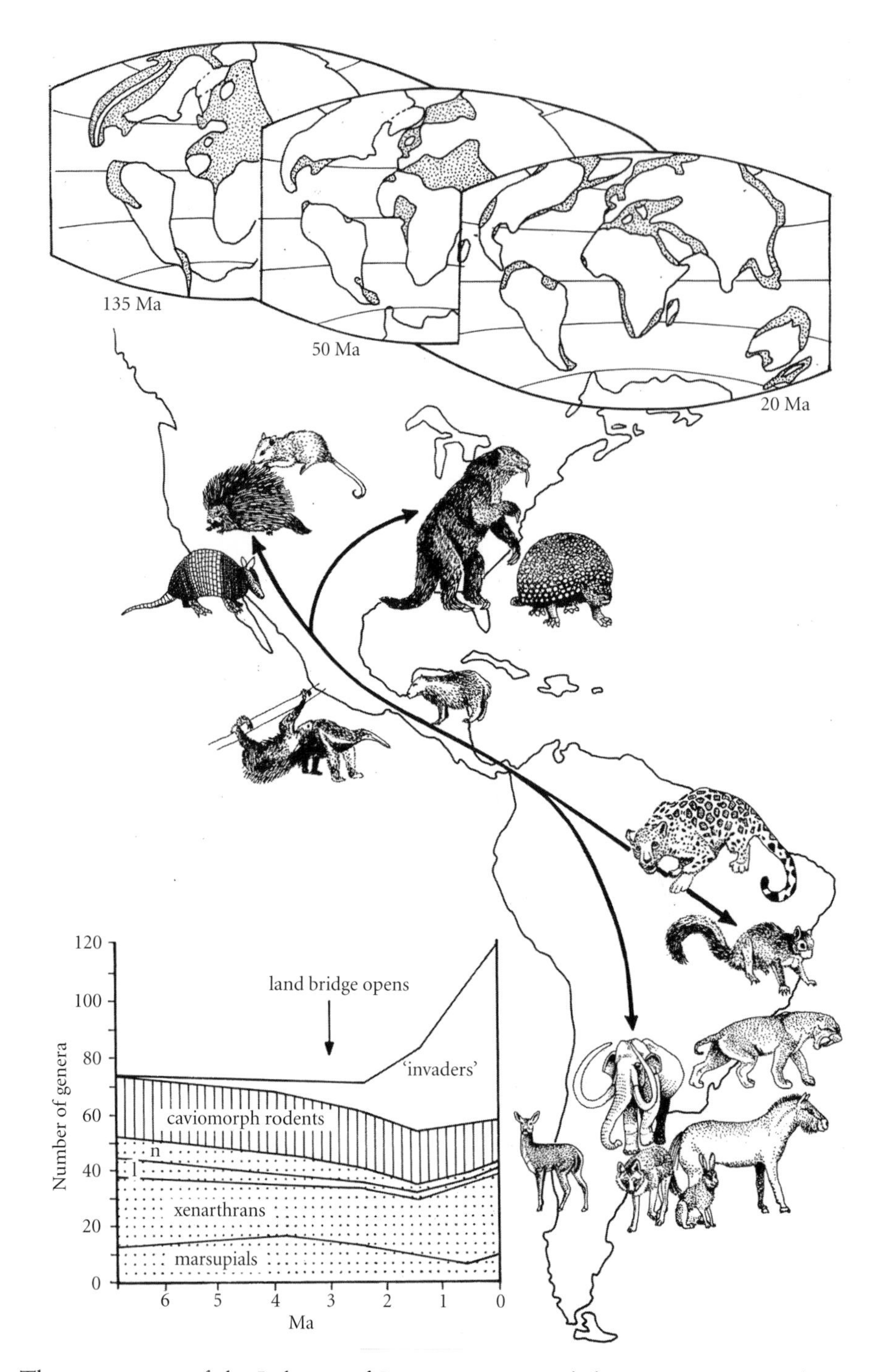

Figure 2.15 The emergence of the Isthmus of Panama promoted the great American biotic interchange (GABI) between North and South American terrestrial vertebrates together with the radiation of the shallow-water marine benthos of the Caribbean Sea. l, litopterns; n, notoungulates. (Based on Benton 2005.)

understand the role of such islands through geological time; by their very nature, being short lived and located in tectonically active areas, they are quickly lost and often destroyed.

Geological and paleontological implications: using the data

Much of the early evidence for **continental drift** was paleontological, although it was

Box 2.6 Analytic methods

Two main types of biogeographic analysis are widely used and are based on either phenetic or classic cladistic methods (Hammer & Harper 2005). Phylogenetic methods are being increasingly used to study past biogeographic patterns (Lieberman 2000). A third technique, **area cladistics**, is rapidly developing and converts a taxon-based cladogram into an area cladogram, independently of geological data; in simple terms geographic areas can be mapped onto the branches of a taxon-based tree. Cladistic methods are based on the assumption that an original province has since fragmented with the creation of subprovinces characterized by new endemics, essentially analogous to apomorphies in taxonomic cladistics (see p. 129). This is not always the case since nodes on the cladogram may equally represent widespread range expansion of taxa, perhaps associated with a marine transgression.

The phenetic methods usually start from a similarity matrix between sites based on the presence and absence of taxa, or more rarely the relative abundance of organisms across the sites (see also Chapter 4). There are a large number of distance and similarity measures to choose from. A few of the commoner coefficients are listed:

$$\text{Dice coefficient} = 2A/(2A + B + C)$$

$$\text{Jaccard coefficient} = A/(A + B + C)$$

$$\text{Simple matching coefficient} = (A + D)/(A + B + C + D)$$

$$\text{Simpson coefficient} = A/(A + E)$$

A is the number of taxa common to any two samples, B is the number in sample 1, C is the number in sample 2, D is the number of taxa absent from both samples, and E is the smaller value of B or C.

On the basis of an intersite similarity or distance matrix, a dendrogram can be constructed linking first the sites with the highest similarities or the closest distances. When the distance or similarity matrix is recalculated to take into account the first clusters, additional sites or genera are clustered until all the data points are included in the dendrogram. Clearly the first clusters, with the highest similarities or lower distances, have the greatest significance and less importance is usually attached to later linkages.

widely derided through the 1940s and 1950s. However, paleontological data are now crucial to an understanding of the fine details of the dance of the continents through time. Wegener suggested that the continents merely ploughed through oceanic crust. But during the 1960s, plate tectonic theory with seafloor spreading, the subduction of ocean crust under the continents and the collision of the continents themselves, provided a mechanism. In the mid-1960s, during the early stages of the **plate tectonic** revolution, Tuzo Wilson (1966) predicted that the remains of an ancient seaway would be found in Lower Paleozoic rocks of the northern hemisphere. North American and European fossil assemblages of brachiopods, trilobites and graptolites were separated by a major suture running the length of the modern Appalachian and Caledonian mountain belts. On this basis, together with a few other lines of evidence, Wilson inferred the existence of a much older ocean, the proto-Atlantic (now termed Iapetus), that separated North America from most of Europe prior to an initial collision of these continents and oceanic closure in the Silurian-Devonian.

Wilson's classic study depicted a two-dimensional ocean with opening and closing between Europe and North America (Fig. 2.16a). The Iapetus Ocean first opened during the Late Precambrian with the breakup of a supercontinent, and developed during the

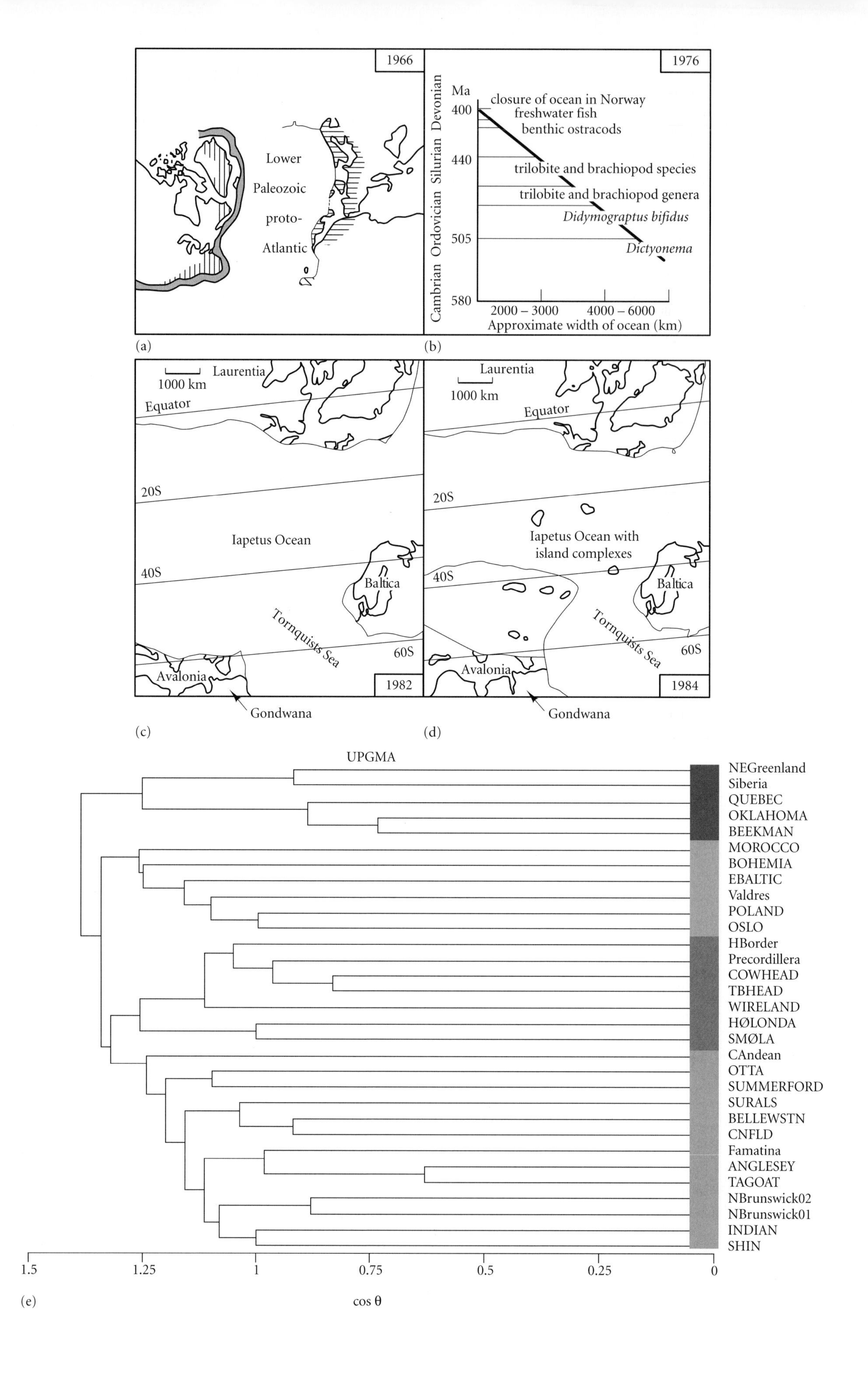
1966
Lower
Paleozoic
proto-
Atlantic
1976
Ma
Cambrian Ordovician Silurian Devonian
400
440
505
580
closure of ocean in Norway
freshwater fish
benthic ostracods
trilobite and brachiopod species
trilobite and brachiopod genera
Didymograptus bifidus
Dictyonema
2000 – 3000
4000 – 6000
Approximate width of ocean (km)
(a)
(b)
Laurentia
1000 km
Equator
20S
Iapetus Ocean
40S
Baltica
Tornquists Sea
60S
Avalonia
Gondwana
1982
Laurentia
1000 km
Equator
20S
Iapetus Ocean with
island complexes
40S
Baltica
Tornquists Sea
60S
Avalonia
Gondwana
1984
(c)
(d)
UPGMA
NEGreenland
Siberia
QUEBEC
OKLAHOMA
BEEKMAN
MOROCCO
BOHEMIA
EBALTIC
Valdres
POLAND
OSLO
HBorder
Precordillera
COWHEAD
TBHEAD
WIRELAND
HØLONDA
SMØLA
CAndean
OTTA
SUMMERFORD
SURALS
BELLEWSTN
CNFLD
Famatina
ANGLESEY
TAGOAT
NBrunswick02
NBrunswick01
INDIAN
SHIN
1.5
1.25
1
0.75
0.5
0.25
0
(e)
cos θ

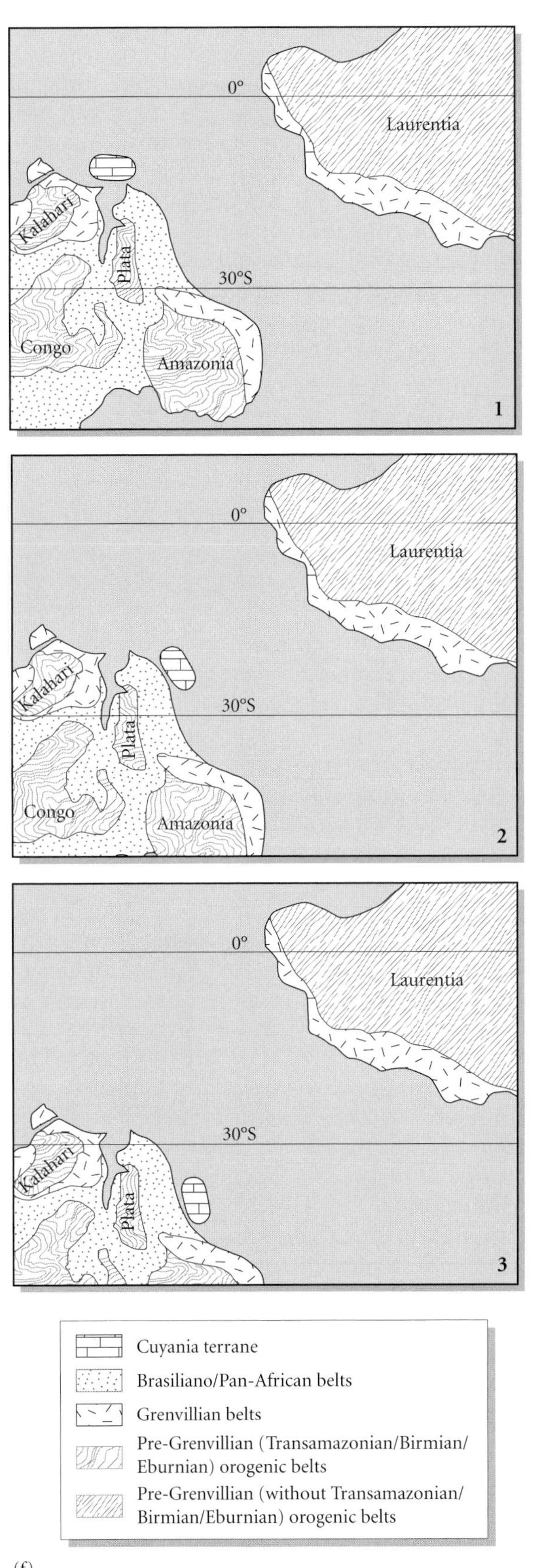

Cambrian. At its widest in the late Cambrian, possibly extending as much as 4000 km across, only floating graptolites were similar on both sides of the Iapetus. But as the ocean closed, swimming organisms such as the conodonts could next cross the seaway (McKerrow & Cock 1976), and later so could the mobile and eventually the fixed benthos, the trilobites and brachiopods (Fig. 2.16b). By the late Silurian, as the Iapetus Ocean narrowed to only a few hundred kilometers, benthic ostracodes scuffled their way across. By the Devonian, when the ocean was almost completely closed, freshwater fishes were similar in Europe and North America. In a refinement to the original model, Cocks and Fortey (1982) described the ocean in terms of a three-plate model with oceans separating Gondwana, Baltica and Avalonia. The smaller Avalonia broke away from Gondwana during the late Cambrian-earliest Ordovician and, together with Baltica, headed north towards Laurentia (Fig. 2.16c, d). Neuman (1984) placed islands within the Iapetus Ocean, small suspect terranes with peculiar faunas, not seen elsewhere. Even more intriguing, Baltica spun anticlockwise as it moved towards the equator picking up these various terranes on the edge of the continent (Torsvik et al. 1991). Both cladistic and phenetic techniques have been used to analyze the large amount of distributional data from within and around the

Figure 2.16 (opposite and this page) Changing ideas on the development of the Early Paleozoic Iapetus Ocean and its faunas: (a, c, d) paleogeographic reconstructions; (b) the mobility of organisms across a closing ocean; (e) a cluster analysis of the Iapetus and related Early Ordovician brachiopod faunas (tinted blocks in descending order indicate low-latitude, high-latitude, low-latitude marginal and high-latitude marginal provinces); and (f) the possible movement of the Precordilleran terrane in three stages, 1–3. A dataset of early Ordovician brachiopod distribution across the Iapetus terranes is available at http://www.blackwellpublishing.com/paleobiology/. These data may be analyzed and manipulated using a range of multivariate techniques including cluster analysis (see also Hammer & Harper 2005). (a–d, from Harper, D.A.T. 1992. *Terra Nova* **4**; f, based on Finney 2007.)

Iapetus Ocean, all confirming in broad terms current paleogeographic reconstructions of this complex ocean system (Harper et al. 1996) (Fig. 2.16e). Finally in this apparent confusion, some terranes, such as the Argentine Precordillera, have faunas that have even switched provinces as their terranes drifted across latitudes (Astini et al. 1995) (Fig. 2.16f). But this evidence has been disputed. The view of fauna switching is not entirely supported by a geochronometric study of detrital zircons that shows that the Precordillera had an origin in Gondwana, where the basement rocks that supplied the zircons probably occur (Finney 2007). Perhaps on this occasion the faunal data require an alternative explanation.

Careful paleogeographic study has shown that some continents have been put together from numerous formerly separated strips of land. Geological mapping may highlight major **fault** zones, lines of disjunction between unmatched rock units on either side, but it is, in fact, the fossils that can pin down where each continental slice, or **terrane**, came from in the first place. A classic example is the North American Cordillera, which is a mosaic of terranes, now plastered onto the west coast of the continent, but probably originating at lower latitudes. Paleontologists have recognized so-called Boreal (northern, low-diversity) and Tethyan (southern, high-diversity) faunas of marine invertebrates in the separate terranes in the Mesozoic. In an east–west traverse across the North American Cordillera, there is a progressive northward displacement of Tethyan-type faunas of Early Jurassic age. Some of the more exotic, far-traveled terranes may have moved over 1300 km (Fig. 2.17).

Biogeography and climatic gradients have driven patterns of changing biodiversity. In broad terms, low latitudes support high-diversity faunas, and biodiversity decreases away from the tropics towards the poles. Studies on modern bivalve, bryozoan, coral and foraminiferan faunas show marked increases in diversity towards the equator, and since many cool-water species breed later in life, polar and temperate-zone animals are sometimes larger than their tropical counterparts. But this is only plausible if the growth rates are the same in both regions; they may not be. What is true today is true in the past (Box 2.7).

Many authors have suggested that changing plate configurations, oscillating between fragmentation and integration, have affected biodiversity through time. For example, the huge Early Ordovician radiation of marine skeletal faunas may be related to the breakup of Gondwana, while the end-Permian extinction event coincides with the construction of Pangaea. More recent diversifications have occurred during the late Mesozoic fragmentation of this supercontinent (Fig. 2.18).

FOSSILS IN FOLD BELTS

One bad fossil is worth a good working hypothesis.

Rudolf Trümpy,
eminent Alpine geologist

Fossils from the deformed zones of mountain belts are rare but important. Relatively few paleontologists study these fossils because they are usually poorly preserved, and are metamorphosed and tectonized; fossils in orogenic or mountain-building zones are also rare and difficult to collect from often hazardous terrains. Nevertheless, fossils are of fundamental importance in the formulation of tectonic models, providing age and geographic constraints, although the fossils themselves are rarely of great morphological significance. The identification of fossiliferous sequences in thrust belts helped identify large-scale horizontal movements of the Earth's crust in the Swiss Alps, the Northwest Highlands of Scotland and in the Scandinavian Caledonides over a century ago (Box 2.8). In many mountain belts fossil data have provided the only reliable dates for rock successions; unlike radiometric clocks, fossils cannot be reset by later thermal and tectonic events.

The Appalachian-Caledonian mountain belt, developed during the Early Paleozoic, contains large pieces of both North America and Europe, but understanding of its complex history and structure is fairly recent. Parts of the belt have been dissected and investigated by paleontological data. For example, Charles Lapworth's studies on the complex structure and stratigraphy of the Southern Uplands of Scotland in the 1870s were based on recognition of the sequence of graptolite faunas.

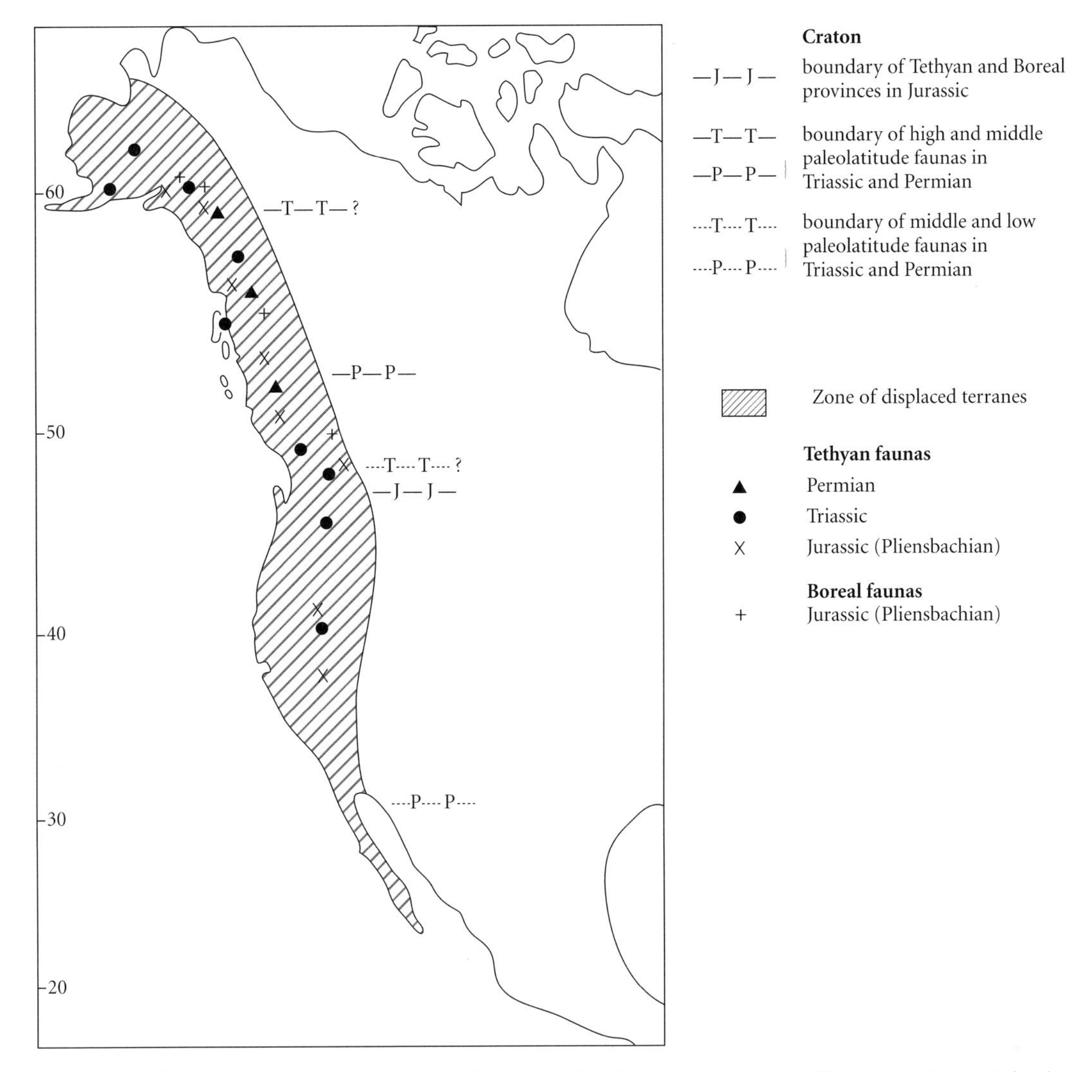

Figure 2.17 Displaced faunas in terranes within the North American Cordillera together with changing provincial boundaries on the craton. Postulated latitudinal boundaries on the craton during the Permian, Triassic and Jurassic are indicated and confirm the northern movement of these displaced terranes. A dataset of Jurassic ammonoid distributions across the cordilleran terranes is available at http://www.blackwellpublishing.com/paleobiology/. These data may be analyzed and manipulated using a range of multivariate techniques including cluster analysis (see also Hammer & Harper 2005). (From Hallam, A. 1986. *J. Geol. Soc.* **143**.)

Much more recently in central Scotland, reliable early Ordovician dates from the Highland Border Complex (based on brachiopods, trilobites and a range of microfossils), previously included as part of the mainly Neoproterozoic Dalradian Supergroup on the continent of Laurentia, suggests that these rocks were deposited in one of a series of basins along the margin of Laurentia. The oceanic terranes, such as volcanic islands and microcontinents, that evolved seaward of the ancient continents are often termed "suspect". In many cases it is not clear to which if any of the continents they were originally attached. The Highland Border Complex was considered a truly suspect terrane. Moreover, the two areas could not have developed together since, firstly, during the Early Ordovician, the Dalradian was deforming and uplifting, and secondly there was a lack of Dalradian clasts in the Highland Border Basin. Some scientists have even suggested the Dalradian was derived

Box 2.7 Latitudinal variation in diversity through time

Today the tropics are teeming with diverse life built around a number of so-called **hotspots**, small areas that have especially high numbers of species. But is this a modern phenomenon? Recent research suggests that latitudinal gradients have intensified dramatically during the past 65 myr and that biotic radiations in the tropics are based on relatively few species-rich groups in both marine and terrestrial environments (Crame 2001). Part of this may have been driven by evolutionary escalation, part by changing climates. In evolution, sometimes predators and prey evolve rapidly in concert – the predators may adopt ever-more deadly means of attacking their prey, but the prey evolves ever-better means of defense. This kind of **escalation**, or **arms race**, has happened in many circumstances (see p. 102), and may have happened in tropical oceans through the past 15 myr. Further, global climate change during this same period probably helped to partition the tropics into a series of diversity hotspots, such as the Indo West Pacific (IWP) center. It is hard to be sure that such hotspots in the geological past will be preserved. How we perceive past diversity may be very much dependent on whether we have or have not properly sampled these hotspots through time.

Other latitudinal diversity gradients tend to confirm current trends. For example, in a study covering the past 100 myr, Markwick (1998) found that crocodilians used to have a wider latitudinal spread than they do today. Modern crocodilians are known primarily from a narrow tropical belt covering the southern United States down to central Brazil, Africa, India and Australasia. Abundant crocodilian fossils from the Cretaceous and Tertiary are known from northern parts of North America and Europe, but the richest finds lie around the paleoequator. So, the tropical, warm-weather part of the world used to be twice as wide as it is today and, in general, global climates have cooled through the last 100 myr. Nevertheless crocodilians are, and were, most abundant round the equator, and their diversity declines the farther one goes away from the tropics.

from Gondwana and has nothing to do with the geological history of North America until later in the Ordovician. This is, however, only one school of thought. New structural data suggest the Highland Border Complex was part of the Dalradian and, indeed, was always intimately linked to the Laurentian craton (Tanner & Sutherland 2007). Elsewhere in the Caledonides, Harper and Parkes (1989) described a series of terranes across Ireland based on paleontological data. While some terranes developed marginal to North America and Avalonia (see above), some smaller terranes in central Ireland almost certainly evolved within the Iapetus Ocean itself, with their own distinctive faunas.

We can thus reassemble ancient mountain belts and trace the origins of their jumbled structure using paleontological data, but can fossils help us understand the rates of these tectonic processes, such as plate movements and the transit of individual thrust sheets within orogenic belts? The Banda Arcs are part of a much younger mountain belt, developed during the Neogene and Quaternary along the continental margin of northern Australia (Harper 1998). A precise stratigraphy based on foraminiferans has allowed the movement of far-traveled thrust complexes to be tracked; thrust sheets were emplaced at rates between 62.5 and 125 mm yr^{-1} whereas the belt as a whole was uplifted at rates of about 15 mm yr^{-1}.

Fossils, surprisingly, can be of great value to structural geologists, not only in understanding the rates and timing of tectonic events. Structural geologists study rocks that have been folded and faulted, and they want to identify how exactly the rocks have been deformed. If they find a fossil that was originally symmetric, but has since been squeezed, or stretched, in particular directions, they have precise evidence of the magnitude of the tectonic forces that have acted. A famous

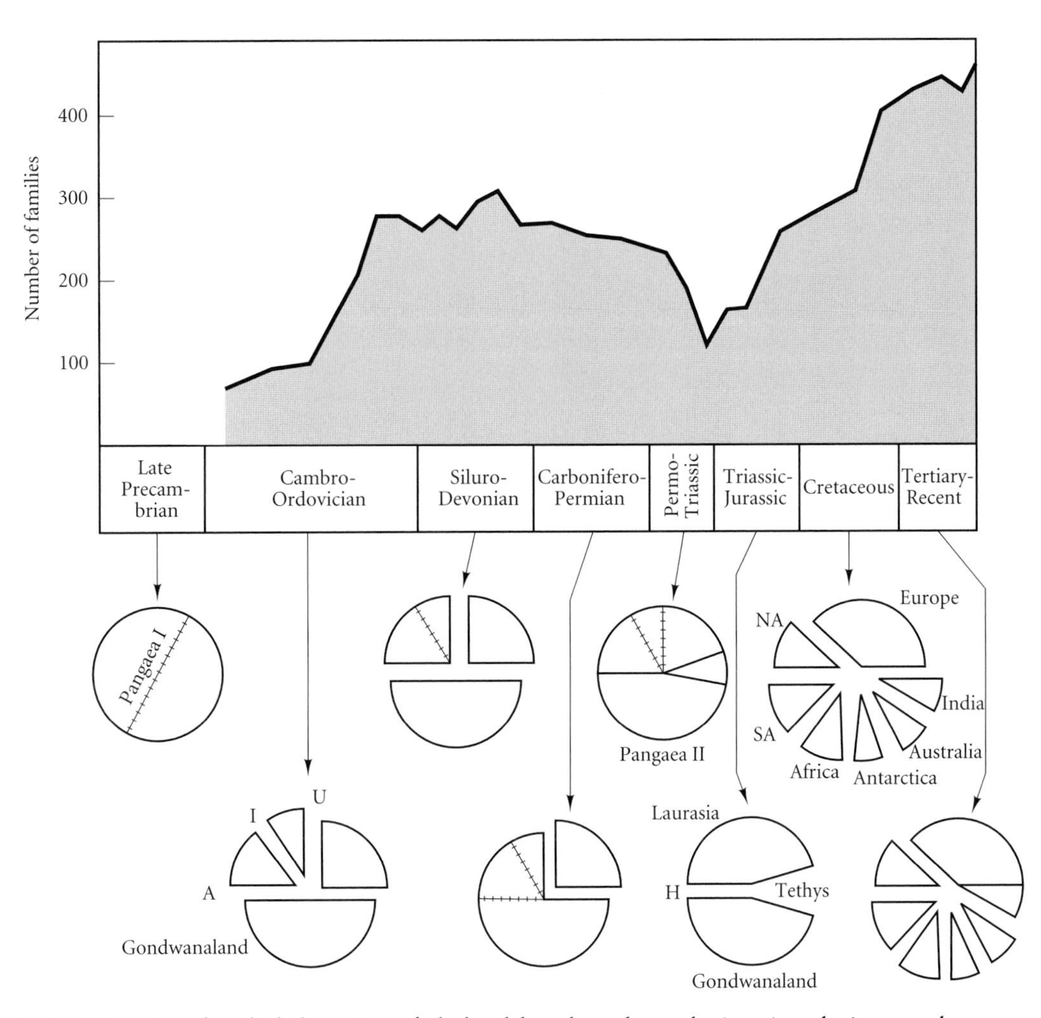

Figure 2.18 Changing familial diversity of skeletal benthos through time in relation to plate configurations: high diversities are apparently coincident with times of greatest continental fragmentation, for example during the Ordovician, Devonian and Cretaceous-Cenozoic. A, pre-Appalachian-Variscan Ocean; H, Hispanic Corridor; I, Iapetus Ocean; U, pre-Uralian Ocean. (Based on Smith, P. 1990. *Geoscience Canada* **15**.)

example is the "Delabole butterfly", so called because quarrymen in the village of Delabole, in Devon (England) thought they were looking at ancient butterflies. In fact, the wide-hinged fossils are spiriferide brachiopods (see p. 306), and they were bent and stretched in all kinds of ways, depending on how they were oriented in the rocks. The fossils are in Devonian sediments that were bent and stretched by the Variscan Orogeny, a great phase of mountain building that affected southern and central Europe during the Carboniferous. By measuring the fossils, these large-scale forces could be reconstructed.

Until fairly recently these and similarly deformed assemblages were of limited value to taxonomic paleontologists; now a range of microcomputer-based graphic techniques are available to "unstrain" specimens. Hughes and Jell (1992), for example, used such techniques to unstrain Cambrian trilobites from Kashmir that had been distorted by earth movements during the uplift of the Himalayan mountain belt (Fig. 2.19). Previous studies had recognized seven species among these trilobites; statistical and graphic removal of the effects of tectonism revealed only one species. The study also allowed Hughes and Jell to identify the trilobites more accurately than before and to understand how they relate to species from India and North China.

Figure 2.19 Strained Cambrian trilobites from Himalaya. (Courtesy of Nigel Hughes.)

Can the actual color of fossils help us understand the geological history of an area? The investigation of **thermal maturation** is now a routine petroleum exploration technique. A number of groups of microfossils change color with changing paleotemperature (Table 2.2). The upper end of the thermally-induced color range has proved useful in mapping metamorphic zones in orogenic belts. Conodonts in particular (see p. 429) are useful thermal indicators. They change color from light amber to gray to black and white, and eventually translucent, on a scale of conodont alteration indices (CAI values) from 1 to 8, through a temperature range from about 60 to 600°C. Carbonaceous organisms, including the graptolites (see p. 412), also show color changes, as does vitrinite derived from plant material. These changes have also been documented in detail for acritarchs (see p. 216), where acritarch alteration indices (AAI values) range from 1 to 5. Spores and pollen have spore color indices (SCI values) ranging from 1 to 10, with colors ranging from colorless to pale yellow through to black. Other groups such as phosphatic microbrachiopods and chitinozoans show similar prospects, but their color changes have yet to be calibrated with precise paleotemperatures. Paleotemperatures can also help predict the oil and gas window, usually located at depths between 2.5 and 3.5 km, and thus have important application to hydrocarbon exploration.

Box 2.8 Scandinavian Caledonides

Mountain belts are a source of all sorts of exciting and significant fossil assemblages. The Scandinavian Caledonides are no exception. This mountain belt stretches for some 1800 km from north to southwest Norway, never exceeding a width of 300 km. It developed during a so-called Wilson cycle (the opening, closing and subsequent destruction of an ancient ocean, named after J. Tuzo Wilson) culminating in the collision of the Baltic plate with those of Avalonia (England, Wales and parts of eastern North America and north central Europe) and then Laurentia (cratonic North America). During its transit from high to low latitudes in the Early Paleozoic, Baltica rotated anticlockwise and first captured terranes adjacent to the craton itself with Baltic faunas, followed by island terranes from within the Iapetus Ocean, with endemic taxa, and finally island complexes that were marginal to the Laurentian plate with North American faunas (Harper 2001). The mountain belt in its pile of thrust sheets thus stores much of the biogeographic history of the Iapetus Ocean and its marginal terranes (Fig. 2.20). Moreover during the Late Silurian-Devonian, as the mountain belt continued to rise, marginal basins contained remarkable marine marginal biotas with spectacular eurypterid faunas. Adjacent basins, for example in Scotland, contain some of the earliest land arthropods and plants. So the collision of plates and the generation of a huge mountain belt was not entirely a destructive process. It has helped preserve key evidence for an ancient ocean with diverse and endemic faunas that helped contribute to the great Ordovician biodiversification event (see p. 253) while its later non-marine basins hold critical information on the early development of life on land (see p. 442).

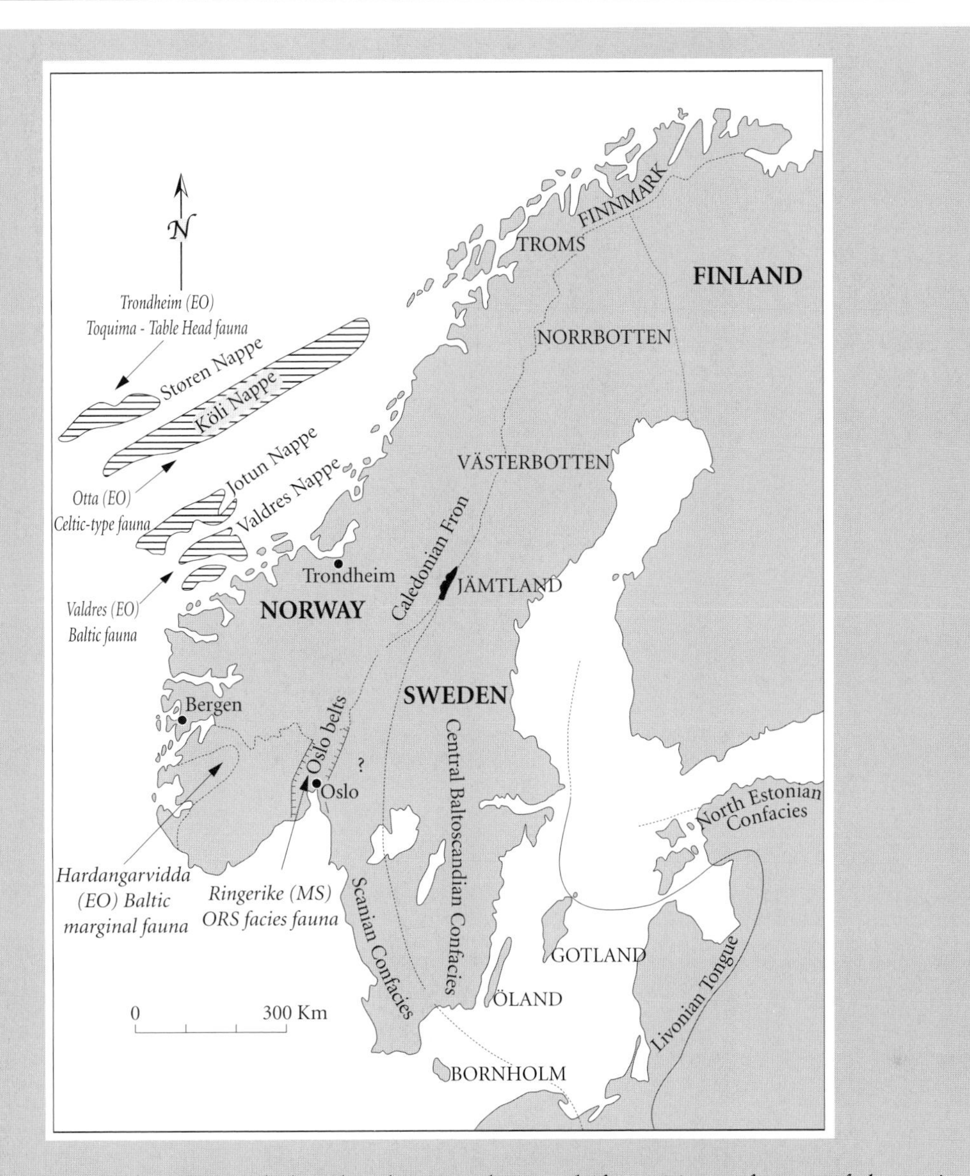

Figure 2.20 The Scandinavian Caledonides showing the pre-drift positions of some of the various thrust sheet complexes. During the Early Ordovician (EO) the most seaward, upper parts of the higher thrust sheets (Støren Nappe) contained North American marginal faunas, whereas the lower parts of these thrust sheets (Köli Nappe) contained Celtic (oceanic) type faunas. The lower parts of the nappe pile (e.g. the Valdres Nappe) have Baltic faunas. The Wenlock-Ludlow (MS) marginal molasse deposits (Old Red Sandstone (ORS) facies), for example at Ringerike, have spectacular marine marginal faunas.

Table 2.2 Various measures of thermal maturation. Color changes recorded in conodonts (CAI), together with corresponding values for vitrinite reflectance and the translucency index of palynmorphs, are related to the oil and gas window and metamorphic grades and zones. (Based on Jones, G.L. 1992. *Terra Nova* **4**.)

CAI	Color	Paleotemperature (°C)	Mean temperature (°C)	Vitrinite reflectance	Palynomorph translucency index	Thermal alteration index approx.	Metamorphic grade	Metamorphic zones
1	pale yellow	50–80	65	0.8	1–5	1.5	↑	Diagenetic zone
						2.0	oil and gas	
1.5	very pale brown	50–90	70	0.7–0.85	5–5ur	2.5	window	
2	brown to dark brown	60–140	100	0.85–1.3	5–6		↓	
2.5		85–180	135			2.7	dry gas	
3	very dark gray-brown	110–200	160	1.4–1.95	5ur–6	3.2		
3.5		150–260	205			3.5		
4	light black	190–300	245	1.95–3.6	6	4.0		
4.5		230–340	285			5.0		
5	dense black	300–400	330	3.6	6ur–7	lower greenschist chlorite/muscovite		Anchizone
5.5	dark gray-black	310–420	365					
6	gray	350–435	400			greenschist meta-argillite		Epizone
6.5	gray-white	425–500	460					
7	opaque white	480–610	550			upper greenschist biotite-garnet		
7.5	semi-translucent	>530						Mesozone
8	crystal clear	>600				garnet		

Review questions

1 The stratigraphic frameworks we use today have been assembled over the last 200 years and are based on litho- and biostratigraphy. Fossils remain our main tool to correlate rock strata. Are they likely to remain as important for correlation over the next 200 years?
2 Cyclostratigraphy is rapidly becoming an important tool for long-distance and precise correlation particularly in Mesozoic and Cenozoic strata. What caused these fine-scale sedimentary rhythms that can sometimes be traced for thousands of kilometers?
3 The past distributions of fossil animals and plants have provided a reliable method to analyze the changing geography of our planet through time. But some fossil groups are more helpful than others. Which types of animals and plants provide the clearest biogeographic signals, and why?
4 Islands are unique ecosystems and some such environments can be recognized in the fossil record. How important were islands for understanding the development of biodiversity and evolution of marine and non-marine biotas?
5 Fossils within mountain belts are hard to find and collect, they often occur in remote, near-inaccessible regions, and are often sheared and poorly preserved. Why is it so important to collect and study these fossils?

Further reading

Ager, D.V. 1993. *The Nature of the Stratigraphical Record*, 3rd edn. John Wiley & Sons, Chichester, UK. (Provocative and stimulating personal view of the stratigraphic record.)

Benton, M.J. (ed.) 1993. *Fossil Record 2*. Chapman & Hall, London. (Massive compilation of diversity change through time at the family level.)

Brenchley, P.J. & Harper, D.A.T. 1998. *Palaeoecology: Ecosystems, Environments and Evolution*. Chapman & Hall, London. (Readable paleoecology text with chapter devoted to paleobiogeography.)

Briggs, D.E.G. & Crowther, P.R. (eds) 1990. *Palaeobiology – A Synthesis*. Blackwell Scientific Publications, Oxford. (Modern synthesis of many aspects of contemporary paleontology.)

Briggs, D.E.G. & Crowther, P.R. (eds) 2003. *Palaeobiology II – A Synthesis*. Blackwell Publishing, Oxford. (Modern and updated synthesis of most aspects of contemporary paleontology; completely revised with new material.)

Bruton, D.L. & Harper, D.A.T. (eds) 1992. Fossils in fold belts. *Terra Nova* **4** (thematic issue). (Collection of papers on the importance and use of fossils in mountain belts.)

Cox, B.C. & Moore, P.D. 2005. *Biogeography. An Ecological and Evolutionary Approach*, 7th edn. Blackwell Publishing, Oxford. (Up-to-date review of biogeography, past and present, and its biological significance.)

Cutler, A. 2003. The Seashell on the Mountaintop. *A Story of Science, Sainthood, and the Humble Genius who Discovered a New History of the Earth*. Heinemann, London. (Accessible account of the life of Steno.)

Doyle, P. & Bennett, M.R. (eds) 1998. *Unlocking the Stratigraphical Record. Advances in Modern Stratigraphy*. John Wiley & Sons, Chichester, UK. (Multi-author text covering all the main areas of modern stratigraphic practice.)

Fortey, R.A. & Cocks, L.R.M. 2003. Palaeontological evidence bearing on global Ordovician-Silurian continental reconstructions. *Earth Science Reviews* **61**, 245–307. (Comprehensive review of the use of paleontological data in early Paleozoic geographic reconstructions.)

Gradstein, F., Ogg, J. & Smith, A. 2004. *A Geologic Time Scale 2004*. Cambridge University Press, Cambridge. (Current in a series of snapshot reviews of the geological time scale.)

Hammer, Ø. & Harper, D.A.T. 2005. *Paleontological Data Analysis*. Blackwell Publishing, Oxford. (Overview of many of the numerical techniques available to paleontologists; linked to software package, PAST.)

Lieberman, B.S. 2000. *Paleobiogeography: Using Fossils to Study Global Change, Plate Tectonics and Evolution*. Plenum Press/Kluwer Academic Publishers, New York. (New, particularly numerical, approaches to the study of paleobiogeography and its wider significance.)

Valentine, J.W. 1973. *Evolutionary Paleoecology of the Marine Biosphere*. Prentice-Hall, Englewood Cliffs, New Jersey. (Visionary study of the marine biosphere through time.)

References

Astini, R.A., Benedetto, J.L. & Vaccari, N.E. 1995. The early Paleozoic evolution of the Argentine Precordillera as a Laurentian rifted, drifted, and collided terrane; a geodynamic model. *GSA Bulletin* **107**, 253–73.

Benton, M.J. 2005. *Vertebrate Palaeontology*. Wiley-Blackwell, Oxford.

Cocks, L.R.M. & Fortey, R.A. 1982. Faunal evidence for oceanic separations in the Palaeozoic of Britain. *Journal of the Geological Society, London* **139**, 465–78.

Crame, J.A. 2001. Taxonomic diversity gradients through geologic time. *Diversity and Distributions* 7, 175–89.

Finney, S.C. 2007. The parautochthonous Gondwanan origin of the Cuyania (greater Precordillera) terrane of Argentina: a re-evaluation of evidence used to support an allochthonous Laurentian origin. *Geologica Acta* **5**, 127–58.

Fortey, R.A. & Cocks, L.R.M. 2003. Palaeontological evidence bearing on global Ordovician-Silurian continental reconstructions. *Earth Science Reviews* **61**, 245–307.

Fortey, R.A., Harper, D.A.T., Ingham, J.K., Owen, A.W. & Rushton, A.W.A. 1995. A revision of Ordovician series and stages from the historical type area. *Geological Magazine* **132**, 15–30.

Gale, A.S., Young, J.R., Shackleton, N.J., Crowhurst, S.J. & Wray, D.S. 1999. Orbital tuning of Cenomanian marly chalk successions: towards a Milankovitch time-scale for the Late Cretaceous. *Philosophical Transactions of the Royal Society, London A* **357**, 1815–29.

Gradstein, F.M. & Ogg, J.G. 2004. Geologic Time Scale 2004 – why, how, and where next? *Lethaia* **37**, 175–81.

Gradstein, F., Ogg, J. & Smith, A. 2004. *A Geologic Time Scale 2004*. Cambridge University Press, Cambridge.

Hammer, Ø. & Harper, D.A.T. 2005. *Paleontological Data Analysis*. Blackwell Publishing, Oxford.

Harper, D.A.T. 1998. Interpreting orogenic belts: principles and examples. *In* Doyle, P. & Bennett, M.R. (eds) *Unlocking the Stratigraphical Record. Advances in Modern Stratigraphy*. John Wiley & Sons, Chichester, UK, pp. 491–524.

Harper, D.A.T. 2001. Fossils in mountain belts. *Geology Today* **17**, 148–52.

Harper, D.A.T., MacNiocaill, C. & Williams, S.H. 1996. The palaeogeography of early Ordovician Iapetus terranes: an integration of faunal and palaeomagnetic constraints. *Palaeogeography, Palaeoclimatology, Palaeoecology* **121**, 297–312.

Harper, D.A.T. & Parkes, M.A. 1989. Palaeontological constraints on the definition and development of Irish Caledonide terranes. *Journal of the Geological Society, London* **146**, 413–15.

Holland, C.H. 1986. Does the golden spike still glitter? *Journal of the Geological Society, London* **143**, 3–21.

Holland, S.M. & Patzkowsky, M.E. 2004. Ecosystem structure and stability: Middle Ordovician of central Kentucky, USA. *Palaios* **19**, 316–31.

Hughes, N.C. & Jell, P.A. 1992. A statistical/computergraphic technique for assessing variation in tectonically deformed fossils and its application to Cambrian trilobites from Kashmir. *Lethaia* **25**, 317–33.

Hutton, J. 1795. *Theory of Earth with Proofs and Illustrations*. William Creech, Edinburgh.

Jackson, J.B.C., Jung, P., Coates, A.G. & Collins, L.S. 1993. Diversity and extinction of tropical American mollusks and emergence of the Isthmus of Panama. *Science* **260**, 1624–26.

Lieberman, B.S. 2000. *Paleobiogeography: Using Fossils to Study Global Change, Plate Tectonics and Evolution*. Plenum Press/Kluwer Academic Publishers, New York.

Markwick, P.J. 1998. Fossil crocodiles as indicators of Late Cretaceous and Cenozoic climates: implications for using palaeontological data in reconstructing palaeoclimate. *Palaeogeography, Palaeoclimatology, Palaeoecology* **137**, 205–71.

McKerrow, W.S. & Cocks, L.R.M. 1976. Progressive faunal migration across the Iapetus Ocean. *Nature* **263**, 304–6.

Neuman, R.B. 1984. Geology and paleobiology of islands in the Ordovician Iapetus Ocean. *Bulletin of the Geological Society of America* **95**, 1188–201.

Peters, S.E. 2005. Geologic contraints on the macroevolutionary history of marine animals. *Proceedings of the National Academy of Sciences, USA* **102**, 12326–31.

Rosenberg, G.D. 2001. An artistic perspective on the continuity of space and the origin of modern geologic thought. *Earth Sciences History* **20**, 127–55.

Signor, P.W. & Lipps, J.H. 1982. Sampling bias, gradual extinction patterns, and catastrophes in the fossil record. *In* Silver, L.T. & Schultz, P.H. (eds) Geological implications of impacts of large asteroids and comets on Earth. *Geological Society of America Special Paper* **190**, 291–6.

Steno, N. 1669. *De solido intra solidum naturaliter contento dissertationis prodomus [Forerunner of a Discourse on Solids Naturally Contained Within Solids]*. Typographia sub signo Stellae, Florence.

Tanner, P.W.G. & Sutherland, S. 2007. The Highland Border Complex, Scotland: a paradox resolved. *Journal of the Geological Society* **164**, 111–16.

Torsvik, T.H., Ryan, P.D., Trench, A. & Harper, D.A.T. 1991. Cambrian-Ordovician paleogeography of Baltica. *Geology* **19**, 7–10.

Valentine, J.W. 1973. *Evolutionary Paleoecology of the Marine Biosphere*. Prentice-Hall, Englewood Cliffs, New Jersey.

Webb, S.D. 1991. Ecogeography and the Great American Interchange. *Paleobiology* **17**, 266–80.

Wegener, A. 1915. *Die Entstehung der Kontinente und Ozeane [Origin of Continents and Oceans]*. Sammlung Vieweg 23, 94 pp.

Wilson, J.T. 1966. Did the Atlantic close and then reopen? *Nature* **211**, 676–81.

Chapter 3

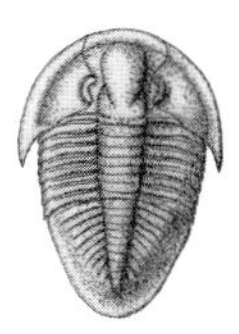

Taphonomy and the quality of the fossil record

Key points

- Plants and animals with hard tissues are most frequently preserved in the fossil record.
- Soft tissues usually decay rapidly, but rapid burial or early mineralization may prevent decay in cases of exceptional preservation.
- Physical and chemical processes may damage hard tissues during transport and compaction.
- Plants may be preserved as permineralized tissues, coalified compressions, cemented casts or as hard parts.
- There has been a longstanding debate about the fidelity and quality of the fossil record.
- The fossil record is clearly affected by the rock record, and apparent rises and falls in biodiversity can mimic rises and falls in sea level, for example.
- Perhaps the parallel patterns of biodiversity and rock record through time are driven by a third factor, such as sea-level change, at least at local and regional scales.
- Quantitative studies suggest that knowledge of the fossil record is improving.
- Paleontologists can use phylogenetic trees *and* fossil records, both largely independent of each other, to establish congruence between the two data sets, and so gain some measure of confidence that the fossil record tells the true history of life.

To examine the causes of life, we must first have recourse to death . . . I must also observe the natural decay and corruption of the human body. Darkness had no effect upon my fancy; and a churchyard was to me merely the receptacle of bodies deprived of life, which, from being the seat of beauty and strength, had become food for the worm. Now I was led to examine the cause and progress of this decay, and forced to spend days and nights in vaults and charnel-houses.

Mary Shelley (1813) *Frankenstein*

The paleontological study of **taphonomy**, which includes all the processes that occur after the death of an organism and before its fossilization in the rock, may seem ghoulish. In fact, many of the analytic approaches used by taphonomists are also used by forensic scientists. A crime scene investigator who is called to inspect a corpse may be asked how long ago the body was buried. The forensic scientist looks at the state of decay – is there any flesh remaining, do the bones still contain fat, what do the remnants of hair and finger nails look like? But now there is a whole armory of analytic techniques. For example, measurement of the chemistry of the bone and, in particular the assessment of the **rare earth elements** (scandium, yttrium and the 15 lanthanides), can help pinpoint the time of death. These forensic science methods are used by archeologists and, stepping back farther in time, also by paleontologists.

A related issue is the quality of the fossil record. Following the decay and loss of fossils, what is actually left? Can paleontologists trust the rock record and use their patchy fossil finds to somehow understand large-scale patterns of evolution? Critics are right to point out that paleontologists should be careful when they attempt to reconstruct a whole plant or animal, and try to understand its biomechanics, when they have just a few bones or bits of twigs. Care is required also in seeking to understand patterns of diversity change and evolution when many fossil species are missing. There is a heated debate about this issue, with some scientists claiming that the fossil record is desperately bad and next to useless, while others claim that the fossils do, in fact, tell us the history of life. We will look at taphonomy first, and the changes that have occurred in typical fossils since they were living organisms, and then consider the wider implications for paleobiology.

FOSSIL PRESERVATION

Fossilization

When a plant or an animal dies, it is likely that it will not end up as a fossil. For those that do, there are several stages that normally occur in the transition from a dead body to a fossil (Fig. 3.1):

1. Decay of the soft tissues of the plant or animal.
2. Transport and breakage of hard tissues.
3. Burial and modification of the hard tissues.

In rare cases, soft parts may be preserved, and these examples of **exceptional preservation** are crucially important in reconstructing past life.

There are two kinds of fossil, **body fossils**, the partial or complete remains of plants or animals, and **trace fossils**, the remains of the activity of ancient organisms, such as burrows and tracks. In most of the book, "fossil" is used to mean "body fossil", which is the usual practice. Trace fossils are treated separately in Chapter 19.

Hard parts and soft parts

Fossils are typically the hard parts – shells, bones, woody tissues – of previously existing plants and animals. In many cases these **skeletons**, materials used in supporting the bodies of the animals and plants when they were alive, are all that is preserved. Skeletons may nonetheless give useful information about the appearance of an extinct animal because they can show the overall body outline and may show the location of muscles, and woody tissues of plants may allow whole tree trunks and leaves to be preserved in some detail. The fossil record is biased in favor of organisms that have hard parts. Soft-bodied organisms today can make up 60% of the animals in certain marine settings, and these would all be lost under normal conditions of fossilization.

There are a variety of hard materials in plants and animals that contribute to their preservation (Table 3.1). These include inorganic mineralized materials, such as forms of calcium carbonate, silica, phosphates and iron oxides. Calcium carbonate ($CaCO_3$) makes up the shells of foraminifera, some sponges, corals, bryozoans, brachiopods, mollusks, many arthropods and echinoderms. Silica (SiO_2) forms the skeletons of radiolarians and most sponges, while phosphate, usually in the form of **apatite** ($CaPO_4$), is typical of vertebrate bone, conodonts and certain brachiopods and worms. There are also organic hard tissues, such as lignin, cellulose, sporopollenin

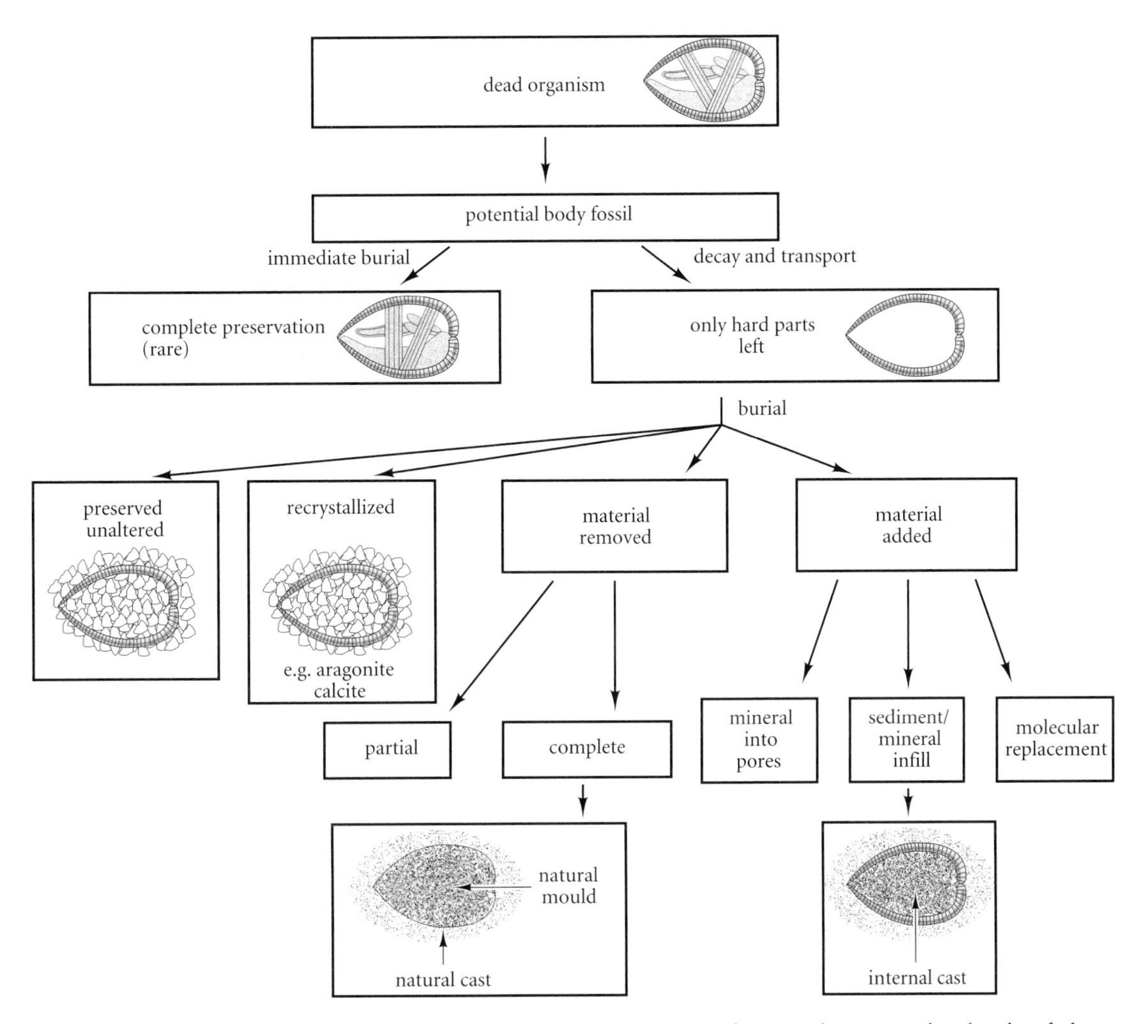

Figure 3.1 How a dead bivalve becomes a fossil. The sequence of stages between the death of the organism and its preservation in various ways.

and others in plants, and chitin, collagen and keratin in animals, which may exist in isolation or in association with mineralized tissues.

Decay

Decay processes typically operate from the moment of death until either the organism disappears completely, or until it is **mineralized**, though mineralization does not always halt decay. If mineralization occurs early, then a great deal of detail of both hard and soft parts may be preserved, so-called exceptional preservation (see below). If mineralization occurs late, as is usually the case, decay processes will have removed or replaced all soft tissues and may also affect many of the hard tissues.

Decay processes exist because dead organisms are valuable sources of food for other organisms. When large animals feed on dead plant or animal tissues, the process is termed **scavenging**, and when microbes, such as fungi or bacteria, transform tissues of the dead organism, the process is termed **decay**. Well-known examples of scavengers are hyenas and vultures, both of which strip the flesh from large animal carcasses. After these large scavengers have had their fill, smaller animals, such as meat-eating beetles, may continue the process of defleshing. In many cases, all flesh is removed in a day or so. Decay is dependent on three factors.

The first factor controlling decay is the supply of oxygen. In **aerobic** (oxygen-rich) situations, microbes break down the organic carbon of a dead animal or plant by convert-

Table 3.1 Mineralized materials in protists, plants, and animals. The commonest occurrences are indicated with XX, and lesser occurrences with X.

Aragonite	Inorganic				Organic				
	Carbonates	Calcite	Phosphates	Silica	Iron oxides	Chitin	Cellulose	Collagen	Keratin
Prokaryotes	XX	X	X		X		X		
Algae	XX	XX		X		X	XX		
Higher plants		X		X	X		XX		
Protozoa		XX		XX	XX	X	X		
Fungi		X	X		X	XX	XX		
Porifera	X	XX		XX	X			XX	
Cnidaria	XX	XX				X		X	
Bryozoa	XX	XX	X			XX		X	
Brachiopoda		XX	XX			XX		X	
Mollusca	XX	X	X	X	X	X		X	
Annelida	XX	XX	XX		X	X		XX	
Arthropoda		XX	XX	X	X	XX		X	
Echinodermata		XX	X	X				XX	
Chordata		X	XX		X		X	XX	XX

ing carbon and oxygen to carbon dioxide and water, according to this equation:

$$CH_2O + O_2 \rightarrow CO_2 + H_2O$$

Microbial decay can also take place in **anaerobic** conditions, that is, in the absence of oxygen, and in these cases nitrate, manganese dioxide, iron oxide or sulfate ions are necessary to allow the decay to occur.

The second set of factors controlling decay, temperature and pH, may be the most important. High temperatures promote rapid decay. Decay proceeds at normal high rates when the pH is neutral, as is the case in most sediments, because this creates ideal conditions for microbial respiration. Decay is slowed down by conditions of unusual pH, such as those found in peat swamps, which are acidic. Fossils preserved in peat or lignite (brown coal) may be tanned, like leather, and many of the soft tissues are preserved. Examples are the famous Neolithic and younger "bog bodies" of northern Europe, in which the skin and internal organs are preserved, and silicified fossils in the lignite of the Geiseltal deposit in Germany (Eocene) that show muscle fibers and skin.

Decay depends, thirdly, on the nature of the organic carbon, which varies from highly **labile** (likely to decay early) to highly decay-resistant. Most soft parts of animals are made from **volatiles**, forms of carbon that have molecular structures that break down readily. Other organic carbons, termed **refractories**, are much less liable to break down, and these include many plant tissues, such as cellulose.

The normal end result of scavenging and decay processes is a plant or animal carcass stripped of all soft parts. In rare cases, some of the soft tissues may survive, and these are examples of exceptional preservation.

Exceptional preservation

There are many famous examples of exceptional preservation (Table 3.2). Certain fossil-bearing formations of different ages, termed **Lagerstätten**, have produced hundreds of remarkable fossil specimens, and in some cases soft parts are preserved. In the most spectacular cases, soft tissues such as muscle, which is composed of labile forms of organic carbon, may be preserved. Usually, however, only the rather more decay-resistant soft tissues, such as chitin and cellulose, are fossilized. Plant and animal tissues decay in a sequence that depends on their volatile content, and the process of decay can only be

Table 3.2 Some of the most famous fossil Lagerstätten (sites of exceptional preservation) in the world.

Lagerstätten	Age	Location
Pre-Cambrian		
Doushantuo Formation	600 Ma	Guizhou Province, China
Ediacara Hills	565 Ma	South Australia
Cambrian		
Maotianshan Shales, Chengjiang	525 Ma	Yunnan Province, China
Emu Bay Shale	525 Ma	South Australia
Sirius Passet	518 Ma	Greenland
House Range	510 Ma	Western Utah, USA
Burgess Shale	505 Ma	British Columbia, Canada
"Orsten"	500 Ma	Sweden
Ordovician		
Soom Shale	435 Ma	South Africa
Silurian		
Ludlow Bonebed	420 Ma	Shropshire, England
Devonian		
Rhynie Chert	400 Ma	Scotland
Hunsrück Slates	390 Ma	Rheinland-Pfalz, Germany
Gilboa	380 Ma	New York, USA
Gogo Formation, Canowindra	360 Ma	New South Wales, Australia
Carboniferous		
Mazon Creek	300 Ma	Illinois, USA
Hamilton Quarry	295 Ma	Kansas, USA
Triassic		
Karatau	213–144 Ma	Kazakhstan
Jurassic		
Posidonienschiefer, Holzmaden	160 Ma	Württemberg, Germany
La Voulte-sur-Rhône	160 Ma	France
Solnhofen Limestone	149 Ma	Bavaria, Germany
Cretaceous		
Yixian Formation	125 Ma	Liaoning, China
Las Hoyas	125 Ma	Cuenca, Spain
Crato Formation	*c.* 117 Ma	Northeast Brazil
Xiagou Formation	*c.* 110 Ma	Gansu, China
Santana Formation	*c.* 100 Ma	Northeast Brazil
Auca Mahuevo	80 Ma	Patagonia, Argentina
Eocene		
Green River Formation	50 Ma	Colorado/Utah/Wyoming, USA
Monte Bolca	49 Ma	Italy
Messel Oil Shale	49 Ma	Hessen, Germany
London Clay	54–48 Ma	UK
Florissant Formation	34 Ma	Colorado, USA
Oligocene-Miocene		
Dominican amber	30–10 Ma	Dominican Republic
Riversleigh	25–15 Ma	Queensland, Australia
Miocene		
Clarkia Fossil Beds	20–17 Ma	Idaho, USA
Ashfall Fossil Beds	10 Ma	Nebraska, USA
Pleistocene		
Rancho La Brea Tar Pits	20,000 ya	California, USA

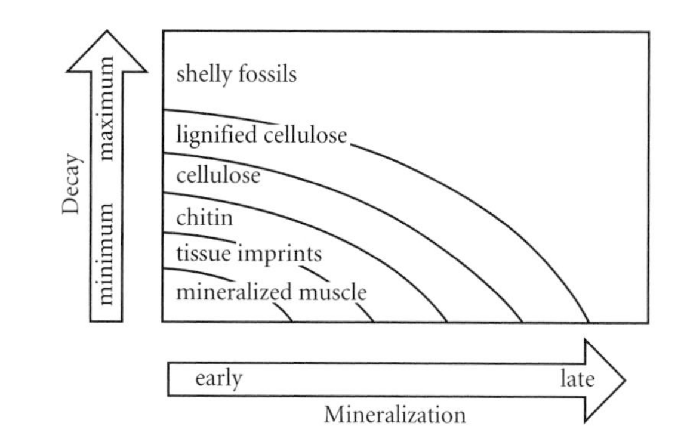

Figure 3.2 The relative rates of decay and mineralization determine the kinds of tissues that may be preserved. At minimum decay rate and with very early mineralization, highly labile muscle tissues may be preserved. When decay has gone to a maximum, and when mineralization occurs late, all that is left are the non-organic tissues such as shells. (Based on Allison 1988.)

halted by mineralization (Fig. 3.2). In the process of fossilization, then, it is possible to think of a race between rates of decay and rates of pre-burial mineralization: the point of intersection of those rates determines the quality of preservation of any particular fossil.

Early mineralization of soft tissues may be achieved in pyrite, phosphate or carbonate, depending on three factors: (i) rate of burial; (ii) organic content; and (iii) salinity (Fig. 3.3a). Physical and chemical effects, such as these, that occur after burial, are termed **diagenesis**. Early diagenetic pyritization (Fig. 3.3b) of soft parts is favored by rapid burial, low organic content and the presence of sulfates in the sediment. Early diagenetic phosphatization (Fig. 3.3c) requires a low rate of burial and a high organic content. Soft-part preservation in carbonates (Fig. 3.3d) is favored by rapid burial in organic-rich sediments; at low salinity levels, siderite is deposited, and at high salinity levels, carbonate is laid down in the form of calcite. In rare cases, decay and mineralization do not occur, when the organism is instantly encased and preserved in a medium such as amber (Fig. 3.3e) or asphalt.

Mineralization of soft tissues occurs in three ways. Rarely, soft tissues may be replaced in detail, or replicated, by phosphates. **Permineralization** occurs very early, probably within hours of death, and may preserve highly labile structures such as muscle fibers (Fig. 3.3b), as well as more refractory tissues such as cellulose and chitin. The commonest mode of mineralization of soft tissues is by the formation of mineral coats of phosphate, carbonate or pyrite, often by the action of bacteria (Box 3.1). The mineral coat preserves an exact replica of the soft tissues that decay away completely. The third mode of soft tissue mineralization is the formation of tissue casts during early stages of sediment compaction. Examples of tissue casts include siliceous and calcareous nodules that preserve the form of the organism and prevent it from being flattened or dissolved.

The mode of accumulation of fossils also determines the nature of fossil Lagerstätten. Fossil assemblages may be produced by **concentration**, the gathering together of remains by normal processes of sedimentary transport and sorting to form fossil-packed horizons (see p. 65), or by **conservation**, the fossilization of plant and animal remains in ways that avoid scavenging, decay and diagenetic destruction (Fig. 3.5). Exceptionally preserved fossil assemblages are produced mainly by processes of conservation. Certain sedimentary regimes, in the sea or in lakes, are stagnant, where sediments are usually anoxic, and are devoid of animals that might scavenge carcasses. In other situations, termed **obrution deposits**, sedimentation rates are so rapid that carcasses are buried virtually instantly, and this may occur in rapidly migrating river channels or at delta fronts and other situations where mass flows of sediment are deposited. Some unusual conditions of instant preservation are termed **conservation traps**. These include **amber**, fossilized resin that oozes through tree bark, and may trap insects, and tar pits and peat beds where plants and animals sink in and their carcasses may be preserved nearly completely.

Breakage and transport

The hard parts left after scavenging and decay have taken their toll may simply be buried without further modification, or they may be broken and transported. There are several processes of breakage (Fig. 3.6), some physical (disarticulation, fragmentation, abrasion)

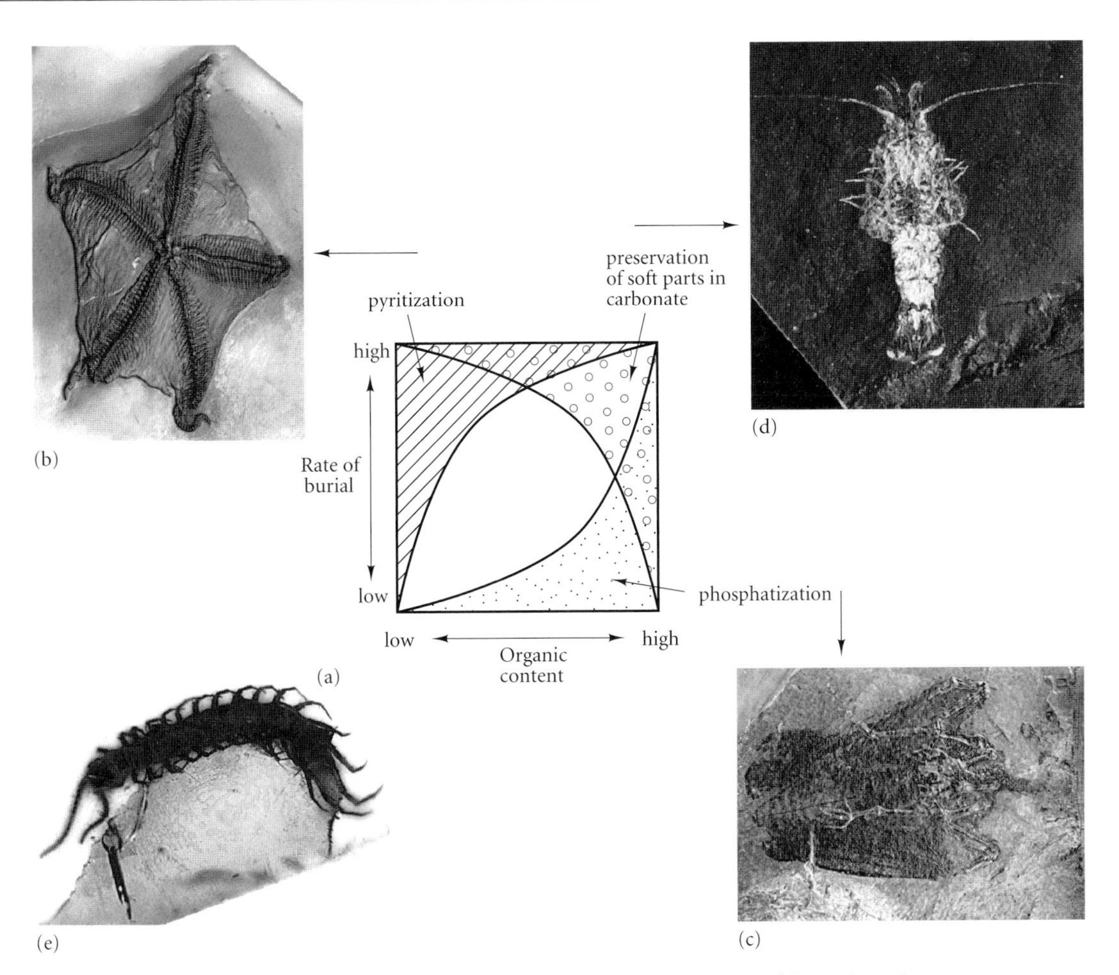

Figure 3.3 The conditions for exceptional preservation. (a) The rate of burial and organic content are key controls on the nature of mineralization of organic matter in fossils. Pyritization (high rate of burial, low organic content) may preserve entirely soft-bodied worms, as in an example from the Early Devonian Hunsrückschiefer of Germany (b). Phosphatization (low rate of burial, high organic content) may preserve trilobite limbs such as this example of *Agnostides* from the Cambrian of Sweden (c). Soft parts may be preserved in carbonate (high rate of burial, high organic content), such as polyps in a colonial coral, *Favosites*, from the Early Silurian of Canada (d). If decay never starts, small animals may be preserved organically and without loss of material, such as a fly in amber from the Early Tertiary of the Baltic region (e). (a, based on Allison 1988; b, courtesy of Phil Wilby; c–e, courtesy of Derek Briggs.)

and some chemical (bioerosion, corrosion and dissolution).

Skeletons that are made from several parts may become **disarticulated**, separated into their component parts. For example, the multielement skeletons of armored worms and vertebrates may be broken up by scavengers and by wave and current activity on the seabed (Fig. 3.6a). Disarticulation happens only after the scavenging or decay of connective tissues that hold the skeleton together. This may occur within a few hours in the case of crinoids, where the ligaments holding the separate skeletal elements together decay rapidly. In trilobites and vertebrates, normal aerobic or anaerobic bacterial decay may take weeks or months to remove all connective tissues.

Skeletons may also become **fragmented**, that is, individual shells, bones or pieces of woody tissue break up into smaller pieces (Fig. 3.6b), usually along lines of weakness. Fragmentation may be caused by predators and scavengers such as hyenas that break bones, or such as crabs that use their claws to

Box 3.1 Exceptional preservation of muscle and microbes

There are now many examples of fossil animals with muscle tissue preserved. These range in age right back to the Cambrian, and there is no diminution in the quality of the specimens with geologic age. A good example is the report of a horseshoe crab from the Upper Jurassic of Germany, presented by Derek Briggs and colleagues (2005) from Yale University and the University of Bristol. The specimen of *Mesolimulus walchi* (Fig. 3.4a) from the Plattenkalk at Nusplingen in Baden-Württemberg

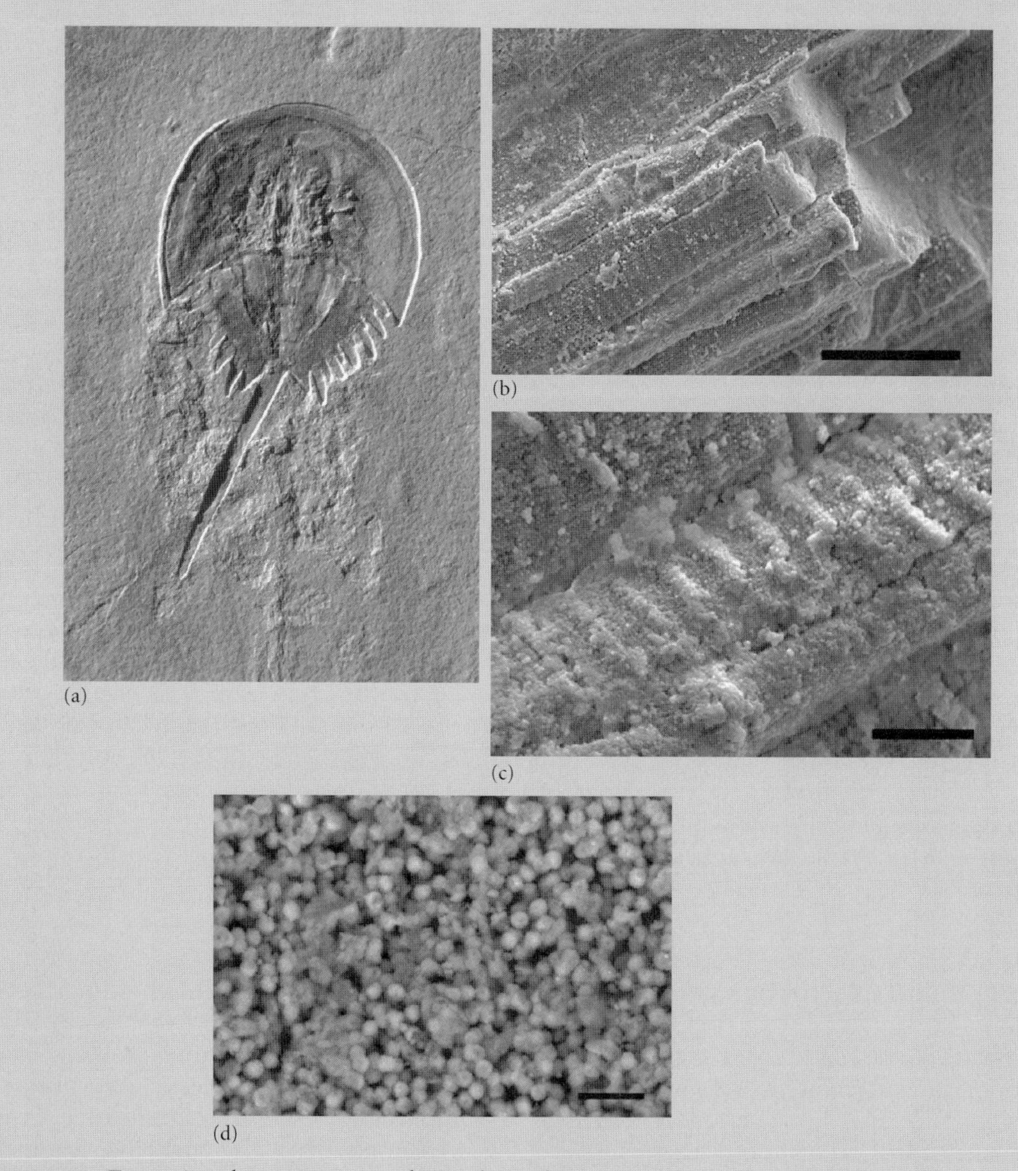

Figure 3.4 Exceptional preservation of muscle in the Jurassic horseshoe crab *Mesolimulus walchi*: (a) the whole specimen showing the rounded headshield (prosoma), with preserved muscle tissues in the middle; (b) muscle fibers; (c) banding across muscle fibers revealed by early decay; and (d) small coccoid microbes associated with the muscle fibers. Scale bars: 20 mm (a), 50 μm (b), 10 μm (c, d). (Courtesy of Derek Briggs.)

looks very like a modern horseshoe crab. The site had been known since 1839 as a source of exquisite fossils of shallow-water marine organisms such as crocodilians, fishes, ammonites and nautiloids with beaks and gut contents, crustaceans and other arthropods, as well as well-preserved land plants washed in from the nearby shore, and pterosaurs that must have fallen in the water.

The specimen was collected during an excavation by the Museum at Stuttgart, and volunteer excavator, Rolf Hugger, who found the specimen, was amazed when he saw that the major muscles of the prosoma, the broad head shield, of this horseshoe crab had survived. Chemical analysis showed that the muscles are preserved as calcium phosphate (apatite). These muscles had a variety of functions: compressing and moving food through the crop, operating the limbs, and bending the body. Under the scanning electron microscope, all the muscle fibers are clear (Fig. 3.4b), and decay had highlighted cross-banding on some of the muscle fibers (Fig. 3.4c). At higher magnification, spherical coccoids (Fig. 3.4d) and spirals could be seen, associated with the preserved muscles. These coccoids and spirals are actually preserved microbes that were presumably feeding on the muscle tissue after the animal died, and formed a so-called **biofilm** over the carcass.

It is well known that muscle tissue breaks down rapidly after an animal dies. Experiments have shown that the muscle here must have been mineralized as apatite within a matter of days, or at most a couple of weeks. The seabed was saturated in calcium carbonate at the time of deposition (the rock is a limestone), and pH has to be lowered slightly to allow calcium phosphate to precipitate. Perhaps the carapace of the dead horseshoe crab acted as a protective roof, inside which microbes began feasting on the muscle tissues and thereby lowered the pH locally enough for apatite to precipitate. The decaying muscle provided some calcium phosphate, but more must have been derived from the surrounding sediment.

Find web references about the Nusplingen fossils at http://www.blackwellpublishing.com/paleobiology/.

snip their way into shelled prey. Much fragmentation is caused by physical processes associated with transport: bones and shells may bang into each other and into rocks as they are transported by water or wind. Wave action may cause such extensive fragmentation that everything is reduced to fine-grained sand.

Shells, bones and wood may be **abraded** by physical grinding and polishing against each other and against other sedimentary grains. Abrasion removes surface details, and the fragments become rounded (Fig. 3.6c). The degree of abrasion is related to the density of the specimen (in general, dense elements survive physical abrasion better than porous

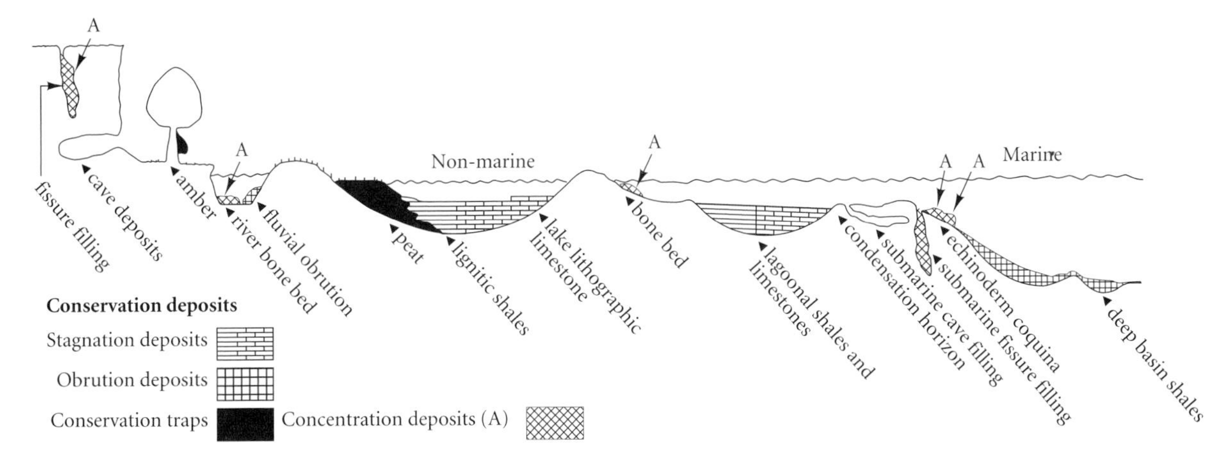

Figure 3.5 An imaginary cross-section showing possible sites of exceptional fossil preservation, most of which are conservation deposits, but a few of which are concentration deposits. (Based on Seilacher et al. 1985.)

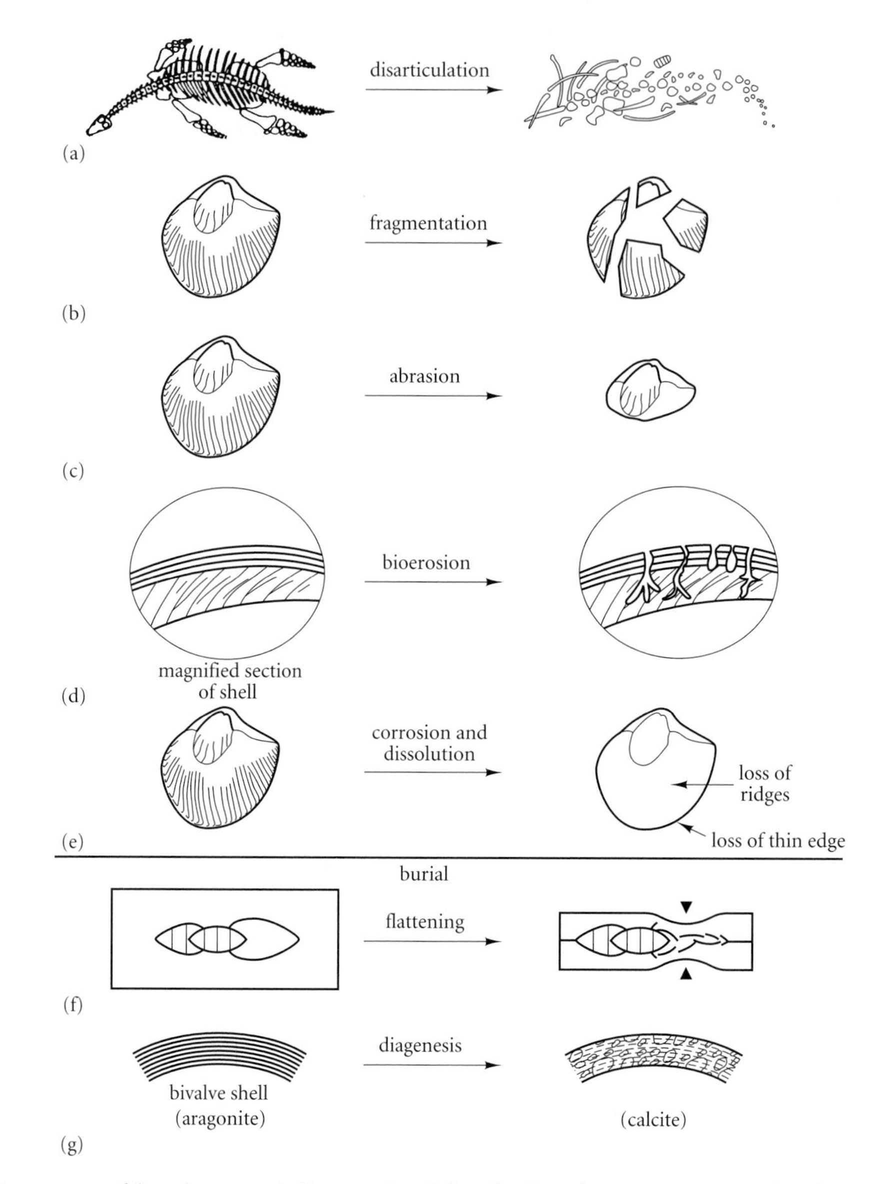

Figure 3.6 Processes of breakage and diagenesis of fossils. Dead organisms may be disarticulated (a) or fragmented (b) by scavenging or transport, abraded (c) by physical movement, bioeroded (d) by borers, or corroded and dissolved (e) by solution in the sediment. After burial, specimens may be flattened (f) by the weight of sediment above, or various forms of chemical diagenesis, such as the replacement of aragonite by calcite (g) may take place.

ones), the energy of currents and grain size of surrounding sedimentary particles (large grains abrade skeletal elements more rapidly than small grains), and the length of exposure to the processes of abrasion.

In certain circumstances shells, bones and wood may undergo **bioerosion**, the removal of skeletal materials by boring organisms such as sponges, algae and bivalves (Fig. 3.6d). Minute boring sponges and algae operate even while their hosts are alive, creating networks of fine borings by chemical dissolution of the calcareous shell material. This process continues after death, and some fossil shells are riddled with borings that may remove more than half of the mineral material of any single specimen. Other boring organisms eat their way into logs, and heavily modify the internal structure.

Before and after burial, skeletal materials are commonly corroded and dissolved by chemical action (Fig. 3.6e). The minerals within many skeletons are chemically unstable, and they break down after death while the specimen lies on the sediment surface, and also for some time after burial. Carbonates are liable to corrosion and dissolution by weakly acidic waters. The most stable skeletal minerals are silica and phosphate.

Burial and modification

Animal and plant remains are typically buried after a great deal of scavenging, decay, breakage and transport. Sediment is washed or blown over the remains, and the specimen becomes more and more deeply buried. During and after burial, the specimen may undergo physical and chemical change.

The commonest physical change is flattening by the weight of sediment deposited above the buried specimen, and this may occur soon after burial. These forces flatten the specimen in the plane of the sedimentary bedding. The nature of flattening depends on the strength of the specimen: the first parts to collapse are those with the thinnest skeleton and largest cavity inside. Greater forces are required to compress more rigid parts of skeletons. Ammonites, for example, have a wide body chamber cavity that would fill up with sand or water after the soft body decayed. This part collapses first (Fig. 3.6f) and, because the shell is hard, it fractures. The other chambers are smaller, fully enclosed and hence mechanically stronger: they collapse later. Plant fossils such as logs are usually roughly circular in cross-section, and they flatten to a more ovoid cross-section after burial. The woody tissues are flexible and they generally do not fracture, but simply distort.

These are examples of diagenesis, and they may occur early, very soon after burial (for example, flattening and some chemical changes), or thousands or millions of years later, as a result of the passage of chemicals in solution through rocks containing fossils. Other examples of late diagenesis include various kinds of deformation by metamorphic and tectonic processes, often millions of years after burial (Box 3.2).

The calcium carbonate in shells occurs in four forms: aragonite, calcite (in two varieties: high magnesium (Mg) calcite, and low Mg calcite), and combinations of aragonite + calcite. The commonest diagenetic process is the conversion of aragonite to calcite. After burial, pore fluids within the sediment may be undersaturated in $CaCO_3$, and the aragonite dissolves completely, leaving a void representing the original shell shape. Later, pore fluids that are supersaturated in $CaCO_3$ allow calcite to crystallize within the void, thus producing a perfect replica of the original shell. This process of replacement of aragonite by calcite occurs commonly, and may be detected by the change of the crystalline structure of the shell (see Fig. 3.6g). The regular layers of aragonite needles have given way to large irregular calcite crystals (sparry calcite) or tiny irregular calcite crystals (micrite).

A common diagenetic phenomenon is the formation of carbonate **concretions**, bodies that form within sediment and concentrate $CaCO_3$ (calcite) or $FeCO_3$ (siderite). Carbonate concretions generally form early during the burial process, and this is demonstrated by the fact that enclosed fossils are uncrushed, having been protected from compaction by the formation of the concretion. Carbonate concretions form typically in black shales, sediments deposited in the sea in anaerobic conditions. Black shales contain abundant organic carbon, and, when this is buried, bacterial processes of anaerobic decay begin. These decay processes reduce oxides in the sediment, and produce bicarbonate ions that may combine with any calcium or iron ions to generate carbonate and siderite concentrations. Such concentrations may grow rapidly to form concretions around the source of calcium and iron ions, usually the remains of an organism.

Another early diagenetic mineral that occurs in anaerobic marine sediments is pyrite (FeS_2). It is also produced as a by-product of anaerobic processes of microbial reduction within shallow buried sediments. Pyrite may replace soft tissues such as muscle in cases of rapid burial, and replaces hard tissues under appropriate chemical conditions. Wood, for example, may be pyritized, and dissolved aragonite or calcite shells may be entirely replaced by pyrite. In both cases, the original skeletal structures are lost.

Phosphate is a primary constituent of vertebrate bone and other skeletal elements. In some cases, masses of organic phosphates are

Box 3.2 Retrodeformation of deformed fossils

Some fossils may be heavily deformed or distorted, so that they do not retain their original shapes. These distortions may be the result of collapse or diagenesis, but they may indicate **metamorphism** – that is, processes connected with **tectonic** activity, faulting, folding and mountain building. If a mudstone is folded and, under high pressure, is changed into a slate, any contained fossils are likely to be stretched and distorted. The deformation is very clear in symmetric fossils (e.g. Fig. 3.7), where the form is stretched in such a way that the original symmetry has been lost. In a slab where numerous fossils lie at different orientations, they will clearly be deformed in different ways, all subject to the same forces in the rocks.

It is possible to restore the original shape of the fossil, a process called **retrodeformation**, meaning "back deformation". The outlines of one, or preferably several, deformed fossils are drawn, usually in two dimensions, and these can be most easily restored to original symmetry in a standard computer drawing software program by manipulating the shape dimensions. This method also allows the analyst to calculate the amount by which the fossil was retrodeformed, and in which direction. This can tell us much about the nature of the tectonic forces that were in operation.

Deformed fossils become commoner the farther back in time one goes, simply because of the greater likelihood than any particular fossiliferous sediment has undergone metamorphism and tectonism.

Find web references about retrodeformation of fossils at http://www.blackwellpublishing.com/paleobiology/.

Figure 3.7 (a) Numerous examples of deformation of the brachiopod *Eoplectodonta*: in a tectonized mudstone from the Silurian of Ireland. (b) A single deformed example (c. 20 mm wide) of a Cambrian *Billingsella* fossil from the Himalayas (Bhutan) and (c) the same example retrodeformed to its original shape.

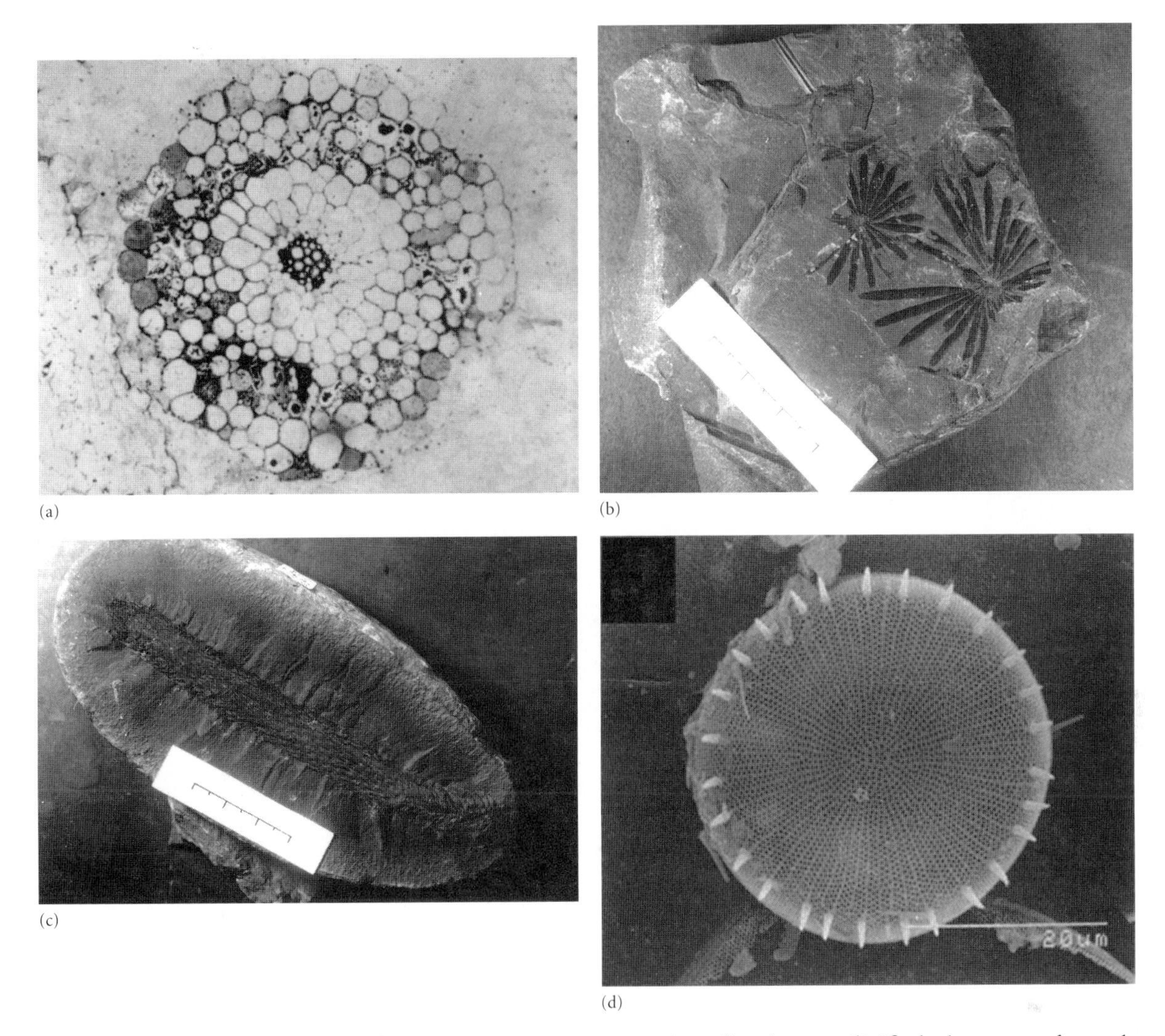

Figure 3.8 Different modes of plant preservation. (a) Permineralization, a silicified plant stem from the Rhynie Chert (Early Devonian, Scotland) (× 50). (b) Coalified compression, leaves of *Annularia* from the Late Carboniferous, Wales (× 0.7). (c) Authigenic preservation, a mold of *Lepidostrobus* from the Late Carboniferous, Wales (× 0.5). (d) Direct preservation of a microscopic fossilized diatom in the original silica (scale bar, 20 µm). (a, courtesy of Dianne Edwards; b, c, courtesy of Chris Cleal; d, courtesy of David Ryves.)

modified by microbial decay, which releases phosphate ions into the sediment. These may combine with calcium ions to form apatite, and this can entirely replace dissolved calcareous shells. In other cases, the microbial processes enable soft tissues, and entirely soft-bodied organisms, to be replaced by phosphate. Coprolites, fossil dung, may also be phosphatized. In these cases, apatite has been liberated from the organisms themselves, and from surrounding concentrations of organic matter, and the replacement destroys most, or all, of the original skeletal structures.

Plant preservation

We deal with plant preservation separately because some modes are different from those seen for fossil animals. Plant parts are usually preserved as compression fossils in fine-grained clastic sediments, such as mudstone, siltstone or fine sandstone, although three-dimensional preservation may occur in exceptional situations. There are four main modes of plant preservation (Schopf 1975): cellular permineralization, coalified compression, authigenic preservation and hard-part preservation (Fig. 3.8).

Plant fossils preserved by cellular permineralization, or **petrifaction**, may show superb microscopic detail of the tissues (Fig. 3.8a), but the organic material has gone. The plant material was invaded throughout by minerals in solution such as silicates, carbonates and iron compounds that precipitated to fill all spaces and replaced some tissues. Examples of cellular permineralization are seen in the Devonian Rhynie Chert and the Triassic wood of the Petrified Forest, Arizona. The most studied examples of permineralized plant tissues are from coal balls. **Coal balls** are irregular masses, often ball-shaped, of concentrated organic plant debris in a carbonate mass, that are commonly found in Carboniferous rocks in association with seams of bituminous coal. Huge collections of coal balls have been made in North America and Europe, and cross-sections of the tissues can reveal astounding detail.

The second common kind of plant preservation is **coalified compression**, produced when masses of plant material lose their soluble components and are compressed by accumulated sediments. The non-volatile residues form a black coaly material, made from broken leaves, stems and roots, and with rarer flowers, fruits, seeds, cones, spores and pollen grains. Coalified compressions may be found within commercially workable coal beds, or as isolated coalified films impressed on siltstones and fine sandstones (Fig. 3.8b).

The third mode of plant preservation, **authigenic preservation** or **cementation**, involves casting and molding. Iron or carbonate minerals become cemented around the plant part and the internal structure commonly degrades. The cemented minerals produce a faithful cast of the external and internal faces of the plant specimen, and the intervening space may be filled with further minerals, producing a perfect replica, or mold, of the original stem or fruit. Some of the best examples of authigenic preservation of plants are ironstone concretions, such as those from Mazon Creek in Illinois and from the South Wales coalfields (Fig. 3.8c).

The fourth typical mode of plant preservation is the direct preservation of hard parts. Some microscopic plants in particular have mineralized tissues in life that survived unchanged as fossils. Examples are coralline algae, with calcareous skeletons, and diatoms, with their silicified cell walls.

QUALITY OF THE FOSSIL RECORD

Incompleteness of the record

From the earliest days of their subject, paleontologists have been concerned about the incompleteness of the fossil record. Charles Darwin famously wrote about the "imperfection of the geological record" in his *On the Origin of Species* in 1859; he clearly understood that there are numerous biological and geological reasons why every organism cannot be preserved, nor even a small sample of every species. In a classic paper in 1972, David Raup explained all the factors that make the fossil record incomplete; these can be thought of as a series of filters that stand between an organism and its final preservation as a fossil:

1 *Anatomic filters*: organisms are likely to be preserved only if they have hard parts, a skeleton of some kind. Entirely soft-bodied organisms, such as worms and jellyfish, are only preserved in rare cases.
2 *Biological filters*: behavior and population size matter. Common organisms such as rats are more likely to be fossilized than rare ones such as pandas. Rats also live for a shorter time than pandas, so more of them die, and more can become potential fossils.
3 *Ecological filters*: where an organism lives matters. Animals that live in shallow seas, or plants that live around lakes and rivers, are more likely to be buried under sediment than, for example, flying animals or creatures that live away from water.
4 *Sedimentary filters*: some environments are typically sites of deposition, and organisms are more likely to be buried there. So, a mountainside or a beach is a site of erosion, and nothing generally survives from these sites in the rock record, whereas a shallow lagoon or a lake is more typically a site of deposition.
5 *Preservation filters*: once the organism is buried in sediment, the chemical conditions must be right for the hard parts to survive. If acidic waters run through the sediment grains, all trace of flesh and

bones or shells might be destroyed. Or if the sediment is constantly being deposited and reworked, for example in a river, any skeletal remains may be worn and damaged by physical movement.

6 *Diagenetic filters*: after a rock has formed, it may be buried beneath further accumulating sediment. Over thousands or millions of years, the rock may be transformed by the passage of mineralizing waters, for example, and these may either enhance the fossils, by replacing biological molecules with mineral molecules, or they may destroy the fossil.
7 *Metamorphic filters*: over millions of years, and the movements of tectonic plates, the fossiliferous rock might be baked or subjected to high pressure. These kinds of metamorphic processes turn mudstones into shales, limestones into marbles. The fossils may survive these terrible indignities, or they may be destroyed.
8 *Vertical movement filters*: nearly all fossils are in sedimentary rocks that have been buried. Burial means the rock has been covered by younger rock, and has gone down to some depth. Tectonic movements must subsequently raise the fossiliferous rock to the Earth's surface, or the fossil remains forever buried and unseen.
9 *Human filters*: the fossil must finally be seen and collected by a human being. Doubtless, the majority of fossils that go through the burial and uplift cycle are lost to erosion, washed away from the foot of a sea cliff or blasted by sand-carrying winds in the desert. Someone has to see the fossil, collect it and take it home. Even then, of course, the fossil has to be registered in a museum before it becomes part of collective human paleontological knowledge. Many that are collected molder in someone's bedroom before they are thrown away with the garbage.

After all this, it's a wonder any fossils survive at all!

The fact that the museums of the world contain so many millions of fossils is a testament to the hard work of paleontologists of all nations. But it also reflects the enormity of geological time and the sheer numbers of organisms that have ever existed.

Bias and adequacy

In his 1972 paper, David Raup argued persuasively that the fossil record is not only incomplete, but also that it is **biased**. This means that the distribution of fossils is not random with respect to time, but that it gets worse in older and older rocks. The evidence is twofold: theoretical and observational. The theoretical evidence is persuasive. The last two or three of the filters just mentioned are time related; the older the rocks, the more substantially they will have removed fossils from the potential record. As times goes by, ancient fossiliferous deposits are ever more likely to have been metamorphosed, buried under younger rocks, subducted into the mantle or eroded. The longer a fossil sits in the rock, the more likely one of these processes is to destroy it. Further, paleontologists are familiar with this steady loss of information. If you try to collect fossils from a Miocene lagoonal deposit, the shells are abundant and beautifully preserved, and you can collect thousands in an hour or two. If you try to collect from a fossiliferous deposit from the same environment in the Cambrian, fossils may be rare, they may be distorted by metamorphism, and they may be hard to get out of the rock.

Others have argued, however, that these biases apply only at certain levels of study. Clearly, in collecting individual shells, you fill your rucksack faster at a Miocene locality than a Cambrian locality. You may also identify many more species based on those collections. But, perhaps if you step back and consider families or genera, rather than species or specimens, and you consider the fossils from whole continents rather than just one quarry, the representation may be relatively uniform. After all, you can recognize the presence of a species or genus from just a single specimen; it does not require a million specimens.

In a study in 2000, Mike Benton and colleagues suggested that the temporal bias identified by Raup might be an issue of scaling. Clearly Raup was right that fossils are steadily lost from the record in older and older rocks. But could the record be *adequate* nonetheless for coarser-scale studies? Benton and colleagues applied clade–stratigraphy measures (Box 3.3) to a sample of 1000 published phylogenetic trees (see p. 129). These trees repre-

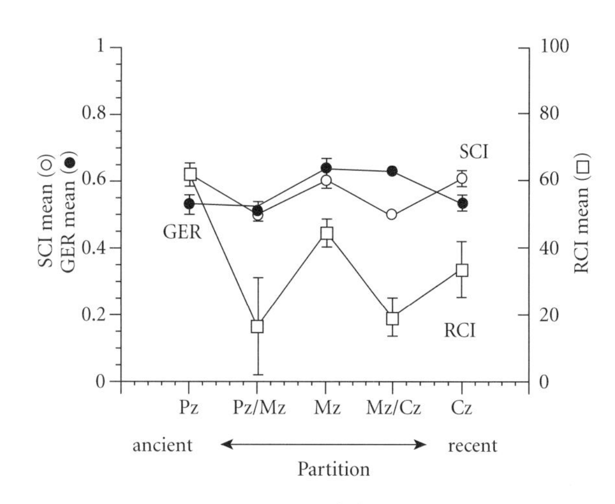

Figure 3.9 Mean scores of the stratigraphic consistency index (SCI), the relative completeness index (RCI) and the gap excess ratio (GER) for five geological time partitions of the data set of 1000 cladograms. Note that the SCI and GER indicate no change through time, while the RCI becomes worse (lower values) from the Paleozoic to Cenozoic – but the RCI depends on total geological time, and so is not a good measure for this study. Pz, cladograms with origins solely in the Paleozoic; Pz/Mz, cladograms with origins spanning the Paleozoic and Mesozoic; Mz, cladograms with origins solely in the Mesozoic; Mz/Cz, cladograms with origins spanning the Mesozoic and Cenozoic; Cz, cladograms with origins solely in the Cenozoic. (Based on Benton et al. 2000.)

sented the branching patterns of different sectors of the tree of life, some of them dating back to the Paleogene, others to the Mesozoic, and yet others to the Paleozoic. These authors divided the 1000 trees into five time bins, each of roughly 200 trees, and they assessed how well the trees matched the fossil record. Using different metrics, the trees showed nearly identical measures of agreement from the Paleozoic to the Cenozoic (Fig. 3.9). Benton and colleagues argued that this confirmed that sampling of the record was equally good (or bad) through the last 500 million years *at a coarse scale*. The cladograms (see p. 129) were generally drawn at coarse taxonomic levels (genera and families, not species) and a coarse time scale was used (stratigraphic stages, average duration 7 million years).

So, paleontologists could breathe a sigh of relief: their studies of the Cambrian might be just as well, or badly, supported by data as their studies of the Carboniferous or Cenozoic. Or could they? What exactly was being measured here, the fossil record or reality?

Preservation bias or common cause?

Many paleontologists have noticed a close linkage between the rock record and the fossil record. Some time intervals, for example, appear to be represented by thick successions of sedimentary rocks that are bursting with fossils, and so the paleontological record of that time interval is especially well documented. What if the fossil record is largely driven by the rock record?

Peters and Foote (2002) noted a close correspondence between the number of named geological formations (standard rock units; see p. 25) and the diversity of named fossils. When they plotted the patterns of appearance and disappearance of marine formations through time (Fig. 3.11a), they noted that this seemed to match the calculated rates of extinction and origination of marine organisms through time. They concluded that perhaps the appearance and disappearance of fossils was controlled by the appearance and disappearance of rocks. If this is the case, then any patterns of diversity, extinction or origination of life through time would really show a geological rather than a biological signal. In other words, the fossil record perhaps shows us little about evolution, and that would be a rather shocking and depressing observation for a paleontologist! This is the **preservation bias hypothesis**, the view that geology controls what we see of the fossil record, as argued by Raup in his classic 1972 paper.

If geology controls the fossil record, what lies behind the appearance and disappearance of formations? Smith (2001) showed that much of the marine rock record relates to relative global sea level. The sea-level curve for the past 600 myr (Fig. 3.11b) shows major rises and falls that reflect phases of seafloor spreading, movements of the tectonic plates, and relative ice volumes (when there are large volumes of polar ice, as at present, global sea levels are low). Smith (2001) noted that many details of the sea-level curve are mimicked by the curves for diversity of marine life (Fig.

Box 3.3 Clade–stratigraphic metrics

Paleontologists have two sources of data about the history of life: the fossils in the rocks and evolutionary trees. If the evolutionary trees are produced using analytic approaches either from molecular or morphological data (see pp. 129–33), there should be no direct linkage between the ages of fossils and the shape of the tree. If that is so, then it should be useful to compare the **congruence** (or agreement) of fossil sequences and phylogenetic trees. If they agree, then perhaps they are both telling the correct story; if they are not congruent, then the fossils, or the tree, or both, could be telling us the wrong story.

There are a variety of metrics for comparing phylogenies and fossil records. The simplest is the Spearman rank correlation coefficient (SRC). This is a **non-parametric** measure that simply compares the order of two series of numbers: if the order is similar enough, the correlation coefficient is statistically significant; if not, the SRC will indicate a non-significant result. So, in the tree in Fig. 3.10a, the **nodes** (branching points) may be numbered 1, 2, 3 and 4 from the bottom to the clade AB or CD (we can not tell whether the node of AB comes before or after that for CD, so can use only one or other in the time series). If the oldest fossils of the clades are in sequence 1, 2, 3, 4, then it is obvious that the two series of numbers (clades and fossils) agree, and the SRC would be +1 indicating a perfect positive correlation. But what if the order of fossils was 1, 2, 4, 3? Is that a good enough agreement or not? With so few digits, the SRC test is inconclusive, but with 10 or more it can give useful outcomes. In an early study, Norell and Novacek (1992) found that 75% of mammal cladograms agreed significantly with the order of fossils. Those that failed the clade versus fossil order SRC test were groups such as primates that are suspected to have a poor fossil record.

Other metrics for comparing cladograms with geological time and fossil occurrences are the stratigraphic consistency index (SCI), the relative completeness index (RCI) and the gap excess ratio (GER).

- The SCI (Huelsenbeck 1994) assesses how well the nodes in a cladogram correspond to the known fossil record. Nodes are dated by the oldest known fossils of either sister group above the node. Each node (Fig. 3.10a) is compared with the node immediately below it. If the upper node is younger than, or equal in age to, the node below, the node is said to be stratigraphically consistent. If the node below is younger, the upper node is stratigraphically inconsistent. The SCI for a cladogram compares the ratio of the sums of stratigraphically consistent to inconsistent nodes. SCI values can indicate cladograms whose nodes are all in line with stratigraphic expectations through to cladograms that imply a sequence of events that is entirely opposite to the known fossil record.
- The RCI (Benton & Storrs 1994) takes account of the actual time spans between nodes, and of implied gaps before the oldest known fossils of lineages. Sister groups, by definition, originated from an immediate common ancestor, and diverged from that ancestor. Thus, both sister groups should have fossil records that start at essentially the same time. In reality, usually the oldest fossil of one lineage will be older than the oldest fossil of its sister lineage. The time gap between these two oldest fossils is the **ghost range** or minimal cladistically-implied gap. The RCI (Fig. 3.10b) assesses the ratio of the ghost range to the known range, and high values imply that ghost ranges are short, and hence that the fossil record is good.
- The GER (Wills 1999) is a modification of the RCI that compares the actual proportion of ghost range in a particular example with the minimum and maximum possible relative amount of ghost range when the cladogram shape is modified to maximize and minimize the ghost range (Fig. 3.10c, d). This then places the result in the context of all possible results, and so assesses the congruence of the tree with the fossil record, taking account of the particular cladogram shape.

These metrics can be used to assess the stratigraphic likelihood of competing cladistic hypotheses that are otherwise equally likely – in other words, if one cladogram implies very little ghost range,

Continued

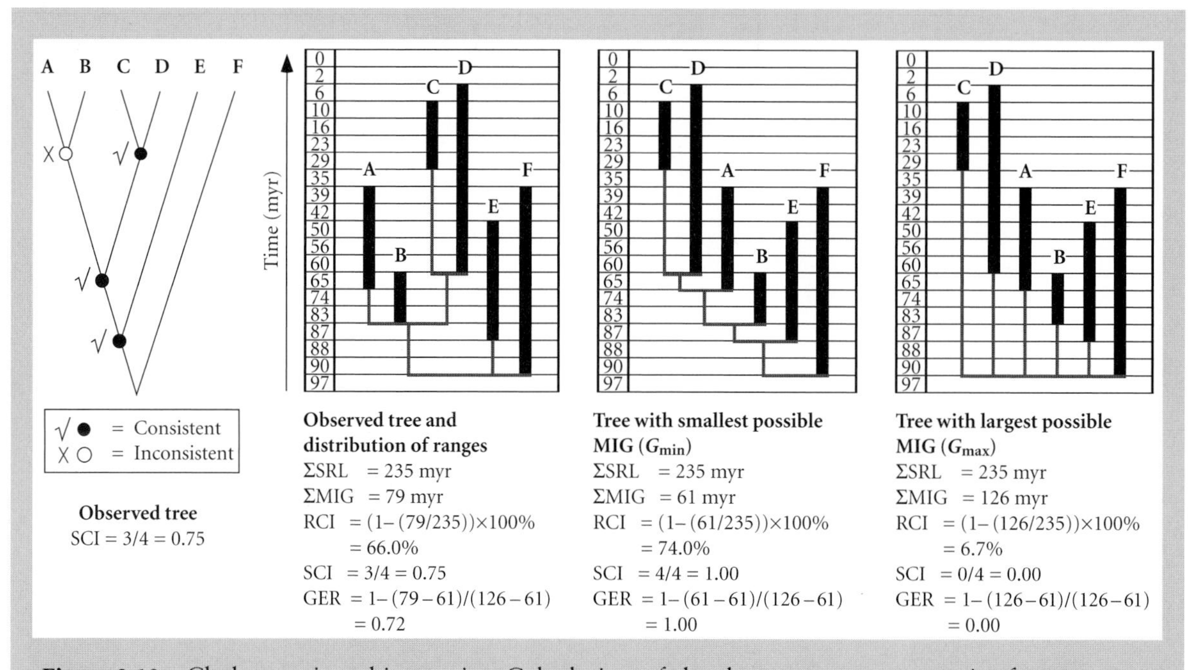

Figure 3.10 Clade–stratigraphic metrics. Calculation of the three congruence metrics for age versus clade comparisons. SCI is the ratio of consistent to inconsistent nodes in a cladogram. RCI is RCI = 1(ΣMIG/ΣSRL), where MIG is minimum implied gap, or ghost range, and SRL is standard range length, the known fossil record. GER is GER = 1(MIG – G_{min})/(G_{max} – G_{min}), where G_{min} is the minimum possible sum of ghost ranges and G_{max} the maximum, for any given distribution of origination dates. (a) The observed tree with SCI calculated according to the distribution of ranges in (b). (b) The observed tree and observed distribution of stratigraphic range data, yielding an RCI of 66.0%. GER is derived from G_{min} and G_{max} values calculated in (c) and (d). (c) The stratigraphic ranges from (b) rearranged on a pectinate tree to yield the smallest possible MIG or G_{min}. (d) The stratigraphic ranges from (b) rearranged on a pectinate tree to yield the largest possible MIG or G_{max}. (Based on Benton et al. 2000.)

and the other implies a huge amount, then the former is probably more likely. Further, large samples of cladograms might give general indications about the preservation and sampling quality of different habitats or fossil groups. For example, Benton et al. (2000) found no overall difference in clade versus fossil matching for marine and non-marine organisms (despite an assumption that marine environments tend to preserve fossils better than non-marine) or between, say, vertebrates and echinoderms. Such comparisons obviously depend on equivalent kinds of cladograms (similar sizes and shapes) within the categories being compared, or the measures become too complex.

Read more in Benton et al. (2000) and Hammer and Harper (2006), and at http://www.blackwellpublishing.com/paleobiology/.

3.11b). Clearly, some drops in biodiversity parallel falls in sea level, and rises in both curves also run in parallel. But, over the past 100 million years, sea level has been falling while diversity has been rising dramatically, so perhaps the pattern can only be read in certain details, but not overall.

What does all this mean? The first conclusion was that geology drives paleontology: the fossil record is closely controlled by sea level and the volume of sedimentary rock being deposited. But what if both are controlled by a third factor? Perhaps times of rare fossils and low rates of deposition really mean some-

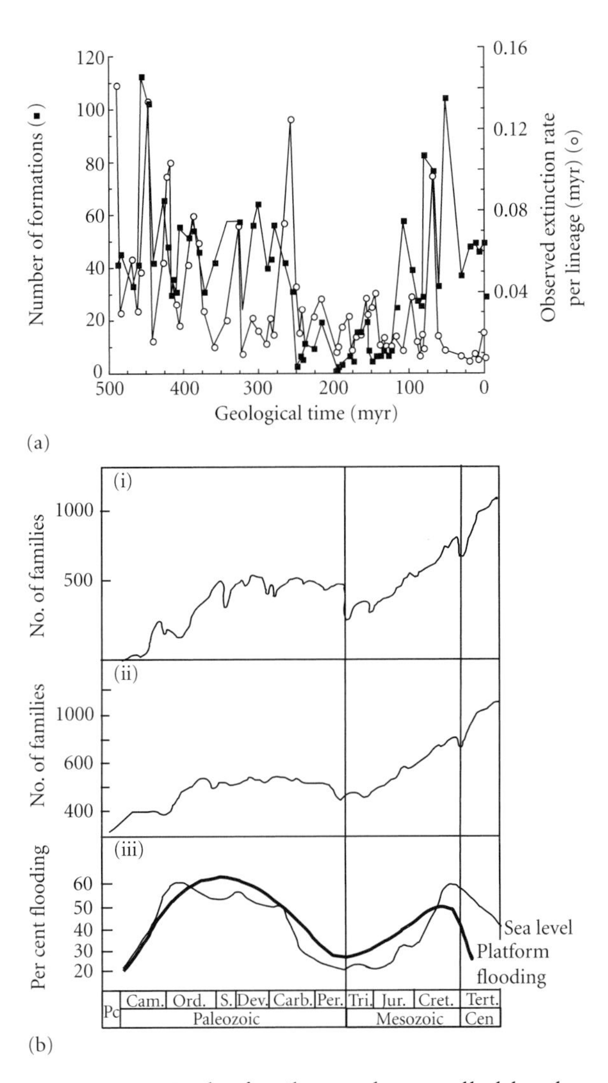

Figure 3.11 Is the fossil record controlled by the rock record? (a) Plot of number of marine geological formations and extinction rate against the last 500 myr of geological time. Note how closely the rock and fossil curves follow each other. (b) Plot of diversification curves for marine families of animals from analyses by Sepkoski (i) and Benton (ii), compared with (iii) the sea-level curve for the Phanerozoic (fine line) and the percentage of platform flooding (heavy line). Note the approximate matching of diversity and sea-level curves until the past 100 myr. (a, based on Peters & Foote 2002; b, based on Smith 2001.)

thing: after a major global catastrophe, for example, rates of shallow marine rock deposition might be low because of a major regression (withdrawal of the sea), and life would also be sparse at the same time. Further, many rocks, most notably certain kinds of limestones, depend on abundant shells and other biological debris for their composition. There is also a human factor – geologists tend to name more formations where fossils are abundant than if they are absent. The fossils provide the basis for biostratigraphy and the discrimination of rock units (see pp. 25–7).

On reflection, many paleontologists and geologists prefer a third option, not that the rocks control the fossils or the fossils control the rocks, but that both are dependent on a third driving factor. This has been termed the **common cause hypothesis** by Peters (2005). The third driving factor is likely to relate to plate tectonic movements and long-term rises and falls in sea level: perhaps marine diversity is high at times of high sea level, and low at times of low sea level. The common cause hypothesis seems to be a better explanation of the apparent correlation between the rock and fossil records than the preservation bias hypothesis (Raup 1972; Smith 2001; Peters & Foote 2002). It is hard to distinguish between the two views, but Peters (2008) shows that, although there is a correlation between fossil and rock records for a comprehensive marine fossil dataset, the agreement breaks down when it is partitioned into a major "Paleozoic" and "modern" division.

Times of crisis in the geological record may provide tests of the common cause and preservation bias hypotheses. Generally, as Peters and Foote (2002) showed, the numbers of geological formations decline after major extinction events. So, for example, there are many fossiliferous geological formations before the Permo-Triassic boundary (PTB) and Cretaceous-Tertiary (KT) mass extinctions, and fossils are abundant and diverse. After both events, the number of formations plummets, as do the numbers of fossils. When studied in detail, some examples appear to weaken the preservation bias hypothesis and support the common cause hypothesis. While fossil diversity and abundance plummet through a mass extinction event, sampling may be constant (i.e. equal numbers of fossiliferous localities in similar rock facies across a time interval). In such cases, the preservation bias hypothesis would predict that fossil abundance and diversity would rise and fall with the numbers of localities or formations sampled. To find the opposite, that fossil diversity falls, while fossil abundance and numbers of localities remain constant, or even rise, suggests that the fossil signal is robust (Wignall & Benton 1999; Benton et al. 2004).

This debate between the preservation bias and common cause hypotheses only reflects the fossils in the rocks, the fossil record as it is recorded. But paleontologists are concerned about a deeper question: do the fossils in the rocks reflect the reality of the past?

Sampling and reality

What are paleontologists doing when they sample the fossil record? Can they build up better and better knowledge of the history of life, or are they simply improving their sampling of a faulty and incomplete record? In a 1994 study, Benton and Storrs showed that sampling is improving through time. Using a clade–stratigraphic metric (see Box 3.3), they compared how paleontological knowledge changed between 1967 and 1993, and they found an apparent improvement of 5% in the 26 years (Fig. 3.12). At least, the congruence between the fossil record as understood in

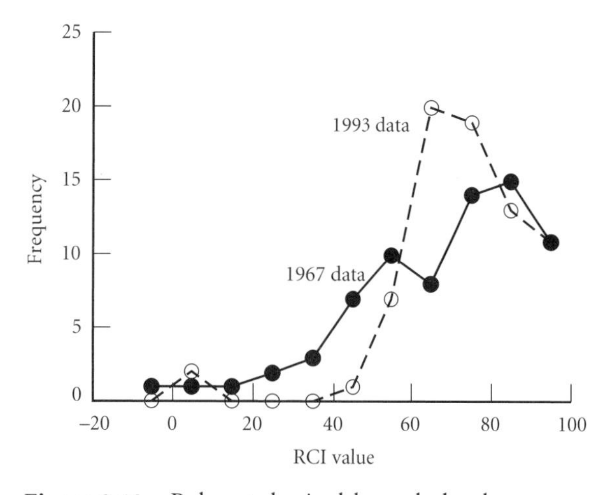

Figure 3.12 Paleontological knowledge has improved by about 5% in the 26-year period between 1967 and 1993. According to 1993 data there is 5% less gap, as assessed by a relative completeness index (RCI), implied in the fossil record of tetrapods than in 1967. This figure was obtained by comparing the order of branching points in cladograms with the order of appearance of fossils in the rocks. Will there be a further 5% shift to the right (i.e. towards 100% completeness) by the year 2019? (Based on Benton & Storrs 1994.)

1993 was 5% better than 1967 when plotted against a set of static cladograms. New fossils were filling the gaps (i.e. reducing the ghost ranges), rather than adding new gaps (i.e. increasing the ghost ranges). One conclusion could be that everything would be known by about the year 2019, but then there is probably a "law of diminishing returns", that ghost ranges will never entirely disappear, and new finds will remove ghost ranges less and less frequently. There is a whole study of ghost ranges, and their markers, the so-called Lazarus taxa (Box 3.4).

All these studies are looking at our knowledge of the fossil record. There are three meanings of the term **fossil record**:

1 Our current knowledge of the fossils in the rocks (the usual meaning).
2 Our ultimate knowledge of the fossils in the rocks (when all fossils have been collected).
3 What actually lived in the past.

As we have seen, many species never left fossils of any kind because they had no hard parts or lived in the wrong place. So, paleontologists can strive to fill the gaps in the fossil record, and that is demonstrably happening (see Fig. 3.12), but how much closer does that bring us to an understanding of what actually lived at any time in the past?

Without supernatural knowledge, that might seem hard to assess. On a good day, paleontologists believe the fossil record (meanings 1 or 2) actually does give us a good outline of the key events in the history of life. On a bad day, it is easy to despair of ever really understanding the history of life (meaning 3) because the fossils we have to hand are such a small remnant of what once existed.

Nonetheless, paleontologists, and other scientists, mostly accept that the fossil record (meaning 1) does give us a broadly correct picture of the history of life (meaning 3). As evidence for this slightly optimistic view, they might point to the lack of surprises. If the fossils were wildly out of kilter with the history of life, we might expect to find human fossils in the Jurassic or dinosaur fossils in the Miocene. We do not (despite Charles Lyell's famous expectation in the 1830s that we might do just that, see p. 13). In fact, new

Box 3.4 Lazarus taxa, Elvis taxa and dead clade walking

There is now a whole terminology for fossils that are absent, or seemingly in the wrong place at the wrong time. David Jablonski of the University of Chicago began the story in 1983 when he invented the term **Lazarus taxa** for species or genera that are present, then seemingly disappear, and then reappear. The name is based on Lazarus in the Bible, who had died, but was brought back to life by Jesus. Clearly species cannot reappear after they have become extinct, so Lazarus taxa identify **gaps** in the record where fossil preservation is poorer than in the beds below and above.

Doug Erwin of the Smithsonian Institution and Mary Droser of the University of California at Riverside then invented the term **Elvis taxa** in 1993 for species or genera that disappear, to be replaced some time later by unrelated by strikingly similar impersonators (i.e. highly convergent species). Elvis taxa can be mistaken for Lazarus taxa if the paleontologist does not study the anatomy carefully.

Not to be outdone, David Jablonski then coined the term **dead clade walking** in 2002 to refer to short-lived survivors of mass extinctions. He had found that many of the organisms that are found after a mass extinction flourish for a while and then go – they had survived the extinction event, but lacked the evolutionary staying power to be a serious part of the recovery.

As Claude Hopkins said in his book *Scientific Advertising* in 1923, "Often the right name is an advertisement in itself".

fossil finds that add to time ranges almost always fill ghost ranges. In other words, new finds, despite the hype in the press ("oldest human fossil rewrites the text books"), almost always fit into expected patterns in time and space.

Perhaps the clade–stratigraphy comparisons (Box 3.3) are the closest to an assessment of the congruence between the fossil record and reality. To put it bluntly, if the fossils fit closely with a phylogenetic tree based on analysis of the DNA of 100 modern species, then perhaps the fossil record (meaning 1) correctly represents reality (meaning 3). This can never be an entirely decisive demonstration, but the more often congruence is found between trees of living organisms and their fossil record, the more confidence perhaps paleontologists might have that the fossils tell the true story of the history of life.

Review questions

1 Summarize the key hard and soft tissues in the human body. Which would decay first (the most labile tissues) and which last (the most refractory tissues)?
2 Which of these groups of fossils are likely to be more completely known, and why: dinosaurs or frogs, mollusks or annelids, birds or bats, land snails or clams?
3 When a tree dies, what might happen step by step to its various parts – leaves, nuts, branches, trunk and roots? How long might each element survive, and where might they end up?
4 Why are Cambrian fossils likely to be less abundant and less well preserved than Miocene fossils?
5 If you were determined to find a new species of fossil, how would you plan your expedition to ensure success?

Further reading

Allison, P.A. & Briggs, D.E.G. 1991. *Taphonomy: Releasing the Data Locked in the Fossil Record.* Plenum Press, New York.

Briggs, D.E.G. 2003. The role of decay and mineralization in the preservation of soft-bodied fossils. *Annual Review of Earth and Planetary Sciences* **31**, 275–301.

Briggs, D.E.G. & Crowther, P.R. 2001. *Palaeobiology; A Synthesis*, 2nd edn. Blackwell Publishing, Oxford.

Donovan, S.K. 1991. *The Processes of Fossilization.* Belhaven Press, London.

Hammer, O. & Harper, D.A.T. 2005. *Paleontological Data Analysis*. Blackwell Publishing, Oxford.

Hopkins, C. 1923. *Scientific Advertising*. Lord & Thomas, New York.

Schopf, J.M. 1975. Modes of plant fossil preservation. *Review of Palaeobotany and Palynology* **20**, 27–53.

References

Allison, P.A. 1988. The role of anoxia in the decay and mineralization of proteinaceous macrofossils. *Paleobiology* **14**, 139–54.

Benton, M.J. & Storrs, G.W. 1994. Testing the quality of the fossil record: paleontological knowledge is improving. *Geology* **22**, 111–14.

Benton, M.J., Tverdokhlebov, V.P. & Surkov, M.V. 2004. Ecosystem remodelling among vertebrates at the Permian-Triassic boundary in Russia. *Nature* **432**, 97–100.

Benton, M.J., Wills, M. & Hitchin, R. 2000. Quality of the fossil record through time. *Nature* **403**, 534–7.

Briggs, D.E.G., Moore, R.A., Shultz, J.W. & Schweigert, G. 2005. Mineralization of soft-part anatomy and invading microbes in the horseshoe crab *Mesolimulus* from the Upper Jurassic Lagerstatte of Nusplingen, Germany. *Proceedings of the Royal Society, London B* **272**, 627–32.

Darwin, C.R. 1859. *On the Origin of Species by Means of Natural Selection, or the Preservation of Favoured Races in the Struggle for Life*. John Murray, London, 502 pp.

Huelsenbeck, J.P. 1994. Comparing the stratigraphic record to estimates of phylogeny. *Paleobiology* **20**, 470–83.

Norell, M.A. & Novacek, M.J. 1992. The fossil record: comparing cladistic and paleontologic evidence for vertebrate history. *Science* **255**, 1690–3.

Peters, S.E. 2005. Geologic constraints on the macroevolutionary history of marine animals. *Proceedings of the National Academy of Sciences USA* **102**, 12326–31.

Peters, S.E. 2008. Environmental determinants of extinction selectivity in the fossil record. *Nature* **453**, in press.

Peters, S.E. & Foote, M. 2002. Determinants of extinction in the fossil record. *Nature* **416**, 420–4.

Raup, D.M. 1972. Taxonomic diversity during the Phanerozoic. *Science* **177**, 1065–71.

Schopf, J.M. 1975. Modes of plant fossil preservation. *Review of Palaeobotany and Palynology* **20**, 27–53.

Seilacher, A., Reif, W.-E., Westphal, F., Riding, R., Clarkson, E.N.K. & Whittington, H.B. 1985. Extraordinary fossil biotas: their ecological and evolutionary significance. *Philosophical Transactions of the Royal Society B* **311**, 5–23.

Smith, A.B. 2001. Large-scale heterogeneity of the fossil record, implications for Phanerozoic biodiversity studies. *Philosophical Transactions of the Royal Society B* **356**, 1–17.

Wignall, P. & Benton, M.J. 1999. Lazarus taxa and fossil abundance at times of biotic crisis. *Journal of the Geological Society of London* **156**, 453–6.

Wills, M.A. 1999. Congruence between phylogeny and stratigraphy: randomization tests. *Systematic Biology* **48**, 559–80.

Chapter 4

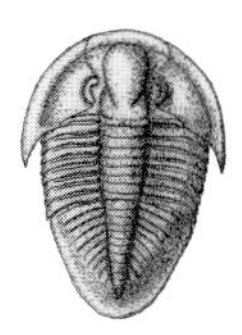

Paleoecology and paleoclimates

Key points

- Fossil organisms provide fundamental evidence of evolution; they also allow the reconstruction of ancient animal and plant communities.
- Paleoecologists study the functions of single fossil organisms (paleoautecology) or the composition and structure of fossil communities (paleosynecology).
- The paleoecology of fossil organisms can be described in terms of their life strategies and trophic modes together with their habitats; virtually all fossil organisms interacted with other fossil organisms and their surrounding environment.
- Populations and paleocommunities may be analyzed with a range of statistical techniques.
- Evolutionary paleoecology charts the changing structure and composition of paleocommunities through time.
- There have been marked changes in the number and membership of Bambachian megaguilds (groups of organisms with similar adaptive strategies), the depth and height of tiering, the intensity of predation, and the composition of shell concentrations through time.
- Ecological events can be classified and ranked in importance; they can be decoupled in significance from biodiversification events.
- Paleoclimates can be described on the basis of climatically-sensitive biotas and sediments together with stable isotopes.
- Climate has been an important factor in driving evolutionary change at a number of different levels.
- Feedback loops between organisms and their environments indicate that the Gaia hypothesis is a useful model for some of geological time.

I do not know what I may appear to the world, but to myself I seem to have been only a boy playing on the sea-shore, and diverting myself in now and then finding a smoother pebble or a prettier shell than ordinary, whilst the great ocean of truth lay all undiscovered before me.

Sir Isaac Newton (shortly before his death in 1727)

PALEOECOLOGY

Pebbles and shells on the beach give us clues about their sources. Paleontologists can reconstruct ancient lifestyles and ancient scenes based on such limited information, and this is the basis of paleoecology. **Paleoecology** is the study of the life and times of fossil organisms, the lifestyles of individual animals and plants together with their relationships to each other and their surrounding environment. We know a great deal about the evolution of life on our planet but relatively little about the ways organisms behaved and interacted. Paleoecology is undoubtedly one of the more exciting disciplines in paleontology; reconstructing past ecosystems and their inhabitants can be great fun. But can we really discover how extinct animals such as the dinosaurs or the graptolites really lived? How did the bizarre animals of the Burgess Shale live together and how did such communities adapt to environmental change?

It is impossible to journey back in time to observe extraordinary ancient communities, so we must rely on many lines of indirect evidence to reconstruct the past and, of course, some speculation. This element of speculation has prompted some paleontologists to exclude paleoecology from mainstream science, suggesting that such topics are better discussed at parties than in the lecture theatre. Emerging numerical and statistical techniques, however, can help us frame and test hypotheses – paleoecology is actually not very different from other sciences.

More recently, too, paleoecology has developed much wider and more serious significance in investigations of long-term planetary change; ecological data through time now form the basis for models of the planet's evolving ecosystem. The influential writings of James Lovelock have extravagantly echoed the suspicions of James Hutton over two centuries ago, that Earth itself can be modeled as a superorganism. The concept of **Gaia** describes the planet as a living organism capable of regulating its environment through a careful balance of biological, chemical and physical processes. Ecological changes and processes through time have been every bit as important as biodiversity changes; these studies form part of the relatively new discipline of evolutionary paleoecology.

Paleoecological investigations require a great deal of detective work. It is relatively easy to work out what is going on in a living community (Fig. 4.1). Ecologists are very interested in the adaptations of animals and plants to their habitats, the interactions between organisms with each other and their environment, as well as the flow of energy and matter through a community. Ecologists also study the planet's life at a variety of levels ranging through populations, communities, ecosystems and the biosphere as a whole. By sampling a living community, ecologists can derive accurate estimates of the abundance and biomass of groups of organisms, the diversity of a community and its trophic structure. But fossil animals and plants commonly are not preserved in their life environments. Soft parts and soft-bodied organisms are usually removed by scavengers, whereas hard parts may have been transported elsewhere or eroded during exposure (see Chapter 3). In a living nearshore community (Fig. 4.1) the soft-bodied organisms, such as worms, would rapidly disappear together with the soft parts of the bony and shelly animals, for example the fishes and the clams; the multiskeletal organisms such as the bony fishes would disaggregate and animals with two or more shells would disarticulate. Fairly quickly there would only be a layer of bones and shells left with possibly some burrows and tracks in the sediment. Moreover, some environments are more likely to be preserved than others; marine environments survive more commonly than terrestrial ones.

Although fossil assemblages suffer from this information loss, paleoecological studies must, nevertheless, have a reliable and sound taxonomic basis – fossils must be properly identified. And although much paleoecological deduction is based on **actualism** or **uniformitarianism**, direct comparisons with living analogs, some environments have changed through geological time as have the lifestyles and habitats of many organisms. For example, some ecosystems such as the "stromatolite world" – sheets of carbonate precipitated by cyanobacteria (see p. 189) – existed throughout much of the Late Precambrian, returning during the Phanerozoic only after some major extinction events and only for a short time (Bottjer 1998). Nevertheless, a few basic principles hold true. Organisms are adapted for,

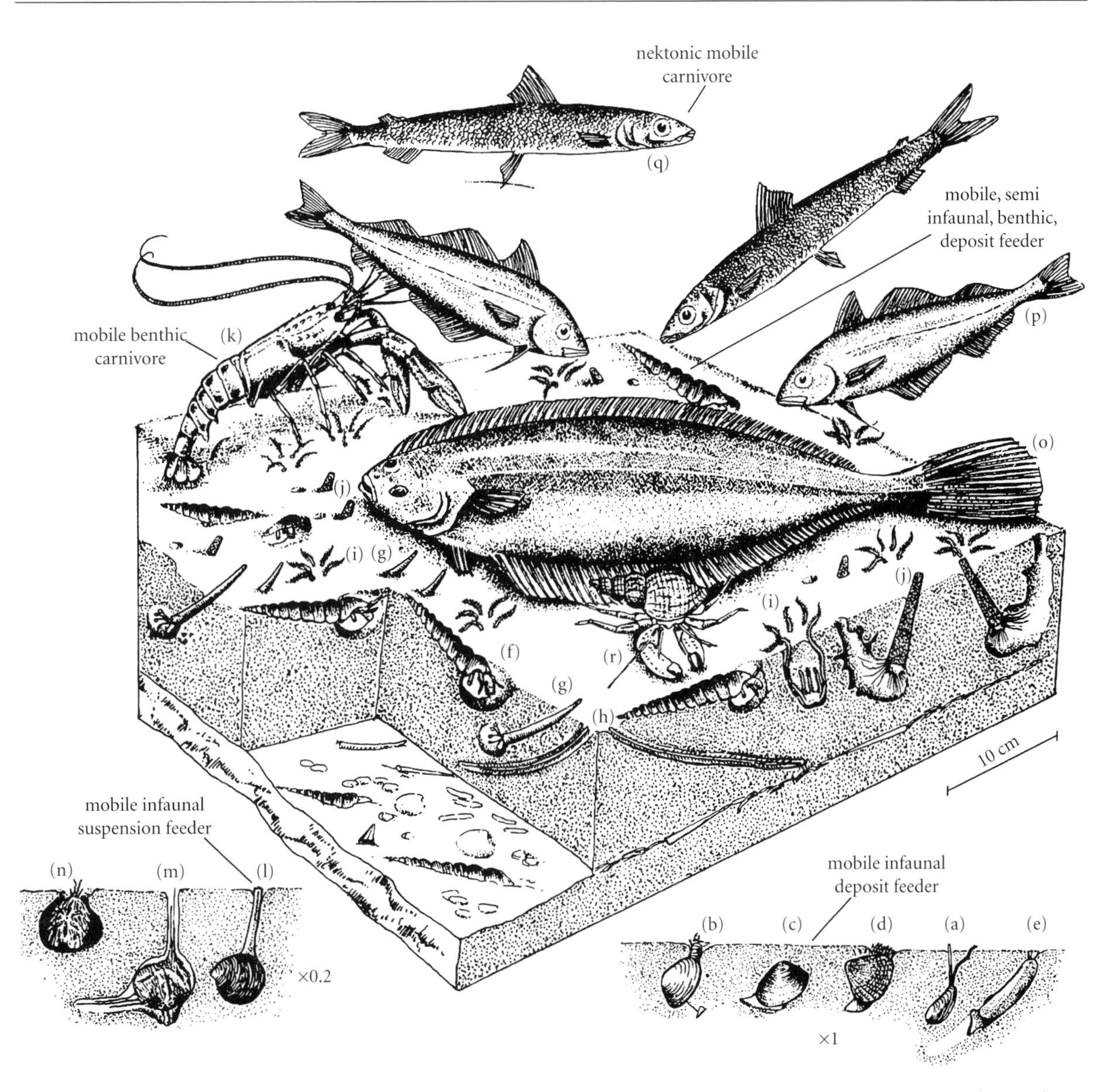

Figure 4.1 Life modes of marine organisms in a living offshore, muddy-sand community in the Irish Sea with a range of bivalves (a–e, l), gastropods (f), scaphopods (g), annelids (h, j), asterozoans (i), crustaceans (k, r), echinoids (m, n) and fishes (o–q). Insets indicate large and small burrowers. (From McKerrow 1978.)

and limited to, a particular environment however broad or restricted; moreover most are adapted for a particular lifestyle and all have some form of direct or indirect dependence on other organisms. These principles are valid also for the study of the ecology of ancient animals and plants.

There are two main areas of paleoecological research: **paleoautecology** is the study of the ecology of a single organism whereas **paleosynecology** looks at communities or associations of organisms. For example, autecology covers the detailed functions and life of a coral species, and synecology might be concerned with the growth and structure of an entire coral reef, including the mutual relationships between species and their relationship to the surrounding environment. The autecology of individual groups is discussed in the taxonomic chapters. In most studies the functions of fossil animal or plants are established through analogies or homologies with living organisms or structures or by a series of experimental and modeling techniques. Geological evidence, however, remains the main test of these comparisons and models. In this chapter we focus on the community aspects of paleoecology (synecology), reviewing the tools available to reconstruct past ecosystems and see how their organisms socialized.

Taphonomic constraints: sifting through the debris

As noted above, most fossil assemblages have been really messed about before being buried and preserved in sediment. The decay and degradation of animal and plant communities after death results in the loss of soft-bodied organisms, while decay removes soft tissue with the disintegration of multiplated and multishelled skeletal taxa (see Chapter 3). If that were not enough, transport and compaction add to the overall loss of information during fossilization. On the other hand, areas occupied by dead communities may be recolonized and animal and plant debris may be supplemented by material washed in from elsewhere. This process of time averaging can thus artificially enhance the diversity of an assemblage over hundreds of years. But can we rely on fossil assemblages to recreate ancient communities with any confidence and accuracy? Paleoecologists know we can, with varying degrees of precision.

The similarity of a death assemblage to its living counterpart, its **fidelity**, can be assessed in different ways. In a series of detailed studies of the living and dead faunas of Copana Bay and the Laguna Madre along the Texas coast, George Staff and his colleagues (e.g. Staff et al. 1986) discussed the paleoecological significance of the taphonomy of a variety of nearshore communities, sampled over a number of years. Most animals in living communities are not usually preserved, nevertheless the majority of animals with preservation potential (mainly shelled organisms) are in fact fossilized. More were actually found in death assemblages than in living assemblages, where the effects of time averaging were clearly significant. Suspension feeders and infaunal organisms were the most likely to be preserved (Fig. 4.2). Measurements of biomass and taxonomic composition rather than those of numerical abundance and diversity are the best estimates of the structures of communities, and counts of the more stable adult populations are the most realistic monitors of community structure.

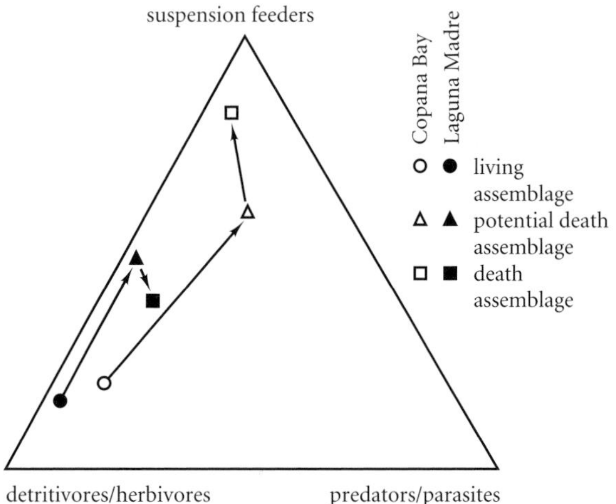

Figure 4.2 The transition from a living assemblage to a death assemblage. Relative proportions of different types of organism change in two living marine assemblages off the Texan coast. Living assemblages are dominated numerically by detritivores and herbivores, death assemblages by suspension feeders. (Based on Staff et al. 1986.)

Another method to estimate taphonomic loss involves a census of an extraordinarily preserved Lagerstätte deposit. Whittington (1980) and his colleagues' detailed reinvestigation of the mid-Cambrian Burgess Shale fauna revealed a community dominated by soft-bodied animals with very few of the more familiar skeletal components of post-Cambrian faunas such as brachiopods, bryozoans, gastropods, bivalves, cephalopods, corals and echinoderms. More importantly, the deep-water Burgess fauna is quite different from more typical Cambrian assemblages with phosphatic brachiopods, simple echinoids and mollusks together with trilobites. Although the Burgess fauna has many other peculiarities (see Chapter 10), the high proportion of, for example, annelid and priapulid worms, adds a different dimension to the more typical reconstructions of mid-Cambrian communities (Fig. 4.3).

These important taphonomic constraints must be addressed and built into any paleoecological analysis and may be partly countered by a careful selection of sampling methods. A variety of methods involving the

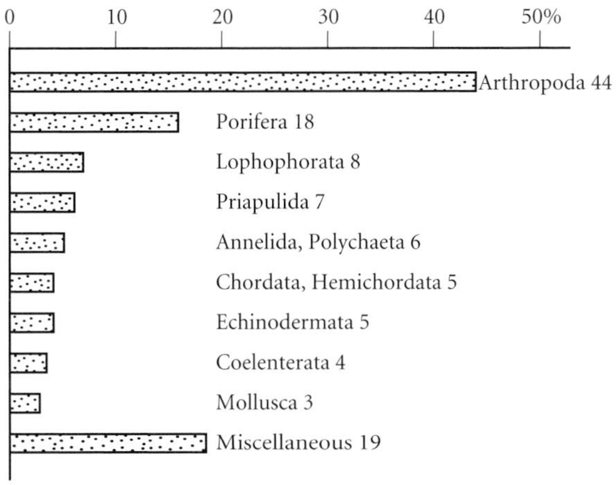

0 10 20 30 40 50%
Arthropoda 44
Porifera 18
Lophophorata 8
Priapulida 7
Annelida, Polychaeta 6
Chordata, Hemichordata 5
Echinodermata 5
Coelenterata 4
Mollusca 3
Miscellaneous 19

Figure 4.3 Census of organisms preserved in the Middle Cambrian Burgess Shale. Many groups, such as the priapulid and annelid worms, together with the diverse arthropod biota, are rarely represented in more typical mid-Cambrian faunas, dominated by phosphatic brachiopods and trilobites. (From Whiltington 1980.)

study of size–frequency histograms (see below), the degree of breakage, disarticulation and fragmentation of individuals, together with the attitude of fossils in sediments, generate useful criteria to separate **autochthonous** (in place) from **allochthonous** (transported) assemblages (see Chapter 4). A number of terms have been developed to describe the fate of a once-living assemblage on its journey to fossilization. The living assemblage, or **biocoenosis**, is transformed into a **thanetocoenosis** after death and decay. The **taphocoenosis** is the end product that is finally preserved. In addition **life assemblages** still retain the original orientations of their inhabitants, **neighborhood assemblages** are still close to their original habitats, whereas **transported assemblages** include broken and abraded bones and shells that have traveled.

Populations: can groups of individuals make a difference?

Populations are the building blocks of communities, and can themselves spark dramatic changes in community and ecosystem structures. A **population** is a naturally occurring assemblage of plants and animals that live in the same place at the same time and regularly interbreed. Within an **ecosystem** – all the populations of species living in association – there may be **keystone species**, species that help shape the ecosystem and that can trigger large-scale changes if they disappear. A classic keystone species is the elephant: it forms the landscape in large parts of Africa by knocking down trees and feeding on certain plants, and the whole scene looks different when it disappears. **Incumbent species** can occupy the same ecological niche for many millions of years, adding stability to many ecosystems. For example, although the dinosaurs and the mammals appeared at roughly the same time, it was the dinosaurs that dominated the land throughout the Mesozoic; mammals had limited niches (insectivores, seed eaters and small omnivores) until after the extinction of the incumbent dinosaurs, when they were able to radiate into vacant ecospaces.

The dynamics and structures of individual populations can provide us with useful clues about how the once-living community functioned and whether the assemblage is actually in place or has been transported. A measurement, such as the length of a brachiopod shell, is chosen as a proxy for the size (and sometimes for the age) of shells. These data, entered into a frequency table, based on discrete class intervals, are plotted as **size–frequency histograms, polygons** or even **cumulative frequency polygons** (Fig. 4.4). Right, positively-skewed curves generally indicate high infant mortality and these are typical of most invertebrate populations. A normal (Gaussian) curve can indicate a steady-state population or transported assemblages whereas a left, negatively-skewed curve indicates high senile mortality. Mortality patterns are, however, best displayed as **survivorship curves**, where the number of survivors at each defined growth intervals is plotted (Fig. 4.5). Size–frequency and survivorship curves store a great deal of information regarding the lifestyle, habitat and life history of an individual organism (Box 4.1). For example, species that mature early and produce small but numerous offspring, many dying before maturity, have been labeled "***r* strategists**". "***K* strategists**", on the other hand, are long-lived species, with low reproduction rates. These two strategies are end members of a spectrum of possibilities described by the following model:

$$dN/dt = rN[(K - N)/K]$$

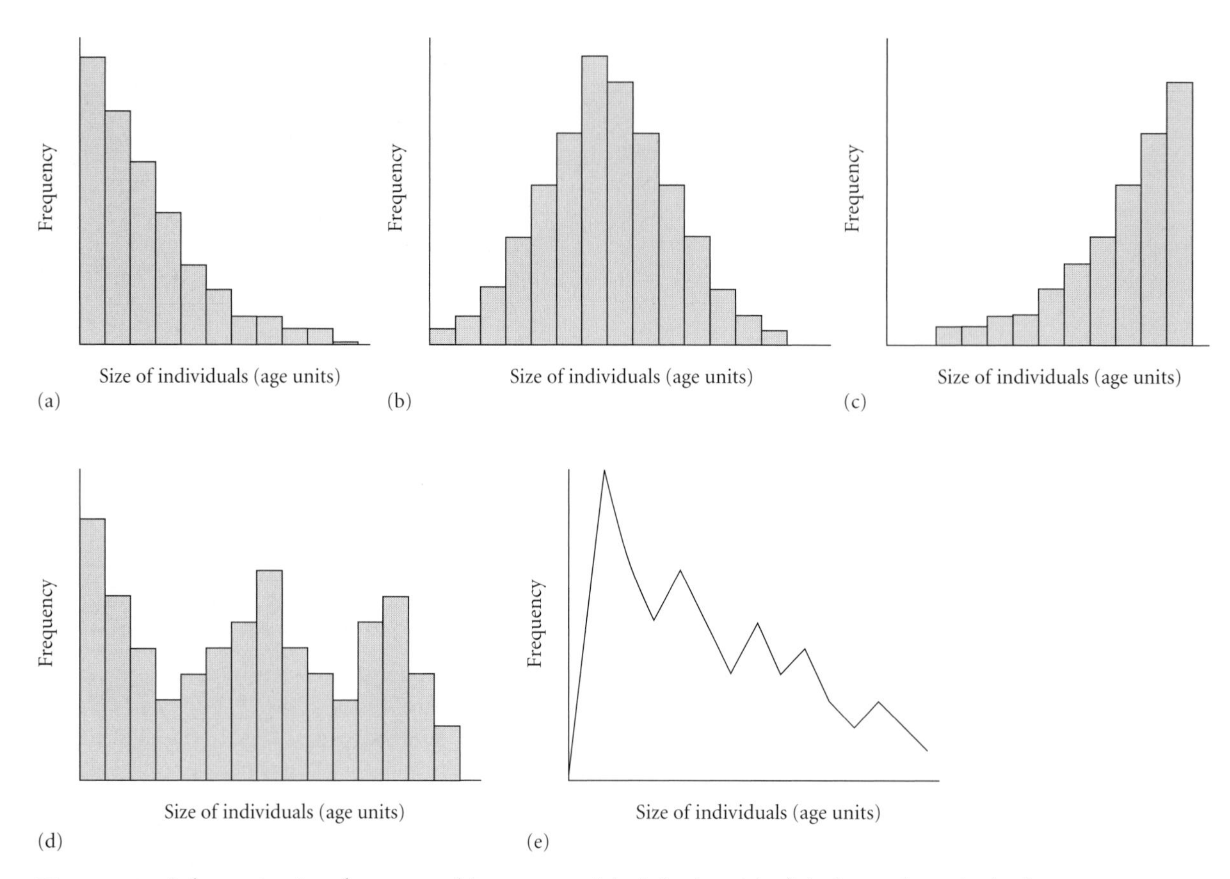

Figure 4.4 Schematic size–frequency histograms: (a) right (positively) skewed, typical of many invertebrate populations with high infant mortality; (b) normal (Gaussian) distribution, typical of steady-state or transported assemblages; (c) left (negatively) skewed, typical of high senile mortality; (d) multimodal distribution, typical of populations with seasonal spawning patterns; and (e) multimodal distribution, with decreasing amplitude, typical of populations growing by molting (ecdysis).

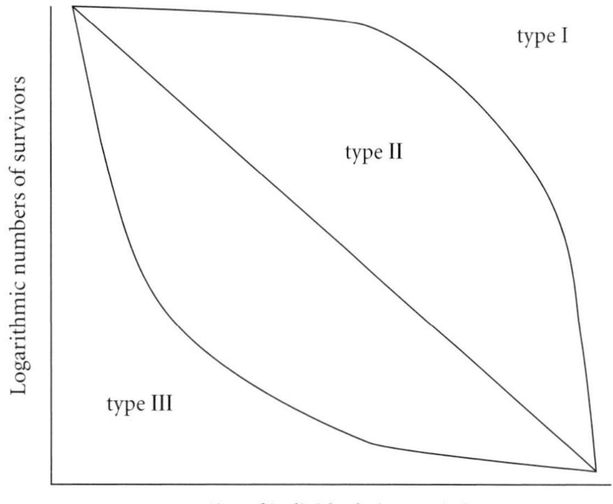

Figure 4.5 Schematic survivorship curves: type I tracks, increasing mortality with age; type II, constant mortality with age; type III, decreasing mortality with age.

where K is the carrying capacity of the population or upper limit of population size, N is the actual population size, r is the intrinsic rate of population increase and t is the unit of time.

Thus, when N approaches K the rate of population growth slows right down and the population will approach a stable equilibrium. Such populations are typical of more stable environments dominated by **equilibrium species** (K strategists). By contrast **opportunistic species** thrive in more adverse, unstable environments, where high growth rates are common (r strategists).

Habitats and niches: addresses and occupations

All modern and fossil organisms can be classified in terms of their **habitat**, where they live

Box 4.1 The terebratulide brachiopod *Dielasma* from the Permian of the Tunstall Hills

The smooth terebratulide brachiopod *Dielasma* is common in the limestones and dolomites associated with the Permian reefs of the Sunderland area in northeast England. Is it possible to use data from simple length measurements of the brachiopod shell to determine the growth strategies of these animals? One sample shows a bimodal pattern suggesting two successive cohorts are present in the population; overall the survivorship curve suggests increasing mortality with age, in possibly a stable, equilibrium environment (Fig. 4.6). But this was not the only environment around these Permian reefs; other samples show different-shaped curves, some demonstrating high infant mortality in possibly less stable environments, whereas a population with a bell-shaped curve suggests that the shells have been transported and sorted prior to burial. A selection of datasets is available by following this link, http://www.blackwellpublishing.com/paleobiology/.

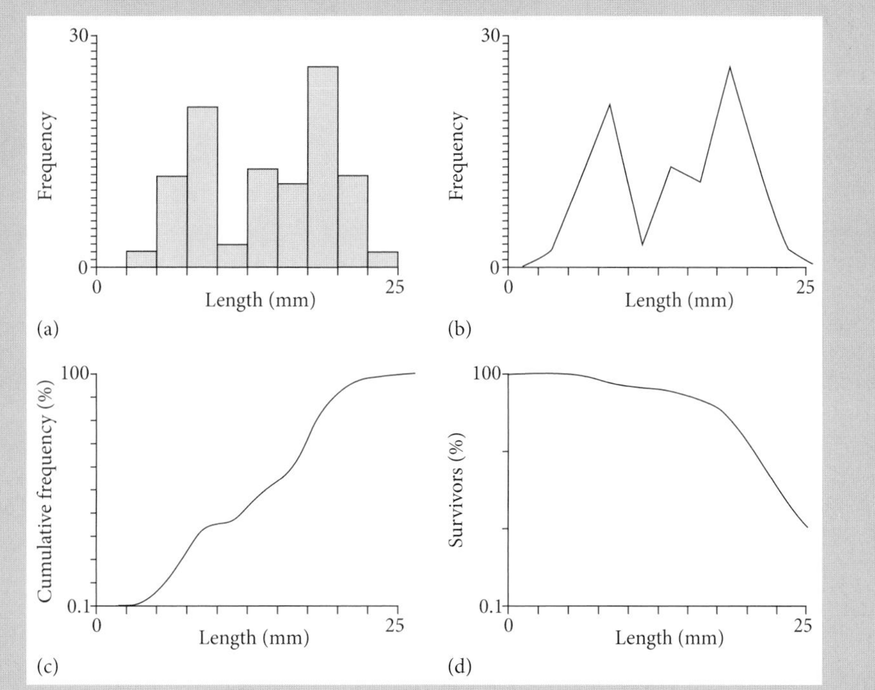

Figure 4.6 Size–frequency histogram (a), polygon (b), cumulative frequency polygon (c) and survivorship curve (d) for a sample of 102 conjoined valves of *Dielasma* from the Permian reef base deposit of the Tunstall Hills, Sunderland. (From Hammer & Harper 2005.)

(their address) or with reference to their **niche**, their lifestyle (their occupation). Modern organisms occupy a range of environments from the top of Mount Everest at heights of nearly 9 km to depths of over 10 km in the Marianas Trench in the Pacific Ocean. Recognition of extremophiles (see p. 205), living in even more bizarre habitats, has considerably extended our understanding of the environmental range of life on Earth. A large number of physical, chemical and biological factors may characterize an organism's environment; unfortunately, few can be recognized in the fossil record.

Some of the most abundant and diverse communities inhabit the littoral zone, where rocky shores hold some of the most varied and extensively studied faunas. For example, nearly 2000 individual organisms have been recorded from a 250 mm^2 quadrat on an

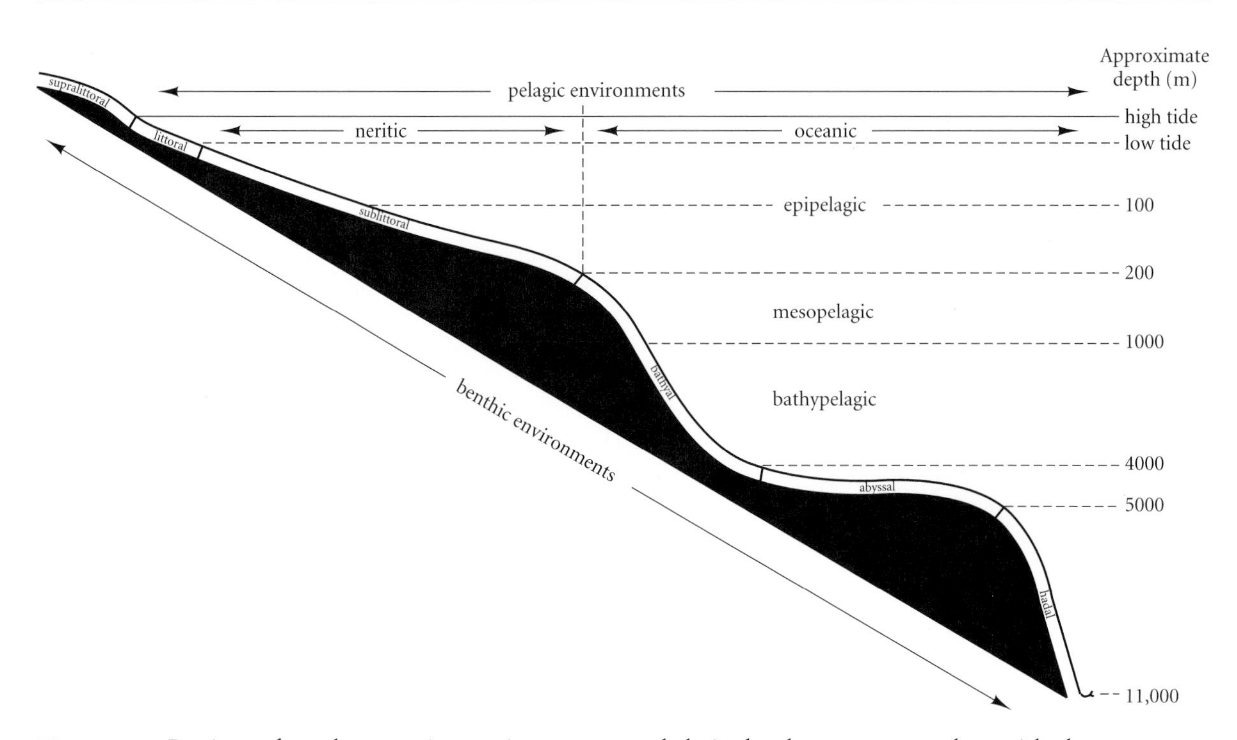

Figure 4.7 Review of modern marine environments and their depth ranges, together with the approximate positions of the main benthic zones. (Based on Ager 1963.)

exposed wave-battered platform around the Scottish island of Oronsay. Unfortunately, few rocky coasts have been recorded from the geological record; where they occur, often associated with paleo-islands, there are exciting and unusual biotas and sediments (Johnson & Baarli 1999).

The majority of fossil animals have been found in marine sediments, occupying a wide range of depths and conditions. The distribution of the marine benthos is controlled principally by depth of water, oxygenation and temperature. The main depth zones and pelagic environments are illustrated on Fig. 4.7. In addition, the **photic zone** is the depth of water penetrated by light; this can vary according to water purity and salinity but in optimum conditions it can extend down to about 100 m. Terrestrial environments are mainly governed by humidity and temperature, and organisms inhabit a wide range of continental environments, ranging from the Arctic tundras to the lush forests of the tropics.

Marine environments host a variety of lifestyles (Fig. 4.8). The upper surface waters are rich in floating **plankton**, and **nektonic** organisms swim at various levels in the water column. Within the **benthos** – the beasts that live in or on the seabed – mobile nektobenthos scuttle across the seafloor and the fixed or sessile benthos are fixed by a variety of structures. **Infaunal** organisms live beneath the sediment–water interface, while **epifauna** live above it.

Members of most communities are involved in some form of competition for food, light and space resources. For example, the stratification of tropical rain forests reflects competition in the upper canopy for light, while vegetation adapted for damp, darker conditions is developed at lower levels. Similar stratification or **tiering** is a feature of most marine communities, becoming higher and more sophisticated through geological time (Fig. 4.9), rather like the skyscrapers in Manhattan seeking to optimize space on a densely populated island. Low-level tiers were typically occupied by brachiopods and corals during the Paleozoic, while the higher tiers were occupied by crinoids. The Mesozoic and Cenozoic faunas, however, are more molluskan-based with the lower tiers occupied by epifaunal bivalves and brachiopods and the upper tiers occupied by bryozoans and crinoids.

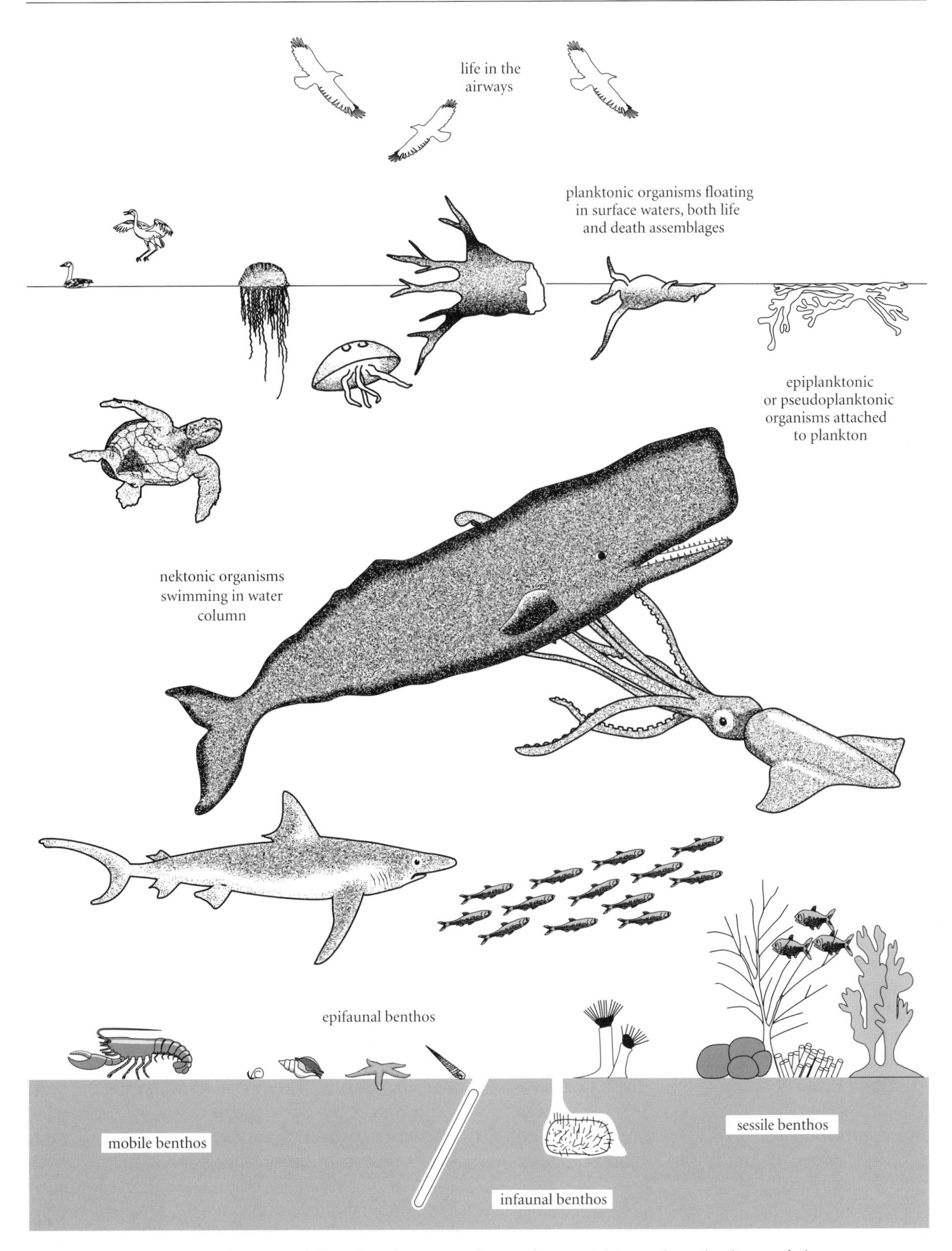

Figure 4.8 Selection of marine lifestyles above, at the surface, within and at the base of the water column. (Based on Ager 1963.)

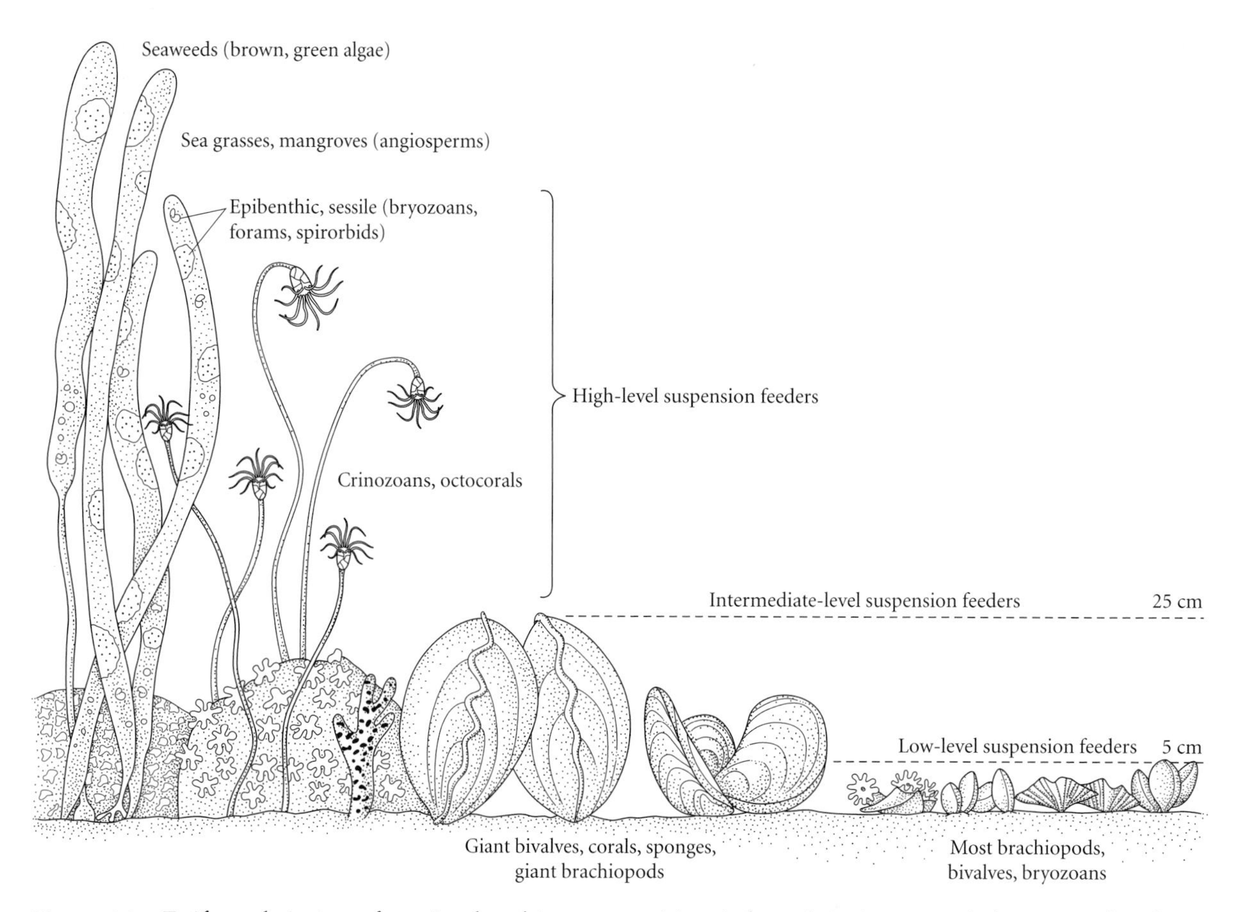

Figure 4.9 Epifaunal tiering of marine benthic communities; infaunal tiering recorded in trace fossil assemblages is discussed on p. 205. (From Copper 1988.)

Trace fossil associations show that burrows may be organized in an infaunal, tiered hierarchy (see Chapter 19). Ausich and Bottjer (1982) defined three levels with increasing depth from the sediment–water interface: 0 to −60 mm, −60 to −120 mm and −120 mm to −1 m. During the earliest Paleozoic, only the first tier was consistently occupied, the second tier was occupied from the Late Silurian and, finally, the third tier was populated in the Carboniferous. Tiering was also selectively affected by extinction events, and tiers deeper than 500 mm are rare after the Late Cretaceous because of predation by bony fishes.

Trophic structures: bottom or top of the food chain?

Food pyramids form the basis of most ecological systems, defining the energy flow through a chain of different organisms from extremely abundant primary producers to relatively few predators. A number of basic trophic or feeding strategies are known (Fig. 4.10). Several marine **food chains** (basically, who eats what) have been documented including those dominated by suspension feeders such as brachiopods, bryozoans and sponges. These fed mainly on phytoplankton and other organic detritus. Suspension feeding was particularly common in Paleozoic benthos; the Mesozoic and Cenozoic faunas were more dominated by detritus feeders, such as echinoids, and food chains were generally longer and more complex (Fig. 4.11).

It might seem rather easy at first sight to reconstruct a food chain for a fossil assemblage, providing you can work out who ate what. But that is easier said than done. One of the most spectacular fossil lake deposits, dominated by amphibians, has been documented from the Upper Carboniferous of

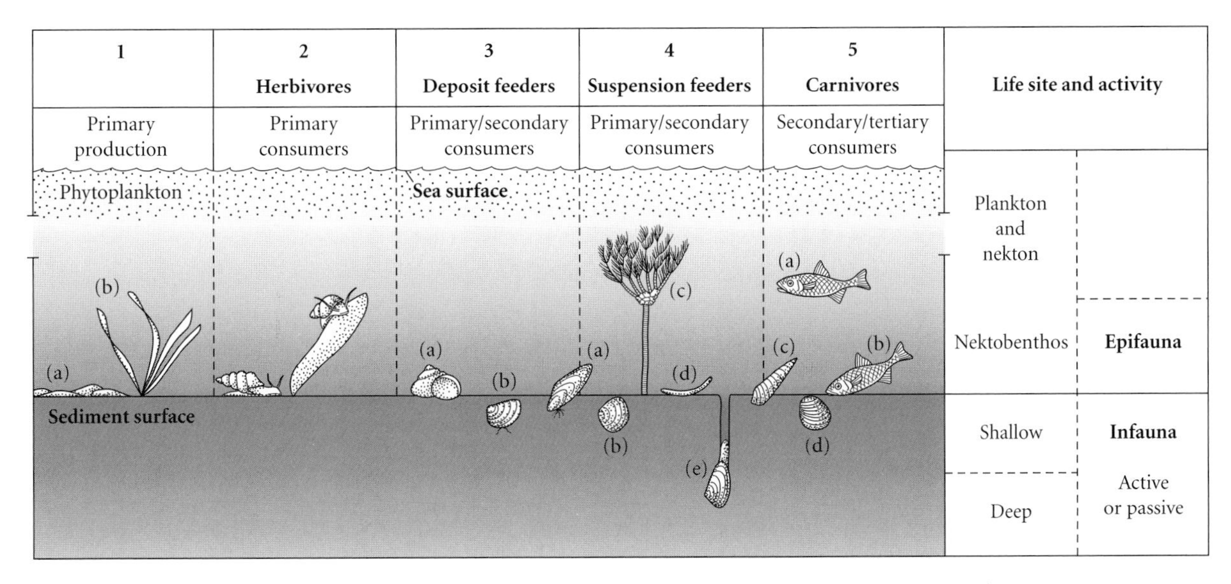

Figure 4.10 Trophic groups, activity of members and their life sites. 1, Primary producers: phytoplankton in surface waters with (a) cyanobacteria and (b) benthic algae. 2, Herbivores: browsing and grazing gastropods. 3, Deposit feeders: (a) deposit-feeding gastropod and (b) shallow infaunal bivalve. 4, Suspension feeders: (a) semi-infaunal, byssally-attached bivalve, (b) shallow infaunal bivalve, (c) crinoid, (d) epifaunal bivalve, and (e) deep infaunal bivalve. 5, Carnivores: (a) nektonic fishes, (b) nekton-benthic fishes, (c) epifaunal gastropod, and (d) infaunal gastropod. (From Brenchley & Harper 1998.)

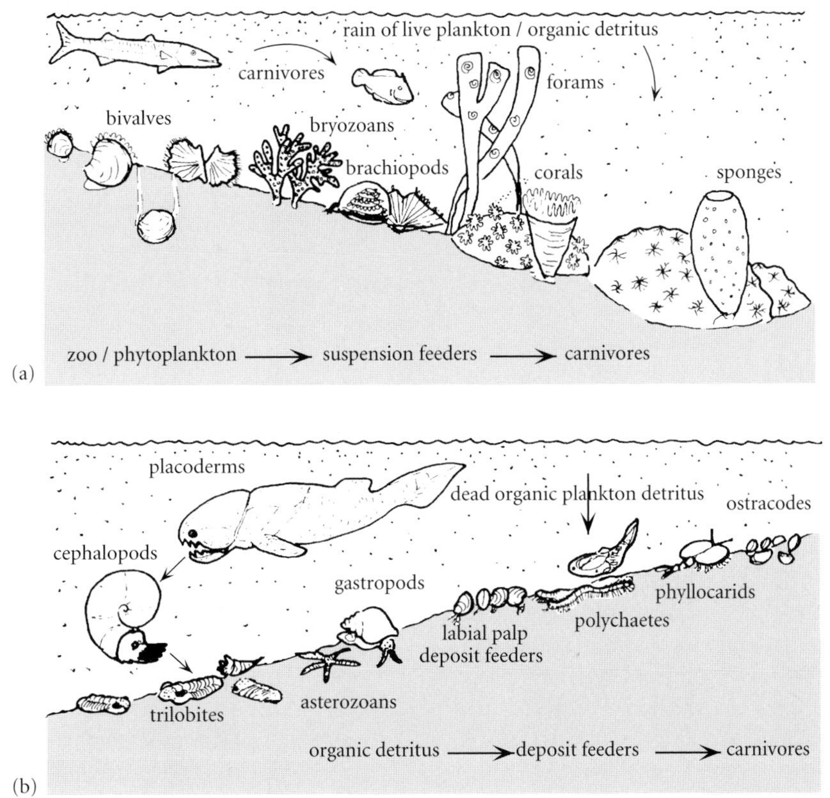

Figure 4.11 Reconstructions of two different food chain communities. (a) A community with a suspension-feeding food chain, displaying a variety of suspension feeders, collecting food in different ways (bivalves with a mucous trap or setae, bryozoans and brachiopods with lophophores, foraminiferans with cilia, corals with tentacles, and sponges with flagellae). (b) A community with a detritus-feeding food chain dominated by various types of bottom-dwelling deposit feeders and nektonic carnivores represented by a cephalopod and placoderm. (From Copper 1988.)

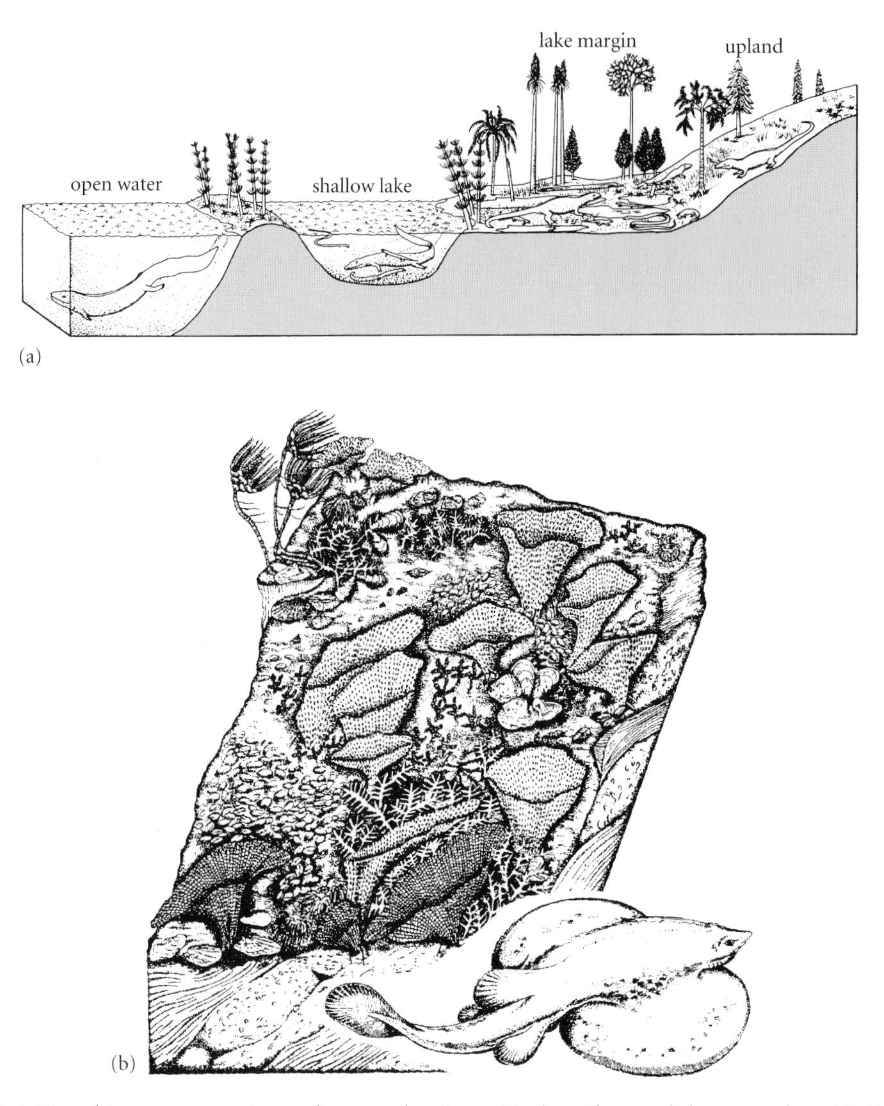

Figure 4.12 (a) Trophic structures in and around a Late Carboniferous lake complex, Nýřany, Czechoslavakia. (b) Trophic structures in a Late Permian reef complex, northeast England. (a, based on Benton 1990; b, from Hollingworth & Pettigrew 1998.)

Czechoslovakia (Fig. 4.12a). The lake ecosystem recreated for the inhabitants of the Nýřany Lake complex has three main ecological communities: an open water and lake association, dominated by fishes together with various larger amphibians; a shallow water and swamp/lake association with amphibians, small fishes, land plants and other plant debris; and finally a terrestrial–marginal association with microsaur (small, primitive) amphibians and primitive reptiles. Food chains have been worked out for each of these associations by careful study of the teeth (was it a herbivore with grinding teeth or a carnivore with slashing teeth?) of each beast, and comparisons with modern relatives. For example, in the open-water environments fishes, such as the spiny acanthodians, fed on plankton but were themselves attacked by the amphibians, presumably at the top of the food chain. In the associated terrestrial environments, plant material was consumed by a variety of invertebrates, including insects, millipedes, spiders, snails and worms; these provided food and nutrients for a range of small amphibians, themselves prey for larger amphibians and reptiles.

A good example of a marine food web comes from the Zechstein Reef facies of northern Europe, dating from the Late Permian

(Fig. 4.12b). The Zechstein benthos was dominated by diverse associations of brachiopods, overshadowed in the higher tiers by fan- and vase-shaped bryozoans (Hollingworth & Pettigrew 1988). Both groups were sessile filter feeders. Stalked echinoderms were rarer and occupied the highest tiers. Mollusks such as bivalves and gastropods were important deposit feeders and grazers. One of the largest predators was *Janassa*, a benthic ray, equipped with a formidable battery of teeth capable of crushing the shells of the sedentary benthos.

Megaguilds

Assignment of organisms to megaguilds provides another way to classify and understand the components of a fossil community. **Guilds** are groups of functionally similar organisms occurring together in a community. **Megaguilds** are simply a range of adaptive strategies based on a combination of life position (e.g. shallow, active, infaunal burrower) and feeding type (e.g. suspension feeder). Some paleontologists have used the term "guild" for these categories; however, these were probably finer ecological divisions within the so-called Bambachian megaguilds, named after the American paleontologist Richard Bambach, who first used the concept (Bambach 1983). Megaguilds have also become an effective tool in assessing long-term ecological change (see p. 105).

Controlling factors

The ecological niche of an organism is determined by a huge range of **limiting factors**, many of which are not recorded in the rock record (Fig. 4.13). Key limiting factors for marine organisms are light, oxygen levels, temperature, salinity, depth and substrate (Pickerill & Brenchley 1991).

Light is the main energy source for primary producers, thus diatoms, dinoflagellates, coccoliths and cyanobacteria are dependent on light and usually occupy the photic zone. Most biological productivity occurs in the top 10–20 m of the water column. Virtually all eukaryotic organisms require oxygen for their metabolic processes, absorbing oxygen by diffusion, in the case of small-bodied organisms, or through gills or lungs in the case of the larger metazoans. There is a well-developed

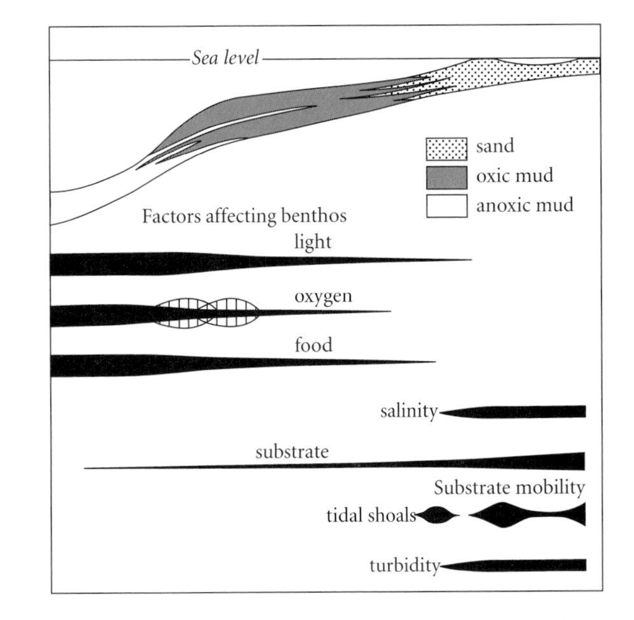

Figure 4.13 Shoreline to basin transect showing the relative importance of different factors on the distribution of organisms. (From Brenchley & Harper 1998.)

oxygen–depth profile in the world's seas and oceans. Oxygen levels generally decrease down to 100–500 m, where the amount of oxygen absorbed by organic matter exceeds primary oxygen production. Here in the **oxygen minimum zone** (OMZ), the lowest oxygen values are reached. The numbers of many organisms, such as corals, echinoderms, mollusks, polychaetes and sponges drop off dramatically in the OMZ.

Levels of oxygen in marine environments are important in determining who lives where. Aerobic (**normoxic**) environments have >1.0 ml L^{-1} concentrations of oxygen, dysaerobic (**hypoxic**) environments have 0.1–1.0 ml L^{-1} and anaerobic (**hypoxic-anoxic**) have <0.1 ml L^{-1}. Although there is marked decrease of biodiversity in oxygen-poor environments, these environments encourage more unusual adaptations such as the flat shells of the "paper pectens" (e.g. the genus *Dunbarella*) and the compressed bodies of the flat worms; the increased surface areas of both presumably helped the diffusion of oxygen.

Temperature is one of the most important limiting factors. Most marine animals are **poikilotherms**, having the same body temperature as their surroundings, and they live within a temperature range of about −1.5 to 30°C.

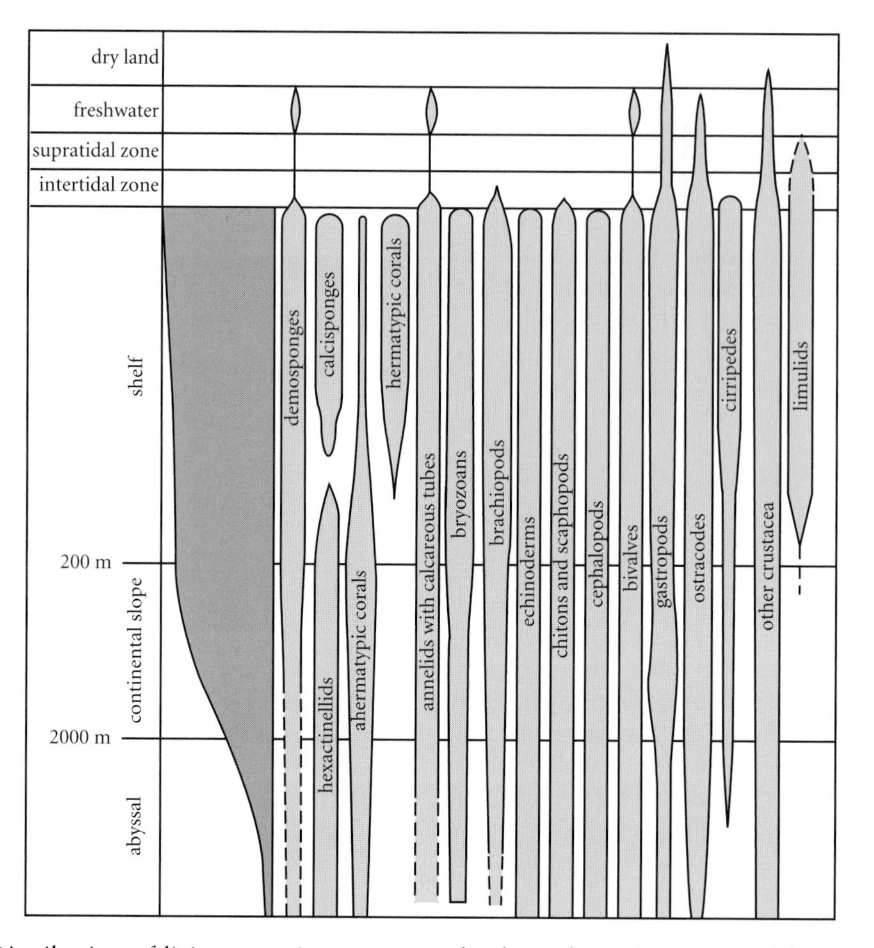

Figure 4.14 Distribution of living organisms across a depth gradient. (From Brenchley & Harper 1998.)

Water temperature in the oceans decreases steadily to the base of the thermocline, the layer within a body of water where the temperature changes rapidly, at around 1000 m depth, where it reaches about 6°C. Temperatures on the ocean floor rarely exceed more than 2°C. Temperature also changes with latitude and obviously affects the broad geographic distribution of organisms; those from the poles are generally quite different from those from the tropics.

Salinity, too, controls the distribution of organisms. Most marine animals are **isotonic** ("same salinity") with seawater and live within narrow (**stenohaline**) rather than wide (**euryhaline**) ranges of salinity, commonly with 30–40‰ dissolved salts in seawater. In broad terms normal marine water is characterized by stenohaline groups such as the ammonites, belemnites, brachiopods, corals, echinoderms and large benthic forams. Brackish waters have mainly low-diversity assemblages with bivalves, crustaceans, ostracodes and small benthic forams, whereas hypersaline assemblages are of very low diversity with just a few bivalves, gastropods and ostracodes.

Depth is one of the most often quoted controls on the distribution of marine organisms (Fig. 4.14). Although the direct affects of depth are related to hydrostatic pressure, many other factors, both chemical and physical, are related to depth; for example, in general terms, the grain size of sediment and water temperature decreases with depth. Although hydrostatic pressure does not usually distort the shells and soft tissues of organisms it can dramatically affect organisms with pockets of gas in their bodies, such as fishes and nautiloids. Apart from the effects of hydrostatic pressure, depth can also control the solubility of calcium carbonate; cold water contains more dissolved carbon dioxide (CO_2) providing a means to corrode carbonate. At given depths in the world's oceans, carbonate material begins to dissolve at so-called compensations depths. Below the **carbonate compensation depth** (CCD) the dissolution of calcium carbonate exceeds supply and at about 4–5 km calcite is not preserved. The depth is shallower for aragonite, with the **aragonite compensation depth** (ACD) placed at 1–2 km. Both the CCD and ACD vary with latitude, being shallower at higher latitudes, and both parameters have varied throughout geological time. Nevertheless, depth alone probably has little effect on biotic distribution, rather the many depth-related factors can be used to reconstruct the water depths of ancient marine communities.

Finally, the state of the substrate, rates of sedimentation and turbidity dramatically affect the distributions of benthic organisms (Brenchley & Pickerill 1993). Organisms have complex ecological requirements, some preferring a particular grain size, a certain type of organic material or they even respond to chemical signals (**chemotaxic**). There are also complex taphonomic feedback processes, where biogenic substrates such as shell pavements can form attachment sites for new communities. In general terms, within near-shore environments, there is a broad correlation between community distribution and grain size. Diversity tends to be highest in muddy sands, moderate in sandy muds, low in pure sands and virtually zero in soft muds. Moreover whether the sediments form soupy muds, loose sands, firmgrounds or hardgrounds will influence faunal distributions.

Paleocommunities

Paleocommunities are recurrent groups of organisms related to some specific set of environmental conditions or limiting factors. Many of the concepts and techniques applied to marine fossil communities are based on the work of biologists such as the Danish scientist Carl Petersen, researching in the late 1800s and early 1900s. Petersen recognized a series of level-bottom benthic communities around the Scandinavian coasts; the major control on community distribution was water depth, although other factors such as the substrate were also influential.

Paleontologists were slow to adopt these insights from modern marine biology. There were a few pioneer studies on Carboniferous assemblages in the 1930s, but it was the classic work by Alfred Ziegler in the 1960s that really brought these methods to the attention of paleontologists. He identified five depth-related, brachiopod-dominated communities in the Lower Silurian rocks of Wales and the Welsh borderlands (see Chapter 12). These communities stretched from the intertidal zone in the east to the deep shelf and continental slope towards the west, matching perfectly the ancient paleogeography. This whole system has been revamped and is now more widely known as the **benthic assemblage zones** (Fig. 4.15). These zones are defined on a wide range of faunal and sedimentological criteria and may be subdivided, internally, on the basis of, for example, substrate type and the degree of turbulence (Brett et al. 1993).

Describing fossil communities

Sometimes the simplest jobs are the hardest to do properly. For over a hundred years, paleontologists have provided lists of species from particular localities, but these are not helpful for ecological work unless the relative abundances of the different species are documented as well. We need to know which species dominate (sometimes one species makes up more than 50% of the sample) and which are rare (i.e. less than 5% of the collection). Now it is more common to document the absolute and **relative abundance** of each organism, illustrated graphically with frequency histograms, and based on data derived from line transects, quadrats or more commonly now from bed-by-bed collecting or bulk samples.

Counting conventions remain a problem. With many organisms it is relatively simple to calculate how many individuals were actually represented in a given assemblage: univalved species (e.g. gastropods) count as one, whereas twin-valved species (e.g. bivalves and brachiopods) may be assessed by adding the most common valve (right or left, dorsal or ventral) to the number of articulated or conjoined shells. Animals that molt, such as ostracods and trilobites, colonial organisms and those

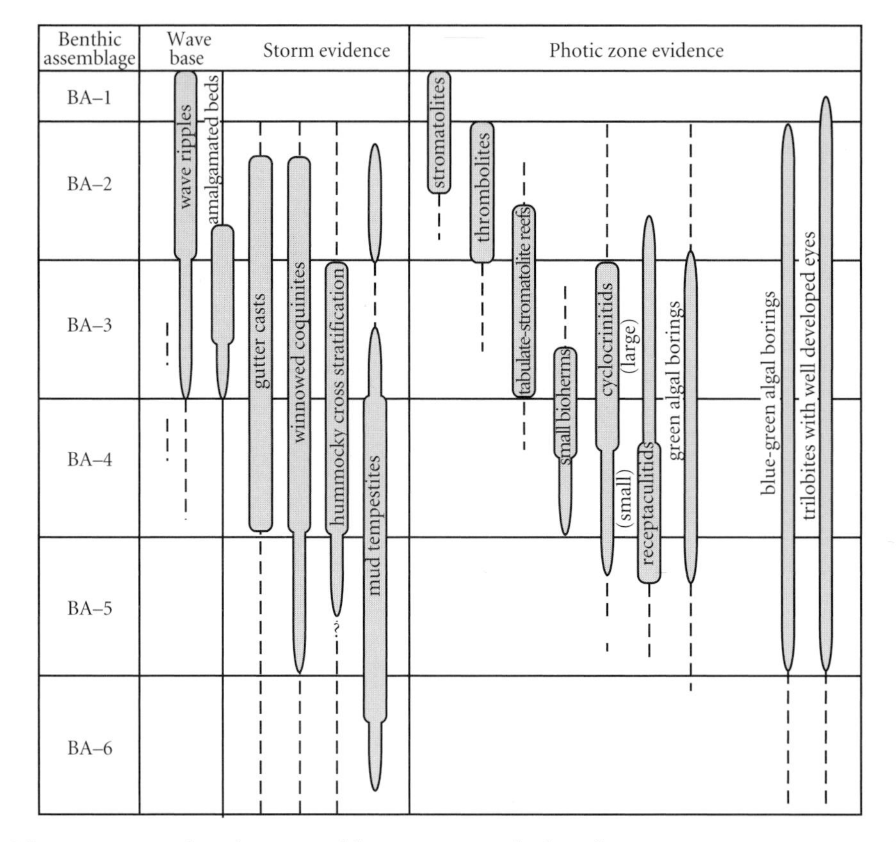

Figure 4.15 Silurian marine benthic assemblage zones and identifying criteria. (From Brenchley & Harper 1998.)

that easily fragment (e.g. bryozoans) and those with multi-element skeletons (e.g. crinoids and vertebrates) require more specialized counting techniques. These basic data are then transformed into a more realistic picture of ancient communities populating past landscapes and seascapes, through histograms and pie charts. Raw numerical data are extremely useful, but these can also be converted to diversity, dominance and evenness parameters, and parameters for taxonomic distinctiveness (Box 4.2). These together can give us a rich overview of the composition and structure of the paleocommunity and allow numerical comparisons with other similar assemblages. Such approaches have become routine in studies of invertebrate paleoecology but it is much more difficult to apply these methods to vertebrate assemblages where sample sizes are generally much smaller.

Detailed analysis of paleocommunity structures has permitted recognition of a number of specific community types. **Pioneer communities** are those that have just entered new ecospace, and they may be dominated by one or two very abundant opportunistic species, in contrast to long-established and rather stable **equilibrium communities** where relatively high diversities of more or less equally abundant animals are present. The ecological relationships between organisms is also an important aspect of community development (Box 4.3).

Paleocommunity development through time

Communities undoubtedly change with time. Factors such as environmental fluctuation, immigration and emigration of animals and plants, evolution and extinction of species and coevolutionary changes will alter the composition and structure of a community. But are the components of communities tightly linked and thus evolve together or is it a rather haphazard random process? Living communities, when first established, show initial high rates of replacement and instability, whereas later stages are more stable with little change,

Box 4.2 Ecological statistics and sampling sufficiency: are you getting enough?

It is often difficult to assess the adequacy of a paleoecological sample. Some authorities have suggested that samples of about 300 give a fairly accurate census of a fossil assemblage. Commonly, investigators plot rarefaction curves (Fig. 4.16). These are produced simply by collecting samples of 10 and identifying the number of species in each. For each sample of 10 plotted along the x-axis, the cumulative number of species is plotted along the y-axis. The curve may level off at the point where no additional species are identified with additional collecting and this fixes the sample size that is adequate to count the majority of species present (Fig. 4.16).

A range of statistics has been used to describe aspects of fossil communities. Although the number of species collected from an assemblage provides a rough guide to the diversity of the association, obviously in most cases the larger the sample, the higher the diversity. Diversity measures are usually standardized against the sample size. Dominance measures have high values for communities with a few abundant elements and low values where species are more or less evenly represented; measures of evenness are usually the inverse of dominance.

$$\text{Margalef diversity} = S - 1/\log N$$

$$\text{Dominance} = \sum (n_i/N)^2$$

$$\text{Evenness} = 1/\sum (p_i)^2$$

where S is the number of species, N is the number of specimens, n_i is the number of the ith species, and p_i is the relative frequency of ith species.

Many numerical techniques have been used to analyze paleocommunities and their distributions. Phenetic methods (see Chapter 2) are based on the investigation of a similarity or distance matrix

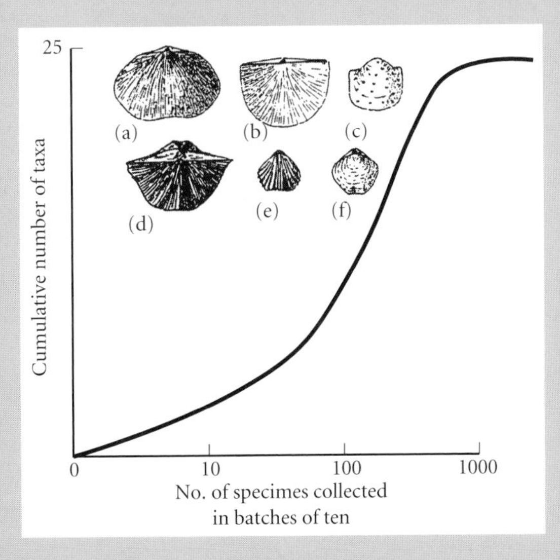

Figure 4.16 Construction of a rarefaction curve based on data collected from a mid-Devonian brachiopod-dominated fauna, northern France. The main types of brachiopod are illustrated: (a) *Schizophoria*, (b) *Douvillina*, (c) *Productella*, (d) *Cyrtospirifer*, (e) *Rhipidiorhynchus*, and (f) *Athyris*. The curve levels off at about 300 specimens, suggesting this sample size is a sufficient census of the fauna. Magnification approximately ×0.5 for all.

Continued

derived from a raw data matrix of the presence or absence or numerical abundance of fossils at each site. Cluster analysis is most commonly used in ecological studies and there is a wide range of both distance and similarity measures, together with clustering techniques, to choose from. R-mode analysis clusters the variables, in most paleoecological studies the taxa, whereas Q-mode analysis clusters the cases, usually the localities or assemblages (Fig. 4.17).

For example, Late Ordovician brachiopod-dominated assemblages from South China have been investigated by cluster analyses (Hammer & Harper 2005) and fall into a number of ecogroups. These data are available at http://www.blackwellpublishing.com/paleobiology/.

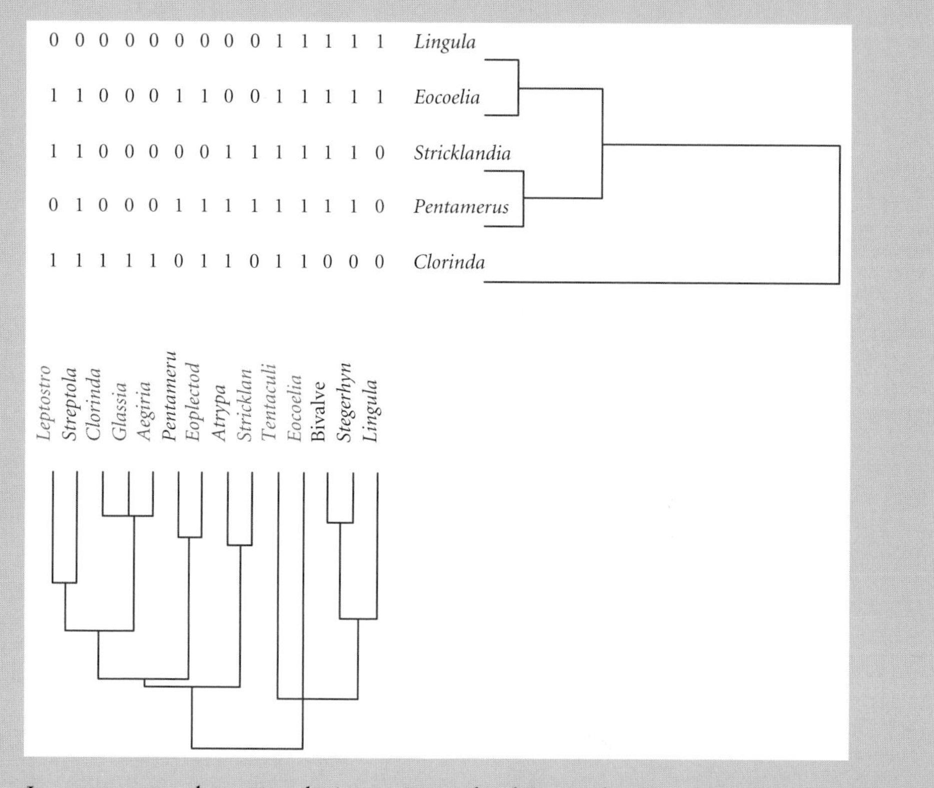

Figure 4.17 In a two-way cluster analysis, an R-mode clusters the genera (bottom) and a Q-mode clusters the community type (right). The original data matrix is in the center of the diagram. The data indicate the reality of a shallow-water biofacies (*Lingula* and *Eocoelia* communities), and mid to deep shelf (*Pentamerus* and *Stricklandia* communities) and outer shelf to slope (*Clorinda* community) assemblages.

building up to a climax community in equilibrium with its environment. There is still some discussion among ecologists about whether communities conform to **Eltonian** models of change (predictable over long periods of time), **Gleasonian** models (short-term, rapid change and instability) or perhaps even both. Evidence from Quaternary, mainly Holocene, communities suggests them to be rather ephemeral (Davis et al. 2005). Species may evolve, become extinct or migrate out of the immediate area during intervals of climate change thus destroying the community structure. They may, however, return and recombine to form the original communities during intervals of more favorable climate (Bennett 1997). Nevertheless, paleocommunities dominated by incumbent taxa such as the dinosaurs during the Jurassic and Cretaceous or pentameride brachiopods during the Silurian

Box 4.3 Ecological interactions

Animals and plants have participated in a wide range of relationships throughout geological time. Ecologists have classified these arrangements in terms of gain (+), loss (–) and neutrality (0). **Antagonistic** arrangements include **antibiosis** (–,0), **exploitation** (0,+) and **competition** (0,0) whereas symbiosis involves both **commensalism** (+,0) and **mutualism** (+,+).

Antibiosis is difficult to demonstrate although mass mortalities of fishes have been ascribed to dinoflagellate blooms. Some paleontologists believe that the twisted skeleton of a Late Cretaceous *Struthiomimus* from Alberta may show the animal died from strychnine poisoning.

Exploitation includes predation and parasitism. There are many records of bite marks, particularly by marine reptiles on mollusk shells, while the stomach contents of Jurassic ichthyosaurs have revealed a diet of belemnites. Moreover a wide variety of nibble marks have been reported from fossil leaves. The relationship between the Devonian tabulate coral *Pleurodictyum* and the worm *Hicetes* fooled many paleontologists. Was this a bizarre compound organism? In fact the worm was probably a parasite; the association is common throughout Europe and virtually every specimen of *Pleurodictyum* has a parasitic worm at its core.

Competition is often difficult to observe directly in the fossil record. Encrusting bryozoans, however, commonly compete for space and food resources on the seabed. Competition between the cyclostome and cheilostome clades (see Chapter 12) may have influenced the post-Paleozoic history of the phylum in favor of the latter. Encrusting bryozoans can also faithfully replicate their substrate, recording the imprint of a soft-bodied animal or aragonitic mollusk. This process of **bioimmuration** ("biological burial") is a useful means of preserving an organism that otherwise may have escaped detection.

Commensalism is one of the most common relationships apparent in the fossil record, where small epifauna or epibionts use larger organisms for attachment and support. Small and immature productoid brachiopods are often attached by clasping spines to crinoid stems while microconchids, previously thought to be *Spirobis* worms (see Chapter 12), are commonly attached near the exhalent currents of Carboniferous non-marine bivalves. Some of the most spectacular examples have been reported from the shells of Devonian spiriferide brachiopods. Derek Ager (University of Wales, Swansea) reported a succession of epifauna, commencing with *Spirobis* (microconchids) followed by *Hederella* and *Paleschara* and finally the tabulate coral *Aulopora*, clustered near the inhalent current of the brachiopod (Fig. 4.18).

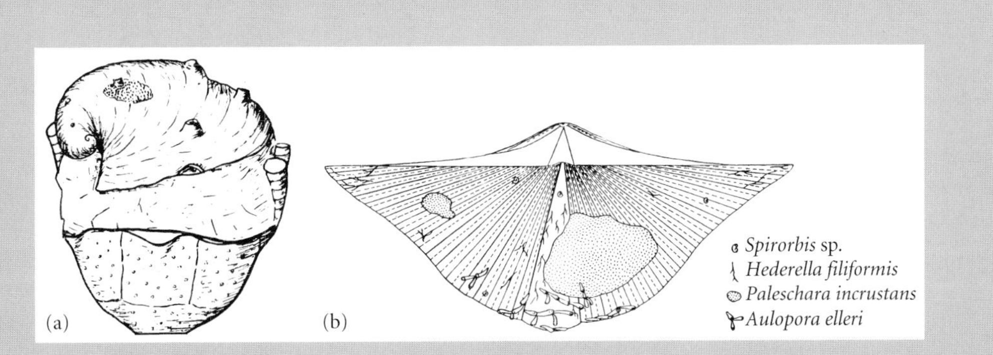

Figure 4.18 Commensalism between (a) the gastropod *Platyceras* and a Devonian crinoid and (b) *Spinocyrtia iowensis* with an epifauna primarily located on the fold of the brachial valve adjacent to inhalant or exhalent currents. (Based on Ager 1963.)

can exist for tens or even hundreds of millions of years (Sheehan 2001).

Despite the fantastic potential to test models for community change through time there have been relatively few rigorous attempts. Some paleontologists have recognized a pattern of **coordinated turnover** followed by **stasis** in which many species disappear over a short time interval and are replaced by other functionally and taxonomically similar species. The new assemblages may retain their structure for 2–8 million years (Brett et al. 1996), although in some cases this appears to occur through repeated reassembly following disturbances. By contrast Ordovician marine assemblages from the Appalachians (see p. 38) show a strong relationship between environmental fluctuations and uncoordinated changes in the composition and dominance of animals in their assemblages. Even life at a small scale shows these patterns. An Ordovician hardground paleocommunity constructed by encrusters, mainly bryozoans and edrioasteroids, on cobbles shows first a low-diversity pioneer community, then a high-diversity association, and finally a monospecific assemblage characterized by a late successional dominant (Wilson 1985). Environmental disturbances, such as the tipping over of the cobbles, allowed a recolonization of the hardground, thus maintaining high diversity within the paleocommunity. Clearly in some cases an Eltonian model may be applicable and in others Gleasonian paradigms rule.

Evolutionary paleoecology

Biodiversity trends through time for the majority of animal and plant groups have been documented in some accuracy and detail since the late 1970s (see p. 534). However, it is clear that there are a series of ecological changes underpinning this incredible taxonomic diversification and such changes were probably decoupled from each other. For example, there have been marked changes in the use of ecospace through the evolution of new adaptive strategies (Bambachian megaguilds), an escalation in the number of guilds and accelerated tiering both above and below

Box 4.4 Chemosynthetic environments

Amazing new discoveries in the modern oceans have revealed some of the most bizarre living creatures that survive in the dark, cold depths, clinging onto the life support provided by hydrocarbon seeps and hydrothermal vents. Uniquely, these bizarre organisms never see the light, and their food chains are not based on sunlight and carbon, but on sulfur from hydrothermal vents. The search is on to find their fossil counterparts. Kathleen Campbell, together with a range of colleagues, has been exploring the distribution of these types of weird communities through time (Campbell 2006). She has identified 40 fossil examples, recognized on the basis of key types of faunas, specific biomarkers and their geochemical and tectonic settings. Such communities associated with Precambrian vents were populated by microbes and it was not really until the Silurian that we find our first groups of metazoan chemosynthetic organisms. These organisms are strange. Gigantic non-articulated brachiopods are associated with large bivalves and worm tubes in a massive volcanic sulfide in the Ural Mountains (Fig. 4.19). In general terms, pre-Jurassic vent faunas were dominated by extinct groups of brachiopods, monoplacophorans, bivalves and gastropods, and post-Jurassic faunas were populated by extant families of bivalves and gastropods. The modern vent-seep fauna is endemic and may either have evolved from a collection of Paleozoic and Mesozoic relics or perhaps some invertebrate groups periodically migrated into the gloom during the Phanerozoic to set up their own communities. Although unusual, the chemosynthetic world was yet another ecosystem with its own set of rules and evolutionary paradigms, contributing to the past and present biodiversity of our planet. Perhaps, also, during times of major and rapid environmental change, this ecosystem provided a stable refugia where at least some organisms could escape fluctuations in those other ecosystems that rely on light and organic nutrients.

Figure 4.19 Selection of fossils from ancient hydrothermal vent sites. All specimens are pyritized and are contained within a matrix of sulfide minerals. (a) Gastropod: *Francisciconcha maslennikovi* from the Lower Jurassic Figueroa sulfide deposit, California. (b) Small worm tubes from the Upper Cretaceous Memi sulfide deposit, Cyprus. (c) Bivalve: *Sibaya ivanovi* from the Middle Devonian Sibay sulfide deposit, Russia. (d, e) From the Lower Silurian Yaman Kasy sulfide deposit, Russia: (d) monoplacophoran, *Themoconus shadlunae* and (e) vestimentiferan worm tube, *Yamankasia rifeia*. Scale bars: 5 mm (a, b), 20 mm (c–e). (Courtesy of Crispin Little.)

the substrate. Through time animals and plants have developed innovative ways to exist in new habitats. The exciting, relatively new field of evolutionary paleoecology seeks to tackle some large-scale ecological patterns and trends through geological time. Why, for example, were Cambrian food chains so short, with few guilds, and why were tiering levels both above and below the sediment so restricted? In marine environments,

Sepkoski's (1981) robust division of Phanerozoic life into his Cambrian (trilobites, non-articulated brachiopods, primitive echinoderms and mollusks), Paleozoic (suspension feeders such as the articulated brachiopods, bryozoans, corals and crinoids) and the Modern faunas (detritus feeders such as the echinoids, gastropods and bivalves together with crustaceans, bony fishes and sharks) has acted as template for much paleoecological research (see p. 538).

Communities and habitats through time

Most paleoecological studies attempt to recreate the dynamism and reality of past communities from environments ranging from mountain lakes (see p. 90) to the strange chemosynthetic environments of the deep sea (Box 4.4). Despite the significant loss of information through taphonomic processes, realistic reconstructions are possible, depicting the main components, their relationships to each other and the surrounding environment. During most of geological time, microbial organisms were the sole inhabitants of Earth. The Ediacara biota appeared at the base of the Ediacaran System, some 630 Ma, but most members had disappeared by the start of the Cambrian. McKerrow (1978) was first to summarize, in broad terms, the development of communities throughout the Phanerozoic (see also Chapter 20). These tableaux were necessarily qualitative but now there are a growing number of more quantitative approaches providing more accurate reconstructions of community change through time. One of the first seascape reconstructions, Sir Henry de la Beche's watercolor of *Duria antiquior* (1830), depicted life in an early Jurassic sea. It was an iconic painting but nevertheless scientific, illustrating, graphically, the relationships between predators and prey in the Modern evolutionary fauna. Today we know much more about the range of environments that existed during the Jurassic Period.

Jurassic Park *and deep-sea worlds*

Jurassic environments provide a wide range of communities and habitats showing the early stages of development of post-Paleozoic faunas. A selection demonstrating environments, life modes and trophic strategies are illustrated (Fig. 4.20). Such tableaux have been criticized for their lack of science. They are, however, based on real case histories and numerical data are now available for many of these reconstructions. For example, two spectacular deposits, the Newark Supergroup and the Posidonia Shales, provide important windows on life in continental and marine environments, respectively, during the early part of the Jurassic.

Major new finds in the Newark Supergroup and equivalent strata in eastern North America, have painted a vivid picture of life on Late Triassic and Early Jurassic arid to humid landscapes of Laurentia swept by occasional monsoons. Olsen and his colleagues (1978, 1987) have described diverse dinosaur communities of both large and small carnivorous theropods, at the top of the food chain, together with large herbivorous sauropods and some early armored forms. Most of the terrestrial tetrapods are preserved in volcaniclastic deposits, but adjacent fluviatile facies contain crocodiles. Lake facies have preserved diverse floras of conifers, cycads, ferns and lycopods. Fast-swimming holostean fishes patrolled the lakes and abundant insects of modern aspect, representing seven orders, populated the forests and shores or may have swum in the shallows together with crustaceans.

The Posidonia Shales crop out near the village of Holzmaden in the Swabian Alps, Germany. The shales are bituminous or tar-like, packed with fossils, generally with echinoderms and vertebrates towards the base and cephalopods at the top. Seilacher (1985) and his colleagues showed how this rock unit with exceptionally preserved fossils, or Lagerstätte, was a stagnation deposit (see p. 60) where fossils accumulated in almost completely anoxic seabed conditions, and so were hardly damaged by decomposers. Benthos is rare, and encrusting and recumbent brachiopods, bivalves, crinoids and serpulids that could not live on the stagnant seabed attached themselves to driftwood, ammonite shells and other floating or swimming organisms to pursue a so-called psedoplanktonic life mode. The dominant animals were nektonic ammonites and coleoids together with the superbly preserved ichthyosaurs and plesiosaurs, now displayed in many European museums. Some horizons are characterized by monotypic assemblages of small taxa such as diademoid echinoids and byssate bivalves, like *Posidonia* itself. These benthic colonizations may have

Figure 4.20 A cocktail of Jurassic environments. Early Jurassic: (a) sand, (b) muddy sand, and (c) bituminous mud communities. Late Jurassic: (d) mud, (e) reef, and (f) lagoonal communities. (From McKerrow 1978.)

been promoted by storms, providing fresher-water conditions for short periods of time.

Ecological patterns and trends through time

During the last 600 myr, both animal and plant communities expanded and diversified (Box 4.5). In simple terms the number of Bambachian megaguilds multiplied through the Cambrian (nine megaguilds), Paleozoic (14) and Modern (20) evolutionary faunas. The focus in the Cambrian was on marine animals that were either attached or mobile with suspension- or deposit-feeding strategies, such as the eocrinoids and trilobites. The morphologies of individual organisms were rather plastic as were their community compositions and structures. Relatively few class-level taxa were included in each ecological box (Fig. 4.21). By the Ordovician, however, the number of megaguilds had expanded, with an overall numerical dominance of suspension feeders, such as the brachiopods, bryozoans, corals and crinoids. The Paleozoic fauna was characterized by sedentary organisms. The Modern fauna, by contrast, was dominated by deposit-feeding, essentially mobile animals bound into a process of **escalation**, or ever-increasing competition, and the first intense arms race on the planet. The term **arms race** is used by ecologists to describe ever-intensifying interactions between predators and prey, for example.

Throughout the Phanerozoic there seems to have been an offshore movement in marine faunas. New communities and taxa may have occurred in nearshore, high-energy environments first, before migrating into deeper water. Thus older, more archaic groups tended to characterize deeper-water habitats. For example during the Ordovician radiation (see p. 253), typical members of the Paleozoic fauna (brachiopods, bryozoans and crinoids) expanded and migrated into deeper-water habitats, while their place in shallow water was taken by components of the Modern fauna (bivalves and gastropods). But why? Are nearshore habitats particularly harsh, driving innovative communities and taxa into deep water, or can innovative organisms arise at any depth and those in shallower-water environments are just more resistant to extinction and can readily migrate into deeper water (Jablonski & Bottjer 1990)? Perhaps it was a combination of both.

In marine environments acceleration of the height, complexity and stratification of benthic tiering was later matched by increases in the depth and sophistication of infaunal tiering as, particularly in Mesozoic and Cenozoic faunas, many more organisms adopted burrowing lifestyles and the benthos switched from filter to deposit feeding with significantly more predators. The Cambrian evolutionary fauna occupied, more or less, only the surface of the seabed, but by the Ordovician crinoids had developed tiers over a meter above the seabed and burrowing had already commenced into the sediment. Terrestrial environments, initially dominated by small green plants, various arthropods and snails, together with diverse amphibian faunas in the Mid to Late Paleozoic, changed significantly during the Mesozoic, with the diversification of vegetation and eventually flowering plants, and terminating, for now, in the high and elaborate canopies we see today in the tropical rain forests (see p. 505).

The Modern fauna was also characterized by something rather special, an arms race (Harper 2006). During the so-called Mesozoic marine revolution, predators, such as bony fishes, crustaceans, marine reptiles and starfishes began to develop better and better ways of crushing or opening shells. The Modern world was a much more dangerous place and in order to survive, potential prey had to develop thicker, more elaborately ornamented shells with smaller apertures (Box 4.6) and devise more cunning evasive strategies such as greater mobility or deeper and deeper burrowing. Unfortunately exposure to intense predation and a much more bioturbated seafloor was no place for many groups of epifaunal animals such as the brachiopods, some groups of bivalves and echinoderms. But as prey developed more armor and better evasive strategies, the hunters developed better weaponry. Together this escalation and increased tiering set the Modern fauna quite apart from those of the Cambrian and Paleozoic. Perhaps the whole ecosystem functioned in a different way, allowing biodiversity to continue to expand way beyond the plateau of the Paleozoic fauna (see p. 541).

Unlike biodiversity change, where we have numbers of taxa to count and monitor, ecological change is much more difficult to describe and quantify. Since some changes are much

Table 4.1 Hierarchical levels of ecological change and their signals.

Level	Definition	Signals
First	Appearance/disappearance of an ecosystem	Initial colonization of environment
Second	Structural changes within an ecosystem	First appearance of, or changes in, ecological dominants of higher taxa Loss/appearance of metazoan reefs Appearance/disappearance of Bambachian megaguilds
Third	Community-type level changes within an established ecological structure	Appearance and/or disappearance of community types Increase and/or decrease in tiering complexity "Filling-in" or "thinning" within Bambachian megaguilds
Fourth	Community-level changes	Appearance and/or disappearance of paleocommunities Taxonomic changes within a clade

more significant than others, one way is to establish a series of levels with key, identifiable characteristics (Droser et al. 2000). Four ranks or paleoecological levels have been identified (Table 4.1) based on, for example, the appearance or disappearance of an entire ecosystem (first), the appearances and disappearances of dominant taxa (second), thickening or thinning of the Bambachian megaguilds (third) or the mere appearance or disappearance of a community (fourth). During the Phanerozoic ecological changes can be charted at all levels: the appearance of the Ediacara biota was clearly a first-order change, whereas the Cambrian explosion and the Ordovician radiation involved changes at the second, third and fourth levels. Recovery after the end-Permian mass extinction event is a textbook example involving the addition to existing Bambachian megaguilds, when the tiering of marine faunas really took off (Twitchett 2006). The major mass extinction events have been ranked ecologically too (Box 4.7).

PALEOCLIMATES

The Greenland ice sheet is likely to be eliminated [within 50 years] unless much more substantial reductions in emissions are made than those envisaged [and changes will] probably be irreversible, this side of a new ice age.

Kofi Annan, Past Secretary General of the United Nations (2004)

Box 4.5 Occupation of ecospace through time

Life through time has increased in taxonomic diversity, but have the number of life modes also increased? One way to investigate this is by mapping the increase in Bambachian megaguilds (Bambach 1983) across the three great evolutionary faunas. The trend is one of not only increasing numbers of megaguilds through time but also one of increased urbanization as more taxa are squeezed into each category (Fig. 4.21a–c). But in order to sustain the increased membership there must have been some fine-tuning and splitting within the megaguilds as new guilds and niches were developed within the Bambachian structure. Can this be tested with new data and why should the number and importance of various life modes change through time? Richard Bambach and his colleagues (Bambach et al. 2007) have reported an increase from about one, in the Late Ediacaran, to over 90 lifestyles in Recent and Neogene faunas (Fig. 4.21d). Between the Paleozoic and Neogene faunas, there has been an increase in motility, infaunalization and predation. Thus the expansion of predation and increased bioturbation may have forced organisms to adjust to new challenges and participate in ever more complex ecosystems.

Continued

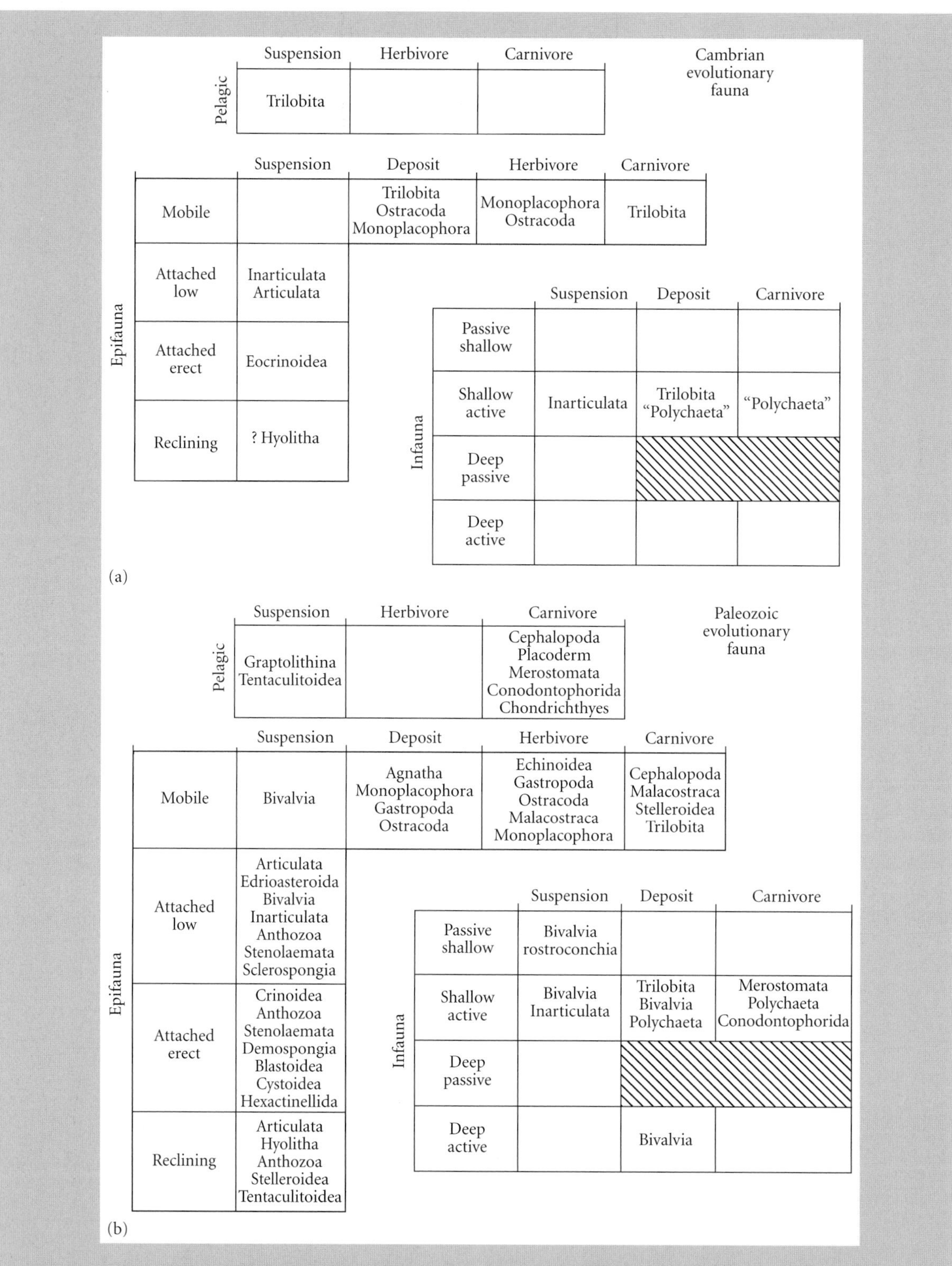

Figure 4.21 Bambachian megaguilds. A near full complement of lifestyles is present in the Modern fauna (c), while fewer are represented in the matrices for the Cambrian (a) and Paleozoic (b) faunas. (d) The numbers of life modes have increased consistently through time.

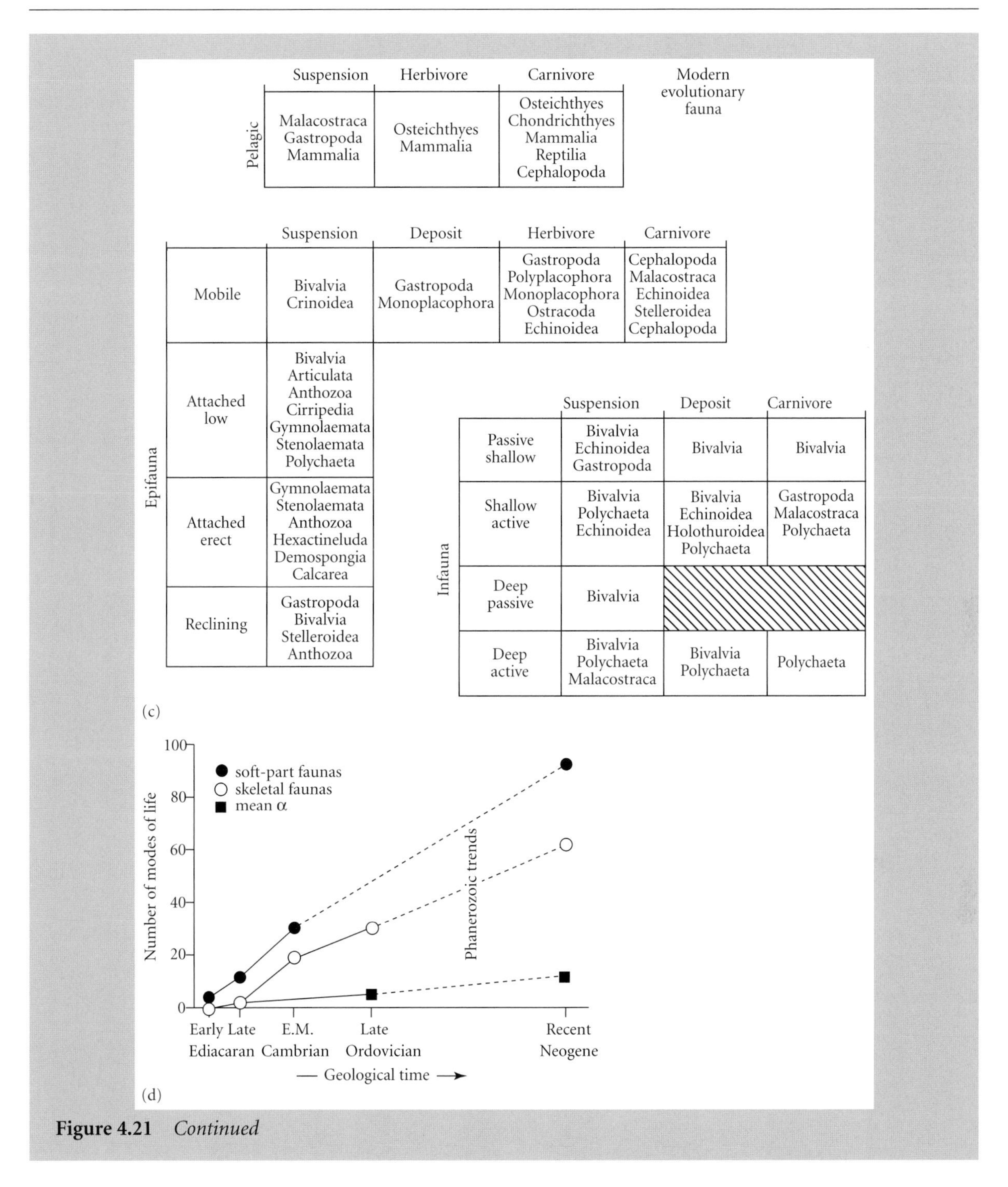

Figure 4.21 *Continued*

During the last 600 million years the Earth has oscillated at least five times between ice-house and greenhouse conditions, spending most of the time in greenhouse climates (Fig. 4.23). For much of the Precambrian the Earth probably endured relatively cool climates. The Earth is generally divided into five climate zones: humid tropical (no winters and average temperatures above 18°C), dry subtropical (evaporation exceeds precipitation), warm temperate (mild winters), cool temperate (severe winters) and polar (no summers and temperatures below 10°C). But can these zones be recognized through deep time and be

Box 4.6 Shell concentrations

Shell concentrations of various types can tell us a huge amount about environments of deposition but also can act as a proxy for biological productivity through time (Kidwell & Brenchley 1994). Moreover, it is possible that evolutionary changes in the diversity and ecology of organisms that produce and destroy calcareous skeletons suggest that the nature of these concentrations may have changed through the Phanerozoic. Data from marine siliciclastic rocks, silicate-based clastic sediments, of Ordovician-Silurian, Jurassic and Neogene ages show a significant increase in the thickness of densely packed bioclastic concentrations, from thin-bedded brachiopod-dominated concentrations in the Ordovician-Silurian to a mollusk-dominated record with more numerous and thicker shell beds in the Neogene (Fig. 4.22). Jurassic shell beds vary in thickness depending on whether they have Paleozoic or modern affinities as the main components. This suggests that the Phanerozoic increase in shell-bed thickness was not controlled by diagenesis or by a shift in taphonomic conditions on the seafloor, but rather by the evolution of biogenic clast producers, themselves – i.e. groups with, firstly, more durable low-organic skeletons, secondly, greater ecological success in high-energy environments, and thirdly higher rates of carbonate production. These results indicate that (i) reproductive and metabolic output has increased in benthic communities over time; and (ii) the scale of time averaging in benthic assemblages has increased owing to greater hard-part durability of modern groups. New data, however, suggest that brachiopods were probably just as durable as mollusks, but their communities simply did not produce so many shells. The frequency and thickness of shell beds through time may simply be down to the relative biological productivity of different groups of organisms.

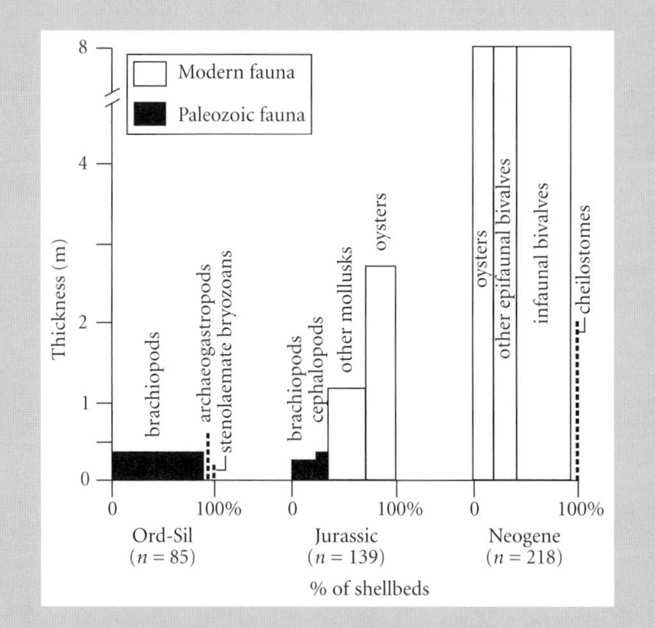

Figure 4.22 Thicknesses of shell concentrations during the Ordovician-Silurian, Jurassic and Neogene. Thick shell beds are a phenomenon of the Modern fauna, mainly generated by bivalves. (From Kidwell & Brenchley 1994.)

Box 4.7 Ecology of extinction events

We now have a massive amount of data across all the big five Phanerozoic extinction events, but are taxon counts a good guide to the severity of each extinction? Probably not! There is a strong ecological dimension to each event. George McGhee and his colleagues (2004) have ranked the ecological severity of each event and the order of severity is in fact different from that established from taxon counts. First, the ecological impacts of the five Phanerozoic biodiversity crises were not all similar (Table 4.2). Second, ranking the five Phanerozoic biodiversity crises by ecological severity shows that the taxonomic and ecological severities of the events are decoupled. Most marked is the end-Cretaceous biodiversity crisis, the least severe in terms of taxonomic diversity loss but ecologically the second most severe. The end-Ordovician biodiversity crisis was associated with major global cooling produced by the end-Ordovician glaciations; it prompted a major loss of marine life, yet the extinction failed to eliminate any key taxa or evolutionary traits, and thus was of minimal ecological impact. The decoupled severities clearly emphasize that the ecological importance of species in an ecosystem is at least as important as species diversity in maintaining an ecosystem. Selective elimination of dominant and/or keystone taxa is a feature of the ecologically most devastating biodiversity crises. A strategy emphasizing the preservation of taxa with high ecological values is the key to minimizing the ecological effects of the current ongoing loss of global biodiversity.

Table 4.2 Classification of the ecological impacts of a diversity crisis.

Impact category	Ecological effects
Category I	Existent ecosystems collapse, replaced by new ecosystems post-extinction
Category II	Existent ecosystems disrupted, but recover and are not replaced post-extinction
Subcategory IIa	Disruption produces permanent loss of major ecosystem components
Subcategory IIb	Disruption temporary, pre-extinction ecosystem organization re-established post-extinction in new clades

used to develop models for both short- and long-term climate change? A range of geological and paleontological criteria has helped identify climatic zones through time (Fig. 4.24). Specific sedimentary rocks such as **calcretes** (soils rich in calcium carbonate) and **evaporites** (evaporated salts) can help identify dry, arid climates whereas **dropstones** (stones that plummet from the bottoms of melting icebergs into seabed sediments) and **tillites** (rocks and sand left behind by an advancing glacier) indicate polar conditions. These criteria have formed the basis for Christopher Scotese and colleagues' reconstruction of climates and paleogeogeography through time (http://www.blackwellpublishing.com/paleobiology/). Global climate change can now be mapped through time with some degree of confidence.

Climatic fluctuations through time

Short-term trends

Many climatic events are short term, occurring within a time span of 100 kyr. Many surface processes respond rapidly to climate change, for example the atmosphere and ocean surface waters can change within days to a few years whereas the deep water of the ocean basins and terrestrial vegetation may take centuries to alter; the buildup of ice sheets and associated sea-level changes, however, occur over millennia. Changes in

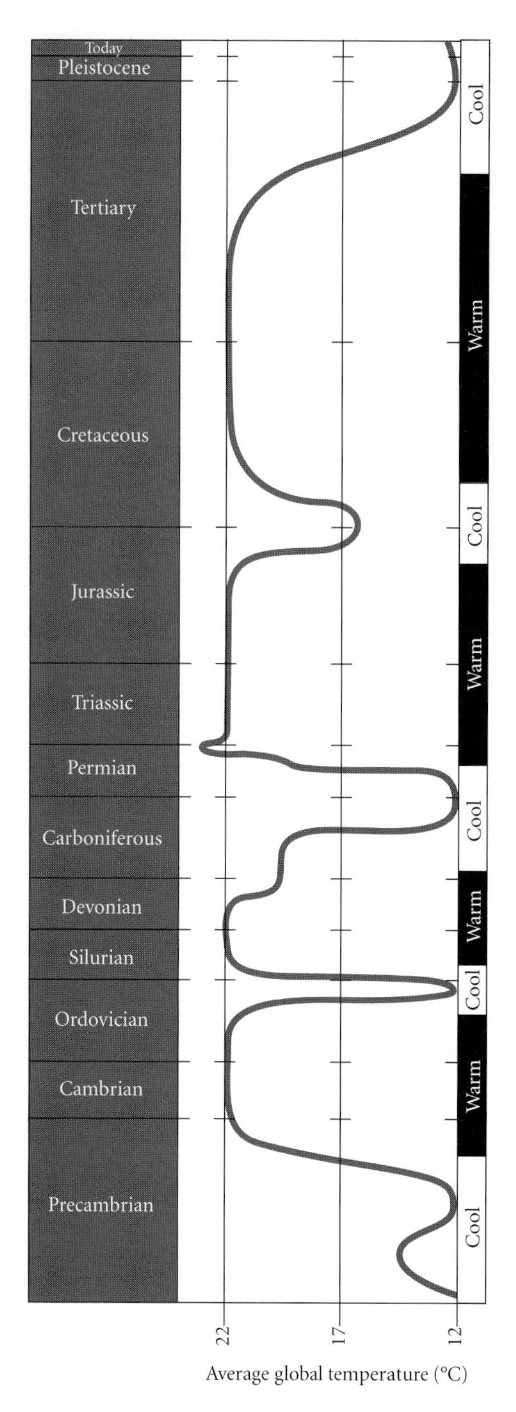

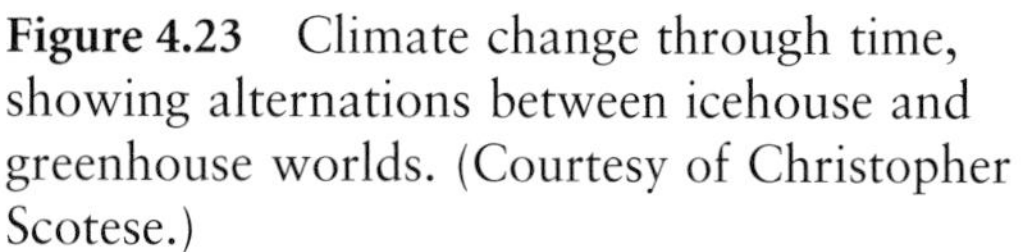

Figure 4.23 Climate change through time, showing alternations between icehouse and greenhouse worlds. (Courtesy of Christopher Scotese.)

precipitation and temperature in the recent past may have influenced the course of human events and almost certainly impacted on the direction of hominid evolution during the Late Pliocene and Pleistocene. Many short-

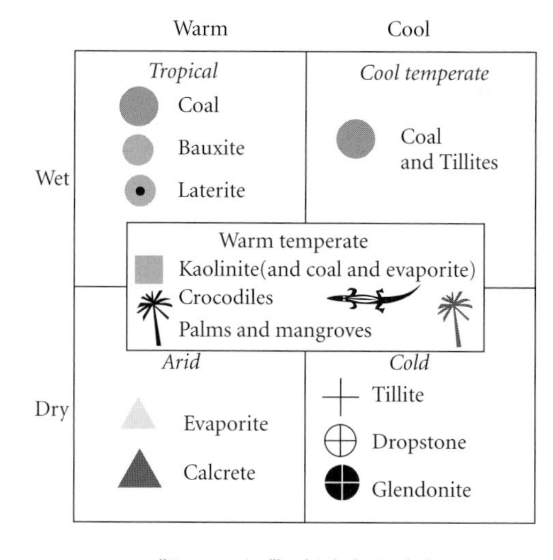

Figure 4.24 Some key indicators of climate and temperature. (Courtesy of Christopher Scotese.)

term climatic fluctuations have been related to Milankovitch cycles (see p. 36), patterns of change in climates and sedimentation patterns that are driven by changes in the eccentricity, obliquity and precession of the Earth's orbit and generally on scales from 20 to 400 kyr. These short-term trends are associated with evolutionary changes at the speciation level and more local regional changes in the composition and structure of ecosystems (Box 4.8).

Long-term trends

As noted above, the Earth has oscillated between greenhouse and icehouse conditions (Box 4.9) at least five times in the past 900 myr (Frakes et al. 1992). These megacycles have been compared with patterns of change in extinctions, sea level and volcanicity (Fig. 4.26). Moreover, there may be a correlation between these variables and the assembly and breakup of the supercontinents. In marine environments two extreme states occurred:

1 The icehouse state involved unstratified, unstable oceans, cool surface waters between 2 and 25°C and bottom waters ranging from 1 to 2°C together with rapidly moving bottom waters, rich in oxygen and with high productivity in areas of upwelling.

Box 4.8 Climate change and fossil size

There is strong evidence that climate and environmental changes have controlled extinctions and speciations, but do they have a direct influence on the size of organisms? Daniela Schmidt and her colleagues (2004) have investigated size changes in planktic foraminiferans during the last 70 myr from well-dated cores furnished by various ocean drilling programs. There was a sharp decrease in size at the Cretaceous–Paleogene boundary with the disappearance of many large taxa, and after this extinction event high-latitude taxa remained consistently small. Fluctuations in size, however, occurred in low-latitude assemblages (Fig. 4.25). A first phase (65–42 Ma) is characterized by dwarfs, a second (42–12 Ma) contains moderate size fluctuations, whereas the third (12 Ma to present) has the relatively large-sized taxa that typify Modern assemblages. Size increases are correlated with intervals of global cooling (Eocene and Neogene), when there were marked latitudinal and temperature gradients and high diversity. More minor size changes in the Paleocene and Oligocene may have been associated with changes in productivity. Cenozoic planktic foraminiferans thus provide strong support for a stationary model of evolutionary change, with size changes being strongly correlated with extrinsic factors such as fluctuations in latitudinal and surface-water temperature gradients.

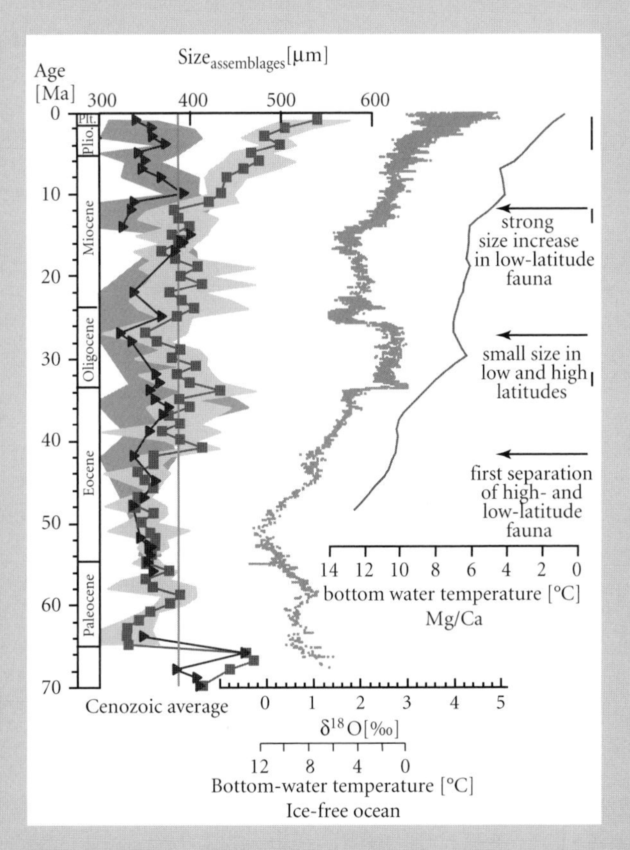

Figure 4.25 Size changes in planktic foraminiferans from high and low latitudes during the last 70 Ma, compared to temperature profiles generated from oxygen isotope data and Mg:Ca ratios. Three phases are recognized, a first (65–42 Ma) with dwarf taxa, a second (42–12 Ma) with moderate-sized taxa, and a third (12 Ma to present) with large-sized taxa. Size increases are correlated with intervals of global cooling. (Courtesy of Daniela Schmidt.)

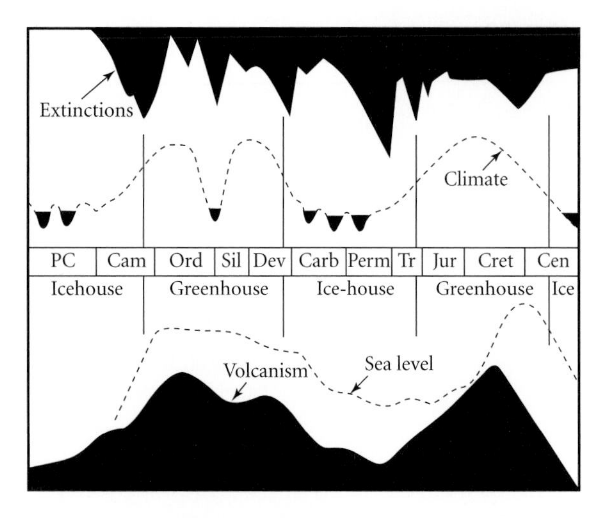

Figure 4.26 Climate change through time illustrated together with changes in sea level and fluctuations in the intensity of volcanicity. (Based on various sources.)

2 On the other hand, greenhouse oceans were more stable, and better stratified with surface waters ranging in temperature from 12 to 25°C with deep-water temperatures between 10 and 15°C. Slow bottom currents carried little oxygen and productivity was generally low.

Extinctions were associated with the transitions between these oceanic states.

In addition to these major climatic fluctuations a series of major extinction events, some associated with extraterrestrial causes, clearly prompted major climate change over several million years. Such events caused major taxonomic extinctions together with major restructuring of the marine and terrestrial ecosystems. Generally, greenhouse biotas were most susceptible to extinction; their species were more specialized and thus more exposed to environmental change.

Consequences for evolution

Microevolution is obvious in many fossil lineages (Benton & Pearson 2001) although the link between speciation events and climatic change is more controversial. Generally, marine plankton show gradual evolution whereas marine invertebrates and vertebrates display a pattern of punctuated equilibria (see p. 121). Moreover, it is probable that narrowly-fluctuating, changing environments host persistent gradualistic evolution whereas widely-fluctuating environments host morphological stasis (Sheldon 1996). This resistance to morphological change is clear in a number of lineages such as Ordovician trilobites and Pliocene mollusks (see p. 123).

Short-term climatic fluctuations, for example those associated with Milankovitch cycles, can clearly disrupt and promote the reassembly of both marine and terrestrial communities. In some cases they can drive local extinctions and radiations, for example in the conodont and graptolite faunas of the Silurian (see p. 434).

Climate surely drives larger-scale aspects of evolution. For example, the Cambrian explosion (see p. 249) – marked by the diversification of skeletal organisms and the appearance of reef-building organisms and the first predators – is associated with increasingly warm climates and higher sea levels. On land the radiation of early terrestrial tetrapods in the Early Carboniferous and the diversification of large flying insects in the first extensive forests, in cooler climates and more exposed land areas, have been correlated with high levels of atmospheric oxygen (Berner et al. 2000).

Some of the largest events of all, such as the appearance of entire new biotas and grades of organization, for example the origin of life itself, the development of photosynthesis and the appearance of the metazoans, may also be associated with climate change. The first two events have been associated with a stable Archaean crust and relatively cooler climates, which are favorable for carbon-based life to evolve. Metazoans appeared and diversified after the decay of the near global ice sheets of "snowball Earth" (Box 4.10), whereas skeletal organisms radiated during the greenhouse climates and higher oxygen levels of the Early Cambrian.

Biological feedbacks

If climate drives evolution, could life itself drive climate change? Few people doubt that humans can affect the climate, and everyone is aware of how the industrialized nations are

Box 4.9 Paleotemperature: isotopes to the rescue?

Is it possible to find out how hot or cold the Earth really was in the past? Stable oxygen isotopes can be extremely useful as paleothermometers but also in assessing the salinity of ancient oceans and the extent of ancient ice caps. Oxygen has three stable isotopes, the lightest being ^{16}O, then ^{17}O, and the heaviest ^{18}O. The ratio of $^{18}O:^{16}O$ is used in most geological investigations. When calcite is precipitated from seawater the ratio of $^{18}O:^{16}O$ increases with temperature. This ratio is also standardized with respect to standard mean ocean water (SMOW) or the Peedee belemnite standard (PDB), *Belemnitella americana* from the Cretaceous Peedee Formation in South Carolina. A shift of 1‰ in $\Delta^{18}O$ values represents a change in temperature of about 4–5°C. Unfortunately, not all shells are precipitated in equilibrium with surrounding seawater; the vital effects of some organisms interfere with the process. Moreover diagenesis can also affect isotope data. For these reasons corals, calcareous algae and echinoderms do not give good results; on the other hand brachiopods, bivalves and foraminiferans have yielded useful data. In addition, the lightest isotope is generally preferentially found in water vapor and thus rainfall. During glacial episodes, snow and ice can act as reservoirs for ^{16}O, thus depleting the world's oceans of that isotope. Thus during ice ages the oceans are characterized by higher amounts of ^{18}O. This simple model has formed the basis for our understanding of climate change over the last 1 myr and the relationships of such changes to Milankovitch cycles (see p. 36).

A dataset of oxygen isotopes is available for time series analysis at http://www.blackwellpublishing.com/paleobiology/.

burning fossil fuels and pumping greenhouse gases into the atmosphere. Global climate warming will affect the plants and animals of the cold temperate and polar regions as climate zones move about 100 km per century towards the poles (Wilson 1992). Nevertheless, a number of models for long-term climatic change have also involved the role of feedbacks from biological organisms. For example, the Gaia hypothesis is an attractive model that treats the Earth as a living system. The constant interaction between the Earth's living organisms, the atmosphere and the oceans helps keep the planet in check. The idea is certainly not new. James Hutton (1726–1797), the father of geology, once described the Earth as a kind of superorganism. But there were times in the Earth's history, the *Day After Tomorrow* ice age of snowball Earth (see p. 112) or the sustained hot climates of the Cretaceous world, when the Earth's climate seemed to be out of control. Nevertheless some climate change can be modeled by Gaia – some of the most marked during the Precambrian (Fig. 4.28). The diversification of photosynthesizers together with consumers from the Early Proterozoic onwards, hiked oxygen levels concomitant with declines in greenhouse gases. Such models promote the vital effects of life as a stabilizing influence on the planet's climate, reducing the otherwise steady rise in the Earth's surface temperatures. In the same way the extensive coal swamps and forests of the later Paleozoic may also have contributed to an interval of cooler climate as diversifying land plants mediated atmospheric oxygen levels, predicting the importance of modern rain forests as a climatic buffer.

There is no doubt that life on planet Earth is resilient and despite the extremes of climate change through deep time may have, through biological feedbacks, been able to conserve and control its own environment.

Box 4.10 Snowball Earth

Strong evidence suggests that a number of Late Neoproterozoic ice ages were of global extent (Hoffman et al. 1998). The occurrence of tillites in close association with carbonates in near-equatorial positions has suggested to Paul Hoffman and his colleagues that during these intervals the Earth was virtually covered by ice. These data supported a model first developed by Brian Harland in the 1960s, subsequently christened "snowball Earth" by Joe Kirschvink in the 1980s. But paleomagnetic data for low-latitude ice is not the only line of evidence for a global snowball. The majority of these glacial deposits are overlain by so-called cap carbonates. These rocks suggest deposition in extreme greenhouse conditions, under an atmosphere of high concentrations of carbon dioxide and seawater supersaturated with calcium carbonate. Such conditions were promoted by the high temperatures required to kick the Earth out of its "snowball" state (Fig. 4.27). The incredible buildup of the greenhouse gas, carbon dioxide, in the atmosphere was a direct consequence of a lack of liquid water and the cessation of weathering processes; this buildup essentially saved the surface of the planet from an eternal frozen state. The glacial deposits and the cap carbonates, however, are also strongly depleted in the ^{13}C isotope; this suggests very little biological productivity was in progress that could have removed the lighter ^{12}C isotope, causing preferential enrichment of the heavier ^{13}C stable isotope. And, finally, banded ironstone formations (BIFs) are a feature of the snowball interval suggesting the existence of an anoxic, stratified ocean system. Some BIFs are even associated with ice-rafted dropstones. Not everyone, of course, agrees with this hypothesis; some have suggested a milder "slushball Earth" and some even deny the possibility of global ice sheets altogether. But, surely these "freeze–fry" episodes had an important influence on the mode of organic evolution. Biological evolution would certainly have continued, not least associated with active volcanic vents deep under the ice and in other extreme environments. However evidence for metazoan life seems to appear directly after snowball Earth.

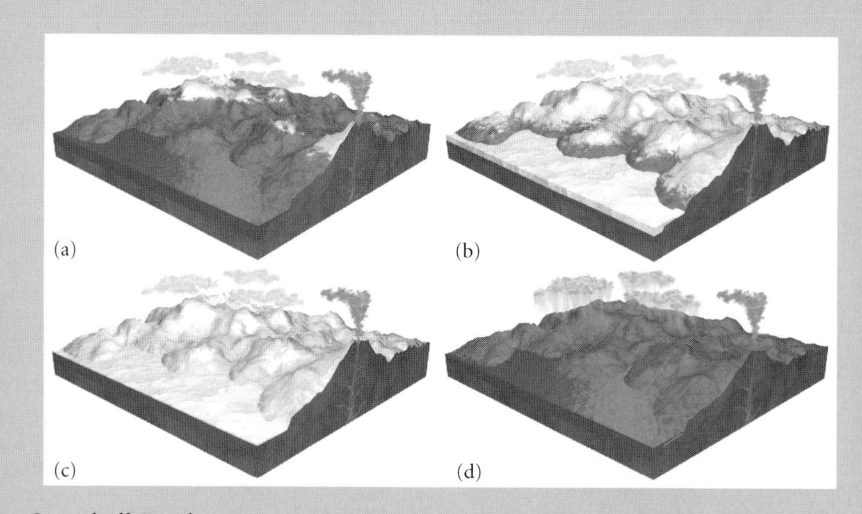

Figure 4.27 Snowball Earth scenario. (a) Continents are near the equator, increasing precipitation removes CO_2 from the atmosphere, and with falling temperatures ice begins to spread from the poles. (b) Ice continues to spread with temperatures further reduced by the albedo (reflection of solar energy) effect. (c) Atmospheric CO_2 increases due to volcanic activity, prompting a reversal in temperatures. (d) Greenhouse conditions return and the ice sheets recede. (Courtesy of Jørgen Christiansen and Svend Stouge.)

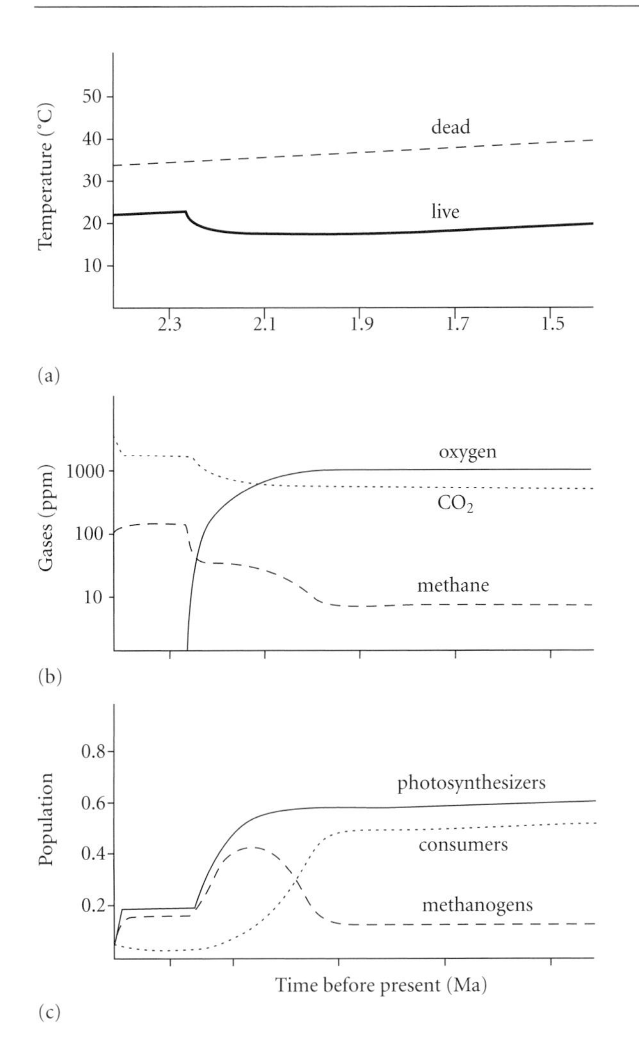

Figure 4.28 Precambrian Gaia and evolution of the biosphere. (a) Changes in climate in a live and lifeless world; there is a sharp fall in temperature when oxygen appears. (b) The changing abundance in atmospheric gases. (c) Changes in the composition of ecosystems: both the photosynthesizers and methanogens increase initially when oxygen appears but the methanogens eventually decline to a much lower level of abundance. (Population is the proportion of the total population in tenths.) (From Lovelock 1998.)

Review questions

1 Living communities contain a very wide variety of lifestyles. Although dominated by members of the Modern fauna, elements of the Paleozoic (suspension-feeding) fauna are still present. Is it possible to predict the sorts of habitats that they prefer?
2 Modern populations can show a variety of different size distributions. Fossil assemblages can also show size–frequency patterns. What sorts of processes can modify the original population polygons of a once-living species?
3 Paleocommunities through time increased their diversity by the expansion of ecospace. Why then did diversity reach a plateau during the Paleozoic but appears to still be increasing in the Modern fauna?
4 Large-scale ecological changes seem to be decoupled from major changes in taxonomic diversity. Is this a valid observation or are our data too crude to actually test this?
5 The Earth has suffered huge extremes in climate through time. Can the Gaia hypothesis help explain these climate swings? Will the planet ever experience a real snowball Earth?

Further reading

Bennett, K.D. 1997. *Evolution and Ecology: The Pace of Life*. Cambridge University Press, Cambridge, UK. (Relationship between ecology and evolution from a Quaternary perspective.)
Brenchley, P.J. & Harper, D.A.T. 1998. *Palaeoecology: Ecosystems, Environments and Evolution*. Routledge. (Readable textbook on most aspects of current paleoecology.)
Briggs, D.E.G. & Crowther, P.R. (eds) 1990. *Palaeobiology – A Synthesis*. Blackwell Scientific Publications, Oxford. (Modern synthesis of many aspects of contemporary paleontology.)
Copper, P. 1988. Paleoecology: paleoecosystems, paleocommunities. *Geoscience Canada* **15**, 199–208. (Concise but informative integration of the main concepts of paleocommunity ecology.)
Cronin, T.M. 1999. *Principles of Paleoclimatology. Perspectives in Paleobiology and Earth History*. Columbia University Press, New York.
Frakes, L.A., Francis, J.E. & Syktus, J.I. 1992. *Climate Modes of the Phanerozoic*. Cambridge University Press, Cambridge, UK. (Overview of ancient climates through time.)
Lovelock, J. 1998. *The Ages of Gaia*. Bantam Books, New York. (Stimulating discussion of the Gaia hypothesis.)
Vermeij, G.J. 1987. *Evolution and Escalation. An Ecological History of Life*. Princeton University Press, Princeton, NJ. (Fundamental text on the influence of predation on the history of life.)
Vrba, E.S. 1996. Climate, heterochrony, and human evolution. *Journal of Anthropological Research* **52**,

1–28. (Important paper relating hominid evolution to climate change.)

Wilson, E.O. 1992. *The Diversity of Life*. Belknap Press, University of Harvard, Cambridge, MA. (Excellent discussion of modern and past biodiversity and the problems that the Earth's ecosystems now face.)

References

Ager, D.V. 1963. *Principles of Paleoecology*. McGraw-Hill, New York.

Ausich, W.I. & Bottjer, D.J. 1982. Tiering in suspension-feeding communities on soft substrata throughout the Phanerozoic. *Science* **216**, 173–4.

Bambach, R.K. 1983. Ecospace utilization and guilds in marine communities through the Phanerozoic. *In* Tevesz, M.J.S & McCall, P.L. (eds) *Biotic Interactions in Recent and Fossil Communities*. Plenum Press, New York, pp. 719–46.

Bambach, R.K., Bush, A.M. & Erwin, D.H. 2007. Autecology and the filling of ecospace: key metazoans radiations. *Palaeontology* **50**, 1–22.

Bennett, K.D. 1997. *Evolution and Ecology: The Pace of Life*. Cambridge University Press, Cambridge, UK.

Benton, M.J. 1990. *Vertebrate Palaeontology*. Chapman and Hall, London.

Benton, M.J. & Pearson, P.N. 2001. Speciation in the fossil record. *Trends in Ecology and Evolution* **16**, 405–11.

Berner, R.A., Petsch, S.T., Lake, J.A. et al. 2000. Isotope fractionation and atmospheric oxygen: implications for Phanerozoic O_2 evolution. *Science* **287**, 1630–3.

Bottjer, D.J. 1998. Phanerozoic non-actualistic paleoecology. *Geobios* **30**, 885–93.

Brenchley, P.J. & Harper, D.A.T. 1998. *Palaeoecology: Ecosystems, Environments and Evolution*. Routledge.

Brenchley, P.J. & Pickerill, R.K. 1993. Animal–sediment relationships in the Ordovician and Silurian of the Welsh Basin. *Proceedings of the Geologists' Association* **104**, 81–93.

Brett, C.E., Boucot, A.J. & Jones, B. 1993. Absolute depths of Silurian benthic assemblages. *Lethaia* **26**, 25–40.

Brett, C.E., Ivany, L.C. & Schopf, K.M. 1996. Coordinated stasis: an overview. *Palaeogeography, Palaeoclimatology, Palaeoecology* **127**, 1–20.

Campbell, K.A. 2006. Hydrocarbon seep and hydrothermal vent paleoenvironments and paleontology: past developments and future research directions. *Palaeogeography, Palaeoclimatology, Palaeoecology* **232**, 362–407.

Copper, P. 1988. Paleoecology: paleoecosystems, paleocommunities. *Geoscience Canada* **15**, 199–208.

Davis, M.A., Thompson, K. & Grime, J.P. 2005. Invasibility: the local mechanism driving community assembly and species diversity. *Ecography* **28**, 696–704.

Droser, M.L., Bottjer, D.J., Sheehan, P.M. & McGhee, G.R. Jr. 2000. Decoupling of taxonomic and ecologic severity of Phanerozoic marine mass extinctions. *Geology* **28**, 675–8.

Frakes, L.A., Francis, J.E. & Syktus, J.I. 1992. *Climate Modes of the Phanerozoic*. Cambridge University Press, Cambridge, UK.

Hammer, Ø. & Harper, D.A.T. 2005. *Paleontological Data Analysis*. Blackwell Publications, Oxford, UK.

Harper, E.M. 2006. Dissecting post-Palaeozoic arms races. *Palaeogeogeography, Palaeoclimatology, Palaeoecology* **232**, 322–43.

Hoffman, P.F., Kaufman, A.J., Halverson, G.P. & Schrag, D.P. 1998. A Neoproterozoic snowball Earth. *Science* **281**, 1342–6.

Hollingworth, N. & Pettigrew, T. 1988. Zechstein reef fossils and their palaeoecology. *Palaeontological Association Field Guides to Fossils* **3**, 75.

Hollingworth, N. & Pettigrew, T. 1998. *Zechstein Reef Fossils and their Palaeoecology*. Palaeontological Association, London.

Jablonski, D. & Bottjer, D.J. 1990. The origin and diversification of major groups: environmental patterns and macro-evolutionary lags. *In* Taylor, P.D. & Larwood, G.P. (eds) *Major Evolutionary Radiations*. Systematics Association Special Volume 42. Clarendon Press, Oxford, UK, pp. 17–57.

Johnson, M.E. & Baarli, B.G. 1999. Diversification of rocky-shore biotas through geologic time. *Geobios* **32**, 257–73.

Kidwell, S.M. & Brenchley, P.J. 1994. Patterns in bioclastic accumulations through the Phanerozoic: changes in input or in destruction. *Geology* **22**, 1139–43.

Lovelock, J. 1998. *The Ages of Gaia*. Bantam Books, New York.

McGhee, G.R. Jr., Sheehan, P.M., Bottjer, D.J. & Droser, M.L. 2004. Ecological ranking of Phanerozoic biodiversity crises: ecological and taxonomic severities are decoupled. *Palaeogeography, Palaeoclimatology, Palaeoecology* **211**, 289–97.

McKerrow, W.S. 1978. *Ecology of Fossils*. Duckworth Company Ltd., London.

Olsen, P.E., Remington, C.L., Cornet, B. & Thomson, K.S. 1978. Cyclic change in Late Triassic lacustrine communities. *Science* **201**, 729–32.

Olsen, P.E., Shubin, N.H. & Anders, M.H. 1987. New Early Jurassic tetrapod assemblages constrain Triassic-Jurassic tetrapod extinction event. *Science* **237**, 1025–8.

Pickerill, R.K. & Brenchley, P.J. 1991. Benthic macrofossils as palaeoenvironmental indicators in marine siliciclastic environments. *Geoscience Canada* **18**, 119–38.

Schmidt, D.N., Thierstein, H.R. & Bollmann, J. 2004. The evolutionary history of size variation of planktic foraminiferal assemblages in the Cenozoic. *Palaeogeography, Palaeoclimatology, Palaeoecology* **212**, 159–80.

Seilacher, A., Reif, W.-E. & Westphal, F. 1985. Sedimentological, ecological and temporal patterns of fossil Lagerstätten. *Philosophical Transactions of the Royal Society B* **311**, 5–23.

Sepkoski, J.J. Jr. 1981. A factor analytical description of the Phanerozoic marine fossil record. *Paleobiology* **7**, 36–53.

Sheehan, P.M. 2001. The history of marine biodiversity. *Geological Journal* **36**, 231–49.

Sheldon, P.R. 1996. Plus ça change – a model for stasis and evolution in different environments. *Palaeogeography, Palaeoclimatology, Palaeoecology* **127**, 209–27.

Staff, G.M., Stanton, R.J. Jr., Powell, E.N. & Cummins, H. 1986. Time-averaging, taphonomy, and their impact on paleocommunity reconstruction: death assemblages in Texas bays. *Bulletin of the Geological Society of America* **97**, 428–43.

Twitchett, R.J. 2006. Palaeoclimatology, palaeoecology and palaeoenvironmental analysis of mass extinction events. *Palaeogeography, Palaeoclimatology, Palaeoecology* **232**, 190–213.

Whittington, H.B. 1980. The significance of the fauna of the Burgess Shale, Middle Cambrian, British Columbia. *Proceedings of the Geologists' Association* **91**, 127–48.

Wilson, E.O. 1992. *The Diversity of Life*. Belknap Press, University of Harvard, Cambridge, MA.

Wilson, M.A. 1985. Disturbance and ecological succession in an Upper Ordovician cobble-dwelling hardground fauna. *Science* **228**, 575–7.

Chapter 5

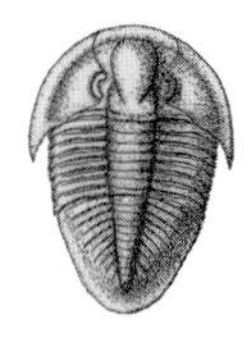

Macroevolution and the tree of life

Key points

- Evolution by natural selection is a core scientific model that was set out by Darwin, and has been confirmed again and again in every branch of biology.
- Creationist attempts to promote their religious beliefs, such as "intelligent design" or belief in a flat Earth, are not testable and therefore are not science.
- Speciation often occurs by the establishment of a barrier, and the isolation of part of a previously interbreeding population.
- Evolution takes place both within species lineages (phyletic gradualism) and at the time of speciation (punctuated equilibrium); the first model is commonest among asexual microorganisms that live in the open oceans, and the latter in sexual organisms that are subject to environmental and geographic barriers.
- There may be a process of species selection, acting independently of natural selection, but examples have been hard to find.
- The evolution of life may be represented by a single branching phylogenetic tree.
- Cladistics is a method of reconstructing phylogeny based on the identification of shared derived characters (homologies).
- Molecular sequencing provides additional evidence for reconstructing and dating the tree of life.
- DNA has been extracted from fossils such as woolly mammoths, but not from truly ancient fossils.

Probably all organic beings which have ever lived on this earth have descended from some one primordial form, into which life was first breathed . . . There is grandeur in this view of life . . . that, whilst this planet has gone cycling on according to the fixed law of gravity, from so simple a beginning endless forms most beautiful and most wonderful have been, and are being evolved.

Charles Darwin (1859) *On the Origin of Species*

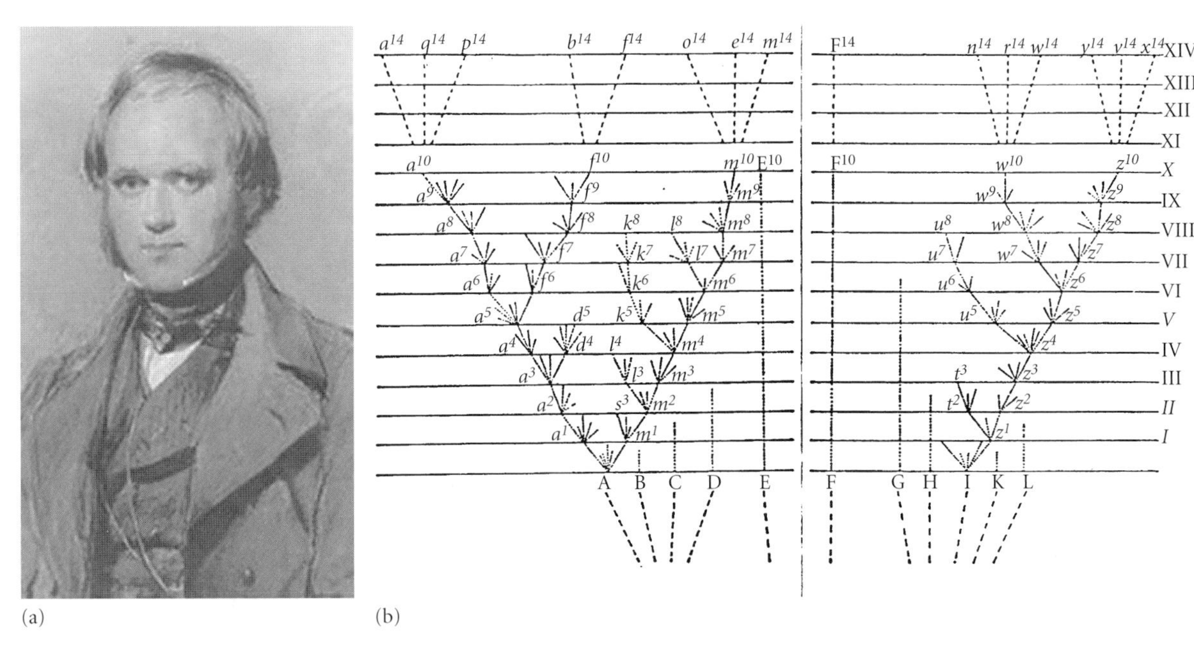

Figure 5.1 (a) Charles Darwin. (b) Branching diagram of phylogeny, the only illustration in *On the Origin of Species* (1859). It shows how two species, A and I, branch and radiate through time. The units I–XIV are time intervals of variable length, and the lower case letters (a, b, c) represent new species.

Darwin laid the framework for evolutionary biology 150 years ago. Despite millions of essays, books and web sites discussing evolution, no one has yet falsified Darwin's theory of evolution by natural selection, and so it stands as the core of modern biology and paleontology, just as Isaac Newton's laws stand at the heart of much of modern physics. And yet, surprisingly, Darwin is quoted, and misquoted, by many special-interest groups who want to use him in support of, or against, their views of politics, sociology and religion. So it is important to understand what Darwin said, how his insights affect science today, and how paleontology relies on modern evolution as its basis.

Charles Darwin's *On the Origin of Species* (1859) is usually remembered as the book that made the case for **natural selection** as the mechanism of evolution, sometimes called "survival of the fittest" (see pp. 118–19). Since the time of Darwin, evolution has been seen in action in the laboratory and in the field, and paleontologists use the principles of evolution to understand how species originate. The origin of species is also core to a second theme in Darwin's writings, namely **phylogeny**, or the pattern of evolution that is often represented as a branching tree diagram. Darwin's idea was that life had diversified to millions of species by the continued splitting of species from a common stem (Fig. 5.1). Indeed, he proposed that all of life, modern and ancient, could be followed back down the phylogenetic tree to a single point of origin: modern evidence confirms this remarkable insight.

Darwin's branching diagram also explained for the first time the meaning of the natural hierarchy of life that Linnaeus had discovered 100 years earlier (Box 5.1). This natural inclusive branching hierarchy is the basis of modern approaches to discovering the **tree of life**, the single great evolutionary tree that links all species living and extinct, from the modern biodiversity of over 10 million species, back to a single hypothetical species 3500 million years ago in the Precambrian. Paleontological aspects of evolution, such as the tree of life and studies of processes over thousands and millions of years, are sometimes called **macroevolution** ("big evolution") to distinguish them from **microevolution** ("small evolution"), all the smaller-scale and shorter-term processes studied by biologists and geneticists in the laboratory or in the field.

Box 5.1 Naming, describing, and classifying fossils

Life is organized in an **inclusive hierarchy**: small things (species) fit in larger categories, and these fit in still larger categories. Early naturalists realized that it was commonplace to be able to identify broad groups, such as wasps, bats, lizards, grasses or snails, and that within each group were many different forms, called species. Life did not consist of a random array of species. The similarity of groups of species suggested two things: first, that a classification system could be drawn up so people could identify and discuss particular forms without confusion, and second, that perhaps the inclusive hierarchy meant something.

Taxonomy is the study of the morphology and relationships of organisms. **Systematics** is the broader science of taxonomy and evolutionary processes, while **classification** refers particularly to the business of naming organisms and identifying the natural hierarchy. When a fossil is described for the first time, the author must name it. Biologists and paleontologists use a modified version of the principles established by the Swedish naturalist and scientist Carl Gustav Linnaeus (1707–1778), often regarded as the founder of systematics. Linnaeus believed that the evident hierarchical order in nature reflected the mind of God. Others at the time were to see things very differently, and to speculate about the possibility of **evolution**, or change through time.

In Linnaean nomenclature a species is given a genus and species name, such as *Homo sapiens*. These names are based on words from ancient Latin and Greek and they are printed in italics, followed by the author's name and date of publication. If, subsequently, another scientist moves a named species to another genus, perhaps because of new observations of similarity, the original author and date must then be placed in parentheses. Where several named species turn out to be the same, the subsequent names are identified as **synonyms**, or aliases, of the first name to have been given to the form.

When a new species is established, a **type specimen** is designated, and it is housed in a major institution, such as a museum or university, accessible to future investigators. The new species is defined by a short **diagnosis**, a few lines emphasizing the distinctive and distinguishing features of the fossil. A fuller description, supported by photographs, drawings and measurements, is also given, together with information on geographic and stratigraphic distribution.

Fossils, like living animals and plants, are classified in a hierarchical system, where species are included in genera, genera in families, and up through orders, classes, phyla, kingdoms and domains.

EVOLUTION BY NATURAL SELECTION

The night of September 28, 1838 was important for Darwin: it was then that he realized the missing piece of the evolutionary puzzle – natural selection. He wrote in his autobiography (Darwin 1859) that,

> *I happened to read for amusement Malthus on Population, and being well prepared to appreciate the struggle for existence which everywhere goes from long-continued observation of the habits of animals and plants, it at once struck me that under these circumstances favorable variations would tend to be preserved and unfavorable ones destroyed.*

On that date he drew a simple branching evolutionary tree in his notebook, and a more elaborate version was the only illustration in the *Origin* (see Fig. 5.1b).

Darwin came to his flash of inspiration by a combination of thoughts and observations:

- He had seen the huge diversity of life during a 5-year long circumnavigation of the world on board the *Beagle*, a British surveying ship; he asked himself why life was *so* diverse – every island he visited had different plants and birds.

- He had seen evidence for relationships in time and space – in South America he saw the bones of giant extinct ground sloths and armadillos, obviously close relatives of living forms; and as he went from island to island in the Galápagos and elsewhere, he saw close similarities between species of plants, reptiles and birds.
- He was aware of the record of fossils in the rocks, and that fossils changed through time, and seemed to progress from simple forms in the oldest rocks towards modern forms in the Pliocene and Pleistocene.
- Thomas Malthus argued in his book *An Essay on the Principle of Population* (1798) that human populations tend to grow far faster than their food supply, and Darwin transferred this concept to the natural world, seeing that reproductive rates are higher than they need to be.

Thus, by September 1838, Darwin understood the concept of evolution, a view that had been expressed by many thinkers before, and that claimed that life had not been static forever, but that species changed and never stopped changing. He had a rich understanding of modern geographic variation. Why, he asked, does every island in the Galápagos archipelago have a different set of species of small birds when the same set would do perfectly well throughout? Further, why did the bird species on neighboring islands look more similar to each other than those on distant islands?

So, Darwin's first key insight was that *life is more diverse than it ought to be if it had been created* and his second was that *all species living and extinct can be linked in a single great evolutionary tree that shows their relationships and that tracks back to a single ancestor*. These are descriptive observations of pattern.

But Darwin is remembered most for his third insight, and this was the principle of natural selection, a process that explains the diversity of life and its branching history of relationships: *only the organisms best adapted to their environment tend to survive and transmit their genetic characteristics in increasing numbers to succeeding generations while those less adapted tend to be eliminated.* Darwin made the case with remorseless logic, and this can be dissected into a series of clear statements:

1 Nearly all species produce far more young than can survive to adulthood (Malthus' principle).
2 The young that survive tend to be those best adapted to survive (larger at birth, faster growing, noisier in the nest, faster to escape predation, less disease, etc.).
3 Characters are inherited from parent to offspring, so the characters that ensure survival (size, aggressiveness, speed, freedom from disease, etc.) will tend to be passed on.
4 These survival characteristics will increase generation by generation. The changes are not inexorable, so cheetahs run fast enough to catch their prey, not at 2000 km per hour, because they do not have to and their bodies would fall to bits if they tried.

Each of these observations can be supported by huge numbers of observations. For point 1, note that most plants and animals produce hundreds, thousands or millions of offspring; if every melon seed grew into a melon plant or every cod egg became an adult, melons and codfish would soon cover the surface of the Earth to a depth of hundreds of meters. For point 2, observe any litter of puppies or nest of fledgling birds and see how siblings compete with each other for their parents' attention. For point 3, observe your parents or children and see the evidence for inherited characters. For point 4, consider how this emerges from points 1–3. Evolution by natural selection is on the one hand rather simple, but also rather complex, and it is frequently misunderstood or misrepresented (Box 5.2).

Darwin's *Origin* (1859) said it all, and he said it so well. In conclusion of this section, Darwin described natural selection in action:

> *It may be said that natural selection is daily and hourly scrutinising, throughout the world, every variation, even the slightest; rejecting that which is bad, preserving and adding up all that is good; silently and insensibly working, whenever and wherever opportunity offers, at the improvement of each organic being in relation to its organic and inorganic conditions of life.*

Box 5.2 The foolishness of intelligent design

Since the earliest days, philosophers have sought to understand the world and where it came from. At one time, many scholars argued that the Earth was flat, while others argued that the Earth was static in space and the sun and planets rotated around the Earth. These views were famously disproved and rejected some 500 years ago.

Most religions have also espoused so-called "creation myths" (see p. 184), often fanciful stories about how the Earth was created, and how it was populated with life. One of the most famous creation myths is the Bible story in Genesis of how God created the first man, Adam, and then the first woman, Eve. For some time, religious fundamentalists – people who believe in the literal truth of every word of the Bible, the Koran or any other religious text – have conducted a campaign against evolution, and often against science and the modern world in general. At present, we see a rise in Christian and Islamic fundamentalism in different parts of the world, and enthusiasts from both religions try to use the political system and the press, and sometimes even violence, to impose their view on others.

Creationism is a belief that the Earth and life were created perhaps 7000 years ago, and that all the areas of science that refer to long time scales (e.g. geology, astronomy, cosmology) and to evolution (all biological and medical sciences) are wrong, and has been particularly prevalent in the United States. After years of ridicule by scientists, creationism has been restyled as **intelligent design** (ID), the view that organisms are so complex that they must have been created by an intelligent being. Proponents of intelligent design range in their beliefs from the hard line (everything you see around you is exactly as it was created, and creation was only a few thousand years ago) to the liberal (the key large groups of plants and animals were created, but perhaps a very long time ago, and perhaps there is some evolution between species). The different branches of creationism, including ID, lack testable hypotheses and they lack evidence, so they are not credible alternatives to evolution.

As an example, many supporters of ID use the flagellum of bacteria as evidence. The **flagellum** (plural, flagella) is a thin structure that beats in a whip-like way to drive the bacterium through the fluid in which it lives. The flagellum is composed of several components, and it is normally driven by a proton pump, the flow of hydrogen ions across a concentration gradient. Supporters of ID have chosen the flagellum as a key piece of evidence that biological structures are so complex they could not have evolved, but must have been created whole. They argue that the flagellum is a good example of **irreducible complexity**, meaning that it can only function as a whole, and if any part is removed it fails to function. In fact this is not true, as has been shown repeatedly, and each component of the flagellum has other functions. So, irreducible complexity, the keystone of ID, has not been demonstrated, and it probably reflects a failure of imagination on the part of the investigator.

Read more about evolution in Darwin (1859), Ridley (1996), Futuyma (2005), Barton et al. (2007) and at http://www.blackwellpublishing.com/paleobiology/, and National Academies Press (2008) for a clear statement about evolution and the lack of evidence for intelligent design.

EVOLUTION AND THE FOSSIL RECORD

Speciation

Species consist of many highly variable individuals, often divided into geographically restricted populations and races. All human beings belong to a single species, *Homo sapiens*, and yet every person is different. The range of genetic and physical variation among humans is enormous, and much of it appears to be associated with geographic distribution. There has also been variation through time, with subspecies of *Homo sapiens*, like *H. s. neanderthalensis*, the Neandertals, being stocky and heavily built, possible adaptations to the cold Ice Age conditions of Europe 30,000 years ago (see p. 473). All species show geographic variation and, where the fossil record is good enough, variation in time as well.

So what is a species? The commonest definition is the **biological species concept** that states, "a species consists of all individuals that naturally breed together and that produce viable offspring". So, all modern humans can breed together and produce fertile (viable) children, and they therefore all belong to one species. Wolves and domestic dogs are also highly variable in external appearance, and yet they interbreed successfully and so they all belong to the one species *Canis lupus*. Domestic dogs belong to the subspecies *C. l. familiaris*, the European wolf to *C. l. lupus*, and there are many other subspecies of wolves from other parts of the world. In other cases, the amount of physical variation may seem much less; there are certain species of frogs and birds, for example, that look identical but are differentiated by their songs and never interbreed with a frog or bird with a different song.

Local populations may be to a great extent autonomous, isolated from other populations of the same species, and with a subtly different **gene pool**, the overall array of genetic material in all the individuals within the population. The cohesion of a species is maintained over its natural range by processes of **gene flow**, the occasional wandering of individuals from one area to another, which interbreed with members of neighboring populations. These processes can be thought of as occurring on many different scales, ranging from the whole Earth for humans, to a tiny patch of forest for some insect species.

If species can show considerable, or little, physical variation, and they can be held together by gene flow, how do they split? The process of splitting of a population to form two species is **speciation**, and there are many models. The most convincing is the **allopatric** ("other homeland") or **geographic model** that was proposed in the 1940s by Ernst Mayr, based on the establishment of geographic barriers. He suggested that populations could be split and gene flow prevented by a barrier, such as a new strip of water, a new mountain chain or even the building of a major road – anything that stops free genetic mixing among populations. The separated populations would then diverge for two reasons:

1 Each population, or set of populations, would start out with a different gene pool, simply because part of the former genetic range of the intact species has now been separated off.
2 Selection pressures would be different, perhaps only subtly, on either side of the barrier.

The separation can cause a divergence in **genotype**, the genetic composition of an individual, population or species, and **phenotype**, the external appearance.

The allopatric model of speciation may take two main forms. The process may be symmetric (Fig. 5.2a), with the ancestral species being divided roughly down the middle of its geographic range, and the two daughter species starting out with similar-sized populations. More dramatic effects may be seen when the split is asymmetric (Fig. 5.2b). Here, a small population, perhaps isolated on an island, evolves independently of the parent species, which may continue roughly unchanged. The smaller population may show unusual and rapid evolution because of what Mayr called the **founder effect**, the fact that its gene pool is a small sample of the overall gene pool, and that new environmental pressures and opportunities may occur.

Speciation and evolution in the fossil record

Biologists generally assumed that speciation happens gradually, with new species branching off from their ancestors slowly. Up to 1970 this view was accepted by most paleontologists, but then everything changed.

Eldredge and Gould (1972) proposed an alternative to the gradual model of evolution, which they termed the punctuated equilibrium model. They argued that the fossil record does not show evolution occurring in species lineages: in fact, they argued, most species lineages show **stasis** ("standing still", i.e. no change) over long spans of time. Change occurs at the time of speciation. Eldredge and Gould contrasted the two evolutionary models in terms of the shape of a phylogeny:

1 In the **phyletic gradualism model** (Fig. 5.3a), with sloping branches, most evolution takes place within species lineages, and speciation events involve no special additional amount of evolution;

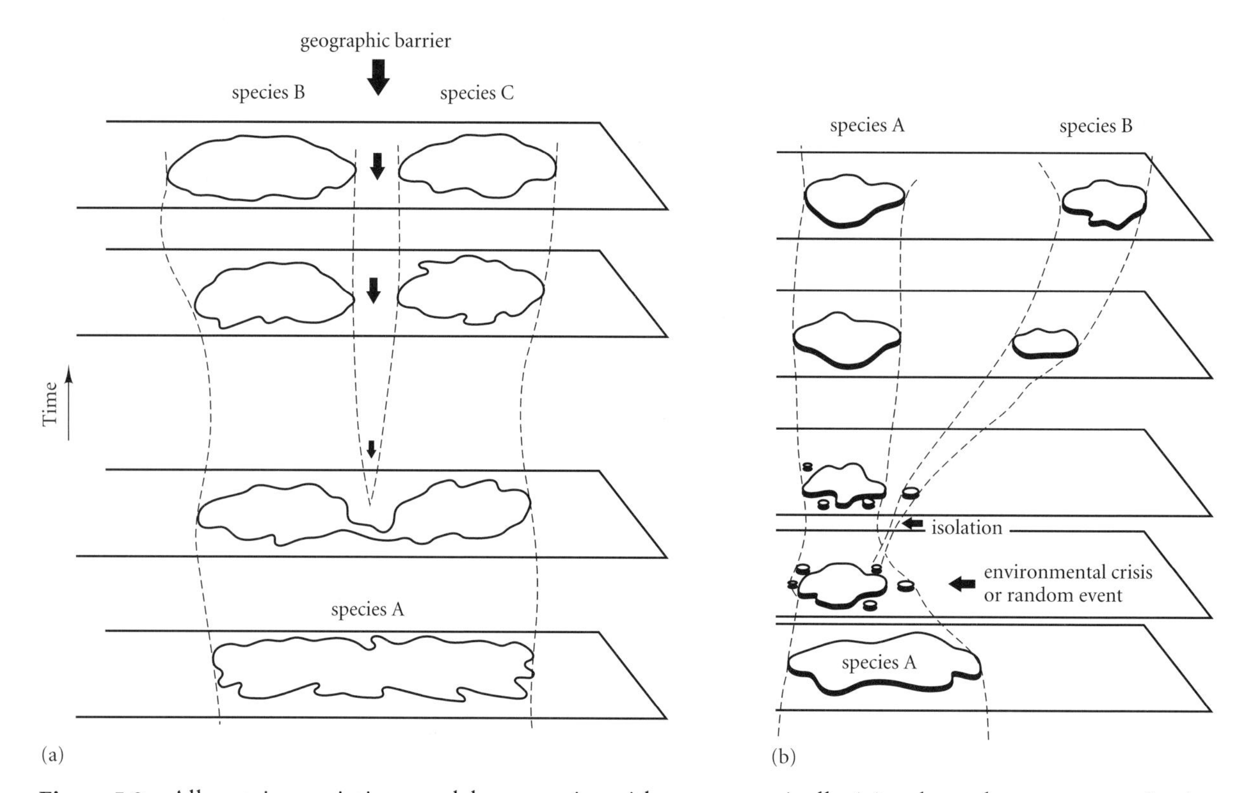

Figure 5.2 Allopatric speciation models, occurring either symmetrically (a), where the parent species is divided into two roughly equal halves by a geographic barrier, or asymmetrically (b), where a small peripheral population is isolated by a barrier. In the first case, two new species may arise; in the second, the parent species may continue unaltered, and the peripheral population may evolve rapidly into a new species.

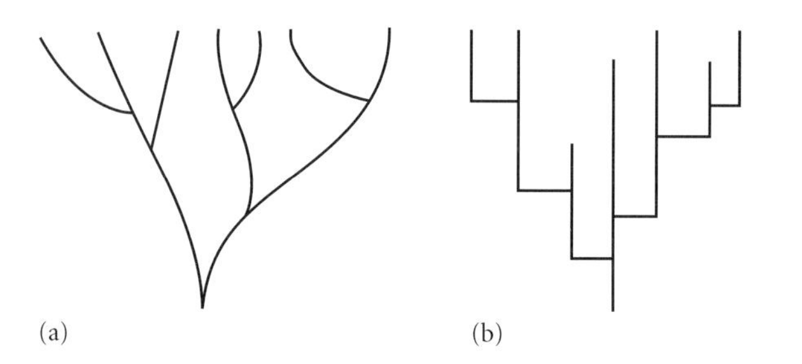

Figure 5.3 Two models of speciation and lineage evolution. (a) Phyletic gradualism, where evolution takes place in the lineages, and speciation is a side effect of that evolution. (b) Punctuated equilibrium, where most evolution is associated with speciation events, and lineages show little evolution (stasis).

2 In the **punctuated equilibrium model** (Fig. 5.3b), with rectangular branches, almost no evolution takes place within species lineages (they show stasis), and evolution is concentrated in the speciation events that coincide with major sideways shifts.

These two models of evolution seem so distinctive, both in the shape of phylogenies, and in their interpretation, that it should be possible to test between them by observations from the fossil record.

Testing punctuated equilibrium: problems

Eldredge and Gould (1972) argued that many test cases of the pattern of evolution at the species level could be studied from the fossil record. These should have the following features:

1 Abundant specimens.
2 Fossils with living representatives, so that species can be identified clearly.
3 Information on geographic variation, so that rapid speciation events (punctuations) could be distinguished from migrations in or out of the area.
4 Good stratigraphic control, in terms of long continuous sequences of rocks

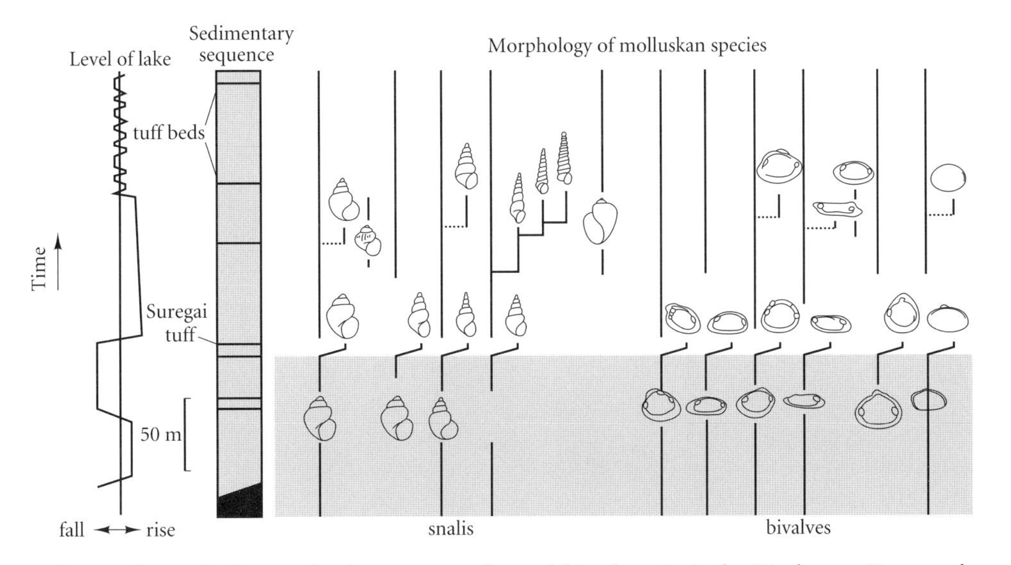

Figure 5.4 Fine-scale evolution in fresh-water snails and bivalves in Lake Turkana, Kenya, through the last 4 myr. The volcanic tuff beds allow accurate dating of the sequence. Major speciation events seem to take place at times of lake-level change: are these examples of punctuational speciation, or merely ecophenotypic shifts? (Based on Williamson 1981.)

without gaps, abundant fossils throughout and good dating.

The problems in testing became evident early on, because sampling was generally not extensive enough. Williamson (1981) attempted to counter this problem in one of the most enormous sampling exercises ever. He studied hundreds of thousands of specimens of snails and bivalves in sediments deposited in the Lake Turkana area of Kenya from 1.3 to 4.5 Ma (Fig. 5.4). Lake Turkana lies in the East African Rift Valley, on a tectonically active line where the continent of Africa is unzipping to form two major plates. Lake muds and sands accumulated in thick deposits as the rift opened, and volcanic ash (tuff) beds occur sporadically throughout the sequence.

Williamson recorded changes in 19 species lineages, and found that stasis was the normal state of affairs, but that rapid morphological shifts had taken place three times, two of which corresponded to substantial lake level rises (Fig. 5.4). He interpreted this as evidence for the punctuated equilibrium model, arguing that rapid environmental changes had caused evolutionary shifts and speciation events. The new species were short-lived, he argued, because the parental stock had survived in neighboring unstressed lakes, and returned to colonize Lake Turkana after the lake-level changes had taken place.

However, even this enormous study aroused controversy. Critics pointed out that the sequence of sediments was not complete enough to be sure that all fossils had been found: there were gaps of 1000 years or more, and a great deal of gradual evolution could take place in that time. Second, Williamson's critics noted that the environmental stress of lake-level change induced only short-term changes in shell shape, and when the stress was over, the shell shapes apparently reverted to normal (Fig. 5.4). Hence, they proposed that speciation had not taken place, and that the shells had merely changed shape **ecophenotypically**. This means that the changes happened during the animals' lifetimes, in response to particular stresses as they grew in size, and these changes were not genetically coded, and hence were not evolutionary.

Most recently, in a thorough re-study of this work, Bert van Bocxlaer and colleagues (2008) have suggested that Williamson got it wrong. They studied mollusks from several of the African great lakes, revised the taxonomy, and argue strongly that the three apparent speciation shifts (Fig. 5.4) are invasion events, when flooding episodes allowed bivalves and gastropods to enter the lakes from nearby rivers. As water levels subsided, the faunas returned more or less to their pre-flooding condition. So, they argue, this classic example of punctuated equilibrium might be better interpreted as an example of repeated climate change and migration. The new study casts serious doubt on a classic case of supposed punctuated equilibrium, but does not, of course, reject the whole concept.

Consensus on speciation in the fossil record

This debate might have led paleontologists to despair of ever finding a convincing case to assess the two models of species evolution. Now, after more than 30 years of debate, and hundreds of case studies, there seems to be a consensus (Benton & Pearson 2001). Both modes of evolution happen in different situations, punctuated equilibrium particularly among sexually reproducing species that live in ever-changing environments where barriers may be established, and phyletic gradualism is seen among asexual organisms, such as microorganisms that live in the surface waters of the oceans, where evolution is slow and barriers non-existent. So, it seems that Williamson (1981) mistook migrations for punctuations, but doubtless his snails were evolving by punctuational speciation, had the evidence been clearer.

The fossil record demonstrates the widespread occurrence of stasis. In a review by Erwin and Anstey (1995) of 58 published studies on speciation patterns in the fossil record, with organisms ranging from radiolaria and foraminifera to ammonites and mammals, and stratigraphic ages ranging from the Cambrian to the Neogene, 41 (71%) showed stasis, associated either with gradualism (15 cases; 37%) or with punctuated patterns (26 cases; 63%). It seems clear then that stasis is common, and that had not been predicted from modern genetic studies.

Microfossil groups such as the single-celled foraminifera, radiolaria and diatoms (see pp. 209, 211 and 229, respectively) commonly show gradual patterns of evolution and speciation. The microscopic skeletons of pelagic (open ocean) plankton can often be recovered in large numbers from sedimentary deposits that can be shown to have accumulated continuously over vast periods of time. A study by Sorhannus and colleagues in 1998 on the diatom *Rhizosolenia* (Box 5.3) is probably the most detailed recent work on speciation in planktonic organisms.

In this case, speciation is evidently **sympatric** (happening on the same spot), because the same splitting event is seen in most of the rock cores from around the equatorial belt of the Pacific. There is no evidence of an invasion of one species from an isolated population elsewhere; indeed, it is difficult to imagine where that population might have hidden and yet remained viable. Second, it is clear that most morphological evolution was not associated with speciation, but occurred afterwards, over about 500,000 years after the morphological distinction first becomes visible. Third, one of the new biological species evolved more rapidly than the other, becoming gradually smaller and evolving a markedly diminished hyaline area, whereas the other retained a morphology more like the ancestral species. Finally, the two species must have evolved

Box 5.3 Gradual speciation in radiolarians

Rhizosolenia is a planktonic diatom that occurs today in huge abundance in the highly productive waters of the equatorial Pacific. The siliceous valves of this genus rain on to the seafloor, where they accumulate in thick piles, mixed with other types of sediment. The morphological evolution of *Rhizosolenia* can be traced by sampling cores of this sediment, which have been taken in several places in the equatorial current system. Relative depths within each core provide a relative chronology, and this chronology can be tied to an absolute age scale using magnetic field reversals in the sediment. Ulf Sorhannus and colleagues from the University of Pennsylvania used this technique to study several million years' worth of evolution of *Rhizosolenia*, which encompasses a well-marked speciation event (Fig. 5.5).

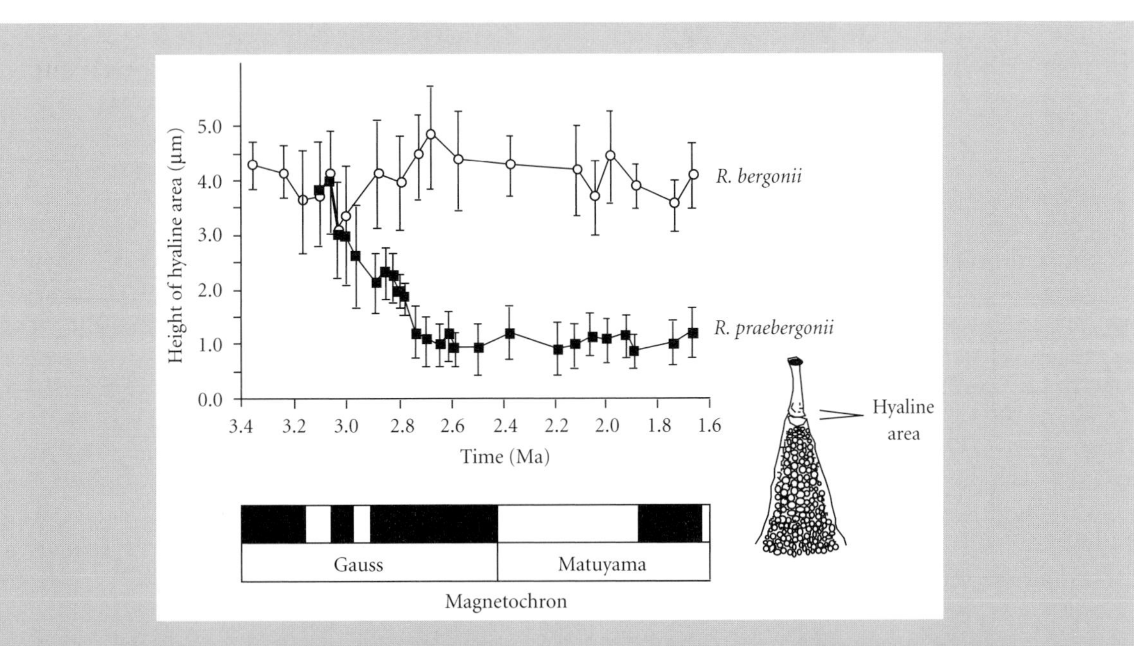

Figure 5.5 Phyletic gradualism and speciation in the planktonic diatom *Rhizosolenia*. Today there are two distinct species, *R. bergonii* and *R. praebergonii*, that do not interbreed and that differ in the height of the hyaline area. When tracked back through the past 3.4 myr, the species can be seen to have diverged through a span of up to 500,000 years, from 3.2 to 2.7 Ma. The plot shows samples taken from deep-sea boreholes in the central Pacific, and each measurement of the height of the hyaline area is based on a large sample of hundreds of individuals; the means and 95% error bars for each sample are shown. The rock succession is dated by reference to the magnetostratigraphic scheme of normal (black) and reversed (white) polarity. (Courtesy of Ulf Sorhannus.)

The valves of *Rhizosolenia* are conical in shape, terminating in an apical process that is rooted in a structure known as the hyaline area. The valves are usually broken at their distal ends, but Sorhannus and his colleagues were able to measure three distinct biometric variables: the length of the apical process, the height of the hyaline area, and the width of the valve at an arbitrary 8 μm from its apex. The first two characters are related to the overall size of the valve; the third is a shape parameter related to both size and the conical angle of the valve. These measurements were conducted on 5000 specimens in a number of populations in eight different cores, spanning 2 million years of evolution and about 60° of longitude.

Planktonic diatoms generally reproduce asexually, but like many predominantly asexual organisms they occasionally produce sexual offspring, probably to counteract the buildup of deleterious mutations (see p. 200). This sexual reproduction means that the large populations of *Rhizosolenia* can be considered as biological species, and speciation must be effected by a permanent barrier to reproduction.

The morphometric data provide convincing evidence that speciation occurred at or before about 3 Ma. Prior to this, there is only one discernible population, but afterwards, two morphologically distinct populations occur, within which there is a range of intergrading variation, but between which there is a morphological gap. The distinction is visible in all three measured parameters. The descendant species (*R. praebergonii*) later invaded the Indian Ocean where it appears abruptly in the sediment record.

Read more about speciation and punctuated equilibrium at http://www.blackwellpublishing.com/paleobiology/.

slightly different environmental tolerances, for although their geographic ranges overlap for all their evolution, one of the two daughter species is entirely absent in one of the cores.

Sympatric speciation and gradual evolution are probably rarer among marine invertebrates and continental vertebrates, where there are many more possibilities for the establishment of physical barriers to interbreeding. Studies of lineage evolution among marine invertebrates from shallow waters suggest punctuated patterns of speciation. Such studies are much harder to make than those of deep-sea microfossils because continental shelf sediments accumulate sporadically, and this makes it harder to acquire information with high sampling precision. Nonetheless, immensely detailed studies have been carried out. For example, in long-term studies Alan Cheetham and Jeremy Jackson of the Smithsonian Institution have sampled various genera of bryozoans in the past 10 million years of sediments in the Caribbean, and their studies suggest punctuational patterns of speciation (Box 5.4).

Current evidence suggests that Eldredge and Gould (1972) were right to challenge the assumption that evolution always had to be slow and gradual; in some cases it seems clear that species can split off rather rapidly, and that is entirely consistent with Darwinian evolution. Paleontological studies have shown that species often remain unchanged for long periods – the new phenomenon of stasis that had not been predicted from genetics. Asexual planktonic microorganisms appear to speciate slowly, perhaps over intervals of 0.5–1 myr in a gradualistic way, and sexually reproducing animals that occupy divided and complex habitats perhaps tend to speciate rapidly, in a punctuated manner.

Species selection

Steven Stanley (1975) argued that a punctuational model of evolution could imply a different kind of process, termed by him **species selection**, that occurred at the same time as, but separate from, natural selection. Stanley envisaged a process that sorted species, and ensured that some parts of the tree of life might diversify rapidly and others more slowly. He emphasized that if there was such a process as species selection, then the species-level characters must be distinct from the individual-level characters involved in natural selection. It is not enough to say, for example, that among African large cats, lions might survive certain kinds of competitive situations because they are larger than the other hunters. Being large is an individual-level character, and selection for size is through natural selection. Species-level characters must be irreducible to the individual level.

Possible species-level characters include the size of the geographic range of a species, the pattern of populations within the overall species' range, characteristic levels of gene flow among the populations of a species, and average species' durations. Some studies have suggested that species-level characters of these kinds may play a part in evolution. Geographically widespread species of gastropods, for example, tend to have longer durations than more localized species, and hence can be said to survive longer because of a species-level character. If species selection is a real force in evolution, then Darwinian evolution would have to be expanded to incorporate a hierarchy, or multilevel array, of processes.

A possible resolution of this issue is the **effect hypothesis** of Vrba (1984). She argued that some species-level characters may be reducible indirectly to the individual level. That is to say, a broadly based feature of the species actually depends on some other character that is under the influence of natural selection. She gave an example from her own work on the evolution of antelope over the past 6 myr (Fig. 5.7). About 5 Ma antelopes branched into two groups, one consisting of long-lived species that never became diverse, and the other of shorter-lived species that radiated widely. Species' duration in the first group was 2–3 myr and total species diversity through the Plio-Pleistocene was two; in the second group species' duration was 0.25–3 myr and 32 species evolved. Surely here, she argued, species selection was taking place: the character of short species' duration in the second group permitted great success, as measured by overall species diversity. Vrba noted, however, that the long-lived antelope had wide ecological preferences, while those in the second group were specialists. Hence, the whole pattern could be explained by natural selection at the level of individual antelope,

Box 5.4 Punctuated speciation in bryozoans

Metrarabdotos is an ascophoran cheilostome bryozoan (see p. 320) that is represented today in the Caribbean by three species. Coastal rocks on Dominica and other islands document the past 10 myr of sedimentation in shallow seas, and they yield abundant fossils of this bryozoan. The fossils show that *Metrarabdotos* radiated dramatically from 8 to 4 Ma, splitting into some 12 species, most of which then died out by the Quaternary. Studies by Cheetham and Jackson have established a variety of protocols for distinguishing species within *Metrarabdotos*, taking into account the genetics of related extant species, and their amount of morphological differentiation, and then extending comparable statistical tests of morphological differentiation to the fossil forms (they demonstrated highly significant correlations between genetic and morphometric differences among the modern forms). Based on 46 morphometric characters, the authors established a mechanism for distinguishing lineages among the fossils (Jackson & Cheetham 1999).

Lineage splitting in *Metrarabdotos* seems to have been rapid and punctuational in character (Fig. 5.6). Speciation was especially rapid in the interval from 8 to 7 Ma, with nine new species appearing. There is some question about sampling quality here, since sampling is poor in the preceding interval, and so some of these nine new species might have appeared earlier. However, the interval from 8 to 4 Ma, represented largely by information from Dominica, has been intensely sampled (DSI, Dominican sampling interval). So, although there are questions over the origins of the nine basal species within this interval, the origins of the remainder (*tenue*, n. sp. 10 and n. sp. 8) are more confidently documented as punctuated. The same kind of punctuated pattern of speciation has been found also in virtually all other studies on fossil marine invertebrates that have been carried out.

Read more about speciation and punctuated equilibrium at http://www.blackwellpublishing.com/paleobiology/.

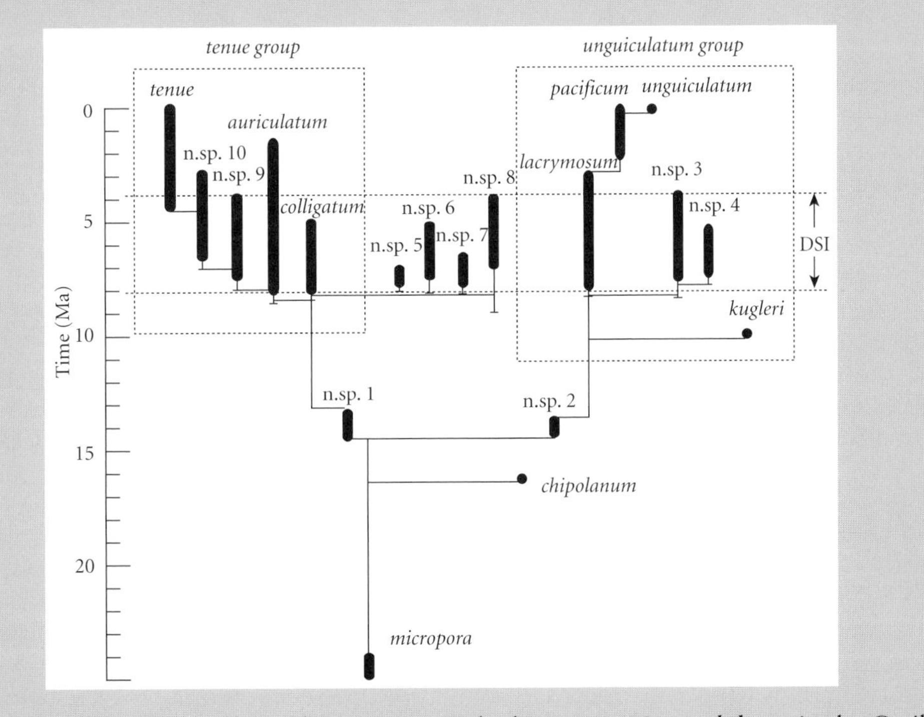

Figure 5.6 Punctuated evolution and speciation in the bryozoan *Metrarabdotos* in the Caribbean. Today, there are three species of this genus, but there have been many more in the past. Careful collecting throughout the Caribbean has shown how the lineages exhibited stasis for long intervals, and then underwent phases of rapid species splitting, especially in the time from 8 to 4 Ma, the Dominican sampling interval (DSI), where records are particularly good. (Courtesy of Alan Cheetham.)

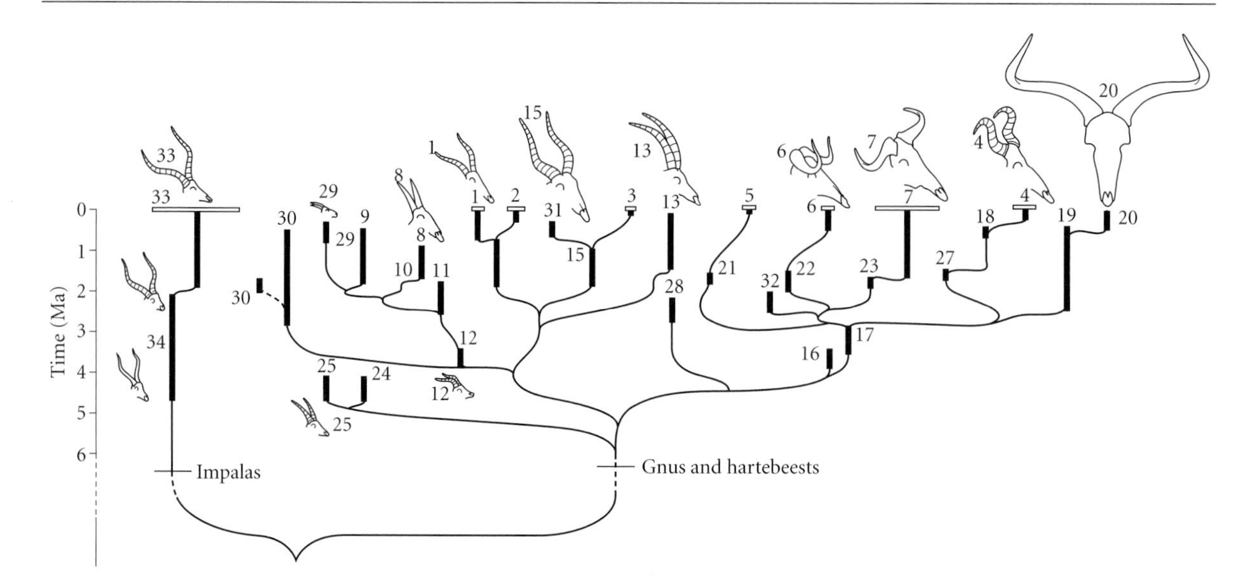

Figure 5.7 Reconstructed phylogeny of African antelopes. Two lineages diverged 6–7 Ma, the slowly evolving impalas and the rapidly speciating gnus and hartebeests. The second group could be said to be evolutionarily more successful than the first, and this might be interpreted as a result of species selection of species-level characters – the rate of speciation. However, the gnus and hartebeests have more specialized ecological preferences than do the species of impalas: perhaps selection has occurred at the individual level (natural selection), and this has had an effect at the species level. Species numbers 14 and 26 are omitted in this study. (Based on Vrba 1984.)

where their ecological tolerances determine their evolutionary rates, and produce a superficial appearance of species selection.

Is evolution hierarchical? And, if so, was Darwin wrong? The case has been overstated by critics: evolution occurs by natural selection, as Darwin said in 1859. Many proposed examples of species selection can be explained by natural selection, coupled with rapid asymmetric geographic speciation and the effect hypothesis. Nonetheless, species selection is a possibility, and convincing examples may be found in the future.

THE TREE OF LIFE

Tree thinking

As noted at the start of this chapter, Darwin was the first to picture evolution as a great branching tree, and to point out that all species had evolved from a single ancestor at the base of the tree. The idea of the **tree of life** has come to the fore recently, with a massive effort by biologists and paleontologists to discover the whole tree. From small-scale questions like "Which species of ape is closest to humans – the gorilla or chimp?", large teams of researchers are hastening to put together complete trees of all species of mammals, angiosperms, amphibians, spiders and many other groups. As each **complete tree**, that is, a tree containing all species, is published, systematists are getting closer to Darwin's ideal of understanding the shape of the whole tree of life.

It is important to distinguish trees from ladders. Many people think that all plants and animals are arranged in a series from simple to complex, or "lower" to higher". The pattern of evolution is then like a ladder, a single long line of progression from one species to the next, an idea that was popular 200 years ago and termed the Scala naturae (see p. 13). But all the evidence shows that evolution is a process of splitting and so the tree is the correct analogy, not the ladder.

Fossils offer fundamental information on the history of life and on large-scale patterns of evolution. There has been a revolution in the ways in which paleontologists interpret evolutionary aspects of the fossil record, and this is true of biologists as well. Nearly all studies of ecology, behavior and evolution are tied to a phylogenetic tree of the organisms involved. Since 1990 phylogenetic trees have been springing up everywhere, both because of new techniques for discovering trees and a realization that nothing in biology means anything without a tree.

Cladistics: reconstructing life's hierarchy

For centuries, biologists have struggled with the search for the true pattern of relationships among organisms: how is the tree of life to be discovered? Debates about whether birds originated from dinosaurs, whether annelids and arthropods are close relatives or not, or whether the gorilla or chimp is the closest relative of humans, all hinge on the need to identify patterns of relationships correctly.

If you had the task of sorting out the relationships among 100 species of parrots, where would you begin? You might note the color of their feathers, and classify them into a blue group, a red group and a green group. But then you might notice that body size or beak shape gives a different classification. If you then looked at the internal anatomy of the 100 parrots, you might find an entirely different classification based on the shape of the skull, the bones of the wing, or the arrangement of muscles or arteries. Up to 1960, systematists had a hard task in seeking to decide which characters were "good" and which were "bad". Good characters are **phylogenetically informative**, that is, indicative of the true phylogeny, but what about the bad, or uninformative, characters?

Phylogenetically uninformative characters fall into two main categories: convergences and plesiomorphies. **Convergence** in evolution is when features, or organisms, evolve to look the same perhaps because they live the same way. The marsupial mole of Australia looks just like the northern hemisphere mole, with great paddle-like limbs, poor eyesight and an excellent sense of smell, because they both burrow and eat worms, and yet they are not closely related. Two species of parrots might have convergently evolved a red patch of feathers on their wings as a signal. **Plesiomorphies** are characters that are shared by the organisms of interest, say parrots, but also by other groups. So, all parrots have beaks, but this is not a helpful character in sorting out the phylogeny of parrots because all other birds have beaks too. True parrots have blue and green feathers that have a special iridescent quality not seen in the feathers of cockatoos. But such special light-reflecting feathers are seen also in many other bird groups, and so are plesiomorphic for parrots.

Phylogenetically informative characters identify **clades, or monophyletic groups**. These are groups that had a single origin and include all the descendants of that common ancestor. A good example of a clade is the Psittaciformes, the parrots, a group that has long been identified as real and distinct from all others by naturalists. Clades are distinguished from two kinds of non-natural groups: (i) **paraphyletic** groups, which had a single common ancestor, but do not include all descendants, such as the Reptilia, which excludes birds and mammals; and (ii) **polyphyletic** groups, which are random assemblages of organisms that arose from more than one ancestor, and so have no place in the search for the tree of life.

Willi Hennig (1913–1976), an eminent German entomologist, realized the difference between phylogenetically informative and uninformative characters, and between monophyletic and paraphyletic/polyphyletic groups. He stressed the need to develop a new, more objective method in systematics, which has come to be called **cladistics**. The fundamental aim of cladistics is to identify clades, and so to discover, or reconstruct, the tree of life. Patterns of relationships are shown as branching diagrams, or **cladograms** (e.g. Fig. 5.8), in

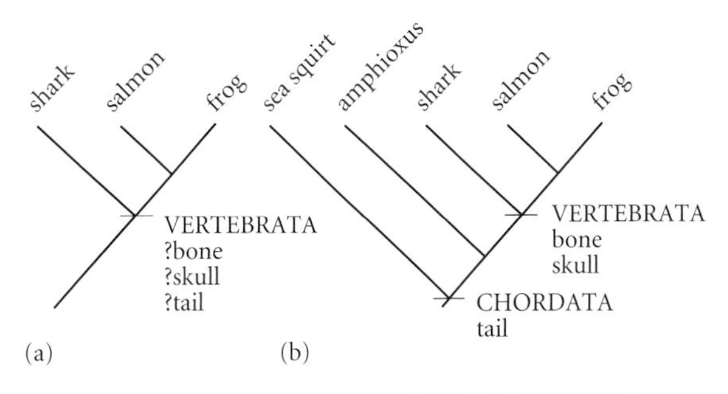

Figure 5.8 Reconstructing the phylogeny of vertebrates by cladistic methods. (a) Are the defining features of vertebrates the possession of bone, a skull and a tail? (b) The tail is found in a wider group, termed the Chordata, but the skull and bone define the Vertebrata.

which the most closely related species are joined most closely. The branches in the cladogram join at branching points, or **nodes**, each of which marks the base of a clade.

Hennig invented a rather complex terminology for cladistics, but some terms are commonly used, and should be mentioned. He called phylogenetically informative characters **apomorphies**, or derived characters (and distinguished them from plesiomorphies, the characters present in wider groups). Apomorphies shared by two or more species are termed **synapomorphies**. Apomorphies are features that arose once only in evolution, and therefore diagnose all the descendants of the first organism to possess that new character. Synapomorphies of parrots, the bird Order Psittaciformes include the deep, hooked beak and the unusual foot in which two toes point forward and two back.

The concept of an apomorphy actually corresponds to an older distinction between **homology** and **analogy** in evolution. The forelimb of vertebrates is a good example of a homology: even though the arm of a human is very different from the wing of a bird or the paddle of a whale, the detailed anatomy of each is the same, and they clearly arose from the same ancestral structure. On the other hand, the swimming limbs of vertebrates differ from group to group: in detail it can be shown that the paddles of ichthyosaurs, whales, plesiosaurs and seals (Fig. 5.9) are not homologs; they are merely **analogs**.

Cladistics might seem relatively straightforward, but it is not in practice (Box 5.5). There

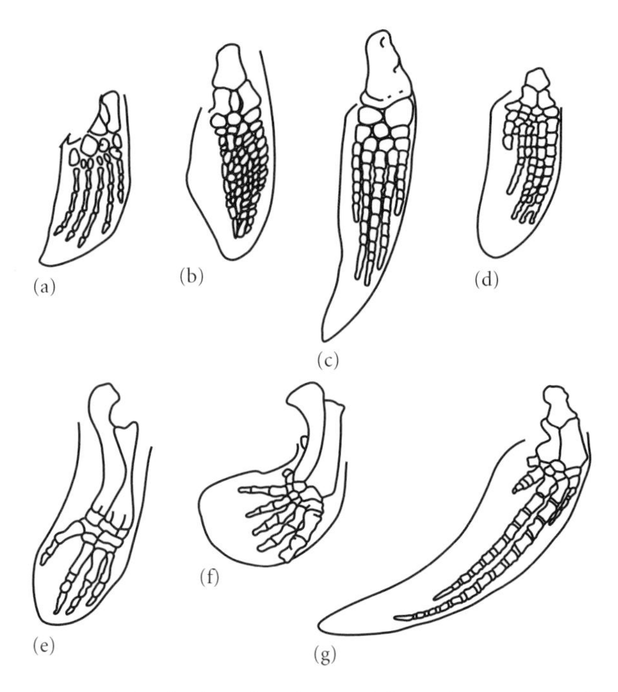

Figure 5.9 Swimming forepaddles of a variety of reptiles (a–d) and mammals (e–g): (a) *Archelon*, a Cretaceous marine turtle; (b) *Mixosaurus*, a Triassic ichthyosaur; (c) *Hydrothecrosaurus*, a Cretaceous plesiosaur; (d) *Plotosaurus*, a Cretaceous mosasaur; (e) *Dusisiren*, a Miocene sea-cow; (f) *Allodesmus*, a Miocene seal; and (g) *Globicephalus*, a modern dolphin. The forelimbs are all homologous with each other, and with the wing of a bird and the arm of a human. However, as paddles, these are all analogs: each paddle shown here represents a separate evolution of the forelimb into a swimming structure.

Box 5.5 Cladistic analysis

There are three steps in drawing a cladogram: character analysis, outgroup comparison and tree calculation. The **character analysis** is the process of listing characters that vary among the organisms of interest, and identifying those that are possible apomorphies. Characters are often coded as presence/absence or 0/1. So, for example, in a character analysis of the apes, a possible apomorphy might be "possession of an opposable thumb, used for gripping". This would be coded as present (or 1) in humans, and as absent (or 0) in chimpanzees and gorillas.

The **outgroup comparison** is the phase when characters and their codings are tested for validity. If the group under study, the **ingroup**, consists of the apes, then the **outgroup** might be monkeys. The outgroup could really be oak trees and worms, but they are so distantly related to the apes that comparisons would be largely meaningless (oak trees and worms do not even have thumbs). Comparison with the outgroup of monkeys shows that monkeys, chimpanzees and gorillas have non-opposable thumbs. So this is coded "0", and the opposable thumb of humans appears to be unique, an apomorphy, and so is coded "1".

The final step is the **tree calculation**. All the characters and their codings are listed in a **data matrix**, a tabulation of the data, listing all the species and all the characters and their codings. The tree calculation is usually run by computer, and a search is carried out, using different methods, to find the single tree, or group of trees, that best explain the data.

We give a simple worked example here, to determine the relationships of six vertebrates: shark, salmon, frog, lizard, chicken and mouse. These are distant relatives, of course, representing cartilaginous fishes, bony fishes, amphibians, reptiles, birds and mammals, respectively, but if we can sort out their relationships, we have a broad outline of the phylogenetic tree of all vertebrates.

Ten (out of many) characters are listed in the data matrix below, and their codings are shown (0 = absent; 1 = present) for each of the six animals. Next, outgroup comparison will help to sort out the phylogenetically informative characters from the plesiomorphies. We will choose two examples: the shark and the salmon have fins (character 1) while the others have legs (character 2), and the chicken and mouse are warm-blooded and the others are cold-blooded (character 3). Outgroup comparison of these two sets of characters suggests that warm-bloodedness is probably an apomorphy (because most members of the outgroup, such as clams, oak trees and bugs, are cold-blooded), but it is harder to tell whether fins or legs are apomorphies or not, so these two are retained.

Character	Shark	Salmon	Frog	Lizard	Chicken	Mouse
1 Fins	1	1	0	0	0	0
2 Legs	0	0	1	1	1	1
3 Warm-bloodedness	0	0	0	0	1	1
4 Bone	0	1	1	1	1	1
5 Diapsid skull	0	0	0	1	1	0
6 Loss of larval stage	0	0	0	1	1	1
7 Lung or swim bladder	0	1	1	1	1	1
8 Amniote egg	0	0	0	1	1	1
9 Elongated neck vertebrae	0	0	0	1	1	0
10 Marginal teeth	0	1	1	1	1	1

Scanning over the data in this table, it is clear that some groupings are indicated by several synapomorphies, but there are contradictions. For example, the diapsid skull (see p. 447) supports a pairing of lizard and chicken, but warm-bloodedness suggests a pairing of chicken and mammal. Both pairings are not possible, and one of these synapomorphies must be wrongly interpreted. The method of testing at this point is to seek the most **parsimonious** pattern of relationships, that is, the one that explains most of the data and implies least mismatch, or incongruence. The data may be run through a computer program, such as PAUP (Swofford 2007) that finds the most parsimonious cladogram (Fig. 5.10a), and highlights the **incongruent** (i.e. probably misinterpreted) characters. The cladogram is of course a best effort, and further study of the specimens, and the discovery of new characters, can confirm or refute it.

The cladogram can be made into a phylogeny by the addition of a time scale (Fig. 5.10b). Here, the fossil evidence for dates of origin of the various groups is used to give a picture of the true shape of this part of the phylogeny of life.

Read more about cladistics in Forey et al. (1998), and about cladistics as applied to fossil organisms in Smith (1994). Cladistic software includes PAUP (Swofford 2007), the most-used program, and basic cladistic routines are available in PAST (Hammer & Harper 2005). Read more at http://www.blackwellpublishing.com/paleobiology/.

Continued

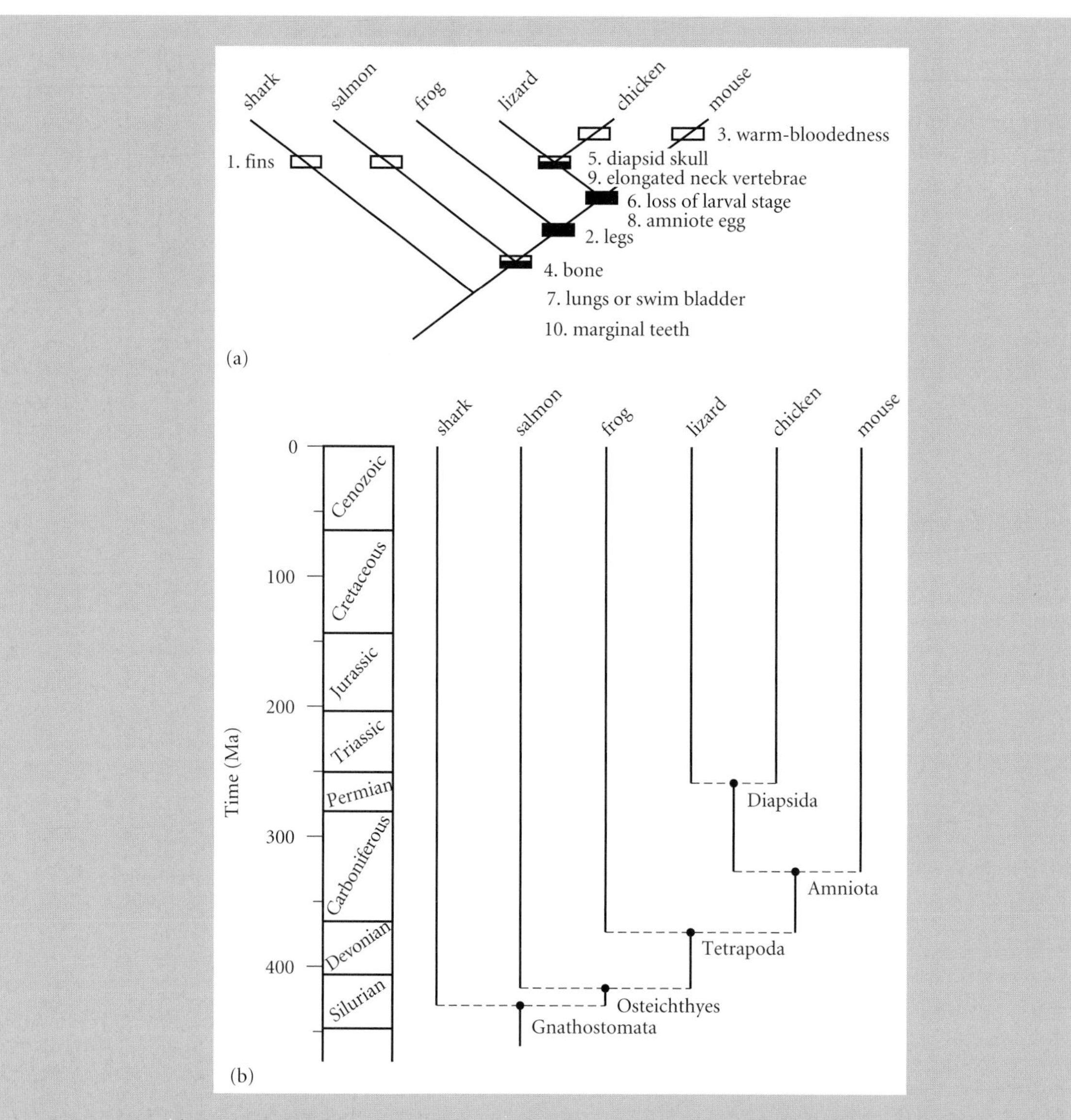

Figure 5.10 The relationships of the major groups of vertebrates, tested using six familiar animals. (a) Postulated relationships, based on the analysis of characters discussed in the text. (b) Phylogenetic tree, showing the cladogram from (a) set against a time scale, and basing the dating of branching points on the oldest known fossil representatives of each group.

are many controversies as systematists try to find agreement on disputed parts of the tree of life. In some cases, the shape of the tree is clear because each node is diagnosed by many apomorphies, but in others the clades and nodes are hard to pinpoint. Perhaps in those cases, evolution happened so fast that apomorphies were not established, or perhaps they have been overwritten in time.

The molecular revolution

The second approach in reconstructing the tree of life is based on comparison of molecules. With the birth of molecular biology in the 1950s and 1960s, it became clear that homologous proteins share similar structures in different organisms. For example, many animals share the molecule hemoglobin, a protein that carries oxygen in the blood, and that makes the blood red. Structurally, the hemoglobin of all organisms that possess it is very similar because it has to perform its oxygen-carrying function – but there are subtle differences. So, the hemoglobin of humans and chimps is identical, but their hemoglobin differs a little from that of a horse or cow, and a great deal from the hemoglobin of a shark or a salmon.

Comparisons of molecules allow analysts to do two things: to draw up trees of relationships and to estimate time. Trees of relationships can be based on a simple comparison of the amount of difference between protein sequences, and a best-fitting **dendrogram**, or branching diagram, is drawn. Identifying specific amino acid changes, and treating them as synapomorphies, allows the dendrogram to be treated as a molecular cladogram.

Time estimation comes from the concept of the **molecular clock**. The amount of difference in the fine structure of a protein between any pair of species is proportional to the time since they last shared a common ancestor. Differences have been documented in the primary structure of proteins, the sequence of **amino acids** from end to end of the unfurled protein backbone. There are some 20 amino acids, and their sequence determines the shape and function of a protein. Small changes in the amino acid sequence of hemoglobin occurred every few million years, somewhat at random, and the rate of change allows a time scale to be calibrated against the molecular tree.

Since 1990, attention has shifted almost entirely from sequencing proteins to sequencing the **nucleic acids** such as DNA and RNA. These are the molecules in the nucleus that comprise the genetic code, and they may be sequenced in a semiautomated manner using a process called the **polymerase chain reaction** (PCR). PCR is a means of **cloning**, or duplicating, small samples of nucleic acid, and then of determining the exact sequence of **base pairs**, the four components of the nucleic acid strand, adenine, cytosine, guanine and thymine (or uracil), abbreviated as A, C, G and T (or U). DNA and RNA may be sequenced from the nucleus or the mitochondria of cells (see p. 186), and molecular biologists generate huge sequences of such information each year. Indeed, the human genome project was one of many examples of international programs to determine the entire DNA sequence of all the chromosomes of a single species. The PCR method has also opened up the possibility of sequencing the genetic material of extinct organisms (Box 5.6).

Box 5.6 Fossil proteins: the real Jurassic Park?

Proteins were extracted from fossils in the 1960s and 1970s, but most of these were decay materials, the proteins of bacteria that decomposed the original tissues. Even in cases of exceptional preservation where soft tissues are preserved (see p. 60), the proteins have usually long vanished. Until 1985, the oldest DNA, recovered in tiny quantities, came from Egyptian mummies, 2400 years old.

Then came *Jurassic Park*! In the book by Michael Crichton (1990), and in the film by Steven Spielberg (1993), a scenario was developed where molecular biologists extracted dinosaur DNA from blood retained in the stomach of a mosquito preserved in amber. The fragments of dinosaur DNA were cloned and inserted into the living cells of a modern frog (an odd choice when the nearest living relatives of dinosaurs are birds), and the whole dinosaur genetic code was somehow reconstructed and living dinosaurs recreated. Amazingly, science then followed the fiction for a time.

Michael Crichton was wise to choose amber as the means of preservation (see p. 63). Insects in amber are trapped instantly, usually overwhelmed by the sticky resin, and no decay takes place; the amber excludes oxygen and water so that no physical or chemical changes should occur during subsequent millennia. A series of scientific reports were published in high-profile journals through the 1990s, announcing original DNA from a termite in Oligocene-Miocene amber, a weevil in Early Cretaceous amber, Miocene leaves, and even supposed dinosaur DNA in 1994. These reports col-

Continued

lapsed like a pack of cards soon after. The "dinosaur" DNA turned out to be human: the PCR technique is so sensitive that the tiniest fragment of DNA, in this case from sweat or sneezed mucus of a lab assistant, can be amplified. Careful study showed that DNA is highly labile and breaks down in even hundreds of years, and is pretty well all gone by 40,000 years, even in the most exceptional preservation.

The most convincing studies of the DNA of fossil species come from ice age mammals such as the cave bear, giant Irish deer, Neandertal man, woolly rhino and woolly mammoth. In a rush of enthusiasm, three labs independently sequenced and published the complete mitochondrial genome of the woolly mammoth in early 2006 (Krause et al. 2006; Poinar et al. 2006; Rogaev et al. 2006). These studies gave conflicting results: it is still not clear whether the closest living relative of the mammoth is the Asian elephant *Elephas maximus* or the African elephant, *Loxodonta africana* (Fig. 5.11). All studies though confirm that modern elephants and the mammoth are about as closely related to each other as humans are to chimps, and that the species split apart 5–6 Ma.

Read more at http://www.blackwellpublishing.com/paleobiology/.

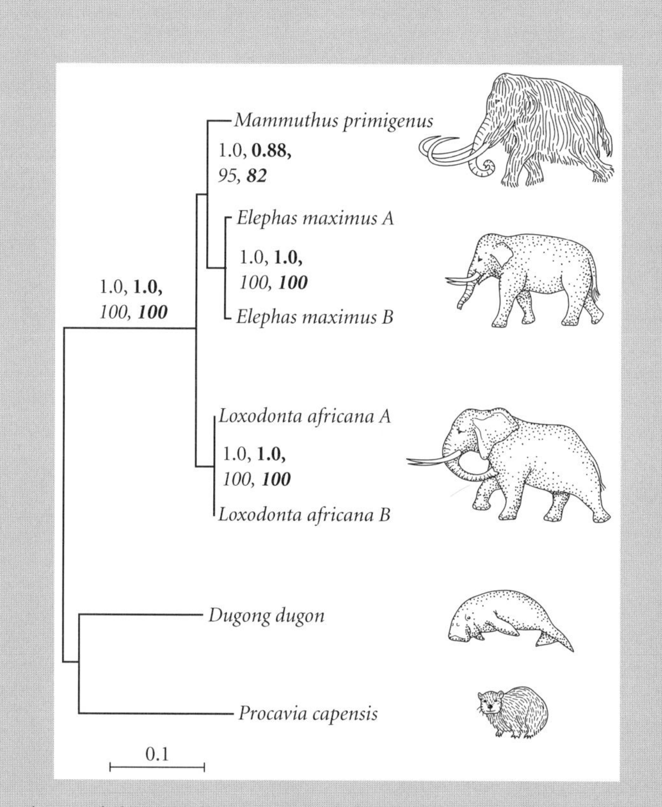

Figure 5.11 Relationships of the woolly mammoth based on mitochondrial DNA (mtDNA). This analysis (Rogaev et al. 2006) places the mammoth *Mammuthus primigenius* closest to the Asiatic elephant *Elephas maximus*, while other analyses of mammoth mtDNA place the mammoth closer to the African elephant *Loxodonta africana*. Either way, the relationship to the modern elephants is close, suggesting all three species diverged in the last 5–6 myr. Two samples of mtDNA for the two modern elephants are included, and the outgroups are the sea cow *Dugong dugon* and the hyrax *Procavia capensis*. The sets of digits at each branching point are various measures of robustness: values range from 0 to 1 and 0 to 100, with 1.0 and 100% indicating maximum robustness of the node. Scale bar is 0.1 base-pair substitutions per site. (Courtesy of Evgeny Rogaev.)

The vast numbers of DNA and RNA sequences that are generated daily from the whole diversity of life are stored as letter codes in open-access databases. Any investigator may download the sequences of any particular gene or chromosome for as many species as are available. There are then two key processes in extracting a phylogenetic tree from such data. First, the nucleic acid sequences must be **aligned**, that is matched, so that the code of a particular gene in one species is lined up with the same sequence in another species. Alignment can be difficult because species do not differ only in the placement of particular base pairs, but sometimes gaps in the sequence are introduced, or whole sections may be repeated. Once the sequences of a number of species are satisfactorily aligned, the phylogenetic analysis is carried out. The base-pair codes are treated like the presence/absence (1/0) codes in a morphological data matrix, and a variety of algorithms are applied to extract the most likely tree that explains the data.

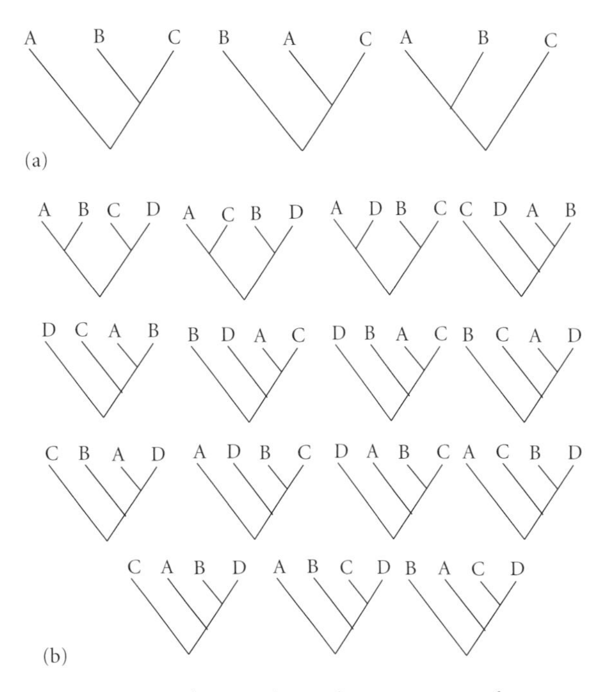

Figure 5.12 The number of unique trees for three (a) and four (b) taxa. These cladograms may be written more simply as (A(BC), (B(AC)) and ((AB)C) for the three-taxon cases, and ((AB)(CD)), ((AC)(BD)), ((AD)(BC)), etc. for the four-taxon cases. Note that (A(BC)) and (A(CB)) are identical trees, and both versions count as one.

The tree of life

Paleontologists and biologists are using morphology and molecules to put together ever-larger sectors of the tree of life. Desktop computers are exploding in labs around the world as analysts ask them to crank out ever-larger trees. Bear in mind that the number of possible trees is $N = (2n - 3)!/(2^{n-2}(n - 2)!)$. So for three species, there are three possible unique trees. For four species, there are 15 possible trees (Fig. 5.12), for eight species 168,210 possible trees, and for 50 species about the same number as there are atoms in the universe. You can do these calculations in table 1.3.1 at http://www.talkorigins.org/faqs/comdesc/section1.html.

And yet 50 species is not a demanding number. Systematists want to know the complete tree for all 240 species of primates, all 4500 species of mammals (now available: Bininda-Emonds et al. 2007), and so on. Mathematicians tinker with the tree-finding software so it finds clever ways to find the best-fitting tree quickly, even though many millions or billions of potential trees are considered and rejected. Another approach is to link existing trees for parts of the group of interest to create a **supertree** that summarizes the information in all existing trees. Researchers are currently using all methods and approaches to draw major sectors of the tree of life, and such huge trees will allow paleontologists and biologists to carry out many novel studies of macroevolution.

Review questions

1 Consider the four logical steps that summarize natural selection (see p. 119), list examples for each of numbers 1–3, and consider how the mechanism could be disproved.
2 Read books and web sites that present evidence for intelligent design (ID). List some specific examples/case studies that are cited by ID supporters, and present these in the form of falsifiable scientific hypotheses.
3 Find 10 paleontological case studies on speciation, by searching the internet with the key words "phyletic gradualism" and

"punctuated equilibrium". Look at the original papers, and consider how well each one fulfils the four criteria for a good study (abundant specimens, fossils with living representatives, information on geographic variation and good stratigraphic control; see p. 122). Which of the 10 studies you have chosen is sufficiently well documented to give a meaningful conclusion?

4 How would you design a study to test whether species selection might have happened?
5 Construct a cladogram of the apes, and identify apomorphies for each node. The likely shape of the tree is (gibbon(orangutan(gorilla(chimp + human)))) (see p. 472) – so read around and find out the cladistic basis for the tree.

Further reading

Barton, N.H., Briggs, D.E.G., Eisen, J.A., Goldstein, D.B. & Patel, N.H. 2007. *Evolution*. Cold Spring Harbor Laboratory Press, Woodbury, New York.

Briggs, D.E.G. & Crowther, P.R. 2001. *Palaeobiology, A Synthesis*, 2nd edn. Blackwell Publishing, Oxford, UK.

Darwin, C. 1859. *On the Origin of Species by Means of Natural Selection, or the Preservation of Favoured Races in the Struggle for Life*. John Murray, London.

Forey, P.L., Kitching, I.J., Humphries, C.J. & Williams, D.M. 1998. *Cladistics*, 2nd edn. Oxford University Press, Oxford, UK.

Futuyma, D. 2005. *Evolution*. Sinauer, Sunderland, MA.

Hammer, Ø. & Harper, D.A.T. 2005. *Paleontological Data Analysis*. Blackwell Publishing, Oxford, UK.

Mayr, E. 2002. *What Evolution Is*. Basic Books, New York.

National Academies Press. 2008. *Science, Evolution and Creationism*. National Academies Press, Philadelphia. Available at http://www.nap.edu/catalog.php?record_id=11876.

Ridley, M. 1996. *Evolution*, 3rd edn. Blackwell, Oxford, UK.

Smith, A.B. 1994. *Systematics and the Fossil Record*. Blackwell, Oxford, UK.

References

Benton, M.J. & Pearson, P.N. 2001. Speciation in the fossil record. *Trends in Ecology and Evolution* **16**, 405–11.

Bininda-Emonds, O.R.P., Cardillo, M., Jones, K.E. et al. 2007. The delayed rise of present-day mammals. *Nature* **446**, 507–12.

Bocxlaer, B. van, Damme, D. van & Feibel, C.S. 2008. Gradual versus punctuated equilibrium evolution in the Turkana Basin molluscs: evolutionary events or biological invasions? *Evolution* **62**, 511–20.

Crichton, M. 1990. *Jurassic Park*. Alfred A. Knopf, New York.

Darwin, C. 1859. *On the Origin of Species by Means of Natural Selection, or the Preservation of Favoured Races in the Struggle for Life*. John Murray, London.

Eldredge, N. & Gould, S.J. 1972. Punctuated equilibria: an alternative to phyletic gradualism. *In* Schopf, T.J.M. (ed.) *Models in Paleobiology*. Freeman, San Francisco, pp. 82–115.

Erwin, D.H. & Anstey, R.L. 1995. Speciation in the fossil record. *In* Erwin, D.H. & Anstey, R.L. (eds). *New Approaches to Speciation in the Fossil Record*. Columbia University Press, New York, pp. 11–39.

Hammer, Ø. & Harper, D.A.T. 2005. *Paleontological Data Analysis*. Blackwell Publishing, Oxford, UK.

Jackson, J.B.C. & Cheetham, A.H. 1999. Tempo and mode of speciation in the sea. *Trends in Ecology and Evolution* **14**, 72–7.

Krause, J., Dear, P.H., Pollack, J.L. et al. 2006. Multiplex amplification of the mammoth mitochondrial genome and the evolution of Elephantidae. *Nature* **439**, 724–7.

Malthus, T.R. 1798. *An Essay on the Principle of Population, as it affects the Future Improvement of Society, with Remarks on the Speculations of Mr Godwin, M. Condorcet and Other Writers, anonymous*. J. Johnson, London.

Poinar, H.N., Schwartz, C., Ji Qi et al. 2006. Metagenomics to paleogenomics: large-scale sequencing of mammoth DNA. *Science* **311**, 392–4.

Rogaev, E.I., Moliaka, Y.K., Malyarchuk, B.A. et al. 2006. Complete mitochondrial genome and phylogeny of Pleistocene mammoth *Mammuthus primigenius*. *PLoS Biology* **4**, e73.

Sorhannus U., Fenster, E.J., Burckle, L.H. & Hoffman, A. 1998 Cladogenetic and anagenetic changes in the morphology of *Rhizosolenia praebergonii* Mukhina. *Historical Biology* **1**, 185–205.

Stanley, S.M. 1975. A theory of evolution above the species level. *Proceedings of the National Academy of Sciences, USA* **72**, 646–50.

Swofford, D.L. 2007. *PAUP, Phylogenetic Analysis Using Parsimony*, version 4.0. Sinauer, Sunderland, MA.

Vrba, E.S. 1984. Patterns in the fossil record and evolutionary processes. *In* Ho, M.-W. & Saunders, P.T. (eds) *Beyond Neo-Darwinism; An Introduction to the New Evolutionary Paradigm*. Academic, London, pp. 115–42.

Williamson, P.G. 1981. Palaeontological documentation of speciation in Cenozoic molluscs from the Turkana Basin. *Nature* **293**, 437–43.

Chapter 6

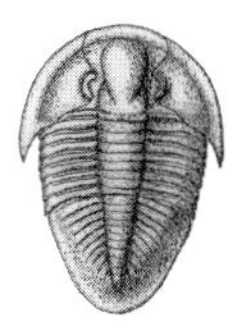

Fossil form and function

Key points

- Fossil species are identified according to their external form; this is termed the morphological species concept.
- Variations in form include normal levels of individual variation between members of a species, as well as variation that results from geographic distribution, sexual dimorphism (different males and females), different growth stages, or ecophenotypic variation (changes in form occurring within the lifetime of an organism as a result of the environment).
- Fossil species may show allometry, or changes in relative proportions during growth; specific organs may show positive (grow faster) or negative (grow slower) allometry.
- The development of an organism may give some evidence about phylogeny.
- Changes in developmental rates and timing (heterochrony) may affect evolution.
- The new "evo-devo" perspective shows how certain developmental genes control fundamental aspects of form, such as symmetry, front–back orientation, segmentation and limb form.
- Inferring function from ancient organisms is difficult. There are various methods of doing this: comparison with modern analogs, biomechanical testing and circumstantial evidence.
- Modern analogs may provide exact parallels with some fossil organisms, but more often they provide only principles or rules.
- Biomechanical models may be used to assess how the design of an ancient organism matches the hypothetical forces acting on it; an example is finite element analysis, a standard engineering technique.
- Biomechanical models of locomotion are easy to produce, but it is important to check that all possible gaits have been considered.
- Circumstantial evidence, such as the enclosing rocks, associated fossils, trace fossils and close study of the fossils themselves, can add considerable information on fossil function. Many such observations are the result of chance preservation.

There is grandeur in this view of life that, whilst this planet has gone cycling on according to the fixed law of gravity, from so simple a beginning endless forms most beautiful and most wonderful have been, and are being evolved.

Charles Darwin (1859) *On the Origin of Species*

Charles Darwin gave us phylogenetic trees (see p. 117) and he gave us biodiversity (see p. 535); he also gave us an evolutionary view of form. He was astonished by the variety of external appearances of plants and animals, and by their wonderful adaptations to life. He discovered many remarkable examples of extraordinary bodily appearance and function in insects, birds and plants. He was intrigued by the specialized beaks and tongues some birds and moths have for feeding on nectar from particular plants. He was fascinated by species where the males and females look utterly different because of the rigors of pre-mating displays. He tried to understand how such remarkable adaptations could be honed and sustained through the generations.

The **form**, or external appearance, of any microbe, plant or animal is shaped by evolution. The wings of birds are adapted for flight; the long beak and tongue of the hummingbird is adapted for feeding on nectar deep in a flower. The amazing tail of the male peacock is adapted through sexual selection to attract a mate. **Sexual selection** is the set of evolutionary processes that depend on interactions between the sexes. The most familiar examples are the astonishing tails and colors of male birds, the antlers and horns of male deer and antelope, the mane of the male lion, and other structures that are there to impress females. These structures generally have little to contribute in protecting the animal from predators or in helping it to find food: in many cases they are a considerable handicap. So sexual selection can act counter to natural selection; but the benefit for a male peacock in finding a willing female, or females, and in mating (sexual selection) is clearly greater than the disadvantage of carrying such a huge tail when trying to avoid a predator (natural selection).

Plants and animals have **adaptations** that function in the context of both natural and sexual selection. An adaptation is an aspect of form that performs a physical or behavioral function. It may not perform that function terribly well, merely well enough. So, human beings are adapted to walking upright, and this has changed many aspects of our body shape. But we are not enormously good at it, and we betray some of our quadrupedal ancestry: humans still get bad backs, arthritic joints and cannot run very fast. In the early years of bipedalism on the African plains 5 million years ago, perhaps speed was not always essential. A human could never outrun a lion or a cheetah, but could perhaps climb a tree or hide in a cave, or act with other humans to distract the predator.

Paleontologists have always been fascinated by the form and function of fossils. Not only are fossil forms often startlingly beautiful, they may be puzzling. So many fossils belong to groups without modern analogs that it becomes an intriguing exercise to determine why they had the forms they had, and what their functions may have been. The form of fossils is important for paleontologists for three reasons:

1 Form is the only evidence we have in fossils for identifying species and wider relationships to reconstruct the tree of life (see below and p. 128).
2 Form can tell us about behavior and ecology (see p. 80).
3 Variations in form are commonplace within a species, and the study of changes in form through time informs us about evolution (see p. 121).

GROWTH AND FORM

Recognizing ancient species

Paleontologists must interpret fossil species, and their ranges of variation, solely from the morphology, or external shape, of the specimens. There are problems in deciding where one species ends and another begins. When there are close living relatives, it may be possible to compare the modern species with the fossils. But how are paleontologists to decide just what is a species of dinosaur or trilobite?

For modern plants and animals, systematists ideally apply the biological species concept (see p. 121). Clearly paleontologists cannot test whether fossil species can or cannot interbreed. So, paleontologists use the **morphological species concept**, judging the bounds of a species entirely on form. The assumption is that all members of a species should look similar, and that a few simple statistical observations should define the mean or average characteristics of members of a

Box 6.1 Phenetics and variation within populations

Frequently, paleontologists are faced with problems that require the simplification of a great mass of measurements. For example, a paleontologist may have a large sample of fossils from a single rock horizon and may wish to determine whether these represent one or more species. It might be sufficient to plot univariate frequency histograms (see p. 16) of particular measures, such as width, length and depth of the shells, as well as the hinge width, the diameter of the pedicle foramen, and the length and width of internal muscle scars. In addition, bivariate plots could be prepared, in which various measures are plotted against each other. However, it might still be difficult to differentiate clusters of points, and this approach means the paleontologist has many separate graphs to compare.

Multivariate techniques can help solve these problems by dealing with all the measured variates together. Two common techniques are cluster analysis and principal components analysis (PCA). In PCA, the maximum direction of variation is determined from the table of raw measurements of many characters, and this direction is termed eigenvector 1. Further eigenvectors are then plotted in sequence perpendicular to the first, representing successively less variation in the sample. The first eigenvector usually reflects growth-related or size-dependent variation, and it is typically ignored in taxonomic studies. Species are usually plotted against the second and third eigenvectors, and tests can then be applied to determine whether there are separate clusters of points.

As an example, a comparison may be made between specimens of two species of brachiopod, *Dicoelosia biloba* from the Early Silurian of Sweden, and *D. hibernica* from rocks of the same age in Ireland. Four measurements were made on samples of both species and a PCA was performed. Both species were then plotted against the second and third eigenvectors (Fig. 6.1). Although both samples overlap, in general the Irish specimens have lower scores on eigenvector 2, showing that *D. hibernica* is wider and less deep than *D. biloba*, and strongly suggesting that there are two species.

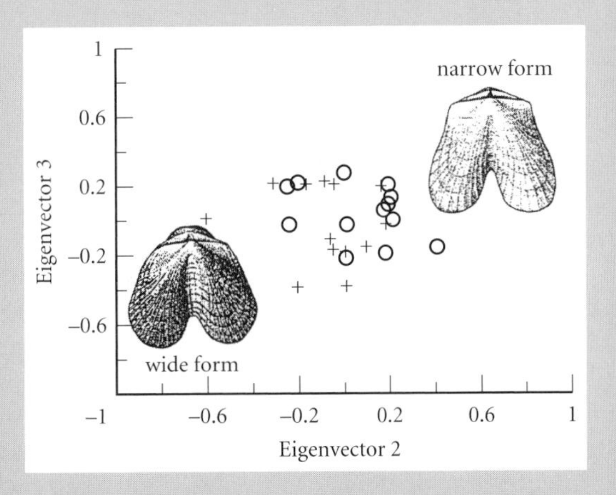

Figure 6.1 Variation in the Early Silurian brachiopod species *Dicoelosia biloba* from Sweden (o) and *D. hibernica* from Ireland (+), based upon numerous measurements. A principal components analysis plot separates wide and narrow forms along eigenvector 2, so there may truly be two species, although there is considerable overlap between the two.

species and the range of variation around that mean. In practice, this usually seems to be the case (Box 6.1). Critics are correct to point out that the morphological species concept suffers from many problems. Not least is the concern that the true species boundaries might be missed: how do you cope with a single species that has many different forms; how do you recognize species that differ in color, song or smell; how can you distinguish cases where the male and female look very different from each other?

Actually paleontologists need not be too downhearted. Where the biological and morphological species concepts have been applied to particular groups, both generally give the same answers. Further, it would be wrong for a systematist of modern organisms to be too smug. Most decisions on the species bounds of living plants and animals are based on assessments of the morphologies of dead specimens in museums: it is impractical to carry out extensive crossbreeding tests even with living organisms.

Problems with fossil species usually arise from the added dimension of time. If a paleontologist finds a long evolving lineage, where should the dividing line be drawn between one species and the next? Decisions are often made easier by gaps in the fossil record that create artificial divisions within evolving lineages. Where gaps are not present, splitting events clearly mark off new species. If there are few of these, an evolving lineage is divided somewhat arbitrarily into **chronospecies** ("time species"), each being defined by particular morphological features.

Variations in form within species

Within a species, there may be a range of morphologies; think of the variation among humans, or more dramatically, among domestic dogs. In naturally occurring species, morphology may vary as a result of several factors:

1 *Individual variation*, the normal differences between any pair of individuals of a species that are not identical twins; this base level variation is the stuff of natural selection, as Darwin stressed.
2 *Geographic variation* and physical differences between populations or subspecies in different parts of the overall species range.
3 *Sexual dimorphism*, in which males and females may show different sizes, and different specialized features (horns, antlers, tail feathers) often related to sexual selection.
4 *Growth stages*, where there may be quite different larval and adult stages, or where body form alters during growth.
5 *Ecophenotypic effects*, where local ecological conditions affect the form of an organism during its lifetime (see p. 123).

Geographic variation may be substantial among members of modern species, particularly those distributed over wide ranges. **Sexual dimorphism** is seen in living animals, particularly in those where males engage in ritualized displays, or where females have special reproductive activities. Sexual dimorphism is also common in fossils, and it has often caused serious problems of identification where males and females look very different. For example, many ammonites show sexual dimorphism, where the postulated females are much larger than the males, and the males possess unusual lappets on either side of the aperture (Fig. 6.2).

Most organisms change substantially in form as they grow from egg to adult, and these growth stages will be explored next. Ecophenotypic variation was introduced in Chapter 5 (see p. 123) and this includes all the changes in form that may occur through an individual's lifetime, but that are not coded genetically. Ecophenotypic variation in a human might include minor features such as the acquisition of powerful arm muscles through work or exercise or the loss of liver function through alcohol abuse. Major ecophenotypic changes might include the loss of a limb in an accident or a carefully maintained Mohican haircut. None of these changes can be passed on genetically to a son or daughter by the legless, muscular or unusually coiffed individual.

Allometry

Changes in form during growth are common. Think of human growth: babies have relatively large heads and eyes, and small limbs. Similar features are found in fossil examples too. Juvenile vertebrates, not just humans, usually have large eyes and heads in proportion to overall body size. A tiny embryo of an ichthyosaur (Fig. 6.3) shows just these features. If measurements of the variable parts (eye diameter, head length) are scaled against a standard measure of the animal (total body length, for example), it is evident that the proportions change as the animal grows older (Fig. 6.4). In the case of the ichthyosaur, the

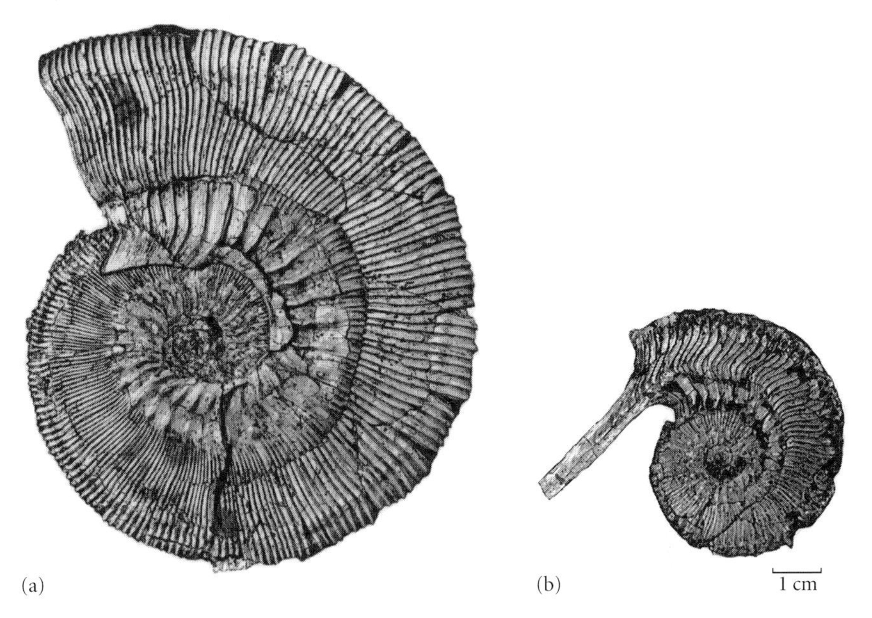

Figure 6.2 Sexual dimorphism in ammonites, the Jurassic *Kosmoceras*. The larger shell (a) was probably the female, the smaller (b) the male. (Courtesy of Jim Kennedy and Peter Skelton.)

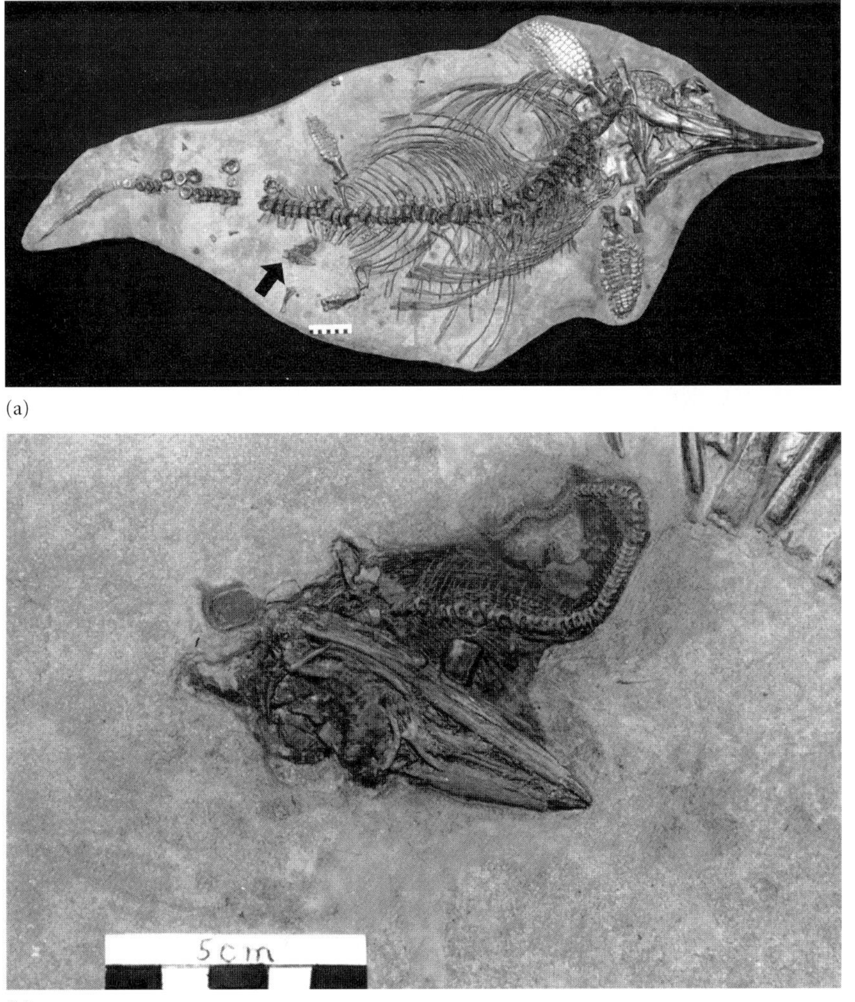

Figure 6.3 Adult female *Ichthyosaurus* (a) from the Lower Jurassic of Somerset, England, showing an embryo that has just been born (arrow), and detail of the curled embryo (b). (Courtesy of Makoto Manabe.)

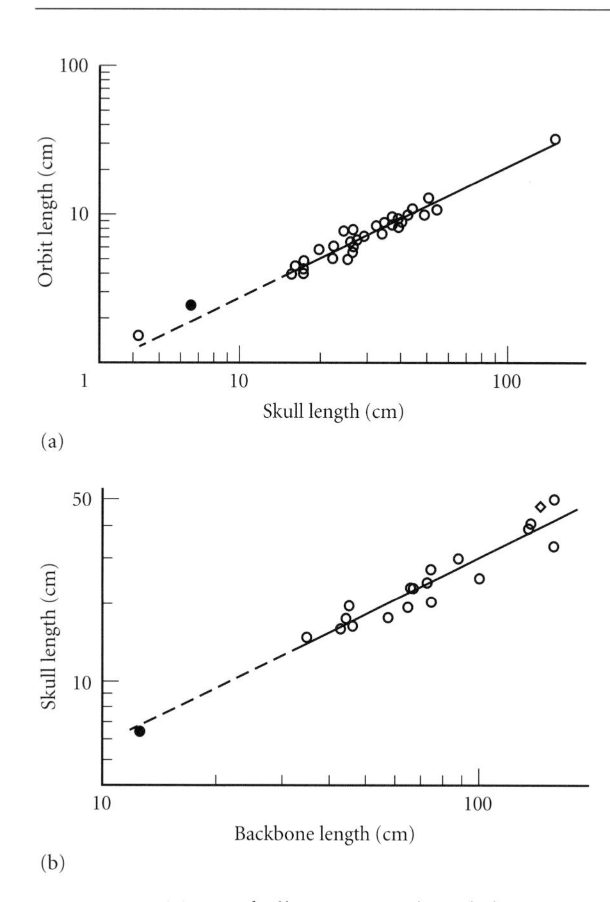

Figure 6.4 Tests of allometry in the ichthyosaur *Ichthyosaurus*. (a) Plot of orbit length against skull length, and (b) plot of skull length against backbone length. The Somerset embryo (Fig. 6.3b) is indicated by a solid circle. Both graphs show negative allometry (orbit diameter = 0.355 (skull length)$^{0.987}$; skull length = 1.162 (backbone length)$^{0.933}$), confirming that embryos and juveniles had relatively large heads and eyes. (Courtesy of Makoto Manabe.)

ratio of eye diameter to body length diminishes as the animal approaches adulthood. This is an example of **allometric** ("different measure") growth. If there is no change in proportions during growth, the feature is said to show **isometric** ("same measure") growth.

Allometric growth is commoner than isometric. Positive allometry is when the organ or feature of interest increases faster than the isometric expectation, and negative allometry is when growth of the structure of interest is slower than isometry. Head and eye size usually show negative allometry, starting relatively large in the juvenile, and becoming relatively smaller in the adult. An example of positive allometry is in the antlers of the Irish deer (Box 6.2), and indeed in many other sexually selected features that are minute or absent in the juvenile, but very large in the adult.

Allometry is commonly considered only in the context of **ontogeny**, the growth from egg or embryo through juvenile to adult. But studies of form may compare species, and shape variation can be accounted for in an evolutionary context too. For example, a comparison of species of antelope would show positive allometry in leg width: scaled against body length, the sturdiness or width of the leg increases positively allometrically. This is because of the well-known **biological scaling principle**: some organs and functions relate to the mass of an animal (a three-dimensional measure), whereas others relate to body length or body outline (one- and two-dimensional measures). As body mass (three-dimensional) increases, the diameter of the legs (two-dimensional) increases in proportion to support the added weight. So, in body outline, small antelope have extremely slender legs, and larger ones have relatively more massive legs.

These aspects of allometry may be understood in terms of the allometric equation,

$$y = kx^a$$

where y is the measurement of interest (e.g. head length, eye diameter), x is the standard of comparison (e.g. body length), k is a constant and a is the allometric coefficient. The constant k is calculated using the allometric equation for each particular case. The allometric coefficient a defines the nature of the slope: if $a = 1$, the slope is at 45° and this defines a case of isometric growth; if $a > 1$, we have positive allometry, and if $a < 1$, we have negative allometry (see Fig. 6.4).

After the nature of any allometric change of parts or organs has been established quantitatively, it is possible to investigate why such changes might occur. The large eyes and small noses of babies are said to make them look cute so their parents will look after them, and feed them. But the fundamental reason is presumably because the eye is complex and is at nearly adult size in the baby for functional reasons, and the relatively large head of a human baby is to accommodate the large

Box 6.2 The Irish deer: too big to survive?

The Irish deer *Megaloceros*, formerly called the Irish elk, is one of the most evocative of the Ice Age mammals (Fig. 6.5a), and one of the most misunderstood. When the first fossils were dug out of the Irish bog, Thomas Molyneux wrote of them in 1697, "Should we compare the fairest buck with the symmetry of this mighty beast, it must certainly fall as much short of its proportions as the smallest young fawn, compared to the largest over-grown buck."

This was a large deer, some 2.1 m tall at the shoulders, and it famously had massive antlers, the largest spanning 3.6 m. The old story was that this deer simply died out because its antlers became too large. Paleontologists understood that the antlers were subject to sexual selection and, as in modern deer, the male with the largest antlers and the scariest display probably gathered the largest harem of females and so passed on his genes most successfully. But can a species really be driven to extinction by sexual selection?

In a classic paper, the young Steve Gould (1974) showed that this was clearly nonsense. He measured the body lengths and antler dimensions of dozens of specimens and showed that they fell precisely on an allometric curve, and that the allometric curve was the same as for other relatives such as the smaller, living red deer and wapiti (Fig. 6.5b). It is clear that sexual selection and natural selection were at odds in this case, as often happens, but the balance was maintained and indeed the Irish deer was successful throughout Europe, existing until 11,000 years ago in Ireland and 8000 years ago in Siberia. It probably died out because of climate change at the end of the Pleistocene and hunting by early humans, rather than by collapsing beneath the weight of its overgrown antlers.

It is worth reading Gould's (1974) classic study of positive allometry in the Irish deer, and a broader review of positive allometry in sexually-selected traits by Kodric-Brown et al. (2006). Read more and see color illustrations at http://www.blackwellpublishing.com/paleobiology/.

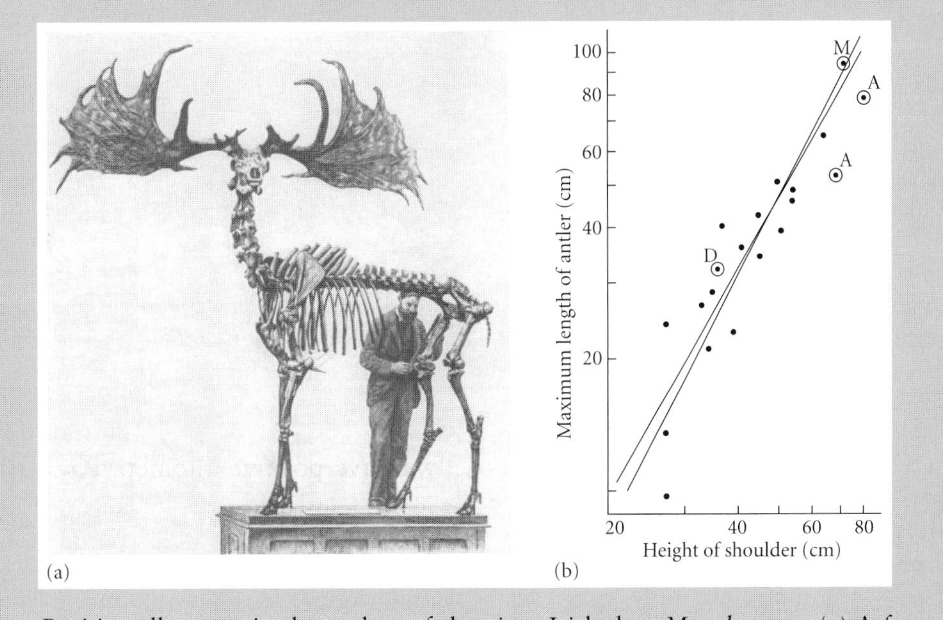

Figure 6.5 Positive allometry in the antlers of the giant Irish deer *Megaloceros*. (a) A famous photograph of an Irish deer skeleton mounted in Dublin in Victorian times. (b) Positive allometry in the antlers of modern deer, showing that *Megaloceros* (M) falls precisely on the expected trend of its closest living relatives. Note that the fallow deer (D) plots above the slope (i.e. antlers are larger than expected from its height), and the European and American moose (A) plot below the line (i.e. antlers are smaller than expected from their height). Two regression lines, the reduced major axis (steeper) and least squares regression, are shown. The allometric equation is antler length = 0.463 (shoulder height)$^{1.74}$. (Based on information in Gould 1974.)

brain, which again is rather well developed at birth.

Ichthyosaurs (see Figs 6.3, 6.4) were born live underwater, as shown by remarkable fossils (see p. 462), and did not hatch from eggs laid onshore, as is the case with most other marine reptiles. Their large head at birth would have allowed them to feed on fishes and ammonites as soon as they were born. The large eyes were perhaps necessary also for hunting in murky water, and had to be near-adult size from the start. Or, perhaps, it made them look cute and encouraged parental care!

Shape variation between species

Within any clade there are many forms. Related plants and animals usually show some common aspects of form, and species and genera vary around a theme. For example, gastropods all have coiled shells and the three-dimensional shape can be thought of as a result of variation in four parameters (see p. 333). When form can be reduced to a small number of parameters like this, then the whole range of possible forms governed by those parameters may be defined – the theoretical **morphospace** for the clade. Studies of the theoretical morphospace for gastropods, ammonoids and early vascular plants show that known species have only exploited a selection of possible morphologies. Some zones of morphospace may represent impossible forms – such as gastropods or ammonoids with a minute aperture, with no room for the living animal – but others have simply not been exploited by chance, or they cannot be reached by normal evolutionary change because of the impossibility of intervening stages.

The range of forms within a clade may also be described as **disparity**, the sum of morphological variation. Disparity may be quantified as the range of values for all possible shape parameters seen in species in a clade. All the measures of shape may be combined in a multivariate analysis that can simplify dozens of shape measures to a smaller number of principal coordinates or eigenvectors (see p. 139) so that size and other general principles may be separated. It is possible to compare the disparity of different clades, or to look at how disparity varies through time. Disparity is generally high early in the history of a clade as the species "try out" all the possibilities of their new body form, and then the disparity of the group remains rather constant for the rest of its history. Changes in disparity through time may roughly mimic changes in diversity (as diversity increases, so too does disparity), but the correlation is usually not perfect, and shape change often goes ahead of diversity increase.

EVOLUTION AND DEVELOPMENT

Ontogeny and phylogeny

Biologists have long sought a link between ontogeny (development) and phylogeny (evolutionary history). In 1866, Ernst Haeckel, a German evolutionist, announced his Biogenetic Law, that "ontogeny recapitulates phylogeny". His idea was that the sequence of embryonic stages mimicked the past evolutionary history of an animal. So, in humans, he argued, the earliest embryonic stages were rather fish-like, with gill pouches in the neck region. Next, he argued was an "amphibian" stage and a "reptile" stage, when the human embryo retained a tail and had a small head, and finally came the "mammal" stage, with growth of a large brain and a pelt of fine hair.

Haeckel's view was attractive at the time, but too simple. Haeckel had drawn on earlier work, including Von Baer's Law, presented in 1828, and this law can be matched with current cladistic models. Von Baer interpreted the embryology of vertebrates as showing that "general characters appear first in ontogeny, special characters later". Early embryos are virtually indistinguishable: they all have a backbone, a head and a tail (vertebrate characters). A little later, fins appear in the fish embryo, legs in the tetrapods. More specialized characters appear later: fin rays in the fish, beak and feather buds in the chick, snout and hooves in the calf, and large brain and tail loss in the human embryo.

"General characters appearing before special characters" has taken on a new meaning with the establishment of a cladistic view of phylogeny (see p. 129). Von Baer's Law draws a parallel between the sequence of development, and the structure of a cladogram. In human development, the embryo passes through the major nodes of the cladogram of vertebrates. The synapomorphies (see p. 130) of vertebrates appear first, then those of tetrapods, then those of amniotes, then those of mammals, of primates, and of the species *Homo sapiens* last.

Three other aspects of development throw light on phylogeny. Certain developmental abnormalities called **atavisms**, or throwbacks, show former stages of evolution, such as human babies with small tails or excessive hair, or horses with extra side toes (Fig. 6.6a), showing how earlier horses had five, four or three toes, compared to the modern one.

Vestigial structures tell similar phylogenetic stories. These are structures retained in living organisms that have no clear function, and may simply be there because they represent something that was once used. So, modern whales have, deep within their bodies, small bones in the hip region that are remnants of their hindlegs (Fig. 6.6b). Whales last had functioning hindlegs over 50 Ma in the Eocene, and the vestigial remnants are still there, even though they serve no further purpose in locomotion, and only support some muscles associated with the penis.

The third aspect of development that forms links with phylogeny is the observation that ontogenetic patterns themselves have evolved. In particular the timing and rate of developmental events has varied between ancestors and descendants, often with profound effects. This phenomenon is termed heterochrony.

Heterochrony: are human adults juvenile apes?

Heterochrony means "different time", and includes all aspects of changes of timing and rates of development. There are two forms of heterochronic change, **pedomorphosis** ("juvenile formation"), or sexual maturity in a juvenile body, and **peramorphosis** ("overdevelopment"), where sexual maturity occurs relatively late. These changes can each occur in three ways, by variation in timing of the beginning of body growth, the timing of sexual maturation or the rate of morphological development (Table 6.1).

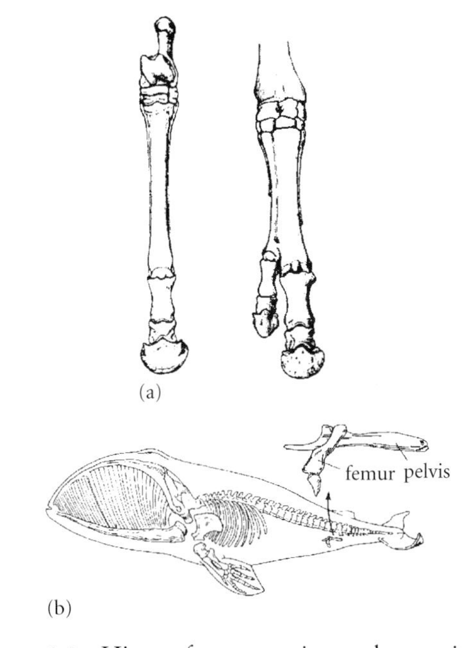

Figure 6.6 Hints of ancestry in modern animals. (a) Extra toes in a horse, an example of an atavistic abnormality in development, or a throw-back, to earlier horses which had more than one toe; normal horse leg (left), extra toes (right). (b) The vestigial hip girdle and hindlimb of a whale; the rudimentary limb is the rudiment of a hindlimb that functioned 50 Ma.

Table 6.1 The processes of heterochrony: differences in the relative timing and rates of development.

	Onset of growth	Sexual maturation	Rate of morphological development
Pedomorphosis			
Progenesis	–	Early	–
Neoteny	–	–	Reduced
Postdisplacement	Delayed	–	–
Peramorphosis			
Hypermorphosis	–	Delayed	–
Acceleration	–	–	Increased
Predisplacement	Early	–	–

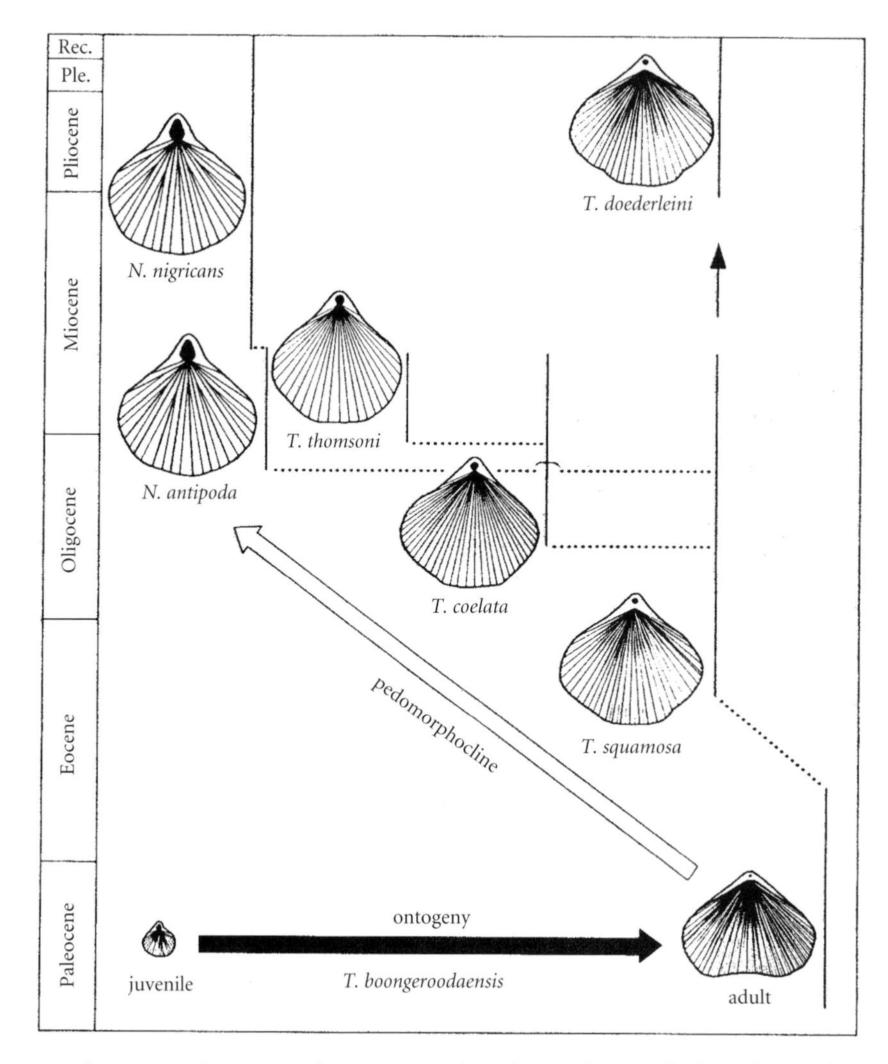

Figure 6.7 Heterochronic evolution in the Cenozoic brachiopods *Tegulorhynchia* and *Notosaria*. Adults of more recent species are like juveniles of the ancestor. Hence, pedomorphosis ("juvenile formation") is expressed in this example. (Based on McNamara 1976.)

In studying heterochrony, it is necessary to have a robust phylogeny of the organisms in question, an adequate fossil record of the group, and a sound set of ontogenetic sequences for each species. This allows the paleontologist to compare juveniles and adults throughout the phylogeny. A classic example is human evolution. It seems obvious that human adults look like juvenile apes, with their flat faces, large brains and lack of body hair. These would imply a pedomorphic change in humans with respect to the human/ape ancestor. However, other characters do not fit this pattern. For example, developmental time in humans is far longer than in apes and ancestral forms, a feature of peramorphosis, and hyperomorphosis in particular (developmental time is longer, but rate of morphological development is not faster). Thus, heterochronic changes can occur in different directions in different characters, a phenomenon called **mosaic evolution.**

In a classic study, McNamara (1976) suggested that species of the Cenozoic brachiopod *Tegulorhynchia* evolved into *Notosaria* by a process of heterochrony (Fig. 6.7). The main changes were a narrowing of the shell, a reduction in the number of ribs in the shell ornament, a smoothing of the lower margin, and an enlargement of the pedicle foramen (the opening through which a fleshy stalk attaches the animal to a rock). These changes

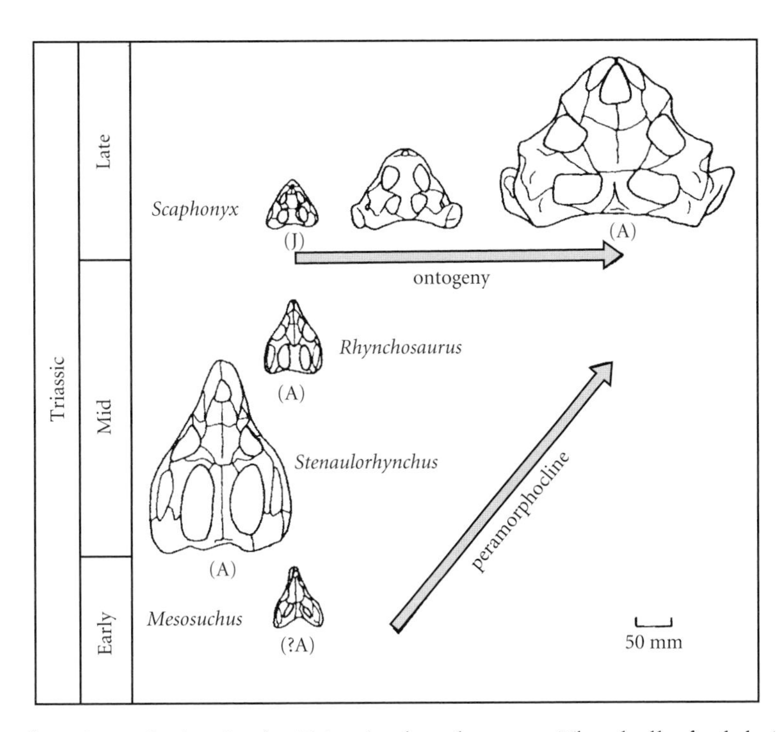

Figure 6.8 Heterochronic evolution in the Triassic rhynchosaurs. The skull of adult (A) Late Triassic forms developed beyond the size and shape limits seen in earlier Triassic adult forms. Here, the juveniles (J) of the descendants resemble the ancestral adults, and this is thus an example of peramorphosis ("beyond formation"). (Based on Benton & Kirkpatrick 1989.)

related to a shift of habitats from deep to shallow high-energy waters: the large pedicle allowed the brachiopod to hold tight in rougher conditions, and the other changes helped stabilize the shell. The developmental sequence of the ancestral species *T. boongeroodaensis* shows that its descendants are like the juvenile stage. Hence, pedomorphosis has taken place along a **pedomorphocline** ("child formation slope"). It is harder here to determine which type of pedomorphosis has taken place; perhaps it was neoteny.

A second example illustrates a peramorphic trend. Rhynchosaurs were a group of Triassic herbivorous reptiles. Later species had exceptionally broad skulls as adults, which gave them vast muscle power to chop tough vegetation. Juvenile examples of these Late Triassic rhynchosaurs retain the rather narrower skulls of the ancestral adult forms (Fig. 6.8). Hence, the evolution of the broad skull is an example of peramorphosis, along a **peramorphocline** ("overdevelopment slope"). The adult Late Triassic rhynchosaurs are larger than earlier forms, which implies that sexual maturation was delayed while the body continued to grow (hypermorphosis) or the rate of morphological development increased in the same duration of ontogeny (acceleration).

Developmental genes

It has been understood since the time of Darwin that the external form, or phenotype, of an organism is controlled by the genotype, the genetic code (see p. 121), but the exact mechanisms have been unclear. At one time people thought there was roughly one gene for each morphological attribute. Some characters seem to be inherited in a unitary manner – you inherit blond, black or red hair from one or the other or both of your parents, and so it might be reasonable to assume that there is a gene variant for each color. But most phenotypic characters are inherited in a much more complex manner, and it is clear that there is no single gene that controls the shape of your nose, the length of your legs or your mathematical ability.

Some clarity has now been shed on how genes control form. There are a number of

developmental genes that are widely shared among organisms and that determine fundamental aspects of form such as symmetry, anteroposterior orientation and limb differentiation. Since the 1980s a major new research field has emerged, sometimes called "**evo-devo**" (short for evolution–development), that investigates these developmental genes. This field is exciting for paleontologists because the developmental genes control aspects of form on a macroevolutionary scale, and so major evolutionary transitions can be interpreted successfully in terms of developmental genes.

The most famous developmental genes are the **homeobox genes**, identified first in the experimental geneticist's greatest ally, the fruit fly *Drosophila*, but since found in a wide range of eukaryotes from slime molds to humans, and yeast to daffodils. Homeobox genes contain a conserved region that is 180 base pairs long (see p. 186) and encodes **transcription factors**, proteins that switch on cascades of other genes, for example all the genes required to make an arm or a leg. In this sense homeobox genes are **regulatory genes**; they act early in development and regulate many other genes that have more specialist functions.

The ***Hox* genes** are a specific set of homeobox genes that are found in a special gene cluster, the *Hox* cluster or complex that is physically located in one region within a chromosome. *Hox* genes function in patterning the body axis by fixing the anteroposterior orientation of the early embryo (which is front and which is back?), they specify positions along the anteroposterior axis, marking where other regulatory genes determine the segmentation of the body, especially seen in arthropods (see p. 362), and they also mark the position and sequence of differentiation of the limbs (Box 6.3).

Box 6.3 *Hox* genes and the vertebrate limb

One of the greatest transitions of form in vertebrate evolution was the remodeling of a fish into a tetrapod, a process that occurred more than 400 Ma in the Devonian (see p. 442). The fossils show how the internal skeleton of a swimming fin was transformed into a walking limb. A crucial part of this repatterning from fin to limb seemed to be the **pentadactyl limb**, the classic arm or leg with five fingers or toes seen in humans and most other tetrapods. But then paleontologists began to find Late Devonian tetrapods with six, seven or eight digits. How could this be explained in a world where there was supposed to be a gene for each digit, and five was the norm?

The tetrapod limb can be divided into three portions that appear in the embryo one after the other, and that appeared in evolutionary history in the same sequence. First is the proximal portion of the limb, the **stylopod** (the upper arm or thigh), then the middle portion of the limb, the **zeugopod** (the forearm or calf), and finally the distal portion, the **autopod** (the hand and wrist or foot and ankle).

This evolutionary sequence is replicated during development of the embryo (Shubin et al. 1997; Coates et al. 2002; Tickle 2006; Zakany & Duboule 2007). At an early phase, the limb is represented simply by a limb bud, a small lateral outgrowth from the body wall. Limb growth is controlled by *Hox* genes. Early in fish evolution, five of the 13 *Hox* genes, numbered 9–13, were coopted to control limb bud development. Manipulation of embryos during three phases of development has shown how this works (Fig. 6.9a). In phase I, the stylopod in the limb bud sprouts, and this is associated with expression of the genes *Hox*D-9 and *HoxD*-10. In phase II, the zeugopod sprouts at the end of the limb bud, and the tissues are mapped into five zones from back to front by different nested clusters of all the limb bud genes *HoxD*-9 to *HoxD*-13. Finally, in phase III, the distal tip of the lengthening limb bud is divided into three anteroposterior zones, each associated with a different combination of genes *HoxD*-10 to *HoxD*-13. Phases I and II have been observed in bony fish development, but phase III appears to be unique to tetrapods.

In the development of vertebrate embryos, there is no fixed plan for every detail of the limb. A developmental axis runs from the side of the body through the limb, and cartilages condense from

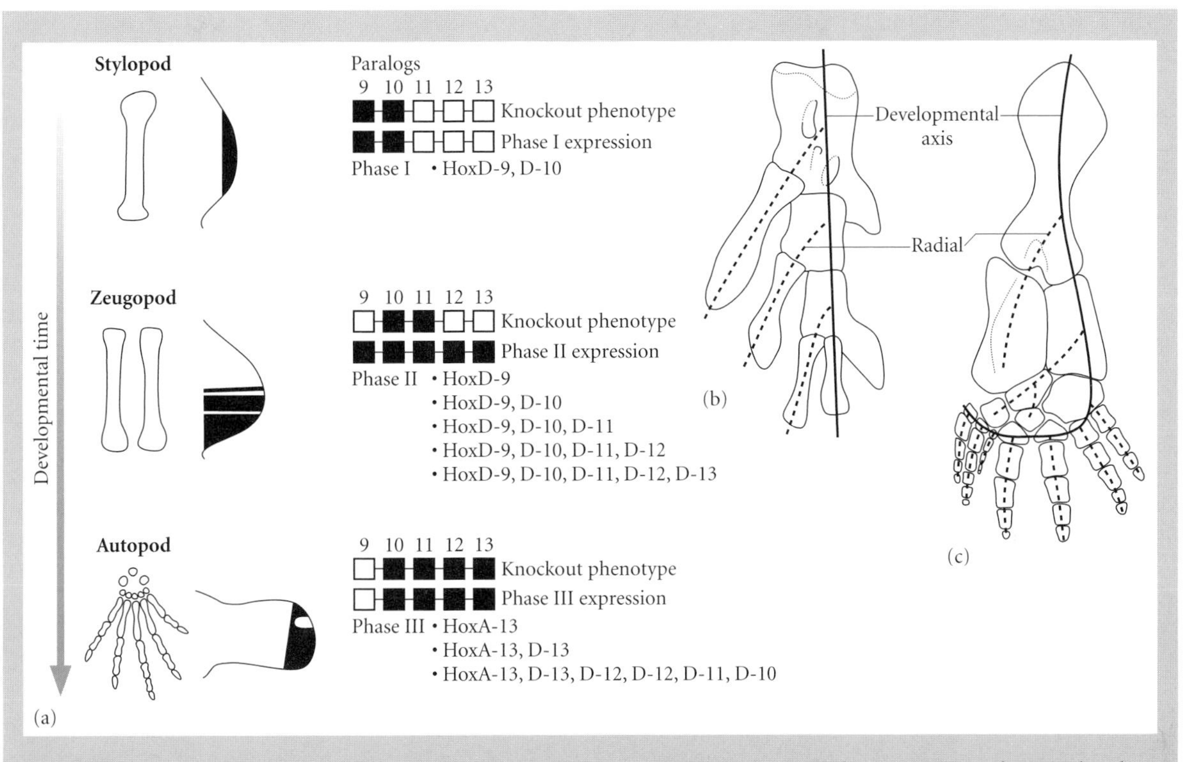

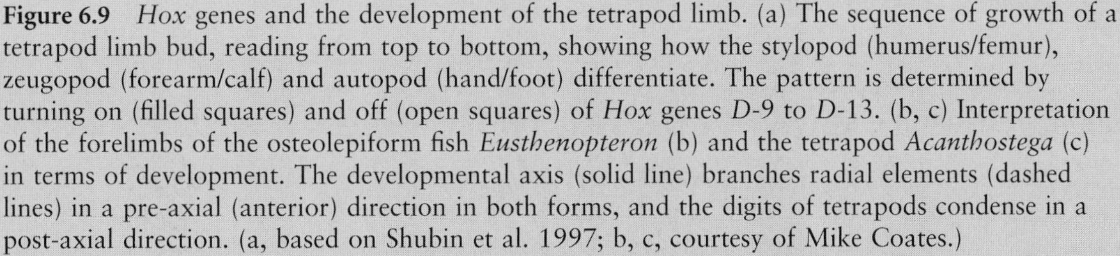

Figure 6.9 *Hox* genes and the development of the tetrapod limb. (a) The sequence of growth of a tetrapod limb bud, reading from top to bottom, showing how the stylopod (humerus/femur), zeugopod (forearm/calf) and autopod (hand/foot) differentiate. The pattern is determined by turning on (filled squares) and off (open squares) of *Hox* genes *D*-9 to *D*-13. (b, c) Interpretation of the forelimbs of the osteolepiform fish *Eusthenopteron* (b) and the tetrapod *Acanthostega* (c) in terms of development. The developmental axis (solid line) branches radial elements (dashed lines) in a pre-axial (anterior) direction in both forms, and the digits of tetrapods condense in a post-axial direction. (a, based on Shubin et al. 1997; b, c, courtesy of Mike Coates.)

soft tissues in sequence from the body outwards to the tips of the fingers. In an osteolepiform fish (Fig. 6.9b), the developmental axis presumably ran through the main bony elements, and additional bones, radials, developed in front of the axis (pre-axial side). In tetrapods (Fig. 6.9c), the axis in the leg (arm) runs through the femur (humerus), fibula (ulna) and ankle (wrist) and then swings through the distal carpals (tarsals). Radials condense pre-axially at first, as in the osteolepiform, forming the tibia (radius) and various ankle (wrist) bones. The developmental process then switches sides to sprout digits post-axially (behind the axis). This reversal of limb-bud growth direction in the hand/foot is matched by a reversal of the expression of the *Hox* genes. In the zeugopod, *HoxD*-9 is expressed in all five zones, *HoxD*-10 in the posterior four zones, down to *HoxD*-13 only in the posterior of the five. In the autopod, on the other hand, *HoxA*-13 is present in all zones, *HoxD*-13 in the posterior two zones, and *HoxD*-10 to *HoxD*-12 only in the posterior zone.

In Late Devonian tetrapods, six, seven or eight digits were freely produced, and it was only at the beginning of the Carboniferous that tetrapods seem to have fixed on five digits fore and aft. Since then, digital reduction has commonly occurred, down to four (frogs), three (many dinosaurs), two (cows and sheep) or one (horses) fingers and toes. Systematists must beware of interpreting such events as unique, however: the new evo-devo perspective suggests that loss of digits has happened many times in tetrapod evolution, and by the same processes of switching *Hox* genes on and off.

Read more about *Hox* genes and limb-bud development at http://www.blackwellpublishing.com/palaeo/, and about evo-devo topics in general in Carroll (2005) and Shubin (2008).

In early studies of the *Hox* genes of *Drosophila*, experimenters were amazed to discover that mutations in particular *Hox* genes might cause the insect to develop a walking leg on its head in place of an antenna. The mutations were not simple changes of the base-pair sequence, but knockouts or deletions of entire functional portions and replacement of their expression domains by more posterior *Hox* genes. Study of such knockouts showed how each *Hox* gene worked; in this case the *Hox* gene acted on the **limb bud**, the small group of cells on the side of the body that appears early in development and eventually becomes a limb. A particular *Hox* gene determines how many limb buds there are and where they are located, and other *Hox* genes determine whether the limb bud becomes a walking leg, a mouthpart or an antenna. If experimenters induce a knockout within a *Hox* gene, it works its magic in the wrong place, giving the fly extra legs or legs in the wrong place. Mutations of *Hox* genes in vertebrates normally do not produce these spectacular effects; the embryo often fails and is aborted.

Such mutations need not always result in damage. Duplication of homeobox genes can produce new body segments, and such duplications may have been important in the evolution of arthropods and other segmented animals. The new evo-devo perspective allows us to understand that an arthropod with numerous body segments and 10 or 100 legs may have evolved by a single evolutionary event, perhaps a relatively straightforward mutation of homeobox genes, rather than an elaborate multistep process of gradual addition of segments and legs through many separate evolutionary events. The evo-devo revolution is beginning to explain some of the most mysterious aspects of evolution.

INTERPRETING THE FUNCTION OF FOSSILS

Functional morphology

Inferring the function of ancient organisms is hard, and yet it is the main reason many people are interested in paleobiology. Just how fast could a trilobite crawl? Why did some brachiopods and bivalves mimic corals? How did that huge seed fern support itself in a storm? How well could pterosaurs fly? Why did sabertoothed cats have such massive fangs? The most fascinating questions concern those fossil organisms that are most different from living plants and animals. This is because it is easy to work out that a fossil bat probably flew and behaved like a modern bat. But what about a pterosaur: so different, and yet similar in certain ways?

There are three approaches to interpreting the function of fossils: comparison with modern analogs, biomechanical modeling and circumstantial evidence. Let us look at some general assumptions first, and then each of those approaches in turn.

The main assumption behind functional morphology is that biological structures are adapted in some way and that they have evolved to be reasonably efficient at doing something. So, an elephant's trunk has evolved to act as a grasping and sucking organ to allow the huge animal to reach the ground, and to gather food and drink. The flower of an angiosperm is colorful to attract pollinating insects, and the nectar is located deep in the flower so the insect has to pick up pollen as it enters. The siphons of a burrowing mollusk are the right length so it can circulate water and nutrients when it is buried at its favored depth.

Fossils can provide a great deal of fundamental evidence of value in interpreting function. For example, the hard skeleton of a fossil arthropod reveals the number and shape of the limbs, the nature of each joint in each limb, perhaps also the mouthparts and other structures relating to locomotion and feeding (see p. 362). Even a fossil bivalve shell gives some functional information in the hinge mechanism, the **pallial line** (which marks the extent of the fleshy mantle) and the muscle scars (see p. 334). Exceptionally preserved fossils may reveal additional structures such as the outline of the tentacles of a belemnite or ammonite (see p. 344), muscle tissue (see p. 64) or sensory organs. The first step in interpreting function then is to consider the morphology, or anatomy, of the fossil.

The vertebrate skeleton can provide a great deal of information about function. The maximum amount of rotation and hinging at each joint can be assessed because this depends on the shapes of the ends of the limb bones. There may be **muscle scars** on the surface of the bone, and particular knobs and ridges (**processes**) that show where the muscles attached, and how big they were. Muscle size is an indicator of strength, and this kind of observation can show how an animal moved.

Comparison with modern analogs

After the basic anatomy of the fossil organism is understood, the logical next step is to identify a modern analog. This can be easy if the fossil belongs to a modern group, perhaps an Eocene crab or a Cretaceous lily plant. The paleontologist then just has to look for the most similar living form, and make adjustments for size and other variations before determining what the ancient organism could do.

But what about ancient organisms that do not have obvious close living relatives? In trying to understand the functional morphology of a dinosaur, for example, should the paleontologist compare the fossil with a crocodile or a bird? In former days, paleontologists might have begun detailed comparisons with a crocodile, but that is not always helpful because crocodiles are different in many aspects of their form and function from dinosaurs. What about birds? After all, we now know that birds are more closely related to dinosaurs than are crocodiles (see p. 460). Again there are problems because birds are much smaller than dinosaurs and they have become so adapted to flying that it is hard to find common ground.

There are two issues here: phylogeny and functional analogs. In phylogenetic terms, it is wrong to compare dinosaurs *exclusively* with crocodiles or with birds. They should be compared with *both*. This is because birds and crocodiles each have their own independent evolutionary histories and there is no guarantee that any of their characters were also present in dinosaurs. However, if both birds and crocodiles share a feature, then dinosaurs almost certainly had it too. This is the concept of the **extant phylogenetic bracket** (EPB) (Witmer 1997): even if a fossil form is distant from living forms, it will be bracketed in the phylogenetic tree by some living organisms. That at least provides a starting point in identifying some unknown characters, especially of soft tissues. The EPB can reveal a great deal about unknown anatomy in a fossil: if crocodiles and birds share particular muscles, then dinosaurs had them too. The same goes for all other normally unpreservable organs. So the EPB has considerable potential to fill in missing anatomy.

But **phylogenetic analogs** may not be much use in determining function. Probably a close study of crocodiles and birds will not solve many problems in dinosaur functional morphology. Dinosaurs were so different in size and shape that a better modern **functional analog** might be an elephant. Elephants are not closely related to dinosaurs, but they are large, and their limb shapes show many anatomic parallels. Watching a modern elephant marching ponderously probably gives the best live demonstration of how a four-limbed dinosaur moved.

The point of using modern analogs is a more general one though. Biologists have learned a great deal about the general principles of **biomechanics**, the physics of how organisms move, from observations across the spectrum. So, the scaling principle mentioned earlier (see p. 142), exemplified by the spindly legs of the antelope and the pillar-like legs of the elephant, is a commonsense observation that clearly applies to extinct forms. And there are many more such commonsense observations: among vertebrates carnivores have sharp teeth and herbivores have blunter teeth; tall trees require broad bases and deep roots so they do not fall over; vulnerable small creatures survive best if they are camouflaged; as animals run faster their stride length increases (see p. 520); fast-swimming animals tend to be torpedo-shaped; and so on. These observations are not "laws" in the sense of the laws of physics, but they are commonsense observations that clearly apply widely across plants and animals, living and extinct. Comparison with modern analogs to learn these general rules is the most important tool in the armory of the functional morphologist (Box 6.4).

Biomechanical modeling

Increasingly, paleobiologists are turning to biomechanical modeling to make interpretations of movements, especially in feeding and locomotion. Such studies use basic principles of biomechanics and engineering to interpret modern and ancient biological structures (Fig. 6.11). A simple example is to consider the vertebrate jaw as a lever, with the jaw joint as the fulcrum (Fig. 6.11c). Simple mechanics shows that the bite will be strongest nearest to the fulcrum, and weakest towards the far end: that is why we bite food off at the front of our jaws but chew at the back. Subtle changes to the positions of the jaw muscles and the relative position of the jaw margin with respect to the fulcrum can then improve the efficiency of the bite. The vertebrate limb can be modeled as a series of cranks, each with a characteristic range of movement at the joints. This kind of model allows the analyst to work out the maximum forwards and backwards bend of the limb and the relative scaling of muscles, for example.

Biomechanical models may be real, three-dimensional models made out of steel rods, bolts and rubber bands. Such models can provide powerful confirmation of the basic principles of movement, clarify the nature of the joint, and the positioning and relative forces of the muscles. Such real-life models may also form the basis for educational demonstrations and museum reconstructions. More commonly now, however, paleobiologists do their modeling on the computer.

Some computer modeling has been very effective in studying the mechanical strength of ancient structures. In particular, paleobiologists have begun looking at the skulls of ancient vertebrates to assess how the structure was shaped by the normal stresses and strains of feeding and head-butting. A useful modeling approach is **finite element analysis** (FEA), a well-established method used by engineers to assess the strength of bridges and buildings before they are built, and now applied to dinosaur skulls (Box 6.5), among other fossil problems. FEA is one of many methods of modeling how forces act on biological struc-

Box 6.4 The Triassic tow-net

For a century or more, fossil hunters had been aware of some astonishing fossils from the Jurassic of Germany that showed long, slender crinoids (see p. 395) attached to driftwood. In life, these crinoids must have dangled beneath the driftwood, and their mode of life was a mystery. Driftwood crinoids have now been identified in many parts of the world, from the Devonian onwards.

Crinoids today can live attached to the seabed, as most of their fossil ancestors did, filtering food particles from the bottom waters. Most living crinoids are free-swimmers, but they do not seem to attach to driftwood. So why did the fossil forms do it, and how did they live?

New discoveries from China (Hagdorn et al. 2007) give some clues. Numerous pieces of driftwood have been identified in the Late Triassic Xiaowa Formation of Guizhou, southwest China, each carrying 10 or more beautiful specimens of the crinoid *Traumatocrinus* (Fig. 6.10a). The juveniles were presumably free-swimming microscopic plankton, as with other echinoderms, and they settled on driftwood logs. Many juveniles have been found on the logs. The crinoids then matured and became very long. Their feeding arms were longer than in seabed crinoids, perhaps to capture more food. This floating mode of life has been termed **pseudoplanktonic**, meaning that the crinoids are living like "fake plankton". They probably fared better up in the oxygenated surface waters than in the black anoxic seabed ooze.

The functional interpretation of a *Traumatocrinus* colony (Fig. 6.10b) is that it worked like a tow-net (Fig. 6.10c), a standard kind of fishing net towed in the open sea. As the boat moves forward, the tow-net hangs passively behind and billows outward. Any fishes encountered are caught. The *Traumatocrinus* colony similarly spread its feeding arms passively as the log moved forward in the gentle Triassic sea currents. Any food particle encountered by the crinoid net would be captured and eaten. Paleontologists have to use their imaginations and intellects in finding plausible functional models for some ancient organisms!

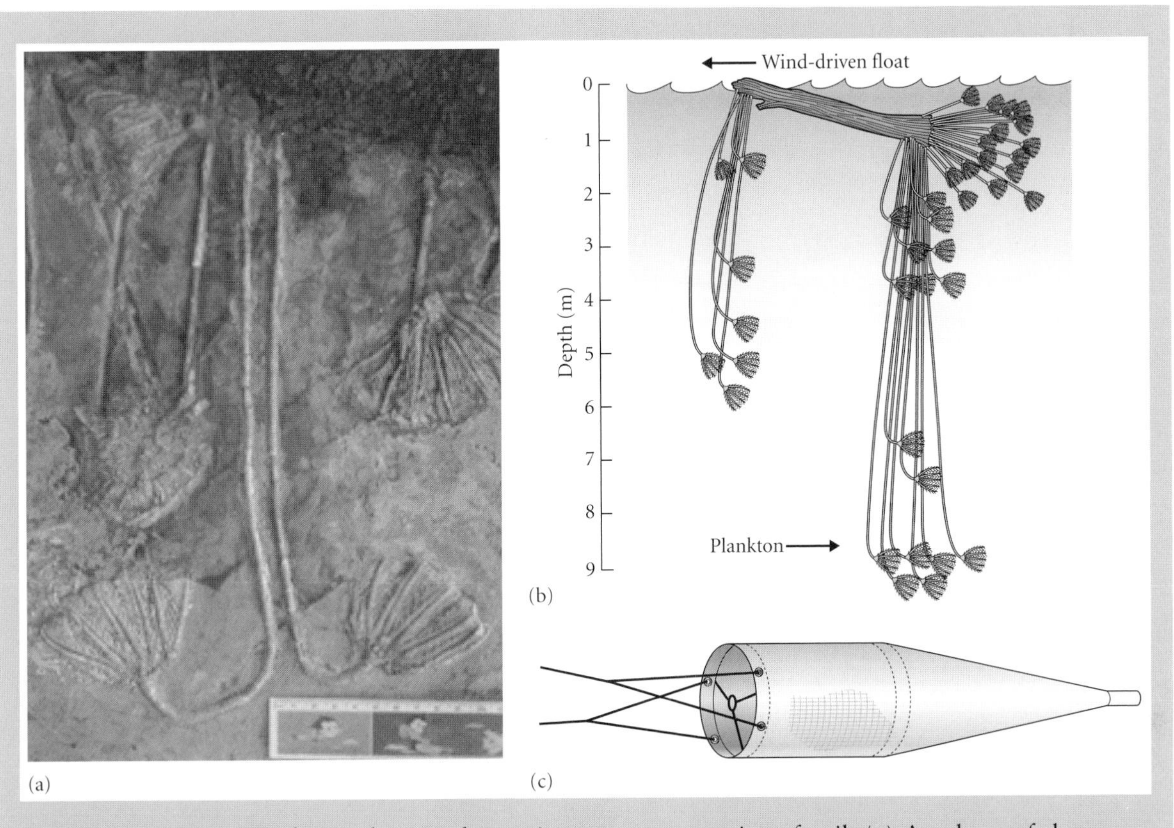

Figure 6.10 The use of a modern analog to interpret a mysterious fossil. (a) A colony of the pseudoplanktonic crinoid *Traumatocrinus* attached to a fossil piece of driftwood, from the Late Triassic of China. (b) Reconstruction of the crinoids in life, showing how the wind pulled the log to the left, and the dangling crinoids captured plankton like a net. (c) A tow-net used to maximize catches of fish, a possible modern analog that explains the feeding mode of the fossil colony. (Courtesy of Wang Xiaofeng.)

tures, while other modeling methods seek to establish how ancient organisms moved.

A number of attempts have been made to understand how dinosaurs ran, and of course everyone focuses on *Tyrannosaurus rex*. At one level, we all know how *T. rex* ran – we have seen it on *Jurassic Park* and *Walking with Dinosaurs*, so what is the problem? The locomotion in those movies was based on study of the limb bones, calculation of their ranges of movement, observation of modern ostriches at speed, and computer animations that rendered a reasonable swing of the leg, and that prevented the animal from falling over. But Hutchinson and Gatesy (2006) have urged caution. They argue that the style of locomotion shown in those films is probably near enough right, but that the animators chose only one out of many possible positions for the limbs.

In the most likely running gait (Fig. 6.13a) the backbone is horizontal and the legs relatively straight and long. Whatever happens, the animal must not fall over, so the first thing in reconstructing locomotion is to determine the **center of mass**, the central point in the core of the body. This can be found crudely by dangling a plastic model from a string and finding the three-dimensional central point of balance – or a more elaborate set of calculations can be done in the computer. In *T. rex* the center of mass lay just in front of the hips, and the tail balanced the body over the hips that acted as a fulcrum.

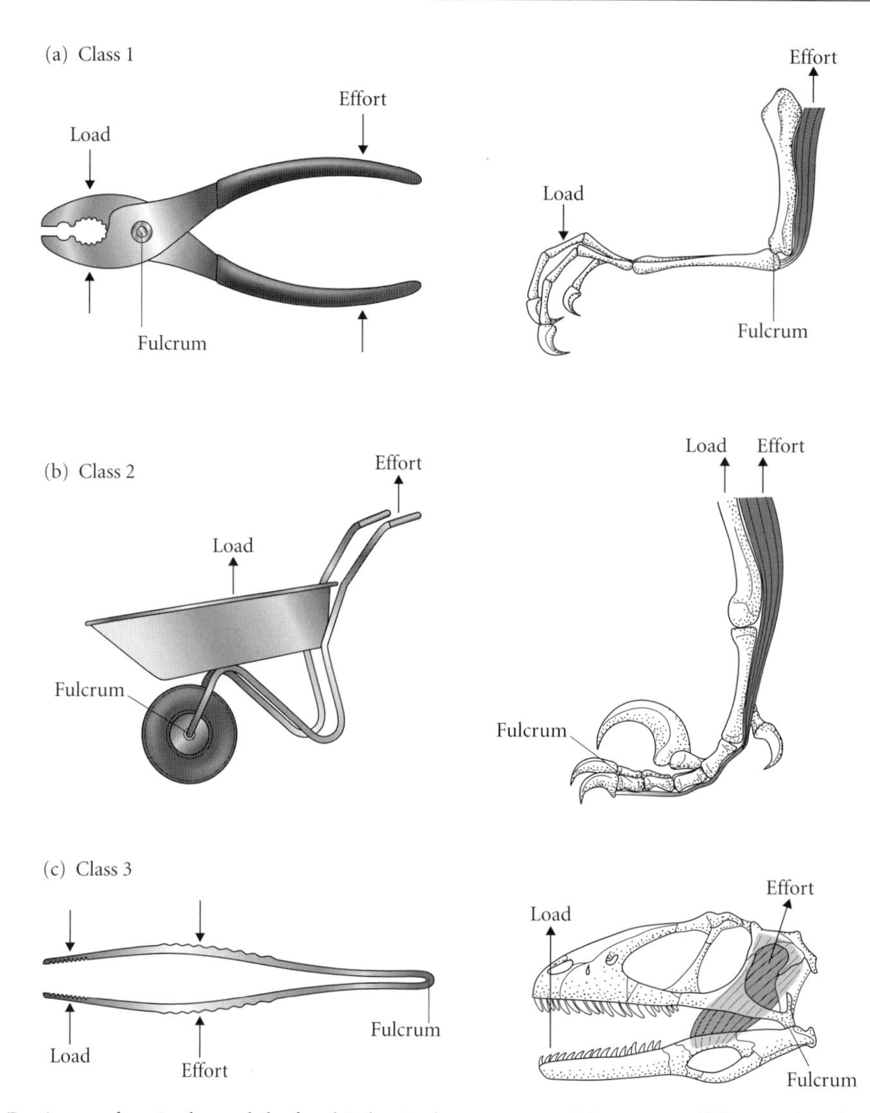

Figure 6.11 Basic mechanical models for biological structures. There are different kinds of levers in use in everyday appliances, and these styles may be seen in biological structures. (a) In a class 1 lever the effort and load are on opposite sides of the fulcrum. (b, c) In class 2 and 3 levers the effort and load are on the same side of the fulcrum, with the effort furthest away in a class 2 lever (b), and closest in a class 3 lever (c).

The running cycle of any animal can be divided into the stance phase, when the foot touches the ground, and the swing phase, when the foot is off the ground (Fig. 6.13a). The limb swings through three extreme postures during the stance phase, from the point at which the foot touches the ground, through mid-stance as the body moves forwards to late stance just before the foot leaves the ground (Fig. 6.13b–d). An animal in contact with the ground produces a ground reaction force (GRF) that is the reaction to its body mass and the force of the limb hitting the ground during movement. The GRF swings its line of action as the limb shifts its position, and the point of maximum stress on the knee is at the mid-stance position (Fig. 6.13c) when the knee is bent, the knee moment arm is longest, and the muscle moment about the knee acting against gravity is at its highest.

Hutchinson and Gatesy (2006) showed that this is only one of many other possible poses for the limbs. Could *T. rex* have run in a high ballet-dancer pose or an extreme crouch (Fig. 6.13e–g)? The ballet-dancer pose is ruled out because the line of the GRF is in front of the knee at mid-stance and this would require the muscles at the back of the leg to act in order to balance the force in front. Living animals do not do this, so there is no reason to assume that extinct ones did. Crouched poses are ruled out too because the knee moment arm would have been too long and the knee muscle moment too high: *T. rex* would have had to have muscles relatively much larger than those of a chicken to cope. So the real *T. rex* probably stood and moved somewhere between these columnar and crouched extremes (Fig. 6.13g), which still leaves a large area of possibilities that cannot be excluded.

Circumstantial evidence

Paleontologists are inquisitive by nature and they gather evidence of all kinds to test their hypotheses. Clues about the lifestyle of an ancient plant or animal may come from the enclosing rocks, associated fossil remains, associated trace fossils and particular features of the body fossils themselves. These can be grouped as circumstantial evidence.

1 Fossils are generally preserved in *sedimentary rocks*, and these record all kinds of features about the conditions of deposition. Fossil plants may be found at certain levels in a cyclical succession that tells a story of the repeated buildup of an ancient delta as it fingers into the sea, the development of soils and forests on top, and its eventual flooding by a particularly high sea level. Marine invertebrates may be found in rocks that indicate deposition in a shallow lagoon, offshore from a reef, on the deep abyssal plain or many other

Box 6.5 Finite element analysis of the skull of *Tyrannosaurus rex*

Emily Rayfield of the University of Bristol (England) had a dream PhD project, to work out how the skulls of the theropod dinosaurs worked, using finite element analysis (FEA). In FEA, the structure is modeled in the computer and its strength characteristics entered. Then the whole three-dimensional shape, however complex, is converted into a network of small triangular or cuboid cells, or elements. When forces are applied (a side wind on a skyscraper, a bite force on a skull or jaw bone) the elements respond and the effect can be seen. In Rayfield's FEA model of a dinosaur skull, as the bite force increases, the zone of element distortion increases and it becomes clear why the skull is shaped the way it is.

In one of her studies, Rayfield (2004) attacked the skull of *T. rex* (Fig. 6.12a). She tried to resolve a paradox that had been noted before: while *T. rex* is assumed to have been capable of producing extremely powerful bite forces, the skull bones are quite loosely articulated. Rayfield applied FEA to assess whether the *T. rex* skull is optimized for the resistance of large biting forces, and how the mobile joints between the skull bones functioned. She studied all the available skulls and constructed a mesh of triangular elements (Fig. 6.12b). Bite forces of 31,000 to 78,060 newtons were applied to individual teeth, and the distortion of the element mesh observed (Fig. 6.12c). The bite forces had been taken from calculations by other paleobiologists, and from observations of tooth puncture marks (a piece of bone bitten by *T. rex* showed the tooth had penetrated the bone to a depth of 11.5 mm, equivalent to a force of 13,400 newtons or about 1.5 tons).

Rayfield's results show that the skull is equally adapted to resist biting or tearing forces and therefore the classic "puncture–pull" feeding hypothesis, in which *T. rex* bites into flesh and tears back, is well supported. Major stresses of biting acted through the pillar-like parts of the skull and the nasal bones on top of the snout, and the loose connections between the bones in the cheek region allowed small movements during the bite, acting as "shock absorbers" to protect other skull structures.

Read about dinosaur feeding behavior in Barrett and Rayfield (2006) and about finite element analysis in Rayfield (2007) and at http://www.blackwellpublishing.com/paleobiology/.

Continued

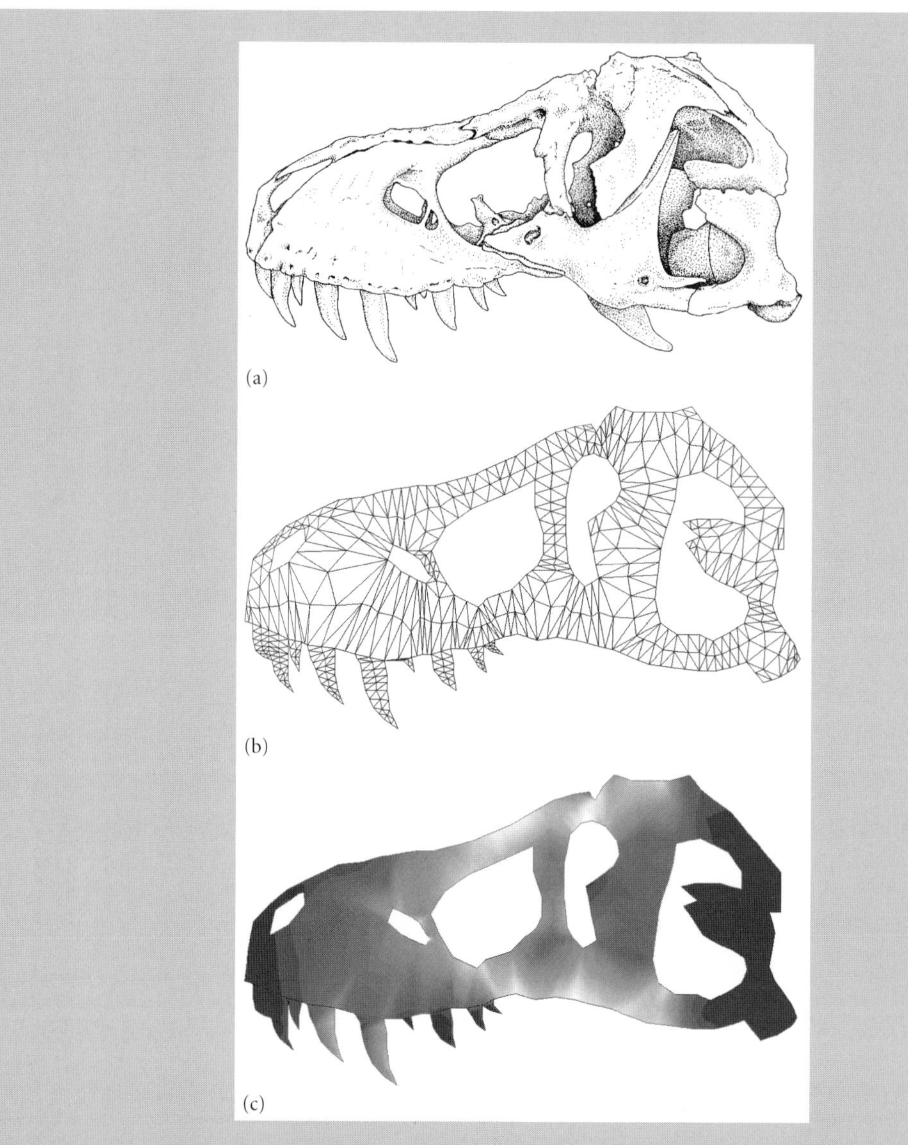

Figure 6.12 Finite element analysis of the skull of *Tyrannosaurus rex*. The skull (a) was converted into a cell mesh (b), and biting forces applied (c). In the stress visualization (c), high stresses are indicated by pale colors, low stresses by black. Each bite, depending on its strength and location, sends stress patterns through the skull mesh and these allow the paleobiologist to understand the construction of the skull, but also the maximum forces possible before the structure fails. (Courtesy of Emily Rayfield.)

situations. Moreover if they had a widespread geographic distribution, perhaps they were planktonic. Dinosaurs or fossil mammals may be found in sandstones deposited in an ancient river or desert. All these clues from sedimentary rocks guide the paleontologist in interpreting the environment of deposition, and in turn can reveal clues about climates and other physical conditions.

2 *Associated fossils* also give clues. They can show where the organism of interest sits in a food web (see p. 88) – who ate it, and what did it eat? Sometimes groups of fossils may be associated in death in such a way that they indicate life habits. For

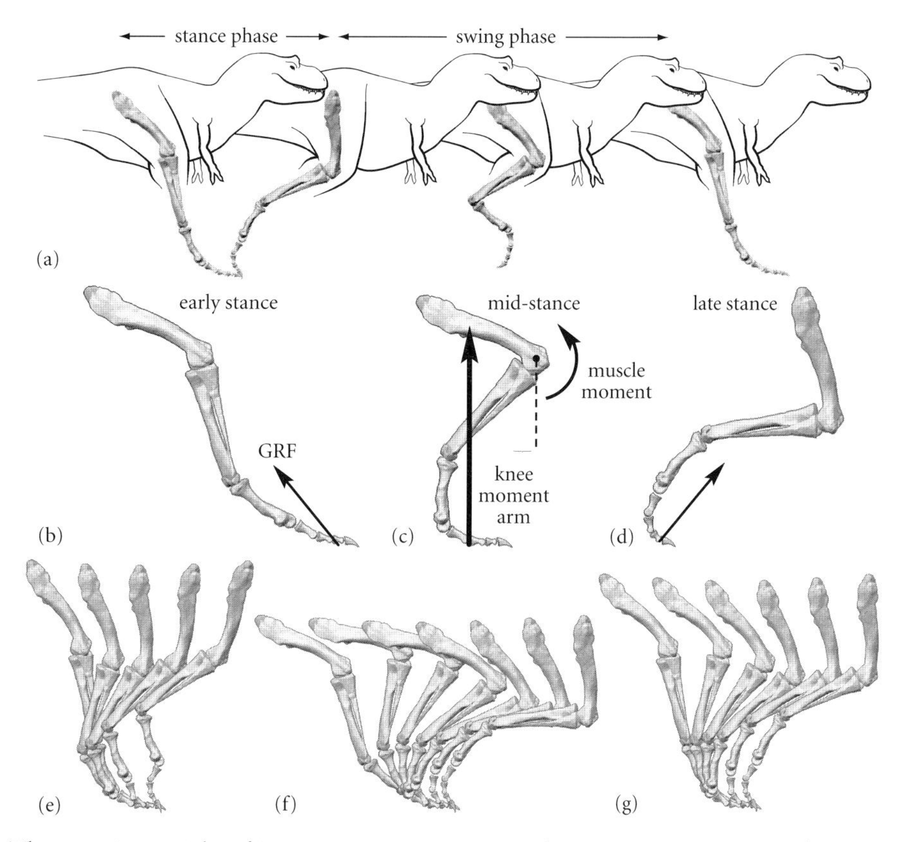

Figure 6.13 The running stride of *Tyrannosaurus rex*. (a) The main components of a stride, showing the stance phase when the foot touches the ground, and the swing phase. (b–d) Three positions of the limb in early stance, mid-stance and late stance, as the body moves forward, and showing the main forces, including the ground reaction force (GRF). (e–g) Three alternative postures for the limb, with the body held high or low. Read more, and see the movies at http://www.rvc.ac.uk/AboutUs/Staff/jhutchinson/ResearchInterests/beyond/Index.cfm. (Courtesy of John Hutchinson.)

example, fossil reefs may be killed off in a particular crisis, and all the organisms that lived together are found in life position; some corals, bryozoans and crinoids may be fixed to the substrate in their normal growth position, and mobile organisms like gastropods or trilobites may be preserved among the thickets of benthic sessile organisms. Similarly, a paleosol (see p. 518) may preserve roots and stems of dozens of plants in life position, together with burrows of insects and worms that lived among them. Associations of fossils can also be more intimate, where for example parasites may be found attached to their hosts, or fossils of one species may be found in the stomach region of another.

3 *Associated trace fossils* can sometimes be linked to their producers, but not always. There are some rare examples of arthropods preserved at the ends of their trails. The link between trace fossil and producer is usually a little less clear: dinosaur footprints may be found at certain levels within a particular geological formation, and the skeletons of likely producers at other levels. The bones of fishes and marine reptiles may be found associated with phosphatic **coprolites** (fossil dung) in certain marine beds – it is likely that the coprolites were dropped by one or other of the associated animals. If a link can be made between a trace fossil and its maker, then a great deal of additional paleobiological information can be established (see Chapter 19).

4 *Close study of the body fossils* themselves is also warranted. Skeletal fossils regularly preserve evidence of soft tissues and other

unpreserved components. Fossil plant stems may be stripped of leaves, but the leaf bases are still there. A fossil trilobite may preserve limbs and other structures under the carapace. Bones often show muscle scars. In conditions of exceptional preservation, of course, skin outlines, muscles, sensory organs and internal organs may also be preserved. For example, the spectacular fossils from Liaoning in China (see p. 463) have confirmed that the fossil birds had feathers and the mammals had hair, as had been expected, but other fossils showed that all the carnivorous dinosaurs had feathers too. That dramatically changes all previous paleobiological interpretations of those dinosaurs because they must have been warm-blooded in some way.

These four kinds of circumstantial evidence have been useful in understanding how ancient rodents fed on nuts (Box 6.6), and also how *T. rex* fed. The biomechanical models of feeding in *T. rex* (see Box 6.5) tell us a great deal. The rocks in which *T. rex* bones are found confirm it lived in hot, lowland, forested areas. Associated fossils include numerous species of plant-eating dinosaurs, and some of these even carry tooth marks likely made by *T. rex*. Tracks of footprints made by *T. rex*, or a relative, show that it trotted along steadily, but not fast. A famous 1 m long coprolite dropped by *T. rex* contained pulverized bone of ornithischian dinosaurs that had been corroded to some extent by stomach acids, but not entirely destroyed. This suggests a relatively rapid transit of food material through the gut. The bones themselves have

Box 6.6 Who ate my nuts?

Animal–plant interactions are often beautifully documented. Paleobotanists have identified marks of chewing, tunneling and munching in leaves, stems and seeds from the Devonian Rhynie Chert (see p. 489) onwards. Margaret Collinson and Jerry Hooker, experts on fossil plants and mammals, respectively, from London, spotted possible feeding damage in small nuts they collected in the Eocene of southern England, and reported in 2000. Some of the tiny seeds of the water plant *Stratiotes* had round holes on one side and the internal contents had been removed, leaving a husk (Fig. 6.14a). *Stratiotes*, sometimes called the water soldier or water aloe, still grows today in the fens and waterways of eastern England, as well as elsewhere in Europe, where it is rooted in the mud or floats on the surface of shallow pools and sends spiky, sword-like leaves up out of the water.

Close study of the holes in the seeds showed that some animal had cut the hole vertical to the outer surface, and that the cut edges of the hole bore numerous parallel grooves. There were more grooves around the hole, as if some creature had been grabbing at the seed to hold it firm while cutting the hole. Collinson and Hooker immediately thought of rodents as the seed eaters – rodents cut straight-sided holes into seeds, and leave parallel grooves formed by their long incisor teeth. The size of the seeds (4–5 mm long) and the size of the holes and grooves suggested a small rodent with incisors at most 1 mm wide, clearly smaller than a squirrel. Today, bankvole, woodmouse and dormouse use different gnawing actions for opening nuts so the authors used modern gnawed nuts for comparison. Of these three, the woodmouse makes the most similar feeding marks (Fig. 6.14b). After gnawing through the surface to make a small hole it grips the outer surface with its upper incisors and vertically chisels the walls of the hole with its lower incisors. It then inserts its lower incisors inside the hole to dig out the kernel.

The Eocene beds of southern England have yielded a variety of rodents, and the two that are the right size to have gnawed the *Stratiotes* seeds are two species of the early dormouse *Glamys*. Eocene *Glamys* may have swum to retrieve the seeds, or patrolled the shores of small ponds looking for any that had been washed up.

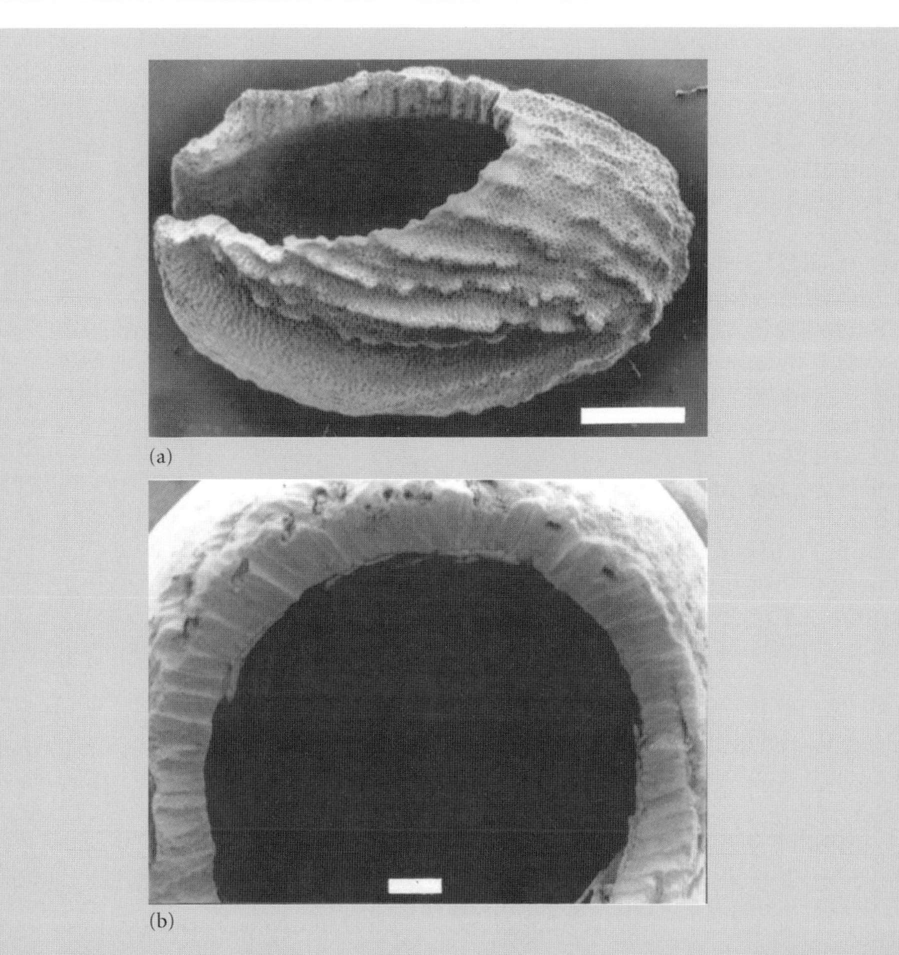

Figure 6.14 Evidence for a rodent–plant interaction from the Eocene. (a) Seed of the water plant *Stratiotes* carrying a neat hole gnawed by a rodent, from the Eocene Bembridge Limestone Formation of the Isle of Wight, southern England. (b) A hole gnawed by a modern woodmouse, showing the same kind of perpendicular narrow grooves made by the tips of the upper incisors. Scale bars, 1 mm. (Courtesy of Margaret Collinson.)

been very revealing. The teeth are sharp and curved, and the edges carry serrations like a steak knife – clear evidence of meat eating. Close study of the teeth also reveals minute scratches that were produced by the bones and other tough food material in the diet (Barrett & Rayfield 2006). Bones of the prey offer clues too: some examples show that *T. rex* could penetrate deep into the bones of its victims, but also that it chomped and tore at the flesh in such a way that it sometimes left dozens of tooth marks as it stripped the bones. All these circumstantial discoveries add to a rich picture of how one fossil animal fed.

Review questions

1 How are fossil species told apart? Look up information on any pair of species within a single genus (such as the human species *Homo erectus* and *Homo sapiens*; the dinosaurs *Saurolophus osborni* from North America and *Saurolophus angustirostris* from Mongolia; or any of the 10 or more species of the trilobite *Paradoxides*), and write down as many distinguishing characters as you can track down. How easy are these morphological characters to observe in the specimens?

2 Make a study of allometry in humans. Select a real baby, several children and an adult (all male or all female), and measure total body length (top of head to base of foot) as a baseline measurement, and then height of head (top of crown to base of chin), length of chin (bottom of lower lip to bottom of chin), arm length (tip of longest finger to armpit) and hand length (tip of longest finger to the line of hinging at the wrist). Which of these show isometric growth, and which are allometric? Are they positively or negatively allometric? If you do not have access to real people of different sizes, use images from books or the web.
3 Read around some recent papers on *Hox* genes, and find out how many are involved in determining the development of the vertebrate hindlimb. What does each gene do?
4 You want to understand how some fossil organisms moved and fed. What would be good modern analogs for trilobites, ichthyosaurs and crinoids? Compare images and descriptions of the fossil and modern groups, and indicate how confident you would be in using each of the modern analogs.
5 Find an image of the skull of the dinosaur *Plateosaurus*. Why is the jaw joint lower than the tooth row? Think of modern analogs, perhaps among common domestic items, and think how the dropped jaw joint might affect the lever performance of the jaw.

Further reading

Barrett, P.M. & Rayfield, E.J. 2006. Ecological and evolutionary implications of dinosaur feeding behaviour. *Trends in Ecology and Evolution* **21**, 217–24.

Briggs, D.E.G. & Crowther, P.R. 2000. *Palaeobiology, A Synthesis*, 2nd edn. Blackwell Publishing, Oxford, UK.

Carroll, S.B., Grenier, J. & Weatherbee, S. 2004. *From DNA to Diversity*, 2nd edn. Blackwell Publishing, Oxford, UK.

Carroll, S.B. 2005. *Endless Forms Most Beautiful: The New Science of Evo devo and the Making of the Animal Kingdom*. W.W. Norton & Co., New York.

Futuyma, D. 1998. *Evolutionary Biology*, 3rd edn. Sinauer, Sunderland, MA.

Gould, S.J. 1974. The origin and function of "bizarre" structures: antler size and skull size in the "Irish Elk," *Megaloceros giganteus*. *Evolution* **28**, 191–220.

Kodric-Brown, A., Sibly, R.M. & Brown, J.H. 2006. The allometry of ornaments and weapons. *Proceedings of the National Academy of Sciences, USA* **103**, 8733–8.

Rayfield, E.J. 2007. Finite element analysis and understanding the biomechanics and evolution of living and fossil organisms. *Annual Review of Earth and Planetary Sciences* **35**, 541–76.

Ridley, M. 2004. *Evolution*, 3rd edn. Blackwell, Oxford, UK.

Shubin, N. 2008. Your Inner Fish: A Journey into the 3.5-Billion-Year History of the Human Body. Pantheon, New York.

References

Barrett, P.M. & Rayfield, E.J. 2006. Ecological and evolutionary implications of dinosaur feeding behaviour. *Trends in Ecology and Evolution* **21**, 217–24.

Benton, M.J. & Kirkpatrick, R. 1989. Heterochrony in a fossil reptile: juveniles of the rhynchosaur *Scaphonyx fischeri* from the late Triassic of Brazil. *Palaeontology* **32**, 335–53.

Coates, M.I., Jeffery, J.E. & Ruta, M. 2002. Fins to limbs: what the fossils say. *Evolution and Development* **4**, 390–401.

Collinson, M.E. & Hooker, J.J. 2000. Gnaw marks on Eocene seeds: evidence for early rodent behaviour. *Palaeogeography, Palaeoclimatology, Palaeoecology* **157**, 127–49.

Gould, S.J. 1974. The origin and function of "bizarre" structures: antler size and skull size in the "Irish Elk," *Megaloceros giganteus*. *Evolution* **28**, 191–220.

Hagdorn, H., Wang, X.F. & Wang, C.S. 2007. Palaeoecology of the pseudoplanktonic crinoid *Traumatocrinus* from southwest China. *Palaeogeography, Palaeoclimatology, Palaeoecology* **247**, 181–96.

Hutchinson, J.R. & Gatesy, S.M. 2006. Dinosaur locomotion: beyond the bones. *Nature* **440**, 292–4.

McNamara, K.J. 1976. The earliest *Tegulorhynchia* (Brachiopoda: Rhynchonellida) and its evolutionary significance. *Journal of Paleontology* **57**, 461–73.

Molyneux T. 1697. A discourse concerning the large horns frequently found under ground in Ireland, concluding from them that the great American deer, call'd a moose, was formerly common in that island: with remarks on some other things natural to the country. *Philosophical Transactions of the Royal Society* **19**, 489–512.

Rayfield E.J. 2004. Cranial mechanics and feeding in *Tyrannosaurus rex*. *Proceedings of the Royal Society of London B* **271**, 1451–9.

Shubin, N., Tabin, C. & Carroll, S. 1997. Fossils, genes and the evolution of animal limbs. *Nature* **388**, 639–48.

Tickle, C. 2006. Making digit patterns in the vertebrate limb. *Nature Reviews: Molecular Cell Biology* **7**, 45–53.

Witmer, L.M. 1997. The evolution of the antorbital cavity of archosaurs: a study in soft-tissue reconstruction in the fossil record with an analysis of the function of pneumaticity. *Journal of Vertebrate Paleontology* **17** (Suppl.), 1–73.

Zakany, J. & Duboule, D. 2007. The role of *Hox* genes during vertebrate limb development. *Current Opinion in Genetics and Development* **17**, 359–66.

Chapter 7

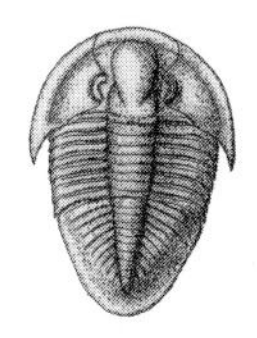

Mass extinctions and biodiversity loss

Key points

- During mass extinctions, 20–90% of species were wiped out; these include a broad range of organisms, and the events appear to have happened rapidly.
- It is difficult to study mass extinctions in the Precambrian, but there seems to have been a Neoproterozoic event between the Ediacaran and Early Cambrian faunas.
- The "big five" Phanerozoic mass extinctions occurred in the end-Ordovician, the Late Devonian, the end of the Permian, the end of the Triassic and the end of the Cretaceous. Of these, the Late Devonian and end-Triassic events seem to have lasted some time and involved depressed origination as much as heightened extinction.
- The end-Permian mass extinction was the largest of all time, and probably caused by a series of Earth-bound causes that began with massive volcanic eruptions, leading to acid rain and global anoxia.
- The end-Cretaceous mass extinction has been most studied, and it was probably caused by a major impact on the Earth.
- Smaller-scale extinction events include the loss of mammals at the end of the Pleistocene, perhaps the result of climate change and human hunting.
- Recovery from mass extinctions can take a long time; first on the scene may be some unusual disaster taxa that cope well in harsh conditions; they give way to the longer-lived taxa that rebuild normal ecosystems.
- Extinction is a major concern today, with calculated species loss as high as during any mass extinction of the past. The severity of the current extinction episode is still debated.

The Dodo never had a chance. He seems to have been invented for the sole purpose of becoming extinct and that was all he was good for.

Will Cuppy (1941) *How to Become Extinct*

Extinction, long studied by paleontologists to inform them of the past, is now a key theme in discussions about the future. Will Cuppy, the famous American humorist, was able to talk about the extinction of dinosaurs, plesiosaurs, the woolly mammoth and the dodo, all of them icons of obsolescence and failure. The dodo is perhaps the most iconic of icons (Fig. 7.1), and it used to be held up as a moral tale for children: here was a large friendly bird, but it was simply too friendly and stupid to survive. The message was: be careful, take care, and don't be as improvident as the dodo! The dodo is now an icon of human carelessness rather than of avian extinction.

The most spectacular extinctions are known as **mass extinctions**, times when a large cross-section of species died out rather rapidly. There may have been only five or six mass extinctions throughout the known history of life, although there were many **extinction events**, smaller-scale losses of species, often in a particular region or involving species with a particular shared ecology.

Figure 7.1 An image of a dodo from another era. Lewis Carroll introduced the dodo as a kindly and wise old gentleman in *Alice Through the Looking Glass*, although at the time most people probably regarded the dodo as rather foolish. Driven to extinction in the 17th century by overhunting, the dodo is now an image of human thoughtlessness.

The serious study of mass extinctions is a relatively new research field, dating only from the 1980s onwards, and it has wide interdisciplinary links across stratigraphy, geochemistry, climate modeling, ecology, conservation and even astronomy. The study of mass extinctions involves careful hypothesis testing (see p. 4) at all levels, from the broadest scale ("Was there a mass extinction at this time? Was it caused by a meteorite impact or a volcanic eruption?") to the narrowest ("How many brachiopod genera died out in my field section? Does their extinction coincide with a negative carbon isotope anomaly? Do the sediments record any evidence for climate change across this interval?"). The excitement of studies of mass extinctions, and smaller extinction events, is that these events were hugely important in the history of life, and yet they are unique paleontological phenomena that cannot be predicted from the modern-day standpoint. In practical terms, the field involves such a broad array of disciplines that research involves teamwork, often groups of five or 10 specialists who pool their expertise and resources to carry out a study.

In this chapter, we will explore what we mean by extinctions and mass extinctions, and whether there are any general features shared by these times of crisis. We shall then explore the two most heavily studied events, the Permo-Triassic mass extinction of 251 million years ago, and the Cretaceous-Tertiary mass extinction of 65 million years ago, in most detail. Finally, it is important to consider how paleobiology informs the current heated debates about extinctions now and in the future.

MASS EXTINCTIONS

Definition

Extinction happens all the time. Species have a natural duration of anything from a few thousand years to a few million, and so they live for a time and then disappear. This means that there is a pattern of normal or **background extinction** that happens without any broad-scale cause. In any segment of time, perhaps 5–10% of species may disappear every million years. In fact, more species have died out during normal times than during the more spectacular mass extinctions.

Nonetheless, mass extinctions fascinate paleontologists and the public because these were times of concentrated misery, and represent perhaps unusually intense environmental catastrophes. But how is a mass extinction to be defined? All mass extinctions share certain features in common, but differ in others. The common features are:

1 Many species became extinct, perhaps more than 30% of plants and animals of the time.
2 The extinct organisms spanned a broad range of ecologies, and typically include marine and non-marine forms, plants and animals, microscopic and large forms.
3 The extinctions were worldwide, covering most continents and ocean basins.
4 The extinctions all happened within a relatively short time, and hence relate to a single cause, or cluster of interlinked causes.
5 The level of extinction stands out as considerably higher than the background extinction level.

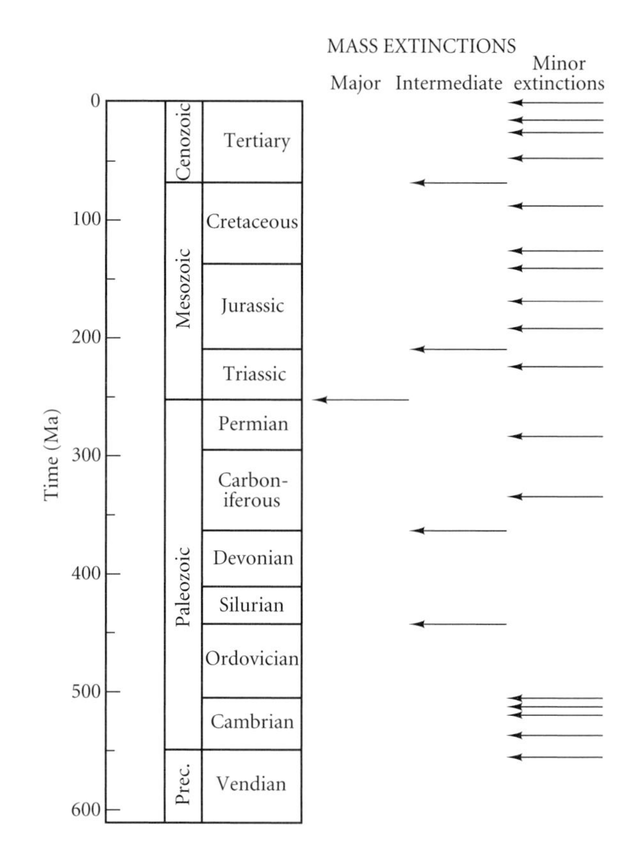

Figure 7.2 Mass extinctions through the past 600 myr include the enormous end-Permian event 251 Ma, which killed two or three times as many families, genera and species (50% of families and up to 96% of species) as the "intermediate" events. These were global in extent, and involved losses of 20% of families and 75–85% of species. Some of the minor mass extinctions were perhaps global in extent, causing losses of 10% of families and up to 50% of species, but many may have been regional in extent, or limited taxonomically or ecologically.

It is hard to define these terms more precisely, first because each mass extinction seems to have been unique, and second because it is sometimes hard to pin down exactly the timing and scale of events.

Paleontologists commonly talk about the "big five" mass extinctions of the last 540 myr, the Phanerozoic, and the current extinction crisis is sometimes called the "sixth extinction". The five mass extinctions (Fig. 7.2) are the end-Ordovician, Late Devonian, end-Permian, end-Triassic, and Cretaceous-Tertiary (KT) events. Study of the Neoproterozoic reveals a further one or two possible mass extinctions, before and after the Ediacaran (see p. 242) so perhaps we should refer to the "big six" or the "big seven" such events.

The notion of five somewhat similar mass extinctions throughout the Phanerozoic has been questioned, however. In a careful statistical survey, Bambach (2006) has shown that there were perhaps only three real mass extinctions, the end-Ordovician, the end-Permian and the KT events. The Late Devonian and end-Triassic events do not stand out so clearly above background extinction rates at those times; each lasted perhaps over 5 myr, and each was caused as much by depressed origination rates as by elevated extinction rates.

In trying to define and scale mass extinctions, the end-Permian event is in a class of its own, because 50% of families disappeared at that time, and this scales to an estimated loss of 80–96% of species. The assumption that a higher proportion of species than families are wiped out is based on the observation that families contain many species, all of which must die for the family to be deemed extinct. Hence, the loss of a family implies the loss of all constituent species, but many families will survive even if most of their contained species disappear. This commonsense observation may be described mathematically as an example of rarefaction (see also p. 95), a useful technique for estimating between scales of observation (Box 7.1). The "intermediate" mass extinctions (Fig. 7.2) are associated with losses of 20–30% of families, scaling to perhaps 50% of species, while the "minor" mass extinctions experienced perhaps 10% family loss and 20–30% species loss.

Pattern and timing of mass extinctions

Good-quality fossil records indicate a variety of patterns of extinction. Detailed collecting of planktonic microfossils based on centimeter-by-centimeter sampling up to, and across, crucial mass extinction boundaries offers the best evidence of the patterns of mass extinctions. In detail, some of the patterns reveal a stepped pattern of decline over a time interval of 0.5–1.5 myr during which 53% of the

Box 7.1 Rarefaction and predicting species numbers from family numbers

Rarefaction is a statistical technique used most commonly by paleontologists to investigate the effect of sample size on taxon counts. So, a common question might be: "How many specimens should I collect in this quarry in order to find all the species?" Ecologists have used this concept, sometimes called the **collector curve** or **accumulation curve**, for decades (see p. 535). By plotting cumulative new species found against the number of specimens collected or observed, you can reconstruct a predictive pattern (Fig. 7.3a). After collecting one specimen, you will have identified one species. The next 10 specimens probably will not add another 10 new species, perhaps only three or four. The next 100 specimens might add another 10 or 15 species. The more you collect, the more you find, but there is a law of diminishing returns. At a certain point, as the species versus effort (that is, specimens or time spent searching) curve approaches an **asymptote**, it is easy to estimate roughly what the final total number of species would be if you just kept on collecting doggedly for days and days.

Rarefaction is a procedure to estimate the completeness of a species list if a smaller sample had been taken. So, if 1000 specimens were collected, it might be of value to know the size of the species count if only 100 specimens, or 10 specimens had been collected at random. The data in the collector curve can be culled or sampled randomly by removing 90% or 99% of records, respectively. In a typical example (Fig. 7.3b), a collection of 750 specimens yielded a species count of 30. If the collection had been half the size, only 20 species would have been identified.

Raup (1979), in a neat example of lateral thinking, applied "reverse rarefaction" to an unknown question: if we know that 50% of families of marine animals were killed off by the end-Permian mass extinction, how many species might that represent? Paleontologists are more confident of their raw data on the numbers of families that existed in the past than the number of species because families are harder to miss (they are bigger, and you only have to find one species to identify the presence of a family). Raup modeled the distribution of species numbers in families – some families contain one species, others contain 200. He then culled at random 50% of families from this distribution, and showed that this equates to a loss of as many as 96% of species. McKinney (1995) criticized Raup's assumption that the 50% of extinct families would be a random cut from all families around at the time. McKinney argued, probably correctly, for the "dodo principle": the extinct families would include a disproportionate number of those that were vulnerable, especially those containing small numbers of species. Highly species-rich families would be less vulnerable, and so the 96% figure might be an overestimate. McKinney (1995) suggested a more likely figure of 80% species loss at the end-Permian event.

Read more about rarefaction in paleobiology in Hammer and Harper (2005) and its use in ecology in Gotelli and Colwell (2001). Implementations may be found through http://www.blackwellpublishing.com/paleobiology/.

Continued

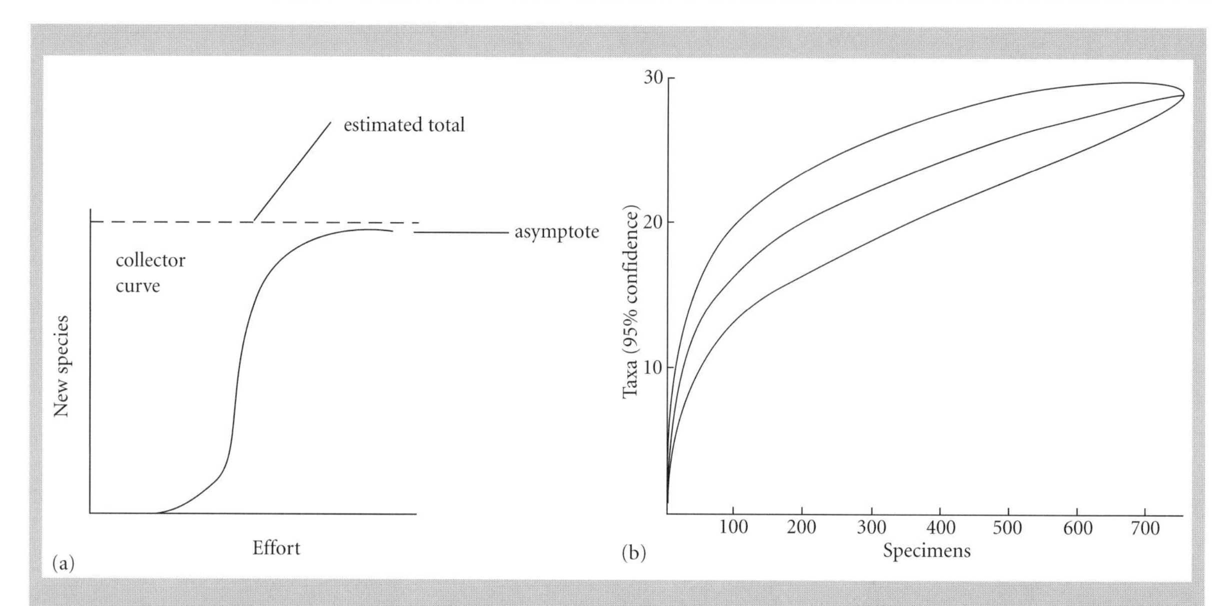

Figure 7.3 (a) The classic collector curve showing the sigmoid (or logistic) shape of the curve of cumulative new species plotted against effort (number of specimens collected/number of days spent looking/number of investigators), with a rapid rise and then a tailing off to an asymptote. (b) Rarefaction curve that shows the number of species likely to be identified from samples of a particular size. (b, based on Hammer & Harper 2005.)

foraminifera species died out (Fig. 7.4). However, should a paleontologist describe this as an example of catastrophic or gradual extinction? A gradualist would argue that the extinction lasts for more than 0.5 myr, too long to be the result of an instant event. A catastrophist would say that the killing lasted for 1–1000 years, and would argue that the stepped pattern in Fig. 7.4 is the result of incomplete preservation, incomplete collecting or reworking of sediment by burrowers. More precise dating and more precise assessment of sampling problems are needed to sharpen the definitions.

The rock record can be misleading (see p. 70), and gradual extinctions might look catastrophic and catastrophic extinctions gradual (Fig. 7.5). If there is a gap in the rock record, especially at a crucial time line such as the KT boundary, species ranges are cut off artificially and the pattern looks sudden (Fig. 7.5a). The opposite effect, an apparently gradual pattern, can happen because paleontologists will never find the very last fossil of a species. Phil Signor and Jere Lipps showed how this backward smearing of the record happens, and it is now termed the **Signor–Lipps effect** in their honor (see also p. 26). The Signor–Lipps effect can make a sudden mass extinction seem gradual (Fig. 7.5b).

These kinds of problems are especially likely for organisms such as dinosaurs. Their bones are preserved in continental sediments, which are deposited sporadically, and specimens are large and rare. Nevertheless, two teams attempted large-scale field sampling in Montana to establish once and for all whether the dinosaurs had drifted to extinction over 5–10 myr, the view of the gradualists, or whether they had survived at full vigor to the last minute of the Cretaceous Period, when they were catastrophically wiped out. Needless to say, one team found evidence for a long-term die-off, and the other team demonstrated sudden extinction.

The problem was not that either team had done their work badly, but that the fossils were still too scattered, and the dating of the rocks was not good enough, to be sure. Geologists work in millions of years, and yet

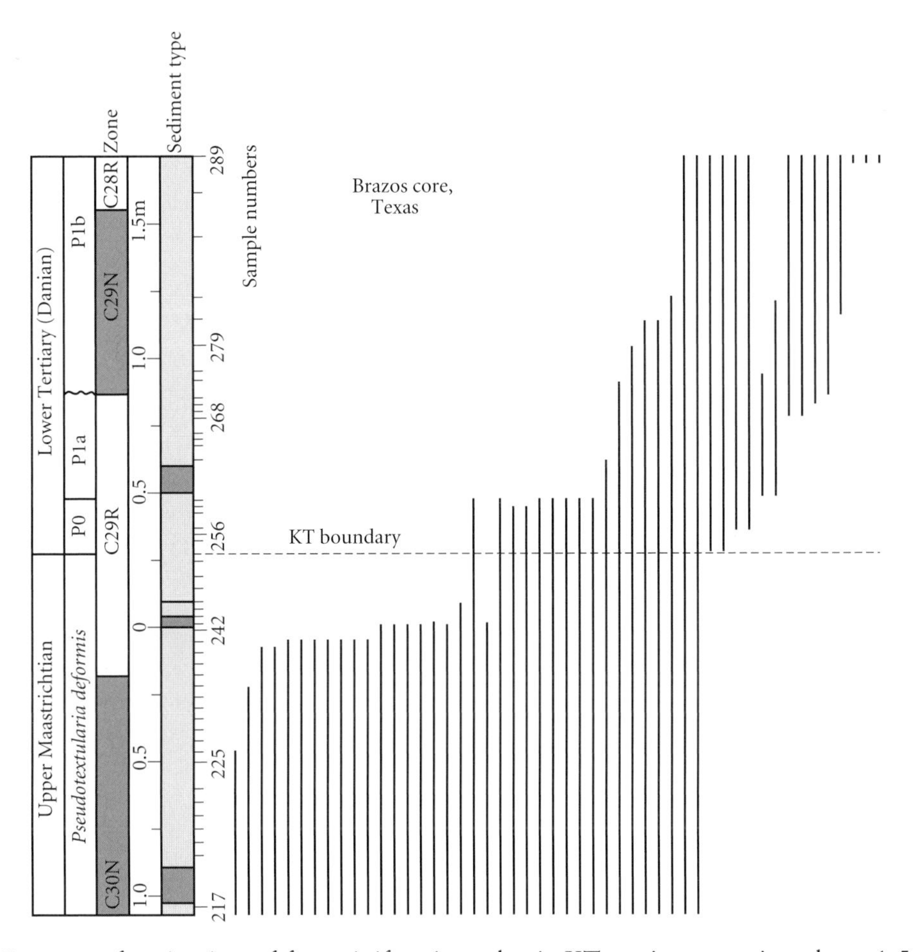

Figure 7.4 Patterns of extinction of foraminifera in a classic KT section spanning about 1.5 myr. A species loss of 53% occurred in two steps close to the KT boundary and iridium anomaly. Dating is based on magnetostratigraphy, and the KT boundary falls in the C29R (reversed) zone. Planktonic zones (P0, P1a, P1b) are indicated; sediment types are mudstones (darker grey) and limestones (pale grey); meter scale bar shows height above and below a particular extinction level, 0. (Based on Keller et al. 1993.)

answers to questions such as these refer to ecological time scales – that is, times of years or decades at most.

It is just as difficult, if not more so, to answer questions of the timing of ancient events from region to region or continent to continent. How can a paleontologist be sure that the supposed KT boundary in Montana is the same as the supposed KT boundary in Mongolia? Perhaps the boundary is marked as the next sedimentary rock layer above the appearance of the last dinosaur fossil. But of course this definition is perfectly circular: the KT boundary is marked by the disappearance of dinosaurs; dinosaurs disappeared just below the KT boundary. Other fossils, such as pollen, may be used to date the boundary, but additional evidence, from magnetostratigraphy (see p. 24) and exact radiometric dating (see p. 38) are also needed.

Selectivity and mass extinctions

The second defining character of mass extinctions (see p. 164) was that they should be ecologically catholic, that there should be little evidence of selectivity. Ecological **selectivity** implies that some organisms might be better able to survive a mass extinction event than others. Mass extinctions do not seem to have been particularly selective, even though it might seem that, for example, large reptiles were specially selected for extinction during the KT event. The dinosaurs and some other large reptiles certainly died out then, but a

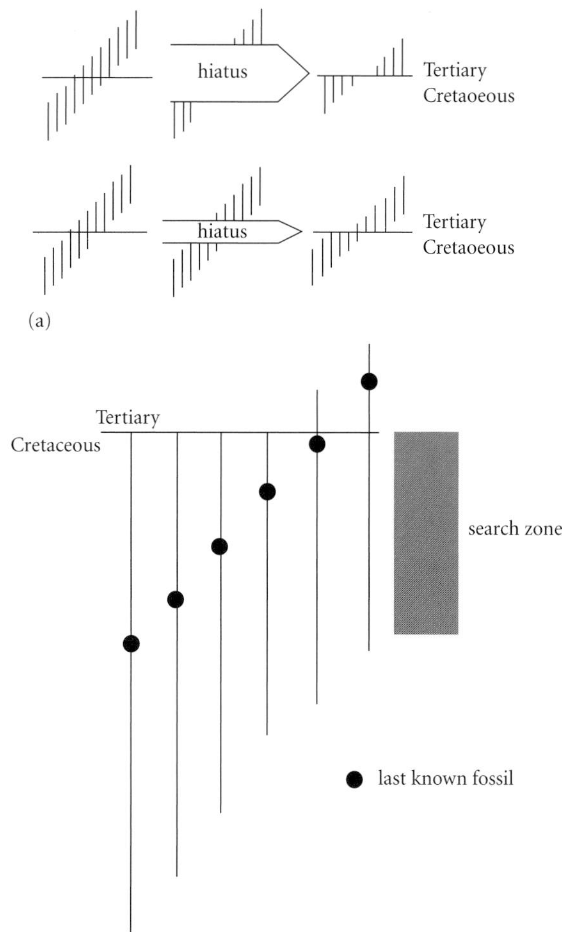

Figure 7.5 Gaps and missing data can make gradual extinction events seem sudden (a) or sudden events seem gradual (b). In both diagrams the vertical lines represent different species. (a) The real pattern of fossil species distribution is shown on the left, and if there is a large or small hiatus, or gap, at the KT boundary (middle diagram), a gradual loss of species might seem artificially sudden (right-hand diagram). (b) It is likely that the very last fossils of a species will not be found, and a sudden extinction might look gradual; this can only be detected by intense additional collecting in the rocks that include the supposed last fossils (shaded gray).

larger number of microscopic planktonic species also died out.

The best evidence of selectivity during mass extinctions has been against genera with limited geographic ranges. Jablonski (2005) could find no evidence for selectivity during the KT event for ecological characters of bivalves and gastropods, such as mode of life, body size or habitat preference. He did find that the probability of extinction for bivalve genera declined predictably depending upon the number of major biogeographic realms they occupied, and the positive survival benefit of a wide geographic range has been found for many other groups during other mass extinctions. Also, genera containing many species survived better than those with few.

Ecological characters that may be important in normal, or background, times often have little influence on survivorship during times of mass extinction. Jablonski (2005), for example, showed that epifaunal bivalves have shorter generic durations than infaunal bivalves in the Jurassic and Cretaceous, suggesting that in evolutionary terms it is better to burrow. However, during the KT event, there was no difference in the pattern of survival and extinction of epifaunal and infaunal bivalves.

This confirms a general principle of mass extinctions, which is that normal evolutionary processes break down. So, if during normal times, it is advantageous to be large, to be secretive, to burrow, to move fast, or to have a particular diet or breeding mode, these positive characters may make no difference at all when the crisis hits. Natural selection hones and shapes the adaptations of species on the scale of generations and normal levels of environmental change; mass extinctions seem to represent a different scale of challenge, much too great for the normal rules to apply. Mass extinctions probably occur too far apart, and too unpredictably, for the normal rules of evolution to apply. As Steve Gould said, mass extinctions re-set the evolutionary clock.

Periodicity of mass extinctions

There are many viewpoints on the causes of mass extinctions, but a fundamental debate has been whether each event had its own unique causes, or whether a unifying principle linking all mass extinctions might be found. If there was a single cause, it might be sporadic changes in temperature (usually cooling) or in sea level, or periodic impacts on the Earth by asteroids (giant rocks) or comets (balls of ice).

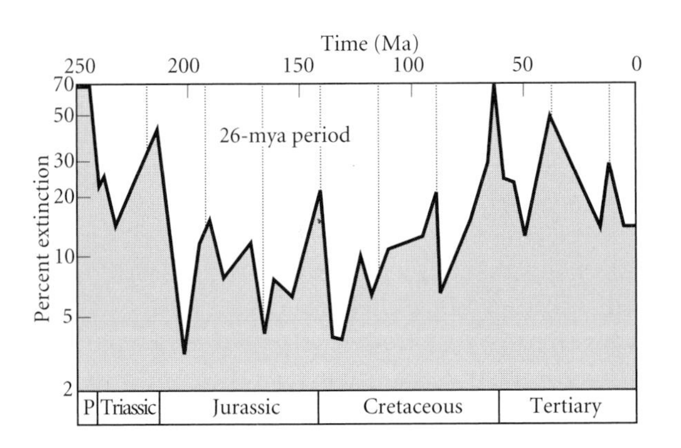

Figure 7.6 Periodic extinctions of marine animal families over the past 250 myr. The extinction rate is plotted as percent extinction per million years. A periodic signal may be detected in a time series like this either by eye, or preferably by the use of time series analysis. There are a variety of mathematical techniques generally termed spectral analysis for decomposing a time series into underlying repeated signals. The techniques are outlined in chapter 7 of Hammer and Harper (2006), and a practical example that repeats the classic Raup and Sepkoski (1984) analysis is given at http://www.blackwellpublishing.com/paleobiology/. (Based on the analysis by Raup & Sepkoski 1984.)

The search for a common cause gained credence with the discovery by Raup and Sepkoski (1984) of a regular spacing of 26 myr between extinction peaks through the last 250 myr (Fig. 7.6). They argued that regular periodicity in mass extinctions implies an astronomical cause, and three suggestions were made: (i) the eccentric orbit of a sister star of the sun, dubbed Nemesis (but not yet seen); (ii) tilting of the galactic plane; or (iii) the effects of a mysterious planet X that lies beyond Pluto on the edges of the solar system. These hypotheses involve a regularly repeating cycle that disturbs the Oört comet cloud and sends showers of comets hurtling through the solar system every 26 myr.

The debate about periodicity of mass extinctions raged through the 1980s. Many geologists and astronomers loved the idea, and they set about looking for Nemesis or planet X – but without success. Some impact enthusiasts found evidence for craters and impact debris associated with the end-Permian and end-Triassic mass extinctions, but not for any of the seven other extinction peaks. And the evidence for impact is frankly rather weak except for the KT event.

Most paleontologists rejected the idea because only three of the 10 supposed mass extinctions were really mass extinctions (end-Permian, end-Triassic and KT) – the seven other high extinction peaks through the Jurassic and Cretaceous were explained away as either too small to signify or as artificial (miscounting of extinctions, mistiming or a major change of rock facies). Re-study of a revised dataset by Benton (1995) did not confirm the validity of any of the seven queried peaks, and with only three out of 10 there is no periodic pattern!

The idea of periodicity of impacts was reawakened by Rohde and Muller (2005) who argued for a 62 myr periodicity in mass extinctions. This cyclicity picks up the end-Ordovician, late Devonian, end-Permian and end-Triassic mass extinctions, but it misses the KT event. It also hints at other intermediate events in the mid-Carboniferous, mid-Permian, Late Jurassic, mid-Cretaceous and Paleogene. Most commentators have been very unhappy with this study, suggesting it does not relate closely to the fossil record, does not replicate the known mass extinctions, and may reflect long-term changes in sea level. So, the search for periodicity in mass extinctions and a single astronomical cause appears to have hit the buffers, but the discovery that perhaps sea level change, or some other forcing factor might itself be periodic, is worth further investigation.

THE "BIG FIVE" MASS EXTINCTION EVENTS

The "big five" or the "big three"?

As noted earlier (see p. 164), there is some debate about whether there were five or three mass extinctions in the past 500 myr. We summarize a few key points about three of the five events, and then concentrated most attention on two of the five.

In the *end-Ordovician* mass extinction, about 445 Ma, substantial turnovers occurred among marine faunas. Most reef-building animals, as well as many families of brachiopods, echinoderms, ostracodes and trilobites died out. These extinctions are associated with evidence for major climatic changes. Tropical-type reefs and their rich faunas lived around the shores of North America and other landmasses that then lay around the equator. Southern continents had, however, drifted over the south pole, and a vast phase of glaciation began. The ice spread north in all directions, cooling the southern oceans, locking water into the ice and lowering sea levels globally. Polar faunas moved towards the tropics, and many warm-water faunas died out as the whole tropical belt disappeared.

The second of the big five mass extinctions occurred during the *Late Devonian*, and this appears to have been a succession of extinction pulses lasting from about 380 to 360 Ma. The abundant free-swimming cephalopods were decimated, as were the extraordinary armored fishes of the Devonian. Substantial losses occurred also among corals, brachiopods, crinoids, stromatoporoids, ostracodes and trilobites. Causes could have been a major cooling phase associated with anoxia (loss of oxygen) on the seabed, or massive impacts of extraterrestrial objects. Perhaps this rather drawn-out series of extinctions is not a clearcut mass extinction, but rather a series of smaller extinction events (Bambach 2006).

The *end-Triassic* event is the fourth of the big five mass extinctions. A marine mass extinction event at, or close to, the Triassic-Jurassic boundary, 200 Ma, has long been recognized by the loss of most ammonoids, many families of brachiopods, bivalves, gastropods and marine reptiles, as well as by the final demise of the conodonts (see p. 429). Impact has been implicated as a possible cause of the end-Triassic mass extinction, but most evidence points to anoxia and global warming following massive flood basalt eruptions located in the middle of the supercontinent Pangea, just at the site where the North Atlantic was beginning to unzip. Perhaps the end-Triassic event is not a clearcut mass extinction either (Bambach 2006): it may have consisted of more than one phase, and it seems to be as much about lowered origination rates as the sudden extinction of many major groups.

The third and fifth of the "big five" were the Permo-Triassic (PT) and Cretaceous-Tertiary (KT) events, and these will now be presented in more detail.

The Permo-Triassic event

The end-Permian, or Permo-Triassic, mass extinction was the most devastating of all time, and yet it was less well understood than the smaller KT event until after 2000. This may seem surprising, but the KT event is more recent and so the rock records are better and easier to study. The KT event is also more newsworthy and immediate because it involved the dinosaurs and meteorite impacts. In the 1990s, paleontologists and geologists were unsure whether the PT extinctions lasted for 10 myr or happened overnight, whether the main killing agents were global warming, sea level change, volcanic eruption or anoxia. The end-Permian mass extinction occurred just below the Permo-Triassic boundary, so is generally termed the PT event.

Since 1995, there have been many additions to our understanding. First, the peak of eruptions by the Siberian Traps was dated at 251 Ma, matching precisely the date of the PT boundary. Further, extensive study of rock sections that straddle the PT boundary, and the discovery of new sections, began to show a common pattern of environmental changes through the latest Permian and earliest Triassic. Fourth, studies of stable isotopes (oxygen, carbon) in those rock sections revealed a common story of environmental turmoil, and this all seemed to point in a single direction, a model of change where normal feedback processes could not cope, and the atmosphere and oceans went into catastrophic breakdown.

The scale of the PT event was huge. Global compilations of data show that more than 50% of families of animals in the sea and on land went extinct. This was estimated by rarefaction (see Box 7.1) to indicate something from 80% to 96% of species loss. Turning these figures round, the PT event saw the virtual annihilation of life, with as few as 4–20% of species surviving. Close study of many rock sections that span the PT boundary has shown the nature of the event at a more local scale (Box 7.2).

The suddenness and the magnitude of the mass extinction suggest a dramatic cause, perhaps impact or volcanism. Evidence for a meteorite impact at the PT boundary has been presented by several researchers: there have been reports of shocked quartz, of supposed extraterrestrial noble gases trapped in carbon compounds, and the supposed crater has been identified – first in the South Atlantic and, in 2005, off the coast of Australia. These proposals of impact have not gained wide support, mainly because the evidence seems much weaker than the evidence for a KT impact (see p. 174).

Most attention has focused on the Siberian Traps, some 2 million cubic kilometers of basalt lava that cover 1.6 million square kilometres of eastern Russia to a depth of 400–3000 m. It is widely accepted now that these massive eruptions, confined to a time span of less than 1 myr in all, were a significant factor in the end-Permian crisis.

The Siberian Traps are composed of basalt, a dark-colored igneous rock. Basalt is generally not erupted explosively from classic conical volcanoes, but emerges more sluggishly from long fissures in the ground; such fissure eruptions are seen today in Iceland. Flood basalts typically form many layers, and may build up over thousands of years to considerable thicknesses. Early efforts at dating the Siberian Traps produced a huge array of dates, from 280 to160 Ma, with a particular cluster between 260 and 230 Ma. According to these ranges, geologists in 1990 could only say that the basalts might be anything from Early Permian to Late Jurassic in age, but probably spanned the PT boundary. More recent dating, using a variety of newer radiometric methods, yielded dates exactly on the boundary, and the range from the bottom to the top of the lava pile was about 600,000 years.

Box 7.2 Close-up view of the mass extinction

Paleontologists have studied PT boundary sections in many parts of the world. One of the best studies so far is by Jin et al. (2000), who looked at the shape of the mass extinction in the Meishan section in southern China. This section has added importance because it was ratified as the global stratotype (see p. 33) for the Permo-Triassic boundary in 1995.

Jin et al. (2000) collected thousands of fossils through 90 m of rocks spanning the PT boundary. They identified 333 species belonging to 14 marine fossil groups – microscopic foraminiferans, fusulinids, radiolarians, rugose corals, bryozoans, brachiopods, bivalves, cephalopods, gastropods, trilobites, ostracodes, conodonts, fishes and algae. In all, 161 species became extinct below the boundary bed (Fig. 7.7a) in the 4 myr before the end of the Permian. Background extinction rates at most levels amounted to 33% or less. Then, just below the PT boundary, at the contact of beds 24 and 25, most of the remaining species disappeared, a loss of 94% of species at that level. Three extinction levels were identified, labeled A, B and C on Fig. 7.7a. Jin and colleagues argued that the six species that apparently died out at level A are probably artificial records, really pertaining to level B (examples of the Signor–Lipps effect; see p. 166). But level C may be real, and this suggests that, after the huge catastrophe at level B, some species survived through the 1 myr to level C, but most disappeared step by step during that interval.

In reconstruction form (Fig. 7.7b, c), the effects of the PT mass extinction are devastating. What was a rich set of reef ecosystems before the event, with dozens of sessile and mobile bottom-dwellers, as well as fishes and ammonoids swimming above, became reduced to only two or three species of paper pectens and the inarticulated brachiopod *Lingula* (which seems to have survived everything; see p. 300). The environment had changed too. Sediments show a well-oxygenated seabed before the event, with masses of coral and shell debris accumulating. After the event, nothing. The sediments are black mudstones containing few or no fossils or burrows. The black color and associated pyrite indicate anoxia (see p. 173). This was the death zone.

Read more about the PT mass extinction in Benton (2003) and Erwin (2006). Benton and Twitchett (2003) is a brief review of current evidence. Web presentations may be read at http://www.blackwellpublishing.com/paleobiology/.

Continued

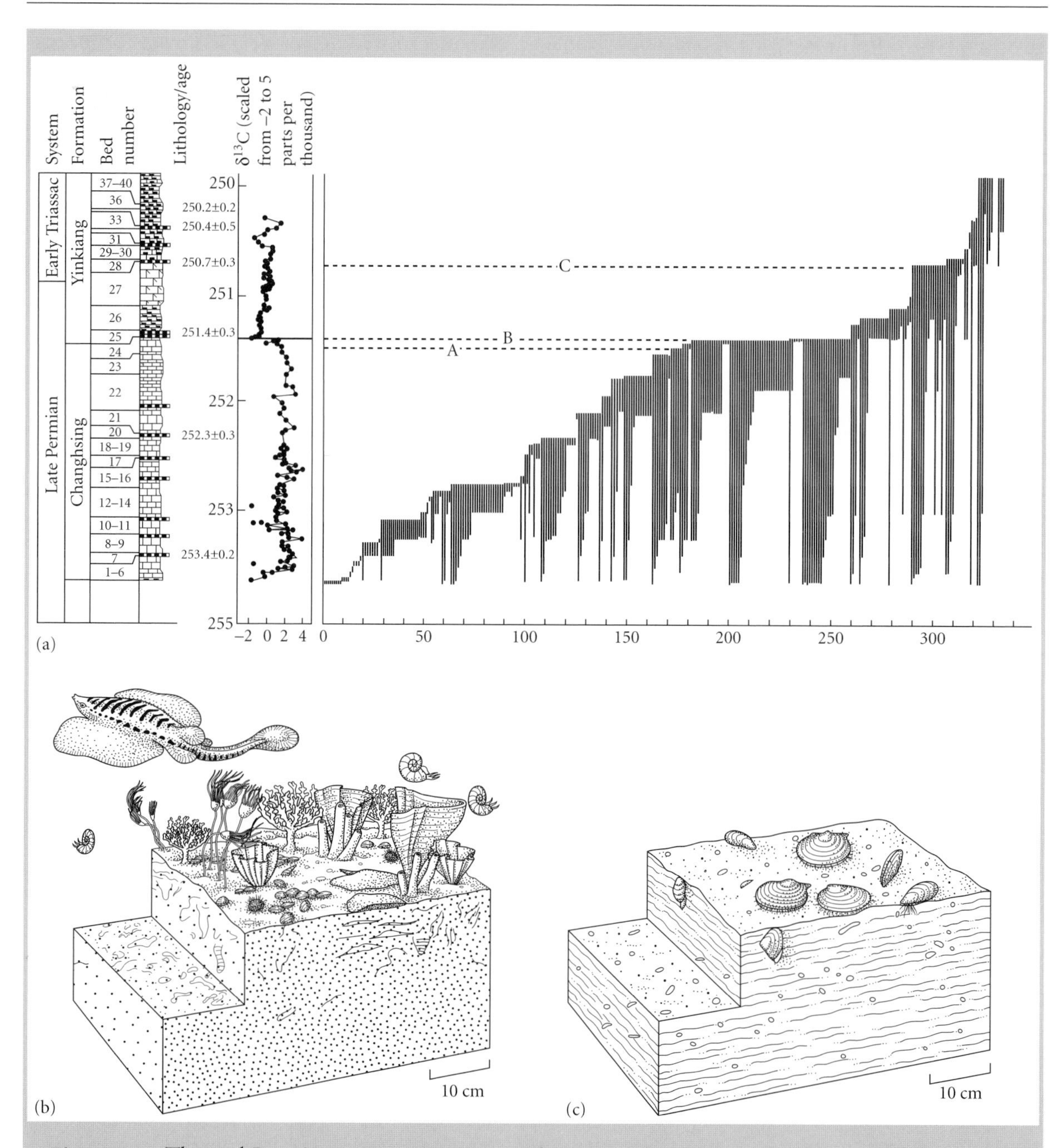

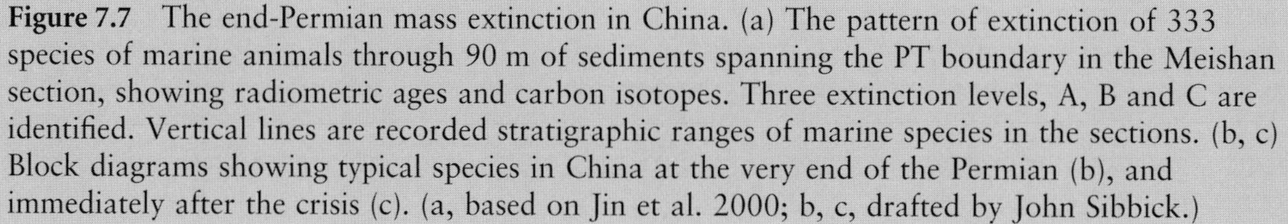
Figure 7.7 The end-Permian mass extinction in China. (a) The pattern of extinction of 333 species of marine animals through 90 m of sediments spanning the PT boundary in the Meishan section, showing radiometric ages and carbon isotopes. Three extinction levels, A, B and C are identified. Vertical lines are recorded stratigraphic ranges of marine species in the sections. (b, c) Block diagrams showing typical species in China at the very end of the Permian (b), and immediately after the crisis (c). (a, based on Jin et al. 2000; b, c, drafted by John Sibbick.)

Studies of sedimentology across the PT boundary in China and elsewhere have shown a dramatic change in depositional conditions. In marine sections, the end-Permian sediments are often bioclastic limestones (limestones made up from abundant fossil debris), indicating optimal conditions for life. Other latest Permian sediments are intensely bioturbated, indicating richly-oxygenated bottom conditions for burrowers. In contrast, sediments deposited immediately after the extinction event, in the earliest Triassic, are dark-colored, often black and full of pyrite. They largely lack burrows, and those that do occur are very small. Fossils of marine benthic invertebrates are extremely rare. These observations, in association with geochemical evidence, suggest a dramatic change in oceanic conditions from well-oxygenated bottom waters to widespread benthic anoxia (Wignall & Twitchett 1996; Twitchett 2006). Before the catastrophe, the ocean fauna was differentiated into recognizably distinct biogeographic provinces. After the event, a cosmopolitan, opportunistic fauna of thin-shelled bivalves, such as the "paper pecten" *Claraia*, and the inarticulated brachiopod *Lingula* spread around the world (see Box 7.2).

Geochemistry gave additional clues. At the PT boundary there is a dramatic shift in oxygen isotope values: a decrease in the value of the $\delta^{18}O$ ratio of about six parts per thousand, corresponding to a global temperature rise of around 6°C. Climate modelers have shown how global warming can reduce ocean circulation, and the amount of dissolved oxygen, to create anoxia on the seabed. A dramatic global rise in temperature is also reflected in the types of sediments and ancient soils deposited on land, and in the plants and reptiles they contain. In many places it seems that soils were washed off the land wholesale. After the event, the few surviving plants were those that could cope with difficult habitats, and virtually the only reptile was the plant-eating dicynodont *Lystrosaurus* (see p. 450). Life was tough in the "post-apocalyptic greenhouse", as it has been called.

So what was the killing model? The key comes from a study of carbon isotopes in marine rocks. They show a sharp negative excursion (see Fig. 7.7a), dropping from a value of +2 to +4 parts per thousand to −2 parts per thousand at the mass extinction level. This drop in the ratio implies a dramatic increase in the light carbon isotope (^{12}C), and geologists and atmospheric modelers have tussled over trying to identify a source. Neither the instantaneous destruction of all life on Earth, and subsequent flushing of the ^{12}C into the oceans, nor the amount of ^{12}C estimated to have reached the atmosphere from the CO_2 released by the Siberian Trap eruptions are enough to explain the observed shift. Something else is required.

That something else might be **gas hydrates**. Gas hydrates are generally formed from the remains of marine plankton that sink to the seabed and become buried. Over millions of years, huge amounts of carbon are transported to the deep oceans around continental margins and the carbon may be trapped as methane in a frozen ice lattice. If the deposits are disturbed by an earthquake, or if the seawater above warms slightly, the gas hydrates may be dislodged and methane is released and rushes to the surface. Because the gas hydrates reside at depth, they are at high pressure, and in the rush to the surface the pressure reduces and they expand sometimes as much as 160 times. The key points are that gas hydrates contain carbon largely in the organic ^{12}C isotopic form, and they may release huge quantities into the atmosphere rapidly.

The assumption is that initial global warming at the end of the Permian, triggered by the huge Siberian eruptions, melted frozen circumpolar gas hydrate bodies, and massive volumes of methane (rich in ^{12}C) rose to the surface of the oceans in huge bubbles. This huge input of methane into the atmosphere caused more warming and this could have melted further gas hydrate reservoirs. So the process continued in a positive feedback spiral that has been termed a "runaway greenhouse" effect. The term "greenhouse" refers to the fact that methane is a well-known greenhouse gas, causing global warming. Perhaps, at the end of the Permian, some sort of threshold was reached, beyond which the natural systems that normally reduce greenhouse gas levels could not operate. The system spiraled out of control, leading to the biggest crash in the history of life.

The current model tracks all the environmental changes back to the eruption of the Siberian Traps (Fig. 7.8). An immediate effect was **acid rain**, as the volcanic gases combined

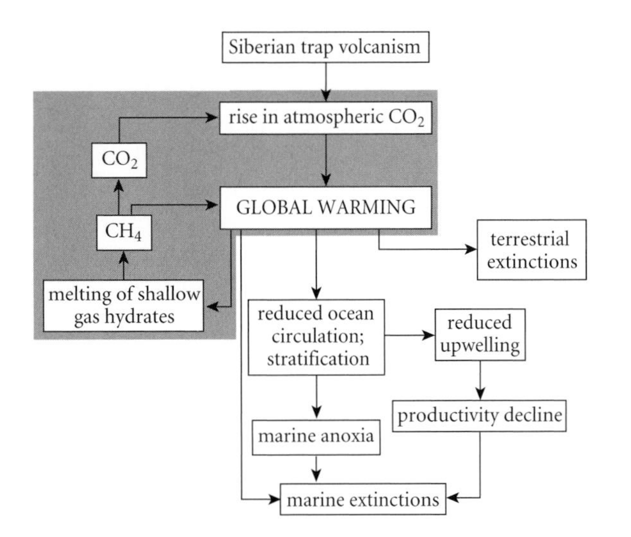

Figure 7.8 The possible chain of events following the eruption of the Siberian Traps, 251 Ma. Volcanism pumps carbon dioxide (CO_2) into the atmosphere and this causes global warming. Global warming leads to reduced circulation and reduced upwelling in the oceans, which produces anoxia, productivity decline and extinction in the sea. Gas hydrates may have released methane (CH_4) which produced further global warming in a "runaway greenhouse" scenario (shaded gray). (Courtesy of Paul Wignall.)

with water in the atmosphere to form a deadly cocktail of sulfuric, carbonic and nitric acids. The acid rain killed the land plants and they were washed away, and this released the soils that were also stripped off the land. With no food, land animals died. The carbon dioxide from the eruptions caused global warming and this perhaps released the gas hydrates, causing further global warming. Warming is often associated with loss of oxygen, and seabeds became anoxic, so killing life in the sea. If this model is correct, it is in some ways more startling than the KT impact because this represents an entirely Earth-bound process when all normal regulatory systems, whether these are part of a Gaia model (see p. 25) or not, broke down. And it all began with global warming . . .

The Cretaceous-Tertiary event

The KT event has been subjected to intense scrutiny since 1980 so much more is known about it than about the PT event. Before 1980, scientists had come up with over 100 theories for what might have happened 65 million years ago. These theories ranged from the reasonable (global climate change, change in plants, impact, plate tectonic movements, sea-level change) to the frankly ludicrous (loss of sexual appetite, increasing stupidity or hormonal imbalance of the dinosaurs, competition with caterpillars for plant food, mammals ate all the dinosaur eggs). A number of serious efforts had been made to document just what happened through the KT interval and to look at environmental and other changes. Then the bombshell struck.

In June 1980, one of the most important papers of the 20th century appeared in *Science*. This paper, by Luis Alvarez and colleagues, made the bold assertion that a 10 km meteorite (asteroid) had hit the Earth, the impact threw up a great cloud of dust that encircled the globe, blacked out the sun, and caused extinction worldwide by stopping photosynthesis in land plants and in phytoplankton. With their plant food gone, the herbivores died out, followed by the carnivores. This simple model was based on limited observational evidence and it was, needless to say, highly controversial.

Luis Alvarez was a physicist who had won a Nobel Prize for his work on subatomic particles. He became involved with his son Walter's geological work in Italy, where a relatively complete rock succession documented the KT boundary in detail. The geological team identified an unusual clay band right at the KT boundary, within a succession of marine limestones. They measured the chemical content of the clay band, and of the rocks above and below, and found an unusual enhancement of the metallic element iridium. This was the famous iridium spike, where the iridium content shot up from normal background levels of 0.1–0.3 parts per billion (ppb) to 9 ppb (Fig. 7.9). Iridium is a platinum-group metal that is rare on the Earth's crust, and reaches the Earth almost exclusively from space, in meteorites. The background low levels represent the results of numerous minor meteorite impacts that go on all the time.

Alvarez proposed that the iridium spike indicated an unusually high rate of arrival of iridium on the Earth's crust, thus a huge meteorite (asteroid) impact. He calculated, working

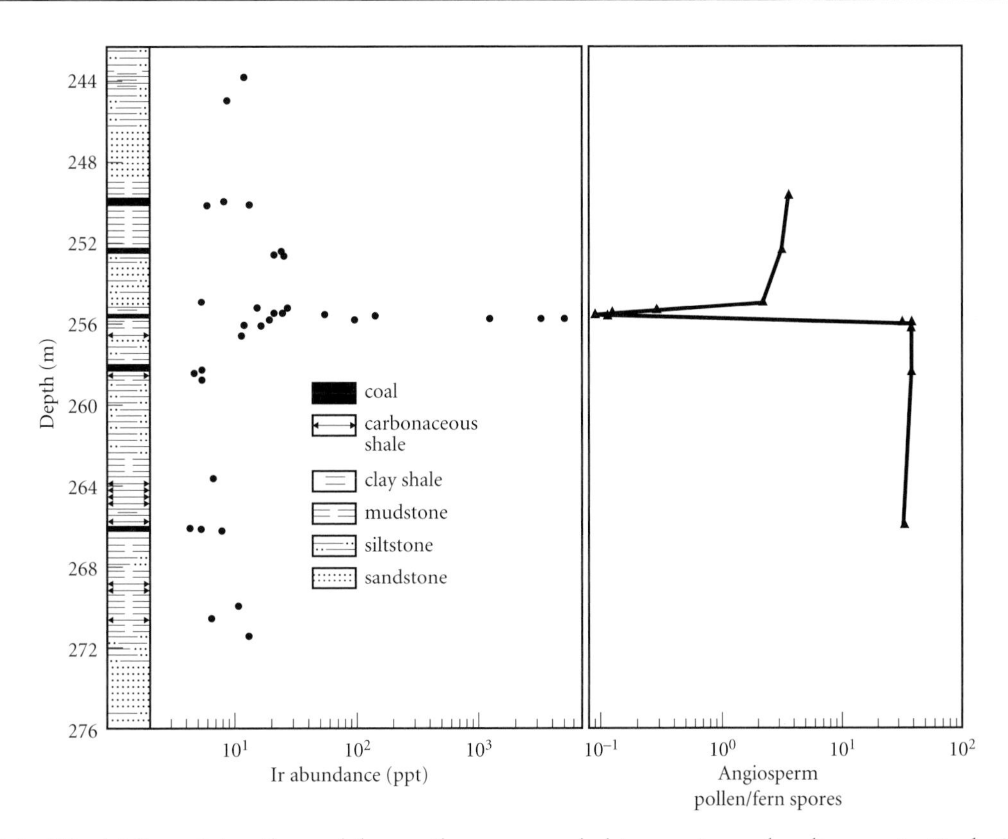

Figure 7.9 The iridium (Ir) spike and fern spike, as recorded in continental sediments in York Canyon, New Mexico. The Ir spike, measured in parts per trillion (ppt), an enhancement of 10,000 times normal background levels, is generally interpreted as evidence for a massive extraterrestrial impact. The fern spike indicates sudden loss of the angiosperm flora, and replacement by ferns. (Based on Orth et al. 1981.)

backwards (Box 7.3), that a killing impact would have to extend its effects worldwide, which meant a dust cloud that encircled the globe. Based on studies of experimental impacts, and on known major volcanic eruptions, he calculated that the crater would have to be 100–150 km across to produce such a large dust cloud, and this implied a meteorite 10 km in diameter. The 1980 *Science* paper attracted instant press coverage on a huge scale, and scientists from all disciplines were alerted to the dramatic new idea immediately.

The Alvarez et al. (1980) paper was hugely controversial, partly because the idea was so outrageous, partly because its chief author was a physicist and not a geologist or paleontologist, and partly because the evidence seemed flimsy in the extreme. But Alvarez and colleagues were vindicated. Since 1980, evidence has piled up that they were right, and indeed in 1991 the crater was identified at Chicxulub in Mexico.

A catastrophic extinction is indicated by sudden plankton and other marine extinctions, and by abrupt shifts in pollen ratios, in certain sections. The shifts in pollen ratios show a sudden loss of angiosperm taxa and their replacement by ferns, and then a progressive return to normal floras. This fern spike (Fig. 7.9), found at many terrestrial KT boundary sections is interpreted as indicating the aftermath of a catastrophic ash fall: ferns recover first and colonize the new surface, followed eventually by the angiosperms after soils begin to develop. This interpretation has been made by analogy with observed floral changes after major volcanic eruptions.

The main alternative to the extraterrestrial **catastrophist model** for the KT mass extinction was the **gradualist model**, in which extinctions were said to have occurred over

Box 7.3 Professor Alvarez's equation

In proposing that the dinosaurs and many other organisms had been killed by an asteroid impact, Luis Alvarez proposed an equation that summarized all the key features of an impact and the blacking-out of the sun. The equation is simple and daring, especially because it is based on limited evidence. This might seem to be a bad thing – surely scientists should be careful? However, sticking your neck out is a good thing for a scientist to do. You have to dare to be wrong; but it helps to be right sometimes as well.

The role of a scientist is to test hypotheses (see p. 4), and that means your own hypotheses have to be open to test by others. The more daring the hypothesis, the easier it would be to disprove. The Alvarez et al. (1980) model for the KT mass extinction was extremely daring and could easily have failed. The fact that it has not been disproved, and indeed that a huge amount of new evidence supports it, makes this a very successful hypothesis.

The Alvarez et al. (1980) formula is:

$$M = \frac{sA}{0.22f}$$

where M is the mass of the asteroid, s is the surface density of iridium just after the time of the impact, A is the surface area of the Earth, f is the fractional abundance of iridium in meteorites, and 0.22 is the proportion of material from Krakatoa, the huge volcano in Indonesia that erupted in 1883, that entered the stratosphere. The surface density of iridium at the KT boundary was estimated as 8×10^{-9} g cm^{-2}, based on the local values at Gubbio, Italy and Stevns Klint, Denmark, their two sampling localities. Measurements of modern meteorites gave a value for f of 0.5×10^{-6}.

Running all these values in the formula gave an asteroid weighing 34 billion tonnes. The diameter of the asteroid was at least 7 km. Other calculations led to similar results, and the Alvarez team fixed on the suggestion that the impacting asteroid had been 10 km in diameter.

Websites about the KT event may be seen at http://www.blackwellpublishing.com/paleobiology/.

long intervals of time as a result of climatic changes. On land, subtropical lush habitats with dinosaurs gave way to strongly seasonal, temperate, conifer-dominated habitats with mammals. Further evidence for the gradualist scenario is that many groups of marine organisms declined gradually through the Late Cretaceous. Climatic changes on land are linked to changes in sea level and in the area of warm shallow-water seas.

A third school of thought is that most of the KT phenomena may be explained by volcanic activity. The Deccan Traps in India represent a vast outpouring of lava that occurred over the 2–3 myr spanning the KT boundary. Supporters of the volcanic model seek to explain all the physical indicators of catastrophe (iridium, shocked quartz, spherules, and the like) and the biological consequences as the result of the eruption of the Deccan Traps. In some interpretations, the volcanic model explains instantaneous catastrophic extinction, while in others it allows a span of 3 myr or so, for a more gradualistic pattern of dying off caused by successive eruption episodes.

The gradualist and volcanic models held sway in the 1980s and 1990s, but increasing evidence for impact has strengthened support for the view expressed in the original Alvarez et al. (1980) paper. The discovery of the Chicxulub Crater, deep in Upper Cretaceous sediments on the Yucatán peninsula, Central America (Fig. 7.10) has been convincing. Melt products under the crater date precisely to the KT boundary, and the rocks around the shores of the proto-Caribbean provide strong support too. For example, sedimentary deposits around the ancient coastline of the proto-Caribbean that consist of massive tumbled

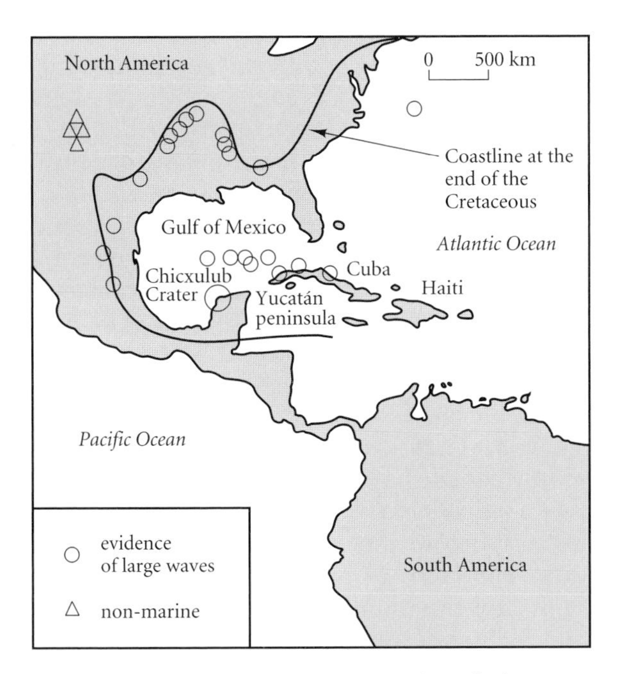

Figure 7.10 The KT impact site identified. Location of the Chicxulub Crater on the Yucatán peninsula, Central America, and sites of tempestite deposits around the coastline of the proto-Caribbean (open circles). Continental KT deposits are indicated by triangles.

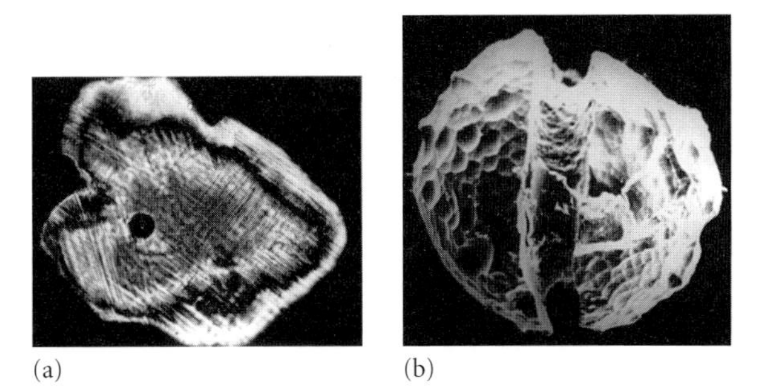

Figure 7.11 Evidence for a KT impact in the Caribbean. (a) Shocked quartz from a KT boundary clay. (b) A glassy spherule from the KT boundary section at Mimbral, northeast Mexico, evidence of fall-out of volcanic melts from the Chicxulub Crater (about 1.5 mm in diameter). (Courtesy of Philippe Claeys.)

and disturbed sedimentary blocks indicate either turbidite (underwater mass flow) or tsunami (massive tidal wave) activity, presumably set off by the vast impact. Further, the KT boundary clays ringing the site also yield abundant shocked quartz (Fig. 7.11a), grains of quartz bearing crisscrossing lines produced by the pressure of an impact. In addition, the KT boundary clays within 1000 km of the impact site also contain glassy spherules (Fig. 7.11b) that have a unique geochemistry. Volcanoes can produce glassy spherules – melt products of the igneous magma – deep in the heart of the volcano. The KT spherules, though, have the same geochemistry as limestones and evaporites, sedimentary rocks that lay on the seafloor of the proto-Caribbean, so the volcanic hypothesis cannot explain them. Sedimentary rocks can be melted only by an unusual process such as a direct hit by an asteroid. Farther afield, the boundary layer is thinner, there are no turbidite/tsunami deposits, spherules are smaller or absent, and shocked quartz is less abundant.

There has been considerable debate about the exact dating of the impact layers. Some evidence suggests that the Chicxulub impact happened up to 300,000 years before the KT boundary and extinction level. This is hotly debated and the idea has been rejected by many paleontologists. But, if the impact happened at a different time from the main pulse of extinction, then the simple KT killing model would have to be revised.

Thus, the geochemical and petrological data such as the iridium anomaly, shocked quartz and glassy spherules, as well as the Chicxulub Crater give strong evidence for an impact on Earth 65 million years ago. Paleontological data support the view of instantaneous extinction, but some still indicate longer-term extinction over 1–2 myr. Key research questions are whether the long-term dying-off is a genuine pattern, or whether it is partly an artifact of incomplete fossil collecting, and, if the impact occurred, how it actually caused the patterns of extinction. Available killing models are either biologically unlikely, or too catastrophic: recall that a killing scenario must take account of the fact that 75% of families survived the KT event, many of them seemingly unaffected. Whether the two models can be combined so that the long-term declines are explained by gradual changes in sea level and climate and the final disappearances at the KT boundary were the result of impact-induced stresses is hard to tell.

EXTINCTION THEN AND NOW

Extinction events

Somewhere between background extinction and mass extinction have been many times when rather large numbers of species have died out, but perhaps only in one part of the world, or perhaps affecting only one or two ecological groups. These medium-sized extinctions are often classed together as extinction events, but clearly each one is different. Many extinction events have been identified (see Fig. 7.2), and some of the better-known ones are noted briefly here.

The first is the *Ediacaran event*, about 542 Ma, which is ill defined in terms of timing, but it marks the end of the Ediacaran animals (see pp. 242–7). Some Ediacaran beasts may have survived into the Cambrian, but the majority of those strange quilted jellyfish-like, frond-like and worm-like creatures disappeared, and the way was cleared for the dramatic radiation of shelly animals at the beginning of the Cambrian. Because of the antiquity of this proposed mass extinction, it is hard to be sure that all species became extinct at the same time, and some would argue that this was not a mass extinction at all. Causes are equally debated, with some evidence for a nutrient crisis or a major temperature change. An older putative mass extinction, at the start of the Ediacaran, some 650 Ma, might have been triggered by global cooling, the "snowball Earth" model (see p. 112), but this is equally debated.

An extinction at the end of the *Early Cambrian* marked the disappearance of previously widespread archaeocyathan reefs (see p. 268).

A series of extinction events occurred during the *Late Cambrian*, perhaps as many as five, in the interval from 513 to 488 Ma. There were major changes in the marine faunas in North America and other parts of the world, with repeated extinctions of trilobites. Following these, animals in the sea became much more diverse, and groups such as articulated brachiopods, corals, fishes, gastropods and cephalopods diversified dramatically during the great Ordovician radiation (see p. 253).

There were many further extinction events or turnover events in the Paleozoic, between the Late Devonian and PT mass extinctions, including a substantial extinction phase between the Middle and Late Permian, some 10 myr before the PT event. This Middle–Late Permian extinction, the *end-Guadalupian event*, may turn out to be a mass extinction in its own right. Numerous marine and nonmarine groups were hard-hit at that time, and it has been hard to identify until recently because its effects were sometimes confused with the end-Permian event, because of lack of clarity about dating.

There were further such events at the end of the Early Triassic and in the Late Triassic. The Late Triassic extinction event, more commonly called the *Carnian-Norian event* (after the stratigraphic stages) occurred some 15–20 myr before the end-Triassic mass extinction. The Carnian-Norian event was marked by turnovers among reef faunas, ammonoids and echinoderms, but it was particularly important on land. There were large-scale changeovers in floras, and many amphibian and reptile groups disappeared, to be followed by the dramatic rise of the dinosaurs and pterosaurs. At this time, many modern groups arrived on the scene, such as turtles, crocodilians, lizard ancestors and mammals. The cause of these events may have been climatic changes associated with continental drift. At that time, the supercontinent Pangaea (see p. 48) was beginning to break up, with the unzipping of the Central Atlantic between North America and Africa.

Extinctions during the *Jurassic* and *Cretaceous* periods were minor. The Early Jurassic and end-Jurassic events involved losses of bivalves, gastropods, brachiopods and ammonites as a result of major phases of anoxia. Free-swimming animals were unaffected, and the events are undetectable on land – they may be partly artificial results of incomplete data recording. Events have been postulated also in the Mid Jurassic and in the Early Cretaceous, but they are hard to determine. The *Cenomanian-Turonian* extinction event some 94 Ma, associated with extinctions of some planktonic organisms, as well as the bony fishes and ichthyosaurs that fed on them, is probably associated with sea-level change.

Extinctions since the KT event have been more modest in scope. The *Eocene-Oligocene* events 34 Ma were marked by extinctions among plankton and open-water bony fishes in the sea, and by a major turnover among mammals in Europe and North America. Later *Cenozoic* events are less well defined. There was a dramatic extinction among mammals in North America in the mid-Oligocene, and minor losses of plankton in the mid-Miocene, but neither event was large. Planktonic extinctions occurred during the Pliocene, and these may be linked to disappearances of bivalves and gastropods in tropical seas.

The latest extinction event, at the end of the *Pleistocene*, while dramatic in human terms, barely qualifies for inclusion. As the great ice sheets withdrew from Europe and North America, large mammals such as mammoths, mastodons, woolly rhinos and giant ground sloths died out. Some of the extinctions were related to major climatic changes, and others may have been exacerbated by human hunting activity. The loss of large mammal species was, however, minor in global terms, amounting to a total loss of less than 1% of species.

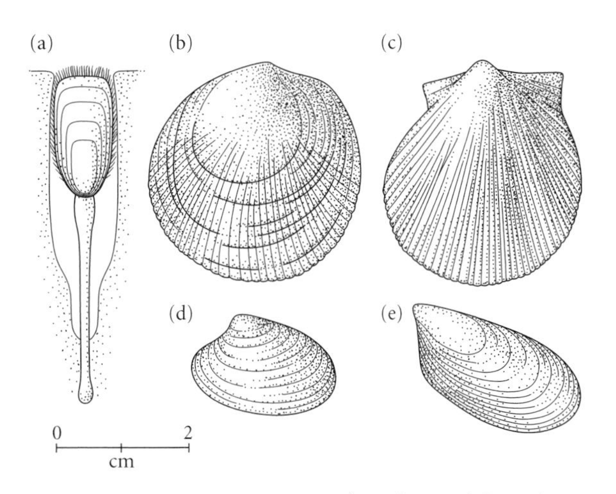

Figure 7.12 Disaster taxa after the end-Permian mass extinction: the brachiopod *Lingula* (a), and the bivalves *Claraia* (b), *Eumorphotis* (c), *Unionites* (d) and *Promyalina* (e). These were some of the few species to survive the end-Permian crisis, and they dominated the black anoxic seabed mudstones for many thousands of years after the event.

Recovery after mass extinctions

After mass extinctions, the recovery time is proportional to the magnitude of the event. Biotic diversity took some 10 myr to recover after major extinction events such as the Late Devonian, the end-Triassic and the KT. Recovery time after the massive PT event was much longer: it took some 100 myr for total global marine familial diversity to recover to pre-extinction levels. Species-level diversity may have recovered sooner, perhaps within 20 or 30 myr, by the Late Triassic. But the deeper diversity of body plans represented by the total number of families took much longer.

It is becoming clear that all the rules change after a profound environmental crisis (Jablonski 2005). **Disaster taxa** prove the point (Fig. 7.12). These are species that, for whatever reason, are able to thrive in conditions that make other species quail. Stromatolites, for example, in marine environments and ferns on land make sudden but brief appearances. After the PT crisis, the inarticulated brachiopod *Lingula* flourished for a brief spell, before retiring to the wings. *Lingula* is sometimes called a "living fossil" because it is a genus that has been known for most of the past 500 myr, and it lives today in low-oxygen estuarine muds. Other post-extinction disaster taxa in the earliest Triassic are the bivalves *Claraia*, *Unionites* and *Promyalina*, found in black, anoxic shales everywhere. These animals could presumably cope with poorly oxygenated waters.

Bivalves and brachiopods diversified slowly in the next 5–10 myr, as did the ammonoids. But other groups had gone forever. The rugose and tabulate corals and other Late Permian reef-builders had been obliterated. The "reef gap" following the PT mass extinction is profound evidence for a major environmental crisis. The rich tropical reefs of the Late Permian had all gone, and nothing faintly resembling a coral reef was seen for 10 myr after the event. When the first tentative reefs reassembled themselves in the Middle Triassic, they were composed of a motley selection of Permian survivors, a few species of bryozoans, stony algae and sponges. It took another 10 myr before corals began to build true structural reefs (see p. 289).

The reef gap in the sea is paralleled by the "coal gap" on land. Coals are formed from dead plants, and there were rich coal deposits formed through the Carboniferous and Permian, indicating the presence of lush forests. After the acid rain had cleared the land of plant life, no coal formed during the first 20–25 myr of the Triassic. It was only in the Late Triassic that forests reappeared. Tetrapods on land had been similarly affected, and ecosystems remained incomplete and unbalanced through the Early and Middle Triassic until they rebuilt themselves in the Late Triassic with dinosaurs and other new groups (see p. 454).

Life recovers slowly after mass extinctions. A flurry of evolution happens initially among disaster taxa, species that can cope with harsh conditions and that can speciate fast. These disaster taxa are then replaced by other species that last longer and begin to rebuild the complex ecosystems that existed before the mass extinction. The mass extinction crisis may have affected life in two ways: conditions after the event may have been so harsh that nothing could live, and the crisis probably knocked out all normal ecological and evolutionary processes.

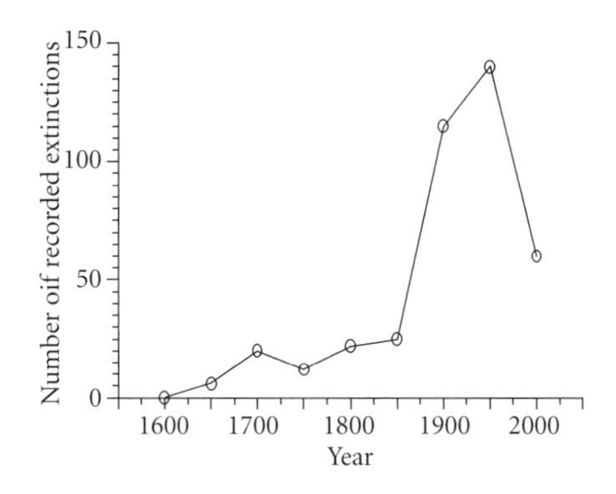

Figure 7.13 The rate of historic extinctions of species for which information exists, counted in 50-year bins. Note the rapid rise in numbers of extinctions in the period 1900–1950; the apparent drop in the period 1950–2000 is artificial because complete counts have not been made for that 50-year period yet.

Extinction today

We started this chapter with the dodo, a representative of how humans cause extinction. There is no question that the extinction of the dodo was regrettable, as is the extinction of any species. But where should we stand on this? Some commentators declare that we are in the middle of an irreversible decline in species numbers, that humans are killing 70 species a day, and that most of life will be gone in a few hundred years. Others declare that extinction is a normal part of evolution, and that there is nothing out of the ordinary happening.

The present rate of extinction can be calculated for some groups from historic records. For birds and mammals, groups that have always been heavily studied, the exact date of extinction of many species is known from historic records. The last dodo was seen on Mauritius in 1681. By 1693, it was gone, prey to passing sailors who valued its flesh, despite the fact that it was "hard and greasie". The last Great auks were collected in the North Atlantic in 1844 – ironically, the last two Great auks were beaten to death on Eldey Island off Iceland by natural history collectors. Some sightings were reported in 1852, but these were not confirmed.

Human activity has not simply caused the extinction of rare or isolated birds. The last Passenger pigeon, named Martha, died at Cincinnati Zoo in 1914. Only 100 years earlier, the great ornithologist John James Audubon, had reported a flock of Passenger pigeons in Kentucky that took 3 days to go by. He estimated that the birds passed him at the rate of 1000 million in 3 h. The sky was black with them in all directions. They were wiped out by a program of systematic shooting, which, at its height, blackened the landscape with Passenger pigeon carcasses as far as the eye could see.

These datable extinctions can be plotted (Fig. 7.13) to show the rates of extinction of birds, mammals and some other groups in historic time. The current rate of extinction of bird species is 1.75 per year (about 1% of extant birds lost since 1600). If this rate of loss is extrapolated to all 20–100 million living species, then the current rate of extinction is 5000–25,000 per year, or 13.7–68.5 per day. With 20–100 million species on Earth, this means that all of life, including presumably *Homo sapiens*, will be extinct in 800–20,000 years. These figures are startling and they are often quoted to compare the present rate of species loss to the mass extinctions of the past.

A reasonable response to this calculation would be to query the annual loss figure and the validity of extrapolating. The birds that have been killed so far are mainly vulnerable species that lived in small populations on single islands (e.g. the dodo) or in extreme conditions (e.g. the Great auk). Perhaps more widespread species such as pigeons, sparrows and chickens will survive such depredations? But recall the Passenger pigeon – it should have been immune to extinction. The other point is to query whether it is right to extrapolate the figures from bird and mammal extinctions to the rest of life. Species of birds and mammal are short-lived (i.e. they evolve fast), and perhaps their extinction rates are not appropriate for insects and plants, for example.

The jury is still out on modern extinction. It is clear that surging human population and increasing tension between development and ecology put pressure on natural habitats and on species. Plants and animals are dying out faster now than at times in the past when the global human population was smaller. Paleontologists and ecologists have an important job to do in seeking to understand just what the threats are and how fast the modern extinction is proceeding.

Review questions

1 How do paleontologists and other earth scientists study mass extinctions? Carry out a census of papers about the Permo-Triassic event published in the last year. Find the first 50 papers using any bibliographic search tool, and classify them by broad theme (paleontology, stratigraphy, geochemistry, atmospheric modeling, volcanology), geographic region (perhaps by continents), sedimentary regime (marine, terrestrial) and key conclusion about the extinction model (eruption of Siberian Traps, gas hydrate release, acid rain, anoxia, meteorite impact). How are our views perhaps biased by limited geographic coverage, a major focus on marine rocks and dominant academic discipline? Are these biases to be expected, and why?

2 Is there any evidence that the media distorts research agendas? Look at news stories about the KT event, and consider the balance of reporting of different aspects: do a census of the animal and plant groups mentioned in the first 50 news reports you encounter.

3 Investigate one of the "other" mass extinctions not covered in detail here: end-Ordovician, Late Devonian and end-Triassic.

4 Calculate the relative magnitudes of the big five events from Jack Sepkoski's database of fossil genera, either through http://strata.ummp.lsa.umich.edu/jack/ or http://geology.isu.edu/FossilPlot/.

5 Why is the current loss of species on Earth sometimes termed the "sixth extinction"?

Further reading

Benton, M.J. 2003. *When Life Nearly Died*. W.W. Norton, New York.

Benton, M.J. & Twitchett, R.J. 2003. How to kill (almost) all life: the end-Permian extinction event. *Trends in Ecology and Evolution* **18**, 358–65.

Briggs, D.E.G. & Crowther, P.R. 2001. *Palaeobiology, A Synthesis*, 2nd edn. Blackwell, Oxford, UK.

Erwin, D.H. 2006. *Extinction: How Life on Earth Nearly Ended 250 Million Years Ago*. Princeton University Press, Princeton, NJ.

Gotelli, N.J. & Colwell, R.K. 2001. Quantifying biodiversity: procedures and pitfalls in the measurement and comparison of species richness. *Ecology Letters* **4**, 379–91.

Hallam, A. & Wignall, P.B. 1997. *Mass Extinctions and their Aftermath*. Oxford University Press, Oxford, UK.

Hammer, Ø. & Harper, D.A.T. 2005. *Paleontological Data Analysis*. Blackwell Publishing, Oxford, UK.

Jablonski, D. 2005. Mass extinctions and macroevolution. *Paleobiology* **31**, 192–210.

Taylor, P. 2004. *Extinctions in the History of Life*. Cambridge University Press, Cambridge, UK, 204 pp.

References

Alvarez, L.W., Alvarez, W., Asaro, F. & Michel, H.V. 1980 Extraterrestrial cause for the Cretaceous-Tertiary extinction. *Science* **208**, 1095–108.

Bambach, R.K. 2006. Phanerozoic biodiversity mass extinctions. *Annual Review of Earth and Planetary Sciences* **34**, 127–55.

Benton, M.J. 1995. Diversification and extinction in the history of life. *Science* **268**, 52–8.

Hammer, Ø. & Harper, D.A.T. 2005. *Paleontological Data Analysis*. Blackwell Publishing, Oxford, UK.

Jablonski, D. 2005. Mass extinctions and macroevolution. *Paleobiology* **31**, 192–210.

Jin, Y.G., Wang, Y., Wang, W., Shang, Q.H., Cao, C.Q. & Erwin D.H. 2000. Pattern of marine mass extinction near the Permian-Triassic boundary in South China. *Science* **289**, 432–6.

Keller, G., Barrera, E., Schmitz, B. & Mattson, E. 1993. Gradual mass extinction, species survivorship, and long-term environmental changes across the Cretaceous-Tertiary boundary in high latitudes. *Bulletin of the Geological Society of America* **105**, 979–97.

McKinney, M.L. 1995. Extinction selectivity among lower taxa – gradational patterns and rarefaction error in extinction estimates. *Paleobiology* **21**, 300–13.

Orth, C.J., Gilmore, J.S., Knight, J.D., Pillmore, C.L., Tschudy, R.H. & Fassett, J.E. 1981. An Ir abundance anomaly at the palynological Cretaceous-Tertiary boundary in northern New Mexico. *Science* **214**, 1341–3.

Raup, D.M. 1979. Size of the Permo-Triassic bottleneck and its evolutionary implications. *Science* **206**, 217–18.

Raup, D.M. & Sepkoski Jr., J.J. 1984. Periodicities of extinctions in the geologic past. *Proceedings of the National Academy of Sciences, USA* **81**, 801–5.

Rohde, R.A. & Muller, R.A. 2005. Cycles in fossil diversity. *Nature* **434**, 208–10.

Twitchett, R.J. 2006. The Late Permian mass extinction event and recovery: biological catastrophe in a greenhouse world. *In* Sammonds, P.M. & Thompson, J. M.T. (eds) *From Earthquakes to Global Warming*. Royal Society Series on Advances in Science No. 2. World Scientific Publishing, Hackensack, NJ, pp. 69–90.

Wignall, P.B. & Twitchett, R.J. 1996 Oceanic anoxia and the end Permian mass extinction. *Science* **272**, 1155–8.

Chapter 8

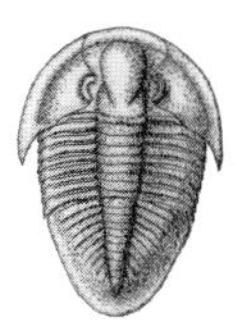

The origin of life

Key points

- Life originated by fusion of organic molecules in the first billion years after the formation of the Earth.
- The precursor to living cells may have been self-replicating RNA; a time before life originated termed "RNA world".
- Photosynthesis by a group of bacteria, called cyanobacteria, generated molecular oxygen (O_2), and the atmosphere became oxygenated at a low level 2.4 Ga. Later, oxygen levels increased further, around 0.8–0.6 Ga.
- The universal tree of life, reconstructed from gene sequencing of modern organisms, shows there are three great domains: Bacteria, Archaea and Eucarya. The first two are prokaryotes, the last eukaryotes.
- The earliest fossils are bacteria in rocks up to 3.2 Ga, indicated by stromatolites, structures built by alternating algal mats and sediment layers.
- Cellular fossils 3.5 Ga are highly controversial; the first widely accepted cellular fossils date from 2.5 Ga.
- Biomarkers, notably lipids, provide evidence for cyanobacteria and eukaryotes 2.7 Ga.
- The oldest eukaryotes, cells with a nucleus and organelles, date back perhaps 1.9 Ga.
- Red algae from 1.2 Ga show that sex had originated – they show mitosis, but also meiosis, which is unique to sexual reproduction.
- Together with sex came multicellularity, the possession of many, often specialized, cells, first seen in 1.2 Ga red algae.

Life is improbable, and it may be unique to this planet, but nevertheless it did begin and it is thus our task to discover how the miracle happened.

Euan Nisbet (1987) *The Young Earth*

Origins are among the deepest questions: Where did humans come from? Where did life itself come from? The most ancient philosophers could see that the world is made of living and non-living things, and they wanted to know where the spark of life came from. How do you go from a non-living thing, like a rock or a glass of water, to a living thing, like a plant or an animal?

These early speculations led to many **creation myths**, stories about how the non-living to living transition might have taken place. Creation myths are common to many religions, and they explain the origin of life by divine intervention. These ideas are not scientific, however, because they cannot be tested. We explored the issue of creationism in Chapter 5.

The current scientific view is that life arose on the Earth some time before 3.5 Ga (Ga = giga years old, or 1000 Ma). In rocks from Australia and South Africa dated at around 3.5 Ga, isotopes of carbon are consistent with the presence of a marine biosphere that preferentially incorporated the carbon-12 (C^{12}) isotope into organic matter relative to C^{13}. The first organisms were simple, single-celled **prokaryotes** similar to modern microbes. More complex cells, **eukaryotes**, arose only later, perhaps 2.7 Ga, and much later than that came the first true plants and animals. This means that the first three-quarters of the history of life passed by in the company of organisms that were neither plant nor animal.

In this chapter, we look first at different ways of explaining origins. Then, we go on to look at the diversity of evidence about when and how life arose. We concentrate on the geological and fossil evidence, of course, but include some necessary molecular biology and biochemistry as well.

THE ORIGIN OF LIFE

Scientific models

There have been many scientific models for the origin of life, some of them now rejected by the evidence, and others still available as potentially valid hypotheses:

1 Spontaneous generation.
2 Inorganic model.
3 Extraterrestrial origins.
4 Biochemical model.
5 Hydrothermal model.

Medieval scholars believed that many organisms sprang into life directly from non-living matter, a form of **spontaneous generation**. For example, frogs were said to arise from the spring dew and maggots were said to come to life in rotting flesh. However, careful tests proved that there was no truth in these ideas. Louis Pasteur in 1861 enclosed pieces of meat in airtight containers, and maggots did not appear. He showed that flies laid their eggs on rotting meat, the eggs hatched as maggots and the maggots then turned into flies. So, the idea of the origin of life by spontaneous generation is a scientific hypothesis because it may be tested, but it turns out to have been wrong. It is important to realize that scientific and non-scientific do not mean "right" and "wrong": science is about testing and rejecting alternate hypotheses until one remains that is not rejected.

The **inorganic model** for the origin of life is that complex organic molecules arose gradually on a pre-existing, non-organic replication platform – silicate crystals in solution. Silicate crystals, clay minerals, were subject to selection pressures on the ancient seabed, and then organic molecules became involved and the inorganic selection became organic. This view has been championed vigorously by Graham Cairns-Smith of Glasgow University, but it has not gained widespread support. The first experiments to test the model were carried out in 2007, but they were not conclusive.

The **extraterrestrial model** is that the building blocks for life were seeded on Earth from outer space. Simple molecules, such as hydrogen cyanide, formic acid, aldehydes and acetylenes are found in certain classes of meteorites called **carbonaceous chondrites**, as well as in comets, and these chemicals might have been delivered to the surface of the Earth during a phase of massive meteorite bombardment about 3.8 Ga. In other, more extreme, forms of this hypothesis, DNA might even exist in space, or life in its entirety might have evolved elsewhere in the universe, and was seeded on the Earth during the Precambrian.

Collectively, these views have sometimes been called "panspermia", meaning "universal seeding". The **panspermia model** received

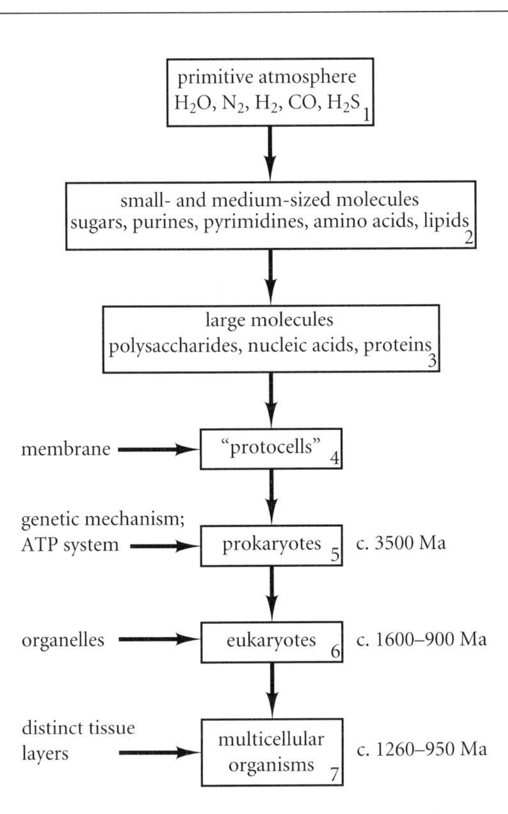

Figure 8.1 The biochemical theory for the origin of life, as proposed by I. A. Oparin and J. B. S. Haldane in the 1920s. Biochemists have achieved steps 1–3 in the laboratory, but scientists have so far failed to create life. ATP, adenosine triphosphate.

a boost in 1996 when David McKay and a team from NASA announced that they had identified fossil bacteria and organic chemical traces of former life in a Martian meteorite. These findings have, however, been disputed vigorously, and the initial excitement has waned. It is hard to see how extraterrestrial/panspermia models for the origin of life could be tested decisively and, in any case, positing the origin of life on another planet still leaves open the question of how that life originated.

The **biochemical model** for the origin of life was developed in the 1920s independently by a Russian biochemist, A. I. Oparin, and a British evolutionary biologist, J. B. S. Haldane. They argued that life could have arisen through a series of organic chemical reactions that produced ever more complex biochemical structures (Fig. 8.1). They proposed that common gases in the early Earth atmosphere combined to form simple organic chemicals, and that these in turn combined to form more complex molecules. Then, the complex molecules became separated from the surrounding medium, and acquired some of the characters of living organisms. They became able to absorb nutrients, to grow, to divide (reproduce) and so on.

The **hydrothermal model** is a recently proposed modification to the Oparin–Haldane biochemical model (Nisbet & Sleep 2001). According to this view, the **last universal common ancestor** of life (sometimes abbreviated as LUCA) was a **hyperthermophile**, a simple organism that lived in unusually hot conditions. The transition from isolated amino acids to DNA (Fig. 8.1) may then have happened in a hot-water system associated with active volcanoes. There are two main kinds of hot-water systems on Earth today, hot pools and fumaroles fed by rainwater that are found around active volcanoes, and black smokers in the deep ocean. Black smokers arise along mid-ocean ridges, where new crust is being formed from magma welling up as major oceanic plates move apart (see p. 42). Seawater leaks down into the crust carrying sulfur as sulfate, mixes with molten magma and emerges as superheated steam, with the sulfur now concentrated as sulfide. As minerals precipitate in the cooler sea bottom waters, they color the emerging hot-water plume black. Black smokers are too hot as a site for the origin of life, but the other kinds of hydrothermal systems are less extreme.

This leaves us the Oparin–Haldane biochemical model as a broad-brush picture of how life might have originated, and the hydrothermal model as a specific aspect. How far have scientists been able to test the biochemical model?

Testing the biochemical model

In cartoons and pop fiction, the white-coated scientist is seen in a laboratory full of mysterious bubbling glass vessels, and he declares, "I've just created life". Could this be true? How far have the experiments gone along the chain of organic synthesis that is postulated in the biochemical model for the origin of life (see Fig. 8.1)?

It took some years before the first laboratory results were obtained. The Oparin–Haldane biochemical model was proposed in the 1920s, but nobody tested it seriously until the 1950s. In 1953, Stanley Miller, then a student at the University of Chicago, made a model of the Precambrian atmosphere and ocean in a laboratory glass vessel. He exposed a mixture of water, nitrogen, carbon monoxide and nitrogen to electric sparks, to mimic lightning, and found a brownish sludge in the bottle after a few days. This contained sugars, amino acids and nucleotides. So, Miller had apparently recreated step 2 in the sequence (see Fig. 8.1). However, nowadays most researchers consider the mixture of gases that Miller used (with high percentage concentrations of H_2 and CH_4) to have been too strongly chemically reducing to represent a likely atmosphere for the early Earth. Atmospheric hydrogen is ultimately replenished from the mixture of gases released from the solid Earth, but the geochemistry of the subsurface means that the mixture generally should contain the oxidized form of hydrogen (i.e. water vapor, H_2O) rather than the large proportion of H_2 in Miller's atmosphere.

Further experiments in the 1950s and 1960s led to the production of polypeptides, polysaccharides and other larger organic molecules (step 3). Sidney Fox at Florida State University even succeeded in creating cell-like structures, in which a soup of organic molecules became enclosed in a membrane (step 4). His "protocells" seemed to feed and divide, but they did not survive for long.

Could scientists ever show how non-living protocells could become living? Did this happen in one jump or was there an intermediate stage?

RNA world

Biochemists and molecular biologists have worried about the transition from non-living to living; it is hard to see how bacterial cells could form from non-living chemicals in one step. What then could have been the transitional form of "precellular" life? The most widely accepted view today is that RNA is the precellular entity, and the time between nonlife and life has been termed the "RNA world".

RNA, or **ribonucleic acid**, is one of the nucleic acids and it has key roles in **protein synthesis**. Proteins are manufactured within the nucleus of eukaryotic cells, and within the cell mass of prokaryotic cells. The **genetic code**, the basic instructions that contain all the information to construct a living organism, is encoded in the DNA (**deoxyribonucleic acid**) strands that make up the chromosomes. There are several different forms of RNA that have different functions: one type acts as the template for the translation of genes into proteins, another transfers amino acids to the **ribosome** (the cell organelle where protein synthesis takes place) to form proteins, and a third type translates the transcript into proteins.

In 1968, Francis Crick (1916–2004), who co-discovered the double-helix structure of DNA in 1953 with James Watson, suggested that RNA was the first genetic molecule. He argued that RNA must have the unique property of acting both as a gene and an enzyme, so RNA on its own could act as a precursor of life. When Harvard molecular biologist Walter Gilbert first used the term "RNA world" in 1986, the concept was controversial. But the first evidence came soon after when Sidney Altman and Thomas Cech independently discovered a kind of RNA that could edit out unnecessary parts of the message it carried before delivering it to the ribosome. Because RNA was acting like an enzyme, Cech called his discovery a **ribozyme**. This was such a major discovery that the two were awarded the Nobel Prize for Chemistry in 1989; Altman and Cech had confirmed part of Crick's prediction.

Since 1990, numerous labs have been chasing evidence for the RNA world. For example, Jack Szostak and colleagues at Massachusetts General Hospital in Boston argued that the first RNA molecules on the **prebiotic** ("before life") Earth were assembled randomly from nucleotides dissolved in rock pools (Szostak et al. 2001). Among the millions of short RNA molecules, there would have been one or two that could copy themselves, an ability that soon made them the dominant RNA on the planet. To take this forward to create a living cell, Szostak identified two stages: (i) the production of a protocell by the combination of an RNA replicase and a self-replicating vesicle; and (ii) the production of a cell by the addition of a living function (Fig. 8.2).

Simply proving that RNA could act as gene and enzyme was one thing; however, a single

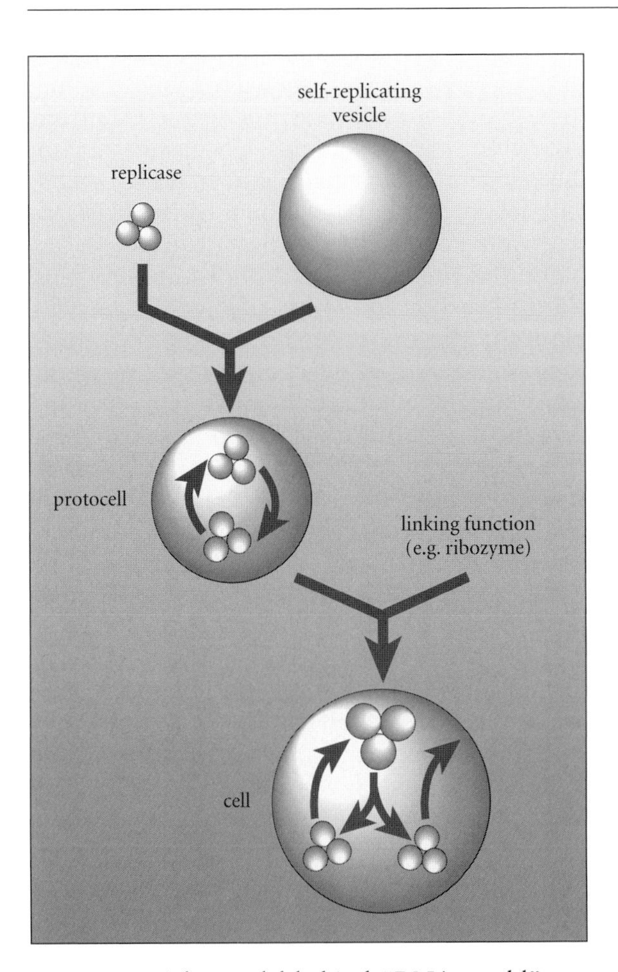

Figure 8.2 The model behind "RNA world", where an RNA replicase and a self-replicating membrane-bound vesicle combine to form a protocell. Inside the vesicle, the RNA replicase functions, and might add a function to improve the production of the vesicle wall through a ribozyme. At this point, the RNA replicase and the vesicle are functioning together, and the protocell has become a living cell, capable of nutrition, growth, reproduction and evolution. Read a general introduction to RNA world at http://www.blackwellpublishing.com/paleobiology/. (Based on information in Szostak et al. 2001.)

molecule cannot both replicate and trigger that replication. The minimum requirement is that two RNA molecules interact, one to act as the enzyme to bring together the components, and the other to act as the gene/template. Together the template and the enzyme RNA combine as an **RNA replicase**. But these components have to be kept together inside some form of compartment or cell, otherwise they would only occasionally come into contact to work together. Szostak and colleagues then proposed there must be a second precellular structure they call a **self-replicating vesicle**, a membrane-bound structure composed mainly of **lipids** (organic compounds that are not soluble in water, including fats) that self-replicates, or grows and divides from time to time. The RNA replicase at some point entered a self-replicating vesicle, and this allowed the RNA replicase to function efficiently.

This is a protocell, but it is not yet living. It is just a self-replicating membrane bag with an independent self-replicating molecule inside. To make the protocell function as an integrated cell, the RNA replicase has to carry out a function that benefits the membrane component. For example, the RNA replicase might generate lipids for the membrane through the medium of a ribozyme. With the membrane keeping the RNA replicase together and so improving its function, and the RNA replicase producing lipids for the membrane, the protocell has become a cell. The two functions are coupled, and the cell can evolve, as vesicles with improved ribozymes can grow and divide, and become more abundant than others. So, we have life and we have evolution. The cell is alive because it has the ability to feed itself, to grow and to replicate. Evolution can happen because the cells show differential survival ("survival of the fittest"), and the genetic information for replication is coded in the RNA.

A number of researchers have carried out experiments to explore all these steps in the hypothetical RNA world model. They have succeeded in evolving ribozymes capable of a broad class of catalytic reactions, including linking components of RNA and lipid molecules, and over time the molecules are selected to perform more efficiently. Much work has yet to be done to show how the whole process could have worked, especially to improve the efficiency and accuracy of copying from the template. The other aspect of the model is the self-replicating vesicle. Experiments here have focused on simple physical models for how oily droplets might incorporate free-floating lipids, and so grow, and then how the droplets or vesicles might divide when they reach a certain size or when external forces are applied, perhaps by the movement of waves in the water. The experiments are complex, and investigators are continuing to explore the behavior of simple RNA replicase, self-replicating vesicles and how the two could come to function together (Szostak et al. 2001).

If the RNA world existed, when was this and for how long? The Earth had to be cool enough for the organic elements to survive being burned off, and the RNA world must pre-date any traces of modern forms of life. Some estimate that this might have been a time of 100–400 myr, somewhere between 4.0 and 3.5 Ga.

EVIDENCE FOR THE ORIGIN OF LIFE

The Early Precambrian world

The Precambrian is divided into the Hadean, Archaean and Proterozoic eons. The Hadean Eon spans from the origin of the Earth, 4.57 to about 4 Ga (Fig. 8.3). At first, the Earth

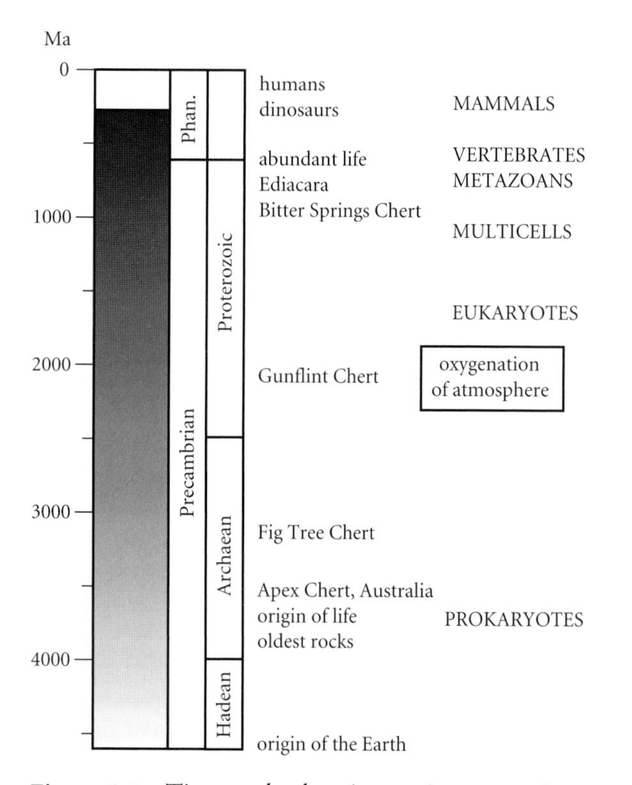

Figure 8.3 Time scale showing major events in the history of the Earth and of life. Most of the time scale is occupied by the Precambrian, whereas the well-known fossil record of the Phanerozoic (Phan.) accounts for only one-seventh of the history of life.

was a molten mass, but it cooled, separating into an outer cool crust and an inner molten mantle and core. Massive volcanic eruptions produced great volumes of gases: carbon dioxide, nitrogen, water vapor and hydrogen sulfide. At the very beginning of the Hadean, temperatures on the Earth's surface were too high, and the crust was too unstable for any form of carbon-based life to exist. During the Hadean, the cratering record on the moon suggests that there were a few ocean-vaporizing impacts on Earth – impacts from large comets or asteroids that would have provided enough energy to turn the ocean into steam. Thus, if life had got started in the early Hadean, it would have been wiped out, only to start afresh. Also, smaller impacts at the end of the Hadean would have destroyed life on the surface of the Earth; only microbes that could stand high temperatures living in the subsurface would have survived.

As the Earth's surface cooled, the **lithosphere**, its rocky crust, began to differentiate as a cooler upper layer above the underlying **asthenosphere**. As the rocky lithosphere formed, magma convection became restricted to the asthenosphere, and the upper crust formed plates that were moved by mantle convection. This marks the beginning of plate tectonics (see p. 42). Heat loss from the Earth now happened mainly round the margins of these early plates, and black smokers, associated with hydrothermal activity, began to form.

The oldest rocks are from Canada and are dated at 3.8–4 Ga, and some mineral grains from Australia have even been dated to 4.4 Ga.

The Archaean Eon lasted from about 4 to 2.5 Ga. The oldest **sedimentary** rocks have been reported from the Isua Group in Greenland, dated at 3.8–3.7 Ga. The rocks are hard to interpret because they have been **metamorphosed** by heat and physical forces, but most geologists accept that some of the Isua Group rocks were originally sediments. Sedimentary rocks prove that the crust had cooled and rivers were flowing and eroding rocks.

The Isua Group rocks have also produced controversial signatures of early life. Nobody would expect to find fossils in these rocks because they have been too metamorphosed, but Rosing and Frei (2004) have reported evidence that photosynthesis was happening then from the carbon isotopes. The carbon atom has two stable isotopes, carbon-12 and carbon-13, usually written as ^{12}C and ^{13}C. The ratio of ^{12}C to ^{13}C, usually written $\delta^{13}C$, can indicate the presence or absence of organic residues of previously living organisms: enrichment in ^{12}C relative to ^{13}C is characteristic of photosynthesizing organisms, and the organisms that eat them. Rosing and Frei (2004) reported values of $\delta^{13}C$ in organic matter from the Isua Group rocks that match those of modern living organic matter, and these might have come from plankton in the oceans that were photosynthesizing. This is a dramatic claim, and it has been disputed, but if true, this is the first evidence for life on Earth.

The Archaean world was anoxic: when did oxygen become a part of the atmosphere, and why?

The "great oxygenation event"

The Proterozoic Eon, from 2.5 Ga to 542 Ma (Fig. 8.3), represents a very different world from the Archaean. Archaean atmospheres contained volcanic gases, but no oxygen. Oxygen levels are maintained in the atmosphere today by the photosynthesis of green plants and cyanobacteria, and the latter were the source of the initial buildup of oxygen during the first part of the Precambrian. Then, 2.4 Ga, atmospheric oxygen levels rose to one-hundredth or one-tenth of modern levels, not much perhaps, but an indicator of a complete change in the global system that has been dubbed the "great oxygenation event" (GOE). What caused this dramatic rise in oxygen?

The first organisms had **anaerobic** metabolisms, that is, they operated in the absence of oxygen. Indeed the first prokaryotes would have been killed by oxygen. This is a shocking fact that is confirmed by living microbes: some can switch from anaerobic to aerobic respiration depending on oxygen levels. Others, though, are **obligate anaerobes** that have to respire anaerobically and cannot survive even the smallest amount of oxygen. Did living things generate sufficient oxygen to change the Earth's atmosphere? Early photosynthetic bacteria did not produce oxygen, and some have argued that modern styles of photosynthesis that liberate oxygen arose about 2.4 Ga. However, there is evidence from biomarkers (see below) that this had happened by 2.7 Ga. Others have proposed that the oxygen built up after a dramatic reduction in volcanic activity; however, there is no compelling evidence for this. Perhaps the secret lies in methane.

David Catling and colleagues at the University of Washington in Seattle proposed that methane was much more abundant in the Archaean atmosphere than today. Methane (CH_4) is a key product of the activities of anaerobic microbes that use a form of anaerobic respiration called **methanogenesis** to breathe. Today, methane is consumed by oxygen in the atmosphere, but in the absence of oxygen Archaean methane levels might have been 100–1500 times as much as today. Methane is also a potent **greenhouse gas**, which would help explain why the early Earth did not freeze over, given that the 4.0 Ga sun was about 25–30% less luminous than today. Methane can diffuse up to the outer fringes of the atmosphere, where it is decomposed by ultraviolet light and the liberated hydrogen atoms are lost into space. In a world without the escape of hydrogen, Catling and Claire (2005) have calculated that oxygen would be mopped up continuously by gases released by volcanism and metamorphism, as well as by soluble metals in hot springs and seafloor vents, and the world would remain forever anaerobic. With high Archaean methane levels, hydrogen atoms were transferred out of the Earth's atmosphere, and the oxygen was not all locked up in water molecules but eventually flooded out as an atmospheric gas. The collapse of the methane greenhouse 2.4 Ga may have triggered glaciation worldwide.

The rise of oxygen in the atmosphere had a profound effect on life and the planet. New aerobic organisms arose that exploited the atmospheric oxygen molecules in their chemical activity. The oxygen also built up a stratospheric **ozone layer** that blocks out solar ultraviolet radiation. The ozone layer has been hugely important since this point in the earliest Proterozoic in blocking solar rays harmful to life, which allowed diverse life to colonize the land surface

After the GOE, oxygen levels remained low, perhaps 1–5% of present levels, for as much as 1 billion years. In the Archaean,

banded iron formations occurred worldwide; these consist of alternating bands of iron-rich (magnetitic/hematitic) chert and iron-poor chert (chalcedony). In the Archaean, iron released from vents in the seafloor was mobile in the deep ocean and welled up onto the continental shelves. This is unlike today, where oxygen extends to the bottom of the sea and iron is immediately deposited as an oxide on the flanks of mid-ocean ridges. The banding in banded iron formations may reflect seasonal plankton blooms that released a great deal of oxygen into the surface ocean, which combined with upwelling iron ions to produce the iron-rich layers. About 1.9 Ga, banded iron formations largely disappear. Continental **red bed** sediments had first appeared at approximately 2.3 Ga, following the rise of oxygen. These red beds indicate higher oxygen levels because the red color comes from weathering of the iron in the rocks in the presence of atmospheric oxygen. A second rise of oxygen around 0.8–0.6 Ga is indicated by increased levels of marine sulfate. Oxygenated rainwater reacts with pyrite on the continents and washes sulfate through rivers to the oceans, so an increase in oceanic sulfate suggests an increase in oxygen.

The two rises in oxygen levels, at the beginning and end of the Proterozoic, respectively, mark the beginning of modern-style **biogeochemical cycles**, in which oxygen and carbon are exchanged continuously between living organisms and the Earth's crust.

The universal tree of life

There used to be a quiz show on British radio called *Animal, vegetable or mineral?* in which a team of scientists had to identify mystery items. Each week, members of the public would send packages of strange tubers, dried internal organs and other revolting fragments for the experts to consider. The division of natural objects into two living (animal, vegetable) and one non-living (mineral) category reflects the common view that life may be divided simply into plants (generally green, do not move) and animals (generally not green, do move). To these two might be added microbes (for all the microscopic critters).

The three-kingdom view was expanded to four by the division of "microbes" into two kingdoms, Protoctista for single-celled eukaryotes and Monera for prokaryotes. Four kingdoms became five in 1969 when Robert Whittaker recognized that Fungi (mushrooms and molds), classed by chefs as plants, are fundamentally different from all other plants.

This five-kingdom picture of life was blown out of the water by a series of revolutionary papers by Carl Woese and colleagues from the University of Illinois from 1977 onwards. Woese and George Fox had been working on molecular phylogenies (see p. 133) of prokaryotes, and they realized that prokaryotes fell into two fundamental divisions, the domains Archaea (named Archaebacteria by Woese and Fox in 1977) and Bacteria (or Eubacteria). The third domain is Eucarya (or Eukaryota), for all eukaryotes. In this view, animals, plants and fungi are then distant twigs within Eucarya. Woese had generated the first **universal tree of life** (UTL). It is likely that the Archaea and Bacteria split first, and then the Eukarya split from the Bacteria, but the root of the UTL is still uncertain.

Further work since 1990 has confirmed Woese's insight, although alternative schemes talk of two domains or six kingdoms, and other subdivisions. With the power of modern gene sequencing, it should have been relatively easy to build the UTL with progressively more detail. One of the largest versions of the UTL consists of 191 organisms for which complete genome sequences have been established (Ciccarelli et al. 2006). However, molecular biologists had not at first contemplated the notion of jumping genes: simple organisms seem to be prone to exchanging genes in a process called **horizontal gene transfer**. Genes can be transferred between eukaryotes, but the process is commoner among prokaryotes. Horizontal gene transfer occurs in bacteria today that take up DNA directly from their surroundings, through infection from a phage virus, or through mating. Jumping genes make the task of the phylogenetic sequencer difficult: parts of the genome may show linkages to one group, while jumping genes may link the organism to another. Once a jumping gene has been identified, however, it may become locked into the genome of all descendants, and so provide evidence for the affiliation of all organisms that possess it.

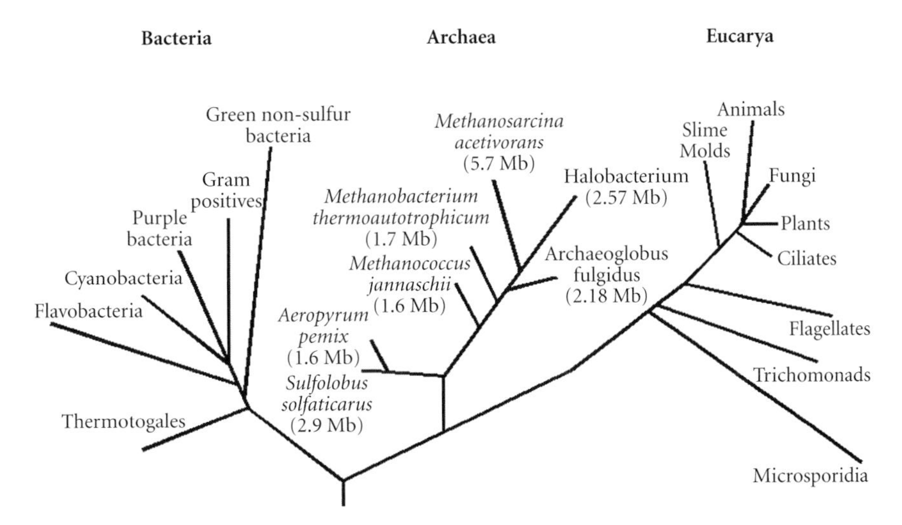

Figure 8.4 The universal tree of life, based on molecular phylogenetic work. The major prokaryote groups are indicated (Bacteria, Archaea), as well as the major subdivisions of Eucarya. Among eukaryotes, most of the groups indicated are traditionally referred to as "algae", both single-celled and multicelled. The metaphytes (land plants), fungi and metazoans (animals) form part of a derived clade within Eucarya, indicated here near the base of the diagram. Mb, megabase (= 1 million base pairs). (Courtesy of Sandie Baldauf.)

The broad patterns of the UTL are not completely resolved (Fig. 8.4) because of jumping genes and other problems: the three domains branch equally, and it is not clear which split came first, between Bacteria and Archaea, or Archaea and Eucarya (Baldauf et al. 2004; Doolittle & Bapteste 2007; McInerney et al. 2008). Until the order of branching is resolved, if it can be, there will be many mysteries about the origin of life. The Domain Bacteria includes Cyanobacteria and most groups commonly called bacteria. The Domain Archaea ("ancient ones") comprises the Halobacteria (salt digesters), Methanobacteria (methane producers) and Eocytes (heat-loving, sulfur-metabolizing bacteria). The Domain Eucarya includes a complex array of single-celled forms that are often lumped together as "algae", a paraphyletic group. Among the "algae" are green algae, flagellates and slime molds, and a crown clade consisting of multicellular organisms. Perhaps the most startling observation is that, within this crown clade, the fungi are more closely related to the animals than to the plants, and this has been confirmed in several analyses. This poses a moral dilemma for vegetarians: should they eat mushrooms or not?

Precambrian prokaryotes

The question of the oldest fossils on Earth has always been controversial. Paleontologists are understandably keen to identify that very first fossil (it is a sure-fire way to attract attention and secure tenure), but that very first fossil is going to be pretty tiny and pretty featureless. How then can the Precambrian paleontologist be sure to identify the fossils correctly, and not be fooled by some whisker or bubble on a microscope slide? The first Archaean fossils were identified only in the 1950s, and over the last decades each new announcement is actively challenged to ensure the specimens are genuine. The latest furor has concerned the reputed microfossils from the 3.5 Ga Apex Chert of Australia (Box 8.1).

The first traces of life occur in rocks dated from 3.5 to 3.0 Ga. These include structures identified as possible **stromatolites** from various parts of the world. Modern stromatolites are constructed by **cyanobacteria** and other prokaryotes (Fig. 8.6). Cyanobacteria live in shallow seawater, and they require good light conditions to enable them to photosynthesize. The cyanobacteria form thin mats on the seafloor in order to maximize their intake of sunlight, but from time to time the mat is overwhelmed by sediment. The microbes migrate towards the light, and recolonize the top of the sediment layer, which may again be swamped by gentle seabed currents. Over time, extensive layered structures may build up. In freshwaters, and sometimes in the sea, stromatolites build up by precipitation of calcite. In most fossil examples, the constructing microbes are not preserved, but the layered structure remains. Many early examples have proved controversial, but the oldest that are generally accepted come from Australia, and are dated as 3.43 Ga (see p. 290).

Perhaps the oldest currently accepted fossils other than stromatolites date from 3.2 Ga. They were found in Western Australia by Birger Rasmussen, and reported in 2000, from

Box 8.1 The Apex Chert: oldest life or hot air?

There was a sensation when Bill Schopf announced the world's oldest fossils in 1987 (Schopf & Packer 1987). He later reported a diverse assemblage of 11 species of bacteria and cyanobacteria from the Apex Chert of the Warrawoona Group in Western Australia, dated as 3465 Ma (Schopf 1993). All specimens are filament-like microbes, ranging in length from 10 to 90 μm; some are circular single cells, while most are filaments consisting of several compartments (Fig. 8.5). These were widely accepted as genuine fossils, and they featured in all the textbooks and web sites as real examples of the earliest cyanobacteria and bacteria.

But their validity was challenged in April 2002. At the second Astrobiology Science Conference held at NASA's Ames Research Center in Moffett Field, California, there was a bombshell. As reported in *Nature*:

> It was the academic equivalent of a heavyweight prizefight. In the red corner, defending his title as discoverer of the Earth's oldest fossils, was Bill Schopf of the University of California, Los Angeles (UCLA). In the blue corner, Martin Brasier of the University of Oxford, UK, who contends that Schopf's "microfossils" are merely carbonaceous blobs, probably formed by the action of scalding water on minerals in the surrounding sediments.

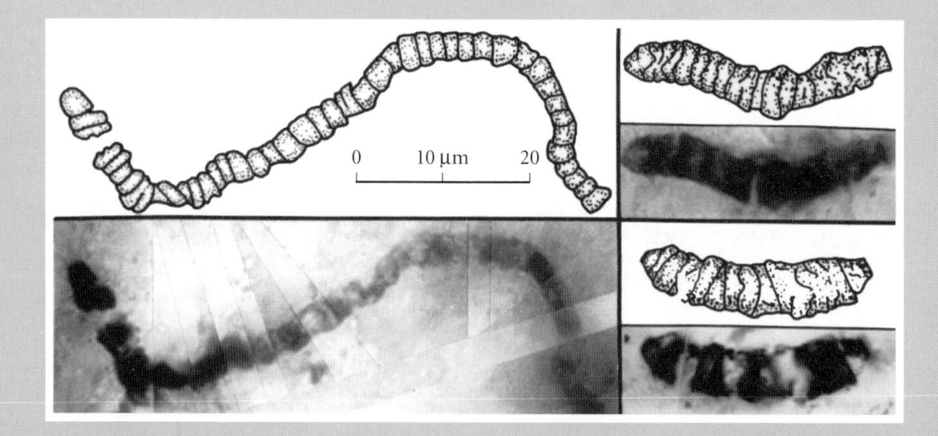

Figure 8.5 Postulated prokaryotes from the Apex Chert of Western Australia (c. 3465 Ma) showing filament-like microbes preserved as carbonaceous traces in thin sections. All are examples of the prokaryote cyanobacterium-like *Primaevifilum*, which measures 2–5 μm wide. (Courtesy of Bill Schopf.)

Brasier and colleagues (2002) had argued the month before that Schopf's "microfossils" were found in a chert that had not formed in shallow seas, but at high temperature in a hydrothermal vein. Any microbes in the solidifying rock would have been roasted. So the "microfossils", said Brasier, must be inorganic structures. Brasier and his colleagues then examined the original specimens, and found that many had been selectively photographed, so that the full complexity of some shapes was not seen in Schopf's published photographs. Many of the "filaments" were extensions of more complex blobs and cavities in the chert, and some showed branching and other features unlikely in a simple prokaryote. Further, the 11 supposed species could not be distinguished, and all kinds of intermediate shapes were found. Brasier believes the "microfossils" are traces of graphite in hydrothermal vein chert and volcanic glass. At high temperature the graphite flowed, forming black, carbon-rich strings and blobs.

Schopf and colleagues (2002) countered that the carbon traces were formed from living material, and they applied a new technique, laser Raman spectroscopy, to prove it. They noted that the spectral bands of the Apex Chert fossils matched signals from known biological materials. But Brasier rebutted this by suggesting that the Raman spectra cannot uniquely identify biological carbon, but simply match color and grain size between areas of a specimen. Their Raman spectra suggested that the "microfossils" and the rock matrix consisted of graphite and silica.

Read more about the dispute at http://www.blackwellpublishing.com/paleobiology/. The debate is renewed in articles by Brasier, Schopf and other commentators in a special issue of the *Philosophical Transactions of the Royal Society* in 2006 (Cavalier-Smith et al. 2006).

Figure 8.6 Stromatolites, a Precambrian example from California, USA (magnification ×0.25). (Courtesy of Maurice Tucker.)

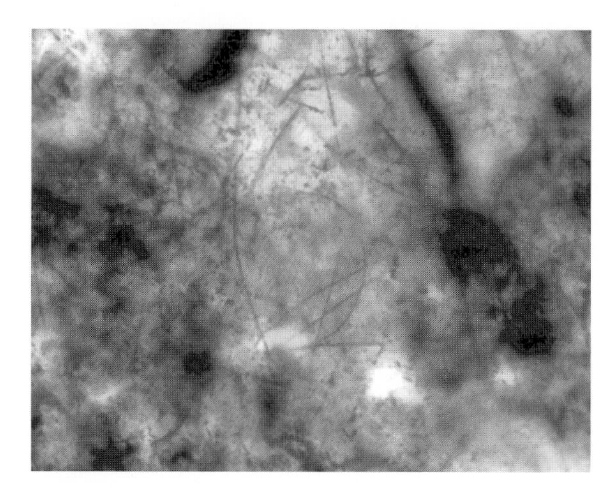

Figure 8.7 The oldest fossils on Earth? A mass of thin thread-like filaments found in a massive sulfide deposit in Western Australia dated at 3.2 Ga. The fact the threads occur in loose groups and in tight masses, and that they are not oriented in one direction, suggests they are organic. The filaments are lined with minute specks of pyrite, showing black, encased in chert. Field of view is 250 μm across. (Courtesy of Birger Rasmussen.)

a massive sulfide deposit produced in an environment like a modern deep-water black smoker, with temperatures up to 300°C. The fossils show evidence of recrystallization by the influx of hydrothermal fluids, and then progressive replacement by later sulfides. The fossils are thread-like filaments (Fig. 8.7) that may be straight, sinuous or sharply curved, and even tightly intertwined in some areas. The overall shape, uniform width and lack of orientation all tend to confirm that these might really be fossils, and not merely inorganic structures. If so, they confirm that some of the earliest life may have been **thermophilic** ("heat-loving") bacteria. Other tubes and filaments of similar age have been reported, but many of these are highly controversial.

There is then a long gap in time until the next generally accepted fossils. These are diverse fossils of cyanobacteria from the Campbellrand Supergroup of South Africa, dated at 2.5 Ga (Altermann & Kazmierczak 2003). The fossils include cell sheaths and capsules that can be identified with modern orders of cyanobacteria. There is then a further long time gap before the next assemblage of prokaryote fossils, from the Gunflint Chert of Ontario, Canada, dated at 1.9 Ga. The Gunflint microorganisms include six distinctive forms, some shaped like filaments, others spherical, and some branched or bearing an umbrella-like structure (Fig. 8.8). These Precambrian unicells resemble in shape various modern prokaryotes, and some were found within stromatolites. Most unusual is *Kakabekia*, the umbrella-shaped microfossil (Fig. 8.8b); it is most like rare prokaryotes found today at the foot of the walls of Harlech Castle in Wales. These modern forms are tolerant of ammonia (NH_3), produced by ancient Britons urinating against the castle walls; so were conditions in Gunflint Chert times also rich in ammonia?

Biomarkers

Even if the oldest fossils are controversial, paleontologists have been able to identify another source of information on early life. These are so-called **biomarkers**, organic chemical indicators of life in general, and of particular sectors of life. Most biomarkers are **lipids**, fatty and waxy compounds found in living cells. For a long time, the oldest accepted biomarkers dated from 1.7 Ga, but Brocks et al. (1999) reported convincing examples from organic-rich shales in Australia dated at 2.7 Ga. The biomarkers they identified were not only 1 billion years older than previous examples, they also proved a wider diversity of life at that time than anyone had suspected.

The 2.7 Ga biomarkers were of two types. First were indicators of cyanobacteria, as might be expected. Brocks and colleagues identified 2-methylhopanes, which are known to be breakdown products of 2-methylbacteriohopanepolyols, specialized lipids that are only found in the membranes of cyanobacteria. The investigators also, unexpectedly, identified C28–C30 steranes, which are sedimentary molecules derived from sterols. Such large-ring sterols are synthesized only by eukaryotes, and not by prokaryotes. Moreover, the biochemical synthesis of such large sterols requires molecular oxygen, so that the eukaryotes likely lived in proximity to oxygen-producing cyanobacteria, strengthening the interpretation of the 2-methylhopanes. So, this biomarker evidence confirms the existence of cyanobacteria at

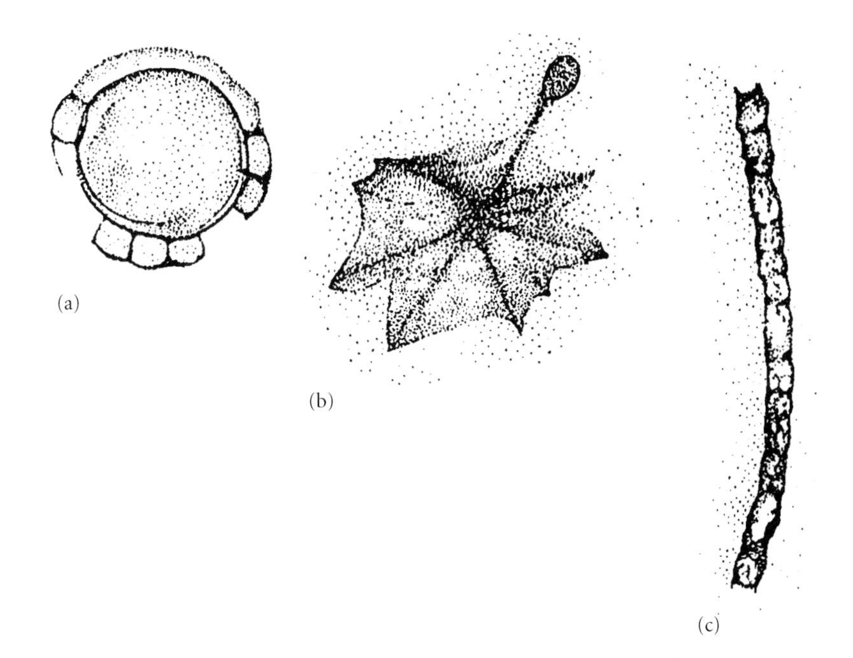

Figure 8.8 Prokaryote fossils from the Gunflint Chert of Ontario, Canada (c. 1.9 Ga): (a) *Eosphaera*, (b) *Kakabekia*, and (c) *Gunflintia*. Specimens are 0.5–10 μm in diameter. (Redrawn from photographs in Barghoorn & Taylor 1965.)

least 2.7 Ga, but it is also the oldest hint of the occurrence of eukaryotes, long before any fossils of that major life domain.

LIFE DIVERSIFIES: EUKARYOTES

Eukaryote characters

Evidence about the earliest evolution of the three domains is scant. It has long been assumed that prokaryotes (i.e. Archaea and Bacteria) were the sole life forms for a billion years or more, and that eukaryotes came much later. This evidence is much more blurred now (Embley & Martin 2006), and the fossils, biomarkers and molecular evidence suggest that eukaryotes might be as old as one or other of the prokaryote domains. The appearance of eukaryotes was important, whenever it happened, because they are complex and include truly multicellular and large organisms.

Eukaryotes are distinguished from prokaryotes (Fig. 8.9a, b) by having a nucleus containing their DNA in chromosomes (prokaryotes have no nucleus, and they have only a circular strand of DNA) and cell organelles, that is, specialized structures that perform key functions, such as **mitochondria** for energy transfer, **flagella** for movement and **chloroplasts** in plants for photosynthesis. There are also many major biochemical differences between prokaryotes and eukaryotes.

The origin of eukaryotes is mysterious because they are in many ways so different from prokaryotes. The most attractive idea for their origin is the **endosymbiotic theory**, proposed by Lynn Margulis in the 1970s. According to this hypothesis (Fig. 8.9c), a prokaryote consumed, or was invaded by, some smaller energy-producing prokaryotes, and the two species evolved to live together in a mutually beneficial way. The small invader was protected by its large host, and the larger organism received supplies of sugars. These invaders became the mitochondria of modern eukaryote cells. Other invaders may have included worm-like swimming prokaryotes (spirochaetes) that became motile flagella, and photosynthesizing prokaryotes that became the chloroplasts of plants.

The endosymbiotic model is immensely attractive, and some aspects have been confirmed spectacularly. Most notable is that the mitochondria and chloroplasts in modern eukaryotes are confirmed as prokaryotes, the mitochondria being closely related to α-proteobacteria and the chloroplasts to

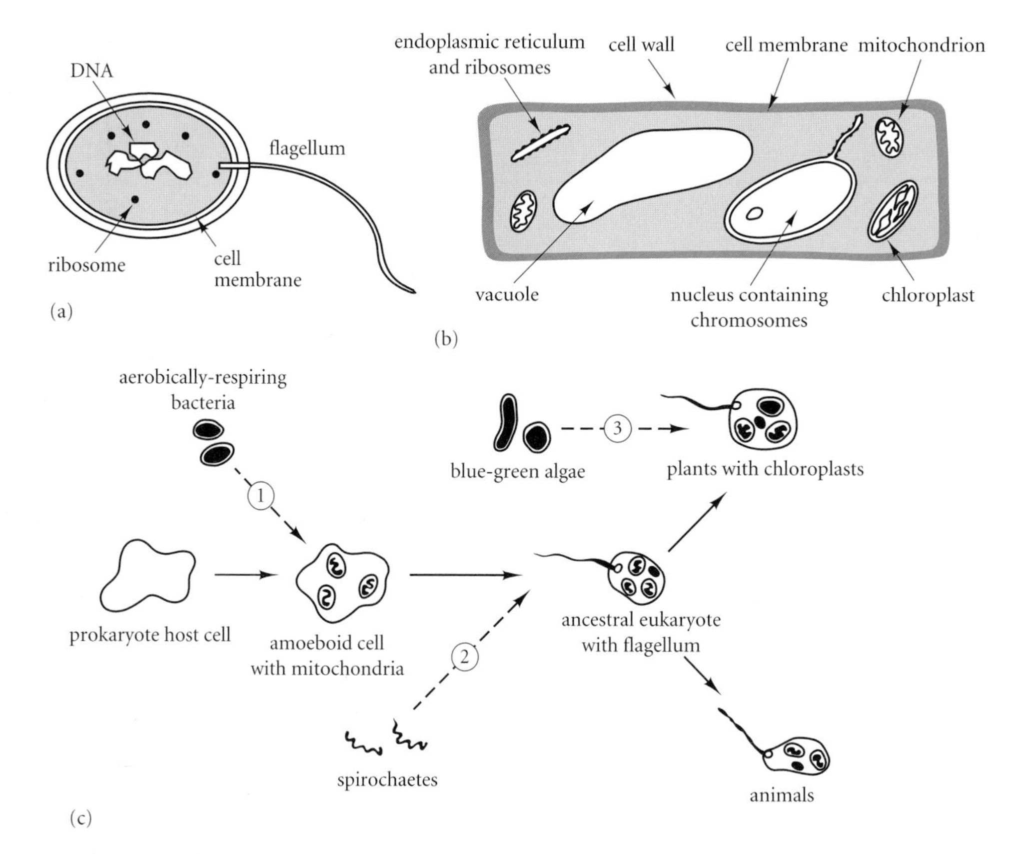

Figure 8.9 Eukaryote characters: a typical prokaryote cell (a) differs from a eukaryote plant cell (b) in the absence of a nucleus and of organelles. (c) The endosymbiotic theory for the origin of eukaryotes proposes that cell organelles arose by a process of mutually beneficial incorporation of smaller prokaryotes into an amoeba-like prokaryote (steps 1, 2 and 3). (Based on various sources.)

cyanobacteria. So, the amazing thing is that a modern eukaryote cell has proven prokaryotic invaders that possess their own DNA and that coordinate their cell divisions with the divisions of the larger host cell.

Many experts reject the endosymbiotic theory, or at least most of it (Poole & Penny 2007). They point out that the only real evidence for engulfment is for the mitochondria. There is no evidence to support the idea that the nucleus was engulfed, nor is it clear what kind of prokaryote did the engulfing, and in fact engulfment is seen today only among eukaryotes, and not among prokaryotes. So, the alternative view, termed the **protoeukaryotic host theory**, is that an ancestral eukaryote, the so-called protoeukaryote, already equipped with a nucleus, indeed did engulf an energy-transferring prokaryote that became the mitochondrion. But this does not tell us where the protoeukaryote itself came from. Further doubt is cast on the classic endosymbiotic theory by the fact that neither Archaea nor Bacteria appear to be ancestral to Eucarya, and that biomarker evidence indicates an unexpectedly ancient origin for eukaryotes.

Which ever model is correct, when did eukaryotes originate? Molecular evidence about dating the universal tree of life (see Fig. 8.3) has been controversial, but current molecular dates for the evolution of basal eukaryotes appear to be roughly in line with the fossils (Box 8.2).

Basal eukaryotes

The oldest eukaryote is controversial. Lipid biomarkers indicate that eukaryotes were around at least by 2.7 Ga (see p. 194). The oldest eukaryote fossil may be *Grypania*, a coiled, spaghetti-like organism that has been reported from rocks as old as 1.85 Ga

(Fig. 8.11a). Slabs are sometimes covered with great loops and coils of *Grypania*, preserved as thin carbonaceous films. It has been identified as a photosynthetic alga, a type of seaweed, based on its overall shape and, if this identification is correct, it is a eukaryote. Many dispute this identification, and would argue that the oldest eukaryotes are micro-

Box 8.2 Dating origins

There was a sensation in 1996 when Greg Wray of Duke University and colleagues announced new molecular evidence that animals had diversified about 1200 Ma. This estimate predated the oldest animal fossils by about 600 myr. In other words, the molecular time scale seemed to be double the fossil age. This proposal suggested three consequences: (i) the Precambrian fossil record of animals (and presumably all other fossils) was even more deficient than had been assumed; (ii) the Cambrian explosion, normally dated at 542 Ma, would shift back deep into the Proterozoic; and (iii) all other splitting dates in the UTL (see Fig. 8.4) would have to be pushed back deeper into the Proterozoic and Archaean.

Wray's view was confirmed by a number of other molecular analyses of basal animal groups, but also of plants, Archaea and Bacteria. Their work is based on gene sequencing from RNA of the nucleus, and it is calibrated against geological time using some fixed points based on known fossil dates. The molecular clock model of molecular evolution (see p. 133) suggests that genes mutate at predictable rates through geological time, so if one or more branching points in the tree can be fixed from known fossil dates, then the others may be calculated in proportion to the amount of gene difference between any pair of taxa.

In Wray's case, mainly vertebrate dates were used, the assumed dates of branching between different groups of fishes and tetrapods in the Paleozoic. So, he had to **extrapolate** his dates from the Paleozoic fixed points back into the Precambrian. Extrapolation (fixing dates outside the range) is tougher than interpolation (fixing dates within a range between a known date and the present day): small errors on those Paleozoic dates would magnify up to huge errors on the Precambrian estimates.

Wray's calculations were criticized by Ayala et al. (1998), who recalculated a date of 670 Ma for the basal radiation of animals, much more in line with the fossil record. In a further revision, Kevin Peterson and colleagues from Dartmouth University (2004) showed that Wray had unwittingly found a very ancient date because vertebrate molecular clocks tick more slowly than those of most other animal groups. So, if vertebrate clocks are slower, it takes longer for a certain amount of genome change to occur than in other animals, and so any calibrations extrapolated from such dates will be much more ancient than they ought to be. Peterson et al. (2004) brought the date of divergence of bilaterian animals down to 573–656 Ma, and so the split of all animals would be just a little older, in line with Ayala et al.'s (1998) estimate.

The reconsideration of molecular clock methods has now opened the way for a great number of studies of the dating of other parts of the UTL (see Fig. 8.4). Most analysts accept a baseline date of 3.5–3.8 Ga for the universal common ancestor, the first living thing on Earth. For example, Hwan Su Yoon and colleagues (2004) from the University of Iowa were able to reconstruct a tree of photosynthetic eukaryotes, the various algal groups, as well as plants (Fig. 8.10), and to date it. They used fixed dates for the origin of life, the oldest bangiophyte red alga (see Box 8.3), the first green plants on land, the first seed plants, and higher branching points among gymnosperms and angiosperms (see pp. 498, 501). These then allowed the team to date splits among marine algae around 1.5 Ga, in line with fossil evidence, and a major radiation of photosynthetic eukaryotes from 1.0 Ga onwards. Their dates also give information on the timing of some events in the endosymbiotic model for the acquisition of organelles by green plant cells (Fig. 8.10).

Read more about the three-domain tree of life at http://www.blackwellpublishing.com/paleobiology/.

Continued

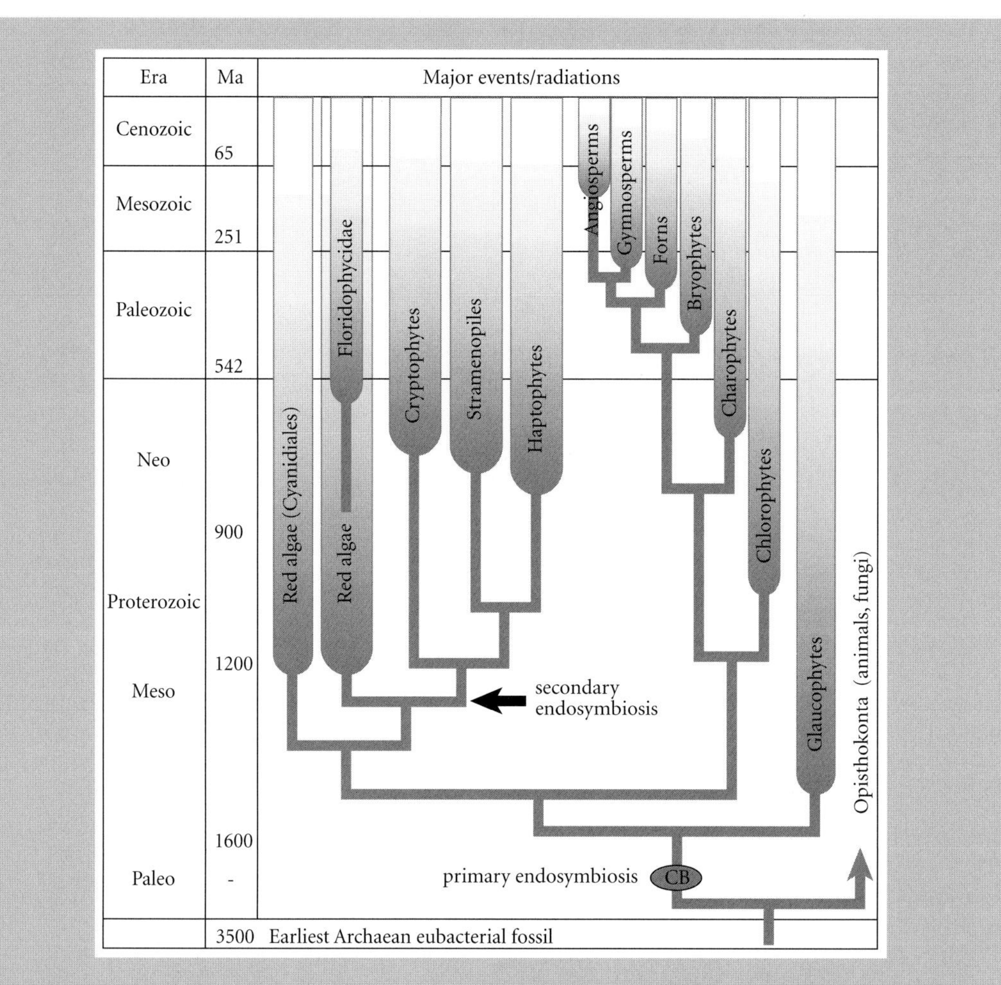

Figure 8.10 Diagram showing the evolutionary relationships and divergence times for the red, green, glaucophyte and chromist algae. These photosynthetic groups are compared with the Opisthokonta, the clade containing animals and fungi. The tree also shows two endosymbiotic events. Some time before 1.5 Ga, the first such event took place, when a photosynthesizing cyanobacterium (CB) was engulfed by a eukyarote. The second endosymbiotic event involved the acquisition of a plastid about 1.3 Ga. Plastids in plants store food and may give plants color (chloroplasts are green). (Courtesy of Hwan Su Yoon.)

scopic acritarchs, marine plant-like organisms (see p. 216) that are known from rocks dated 1.45 Ga.

Eukaryotes may be identified by their nuclei, and paleontologists have hoped to find such clinching evidence in the fossils. For a time, many believed that nuclei had been identified in the diverse eukaryotes from the much younger Bitter Springs Cherts of central Australia, dated at about 800 Ma. Some cells show apparent nuclei (Fig. 8.11b), but the dark areas probably only represent condensations of the cell contents. The Bitter Springs fossils also show evidence of cell division, but what kind of cell division?

Normal cell divisions in growth are called **mitosis**, where all the cell contents, including the DNA, are shared. Mitosis is seen in asexual and sexual organisms. The globular *Glenobotrydion* from the Bitter Springs Chert shows cells in different stages of mitotic division (Fig. 8.11b), where one cell divides into two, and then the two divide into four. *Eotetrahedrion* (Fig. 8.11c), once described as a reproducing eukaryote, is now interpreted as a cluster of cyanobacteria. Other fossils include

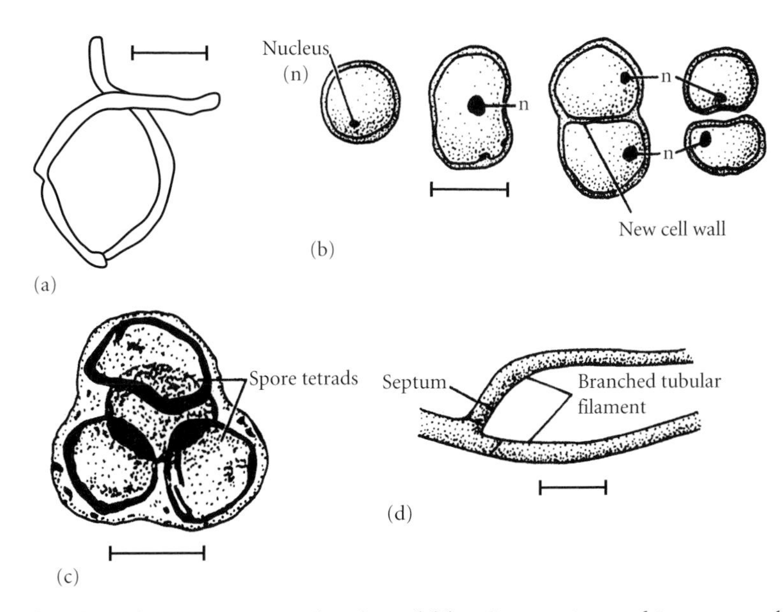

Figure 8.11 Early fossil "eukaryotes". (a) The thread-like *Grypania meeki*, preserved as a carbonaceous film, from the Greyson Shale, Montana (c. 1.3 Ga). (b, c) Single-celled eukaryotes from the Bitter Springs Chert, Australia (c. 800 Ma): (b) *Glenobotrydion* showing possible mitosis (cell division in growth), and (c) *Eotetrahedrion*, probably a cluster of individual *Chroococcus*-like cyanobacteria. (d) Branching siphonalean-like filament. Scale bars: 2 mm (a), 10 μm (b–d). (Courtesy of Martin Brasier, based on various sources.)

branched filaments that look like modern siphonalean green algae (Fig. 8.11d).

Older fossils too look like algae. For example, in the Lakhanda Group of eastern Siberia, 1000–950 Ma, five or six metaphyte species have been found (Fig. 8.12), as well as a colonial form that forms networks rather like a slime mold. But the key fossil in understanding early eukaryote evolution is *Bangiomorpha* (Box 8.3).

Multicellularity and sex

As eukaryotes ourselves, multicellularity and sex seem obvious. Prokaryotes are single-celled organisms, although some form filaments and loose "colonial" aggregations. True **multicellular** organisms arose only among the eukaryotes. These are plants and animals that are composed of more than one cell, typically a long string of connected cells in early forms. Multicellularity had several important consequences, one of which was that it allowed plants and animals to become large (some giant seaweeds or kelp, forms of algae, reach lengths of tens of meters). Another consequence of multicellularity was that cells

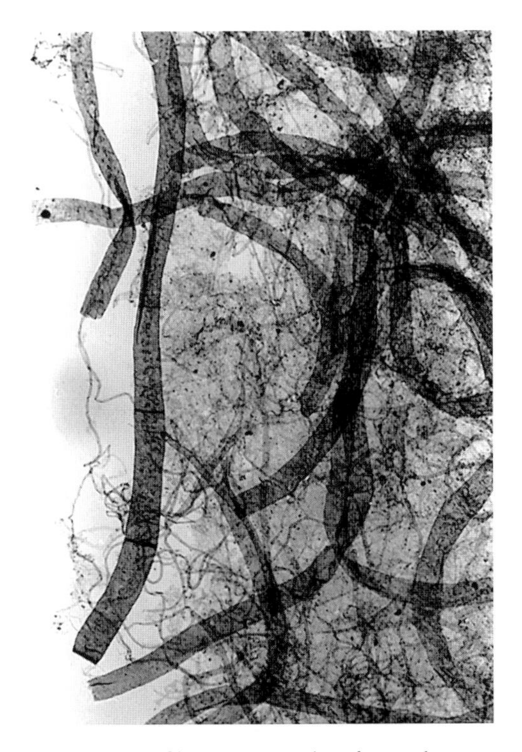

Figure 8.12 A filamentous alga from the Lakhanda Group, Siberia (c. 1000 Ma), 400 μm wide. (Courtesy of Andy Knoll.)

could specialize within an organism, some being adapted for feeding, others for reproduction, defense or communication.

But it seems that multicellularity required sex as well. The first organisms almost certainly reproduced **asexually**, that is, their cells divided and split. Asexual reproduction, or budding as it is sometimes called, is really just a form of growth: cells feed and grow in size, and when they are big enough they split by mitosis to form two organisms. The DNA splits at the same time and is shared by the two new cells. **Sexual** reproduction, on the other hand, involves the exchange of **gametes** (sperms and eggs) between organisms. Typically, the male provides sperm that fertilize the egg from the female. Gametes have half the normal DNA complement, and the two half DNA sets zip together to produce a different genome in the offspring, but clearly sharing features of father and mother. In eukaryotes, the DNA exists as two copies, each strand forming one half of the double-helix structure. Cell divisions in sexual reproduction are called **meiosis**, where the DNA unzips to form two single copies, one going into each gamete, prior to fusion after fertilization.

Box 8.3 *Bangiomorpha*: origin of multicellularity and sex

Red algae (rhodophytes) today range from single cells to large ornate plants, and they may be tolerant of a wide variety of conditions. The modern red alga *Bangia*, for example, can survive in a full range of salinities, from the sea to freshwater lakes. The oldest fossil red alga was announced in 1990, and described in detail by Nick Butterfield from the University of Cambridge in 2000. The specimens are preserved in silicified shallow marine carbonates of the Hunting Formation, eastern Canada, dated at 1.2 Ga, together with a variety of other fossils, both prokaryote and eukaryote.

In his 2000 paper, Butterfield quippishly named the new form *Bangiomorpha pubescens*, the species name *pubescens* chosen "with reference to its pubescent or hairlike form, as well as the connotations of having achieved sexual maturity". The name *Bangiomorpha pubescens* has even made it into the dictionaries of bizarre and cheeky names; one web site notes "The fossil shows the first recorded sex act, 1.2 billion years ago. The 'bang' in the name was intended as a euphemism for sex." The fossils do not show sex acts, and the commentators surely exaggerate: Nick Butterfield may be based at the University of Cambridge in England, home of smutty humor since medieval times (if not before), but he is Canadian by birth!

Bangiomorpha grew in tufts of whiskery strands attached to shoreline rocks by holdfast structures made from several cells (Fig. 8.13a). The individual filaments are up to 2 mm long, and the cells are less than 50 μm wide. The cell walls are dark and enclose circular to disk-like cells, and the whole plant is enclosed in a further thick external layer. The individual filaments may be composed of a single series of cells, or of several series running side by side, or a combination of the two (Fig. 8.13b). Multiple-series filaments are composed of sets of wedge-shaped cells that radiate from the midline of the strand, a diagnostic feature of the modern *Bangia* and of all so-called bangiacean red algae.

Many dozens of specimens of *Bangiomorpha* have been found, and these show how the filaments developed. Starting with a single cell, the filament grew by division of cells (mitosis) along the filament axis. One cell divided into two, then two into four, and so on. Along the filaments (Fig. 8.13b), disk-shaped cells occur in clusters of two, four or eight, and these reflect further cell divisions within the filament. Some broader filaments show clusters of spherical, spore-like structures at the top end; if correctly identified, these prove that sexual reproduction and meiosis were taking place. Close study of the filaments, and of series of developmental stages, shows that *Bangiomorpha* was not only multicellular but that it showed differentiation of cells (holdfast cells versus filament cells), multiple cycles of cell division, differentiated spores and sexually differentiated whole plants.

Read more about *Bangiomorpha* in Butterfield's (2000) paper and at http://www.blackwellpublishing.com/paleobiology/.

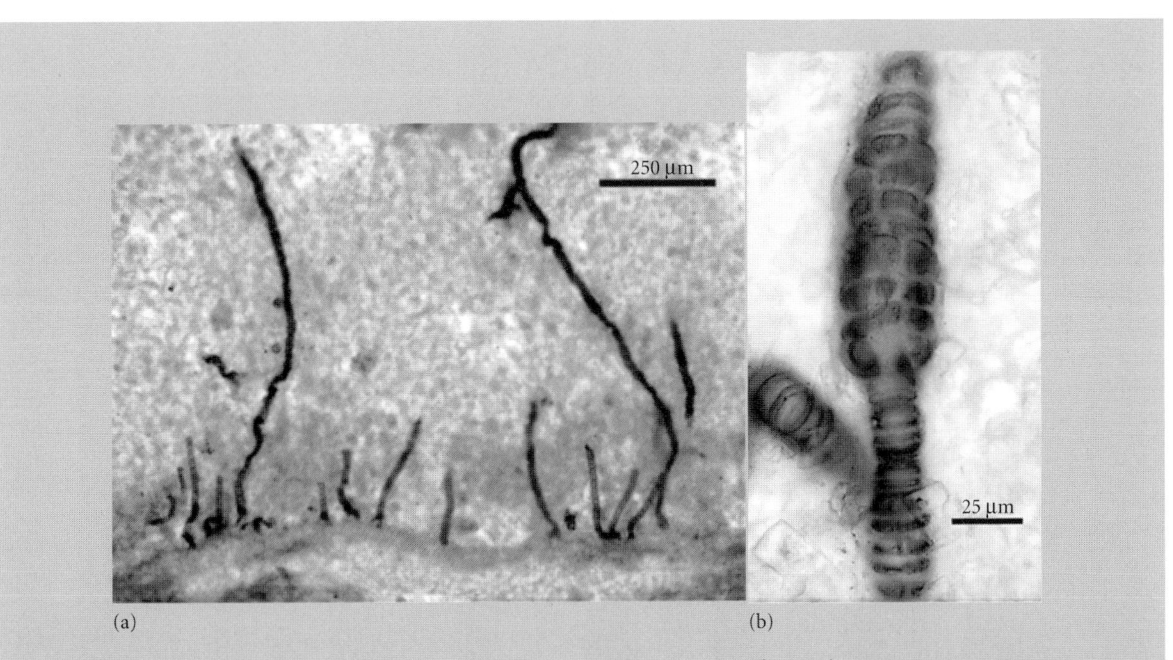

Figure 8.13 The oldest multicellular eukaryote, *Bangiomorpha*, from the 1.2 Ga Hunting Formation of Canada. (a) A colony of whiskery filaments growing from holdfasts attached to a limestone base. (b) A single filament showing a single-series filament making a transition to multiple series, with sets of four wedge-shaped cells; note the sets of four disk-shaped cells in the single-series part of the strand. (Courtesy of Nick Butterfield.)

Whereas some of the Bitter Springs Chert fossils were once supposed to show meiotic cell division, and so sex, this is now doubted. Must paleontologists find fossils of early eukaryotes actually engaged in sexual reproduction in order to prove the origin of sex? The answer is no, and a phylogenetic argument is enough. If we know that all species in a modern clade show sexual reproduction, then their ancestors probably did too. Many modern algae show sexual reproduction, and the oldest member of a sexually reproducing group is a 1.2 Ga red alga (see p. 200), so that provides a minimum date for the origin of sex.

One of the oldest multicellular organisms is *Bangiomorpha* (Box 8.3), obviously multicellular and a member of a modern group that engages in sex. Multicellularity allowed many new forms to appear. The term "algae" refers to a paraphyletic assortment of single-celled and multicelled organisms, all of them eukaryotes, and most of them **photosynthetic**. The major groups are distinguished by their color, morphology and biochemical properties. Molecular phylogenies (see Fig. 8.4) show that many lines of eukaryotes have traditionally been termed "algae". Several algal groups now seem to be closely related to true plants (see p. 483). The fossil record of algae is patchy, but exceptions are the biostratigraphically useful dinoflagellates, coccoliths and diatoms, and calcareous algae such as dasycladaceans, charophytes and corallines (see p. 221).

Why have sex? Budding seems to be efficient enough, and it is what Bacteria, Archaea and many simple eukaryotes have always done, and continue to do today. The benefits are that the process is quick and efficient: what could be better for a successful organism than to replicate identical **clones** of itself? Sex, on the other hand, is a messy and complex business. Many simple organisms, and even fishes and amphibians, produce vast numbers of eggs, sometimes millions that are shed into the water, where most are wasted. Sperm of course is also produced in vast quantities, and most goes to waste. Nonetheless, the invention of sex is usually seen as one of the great milestones in biological evolution (see p. 546).

The reason for its origin may be obscure, but its consequences are manifest. Sex allows rapid evolution and diversification of species because genetic material is swapped and changes during each reproductive cycle. Sexual organisms vary more than asexual organisms, and they can adapt and specialize more readily. Finally, sexual organisms can be multicellular.

The Late Neoproterozoic

The last 100 myr of the Proterozoic, the Late Neoproterozoic, is marked by a dramatic increase in fossil diversity. Sexual reproduction and multicellularity opened the door for more complex, and larger, organisms. Algal groups, including relatives of plants, appeared. In addition, multicellular animals or metazoans, also appeared later in the Proterozoic, and these included the complex Ediacaran animals.

Review questions

1 Find out how many distinct creation myths you can track down on the internet. Arrange them in a classification that links major features of the myths, and match them to their appropriate religions and time span of general acceptability.
2 Many claims have been made over the years about the oldest fossils of life. Look back through the literature to find what was the oldest acceptable record in 1960, 1970, 1980, 1990 and 2000. Read about why many of these claimed oldest finds were eventually doubted or rejected, and list the reasons why.
3 Read around the debate about the universal tree of life, and consider whether it will ever be possible to determine which branched first – Archaea, Bacteria or Eucarya – and give reasons why some analysts believe that this will never be resolved.
4 What are the advantages and disadvantages of sex and of multicellularity? Catalog as many arguments as you can find for and against each of these biological attributes, and describe the possible world today if sex and multicellularity had never arisen.
5 Why are fossils so rare in the Precambrian?

Further reading

Butterfield, N.J. 2000. *Bangiomorpha pubescens* n. gen., n. sp.: implications for the evolution of sex, multicellularity, and the Mesoproterozoic/Neoproterozoic radiation of eukaryotes. *Paleobiology* **26**, 386–404.

Cavalier-Smith, T., Brasier, M. & Embley, T.M. (eds) 2006. How and when did microbes change the world? *Philosophical Transactions of the Royal Society B* **361**, 845–1083.

Cracraft, J. & Donoghue, M.J. (eds) 2004. *Assembling the Tree of Life*. Oxford University Press, Oxford, UK.

Hazen, R. 2005. *Genesis: The Scientific Quest for Life's Origin*. Joseph Henry Press, Washington. http://darwin.nap.edu/books/0309094321/html/.

Knoll, A.H. 1992. The early evolution of eukaryotes: a geological perspective. *Science* **256**, 622–7.

Knoll, A.H. 2003. *Life on a Young Planet: The First Three Billion Years of Evolution on Earth*. Princeton University Press, Princeton, NJ.

Tudge, C.T. 2000. *The Variety of Life*. Oxford University Press, Oxford, UK.

References

Altermann, W. & Kazmierczak, J.2003. Archean microfossils: a reappraisal of early life on Earth. *Research in Microbiology* **154**: 611–17.

Ayala, F.J., Rzhetsky, A.& Ayala, F.J. 1998. Origin of the metazoan phyla: molecular clocks confirm paleontological estimates. *Proceedings of the National Academy of Sciences, USA* **95**, 606–11.

Barghoorn, E.S. & Taylor, S.A. 1965. Microorganisms from the Gunflint Chert. *Science* **147**, 563–77.

Baldauf, S.L., Bhattacharya, D., Cockrill, J., Hugenholtz, P., Pawlowski, J. & Simpson, A.C.B. 2004. The tree of life, an overview. *In* Cracraft, J. & Donoghue, M.J. (eds) *Assembling the Tree of Life*. Oxford University Press, Oxford, UK, pp. 43–75.

Brasier, M.D., Green, O.R., Jephcoat, A.P. et al. 2002. Questioning the evidence for earth's oldest fossils. *Nature* **416**, 76–81.

Brocks, J.J., Logan, G.A., Buick, R. & Summons, R.E. 1999. Archean molecular fossils and the early rise of Eukaryotes. *Science* **285**, 1033–6.

Butterfield, N.J. 2000. *Bangiomorpha pubescens* n. gen., n. sp.: implications for the evolution of sex, multicellularity, and the Mesoproterozoic/Neoproterozoic radiation of eukaryotes. *Paleobiology* **26**, 386–404.

Catling, D.C. & Claire, M. 2005. How Earth's atmosphere evolved to an oxic state: a status report. *Earth and Planetary Science Letters* **237**, 1–20.

Ciccarelli, F.D., Doerks, T., von Mering, C. et al. 2006. Toward automatic reconstruction of a highly resolved tree of life. *Science* **311**, 1283–7.

Crick, F.H.C. 1968. The origin of the genetic code. *Journal of Molecular Biology* **38**, 367–9.

Doolittle, W.F. & Bapteste, E. 2007. Pattern pluralism and the Tree of Life hypothesis. *Proceedings of the National Academy of Sciences, USA* **104**, 243–9.

Embley, T.M. & Martin, W. 2006. Eukaryotic evolution, changes and challenges. *Nature* **440**, 623–30.

Gilbert, W. 1986. The RNA world. *Nature* **319**, 618.

McInerney, J.O., Cotton, J.A. & Pisani, D. 2008. The prokaryotic tree of life: past, present . . . and future? *Trends in Ecology and Evolution* **23**, 276–81.

Nisbet, E.G. & Sleep, N.H. 2001. The habitat and nature of early life. *Nature* **409**, 1083–91.

Peterson, K.J., Lyons, J.B., Nowak, K.S. et al. 2004. Estimating metazoan divergence times with a molecular clock. *Proceedings of the National Academy of Sciences, USA* **101**, 6536–41.

Poole, A.M. & Penny, D. 2007. Evaluating hypotheses for the origin of eukaryotes. *BioEssays* **29**, 74–84.

Rasmussen, B. 2000. Filamentous microfossils in a 3,235-million-year-old volcanogenic massive sulfide. *Nature* **405**, 676–9.

Rosing, M.T. & Frei, R. 2004 U-rich Archean sea-floor sediments from Greenland – indications of >3700 Ma oxygenic photosynthesis. *Earth and Planetary Science Letters* **217**, 237–44.

Schopf, J.W. 1993. Microfossils of the Early Archean Apex Chert: new evidence of the antiquity of life. *Science* **260**, 640–6.

Schopf, J.W., Kudryavtsev, A.B., Agresti, D.G., Wdowiak, T.J. & Czaja, A.D. 2002. Laser-Raman imagery of Earth's earliest fossils. *Nature* **416**, 73–6.

Schopf, J.W. & Packer, B.M. 1987. Early Archean (3.3-billion to 3.5 billion-year-old) microfossils from Warrawoona Group, Australia. *Science* **237**, 70–3.

Szostak, J.W., Bartel, D.P. & Luisi, P.L. 2001. Synthesizing life. *Nature* **409**, 387–90.

Whittaker, R. 1969. New concepts of kingdoms or organisms: evolutionary relations are better represented by new classifications than by the traditional two kingdoms. *Science* **163**, 150–60.

Woese, C.R. & Fox, G.E. 1977. Phylogenetic structure of the prokaryotic domain: the primary kingdoms. *Proceedings of the National Academy of Sciences, USA* **74**, 5088–90.

Wray, G.A., Levinton, J.S. & Shapiro, L. 1996. Molecular evidence for deep pre-Cambrian divergences among the metazoan phyla. *Science* **274**, 568–73.

Yoon, H.S., Hackett, J.D., Ciniglia, C., Pinto, G. & Bhattacharya, D. 2004. A timeline for the origin of photosynthetic eukaryotes. *Molecular Biology and Evolution* **21**, 809–18.

Chapter 9

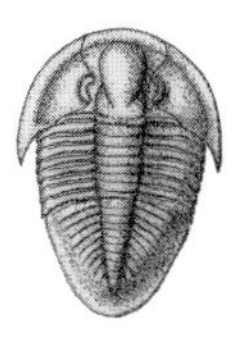

Protists

Key points

- Micropaleontology is a multidisciplinary science, focused on the study of microorganisms or the microscopic parts of larger organisms.
- Prokaryotes, unicellular microbes lacking nuclei and organelles, include the carbonate-producing cyanobacteria, the oldest known organisms; their radiation during the mid-Precambrian promoted an oxygen-rich atmosphere.
- Protists, unicellular organisms with nuclei, include a large variety of organisms with external protective coverings (tests and cysts) assigned to the kingdoms Protozoa and Chromista.
- Fossilized protists can also be split into organisms with organic (acritarchs, dinoflagellates, chitinozoans), calcareous (coccolithophores, foraminiferans) or siliceous (diatoms, radiolarians) skeletons.
- Foraminifera, single-celled animal-like protozoans, contain both benthic and planktonic forms with chitinous, agglutinated, but most commonly calcareous (hyaline and porcellaneous), tests occurring throughout the Phanerozoic.
- Radiolarians, animal-like protozoans with siliceous tests, and diatoms, plant-like protozoans with silicic skeletons, are both important rock formers.
- Acritarchs, dinoflagellates and chitinozoans are palynomorphs, most commonly preserved as cysts, with important biostratigraphic applications. The first two are assigned to the protozoans, the third is currently difficult to classify.
- Coccolithophores and diatoms are assigned to the chromistans.

It has long been an axiom of mine that the little things are infinitely the more important.

Arthur Conan Doyle (1891) *A Case of Identity*

The world of microbes is more bizarre than the most contrived science fiction novel. The Earth is host to creatures that ingest iron and uranium, thrive in environments akin to boiling sulfuric acid or even live within solid rock itself (Box 9.1). These amazing organisms have a huge variety of shapes, belong to a multitude of groups living in many different environments while pursuing a wide range of lifestyles with often apparently alien metabolisms. Microbes such as bacteria and viruses are by far the most abundant life forms on the planet, a situation undoubtedly true of the geological past. Microfossils are the microscopic remains, commonly less than a millimeter in size, of either microorganisms or the disarticulated or reproductive parts of larger organisms. They thus include not only microbes themselves but also the microscopic parts of animals and plants.

In his famous book *Small is Beautiful*, Schumacher argued for small-scale economics in the world. Among paleontologists, micropaleontologists are obsessed with microscopic fossils. Until you have screwed up your eyes and peered down a binocular microscope, you can have no idea of the exquisite beauty of microfossils, their tiny shapes showing infinite detail in their sculpture, spines and plate patterns. And they are not only beautiful, but useful too! Micropaleontology has thus attracted the attentions of botanists, zoologists, biochemists and microbiologists together with, of course, paleontologists and geologists. The disparate taxonomic groups included as microfossils are, nonetheless, united by their method of study – all require the use of an optical microscope, although more recently both scanning and transmission electron microscopes have taken microfossil studies to new, amazing levels. The majority of microfossils are indeed small and perfectly formed; but they display often the most complex and intricate of organic morphologies.

Microfossils thus include material derived from most of the major groups of life, Bacteria, Protozoa, Chromista, Fungi, Plants and Animals, although Fungi are rarely found as fossils. The broad classification adopted by most textbooks is both conventional and operational: microfossils are usually divided into the prokaryotes (mainly bacteria), pro-

Box 9.1 Microbes in extreme environments: the extremophiles

We are aware that microbes are everywhere, but are they as widespread as we believe? Yes, and probably more so. Scientists have been investigating a range of microbes, the **extremophiles** ("lovers of extremes"), that appear to be adapted, with specific enzymes, to some of the most extreme environments on Earth. Thus acidophiles (acid environments), alkaliphiles (alkaline environments), barophiles (high pressure), halophiles (saline environments), mesophiles (moderate temperatures), thermophiles (high temperatures), psychrofiles (cool temperatures) and xerophiles (arid environments) have now been identified. Extremophiles are spread across both the prokaryotes and eukaryotes, although most belong to the Archaea and Bacteria and some scientists have argued they should be included in a separate domain on the basis of their unique metabolic processes. Thus if modern microbes can function in both frozen and geothermal habitats, both acid and alkaline ponds and even deep within the crust, the extreme environments of the Early Precambrian and perhaps even space were probably not a great challenge to evolving life of this type. Moreover such groups of organisms could clearly survive the extreme environments of great extinction events. But it remains a challenge to identify such groups in the fossil record. One group of ingenious algae, the acritarchs (see p. 216), made it through one of the most extreme series of ice ages our planet has experienced. The "snowball Earth" hypothesis (see p. 112) suggests that the planet's oceans froze over during the Late Proterozoic, with life coming to a virtual standstill. Acritarch diversity was maintained through the crises (Corsetti et al. 2006). Have we identified a group of extremophiles, or was the climate not so harsh as suggested by the snowball Earth hypothesis?

tists (unicellular eukaryote organisms with a variety of **tests** (external shells) and **cysts** (enclosed resting stages)), microinvertebrates (mainly the ostracodes, see p. 383), microvertebrates (mainly the conodonts and various other microscopic parts of fishes, see p. 441) and spores and pollen (microscopic reproductive organs of plants, see p. 493). We devote this chapter, however, to the more advanced microbes themselves, represented by the second group. The protists are most probably derived from within the Archaea, splitting from them between 4.2 to 3.5 Ga, but the group is almost certainly polyphyletic. The prokaryotic Archaea and Bacteria are intimately tied to the origin of life and the limited Precambrian fossil record (see pp. 191–4); this is all the evidence of life in rocks over 1 billion years old!

The abundance and durability of many microfossil groups makes them invaluable for biostratigraphic correlation (see p. 25). Sequences of samples can be collected from rock outcrops and even from the very small samples available from drill cores and drilling muds. Consequently they are very widely used in geological exploration by petroleum and mining companies. In addition, many microfossils are produced by planktonic organisms with very wide biogeographic distributions, making them invaluable for reliable long-distance correlation. Microfossils in oceanic sediments also provide a continuous record of environmental change and paleoclimate, and study of changing assemblages and the geochemistry of microfossil shells provide the fundamental data for paleoceanographic research. Moreover, consistent color changes through thermal gradients have made microfossils, particularly conodonts and palynomorphs, invaluable for assessments of thermal maturation and the prediction of hydrocarbon windows.

Microorganisms have made a phenomenal contribution to the evolution of the planet as a whole. Many, such as the coccolithophores, diatoms, foraminiferans and radiolarians, are rock-forming organisms. The prokaryotic cyanobacteria fundamentally changed the planet's atmosphere from anoxic to aerobic during the Precambrian, and probably continued to mediate atmospheric and hydrosphere systems. For example, recent research suggests that carbonate mudmounds – such as the Late Ordovician mudbanks in central Ireland, the north of England and Sweden, the Early Carboniferous Waulsortian mounds in Ireland and elsewhere, together with the Early Cretaceous mudmounds in the Urgonian limestones of the Alpine belt – were precipitated by microbes. The influence of microorganisms may also be more subtle. Coccolith-producing organisms, for example *Emiliania*, can, during blooms, manufacture massive amounts of calcium carbonate; this material is much more readily subducted than shelf carbonates and it is then recycled through volcanoes as carbon dioxide (CO_2). The buildup of this greenhouse gas probably maintained warmer climates during the last 200 million years.

The extraction and retrieval of microfossils from rocks and sediments requires a range of preparation techniques, some of which can only be attempted in purpose-built laboratories. For many groups, preparation consists essentially of disaggregation of the rock in water or more potent solvents followed by sieving to remove the clay fraction. The silt- and sand-sized residue is then hand picked under a microscope to collect microfossils such as foraminiferans and ostracods. For other groups such as radiolarians, diatoms and conodonts, acetic or hydrochloric acid is used to remove the carbonate fraction and concentrate the fossils. For palynomorphs, the silicate minerals are removed with hydrofluoric acid, an extremely dangerous chemical that requires special facilities. Finally, microfossils may be concentrated by settling in heavy liquids or by electromagnetic separation. Many groups, such as algae and foraminiferans, may also be studied in thin section.

PROTISTA: INTRODUCTION

The protists are predominantly single-celled organisms with nuclei and organelles, including both **autotrophs**, organisms that convert inorganic matter such as CO_2 and water into food, and **heterotrophs**, organisms that eat organic debris or other organisms. The Protista is a convenient grouping but it is not well defined. Essentially it consists of all eukaryotes once the multicellular animals, fungi and vascular plants are removed. Consequently it is a paraphyletic collection of rather disparate organisms. Most are microscopic and unicellular but multicellularity has evolved numerous times and the multicellular algae (seaweeds) are conventionally included in the Protista too

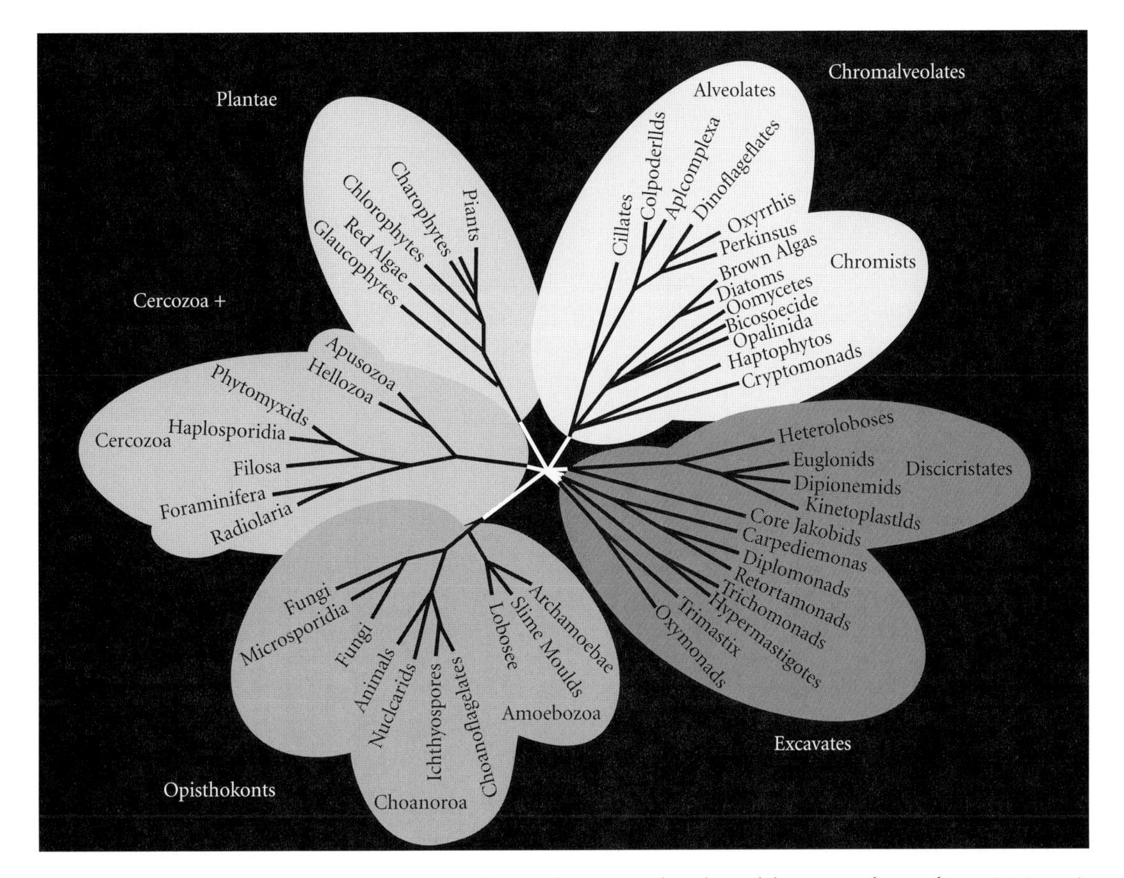

Figure 9.1 Protist positions on the tree of life. In this tree, developed by Patrick Keeling, University of British Columbia, the protozoans (foraminiferans and radiolarians) lie within the Cercozoa far divorced from the chromists (diatoms and dinoflagellates) within the Chromalveolates. (From Keeling et al. 2005.)

(Fig. 9.1). Subdividing the diversity of protists is equally problematic. The division into autotrophic protozoans and heterotrophic algae (chromistans) is important ecologically, but phylogenetically almost meaningless as both groups are polyphyletic. The first protists were almost certainly heterotrophs, but chloroplasts were acquired separately in at least six lineages, producing heterotrophs, and lost secondarily even more often: for example, the classic protozoan ciliates almost certainly evolved from algae. Protists are also often subdivided according to their means of locomotion, most simply into flagellates and amoebans. Again, however, these are polyphyletic groups. So simplisitic attempts at classifying protists do not really work and they are perhaps better regarded as a loose grouping of 30 or 40 disparate phyla with diverse combinations of trophic modes, mechanisms of motility, cell coverings and life cycles. Modern molecular genetic and cytologic research is slowly making sense of this diversity but this is not the place to go into the rapidly changing details of this research. Instead, we should simply note that groups with microfossil records are widely scattered across the diversity of protists. Here, following Cavalier-Smith (2002) and others, the protists are grouped into protozoans (foraminiferans, radiolarians, acritarchs, dinoflagellates and ciliophorans) and chromistans (coccolithophores and diatoms); chitinozoans are difficult to classify in this scheme and are thus treated separately.

EUKARYOTES ARRIVE CENTER STAGE

So when did the eukaryotes first appear? Unicellular eukaryotes, with nuclei and organelles, represented by acritarch cysts are known from rocks dated at about 1.45 Ga. Spiral ribbons of *Grypania*, however, have been reported from rocks as old as 1.85 Ga. One of the oldest multicellular organisms is a bangiophyte red alga (rhodophyte) preserved in silicified car-

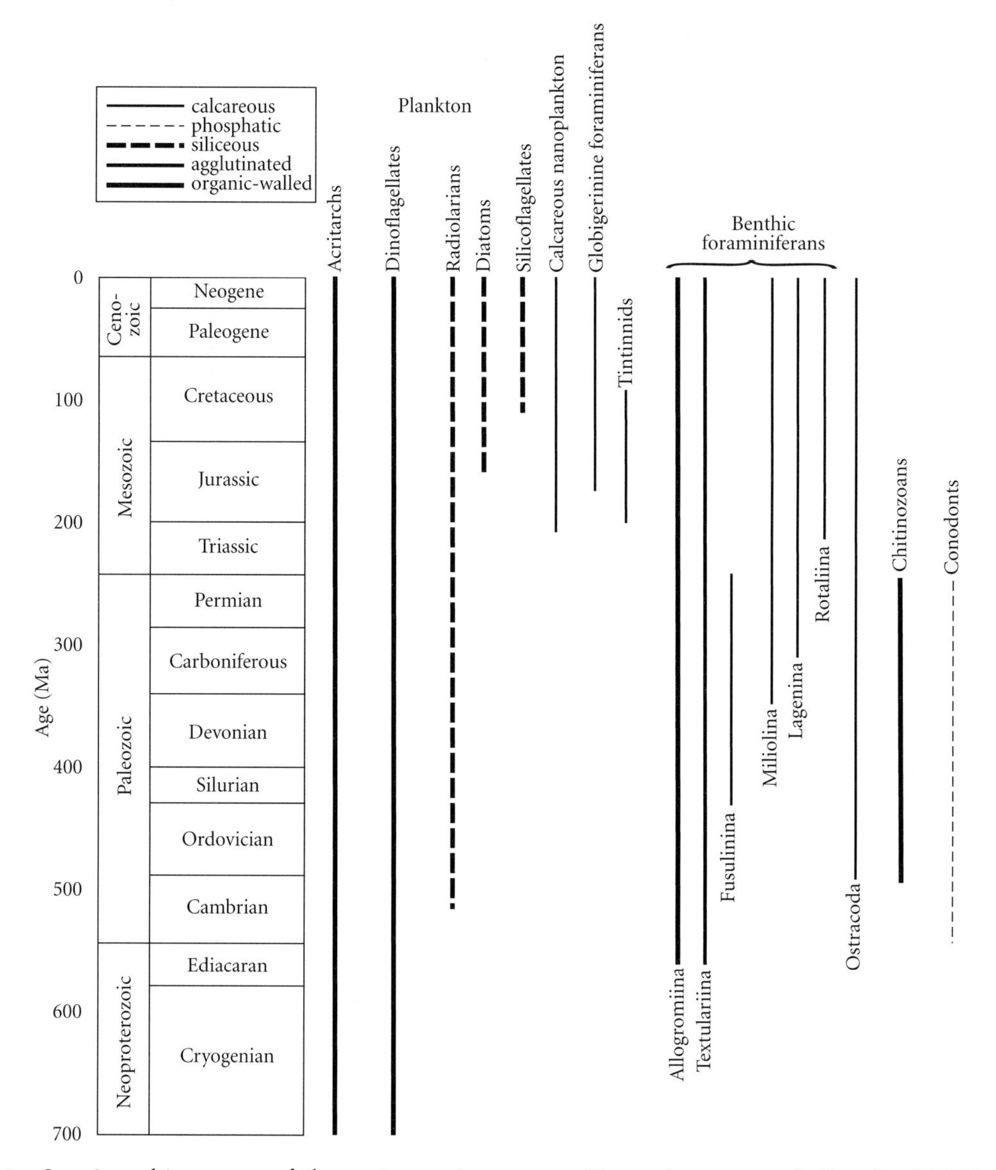

Figure 9.2 Stratigraphic ranges of the main protist groups. (From Armstrong & Brasier 2005.)

bonates of the Hunting Formation, eastern Canada, dated at 1.2 Ga (see p. 200). After 1 Ga, algae are reported from a range of localities around the world. A range of protists such as the acritarchs, chitinozoans, coccolithophores and diatoms dominated the phytoplankton at various stages from the Late Precambrian to the present, whereas the foraminiferans and radiolarians were important parts of the zooplankton (Fig. 9.2). Apart from a role as a primary food source, the marine phytoplankton function as a major carbon sink, initially removing CO_2 from the atmosphere as carbonate ions. These cycles may already have been in place throughout much of the later Proterozoic, anticipating more modern oceanic biological and chemical systems.

PROTOZOA

Protozoans are neither animal nor plant, but single-celled eukaryotes that commonly show animal characteristics such as motility and heterotrophy; some groups are able to form cysts. Most are about 50–100 μm in size and are very common in aquatic environments and in the soil. They can occupy various levels in the food chain ranging from primary producers to predators and some groups function as parasites and symbionts.

Foraminifera

Foraminifera are shelled, heterotrophic protozoans, common in a wide variety of Phanerozoic sedimentary rocks and of considerable biostratigraphic and paleoenvironmental value. The foraminiferans are characterized by a complex network of granular pseudopodia. The foraminiferans were traditionally included in the phylum Sarcodina together with the Radiolaria and a range of other non-flagellate protozoans. In modern classifications the foraminiferans are usually regarded as a discrete phylum, the Granuloreticulosa. Cavalier-Smith (2002), for example, regarded the Foraminifera as a member of the infrakingdom Rhizaria and placed them within the phylum Retaria together with the Radiolaria.

Foraminifera are easily the most abundant of microfossils and can be studied with simple preparation techniques and low-power microscopes. Consequently, pioneer studies in micropaleontology were based on the foraminiferans and techniques established for the study of this group were extended to many other microfossil taxa. Foraminifera have proved extremely useful in the petroleum industry, where detailed biostratigraphic schemes, particularly for Cenozoic rocks, have helped correlate oil field data. Moreover stable isotopes extracted from foraminiferan tests have provided valuable data on ancient sea temperatures through the Mesozoic and Cenozoic.

Morphology and classification

Although many different classifications have been published, shell morphology and mineralogy form the prime basis for identification of species and higher categories of Foraminifera. Most have a shell or test comprising chambers, interconnected through holes or foramina. The test may be composed of a number of materials and three main categories have been documented, organic, agglutinated and secreted calcareous:

1 Organic tests consist of tectin, which is a protinaceous or pseudochitinous substance.
2 **Agglutinated** ("glued") tests comprise fragments of extraneous material bound together by a variety of cements. The debris may be siliciclastic, such as quartz, mica grains or sponge spicules, or calcareous, recycling fragments of coccoliths or other forams.
3 Secreted calcareous tests may be subdivided into three categories: porcellaneous, hyaline and microgranular (Fig. 9.3a). Porcellaneous tests are formed of small, randomly oriented crystals of high-magnesium calcite giving a smooth white shell. Hyaline tests are formed of larger crystals of low-magnesium calcite and have a glassy appearance when well preserved. Hyaline tests have two main modes: the radial tests are made up of minute calcite crystals with their c-axes normal to the test surface, whereas granular forms consist of microcrystals of calcite with variable orientations. Both modes usually have a multilayered structure (Fig. 9.3b) and perforations. Hyaline aragonitic tests occur but are much rarer than calcitic tests. Finally, microgranular tests consist of tightly packed, similar-sized grains of crystalline calcite. Most members of this group are known from the Upper Paleozoic.

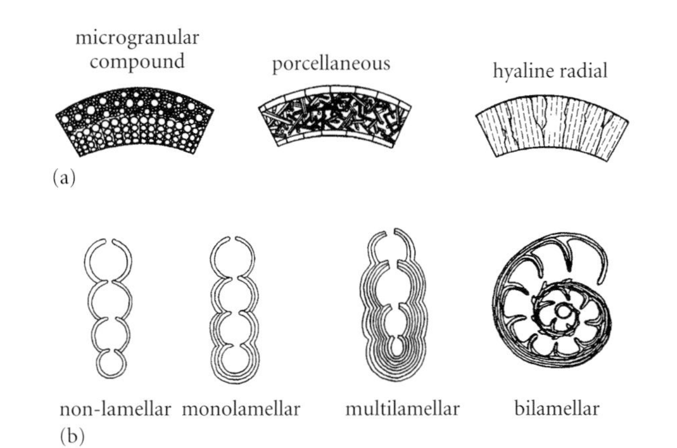

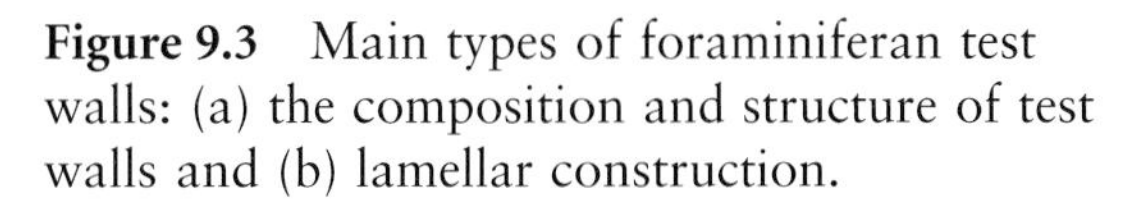

Figure 9.3 Main types of foraminiferan test walls: (a) the composition and structure of test walls and (b) lamellar construction.

The gross morphology of a foraminiferan test is governed by the shape and arrangement of the chambers. The group has evolved a wide range of test symmetries (Fig. 9.4) from simple uniserial and biserial forms to more complex planispiral and trochospiral shapes (Fig. 9.5). Chambers also come in a wide

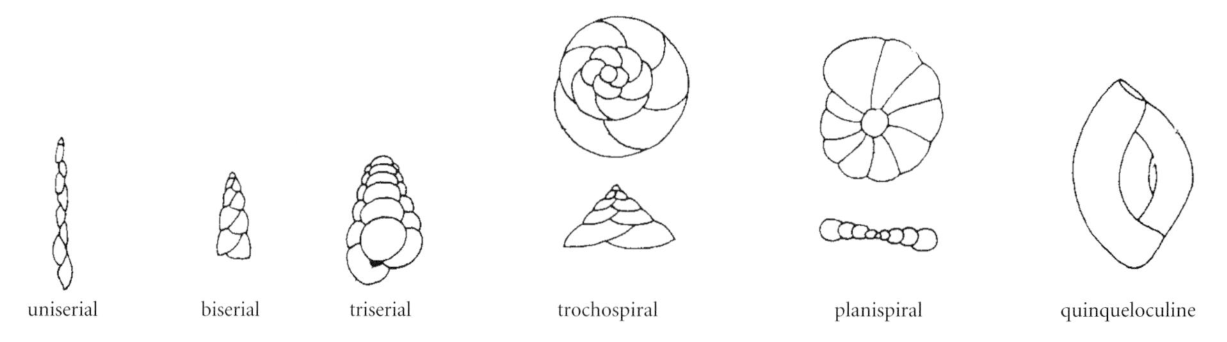

Figure 9.4 Main types of foraminiferan chamber construction.

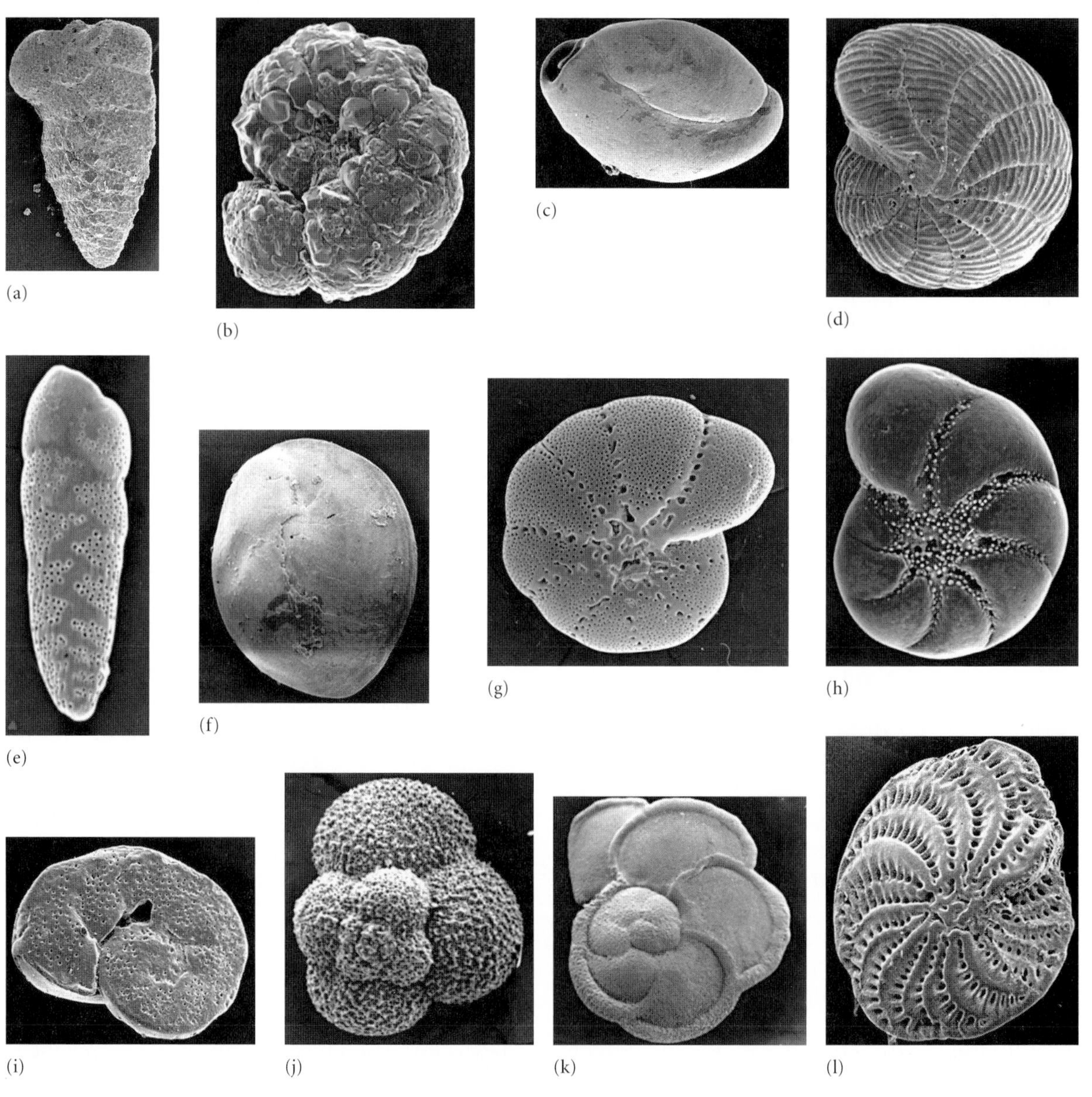

Figure 9.5 Some genera of foraminiferans: (a) *Textularia*, (b) *Cribrostomoides*, (c) *Milionella*, (d) *Sprirolina*, (e) *Brizalina*, (f) *Pyrgo*, (g) *Elphidium*, (h) *Nonion*, (i) *Cibicides*, (j) *Globigerina*, (k) *Globorotalia*, and (l) *Elphidium* (another species). Magnification ×50–100 for all. (Courtesy of John Murray (b, d, e, g, h, j, k) and Euan Clarkson (a, c, f, i, l).)

spectrum of shapes, from simple spherical compartments through tubular to clavate forms. Moreover, the shape and position of the aperture may vary. Surface ornament may include ribs and spines or be merely punctate or rugose. Foraminifera are classified according to test type and ornamentation (Box 9.2).

Life modes

The foraminiferans have adopted two main life modes, benthic and planktonic. The majority are benthic, epifaunal organisms; they are either attached or cling to the substrate or crawl slowly over the seabed by extending their protoplasmic pseudopodia. Infaunal types live within the top 15 cm of sediment. Most benthic forms have a restricted geographic range. Planktonic foraminiferans are most diverse in tropical, equatorial regions and may be extremely abundant in fertile areas of the oceans, particularly where upwelling occurs.

The functional morphology of these groups can now be modeled mathematically (Box 9.3) and potentially can be related to different life modes in the group. Moreover their relationships to different environments, past and present, are well established (Box 9.4).

Evolution and geological history

The earliest foraminiferans are known from the Lower Cambrian, represented by simple agglutinated tubes assigned to *Bathysiphon*, a living benthic genus (Fig. 9.8). More diverse agglutinated forms appeared during the Ordovician while microgranular tests evolved during the Silurian; however, it was not until the Devonian that multichambered tests probably developed. Nevertheless, Carboniferous assemblages have a variety of uniserial, biserial, triserial and trochospiral agglutinated tests. Around the Devonian–Carboniferous boundary the first partitioned tests displaying multilocular growth modes (the addition of new chambers in series) appeared. Two families, the Endothyridae and Fusulinidae, dominated Carboniferous assemblages and the porcellaneous Miliolinidae achieved importance in the Permian. The Fusulinidae were generally large, specialized foraminiferans, adapted to carbonate and reef-type facies during the Late Carboniferous and Permian. Despite a high diversity during the Late Permian, they became extinct at the end of the Paleozoic, and the Endothyridae and the Miliolinidae were very much reduced in diversity.

Although Triassic assemblages were generally impoverished, the stage was set for a considerable radiation during the Jurassic. Two hyaline groups, the benthic Nodosariidae and planktonic Globigerinidae, diversified, while the agglutinates, Lituolitidae and Orbitolinidae, continued. The planktonic foraminiferans diversified in the Cretaceous, culminating in the near extinction of the group during the Cretaceous–Tertiary (KT) mass extinction. Two further periods of diversification took place during the Paleocene-Eocene and the Miocene.

Radiolaria

The radiolarians are marine, unicellular, planktonic protists with delicate skeletons usually composed of a framework of opaline silica (Fig. 9.9). Their name is derived from the radial symmetry, commonly marked by radial skeletal spines, characteristic of many forms. Many others, however, lack radial symmetry. Most radiolarians feed on bacteria and phytoplankton, but also on copepods and crustacean larvae and occupy levels in the water column from the surface to the abyssal depths, although most live in the photic zone commonly associated with symbiotic algae. The radiolarian ectoplasm covers the test and holds symbiotic **zooxanthellae**, microorganisms enclosed within the cell mass, and perforations, providing some nourishment. The radiolarian **endoplasm** (surrounded by the capsular membrane) contains the nucleus and other inclusions. The group has two types of pseudopodia: the axopodia are rigid and not ramified, whereas the filipodia are thin, ramified extensions of the ectoplasm.

Morphology and classification

The radiolarian skeleton or test consists of isolated or networked spicules, composed of opaline silica and forming sponge-like structures or trabeculae. Three of the main groups are recognized (Box 9.5) on the basis of skeletal structure and arrangement of

Box 9.2 Classification of Foraminifera

Suborder ALLOGROMIINA

- Organic tests, usually unilocular, occurring in fresh, brackish and marine conditions. Not usually fossilized
- Cambrian (Lower) to Recent

Suborder TEXTULARIINA

- Agglutinated tests consist of debris bound together with cement; both septate and non-septate
- Cambrian (Lower) to Recent

Suborder FUSULININA

- Microgranular tests, some with two or more laminae; septate and non-septate forms
- Ordovician (Llandeilo) to Permian (Changhsingian)

Suborder INVOLUTININA

- Aragonitic hyaline tests
- Permian (Rotliegendes) to Cretaceous (Cenomanian)

Suborder SPIRILLININA

- Calcitic hyaline tests, planispiral to conical
- Triassic (Rhaetic) to Recent

Suborder CARTERININA

- Tests comprise calcareous spicules in calcareous cement
- Tertiary (Priabonian) to Recent

Suborder MILIOLINA

- Porcellaneous tests, imperforate, both septate and non-septate, which are often large and complex
- Carboniferous (Viséan) to Recent

Suborder SILICOLOCULININA

- Imperforate tests of opaline silica
- Tertiary (Miocene) to Recent

Suborder LAGENINA

- Calcitic monolamellar tests, hyaline radial
- Silurian (Prídolí) to Recent

Suborder ROBERTININA

- Aragonitic, hyaline radial tests; both septate and finely perforate
- Triassic (Anisian) to Recent

Suborder GLOBIGERININA

- Calcitic, hyaline tests; finely perforate planktonic forms
- Jurassic (Bajocian) to Recent

Suborder ROTALIINA

- Calcitic, hyaline radial; perforate multilocular forms
- Jurassic (Aalenian) to Recent

Box 9.3 Modeling of foram tests

David Raup's theoretical work on the modeling of mollusk morphospace created a paradigm shift in our understanding of shell ontogeny (see p. 332). The skeletons of many groups of organisms can now be generated, mathematically, according to a simple set of equations in each case. The shapes of microfossils can also be modeled in this way, with a set of rules based on the angle of deviation, a translation factor and a growth factor (Tyszka 2006). By varying these, a huge range of possible and impossible tests can be homegrown on the computer (Fig. 9.6). The forms illustrated here are only a subset of the total number of possibilities. Interestingly, these sorts of computer models always generate some bizarre forms. The dysfunctional forms, for example, are geometrically possible but the shapes and volumes of the chambers could simply not function; vacant ranges on the other hand contain fully functional morphologies but these forms have not yet been found in the fossil record. Why not?

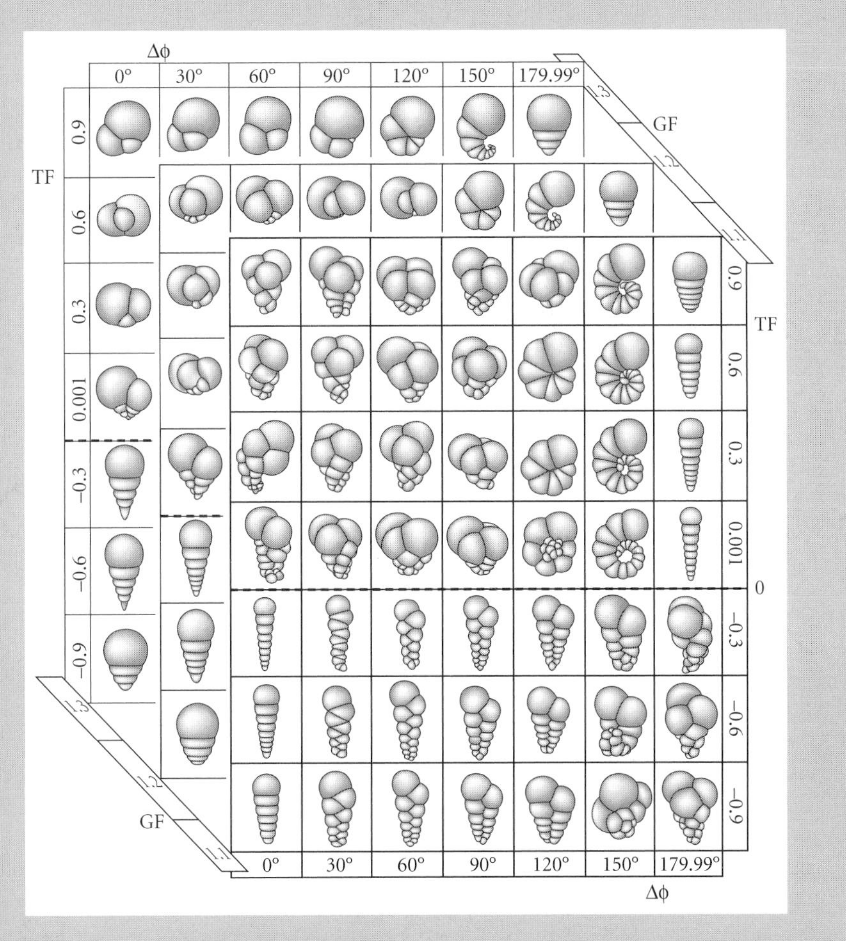

Figure 9.6 Modeling foraminiferan tests: part of a theoretical three-dimensional morphospace for foraminiferans. GF, growth factor; TF, translation factor; Δϕ, deviation factor. (From Tyszka 2006.)

perforations: the nassellarians (Fig. 9.10) and entactinarians develop a lattice from bar-like spicules, each end having a bundle of spicules. The initial nassellarian spicule is enclosed in the cephalis, and the skeleton develops further by the addition of segments following axial symmetry. By contrast, the initial entactinarian spicule is enclosed in a latticed or spongy test with radial symmetry based on a spherical body plan; this is similar to those of the spumellarians (Fig. 9.10), which however have a microsphere (instead of a spicule) internally.

Evolution and geological history

Although some records suggest an origin in the Mid Cambrian or earlier, the radiolarians became common in the Ordovician, and they are often found in deep-sea cherts associated with major subduction zones. The albaillellarians together with the entactinarians were the dominant forms, although after the Devonian, spumellarians with sponge-like tests were more prominent (Fig. 9.10).

Spumellarians remained important during the Triassic, with genera such as *Capnuchosphaera*, although the nassellarians had appeared; they continued as the major group through the Jurassic, Cretaceous and Early Tertiary. Late Tertiary forms evolved thinner skeletons, perhaps because of increased competition with the diatoms for mineral resources.

Radiolarian oozes cover about 2.5% of the ocean floors, accumulating at rates of 4–5 mm per 1000 years. Radiolarians are useful in paleo-oceanographic investigations, and they are particularly useful in dating the formation of deep-water sediments accumulating beneath the carbonate compensation depth (CCD), where carbonate-shelled organisms such as foraminiferans cannot survive. Radiolarian cherts and radiolarites commonly occur in oceanic facies preserved in mountain belts and are commonly associated with **ophiolites**, sections of the ancient ocean crust and upper mantle that have been uplifted (see p. 48), so they are very important in deciphering the origins and destruction of ancient ocean systems such as Tethys.

But the beauty of the radiolarian skeleton has also assured the group's place in the history of art (Box 9.6).

Box 9.4 Forams and environments

The ratio of agglutinated : hyaline : porcellaneous foram tests has been used extensively to differentiate among a range of modern environments. Ternary plots of the relative frequencies of test type distinguish fields for hypersaline and marine lagoons, estuaries and open shelf seas (Fig. 9.7). Fossil faunas may be plotted on these templates, and these allow paleontologists to estimate the salinity of ancient environments.

The ratio of infaunal : epifaunal benthic foraminiferans has also been widely used to determine the relative content of dissolved oxygen and/or organic carbon on the seafloor. Epifaunal and infaunal foraminiferans can be distinguished by their test morphologies, where epifaunal forms occur mainly in aerobic conditions with low amounts of organic carbon, and infaunal forms occur in more oxygen-deficient conditions with higher organic carbon content.

Measures of the ratio of benthic : planktonic foraminiferans are also useful in environmental studies. In general terms, the percentage of benthic taxa declines rapidly below depths of about 500 m in modern seas and oceans. Data from living assemblages have been used to interpret paleoenvironments with diverse fossil foraminiferan faunas. For example, microfossil analysis of the upper part of the Late Cretaceous chalk of the Anglo-Paris basin has suggested water depths of between 600 and 800 m during the Turonian on the basis of the high proportions of planktonic foraminiferans; however, by the Campanian, water depths of about 100 m are suggested by the rich benthic fauna.

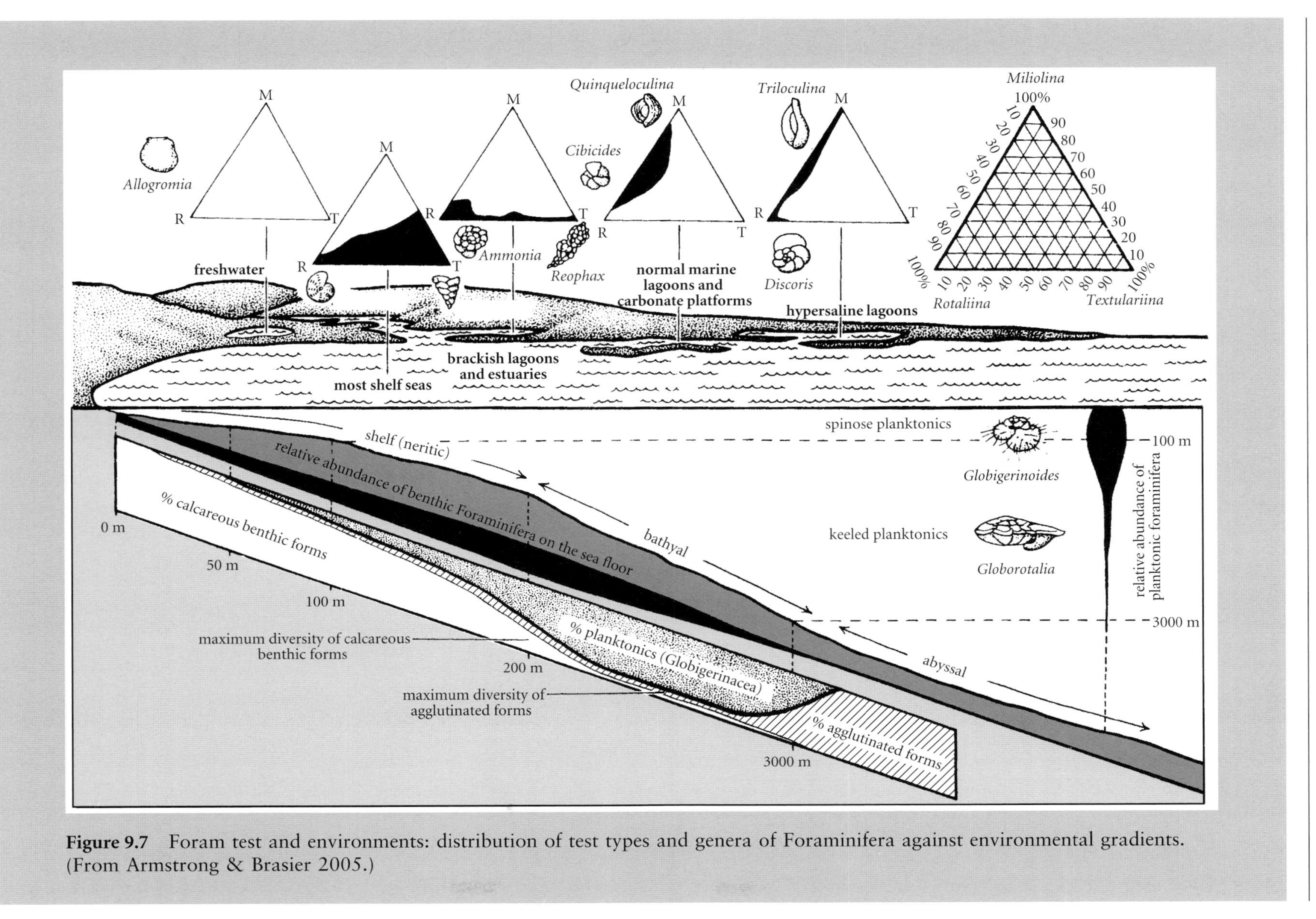

Figure 9.7 Foram test and environments: distribution of test types and genera of Foraminifera against environmental gradients. (From Armstrong & Brasier 2005.)

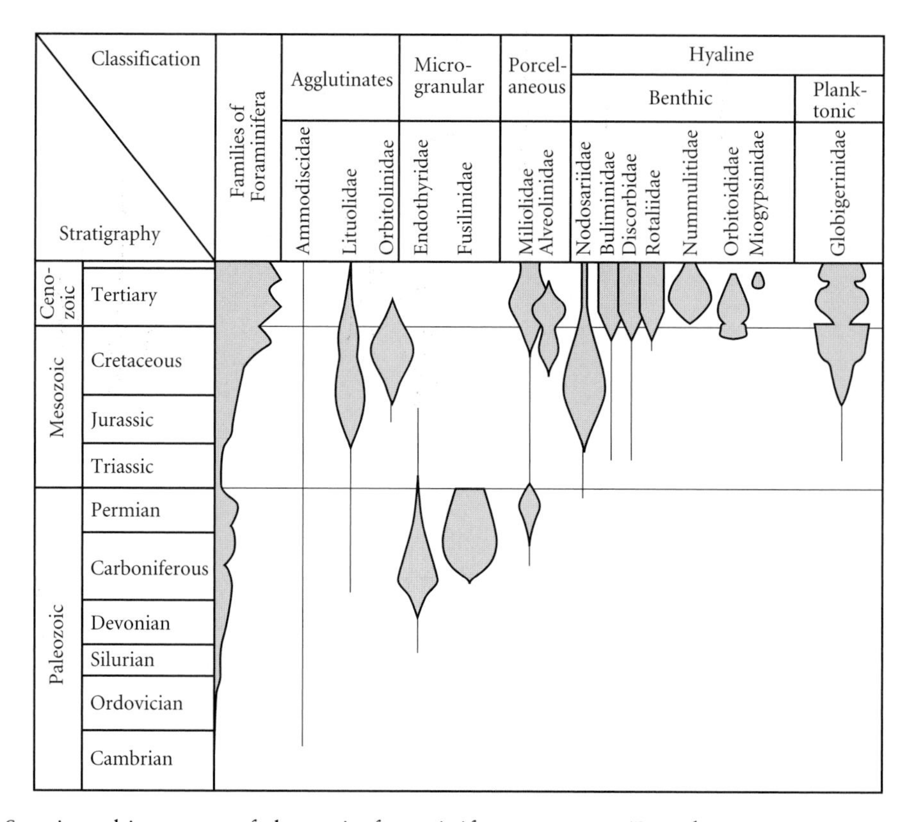

Figure 9.8 Stratigraphic ranges of the main foraminiferan groups. (Based on various sources.)

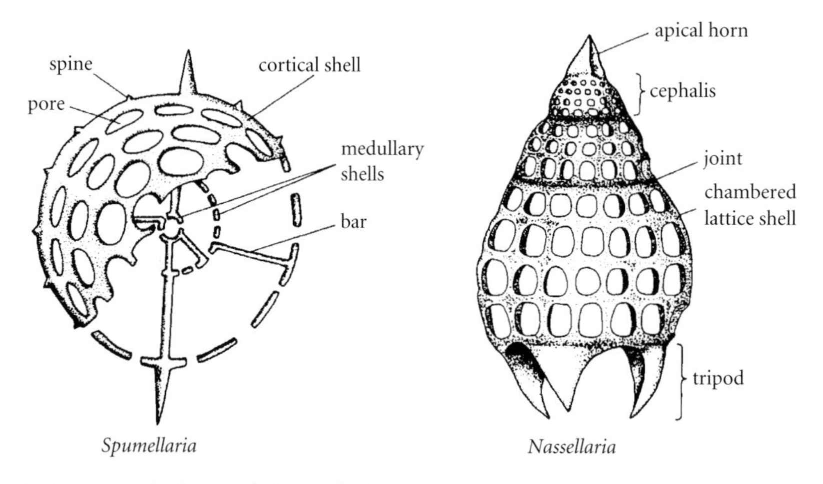

Figure 9.9 Descriptive morphology of the radiolarians.

Acritarchs

The acritarchs are a mixed bag of entirely fossil, hollow, organic-walled microfossils that are impossible to classify. The acritarchs are probably polyphyletic; they include a wide range of forms, probably representing the **cyst stages** or resting phases in the life cycles of various groups of planktonic algae. Fundamental work on the group by Alfred Eisenack (1891–1982) initially suggested that these tiny fossils were the eggs of planktonic invertebrates; however, later he considered the group to be fossil members of the phytoplankton, plants rather than animals. William Evitt of Stanford University, in establishing the scope of the group in the early 1960s, noted that his term "acritarch" (meaning "uncertain origin")

Box 9.5 Classification of the radiolarians

The classification of the radiolarians is currently in a state of flux. Six orders are recognized (De Wever et al. 2001).

Order ARCHAEOSPICULARIA

- Mid Cambrian to Silurian

Order ALBAILLELLARIA

- Late Ordovician to Late Silurian or ?Devonian

Order LATENTIFISTULARIA

- Early Carboniferous (or earlier?) to Permian

Order SPUMELLARIA

- Paleozoic (precise age uncertain) to Recent

Order ENTACTINARIA

- (?Cambrian), Ordovician to Recent

Order NASSELLARIA

- (?Late Paleozoic), Triassic to Recent

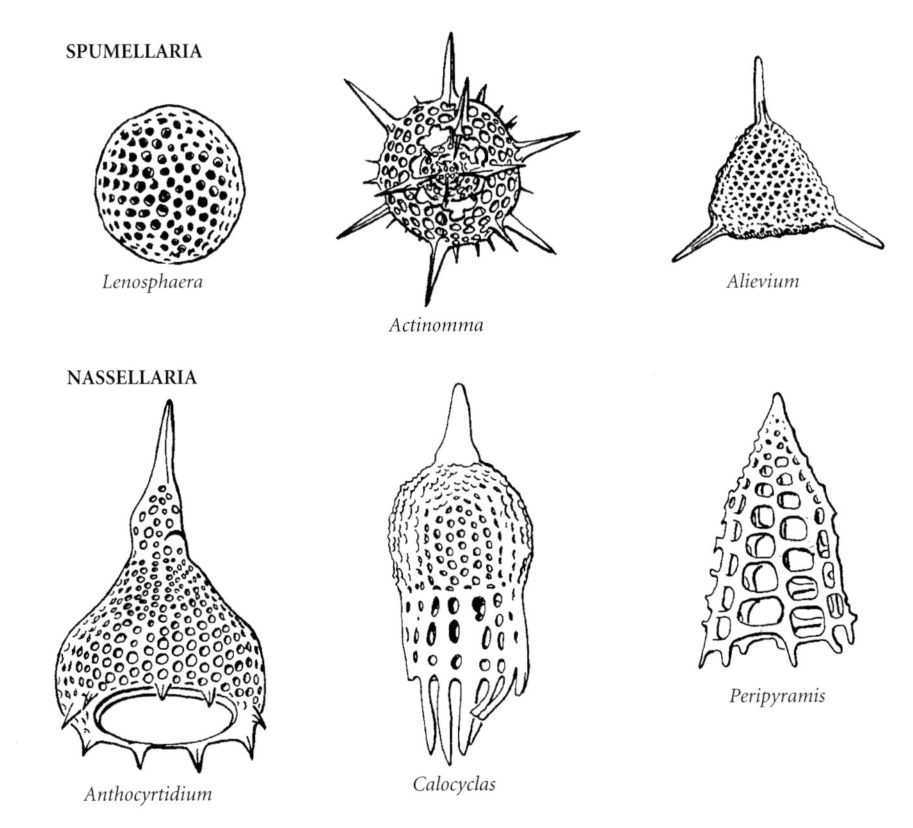

Figure 9.10 Some radiolarian morphotypes: *Lenosphaera* (×100), *Actinomma* (×240), *Alievium* (×180), *Anthocyrtidium* (×250), *Calocyclas* (×150) and *Peripyramis* (×150).

was a monument to our ignorance. Although many more taxa have been described since, and their value in biostratigraphic correlation has been proved, uncertainty still surrounds the origin and affinities of the group. Similarity, however, with the cyst stages of modern **prasinophytes** and **dinoflagellates** suggests a relationship to primitive green algae. However, because they are very useful in hydrocarbon exploration, perhaps the minor issue of their identity can be left for future generations!

Morphology and classification

The composition and broad morphology of the acritarchs suggest similarities with the dinocysts; like the dinocysts, acritarchs are also often found in clusters. The group probably had a similar life cycle to that of the dinoflagellates, single-celled protists that mainly live in the marine plankton today. Acritarchs seem to show encystment structures, or cysts – protective devices similar to those of modern dinoflagellates, in which the organism can survive drying out or lack of food for long periods. When conditions return to normal, usually when the cyst is covered with water again, the organism "escapes" by bursting through the watertight skin of the cyst, and resumes feeding and reproducing. A number of escape structures have been described including median splits, **pylomes** and **cryptopylomes**, that would have allowed material to seep out.

Acritarchs consist of **vesicles** composed of various polymers combined to form sporopollenin (Fig. 9.12). They range in shape from spherical to cubic and in size from usually 50 to 100 μm, although some specimens from the Triassic and Jurassic are as small as 15–20 μm. Many lose these morphological details when preserved as flattened films in black shales. There is a huge variety of basic shapes (Fig. 9.13). Acritarchs can have single- or double-layered walls; the wall structure is often useful taxonomically. The central cavity or chamber can be closed or open externally through a pore or slit called the pylome. The opening or **epityche** presumably allowed the escape of the motile stage and may be modified with a hinged flap.

On the outside the acritarch may be smooth or, for example, have granulate or microgranulate ornament. Moreover, the vesicle may be modified by various extensions or processes projecting outwards from the vesicle wall. If an acritarch has a set of similar processes, they are termed **homomorphic**, and if it has a variety of different projections it is **heteromorphic**.

Over 1000 genera of acritarchs are known, defined mainly on shape characteristics (Box 9.7). All acritarchs were aquatic with the vast majority found in marine environments. The classification of the group is based on the wall structure, the shape of the body vesicle, pylome type and the nature of the extensions and processes.

Box 9.6 Ernst Haeckel, art and the radiolarians

The link between art and paleontology has always been strong, with many images finding their inspiration in the beauty of the fossil form. Ernst Haeckel (1834–1919), the German evolutionary biologist, responsible for such terms as "Darwinism" and "ecology", the phrase "ontogeny recapitulates phylogeny" and the first detailed tree of life (see p. 128) was also an accomplished artist; he believed in the esthetic dimension of morphology (Fig. 9.11). His giant opus *Art Forms in Nature* (1899–1904) is considered to be one of the most elegant, artistic works of the 19th century, his illustrations being a paleontological precursor to the Art Nouveau movement. His style is nowhere better presented than in his monograph on the Radiolaria (Haeckel 1862). Unfortunately his attempts to associate science with art may have damaged his career, but current interest in the tree of life has generated a Haeckel renaissance. His illustrations are even available now as an attractive screensaver!

You can see these beautiful images at http://www.blackwellpublishing.com/paleobiology/.

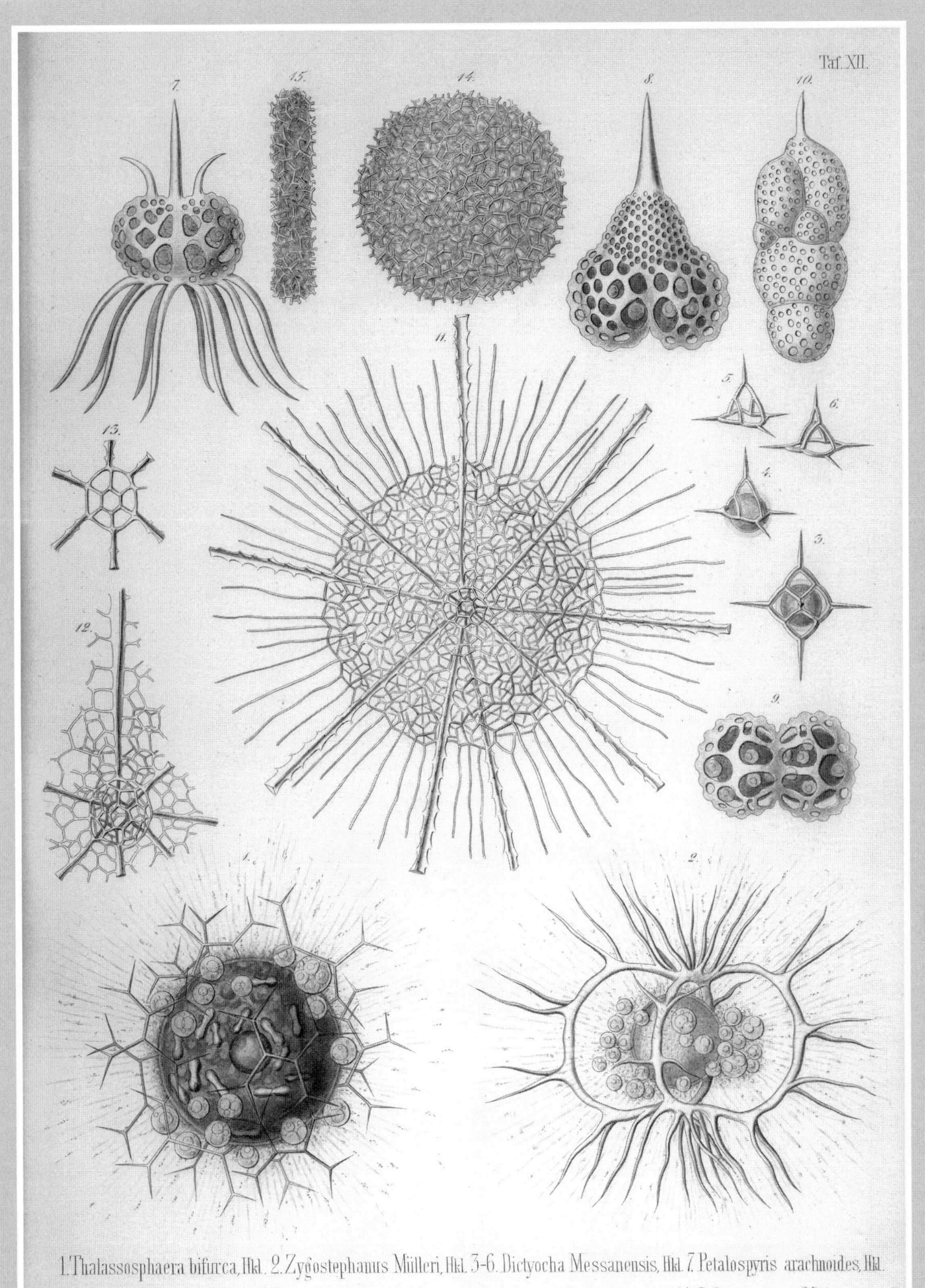

Figure 9.11 Haeckel's radiolarians: plate 12 from *Die Radiolarien (Rhizopoda Radiaria)* by Ernst Haeckel (1862).

Box 9.7 Classification of the main organic-walled groups: acritarch form classification

The classification of acritarchs is based entirely on morphology and as such is merely a set of shape and ornament categories with no phylogenetic status. The names of the main groups thus are also used as morphological terms to define the variation in shape (see Fig. 9.12). Clearly such a classification is rife with convergent morphotypes that may never be properly classified. Recent studies, however, suggest that understanding the mode of encystment may be a step towards the development of a more phylogenetic classification.

ACRITARCHS WITHOUT PROCESSES OR FLANGES

Sphaeromorphs

- Spherical forms lacking processes but with ornamented walls. These morphs are often variably ornamented
- Precambrian (Animikean) to Recent

ACRITARCHS WITH FLANGES BUT LACKING PROCESSES

Herkomorphs

- Subpolygonal or spherical with polygonal ornament defined by crests
- Cambrian (Lower) to Recent

Pteromorphs

- Forms equipped with an equatorial flange
- Ordovician (Caradoc) to Recent

ACRITARCHS WITH PROCESSES BUT WITH FLANGES

Acanthomorphs

- Spherical forms lacking an inner body and crests, with simple or branching processes

Polygonomorphs

- Polygonal forms with simple processes

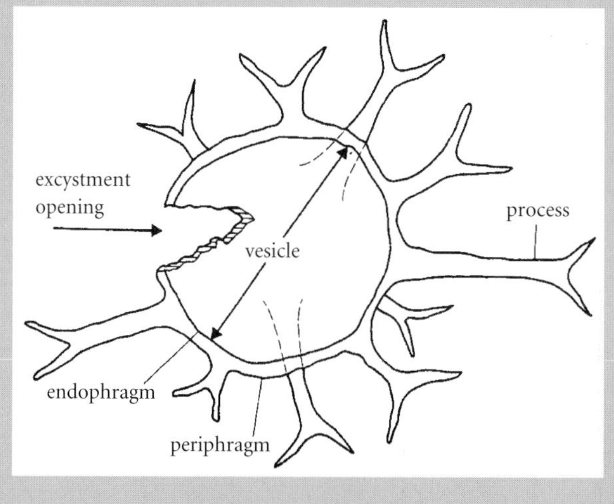

Figure 9.12 Descriptive morphology of the acritarchs.

Netromorphs
- Elongate, commonly fusiform morphs with poles variably developed as processes or spines

Diacromorphs
- Spherical to ellipsoidal, with ornament restricted to around the poles

Prismatomorphs
- Polygonal or prismatic, with edges commonly extended as flanges

Oomorphs
- Egg-shaped forms, one end smooth and the other highly ornamented

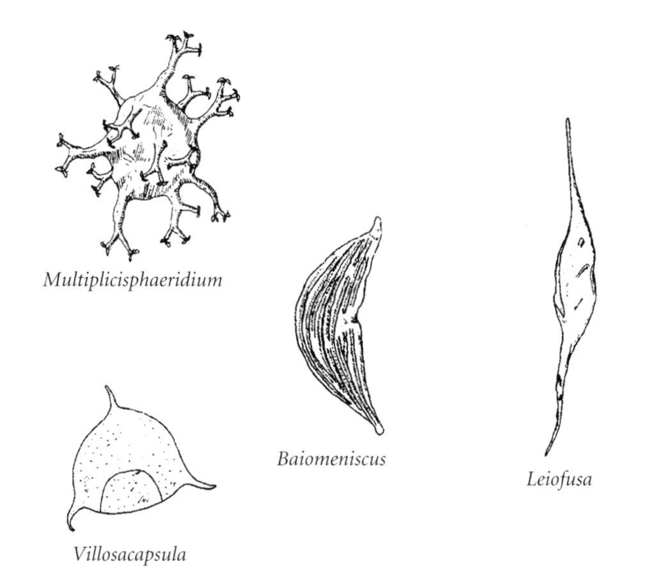

Figure 9.13 Some acritarch morphotypes: *Multiplicisphaeridium* (×800), *Baiomeniscus* (×200), *Leiofusa* (×400) and *Villosacapsula* (×400).

Evolution and geological history

Acritarchs had a wide geographic range, apparently mainly controlled by latitude; the entire group ranged from the poles through the tropics. The wide distribution of the group is similar to that of the dinoflagellates and strongly suggests that acritarchs were also members of the phytoplankton. Biogeographic provinces have been established for the Ordovician, Silurian and Devonian periods and have helped reconstruct ancient climate belts and oceanic currents. Acritarchs have also been of considerable value in regional correlations, particularly during the Ordovician and Silurian.

The acritarchs are some of the oldest documented fossils with a history of over 3000 myr, although the group was not common until some 1 Ga, when the first major diversification of the group, predating the Ediacara biota (p. 242), was marked by large spheromorphs, acanthomorphs and polygonomorphs. During the important Early Cambrian radiation of the group, spinose morphs such as *Baltisphaeridium* and *Micrhystridium*, together with the crested *Cymatiosphaera*, appeared. Significantly these armored vesicles evolved during the expansion of marine predators: Was this a form of arms race or merely a coincidence? By the Late Cambrian to Early Ordovician, acritarch palynofacies (pollen and spore assemblages) were dominated by three main groupings: the *Acanthodiacrodium*, *Cymatiogalea* and *Leiofusa* groups (Box 9.8). The acritarchs declined during the Devonian, and are rare in Carboniferous-Triassic rocks. Nevertheless the group staged a weak recovery during the Jurassic and continued through the Cretaceous and Tertiary.

Dinoflagellates

The dinoflagellates, or "whirling whips", comprise a group of microscopic algae with organic-walled cysts. The life history of these organisms thus oscillates between a **motile** (swimming) and a cyst (resting) stage; the cysts usually range in size from 40 to 150 μm. The motile phase is either flexible and unarmored, or rigid and armored with a network of plates, the **theca**; the arrangement of the

Box 9.8 Acritarchs and the food chain

Groups such as the acritarchs formed a prominent base to the relatively short, suspension-feeding Early Paleozoic food chains, yet it is virtually impossible to quantify the abundance of microfossils in sediments because many factors such as cyst production, hydrodynamic sorting and taphonomy come into play. Unfortunately, diversity cannot be used as a proxy for abundance, so there is no direct evidence in the fossil record of just how densely packed the water column was with phytoplankton, say during the Ordovician. However, it may be possible to speculate that primary production increased rapidly during the Ordovician: This period was marked by the appearance and radiation of the graptolites, phyllocarids, some groups of echinoderms and the radiolarians. Huge bursts in diversity are seen among the brachiopods, mollusks and trilobites, while there was increasing complexity in benthic and reef communities. Yet little is known about the cause of this phenomenal diversification. Marco Vecoli and his colleagues (2005) have now suggested that these massive metazoan radiations probably signal a cryptic explosion in primary production in the world's oceans (Servais et al. 2008) that may have been one of the main triggers for the great Ordovician biodiversification (see p. 253). The diversity curves of these protistan groups appear to match perfectly those of the metazoans (Fig. 9.14).

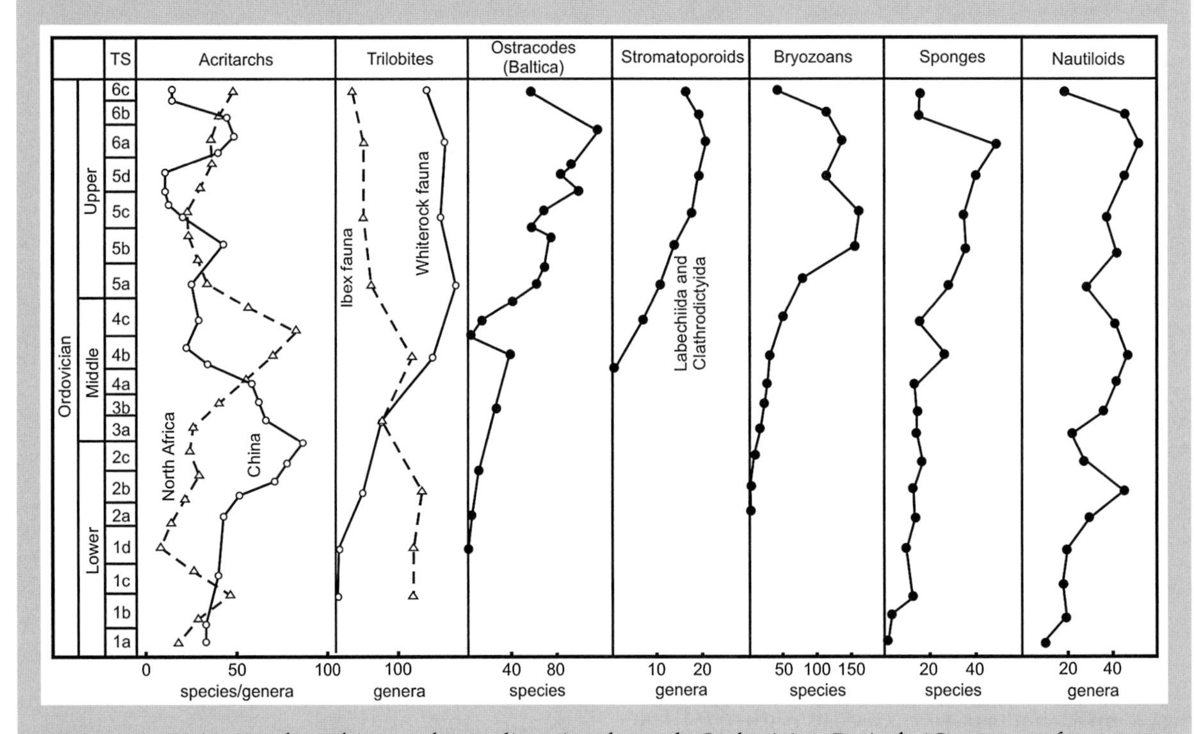

Figure 9.14 Acritarch and invertebrate diversity through Ordovician Period. (Courtesy of Thomas Servais.)

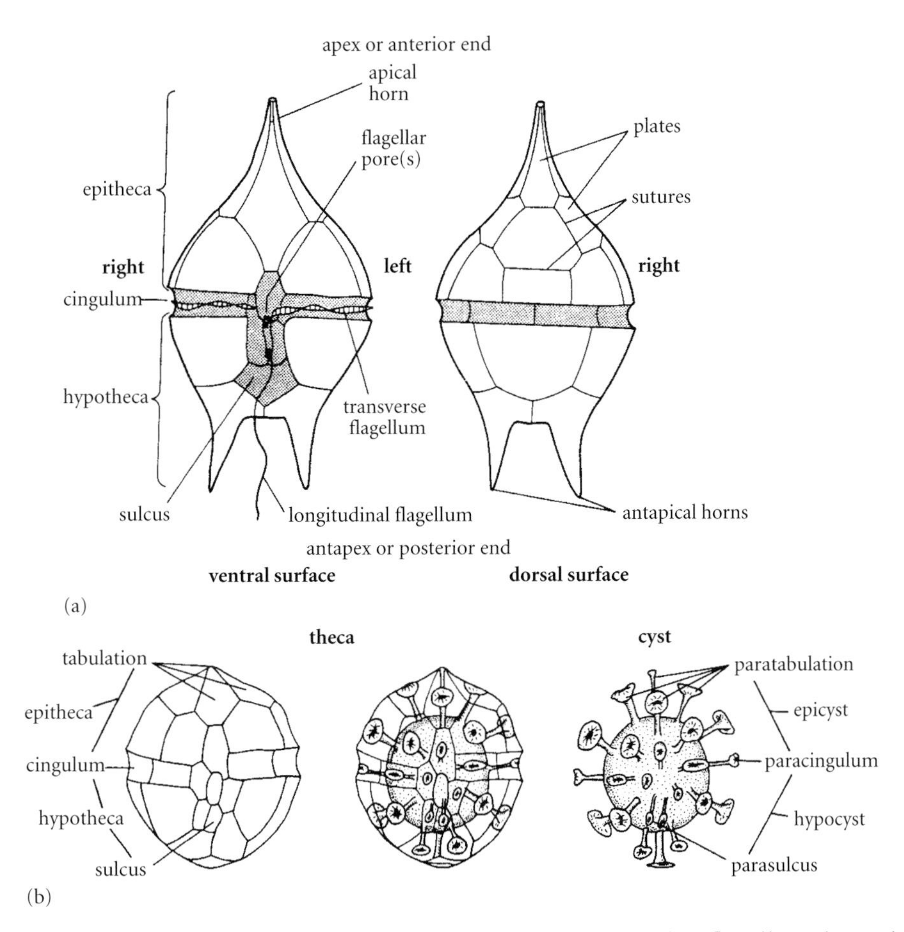

Figure 9.15 Descriptive morphology of (a) a dinoflagellate, and (b) a dinoflagellate theca (left), unpeeled (middle) to reveal the corresponding cyst (right).

thecal plates comprises the **dinoflagellate tabulation**.

Morphology and classification

The plates of a dinoflagellate theca are arranged from the **apex** to **antapex** as follows: apical, precingular, cingular, postcingular and antapical; the first two are part of the eipitheca and the last two, the hypotheca (Fig. 9.15). There are a number of other plates with further specialized terms and together the plates are commonly labeled and numbered in sequence. The motile phase is rarely fossilized. In contrast, the cysts are chemically resistant and relatively common. The morphology of a motile dinoflagellate is crudely similar to its theca and comparable structures in the motile form are prefixed by the term "para".

Cysts have a paratabulation that is useful taxonomically (Box 9.9): for example, the cysts of peridiniaceans have seven **precingular** and five **postcingular** paraplates, whereas the gonyaulacaceans have six precingular and six postcingular paraplates.

Dinoflagellates are abundant and diverse members of the living and more recent fossil phytoplankton (Fig. 9.16), forming an important part of the base of the food chain of the oceans; they may in fact be second only in abundance to diatoms as primary producers. However, dinoflagellate **blooms** or red tides, when there is huge population explosion, can lead to asphyxiation of other marine groups. Mass mortalities of Cretaceous bivalves in Denmark and of Oligocene fishes in Romania have been blamed on fossil red tides.

There are three main cyst types. The proximate cyst is developed directly against the theca itself and has a similar configuration. A

Box 9.9 Classification of the main organic-walled groups: dinoflagellate form classification

Class DINOPHYCEAE

Most dinoflagellates belong to this class, which includes fossil representatives. They are free-living cells with a large nucleus and numerous chromosomes; some are parasites and symbionts.

Order GYMNODINIALES

- Cretaceous to Recent

Order PTYCHODISCALES

- Cretaceous

Order SUESSIALES

- Triassic to Recent

Order NANNOCERATOPSIALES

- Jurassic

Order DINOPHYSIALES

- Jurassic?

Order DESMOCAPSALES

- Recent

Order PHYTODINIALES

- Recent

Order GONYAULACALES

- Jurassic to Recent

Order PERIDINIALES

- Triassic to Recent

Order THORACOSPHAERALES

- Triassic to Recent

Class BLASTODINIPHYCEAE

Parasites on or in copepods and other animals.

Order BLASTODINIALES

- Recent

Class NOCTILUCIPHYCEAE

Very large, naked cells lacking chloroplasts.

Order NOCTILUCALES

- Recent

Class SYNDINIOPHYCEAE

Symbionts or endoparasites lacking chloroplasts

Order SYNDINIALES

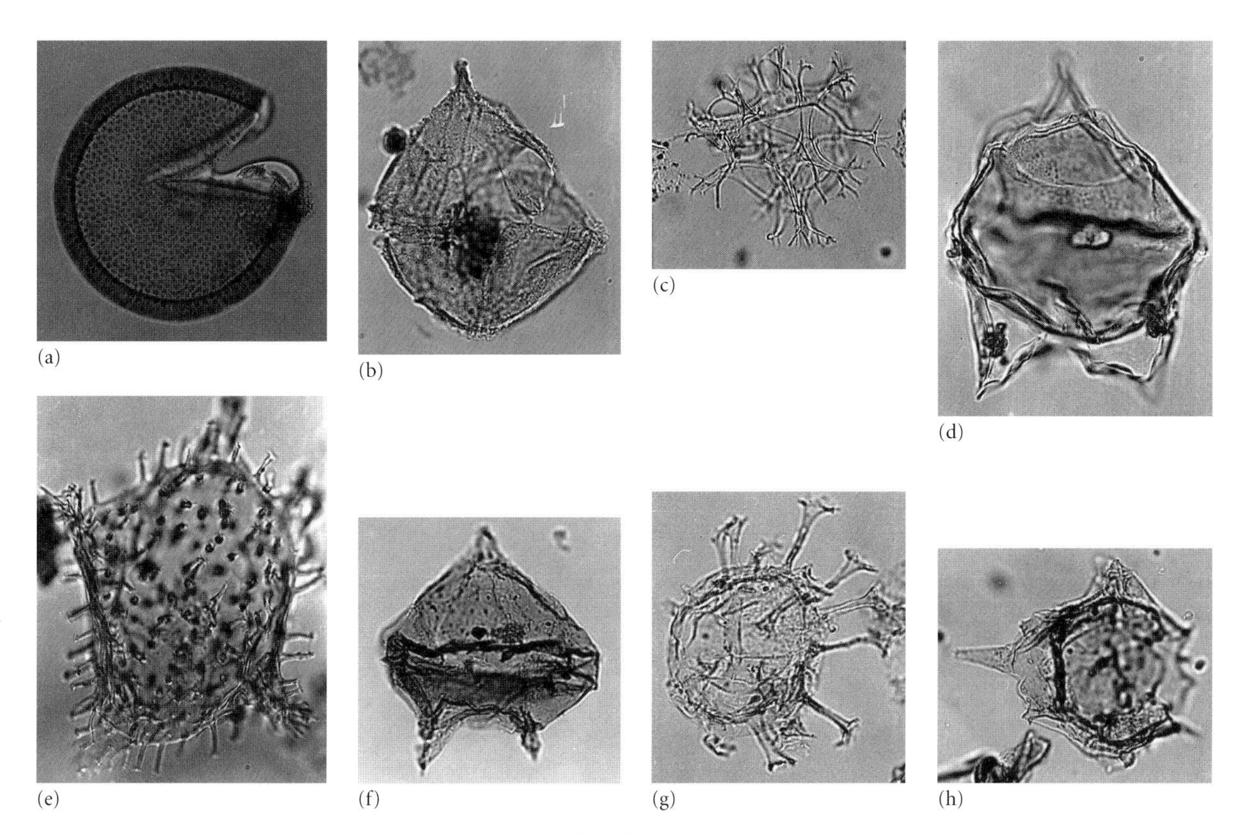

Figure 9.16 A prasinophyte (a) and some dinoflagellate taxa (b–h): (a) *Tasmanites* (Jurassic), (b) *Cribroperidinium* (Cretaceous), (c) *Spiniferites* (Cretaceous), (d) *Deflandrea* (Eocene), (e) Wetzeliella (Eocene), (f) *Lejeunecysta* (Eocene), (g) *Homotryblium* (Eocene), and (h) *Muderongia* (Cretaceous). Magnification ×250 (a, d, e), ×425 (b, c, f, g, h). (Courtesy of Jim Smith.)

chorate cyst is smaller than the theca and the cysts are contained within the theca, interconnected by various appendages and spines, which are related to the external tabulation of the theca. In cavate morphs there is a gap between the cyst and the theca at the two poles.

Evolution and geological history

Dinoflagellate biomarkers have been identified in Upper Proterozoic and Cambrian rocks. Moreover the Late Precambrian and Paleozoic diversifications of the acritarchs may mark an early phase in dinoflagellate radiation, involving non-tabulate forms. To date, however, the oldest dinoflagellate cyst is probably *Arpylorus* from the Ludlow (Upper Silurian) rocks of Tunisia; the cyst has feeble paratabulation and a precingular archeopyle. Oddly, there is a long gap after this record until the Early Triassic, when *Sahulidinium* appears off northwest Australia. Some authors have suggested that a number of Paleozoic acritarch taxa may in fact be dinoflagellates. Multiplated forms such as *Rhaetogonyaulax* and *Suessia* appearing in the Late Triassic characterize dinocyst floras ranging from Australia to Europe. *Nannoceratopsis* cysts with characteristic archeopyles and tabulation are common in Early Jurassic floras, while *Ceratium*-like forms appeared first during the Late Jurassic and diversified in the Cretaceous. Many precise zonation schemes for Mesozoic and Cenozoic strata are based on dinocyst distributions. However, during the Eocene the global biodiversity of the group began a steady decline.

Ciliophora

The Ciliophora today consist of some 8000 species of single-celled organisms that swim by beating their **cilia**, minute hair-like organs. Two fossil groups, the calpionellids and tintinnids, may belong here. Calpionellids are a

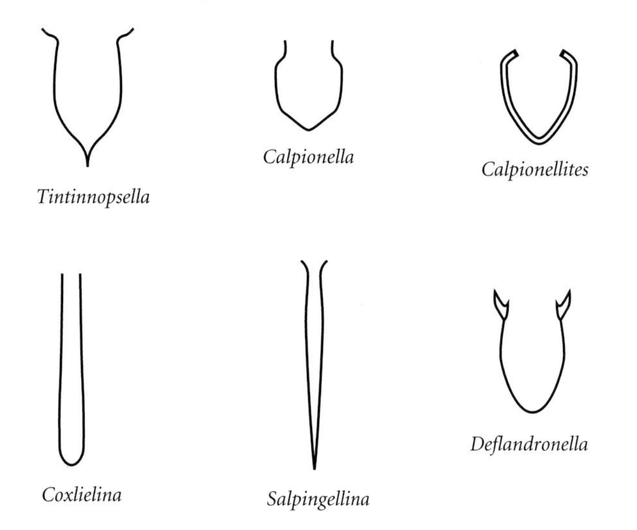

Figure 9.17 Morphology of some tintinnids in cross-section from limestones (×100–200).

group of extinct, cup-shaped, calcareous microfossils that were abundant in Late Jurassic and Early Cretaceous pelagic sediments, especially in the Tethyan realm. As an extinct group with no complex characters, no definitive evidence of their affinities has been found; however, they are strikingly similar in shape and size to an important group of ciliates, the tintinnids.

Tintinnids are part of the zooplankton, grazing on phytoplankton and providing a food source for larger members of the plankton. The cell is enclosed within a cup-shaped test or lorica, often 10 times larger than the cell itself. Modern tintinnids have an organic lorica with, in some cases agglutinated mineral grains or coccoliths, but without biomineralization, whereas the fossil calpionellids had a primary calcareous test (Fig. 9.17).

Two families of fossil tintinnid have been recorded, together ranging in age from the Tithonian (Upper Jurassic) to the Albian (Middle Cretaceous).

CHROMISTA

The chromistans are probably a paraphyletic group of eukaryotes that usually contains chloroplasts with chlorophyll c, which is absent from all known plant groups. The group includes various algae, the coccolithophores and the diatoms and the majority are primary producers, functioning as part of the phytoplankton.

Coccolithophores

Nannoplankton, are defined as plankton less than 63 μm across, the smallest standard mesh size for sieves. Although the nannoplankton includes organic-walled and siliceous forms, the calcareous groups are most prominent in living floras and dominate the fossil record. Coccolithophores are the dominant members of the fossil calcareous nannoplankton, and the calcareous plates they produce, **coccoliths**, dominate nannofossil assemblages. Many calcareous nannofossils lack obvious shared characters with coccoliths and so are excluded from the coccolithophores and instead are termed **nannoliths**. These nannoliths may be related to coccolith-bearing organisms, but in view of their diversity in form, the group may contain calcareous structures produced by quite unrelated microbes. As a whole, calcareous nannoplankton first appeared during the Late Triassic, increased in abundance and diversity through the Jurassic and Cretaceous, reaching an acme of diversity in the Late Cretaceous. They were severely affected by the KT mass extinction, but subsequently radiated in the Early Paleogene and remained a major component of the calcifying plankton throughout the Cenozoic. They are extremely abundant in the surface waters of modern oceans.

Morphology and classification

Coccolithophores are unicellular algae, predominantly autotrophic in dietary mode, usually ranging in size from 5 to 50 μm, and globular, fusiform or pyriform in shape. The group constitutes the Phylum Haptophyta, within the Kingdom Chromista, together with various closely related non-calcifying algae; they have golden-brown photosynthetic pigments and, in motile phases, two smooth flagella together with a third flagellum-like structure, the **haptonema**. Coccolithophores are almost exclusively marine (there is just one, rather rare, freshwater species), usually open marine, occupying the photic zone where they photosynthesize. The group today is most diverse and has its highest relative abundances in the tropics although coccolithophores occur at all latitudes. The shell is composed of distinctive calcitic platelets or coccoliths. These are produced intracellularly;

Box 9.10 Atomic force microscopy of coccolithophores

Coccolithophores, despite their small size, are attractive and sophisticated organisms. A number of plate morphs, emphasizing the diversity of form within the group, have been described (Fig. 9.18c): **asterolith**, star-shaped plates; **cyclolith**, open rings; **lopadolith**, vase-shaped morphs with elevated edges; **placolith**, two disks fused by the median tube; **stetolith**, column-shaped plates; **zygolith**, elliptical ring with arches applied to holococcoliths. Apart from the term placolith most are not in routine use. Additionally, **helioliths**, composed of a large number of small radially arranged crystals, and **ortholiths**, with only a few crystals, have been recognized.

The coccolithophore is precipitated within the cell from the coccolith vesicle or Golgi body with tightly regulated crystal growth, allowing the crystals to integrate as the complex and exquisite networks that comprise a complete skeleton. Karen Henriksen, a former graduate student at the University of Copenhagen, applied atomic force microscopy (AFM) to the surface of three coccolith species, a technique that allows investigation at higher orders of magnitude than even scanning electron microscopy (SEM) and transmission electron microscopy (TEM) equipment. Henriksen and colleagues (2004) established key differences among these taxa suggesting that subtle changes in the mechanisms of biomineralization can drive significant changes in morphology that have knock-on effects for the adaptability, lifestyle and distribution of the coccolith species. The large morphological disparity seen in this remarkable group is thus a function of the mode and orientation of crystal growth at the atomic level and where the organism ultimately lived depended on the whims of a crystal lattice.

they then migrate to the cell surface and are expelled to form a composite exoskeleton, the **coccosphere**. Commonly the coccosphere consists of 10–30 discrete coccoliths, although some forms have many more (Box 9.10). Many taxa produce coccospheres formed of only one type of coccolith, but others show a variety of coccolith morphologies (Fig. 9.18); in particular there are often specialized coccoliths around the flagellar pole of the cell. There are two fundamentally different types of coccoliths: heterococcoliths have a radial array of relatively few (typically 20–50) complex-shaped crystal units, whereas holococcoliths are formed of planar arrays of hundreds of minute uniform-sized (typically c. 0.1 μm) rhombohedral crystallites.

Haptophyte life cycles were very poorly known until recently; research has now shown that cocolithophores, and possibly most haptophtes, typically have alternating haploid and diploid stages that are both capable of asexual reproduction. Coccolithophores usually have life cycles consisting of two main phases producing radically different coccoliths that were often described initially as two different species. The **haploid phase** (with half the complement of chromosomes) is always flagellate, and is usually coated by minute holococcoliths; the **diploid phase** (with full complement of chromosomes) is usually non-flagellate, and is coated by heterococcoliths. Both phases are capable of indefinite asexual reproduction and it appears likely that the two-phase life cycle is an adaptation allowing coccolithophores to survive challenging ecological conditions. The haploid (holoccolith-producing) phase is thought to be adapted to oligotrophic conditions (when nutrients are scarce) whilst the diploid (heterococcolith-producing) phase is thought to be adapted to more eutrophic conditions (when nutrients are abundant).

The classification of extant coccolithophores is based largely on coccosphere morphology and coccolith structure because the intricate and distinctive form of coccoliths makes them ideal for morphological classification. Cell characters can only be studied with transmission electron microscopy and have generally proved rather invariant. Data from cytology and molecular genetics have strongly supported the classification based on morphological criteria. The reliance on coccoliths in the extant classification also means that there are relatively few problems in align-

Figure 9.18 Some coccolith morphotypes: (a) coccospheres of the living *Emiliana huxleyi*, currently the most common coccolithophore (×6500), and (b) Late Jurassic coccolith limestone (×2000). (c) Coccolith plate styles: 1 and 2, *Coccolithus pelagus*; 4 and 5, *Oolithus fragilis*; 5 and 6, *Helicosphera carteri*. In *C. pelagus* and *H. carteri* growth was upwards and outwards with the addition of layer upon layer of calcite; in *O. fragilis* growth was different with curved elements, in non-parallel to crystal cleavage directions. (a, b, courtesy of Jeremy Young; c, courtesy of Karen Henriksen.)

ing modern and fossil taxonomies. Some modern coccolithophores are polymorphic, producing several different types of coccoliths, but these are mostly in taxa with a limited fossil record. More interesting problems are posed by the alternation of holococcolith-bearing and heterococcolith-bearing phases in the life cycle of a single species. In modern coccolithophores the taxonomy is being adjusted to reflect this as data become available.

Together with diatoms, dinoflagellates and picoplankton (tiny, single-celled plankton 0.2–2.0 µm in size), coccolithophores are the most abundant phytoplankton in modern oceans. The greatest diversity is developed in the tropics. Dependence on sunlight for photosynthesis restricts the group to the photic zone, with a depth range of 0 m to about 150 m. Within wave-mixed surface waters there is normally only a slight vertical stratification of assemblages, but a quite different assemblage is often developed beneath the thermocline.

Evolution and geological history

Rare coccoliths first appeared in the Late Triassic and increased in numbers during the Jurassic and Cretaceous; the group peaked in the Late Cretaceous, and chalk from that interval is almost entirely composed of these nannofossils. Only a few species survived the end-Cretaceous extinction event but they radiated again during the Cenozoic, recovering their numbers and abundance. However, in the last 4–5 myr there has been a marked decline in the abundance of larger coccoliths and, as a result, they have become less abundant in oceanic sediments, typically forming only 10–30% of modern calcareous oozes. Biostratigraphic zonal schemes using coccolithophores have been established from the Jurassic to the present day, and these are widely applied because they are reliable and operate over great distances. Moreover, basic biostratigraphic analyses of coccolithophore samples can be carried out rapidly, typically requiring less than an hour per sample. This is because nannofossils are abundant enough to be studied in simple strew mount preparations and can be reliably identified in cross-polarized light. Nannofossils only occur in low-energy marine sediments and are easily destroyed by diagenesis, but when they are present they provide an ideal means of rapidly dating sediments.

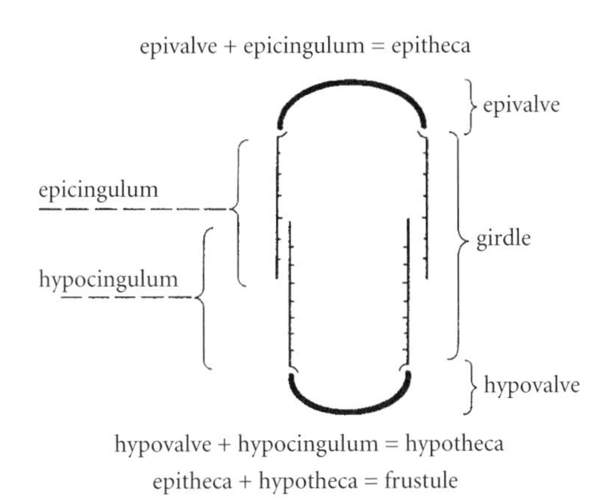

Figure 9.19 Descriptive morphology of the diatoms.

Diatoms

Diatoms are unicellular autotrophs that are included among the chrysophyte algae; they are characterized by large green-brown chloroplasts. Both individuals and loosely integrated colonies of diatoms occur in a range of aquatic environments from saline to freshwater and across a range of temperatures, being particularly common in the Antarctic plankton. Both benthic and planktonic life modes occur, although within the plankton one group – the Centrales – prefer marine environments; the Pennales, on the other hand, are more common in freshwater lakes (Box 9.11).

Morphology and classification

The diatom cell is contained within a siliceous skeleton or **frustule** comprising two unequally-sized valves or **thecae** (Fig. 9.19). The smaller hypotheca fits into the larger epitheca; the valve plates and congula of both valves interface with the congulum of the epitheca covering that of the hypotheca to form a connective seal.

During reproductive fission, both the parent valves are used as the epitheca by the offspring, which then constructs its own hypotheca. This process occurs a number of times each day, progressively reducing the size of the fustule. A stage of sexual reproduction kicks in to restore the growth momentum of the individual.

Classification of the group is based on shell morphology (Box 9.11).

Evolution and geological history

Both diatom frustules and, more commonly, endospores are preserved in the fossil record. A Late Jurassic assemblage from western Siberia that includes *Stephanopyxis* may be the oldest known diatom flora. The first diverse floras appeared during the mid-Cretaceous with almost 10 families recorded from Aptian rocks; the group further diversified after the Turonian. Nearly 100 genera of centric diatoms are recorded from the Upper Cretaceous. Some of the first pennate diatoms appeared during the Paleogene, colonizing freshwater environments for the first time; the group reached an acme during the Miocene.

Remarkably, diatom frustules can accumulate as thick deposits of diatomite (sometimes up to 500 m thick), which is a very porous sediment, often with 80% as spaces, and permeable with a density of about 0.5 g cm^{-1}. These **diatomites**, also termed kieselguhrs and tripolis, are widely used as purifiers for filtering drinks, medicines and water. Over 2 million tons are extracted each year for commercial use. Modern sedimentation rates suggest that 4–5 mm of diatomaceous ooze is deposited over 1000 years; such an ooze currently occupies over 10% of the ocean floor today. Major commercial deposits occur in the Miocene of the Ardèche, France and in the Pliocene and Pleistocene of Cantal, France are some of the main suppliers, although other deposits occur in Spain, Germany and Russia. The Miocene Monterey Formation in California is particularly widespread, occurring in both onshore and offshore basins; this diatomaceous mudstone is also the source and reservoir rock for most of California's petroleum.

Chitinozoans

Chitinozoans are most common in fine-grained sediments, usually those deposited in anoxic environments, and are associated with pelagic macrofauna such as graptolites and nautiloids together with acritarchs. In some lithologies, such as black slates, chitinozoans are the only fossils preserved. These associations, together with their widespread geographic range, suggest that chitinozoans were at least pelagic. The group has proved extremely useful for both regional and global

Box 9.11 Classification of siliceous-walled groups: diatom classification

Two main divisions are recognized based on their shell morphologies: the Centrales, as the name suggests, have round valves with pores radiating in concentric rows from the valve center; the Pennales have more elliptical valves with the pores arranged in pairs (Fig. 9.20). The latter are usually characterized by a median gash or **raphe**.

Order CENTRALES

Suborder COSCINODISCINEAE

- Valves with a ring of marginal processes

Suborder RHIZOSOLENIINEAE

- Valves are unipolar

Suborder BIDDULPHIINEAE

- Valves are bipolar

Order PENNALES

Suborder ARAPHIDINEAE

- Valves without a raphe

Suborder RAPHIDINEAE

- Valves with a raphe

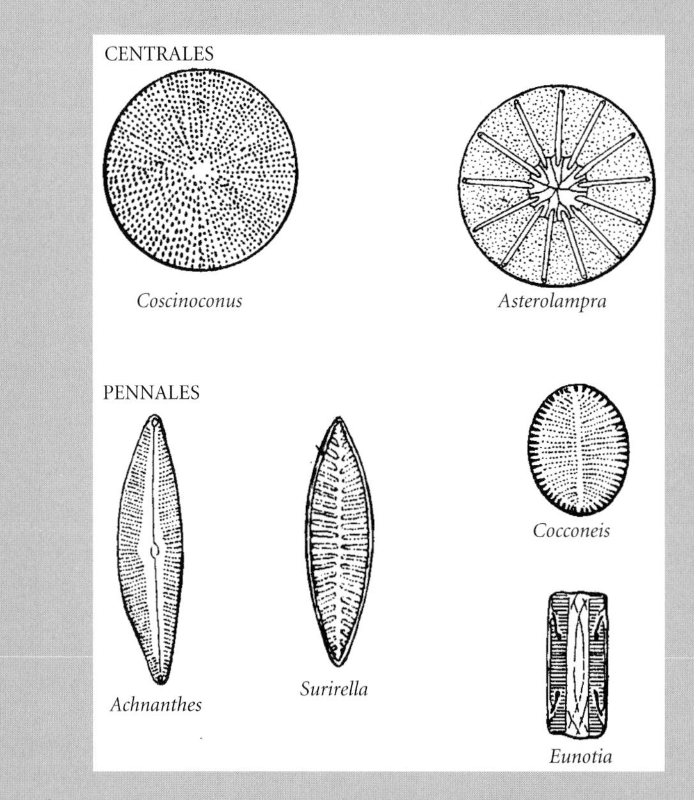

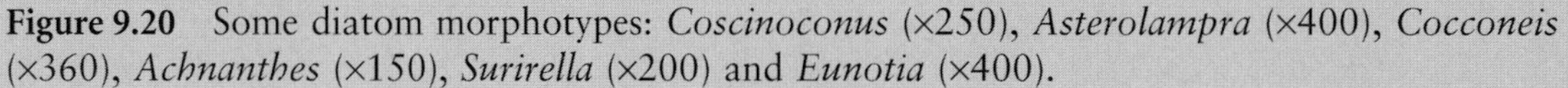

Figure 9.20 Some diatom morphotypes: *Coscinoconus* (×250), *Asterolampra* (×400), *Cocconeis* (×360), *Achnanthes* (×150), *Surirella* (×200) and *Eunotia* (×400).

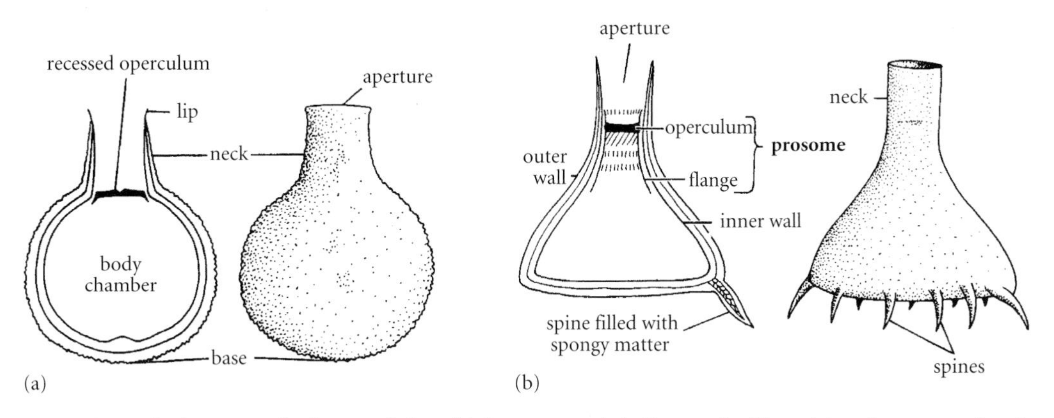

Figure 9.21 Descriptive morphology of the chitinozoans: (a) Operculatifera (simplexoperculate), *Lagenochitina*, and (b) Prosomatifera (complexoperculate), *Ancyrochitina*.

correlations and is a key part of global biostratigraphic schemes for the Ordovician and Silurian systems.

Morphology and classification

Chitinozoans are small (between 50 and 2000 μm), flask- to vase-shaped, hollow vesicles with smooth or ornamented surfaces (Fig. 9.21). The vesicles were thought to have consisted of a protein called **pseudochitin** similar in composition to the graptolite rhabdosome, but recent research suggests that they are actually composed of networks of kerogen, and chitin is in fact absent from the pyrolysates vaporized from the vesicle (Jacob et al. 2007). The vesicle encloses a chamber that ranges in shape from spherical through ovoid to cylindrical and conical forms. The chamber opens through an **aperture** at the oral end, either directly or at the end of a neck with a collar. The aperture is closed by an operculum that may be supported by the prosome. The base of the vesicle may be flat or extended as a variety of structures, for example a **copula** (long hollow tube), **mucron** (short hollow tube), **siphon** (bulb-like process) or **peduncle** (solid process). There are nearly 60 genera of chitinozoans.

The precise affinity of the group remains uncertain (Box 9.12). Chitinozoan vesicles were probably tightly sealed, and they occur as chains and clusters that suggest they may have been eggs or egg capsules or even dormant cysts. Chitinozoans have, in fact, been interpreted in the past as egg cases of a huge range of invertebrates, such as annelids, echinoderms, gastropods and graptolites, but they were probably the products of some soft-bodied, worm-like animal during a pelagic life stage.

Two main groups of chitinozoans have been established based on the way the vesicle is sealed, and are further subdivided according to the outline or silhouette of the vesicle together with modifications of the neck. The **Operculatifera** have a relatively simple operculum and they lack a neck (including the Desmochitinidae with small subspherical vesicles), whereas the **Prosomatifera** have a more complex opercula with a prosoma and a well-developed neck (including the Conochitinida and the Lagenochitinidae, the second with a recessed operculum) (Fig. 9.23).

Evolution and distribution

Possible chitinozoans, in the form of *Desmochitina*-like sacs, have been reported from the Upper Proterozoic of Arizona, but the first true chitinozoans appeared during the Tremadocian (Early Ordovician) and subsequently diversified rapidly during the Early Ordovician, evolving hundreds of different species spread across at least 50 genera. This diversity continued through the Silurian with all the three main groups represented. They declined during the Devonian, disappearing finally at the top of the Famennian, when the last remaining lagenochitinid went extinct. Through time the group developed smaller, self-contained chambers with an increased complexity of ornament and a greater degree of apparent coloniality (Fig. 9.23).

Box 9.12 The chitinozoan Rosetta Stone

But what really were chitinozoans? Material from the Ordovician of Estonia, described as the chitinozoan "Rosetta Stone", may have partially solved the problem. Individual vesicles are linked together in a coiled, chain-like structure; each vesicle belongs to the same species, *Desmochitina nodosa* Eisenack (Fig. 9.22). It is unlikely that these were eggs of a metazoan, because larvae would be unable to escape from the tightly sealed and connected chambers. Therefore Paris and Nolvak (1999) postulated that the coiled, chain-like structure represents an intermediate, immature stage, perhaps an intra-oviduct phase, prior to the final egg-laying event. Unfortunately, distribution in time and space of chitinozoans does not match any skeletonized metazoan group. So we are back to speculation. Possibly chitinozoans are related to a soft-bodied "chitinozoan animal" and the search is on to find this animal in one or more of the Paleozoic Lagerstätten.

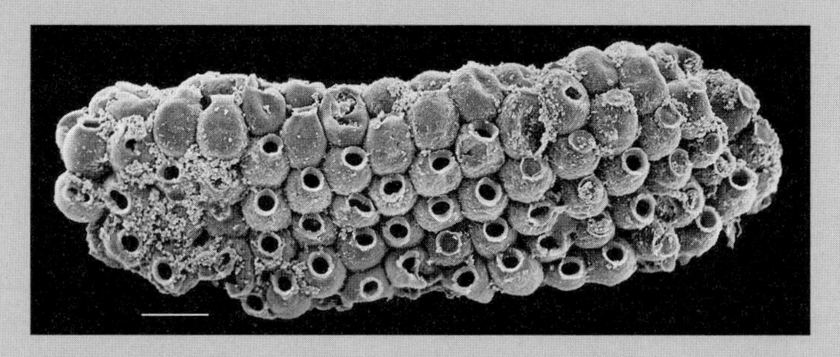

Figure 9.22 Chitinozoan apparatus: a large cluster of *Desmochitina nodus* interpreted as an egg clutch of the chitinozoan animal; the opercula are not present suggesting that the animals had already hatched (×70). (Courtesy of Florentin Paris.)

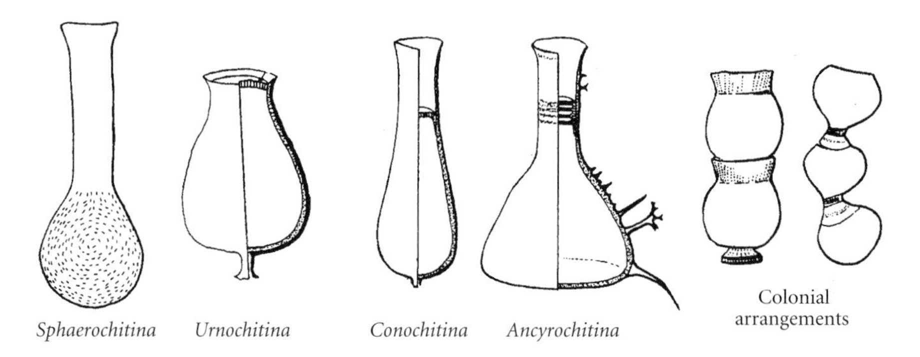

Figure 9.23 Some chitinozoan morphotypes: *Sphaerochitina* (×160), *Urnochitina* (×160), *Conochitina* (×80), *Ancyrochitina* (×240) and colonial arrangements (×40).

Review questions

1 The move from prokaryote to eukaryote cell types was a major evolutionary jump. How was this transition achieved and what sorts of implications did it have for life on Earth?
2 Foraminifera have been widely used by oil and gas companies in exploration. Why are they so useful?
3 Radiolarians have proved very useful in sorting out the stratigraphy of mountain belts. Why are they superior to other micro- and macrofossil groups in these types of studies?
4 Chromistan groups such as the coccolithophores and diatoms have a fundamental effect on the stability of atmospheric and oceanic systems on our planet. But such nannoplankton appeared rela-

tively late in the geological record. Is there any evidence for a Paleozoic nannoplankton?

5 The identity of the chitinozoans may have been solved but how should these fossils be classified?

Further reading

Armstrong, H.A. & Brasier, M.D. 2005. *Microfossils*, 2nd edn. Blackwell Publishing, Oxford, UK.

Bignot, G. 1985. *Elements of Micropalaeontology*. Graham and Trotman, London. (Useful overview of all the main microfossil groups.)

De Wever, P., Dumitrica, P., Caulet, J.P., Nigrini, C. & Caridroit, M. 2001. *Radiolarians in the Sedimentary Record*. Gordon and Breach Science Publishers, the Netherlands. (Key reference on radiolarian paleontology.)

Haeckel, E. 1862. *Die Radiolarien (Rhizopoda Radiaria). Eine Monographie*. Reimer, Berlin. (Classic reference on Radiolaria, beautifully illustrated.)

Jenkins, D.G. & Murray, J.W. 1989. *Stratigraphical Atlas of Fossil Foraminifera*, 2nd edn. British Micropaleontology Association and Ellis Horwood Ltd, London. (Well-illustrated account of the foraminiferans.)

Lipps, J.H. (ed.) 1993. *Fossil Prokaryotes and Protists*. Blackwell Scientific Publications, Oxford, UK. (Multiauthor compilation of the prokaryote and protist microfossil groups.)

References

Armstrong, H.A. & Brasier, M.D. 2005. *Microfossils*, 2nd edn. Blackwell Publishing, Oxford, UK.

Cavalier-Smith, T. 2002. The phagotrophic origin of eukaryotes and phylogenetic classification of protozoa. *International Journal of Systematic and Evolutionary Microbiology* **52**, 297–354.

Corsetti, F.A., Olcott, A.N. & Bakermans, C. 2006. The biotic response to Neoproterozoic snowball Earth. *Palaeogeography, Palaeoclimatology, Palaeoecology* **232**, 114–30.

De Wever, P., Dumitrica, P., Caulet, J.P., Nigrini, C. & Caridroit, M. 2001. *Radiolarians in the Sedimentary Record*. Gordon and Breach Science Publishers, the Netherlands.

Haeckel, E. 1862. *Die Radiolarien (Rhizopoda Radiaria). Eine Monographie*. Reimer, Berlin.

Haeckel, E. 1904. *Kunstformen der Natur*. Verlag des Bibliographischen Institut, Leipzig.

Henriksen, K., Young, J.R., Bown, P.R. & Stipp, S.L.S. 2004. Coccolith biomineralisation studied with atomic force microscopy. *Palaeontology* **47**, 725–43.

Jacob, J., Paris, F., Monod, O., Miller, M.A., Tang, P., George, S.C. & Bény, J.-M. 2007. New insights into the chemical composition of chitinozoans. *Organic Geochemistry* **38**, 1782–8.

Keeling, P.J., Burger, G., Durnford, D.G. et al. 2005. The tree of eukaryotes. *Trends in Ecology and Evolution* **20**, 670–6.

Paris, F. & Nolvak, J. 1999. Biological interpretation and palaeobiodiversity of a cryptic fossil group: the "chitinozoan animal". *Geobios* **32**, 315–24.

Servais, T., Lehnert, O., Li, J., Mullins, G.L., Munnecke, A., Nützel, A. & Vecoli, M. 2008. The Ordovician biodiversification: revolution in the oceanic trophic realm. *Lethaia* **41**, 99–110.

Tyszka, J. 2006. Morphospace of foraminiferal shells: results from the moving reference model. *Lethaia* **39**, 1–12.

Vecoli, M., Lehnert, O. & Servais, T. 2005. The role of marine microphytoplankton in the Ordovician biodiversification event. *Notebooks on Geology, Memoir* **2005/2**, 69–70.

Chapter 10

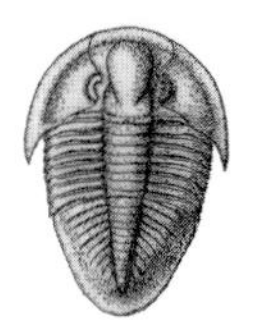

Origin of the metazoans

Key points

- Relatively few basic body plans have appeared in the fossil record; most animals have a triploblastic architecture, with three fundamental body layers.
- Molecular data show there are three main groupings of animals: the deuterostomes (echinoderm–hemichordate–chordate group), the spiralians (mollusk–annelid–brachiopod–bryozoans–most flatworms–rotifers (platyzoans) group) and the ecdysozoans (arthropod–nematode–priapulid plus other taxa group). Together, the spiralians and ecdysozoans are usually called the protostomes.
- Five lines of evidence (body fossils, trace fossils, fossil embryos, the molecular clock and biomarkers) suggest that the metazoans had originated prior to the Ediacaran, 600 Ma.
- Snowball Earth by coincidence or design was a pivotal event in metazoan history; bilaterians evolved after the Marinoan glaciation.
- The first metazoans were probably similar to the demosponges, occurring first before the Ediacaran.
- The Ediacaran biota was a soft-bodied assemblage of organisms largely of uncertain affinities, reaching its acme during the Late Proterozoic, which may represent the earliest ecosystem dominated by large, multicellular organisms.
- The Tommotian or small shelly fauna was the first skeletalized assemblage of metazoans; this association of Early Cambrian microfossils contains a variety of phyla with shells or sclerites mainly composed of phosphatic material.
- The Cambrian explosion generated a range of new body plans during a relatively short time interval.
- The Ordovician radiation was marked by accelerations in diversification at the family, genus and species levels together with increased complexity in marine communities.

Consequently, if my theory be true, it is indisputable that before the lowest Silurian [Cambrian of modern usage] stratum was deposited, long periods elapsed, as long as, or probably far longer than, the whole interval from the Silurian age to the present day; and that during these vast, yet quite unknown, periods of time, the world swarmed with living creatures.

Charles Darwin (1859) *On the Origin of Species*

ORIGINS AND CLASSIFICATION

When did the first complex animals, the metazoans, appear on Earth and what did they look like? How could complex, multicelled animals evolve from the undifferentiated single-celled organisms of most of the Precambrian? Why did they take almost 4 billion years to appear? These questions have puzzled scientists, including Charles Darwin, for over two centuries. In the last few decades a range of multidisciplinary techniques, from molecular biology to X-ray tomography, has helped generate new testable hypotheses regarding the origins of our early ancestors. Apart from the fossil evidence of metazoan body and trace fossils, the investigation of minute fossil embryos, carefully calibrated molecular clocks and more recently biomarkers have placed the investigation of Precambrian life at the top of many scientific agendas.

The first metazoans: when and what?

Life on our planet has been evolving for nearly 4 billion years. Molecular data suggest metazoans have probably been around for at least 600 myr (Fig. 10.1), during which time, according to some biologists, as many as 35 separate phyla have evolved. Five lines of evidence have figured prominently in the search for the earliest metazoans: body fossils, trace fossils, fossil embryos, the molecular clock and biomarkers.

Much controversy still surrounds the timing of their origin. Was there a long cryptic interval of metazoan evolution prior to the Ediacaran – a time when we do not find fossils preserved, either because the animals lacked preservable bodies, or they were small, or perhaps a combination of both? Or, as the recalibrated molecular clocks suggest, can animal origins be tracked back only to the Ediacaran, when there was also a sudden rise

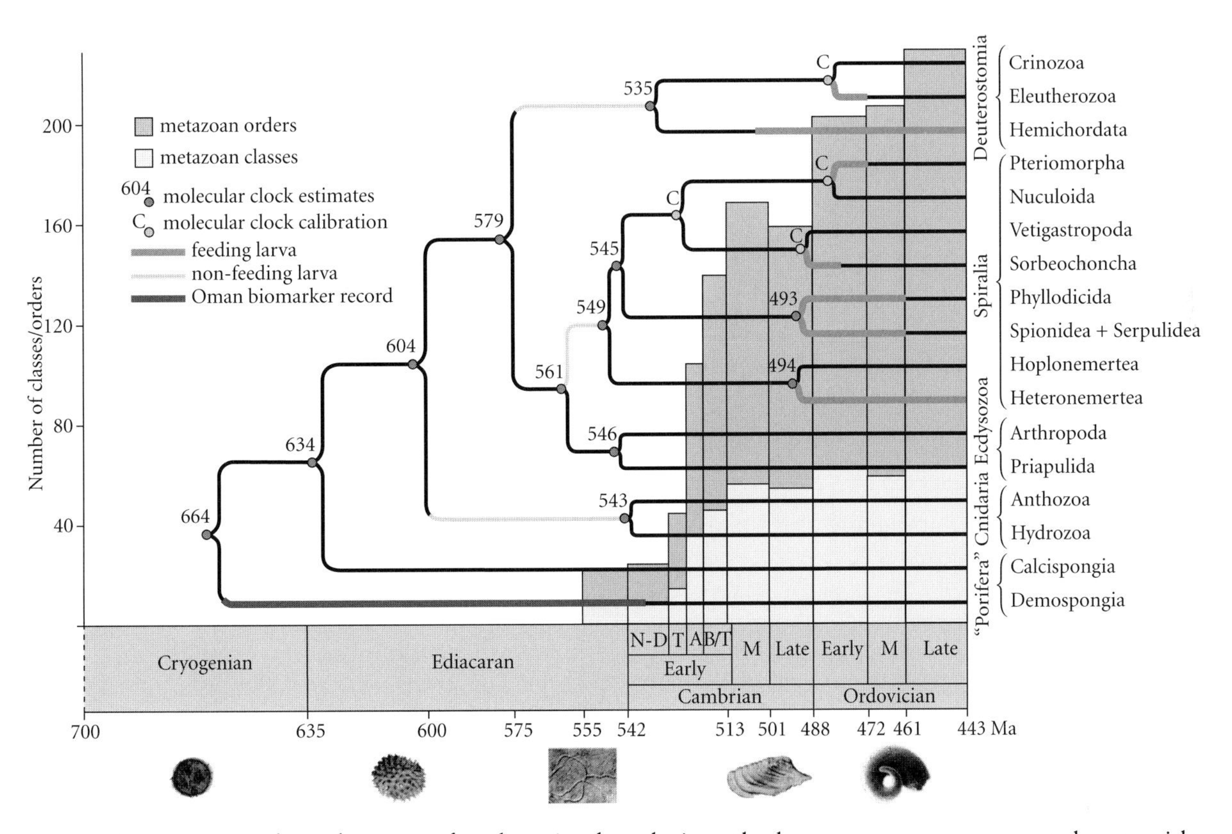

Figure 10.1 Time scale and tempo of early animal evolution: the key metazoan groups are shown with the putative age of their last common ancestor, together with an estimate of the respective numbers of classes and orders indicated against a stratigraphy indicating key biological and chemical events. N–D, Nemakit-Daldynian; T, Tommotian; A, Atdabanian; B/T, Botomian. (Courtesy of Kevin Peterson.)

in oxygen levels in the deep ocean (Canfield et al. 2007).

Body fossil evidence

Body fossils of basal metazoans in the Ediacaran Period are few and far between. The morphology of an early metazoan fossil must be clearly described and convincingly illustrated, different organs and tissues identified, and comparisons drawn with other extant and fossil organisms. Many Upper Precambrian successions have been subjected to intense metamorphism and tectonism (see p. 48) and are now located in some of the Earth's mountain belts. The chances of finding adequately preserved fossils are slight. Nevertheless, the earliest undoubted metazoans occur within the widespread Ediacara biota (see p. 242) dated at approximately 600–550 Ma. Moreover the fact that a relatively advanced metazoan, the mollusk *Kimberella*, possibly equipped with a foot and radula (see p. 330), occurs within the Ediacara biota from southern Australia and Russia could suggest a history of metazoan evolution prior to the Ediacaran. But although a strong case can be made for a significant Proterozoic record for the cnidarians and sponges and perhaps some other metazoans, the Cambrian explosion still marks the arrival, center stage, of the bilaterians (Budd 2008).

Trace fossil evidence

Trace fossils are the behavior of organisms recorded in the sediment (see p. 510). By their very nature they occur in place and thus cannot be transported or reworked by currents. Nevertheless these too must be convincingly demonstrated as biogenic and the age of their enclosing sediments accurately determined. If and when metazoans developed locomotory organs, such as the molluskan foot, and digestive systems, we might expect to find burrows and trails together with fecal pellets. Records of trace fossils from rocks older than 1 Ga in India (Seilacher et al. 1998) and over 1.2 Ga in the Stirling biota of Australia (Rasmussen et al. 2002) generated considerable excitement (Fig. 10.2). Both suggested metazoan life older than 1 Ga but both are now considered questionable (Jensen 2003). The oldest undoubted locomotory trace fossils are from about 550 Ma (Droser et al. 2002) from northwest Russia, whereas fecal strings have been reported from rocks some 600 Ma (Brasier & McIlroy 1998) suggesting the existence of an ancient digestive system. In fact no convincing trace fossils are known from successions older than the **Marinoan glaciation** (635 Ma), the second main icehouse event associated with snowball Earth (see p. 112).

Figure 10.2 Putative trace fossils from the Precambrian of Australia, showing *Myxomitodes*, a presumed trail of a mucus-producing multicellular organism about 1.8–2 billion years old from Stirling Range, Western Australia. (Photo is approximately 65 mm wide.) (Courtesy of Stefan Bengtson.)

Embryo fossil evidence

Fossil Neoproterozoic embryos are now known from a number of localities, although claims that they represent sulfur-oxidizing bacteria or that they are not embryos at all have their advocates. Some of the best studied examples are from the Doushantuo Formation, South China. The part of the formation yielding the embryos was first dated at approximately 580 Ma, predating much of the Ediacaran but postdating the Marinoan glaciation. Revised dates seem to suggest that the faunas are younger and that they overlap with the older Ediacaran assemblages. Cell division and cleavage patterns are obvious although it is difficult to assign the material to distinct metazoan groups in the absence of juvenile and adult forms. There are, however, a lack of epithelia even in clusters of over 1000 cells suggesting that the embryos examined are those, at best, of stem-group metazoans (Hagadorn et al. 2006); they could equally well be fungi or rangeomorphs (enigmatic frond-like fossils). Nonetheless the Doushantuo embryos, although unplaced taxonomically, provide our earliest body fossil evidence for probable metazoan life, albeit very basal, and a fascinating insight into embryologic processes in deep time (Donoghue 2007) (Box 10.1).

Molecular evidence

Not only have the morphologies of organisms evolved with time, but so too have their molecules. This forms the basis of the concept of the **molecular clock** (see p. 133). The molecular clock has opened up tremendous possibilities to date, independently of direct fossil evidence, the times of divergence of say the mammals from the reptiles or the brachiopods from the mollusks. Nevertheless, attempts to date the divergences of the various groups of metazoans have proved controversial. For example, the last common ancestor of the bilaterians, the metazoan clade excluding the sponges and cnidarians, has been variously placed at anywhere between 900 and 570 Ma. Why is there such a spread of ages in a seemingly exact science? The rates of molecular evolution in various groups are unfortunately not constant. The vertebrates appear to have reduced their rates of molecular change through time. So, using the slow vertebrate rates of molecular evolution to calibrate the date of origin of Bilateria gives dates that are too ancient (900 Ma). On the other hand, using mean bilaterian rates of molecular evolution gives a date (570 Ma) that is more in keeping with evidence from the fossil record (e.g. Budd & Jensen 2000) and thus makes the Cambrian explosion much more of an explosion of animals rather than fossils (Peterson et al. 2004). Nevertheless the most recent molecular clock data (Peterson et al. 2008) suggest a major phase of metazoan radiation within the Ediacaran, prior to that in the Cambrian. This radiation probably set the agenda for metazoan macroevolution for the rest of geological time.

Box 10.1 Synchrotron-radiation X-ray tomographic microscopy

Fossil embryos from the Upper Neoproterozoic and Cambrian are providing some important clues about the origin and early evolution of the metazoans. They are, however, tiny and notoriously hard to study. Nevertheless Phil Donoghue and his colleagues (2006) are beginning to accumulate a large amount of new information on the composition, structure and cell division within these minute organisms together with their modes of preservation. Synchrotron-radiation X-ray tomographic microscopy (SRXTM) has provided a whole new way of scanning embryos without actually destroying them (Fig. 10.3). The embryos, most of them 1 mm across or smaller, are held steady in a high-energy beam of photons, and multiple "slices" are produced, spaced a few microns apart. Using imaging software, these slices can be combined to create a detailed three-dimensional model of the internal structure of the fossil. Embryos assigned to the bilaterian worm, *Markuelia*, together with *Pseudooides*, variously show the process of cell cleavage and development of possible blastomeres, clusters of cells produced by cell division after fertilization, rather than yolk pyramids, which are more typical of the arthropods. This high-tech methodology has already demonstrated a real prospect for identifying the animals themselves and charting their early stages of development, some 600 Ma. It also can reject the claims that such fossils were the planula larvae of cnidarians, minute bilaterians or the early stages of gastrulation (see p. 240) of hydrozoans or bilaterians. It has, however, been recently suggested that many of these embryonic structures were created by bacteria (see p. 190). But not all.

Read more about this topic at http://www.blackwellpublishing.com/paleobiology/.

Continued

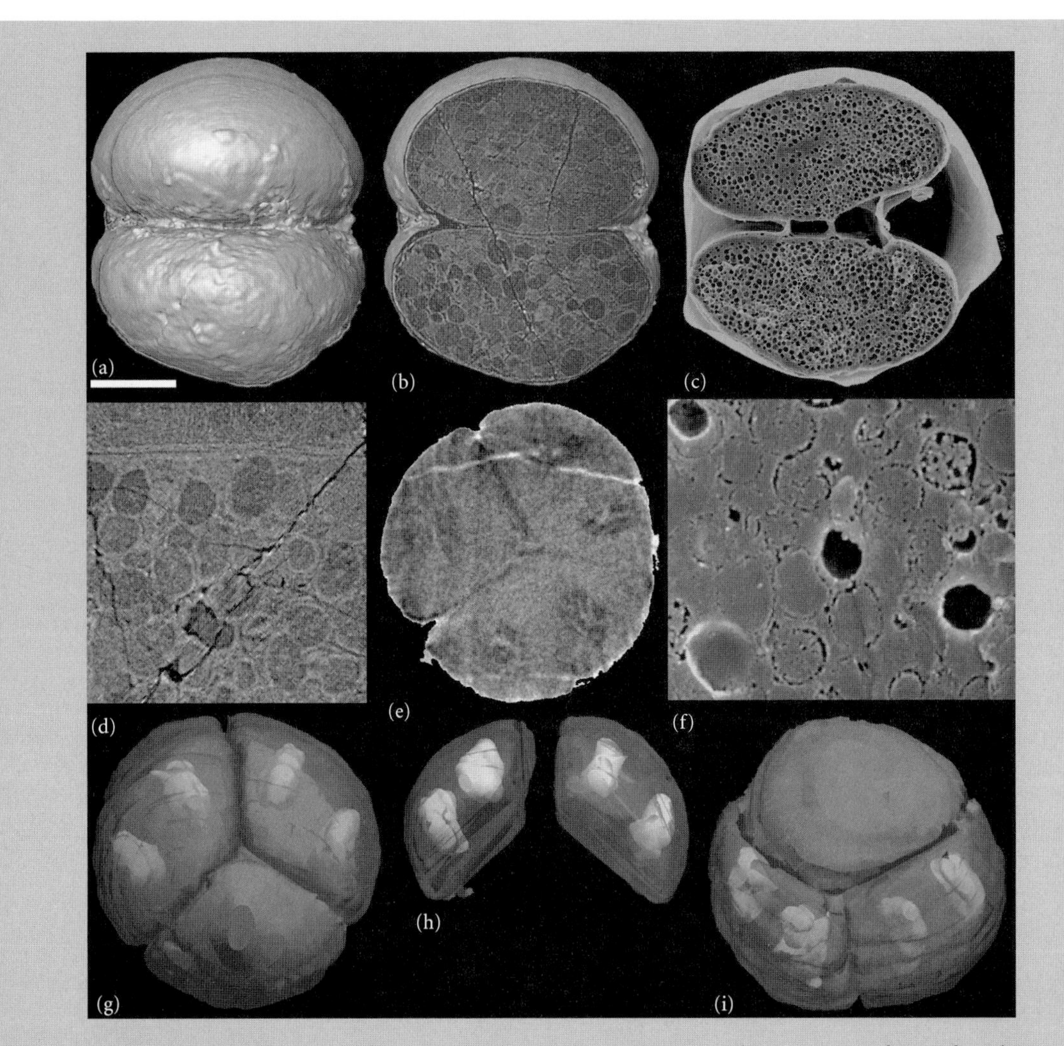

Figure 10.3 Animal embryos from the Doushantou Formation, China. (a) Surface of embryo based on tomographic scans together with (b) an orthoslice revealing subcellular structures analogous to modern lipids and (c) an orthoslice at the boundary between two cells. (c, f) Two-cell embryo of the sea urchin *Heliocidaris* showing lipid vesicles for comparison. (e) Orthoslice rendering of a possible embryo revealing internal structures. (g–i) Models of tetrahedrally arranged cells. Relative scale bar (see top left): 170 μm (a–d, f), 270 μm (e), 150 μm (g–i). (Courtesy of Philip Donoghue.)

Biomarker evidence

Biomarkers, essentially the biochemical fingerprints of life, have become increasingly important in astrobiology, where they have been sought in the quest for extraterrestrial life. But they are also of considerable importance in the investigation of Precambrian life (see p. 188), where other lines of evidence are lacking. Thus amino acids, hopanes, some types of hydrocarbons, evidence of isotopic fractionation in carbon (^{12}C) and biofilms are strong indicators of life forms. More exciting is the fact that specific biomarkers may be related to particular groups of organisms. Significantly, biomarkers associated with metazoan demosponges (see p. 262) have now been reported from rocks older than the Ediacaran, confirming the presence of basal metazoans at this time. But since the sponges are paraphyletic, biomarkers from the homoscleromorph sponges (see p. 262) would also have to be present to prove the presence of the eumetazoans.

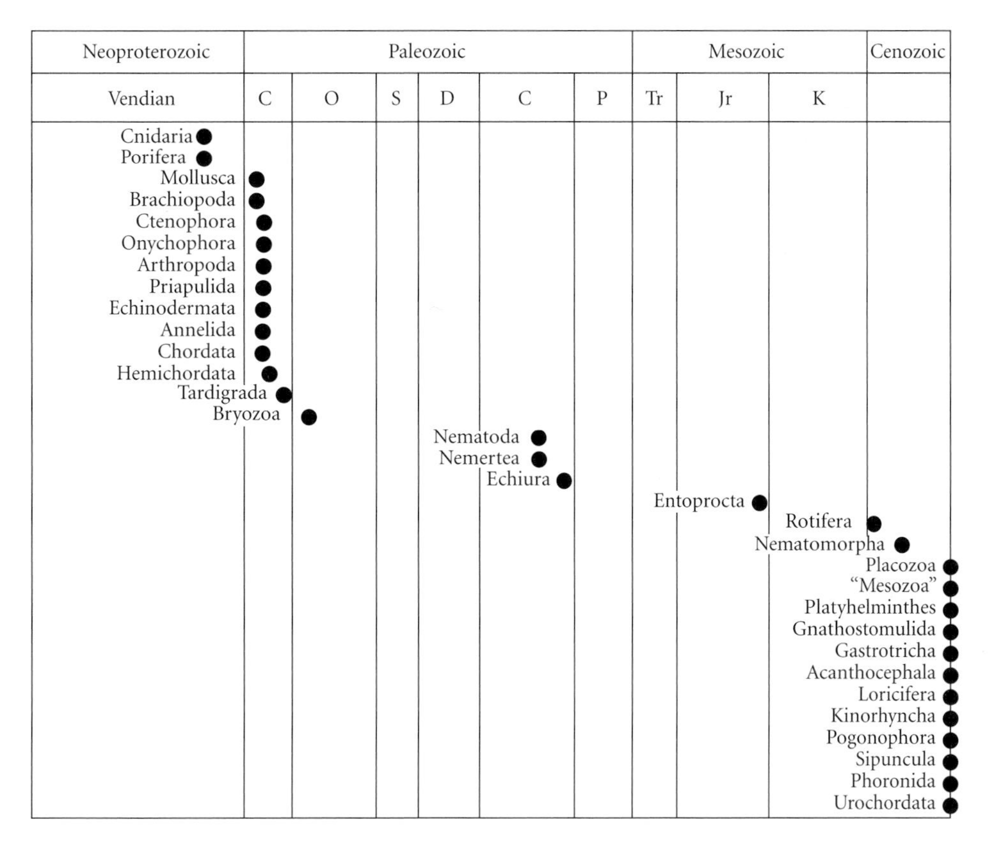

Figure 10.4 Appearance of the main animal phyla and some other high-level taxonomic groups. Geological period abbreviations are standard, ranging from Cambrian (C) to Cretaceous (K). (Based on Valentine 2004.)

Invertebrate body and skeletal plans

Life on our planet has been evolving for nearly 4 billion years. Molecular data suggest metazoans have probably been around for at least at 550 myr, during which time, according to some biologists, as many as 35 separate phyla have evolved. In recent years, new molecular phylogenies have completely changed our views of animal relationships and thus the importance of invertebrate body and skeletal plans. They are important from a functional point of view, but are potentially highly misleading if simply read as telling an evolutionary story. Despite the infinite theoretical possibilities for invertebrate body plans, relatively few basic types have actually become established and many had evolved by the Cambrian (Fig. 10.4). These body plans are usually defined by the number and type of enveloping walls of tissue together with the presence or absence of a **celom** (Fig. 10.5). The basic unicellular grade is typical of protist organisms and is ancestral to the entire animal kingdom. The first metazoans were multicellular with one main cell type and peripheral collar cells or **choanocytes**, equipped with a whip or **flagellum** (Nielsen 2008). There are three main body plans (Table 10.1).

The **parazoan** body plan, seen in sponges, is characterized by groups of cells usually organized in two layers separated by jelly-like material, punctuated by so-called wandering cells or amoebocytes; the cell aggregates are not differentiated into tissue types or organs. In fact molecular phylogenetic studies have suggested that sponges are paraphyletic (see p. 262) so this is only a grade of organization.

The **diploblastic** grade or body plan, typical of cnidarians and the ctenophorans, has two layers – an outer ectoderm and an inner endoderm and epithelia. These two layers are separated by the acellular, gelatinous **mesogloea.**

The **triploblastic** body plan, seen in most other animals, has three layers of tissues from the outside in: the ectoderm, mesoderm and endoderm. Superimposed on this body plan is the bilateral symmetry that defines the bilate-

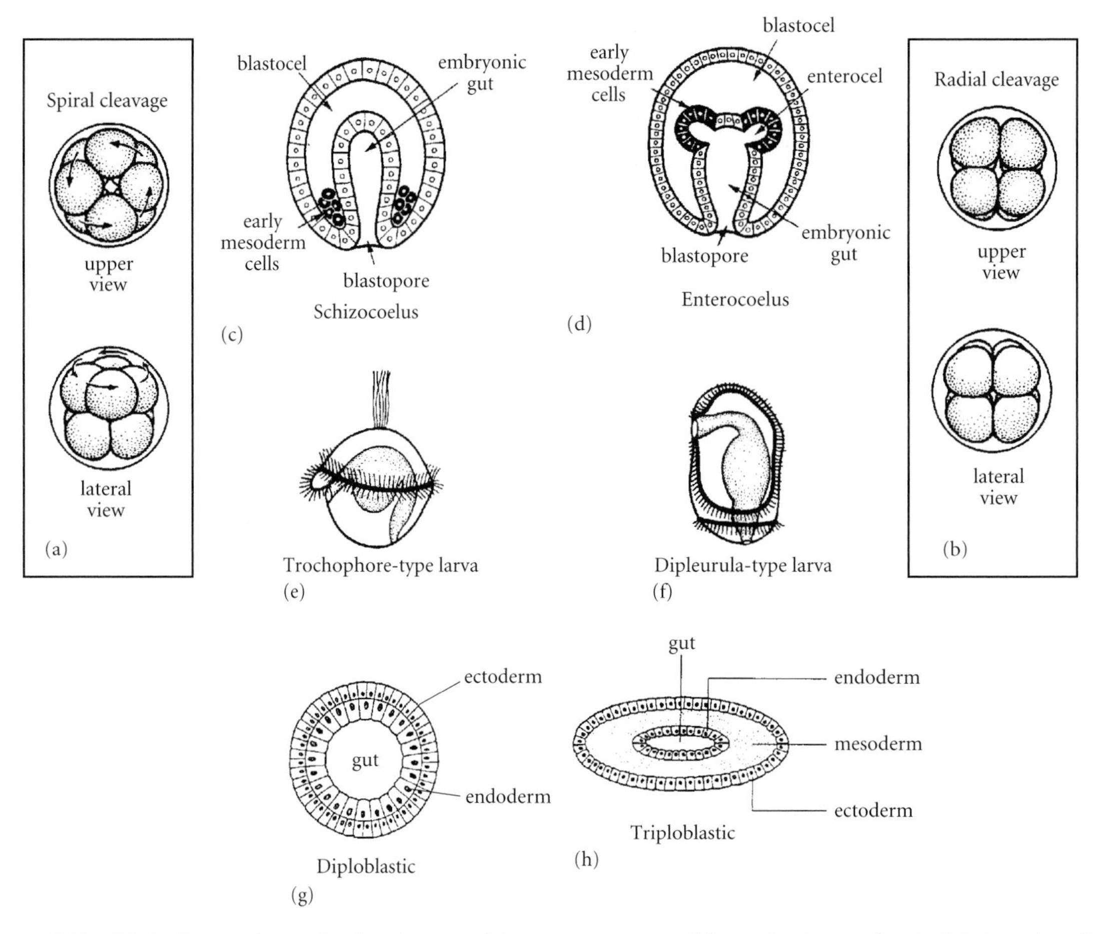

Figure 10.5 Main invertebrate body plans and larvae: upper and lateral views of spiral (a) and radial (b) patterns of cell cleavage; development of the mesoderm in the spiralians (c) and radialians (d); diploblastic (g) and triploblastic (h) body plans and trochophore-type (e) and dipleurula-type (f) larvae.

Table 10.1 Key characteristics of the three main groups of animals.

Group	Grade	Symmetry	Key character	Larvae
Porifera	Parazoan	Bilateral and radial symmetry	Collar cells	Blastula larva
Cnidaria	Diploblastic	Radial symmetry	Cnidoblasts	Planula larva
Bilateria	Triploblastic	Bilateral symmetry	Digestive tract	Various types

rians. And finally the development of the celom or body cavity characterizes most of the animal groups found as fossils. The celom usually functions as a hydrostatic skeleton and is related to locomotion. But the presence and organization of the celom is not phylogenetically significant; the celom has evolved several times and in some groups, such as the flatworms, there are at least two types of celomic cavities.

The annelid worms and the arthropods have a celom divided along its length into segments; each segment possesses identical paired organs such as kidneys and gonads together with appendages. The mollusks, on the other hand, have an undivided celom situated mesodermally and irregularly duplicated organs.

The remaining bilaterians, such as the phoronids, brachiopods, bryozoans, echinoderms and hemichordates have a celom that is divided longitudinally into two or three zones each with different functions. Based around this plan, animals with a specialized feeding and respiratory organ, the **lophophore**, are characterized by sac-like bodies; but this is no guarantee that these so-called "lophophorates", brachiopods and bryozoans, are in fact closely related. The hemichordates possess a crown of tentacles and some have paired gill slits. The echinoderms have an elaborate water vascular system that drives feeding, locomotion and respiration.

The identification of invertebrate body plans is a useful method of grouping organisms according to their basic architecture. However, similarities between grades of construction unfortunately do not always mean a close taxonomic relationship. Be aware that certain body plans have evolved more than once in different groups, Skeletons too, for example, have evolved a number of times in a variety of forms.

The skeleton is an integral part of the body plan of an animal, providing support, protection and attachment for muscles. Many animals such as the soft-bodied mollusks (slugs) possess a hydraulic skeleton in which the movement of fluid provides support. Rigid skeletons based on mineralized material may be external (**exoskeleton**), in the case of most invertebrates, or internal (**endoskeleton**) structures, in the case of a few mollusks (e.g. belemnites), echinoderms and vertebrates. Growth is accommodated in a number of ways. Most invertebrate skeletons grow by the addition of new material, a process termed **accretion**. Arthropods, however, grow by periodic bursts between intervals of **ecdysis** or molting; echinoderms grow by both accretion to existing material and by the appearance of new calcitic plates.

Classification and relationships

Classifications based on purely morphological data and embryology have met with problems. Difficulties in establishing homologous characters and homoplasy (see p. 129) have contributed to a number of different phylogenies. The locator tree (Fig. 10.6), however, outlines some of the main features of animal evolution. From the base of the metazoan tree, the demosponges and calcisponges are the simplest animals whereas the cnidarians are the most basal eumetazoans. Three robust bilaterian groupings are recognized mainly on molecular data: the ecdysozoans, the spiralians and the deuterostomes. The ecdysozoans and the spiralians comprise the **protostomes** ("first mouth") where the mouth develops directly from the first opening, the blastopore, resulting from cell growth and migration. The **deuterostomes** ("second mouth"), however, have a mouth arising from a secondary opening; the true blastopore often develops as an anus. Not all phyla fit simply into these two major divisions, but using a consensus based on comparative morphology, two main streams emerge: the echinoderm–hemichordate–chordate (deuterostomous) and the mollusk–lophophorate–annelid–arthropod (protostomous) groupings (Box 10.2).

Other studies have laid emphasis on the similarities between the larval stages of organisms to investigate phylogenetic relationships.

Most invertebrates develop first a larval stage that may be either **planktotrophic**, free-living and feeding on plankton, or **lecithotrophic**, essentially benthic and feeding on yolk sacs. There is a range of different larval types. For example the nauplius larva is most typical of crustaceans, the planula characterizes the cnidarians, the trochophore larva occurs in the mollusks and the polychaetes whereas the shelled veliger also characterizes the mollusks. Thus those groups (annelids and mollusks) with trochophores may have shared a common ancestor. Invertebrate larvae are occasionally identified in the fossil record. With the availability of more advanced preparatory and high-tech investigative techniques, studies of fossil larvae may yet become a viable part of paleontology.

FOUR KEY FAUNAS

The three great evolutionary faunas of the Phanerozoic, the Cambrian, Paleozoic and Modern (see p. 538), developed during a timeframe of some 550 myr. Nevertheless, in the 100 myr that include the transition between Precambrian and Phanerozoic life, there were a number of distinctive groups of animals that together paved the way to the spectacular diversity we see today in marine and terrestrial communities. The **Ediacara biota** and **small shelly faunas**, together with those that developed during the **Cambrian explosion** and **Ordovician radiation**, set the scene for life on our planet.

Box 10.2 Molecular classification

Can molecular data help? Kevin Peterson and his colleagues (2004, 2005) have presented a minimum evolution analysis (see p. 129) based on amino acid data derived from housekeeping genes (Fig. 10.6). The cladogram separates the Deuterostomia (echinoderms + hemichordates) from the Protostomia, which includes the Spiralia (mollusks + annelids + nemerteans + platyhelminthes) and the Ecdysozoa (arthropods + priapulids). Both are united within the Triploblastica that, together with the cnidarians, forms the Bilateria; the Eumetazoa comprise the Bilateria + Cnidaria and the metazoan clade is completed with the addition of the calcisponges and demosponges. Thus the last common ancestor of the Metazoa was probably rather like a modern sponge. The tree, however, lacks data from a number of problematic groups such as the Bryozoa and Brachiopoda, both commonly united on the basis of their lophophores. Moreover to date it has proved impossible to resolve polychotomies such as that including the mollusks, annelids and brachiopods (see also Aguinaldo & Lake 1998).

These molecular results are being increasingly accepted by zoologists as analysis of different gene datasets produce the same results. The hunt is now on for morphological characters of some of the major clades discovered by molecular means. A good example is the shedding of the exoskeleton (ecdysis) by the Ecdysozoa, a strong morphological synapomorphy that had once been thought to have evolved convergently in arthropods, nematodes and the others.

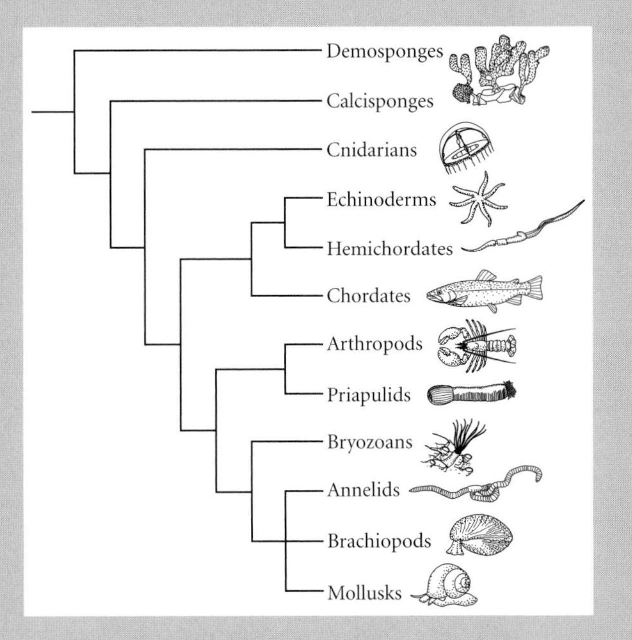

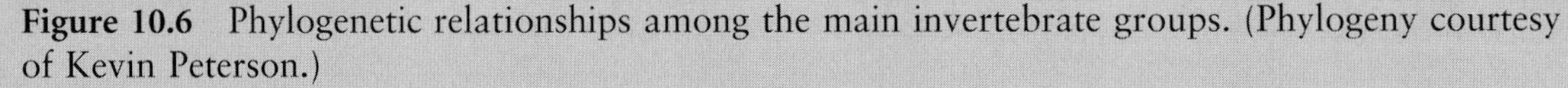

Figure 10.6 Phylogenetic relationships among the main invertebrate groups. (Phylogeny courtesy of Kevin Peterson.)

Ediacara biota

Since the first impressions of soft-bodied organisms were identified in the Upper Proterozoic rocks of Namibia and in the Pound Quartzite in the Ediacara Hills, north of Adelaide in southern Australia in the late 1940s, this remarkable assemblage has now been documented from 30 localities on five continents (Fig. 10.7). More than 100 species of these unique organisms have been described on the basis of molds usually preserved in

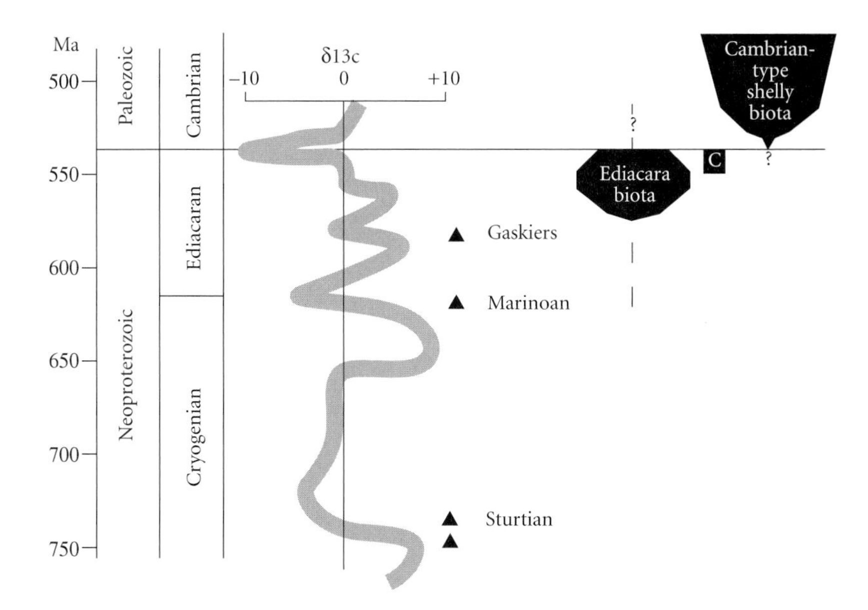

Figure 10.7 Stratigraphic distribution of the Ediacara biota. Solid triangles, glaciations; C, calcified metazoans; T, position of the Twitya disks. (Based on Narbonne 2005.)

shallow-water siliciclastic sediments, consisting of clasts of silicic-rich rocks, or volcanic ash, more rarely carbonates or even turbidites. The sediments were deposited during specific events, such as a storm, and are usually termed event beds. Deep-water biotas are also known such as those from Mistaken Point in Newfoundland. The style of preservation plays an important role in understanding these organisms (Narbonne 2005). The widespread development of algal mats, prior to the Cambrian substrate revolution (see p. 330), suggests that these too aided preservation, sometimes providing "death masks", of these non-skeletal organisms.

Although morphologically diverse, the Ediacaran organisms have many features in common. All were soft-bodied, with high surface to volume ratios and marked radial or bilateral symmetries. These thin, ribbon-shaped animals may have operated by direct diffusion processes where oxygen entered through the skin surface, so gills and other more complex internal organs were perhaps not required. Most Ediacaran organisms have been studied from environments within the photic zone; many collected from deeper-water deposits are probably washed in. Provincialism among these Upper Proterozoic biotas was weak with many taxa having a nearly worldwide distribution. It is possible that the flesh of the Ediacaran organisms littered areas of the Late Precambrian seafloor; predators and scavengers had yet to evolve in sufficient numbers to remove it.

Morphology and classification

Traditionally the Ediacaran taxa, a collection of disks, fronds and segmented bodies, have been assigned to a variety of Phanerozoic invertebrate groups on the basis of apparent morphological similarities. In many cases considerable speculation is necessary and many assumptions are required to classify these impressions. Most of the species have been assigned to coelenterate groups, although some taxa have been identified as, for example, arthropods or annelids. Michael Fedonkin (1990), however, suggested a form classification based on the morphology and structure of these fossils. Key areas of his classification are summarized in Box 10.3 and typical examples illustrated in Figure 10.8. The bilateral forms were probably derived from an initial radial body plan. The concept and classification of the Ediacara biota is in a state of flux and Fedonkin's classification is one of a number of attempts to rationalize the group, assuming the majority are in fact animals. Some have argued, nevertheless, that the Ediacarans are organisms unrelated to modern metazoans (Box 10.4), or are even Fungi.

Box 10.3 The Ediacaran animals: a form classification

RADIATA (RADIAL ANIMALS)

Three main classes are defined. Most colonial organisms in the fauna, for example *Charnia*, *Charniodiscus* and *Rangea*, are assigned to coelenterates and were part of the sessile benthos. The affinities of these animals have been debated in detail, but their close similarity to the sea-pens suggests an assignment to the pennatulaceans.

Class CYCLOZOA

- These animals have a concentric body plan with a large disk-shaped stomach and the class includes mostly sessile forms such as *Cyclomedusa* and *Ediacaria*. About 15 species of jellyfish-like animals have been described and in some, for example, *Eoporpita* tentacles are preserved

Class INORDOZOA

- Medusa-like animals with more complex internal structures, for example *Hielmalora*

Class TRILOBOZOA

- Characterized by a unique three-rayed pattern of symmetry. *Tribrachidium* and *Albumares* are typical members of the group

BILATERIA (BILATERAL ANIMALS)

This division contains both smooth and segmented forms.

Smooth forms

- These morphotypes are rare. They include *Vladimissa* and *Platypholinia*, which may be turbellarians, a type of platyhelminthes worm

Segmented forms

- Much of the Ediacara fauna is dominated by segmented taxa inviting comparisons with the annelids and arthropods. *Dickinsonia*, for example, may represent an early divergence from the radial forms whereas *Spriggina*, although superficially similar to some annelids and arthropods, possesses a unique morphology

Ecology

There is little doubt that the Ediacara biotas dominated the latest Precambrian marine ecosystem, occupying a range of ecological niches and pursuing varied life strategies probably within the photic zone (Fig. 10.10). There is no evidence to suggest that any of the Ediacaran organisms were either infaunal or pelagic, thus in contrast to the subsequent Cambrian Period, life was restricted to the seabed. It is also possible that these flattened organisms hosted photosymbiotic algae, maintaining an autotrophic existence in the tranquil "garden of Ediacara" as envisaged by Mark McMenamin (1986), although this model has its opponents. McMenamin considered that the ecosystem was dominated by medusoid pelagic animals, and that attached, sessile benthos and infaunal animals were sparse; the medusoids have been reinterpreted as bacterial colonies or even holdfasts. Food chains were thus probably short and the trophic structure was apparently dominated by suspension and deposit feeders.

Biogeography

Although provincialism was weak among the Ediacara biotas, three clusters have been rec-

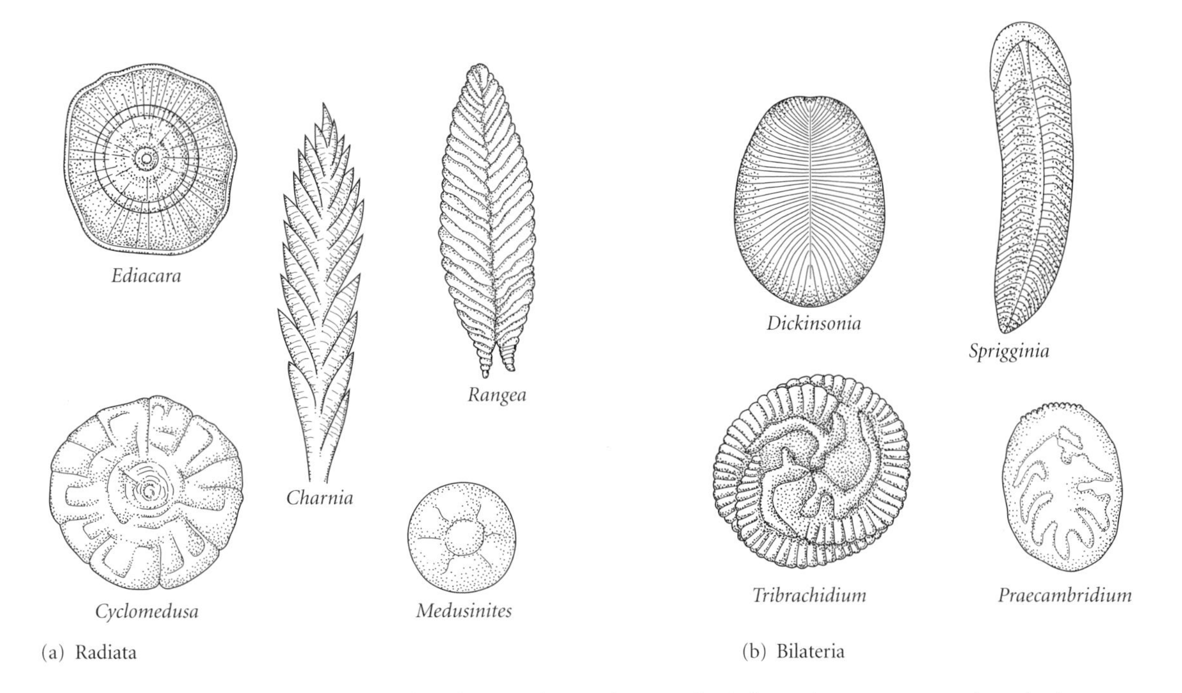

Figure 10.8 Some typical Ediacara fossils: (a) the Radiata, which have been associated with the cnidarians, and (b) the Bilateria, which may be related to the annelids and arthropods. *Ediacaria* (×0.3), *Charnia* (×0.3), *Rangea* (×0.3), *Cyclomedusa* (×0.3), *Medusinites* (×0.3), *Dickinsonia* (×0.6), *Spriggina* (×1.25), *Tribrachidium* (×0.9) and *Praecambridium* (×0.6). (Redrawn from various sources by Anne Hastrup Ross.)

ognized based on multivariate biogeographic analysis (see p. 45) by Ben Waggoner (2003): (i) the Avalon assemblage is from deep-water, volcaniclastic settings in eastern Newfoundland; (ii) the White Sea assemblage represents the classic Vendian section in the White Sea, Russia; and (iii) the Nama assemblage is a shallow-water association from Namibia, West Africa. Unfortunately the distribution of these assemblages does not match any paleogeographic models for the period and the clusters may rather represent a mixture of environmental and temporal factors (Grazhdankin 2004).

Extinction of the Ediacarans

The Ediacara biota, as a whole, became extinct about 550 Ma. Nevertheless, in terms of longevity, the ecosystem was very successful and a few seem to have survived into the Cambrian. The rise of predators and scavengers together with an increase in atmospheric oxygen may have at last prevented the routine preservation of soft parts and soft-bodied organisms. More importantly, the Ediacara body plan offered little defense against active predation. There is abundant evidence for Cambrian predators: damaged prey, actual predatory organisms and the appearance of defense structures, such as trilobite spines and multielement skeletons. All suggest the existence of a predatory life strategy that was probably established prior to the beginning of the Cambrian Period. The Proterozoic–Cambrian transition clearly marked one of the largest faunal turnovers in the geological record, with a significant move from soft-bodied, possibly photoautotrophic, animals to heterotrophs relying on a variety of nutrient-gathering strategies. It is, however, still uncertain whether a true extinction, or the slamming shut of a taphonomic window, accounted for the disappearance of the Ediacara biota from the fossil record.

Cloudina *assemblages*

Although the Ediacara biotas were overwhelmingly dominated by soft-bodied organisms,

Box 10.4 Vendobionts or the first true metazoans

The apparently unique morphology and mode of preservation of the Ediacara biota has led to much debate about the identity and origins of the assemblage. Adolf Seilacher (1989) argued that these organisms were quite different from anything alive today in terms of their constructional and functional morphology (Fig. 10.9). Apart from a distinctive mode of preservation, the organisms all share a body form like a quilted air mattress: they are rigid, hollow, balloon-like structures with sometimes additional struts and supports together with a significant flexibility. Seilacher termed the Ediacaran organisms **vendobionts**, meaning organisms from the Vendian, and he speculated about their unique biology. Reproduction may have been by spores or gametes. The skin must have been flexible, although it could crease and fracture, and it must have acted as an interface for diffusion processes. This stimulating and original view of the Ediacarans, however, remains controversial. Several members of the Vendobionta have been interpreted as regular metazoans, suggesting a less original explanation for the Ediacara group.

Leo Buss and Adolf Seilacher (1994) suggested a compromise. Their phylum Vendobionta includes cnidarian-like organisms lacking cnidae, the stinging apparatus typical of the cnidarians. Vendobionts thus comprise a monophyletic sister group to the Eumetazoa (ctenophorans + bilaterians). This interpretation requires the true cnidarians to acquire cnidae as an apomorphy for the phylum.

The vendobiont interpretation has opened the doors for a number of other interpretations and the understanding of Ediacaran paleobiology is as open as ever: some authors have suggested the Ediacarans are giant protists, lichens, prokaryotic colonies or fungus-like organisms. However most agree that the Ediacara assemblage includes some crown- and stem-group sponges and cnidarians, a conclusion proposed by Sprigg in the late 1940s. This is supported by biomarker and molecular clock data.

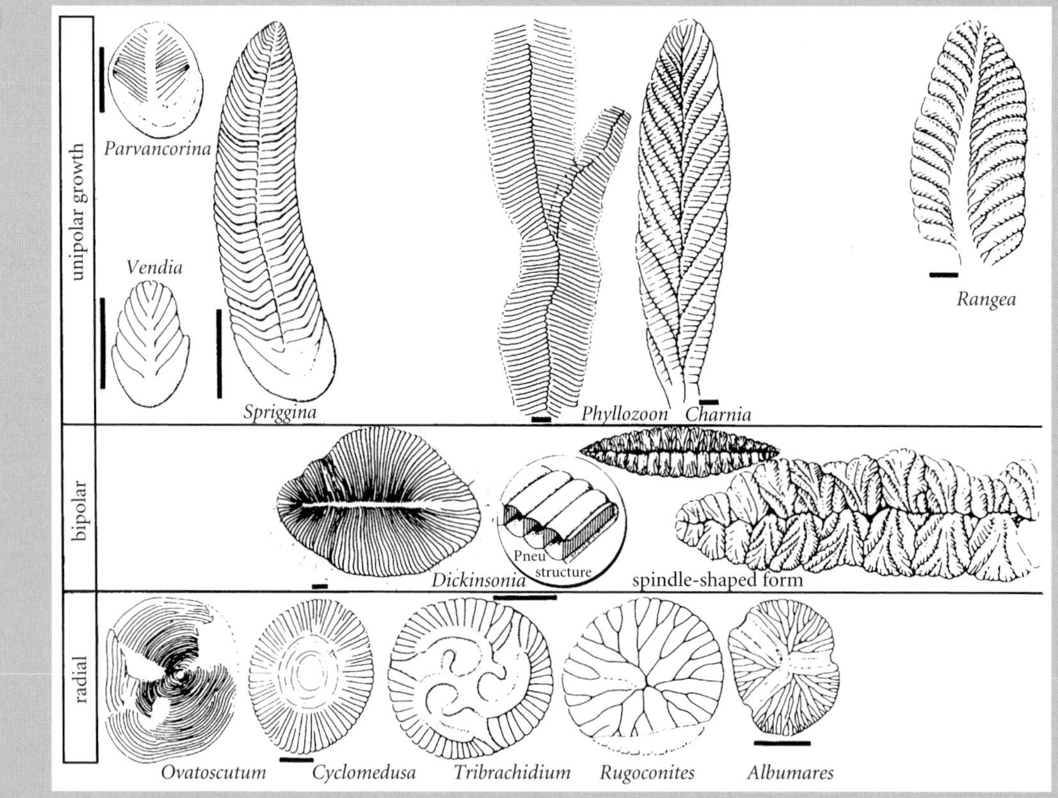

Figure 10.9 Vendozoan constructional morphology, recognizing unipolar, bipolar and radial growth modes within the Ediacara-type biota. Scale bars, 10 mm. (From Seilacher 1989.)

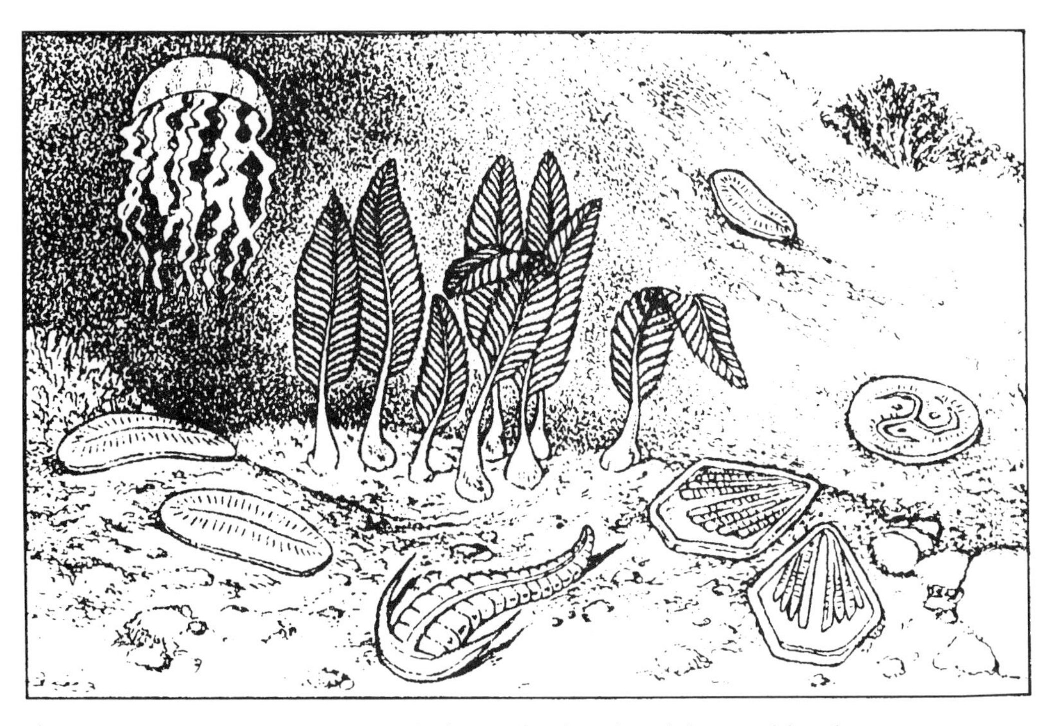

Figure 10.10 An Ediacara community including a fixed and mobile tiered benthos.

minute conical shells were also present in some Ediacaran successions, including localities in Brazil, China, Oman and Spain. *Cloudina* was possibly a cnidarian-type organism with a unique shell structure having new layers forming within older layers. Moreover it was probably related to a suite of similar shells such as *Sinotubulites*, *Nevadatubulus* and *Wyattia* that also occurred close to the Precambrian–Cambrian boundary. In addition to complex multicellularity, modularity, locomotion and predation, biomineralization was already far advanced in the Late Proterozoic, providing a link with what was to follow in the Nemakit-Daldynian assemblages of the earliest Cambrian. Some of the shells of *Cloudina* are bored, suggesting the presence of predators (Fig. 10.11), although it is not certain the animals were still living when bored.

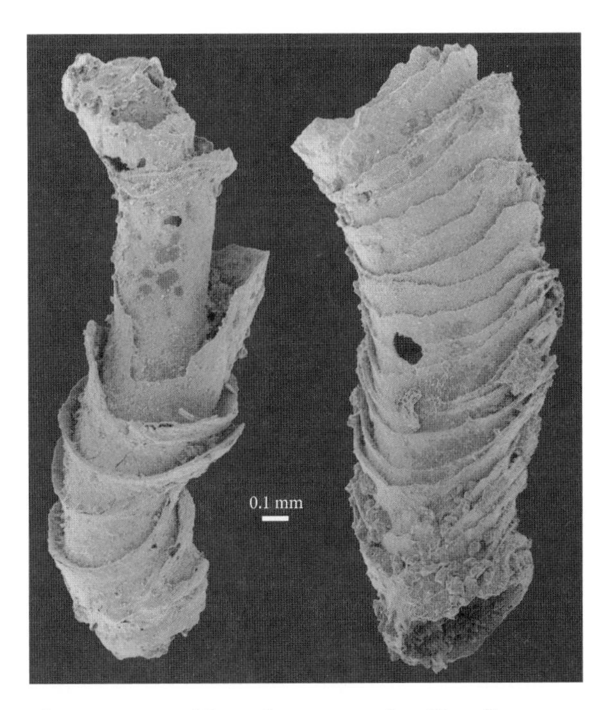

Figure 10.11 The calcareous tube *Cloudina* displaying indications of predation. (Courtesy of Stefan Bengtson.)

Small shelly fauna

A distinctive assemblage of small shelly fossils has now been documented in considerable detail from the Precambrian–Cambrian transition; the assemblage is most extravagantly developed in the lower part of the Cambrian defined on the Siberian platform, traditionally called the **Tommotian**, which gives its name to the fauna. A great deal is now known about the stratigraphic distribution and paleobiogeography of these organisms through current

Figure 10.12 Elements of the Tommotian-type or small shelly fauna. Magnification approximately ×20 for all, except *Fomitchella* which is about ×40. (Based on various sources.)

interest in the definition of the base of the Cambrian System. Nevertheless, the biological affinities of many members of the Tommotian fauna have yet to be established. The assemblage, although dominated by minute species, together with small sclerites of larger species, represents the first major appearance of hard skeletal material in the fossil record, some 10 myr before the first trilobites evolved (see p. 363).

This type of fauna is not restricted to the Tommotian Stage; small shelly fossils are also common in the overlying Adtabanian Stage (see below) and similar assemblages of mainly phosphatic minute shells have been reported from younger condensed sequences in the Paleozoic. The shell substance of the carbonate skeletons within the fauna seems to have been controlled by the ambient seawater chemistry; Nemakit-Daldynian assemblages were mainly aragonite, whereas younger shells were mainly calcitic (Porter 2007). Tommotian-type faunas probably finally disappeared with the escalation of predation during the Mesozoic.

Some scientists such as Stephen Jay Gould suggested the less time-specific term, small shelly fossils to describe these assemblages. The fauna is now known to include a variety of groups united by their minute size and sudden appearance near the base of Cambrian. The small shelly fauna probably dominated the earliest Cambrian ecosystems when many metazoan phyla developed their own distinctive characteristics, initially at a very small scale. Nevertheless, some of this small size may be a preservational artifact, since phosphatization only works at a millimeter scale.

Composition and morphology

Many of the Tommotian skeletons (Fig. 10.12) were retrieved from residues after the acid etching of limestones; thus there is a bias towards acid-resistant skeletal material in any census of the group as a whole. Moreover, there is currently discussion concerning whether the acid-resistant skeletons of the Tommotian-type animals were primary constructions or secondary replacement fabrics. Or perhaps these shells survived in the sediments because of particular chemical conditions in the oceans at the time that allowed phosphatic fossils to survive (Porter 2004). The Tommotian animals had skeletons composed of a variety of materials. For example, *Cloudina* and the anabaritids were tube-builders that secreted carbonate material, whereas *Mobergella* and *Lapworthella* consisted of sclerites comprising organisms that secreted phosphatic material; *Sabellidites* is an organic-walled tube possibly of an unsegmented worm.

Many of the Tommotian animals are **form taxa** (that is, named simply by their shapes) because the biological relationships of most cannot be established and often there are few clues regarding the function and significance of each skeletal part. Most are short-lived and have no obvious modern analogs. Two groups are common – the hyolithelminthids have phosphatic tubes, open at both ends, whereas the tommotiids are usually phosphatic, cone-shaped shells that seem to belong in bilaterally symmetric sets.

Discoveries of near-complete examples of *Microdictyon*-like animals from the Lower Cambrian of China have helped clarify the status and function of some elements of the Tommotian fauna. These worms have round to oval plates arranged in pairs along the length of the body, which may have provided a base for muscle attachment associated with locomotion. As noted previously, many of the small shelly fossils are probably the sclerites of larger multiplated worm and worm-like animals (Box 10.5).

The Meishucunian biota

The Meishucunian Stage of South China has yielded some of the most diverse Tommotian-type assemblages in strata of Atdabanian age (see Appendix 1). Qian Yi and Stefan Bengtson (1989) have described nearly 40 genera that belong to three largely discrete, successive assemblages through the stage. First, the *Anabarites–Protohertzina–Arthrochites* assemblage is dominated by tube-dwelling organisms such as *Anabarites*; the *Siphonguchites–Paragloborilus* assemblage contains mobile mollusk-like and multiplated organisms together with some tube-dwellers and possible predators; whereas the *Lapworthella–Tannuolina–Sinosachites* association has mainly widespread multiplated animals.

Many of these fossils are known from Lower Cambrian horizons elsewhere in the world, highlighting the global distribution of many elements of the fauna. However, the three "community" types are rather mysterious, and probably represent different ecosystems, but it is hard to speculate further.

Distribution and ecology

Although it is still unclear whether many of the Tommotian skeletons are single shells or single sclerites and the autecology of most groups is unknown, the assemblage was certainly the first example in evolution of a skeletalized benthos. Very few of the Tommotian skeletal parts exceed 1 cm; nevertheless many shells were the armored parts of larger worm-like animals. And both mobile and fixed forms occurred together with archaeocyathans and non-articulate brachiopods. The microbenthos of the Tommotian was succeeded by a more typical Cambrian fauna, dominated by trilobites, non-articulate brachiopods, monoplacophoran mollusks and primitive echinoderms together with the archaeocyathans during the Atdabanian Stage (Fig. 10.14).

Cambrian explosion

The Cambrian explosion suddenly generated many entirely new and spectacular body plans (Box 10.6) and coincides with the appearance of the Bilateria over a relatively short period of time (Conway Morris 1998, 2006). This rapid diversification of life formed the basis for Stephen Jay Gould's bestseller, *Wonderful Life* (1989), which took its title from the Frank Capra 1946 film *It's a Wonderful Life*. The rapid appearance of such a wide range of apparently different animals has suggested two possible explanations. The "standard" view is that the diversification of bilaterians happened just as fast as the fossils suggest, and that some reasons must be sought to explain why many different animal groups apparently acquired mineralized skeletons at the same time. An alternative view arose after initial molecular studies had suggested that animals diverged some 800 myr before the beginning of the Cambrian (e.g. Wray et al. 1996). If these molecular views were correct, then the absence of fossils of modern animal phyla through the Proterozoic would have to be explained by an interval of cryptic evolution of probable micro- and meioscopic organisms, living between grains of sand, operating beneath the limits of detection prior to the explosion (Cooper & Fortey 1998). Greater refinement of Cambrian stratigraphy, the taxonomy and phylogeny of key Cambrian taxa and their relative appearance in the fossil

Box 10.5 Coelosclerites, mineralization and early animal evolution

The coeloscleritophorans are an odd group of animals based on the unique structure of their sclerites that appeared first in the Tommotian (Fig. 10.13). The sclerites are made of thin mineralized walls surrounding a cavity with a small basal opening. Once formed, the sclerites did not grow and were secreted by the mineralization of organic material occupying the cavity. The sclerites have longitudinal fibers and overlapping platelets within the mineralized wall. These animals may be extremely important in understanding the origin of biomineralization and the fuse for the Cambrian explosion, as argued by Stefan Bengtson (2005). Coelosclerites may be structures that are not known in any living animal but that were shared by both the bilaterians and non-bilaterians and probably characterized both ecdysozoans and spiralians. Coelosclerites may then have been lost, possibly by progenesis (see p. 145) from the larval to juvenile stages. If these features were developed in larger bilaterians then it is possible that within the Ediacara fauna giant forms – tens of centimeters in length – lurked, adorned by spiny and scaly sclerites. This is a controversial but nonetheless stimulating view that adds even more variety to our interpretations of early metazoan evolution.

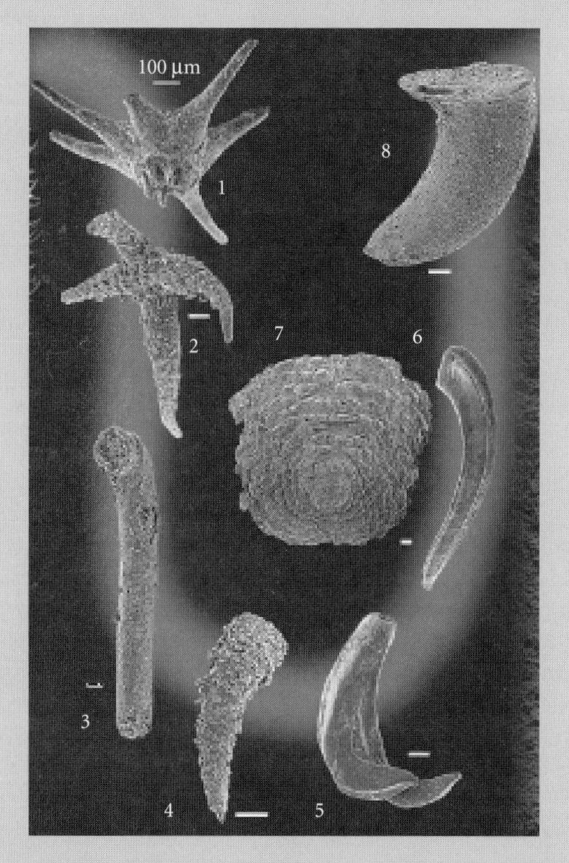

Figure 10.13 Coelosclerites. Chancelloriids: 1 and 2, *Chancelloria*; 3, *Archiasterella*; 4, *Eremactis*. Sachitid: 5, *Hippopharangites*. Siphonoguchitids: 6, *Drepanochites*; 7, *Siphogonuchites*; 8, *Maikhanella*. Scale bars, 100 µm. (Courtesy of Stefan Bengtson.)

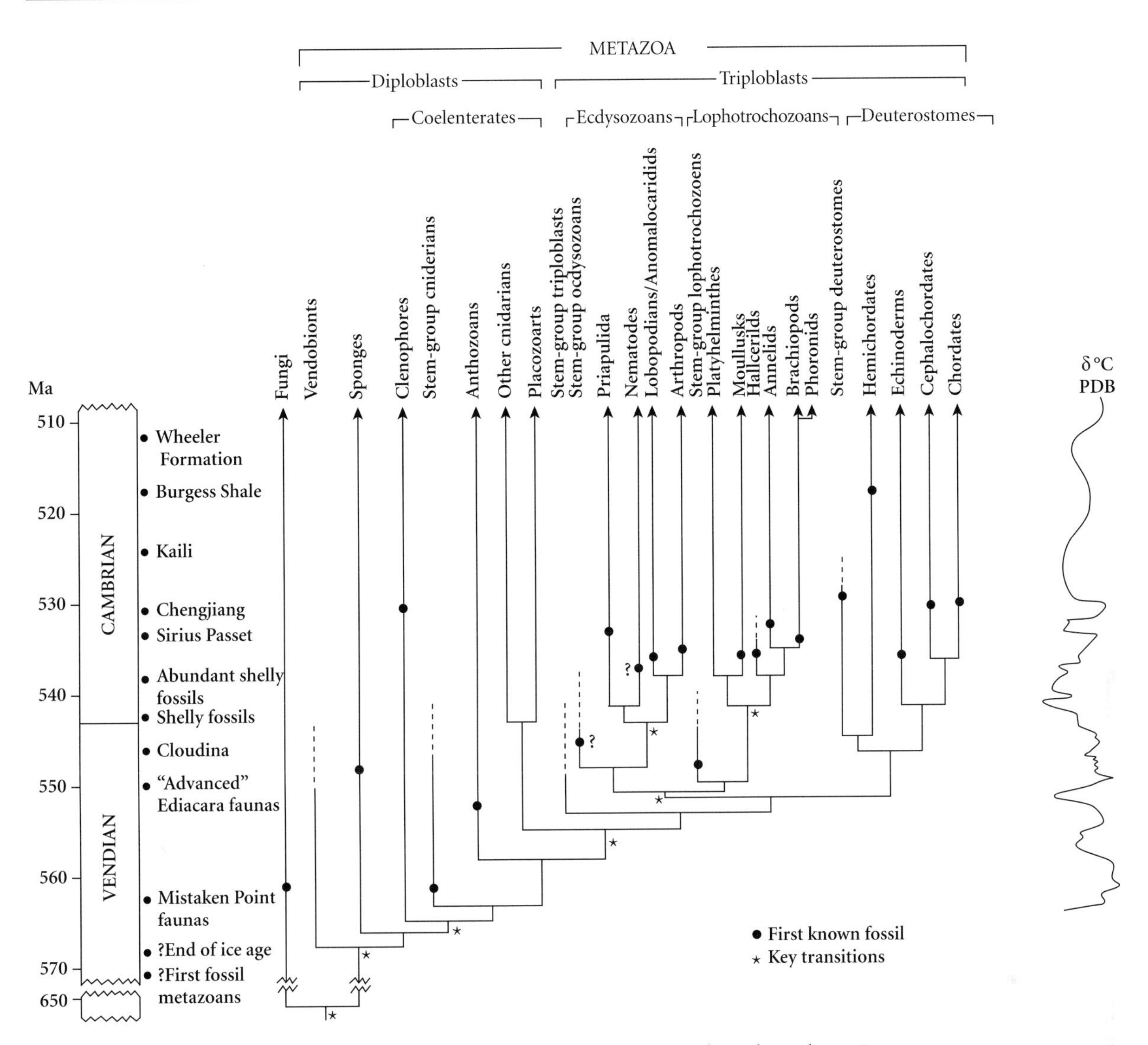

Figure 10.14 Stratigraphic distribution of Late Precambrian and Early Paleozoic metazoan taxa, some key morphological transitions and the carbon isotope record ($\delta^{13}C$). PDB, Vienna Pee Dee beleminite, the standard material for relative carbon isotope measurements. (Based on various sources.)

record, together with a revised molecular clock (see p. 133), have suggested an alternative hypothesis. The current Lower to Middle Cambrian fossil record displays the sequential and orderly appearance of successively more complex metazoans (Budd 2003), albeit rather rapidly (Fig. 10.16), and the timing is closely matched by revised molecular time scales (see p. 235; Peterson et al. 2004). Nevertheless there is some suggestion from the biogeographic patterns of trilobites that the divergence of many metazoan lineages may have already begun 30–70 myr earlier (Meert & Lieberman 2004) and speciation rates during the explosion were not in fact so incredible compared with those of other diversifications preserved in the fossil record (Lieberman 2001).

Much of our knowledge of the Cambrian explosion is derived from three spectacular, intensively-studied Lagerstätte assemblages: Burgess (Canada), Chengjiang (China) and Sirius Passet (Greenland). The diversities of the Cambrian "background" faunas are generally much lower and arguably contain less morphologically different organisms. Reconstructions of these seafloors are possible (Fig. 10.17). But whereas the Cambrian explosion provided higher taxa, in some diversity, the Ordovician radiation generated the sheer biomass, biodiversity and biocomplexity that would fill the world's oceans.

Box 10.6 Roughness landscapes

There have been a number of explanations for the rapid explosion of life during the Early and Mid Cambrian involving all sorts of developmental (genetic), ecological and environmental factors. Why, too, was this event restricted to the Cambrian? Was there some kind of developmental limitation, an ecological saturation, or were there simply no further ecological opportunities left to exploit? One interesting model that may help explain the ecological dimension of the event involves the use of fitness landscapes. The concept is taken from genetics but can be adapted to morphological information (Marshall 2006). Biotas can be plotted against two axes, each representing morphological rules that can generate shapes. The Ediacara fauna has only three recognizable bilaterians, so the landscape is relatively smooth with only three peaks. On the other hand the Cambrian explosion generated at least 20 bilaterian body plans and a very rough landscape rather like the Alps or the Rockies (Fig. 10.15). What roughened the landscape, or why were there more bilaterians in the Cambrian fauna? Much of the bilaterian genetic tool kit was already in place in the Late Proterozoic and the environment was clearly conducive to their existence. The "principle of frustration" (Marshall 2006), however, suggests that different needs will often have conflicting solutions, ensuring that the best morphological design is rarely the most optimal one. Is it possible that, with the rapid development of biotic interactions such as predation, many morphological solutions were developed, some less than optimal but nevertheless driving a roughening of the fitness landscape. Thus "frustration", the multiplication of attempted solutions to new opportunities, led to the roughening of the Cambrian landscape and may have been an important factor in the Cambrian explosion.

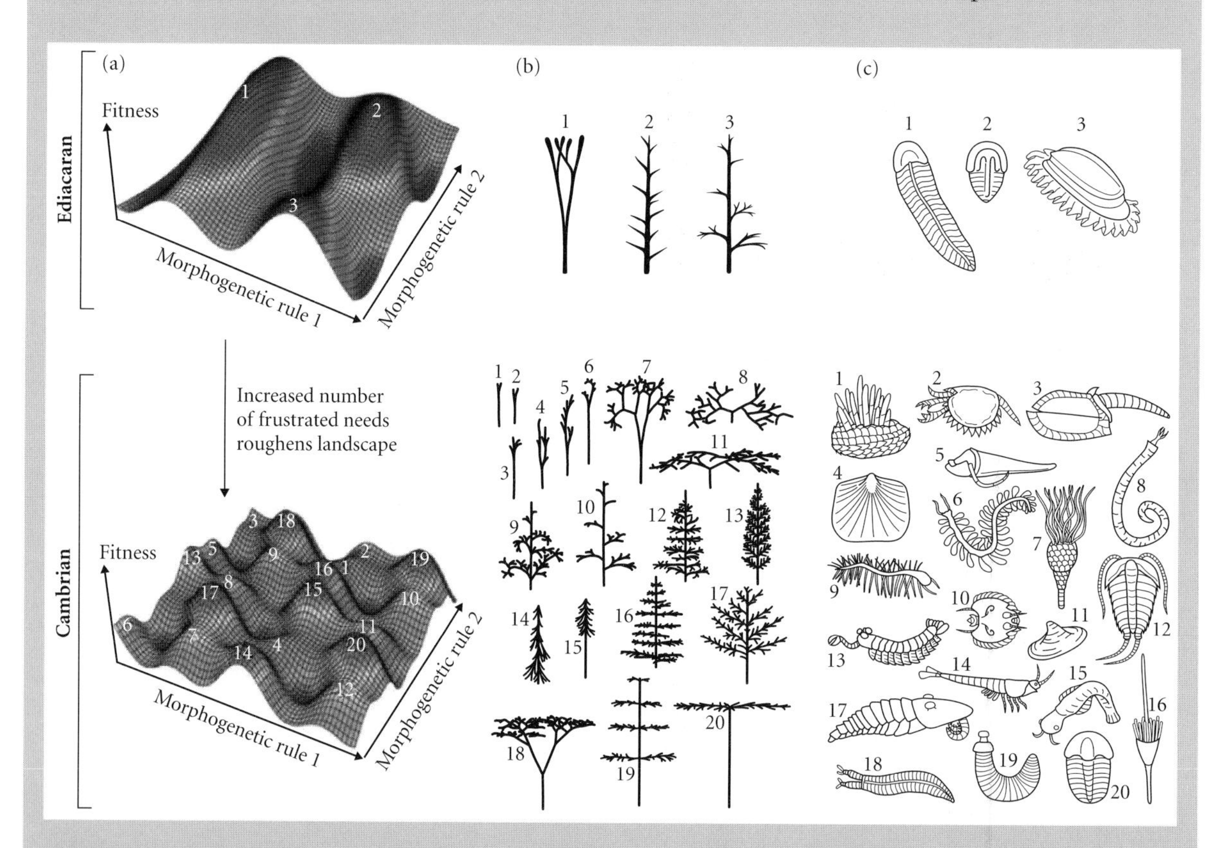

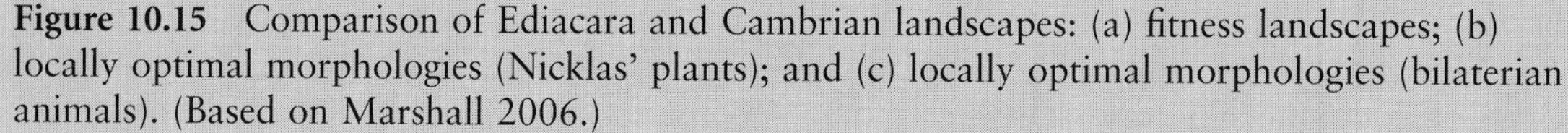

Figure 10.15 Comparison of Ediacara and Cambrian landscapes: (a) fitness landscapes; (b) locally optimal morphologies (Nicklas' plants); and (c) locally optimal morphologies (bilaterian animals). (Based on Marshall 2006.)

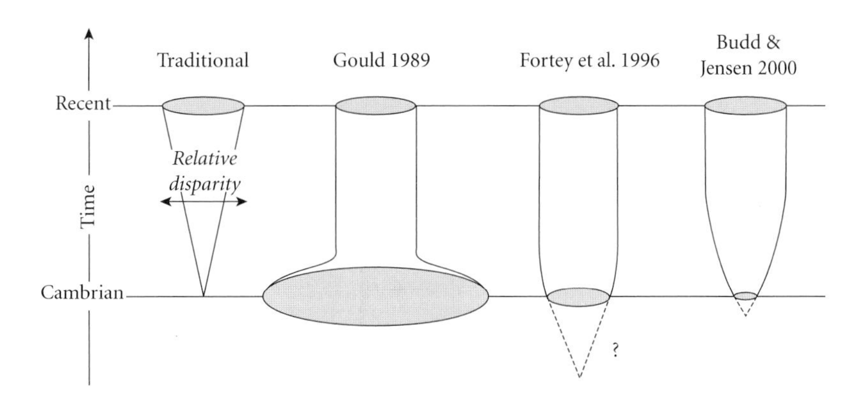

Figure 10.16 Modes of the Cambrian explosion. (Based on Budd & Jensen 2000.)

Ordovician radiation

During an interval of some 25 myr, during the Mid to Late Ordovician, the biological component of the planet's seafloors was irreversibly changed. A massive hike in biodiversity was matched by an increase in the complexity of marine life (Harper 2006). The event witnessed a three- to four-fold increase in, for example, the number of families, leveling off at about 500; these clades would dominate marine life for the next 250 myr. Nevertheless the majority of "Paleozoic" taxa were derived from Cambrian stocks. With the exception of the bryozoans (see p. 313), no new phyla emerged during the radiation, although more crown groups emerged from the stem groups generated during the Cambrian explosion.

The great Ordovician radiation is one of the two most significant evolutionary events in the history of Paleozoic life. In many ways the Ordovician Period was unique, enjoying unusually high sea levels, extensive, large epicontinental seas, with virtually flat seabeds, and restricted land areas, many probably represented only by archipelagos. Magmatic and tectonic activity was intense with rapid plate movements and widespread volcanic activity. Island arcs and mountain belts provided sources for clastic sediment in competition with the carbonate belts associated with most of the continents. Biogeographic differentiation was extreme, affecting plankton, nekton and benthos, and climatic zonation existed, particularly in the southern hemisphere. Finally, during the Mid Ordovician, the Earth was bombarded with asteroids that appear in some way also to be linked to the biodiversification (Schmitz et al. 2008). Taken together, these conditions were ideal for all kinds of speciation processes and the evolution of ecological niches. Most significant was the diversification of skeletal organisms, including the brachiopods, bryozoans, cephalopods, conodonts, corals, crinoids, graptolites, ostracodes, stromatoporoids and trilobites that we will read about later.

Whereas the Cambrian explosion involved the rapid evolution of skeletalization and a range of new body plans, together with the extinction of the soft-bodied Ediacara biota and the appearance of the Bilateria, the Ordovician diversification generated few new higher taxa, for example phyla, but witnessed a staggering increase in biodiversity at the family, genus and species levels. This taxonomic radiation, which included members of the so-called "Cambrian", "Paleozoic" and "Modern" evolutionary biotas (see p. 538), set the agenda for much of subsequent marine life on the planet against a background of sustained greenhouse climates. Although many outline analyses have been made, there are relatively few studies of the ecological and environmental aspects of the Ordovician diversification (Bottjer et al. 2001). Moreover the causes of the event, and its relationship to both biological and environmental factors, are far from clear. Evolution of the plankton, however, may have been a primary factor (Box 10.7).

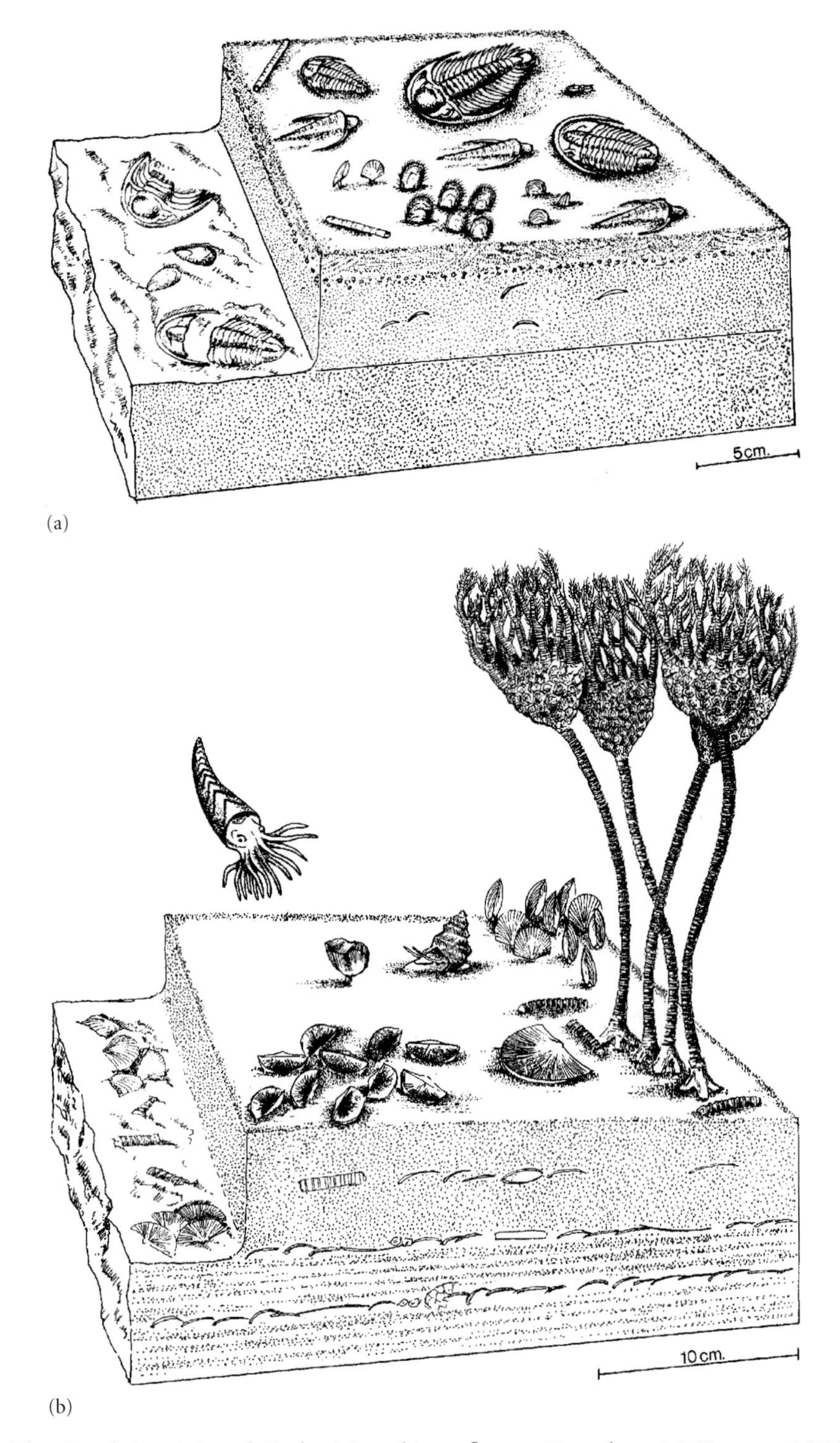

Figure 10.17 The Cambrian (a) and Ordovician (b) seafloors. (Based on McKerrow 1978.)

Box 10.7 Larvae and the Ordovician radiation

Many factors, mainly ecological and environmental, have been invoked to explain the great Ordovician biodiversification or Ordovician radiation. Did the diversification have its origins in the plankton? Most early bilaterians probably had benthic lecithotrophic larvae (see p. 241). But the Cambrian oceans, relatively free of pelagic predators, offered great possibilities. Exploitation of the water column by larvae occurred a number of times independently, turning the clear waters of the Early Cambrian into a soup of planktonic organisms in the Ordovician. The fossil record and molecular clock data suggest that at least six different feeding larvae developed from non-feeding types between the Late Cambrian and Late Silurian (Peterson 2005). In addition to planktotrophic larvae, the oceans were rapidly colonized by diverse biotas of other microorganisms such as the acritarchs (see p. 216). The dramatic diversification of the suspension-feeding benthos coincides with the evolution of planktotrophy in a number of different lineages (Fig. 10.18). These factors had an undoubted effect on the diversification of Early Paleozoic life, which reached a plateau of diversity during the Ordovician.

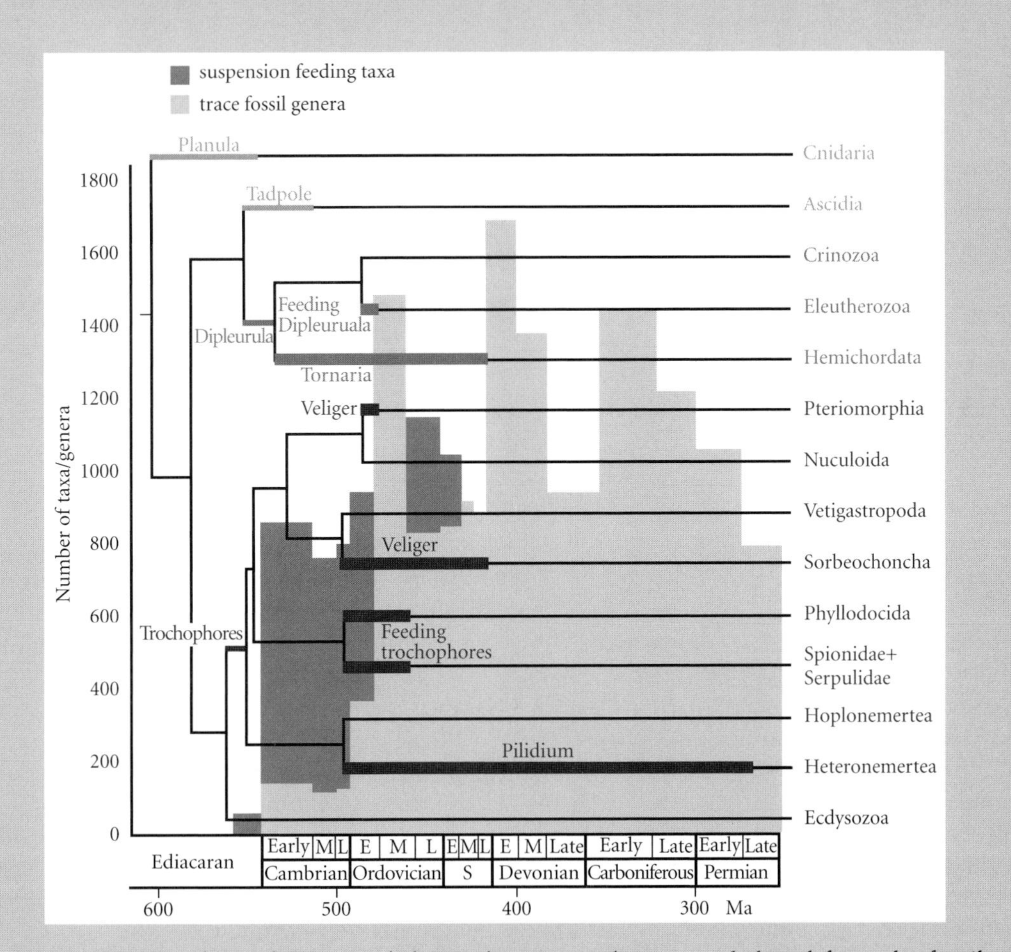

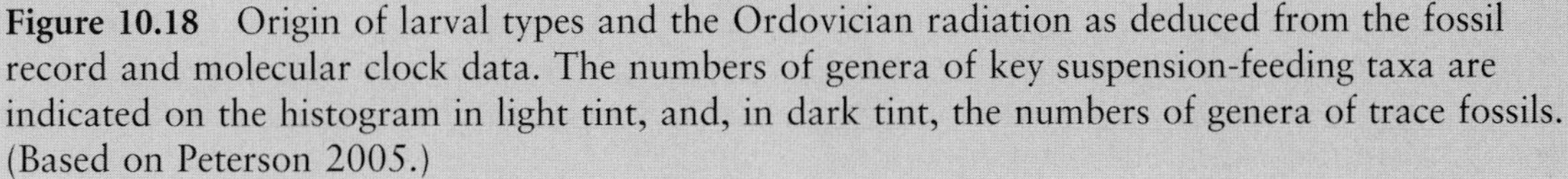

Figure 10.18 Origin of larval types and the Ordovician radiation as deduced from the fossil record and molecular clock data. The numbers of genera of key suspension-feeding taxa are indicated on the histogram in light tint, and, in dark tint, the numbers of genera of trace fossils. (Based on Peterson 2005.)

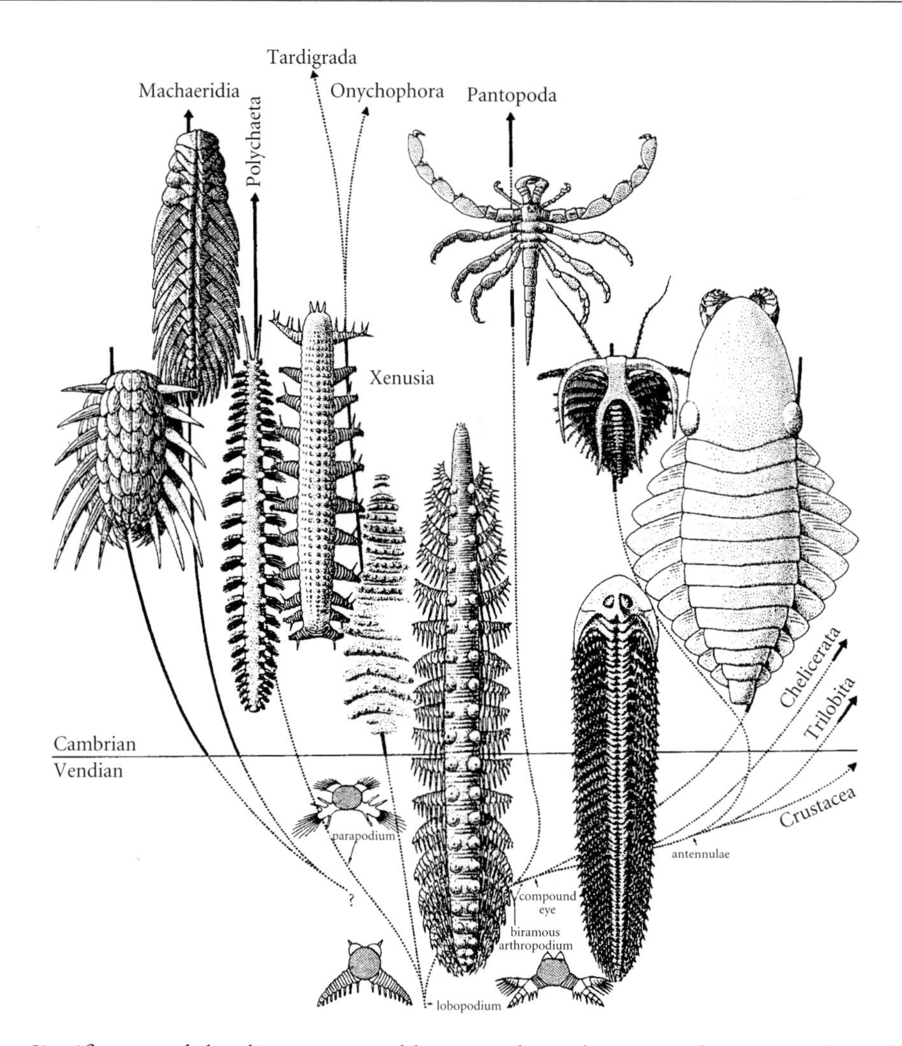

Figure 10.19 Significance of the diverse worm-like animals at the Precambrian–Cambrian boundary and the postulated origins of some major clades. (Based on Dzik, J. & Krumbiegel, G. 1989. *Lethaia* 22.)

SOFT-BODIED INVERTEBRATES

Of the 25 or so commonly recognized animal phyla, fewer than nine (35%) have an adequate fossil record. Many are small phyla represented by relatively few species. However, there are a number of larger phyla whose poor fossil record reflects the lack of a preservable skeleton, although a number of these soft-bodied forms are preserved in fossil Lagerstätten. Most are worms or worm-like organisms (Fig. 10.19). But in spite of unspectacular fossil records, there is considerable interest in these poorly represented invertebrates. The origins of many higher taxa must be sought within the plexus of worm-like organisms. Moreover, the evidence from the Burgess Shale and other such exceptionally preserved faunas suggests that many of these soft-bodied groups dominated certain marine paleocommunities in terms of both numbers and biomass and additionally contributed to associated trace fossil assemblages.

The platyhelminths or the flatworms are bilateral animals with organs composed of tissues arranged into systems. Most are parasites, but the turbellarians are free-living carnivores and scavengers. The Ediacaran animals *Dickinsonia* and *Palaeoplatoda* have been assigned to the turbellarian flatworms by some authors; similarly *Platydendron* from the Middle Cambrian Burgess Shale has been ascribed to the platyhelminthes.

The ribbon worms, or nemertines, are characterized by a long anterior sensory proboscis. The majority are marine, although some inhabit soil and freshwater. Although the bizarre *Amiskwia* from the Middle Cambrian Burgess Shale was assigned to this group, recent opinion suggests it is merely convergent on the nemertine body shape. Some of the Tommotian animals may also be nemertine worms. The nematodes or roundworms are generally smooth and sac-like.

The priapulid worms are exclusively marine, short and broad with **proboscces** ("noses"; singular, proboscis) covered in spines and warts. The Middle Cambrian Burgess Shale contains seven genera assigned to at least five families. The Burgess forms are all characterized by priapulid proboscces, and most have little in common with modern forms. Nevertheless the most abundant taxon, *Ottoia*, is very similar to the living genus *Halicryptus*. Elsewhere in the fossil record the Upper Carboniferous Mazon Creek fauna has yielded *Priapulites*, which has a distinctly modern aspect.

The annelid worms, such as the common earthworm and lugworm, have ring-like external segments that coincide with internal partitions housing pairs of digestive and reproductive organs; the nervous system is well developed and the head has distinctive eyes. The annelid body is ornamented by bristles that aid locomotion and provide stability. Most are predators or scavengers living in burrows. The polychaetes or paddle worms have the most complete fossil record; the record is enhanced by the relatively common preservation of elements of the phosphatic jaw apparatus known as **scolecodonts** (see p. 359). Although some Ediacaran animals, such as *Spriggina*, have been associated with the polychaetes, the first undoubted paddle worms are not known until the Cambrian. A diverse polychaete fauna has been described from the Burgess Shale; it even contains *Canada spinosa*, similar to some living polychaetes.

Review questions

1 Traditional methods of reconstructing the phylogeny of the early metazoans based on morphology have encountered problems. Is the concept of body plans still useful and if so, for what?
2 Interpretations of Ediacaran biotas are as far from a consensus as ever. Why are the Ediacara organisms so difficult to classify and understand?
3 The identification of embryos and trace fossils are both important evidence of animal life. How can both be used to indicate the presence of metazoan life?
4 Was the Cambrian explosion one of animals or fossils? How large was the role of taphonomy in the manifestation of the Cambrian explosion?
5 Within an interval of 100 million years the planet's seafloors were changed for ever. Briefly compare and contrast the changing seascapes through the Ediacaran, Cambrian and Ordovician periods.

Further reading

Briggs, D.E.G. & Fortey, R.A. 2005. Wonderful strife: systematics, stem groups, and the phylogenetic signal of the Cambrian radiation. *Paleobiology* **31** (Suppl.), 94–112.

Brusca, R.C. & Brusca, G.J. 2002. *Invertebrates*, 2nd edn. Sinauer Associates, Sunderland, MA.

Conway Morris, S. 2006. Darwin's dilemma: the realities of the Cambrian explosion. *Philosophical Transactions of the Royal Society B* **361**, 1069–83.

Gould, S.J. 1989. *Wonderful Life. The Burgess Shale and the Nature of History*. W.W. Norton & Co., New York.

Nielsen, C. 2003. *Animal Evolution. Interrelationships of the Living Phyla*, 2nd edn. Oxford University Press, Oxford, UK.

Valentine, J.W. 2004. *On the Origin of Phyla*. University of Chicago Press, Chicago.

References

Aguinaldo, A.M.A. & Lake, J.A. 1998. Evolution of multicellular animals. *American Zoologist* **38**, 878–87.

Bengtson, S. 2005. Mineralized skeletons and early animal evolution. *In* Briggs, D.E.G. (ed.) *Evolving Form and Function*. New Haven Peabody Museum of Natural History, Yale University, New Haven, CT, pp. 101–17.

Bottjer, D.J., Droser, M.L., Sheehan, P.M. & McGhee, G.R. 2001. The ecological architecture of major events in the Phanerozoic history of marine life. *In* Allmon, W.D. & Bottjer, D.J. (eds) *Evolutionary Paleoecology*. Columbia University Press, New York, pp. 35–61.

Brasier, M.D. & McIlroy, D. 1998. *Neonereites uniserialis* from c. 600 Ma year old rocks in western

Scotland and the emergence of animals. *Journal of the Geological Society, London* **155**, 5–12.

Budd, G.E. 2003. The Cambrian fossil record and the origin of the phyla. *Integrative Comparative Biology* **43**, 157–65.

Budd, G.E. 2008. The earliest fossil record of the animals and its significance. *Philosphical Transactions of the Royal Society B* **363**, 1425–34.

Budd, G.E. & Jensen, S. 2000. A critical reappraisal of the fossil record of bilaterian phyla. *Biological Reviews* **75**, 253–95.

Buss, L.W. & Seilacher, A. 1994. The phylum Vendobionta: a sister group of the Eumetazoa? *Paleobiology* **20**, 1–4.

Canfield, D.E., Poulton, S.W. & Narbonne, G.M. 2007. Late-Neoproterozoic deep-ocean oxygenation and the rise of animal life. *Science* **315**, 92–5.

Conway Morris, S. 1998. The evolution of diversity in ancient ecosystems: a review. *Philosophical Transactions of the Royal Society B* **353**, 327–45.

Conway Morris, S. 2006. Darwin's dilemma: the realities of the Cambrian explosion. *Philosophical Transactions of the Royal Society B* **361**, 1069–83.

Cooper, A. & Fortey, R.A. 1998. Evolutionary explosions and the phylogenetic fuse. *Trends in Ecology and Evolution* **13**, 151–6.

Donogue, P.C.J. 2007. Embryonic identity crisis. *Nature* **445**, 155–6.

Donoghue, P.C.J., Bengtson, S., Dong Xi-ping et al. 2006. Synchotron X-ray tomographic microscopy of fossil embryos. *Nature* **442**, 680–3.

Droser, M.L., Jensen, S. & Gehling, J.G. 2002. Trace fossils and substrates of the terminal Proterozoic-Cambrian transition: implications for the record of early bilaterians and sediment. *Proceedings of the National Academy of Sciences, USA* **99**, 12572–6.

Fedonkin, M.A. 1990. Precambrian metazoans. *In* Briggs, D.E.G. & Crowther, P.R. (eds) *Palaeobiology, A Synthesis*. Palaeontological Association and Blackwell Scientific Publications, Oxford, UK, pp. 17–24.

Fortey, R.A., Briggs, D.E.G. & Wills, M.A. 1996. The Cambrian evolutionary "explosion": decoupling cladogenesis from morphological disparity. *Biological Journal of the Linnaean Society* **57**, 13–33.

Gould, S.J. 1989. *Wonderful Life. The Burgess Shale and the Nature of History*. W.W. Norton & Co., New York.

Grazhdankin, D. 2004. Patterns of distribution in the Ediacaran biotas: facies versus biogeography and evolution. *Paleobiology* **30**, 203–21.

Hagadorn, J.W., Xiao Shuhai, Donoghue, P.C.J. et al. 2006. Cellular and subcellular structure of Neoproterozoic animal embryos. *Science* **314**, 291–4.

Harper, D.A.T. 2006. The Ordovician biodiversification: setting an agenda for marine life. *Palaeogeography, Palaeoclimatology, Palaeoecology* **232**, 148–66.

Jensen, S. 2003. The Proterozoic and earliest Cambrian trace fossil record: patterns, problems and perspectives. *Integrative Comparative Biology* **43**, 219–28.

Lieberman, B.S. 2001. A probabilistic analysis of rates of speciation during the Cambrian radiation. *Proceedings of the Royal Society, Biological Sciences* **268**, 1707–14.

Marshall, C.R. 2006. Explaining the Cambrian "Explosion" of animals. *Annual Reviews of Earth and Planetary Science* **33**, 355–84.

McKerrow, W.S. 1978. *Ecology of Fossils*. Duckworth Company Ltd., London.

McMenamin, M.A.S. 1986. The garden of Ediacara. *Palaois* **1**, 178–82.

Meert, J.G. & Lieberman, B.S. 2004. A palaeomagnetic and palaeobiogeographic perspective on latest Neoproterozoic and Cambrian tectonic events. *Journal of the Geological Society, London* **161**, 1–11.

Narbonne, G.M. 2005. The Ediacara biota: Neoproterozoic origin of animals and their ecosystems. *Annual Reviews of Earth and Planetary Science* **33**, 421–42.

Nielsen, C. 2008. Six major steps in animal evolution: are we derived sponge larvae? *Evolution and Development* **10**, 241–57.

Peterson, K.J. 2005. Macroevolutionary interplay between planktic larvae and benthic predators. *Geology* **33**, 929–32.

Peterson, K.J., Cotton, J.A., Gehling, J.G. & Pisani, D. 2008. The Ediacaran emergence of bilaterians: congruence between genetic and the geological fossil records. *Philosphical Transactions of the Royal Society B* **363**, 1435–43.

Peterson, K.J., Lyons, J.B., Nowak, K.S., Takacs, C.M., Wargo, M.J. & McPeek, M. 2004. Estimating metazoan divergence times with a molecular clock. *Proceedings of the National Academy of Sciences, USA* **101**, 6536–41.

Peterson, K.J., McPeek, M.A. & Evans, D.A.D. 2005. Tempo and mode of early animal evolution: inferences from rocks, Hox, and molecular clocks. *Paleobiology* **31** (Suppl.), 36–55.

Porter, S.M. 2004. Closing the phosphatization window: testing for the influence of taphonomic megabias on the patterns of small shelly fauna decline. *Palaios* **19**, 178–83.

Porter, S.M. 2007. Seawater chemistry and early carbonate biomineralization. *Science* **316**, 1302.

Qian Yi & Bentson, S. 1989. Palaeontology and biostratigraphy of the Early Cambrian Meishucunian Stage in Yunnan Province, South China. *Fossils and Strata* **24**, 1–156.

Rasmussen, B., Bengtson, S., Fletcher, I.R. & McNaughton, N.J. 2002. Discoidal impressions and trace-like fossils more than 1200 million years ago. *Science* **296**, 1112–15.

Schmitz, B., Harper, D.A.T., Peucker-Ehrenbrink, B. et al. 2008. Asteroid breakup linked to the Great

Ordovician Biodiversification Event. *Nature Geoscience* **1**, 49–53.
Seilacher, A. 1989. Vendozoa: organismic construction in the Proterozoic biosphere. *Lethaia* **22**, 229–39.
Seilacher, A., Bose, P.K. & Pflüger, F. 1998. Triploblastic animals more than 1 billion years ago: trace fossil evidence from India. Science **282**, 80–3.
Valentine, J.W. 2004. *On the Origin of Phyla*. University of Chicago Press, Chicago.
Waggoner, B. 2003. The Ediacara biotas in space and time. *Integrative Comparative Biology* **43**, 104–13.
Wray, G.A., Levinton, J.S. & Shapiro, L.M. 1996. Molecular evidence for deep pre-Cambrian divergences amongst metazoan phyla. *Science* **214**, 568–73.

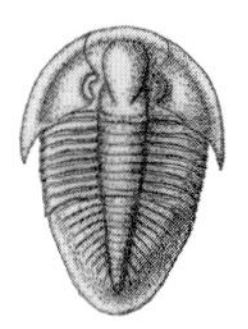

Chapter 11

The basal metazoans: sponges and corals

Key points

- Parazoans are a grade of organization within the metazoans composed of multicellular complexes with few cell types and lacking variation in tissue or organs; the sponges (Phylum Porifera) are typical parazoans that lack a gut.
- Sponges are almost entirely filter-feeding members of the sessile benthos. The group contains a variety of grades of functional organization that cut across the traditional classification of the phylum.
- Sponge reefs were dominated, during most of the Phanerozoic, by calcareous grades developed convergently across the phylum; siliceous sponges were important reef builders mainly during the Mesozoic.
- Stromatoporoids are a grade of organization within the Porifera with a secondary calcareous skeleton, important in reefs during the mid-Paleozoic and mid-Mesozoic.
- Archaeocyaths are Cambrian organisms of sponge grade. They were mainly solitary but developed a branching, modular growth mode and successfully built reefs in often turbulent and unstable environments.
- Reef-type structures were already present in the Late Precambrian hosting large, robust, colonial organisms.
- The cnidarians are the simplest of the higher metazoans with a radial diploblastic body plan and stinging cells or cnidoblasts. The phylum includes sea anemones, jellyfish and hydra together with the corals.
- The Paleozoic rugose and tabulate corals displayed a wide range of growth modes often related to environments; neither group was a dominant reef builder.
- The scleractinians radiated during the Mesozoic with zooxanthellate forms dominating biological reefs. Scleractinian-like morphs in Paleozoic faunas arose several times independently from anemones with scleractinian-type polyps.
- Reef development through time has waxed and waned, dominated at different times by different groups of reef-building organisms.
- Coloniality within the metazoans has evolved many times; one hypothesis suggests that a Precambrian colonial organism may have been a source for the bilaterians.

Until nearly the end of the Neoproterozoic, some 80% of geological time, the oceans of the world were mainly occupied by rather simple, usually unicellular, organisms. By the Ediacaran it was clear that this simple existence was not enough, and more complex body plans were soon to develop their own ecosystems. Two groups, the Porifera and the Cnidaria, form the basal parts of the metazoan tree, diverging during the Neoproterozoic. Despite their origins in deep time and their relative simplicity, both maintained high diversity, notably as colonial organisms, throughout the Phanerozoic, frequently becoming important parts of the planet's reef ecosystems.

PORIFERA

So he dissected sea sponges by night, winter night after winter night . . . adult and embryo human body parts by day, adult and larval sponge body parts by night.

Rebbeca Stott (2003) *Darwin and the Barnacle*, on the sponge doctor, Robert Grant

Most of us have used a bath sponge, probably a synthetic replica of the real thing. But ancient peoples used sponge skeletons as an aid to bathing and possibly exfoliation in some of the world's earliest and most exclusive health farms. Most considered they were some form of plant until proper biological study in the mid and late 1700s suggested they were animals – and at first they were classified as corals. It was in fact Dr Robert Grant (1793–1874), one time mentor to Charles Darwin, who later established the Porifera as a unique group in its own right. The poriferans or sponges have a unique porous structure and a body plan based at the cellular level of organization; they are said to lack true tissues. Most lack symmetry, true differentiated tissues, and organs, although their cells, like those of the protists, can switch function. They reproduce both asexually (by budding) and sexually with different cells expelling clouds of eggs and sperm out through an opening; some are even viviparous, with the eggs hatching within the parent sponge, and larvae released into the water.

There are over 10,000 species of sponge. All are aquatic, and most are marine. Sponges are part of the sessile benthos, fixed to the seabed, pumping large volumes of water – in extreme cases over 1000 L per day – through their fixed but commonly flexible bodies, which act as filters for nutrients. The group has a remarkable range of morphologies; the more specialized, stalked forms live in deep-water environments and flattened, dumpy forms prefer shallower-water, high-energy environments. Despite the apparent simplicity of the sponges, the classification of the phylum has recently undergone considerable revision (Box 11.1). Some well-established calcified groups, such as the "chaetetids" and "sphinctozoans", are probably polyphyletic, merely representing convergence towards common grades of organization. The well-established and diverse Demospongea, the common sponges, may too be polyphyletic. Despite their relative simplicity, the complex relationships of "sponge-grade" animals have yet to be resolved.

Morphology: examining a typical sponge

A typical sponge individual is not particularly complex or intellectually demanding to understand; it is nonetheless a remarkable organism. It is sac-shaped with a central cavity or **paragaster**, which opens externally at the top through the **osculum** (Fig. 11.1). The sponge is densely perforated by **ostia**, small holes marking the entrances to minute canals through which pass the inhalant currents. In simple terms, there are three main cell types: (i) flattened epithelial cells; (ii) collar cells or **choanocytes**, which occupy the internal chambers and move water along by beating their **flagella**; and (iii) **amoeboid** cells, which have digestive, reproductive and skeletal functions. Amoeboid cells can actually irreversibly change into other cell types with other functions. Nutrient-laden water is thus sucked through the ostia, flagellated by the choanocytes and processed by the amoeboids. Waste products and spent water, together with reproductive products when in season, are ejected upwards through the paragaster into the water column.

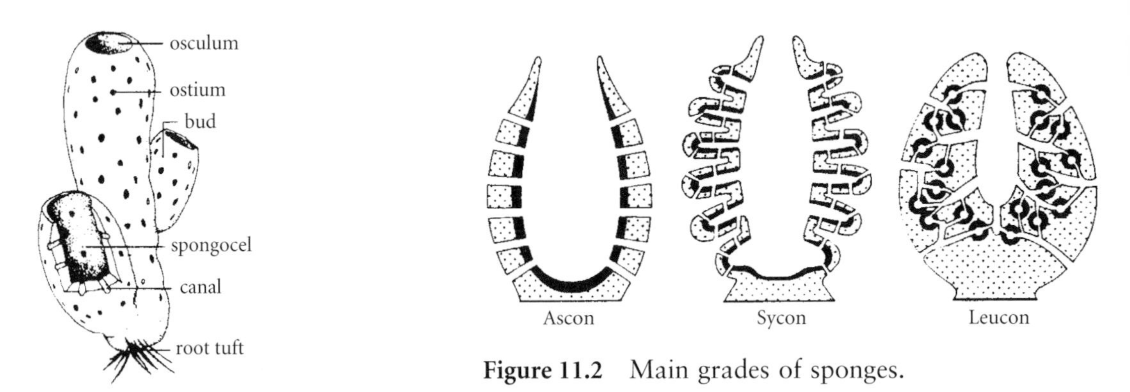

Figure 11.2 Main grades of sponges.

Figure 11.1 Basic sponge morphology.

Box 11.1 Classification and spicule morphology of the sponges

CLASSIFICATION OF THE SPONGES

The phylum Porifera was traditionally subdivided into four classes, the Demospongea, Calcarea, Sclerospongea and Hexactinellida, based mainly on the composition of the skeleton and type of spicules. Higher-level taxonomy is based exclusively on soft-tissue morphology. Some workers have suggested the exclusion of the glass sponges from the Porifera but this is poorly supported; rather they are closely related to the demosponges. However, the sclerosponges, with some additional calcareous skeletons, are now placed within the Demospongea. Thus three classes now comprise the phylum (Fig. 11.3).

Class CALCAREA (calcareous sponges)

- Sponges with calcitic spicules, usually simple, and/or porous calcareous walls. Marine environments
- Cambrian to Recent

Class DESMOSPONGEA (common sponges)

- Sponges with skeletons of spongin, a mix of spongin and siliceous spicules or only siliceous spicules. The spicules may be of two different sizes and the larger are represented by monaxons and tetraxons. Marine, brackish and freshwater environments. Living sponges previously assigned to the Sclerospongiae (coralline sponges) – sponges with a compound skeleton of siliceous spicules, spongin and an additional basal layer of laminated fibrous aragonite or calcite – are now also included here
- Cambrian to Recent

Class HEXACTINELLIDA (siliceous sponges)

- These are the glass sponges with complex siliceous spicules having six rays directed along three mutually perpendicular axes. Deep-water marine environments
- Precambrian (?) and Cambrian to Recent

However, two form-groups of sponge, the sphinctozoans (with a segmented chambered skeleton) and the chaetetids (with microscopic tubules) have representatives within the Calcarea and Demospongea; both were important reef builders.

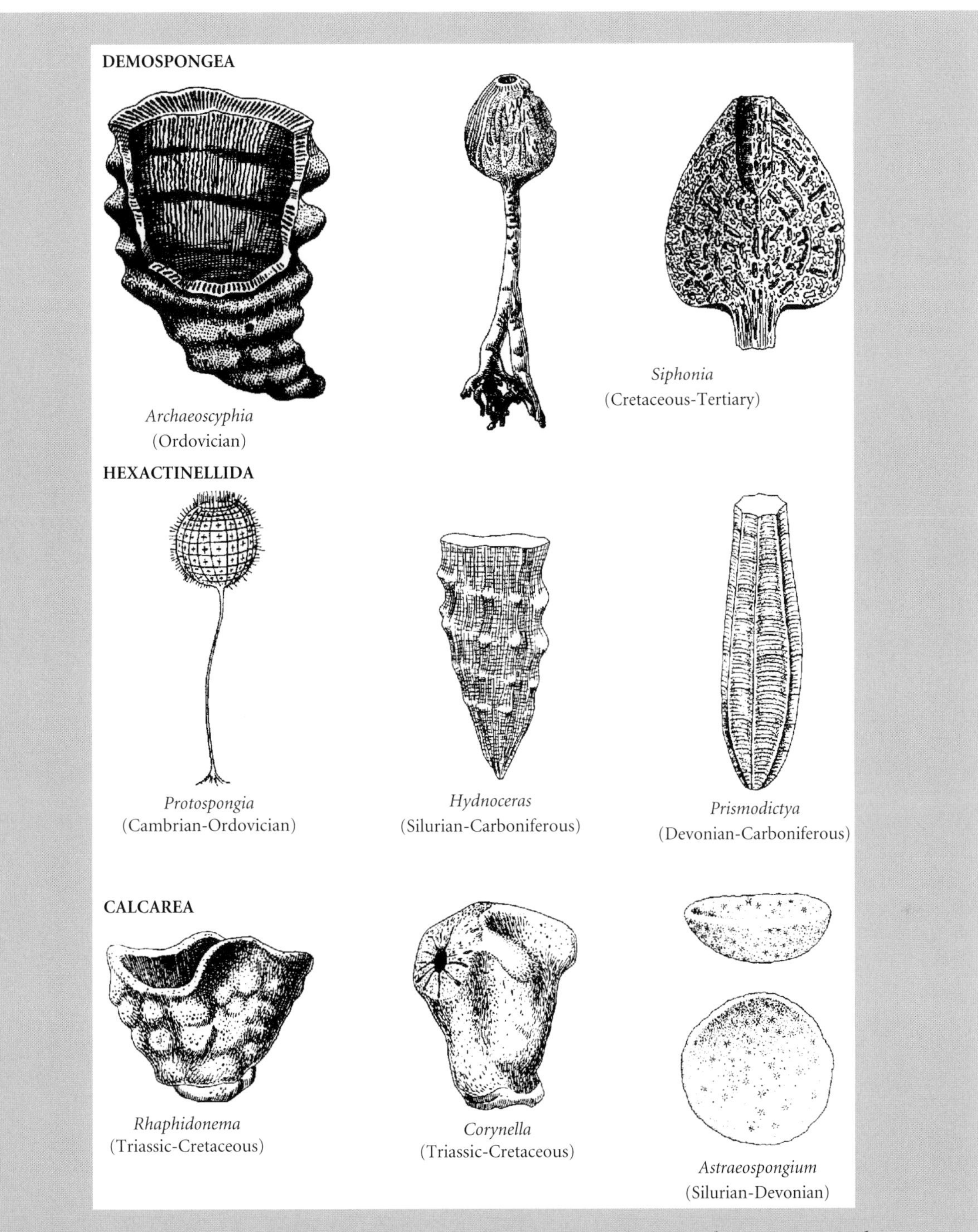

Figure 11.3 Some examples of the main groups of sponges: *Archaeoscyphia* (×0.25), *Siphonia* (×0.4 and 0.8), *Protospongia* (×0.4), *Hydnoceras* (×0.25), *Prismodictya* (×0.6), *Rhaphidonema* (×0.8), *Corynella* (×0.8) and *Astraeospongium* (×0.4).

The relationships among the three major groups of Porifera are obscure. Analyses of poriferan morphology and structure, cytology and molecular biology suggest that, first, sponges are a paraphyletic grouping (Sperling et al. 2007) and, second, that the Calcarea and Hexactinellida form monophyletic groups that are close to the base of the Eumetazoa. The Demospongea is more basal (Fig. 11.4).

Continued

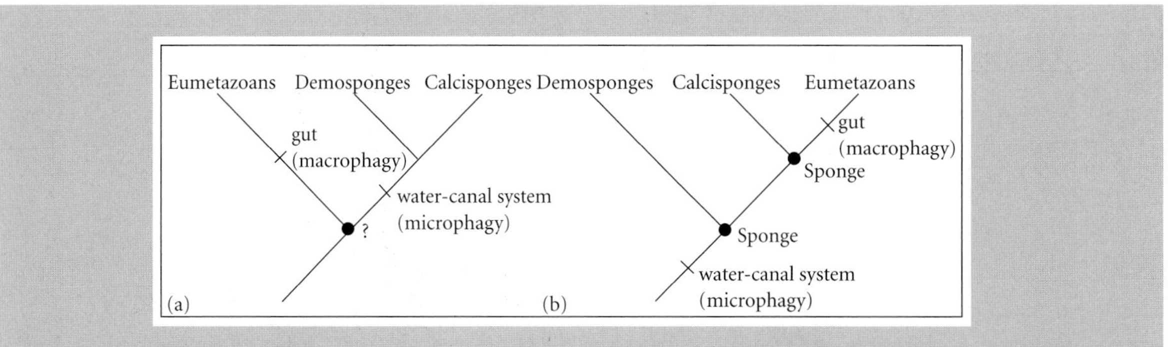

Figure 11.4 Sponge paraphyly. (a) The more traditional view presenting both the eumetazoans and poriferans as monophyletic groups; feeding strategies cannot be polarized since all the outgroups are non-metazoan. (b) If, however, poriferans are paraphyletic and calcisponges are more closely related to eumetazoans then the water canal system is a primitive character and the gut is more derived.

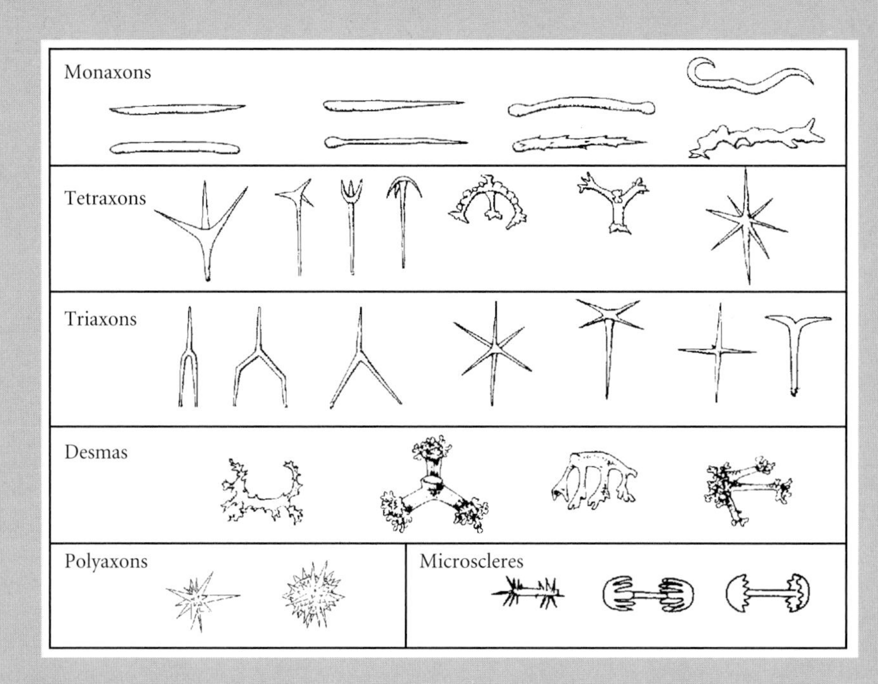

Figure 11.5 Main categories of spicule morphology. Magnification approximately ×75 for all, except microscleres which are about ×750.

SPICULE MORPHOLOGY

Commonly the spongin skeletons decay and unfused spicular skeletons disintegrate shortly after death leaving only a selection of hard parts, such as spicules (Fig. 11.5). Spicule morphology is thus a fundamental means of identification of those spiculate forms. Spicules may be large (**megascleres**), acting as part of the skeleton, or small (**microscleres**), scattered throughout the sponge and rarely preserved. Five basic types of spicule have been recognized:

1 **Monaxons:** single axial forms that may grow in one (monactinal) or two (diactinal) directions.
2 **Tetraxons (hexactines):** four-rayed forms that may have axes of equal length (calthrop).
3 **Triaxons:** six-rayed forms that form regular networks within the Hexactinellida or glass sponges.
4 **Desmas:** irregular-shaped forms with ends modified to articulate with one another.
5 **Polyaxons:** multirayed forms including spherical or star-shaped spicules.

Despite the flexibility of a typical bath sponge, sponges are skeletal organisms. Skeletons are composed of a colloidal jelly or **spongin**, a horny organic material; calcareous or siliceous **spicules** may occur with or without spongin. These structures support the body shape and provide a framework for the rather disparate cells of the sponge. In simple terms, the sponge animal functions as a colonial, loosely-integrated protist, but with a higher degree of physiological integration.

Three basic levels of chamber organization have been recognized among the sponges (Fig. 11.2), and these provide a useful guide to their shape. The simple **ascon** sponges are sacs with a single chamber lined by flagellate cells, whereas the **sycon** grade has a number of simple chambers with a single central paragaster. The **leucon** grade is the most common where a series of sycon chambers access a large central paragaster.

Autecology: life as a sponge

Sponges are part of the sedentary benthos, with large exhalant openings, communicating upwards with the water column. When not resting, the sponge sucks in water through its upward-facing ostia, forming inhalant currents; material is then pumped out of the animal though the exhalant opening. The group is entirely aquatic, living attached in a range of environments from the abyssal depths of oceans to the moist barks of trees in the humid tropics. Most Paleozoic and early Mesozoic forms have been collected from shallow-water environments, although like many other groups they expanded into deep-water environments during the Ordovician where they remained an important part of the benthos.

Today, sponges occupy a wider range of environments than in the past. Modern hexactinellids prefer depths of 200–600 m, probably extending down onto the abyssal plains and into submarine trenches, whereas the calcareous sponges are most common in depths of less than 100 m. The modern calcified sponges are either deep-reef or, more often, cave dwellers, lurking in the shadows of submarine crevices at depths of 5–200 m, mainly in the Caribbean although the group occurs elsewhere, including the Mediterranean. The meadows of Antarctic cold-water sponges can comprise up to 75% of the living benthos in seas under the ice sheets.

Sponges use a variety of substrates. Clionid sponge borings, producing the trace fossil *Entobia* (see p. 523) in mollusk shells, have a long geological history and today *Cliona* is commonly associated with many oyster beds. Spicules themselves can form mat-like substrates that when colonized form local pockets of biodiversity. Although almost all sponges are fixed filter feeders, some deep-water forms are carnivorous: their long barbed spicules entangle fish and arthropods, and the sponge tissue rapidly grows over the prey to digest it. Moreover some encrusting sponges can crawl slowly over the surface in search of food. Few predators attack sponges, although some fishes, snails, starfish and turtles have been observed eating their soft tissues in the tropics; and some organisms have used sponges as a refuge, including hermit crabs, while dolphins sometimes use sponges to protect their snouts when investigating crevices.

Synecology: sponges and sponge reefs through time

Sponges and corals are the major components of modern and ancient reefs (Wood 1990). The first sponges probably appeared in the Late Proterozoic as clusters of flagellate cells. But the evolution of the main groups of fossil sponges is intimately related to their participation in reef ecosystems (Fig. 11.6). Particular grades of organization were suited to special environmental conditions and sponges can possess a rigid, reef-building skeleton by the fusion of strong spicules or by the development of an additional basal calcareous skeleton. The Cambrian sponge fauna, of thin-walled and weakly-fused spiculate demosponges and hexactinellids together with early calcisponges, is mainly cosmopolitan, having a wide geographic distribution. In contrast, Ordovician sponge faunas are characterized by the heavier, thick-walled demosponges that continued to dominate Silurian faunas in carbonate environments; siliciclastic facies were dominated by the hexactinellids. The demosponges, however, became less important as the stromatoporoids together with rugose and tabulate corals began to sneak into these sorts of niches. Hexactinellids were locally abundant during the Late Devonian, and in the

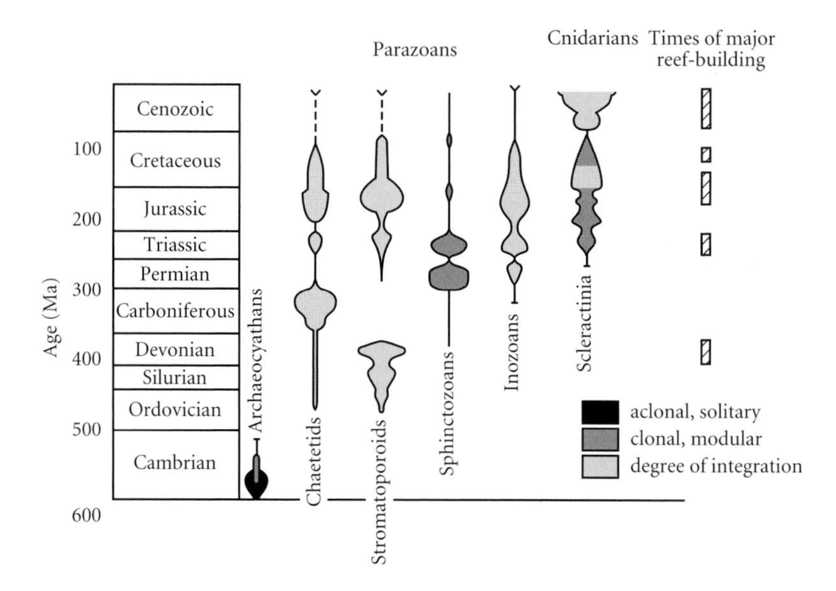

Figure 11.6 Stratigraphic distribution of reef-building sponges and related parazoans, together with the scleractinian corals.

Late Carboniferous the chaetetid calcified sponges were important reef builders. In the Permian and mid-Triassic, structures involving sphinctozoans were common and the Mid to Late Jurassic was marked by bioherms of lithistid demosponges, while the hexactinellids migrated into deeper-water environments. Jurassic sponge reefs dominated by hexactinellids and lithistids have been documented throughout the Alpine region. Cup-shaped and discoidal morphotypes dominated hard and soft substrates, respectively, and these developed a substantial topography above the seafloor, and modern analogs of these hexactinellid reefs are now known from off the coast of Canada.

As noted earlier, the acquisition of a calcareous skeleton was not confined to any one class; the calcareous skeleton was developed a number of times, convergently, across the phylum, with a few basic plans superimposed on pre-existing sponge morphology. Consequently, various groups have been recognized on the basis of the calcareous skeleton, but components of each group arose independently in different clades. In broad terms, the chaetetids and sphinctozoans, together with the archaeocyaths and stromatoporoids, were the most important calcareous reef builders. However, the decline of the calcareous sponges in reef ecosystems during the Mesozoic is often correlated with the rise of the scleractinian corals, equipped with a superior nutrition-gathering system, associated with symbiotic zooxanthellae (see p. 285).

Stromatoporoidea

The stromatoporoids were mound and sheet-like marine, modular organisms that appeared in the Mid Ordovician. These animals were common components of Late Ordovician, Silurian and Early to Mid Devonian shallow-water marine communities, forming irregular mounds on the seabed, associated with calcareous algae and corals. They have a superficial resemblance to some tabulate corals. The group reached an acme during the Mid Devonian but declined during the later Paleozoic and Mesozoic. Although stromatoporoids have been understandably classified with the cnidarians, their similarity to the modern calcified sponges and the discovery of spicules within the skeleton suggest that these, too, are poriferans and may well be a grade of organization within the Demospongea. In common with a number of other poriferans, the group is polyphyletic, with stromatoporoid taxa showing gross morphological convergence towards a common body plan or grade of organization. Because most stromatoporoids

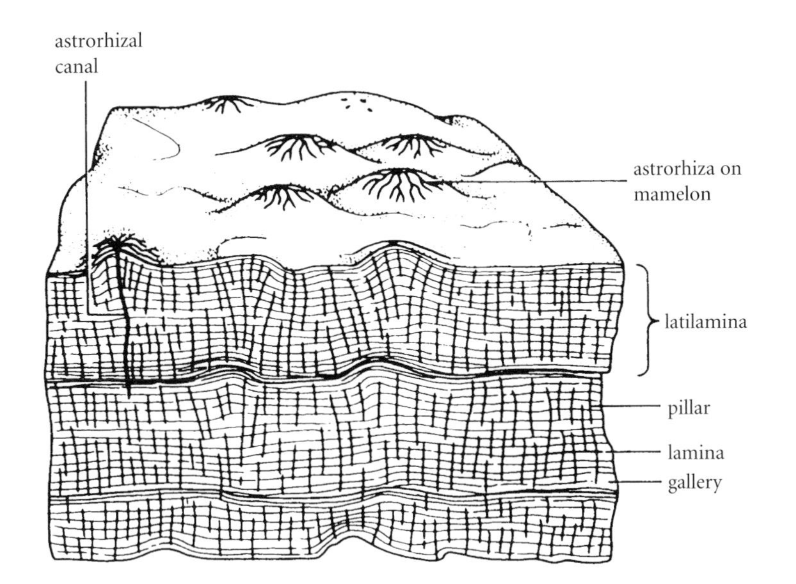

Figure 11.7 Stromatoporoid morphology.

look like solidified cow pats, and are all superficially very similar, paleontologists must use thin sections to describe the microstructure and classify the species.

Morphology and classification

Typical stromatoporoids have a calcareous skeleton with both horizontal and vertical structures and often a fibrous microstructure (Fig. 11.7). The skeleton is constructed from undulating layers of calcareous laminae punctuated perpendicularly by vertical pillars. The surfaces of some forms are modified by small swellings or **mamelons** together with **astrorhizae**, radiating **stellate canals**, which are the traces of the exhalant current canal system. Siliceous spicules have been identified in some Carboniferous and Mesozoic taxa, suggesting that the primary skeleton was in fact spiculate; the calcareous casing is secondary with probably low magnesium calcite precipitated within a framework of spongin.

Some authors have included the extinct stromatoporoids within the sclerosponges, a small group of enigmatic sponges with siliceous spicules embedded in aragonite, commonly found today in cryptic environments in the tropics. Others have classified them as cyanobacteria, foraminiferans or even as a separate phylum. But these assignments are probably only of historical interest because most morphological evidence places them firmly in the sponges.

Autecology and synecology: stromatoporoid life and times

Stromatoporoids were marine organisms usually associated with shallow-water carbonate sediments often deposited in turbulent environments. Many genera were important constituents of reefs, particularly during the Silurian and Devonian. For example, the spectacular Silurian reefs on the Swedish island of Gotland are characterized by a variety of stromatoporoid growth forms (Kershaw 1990), whereas throughout North America and northern Europe Devonian reef complexes and bioherms are dominated by stromatoporoids. These animals had complex water systems and grew in a variety of different ways: columnar, dendroid, encrusting and hemispherical forms were associated with specific energy and turbulence levels (Fig. 11.8).

Stromatoporoids were also associated with their own diverse microecosystems; those preserved in the Silurian of Gotland provided habitats for communities with over 30 epibiont species (see p. 97) that lived attached to the animals. Boring, encrusting and epifaunal organisms made good use of the cavities and substrates available in and on the stromato-

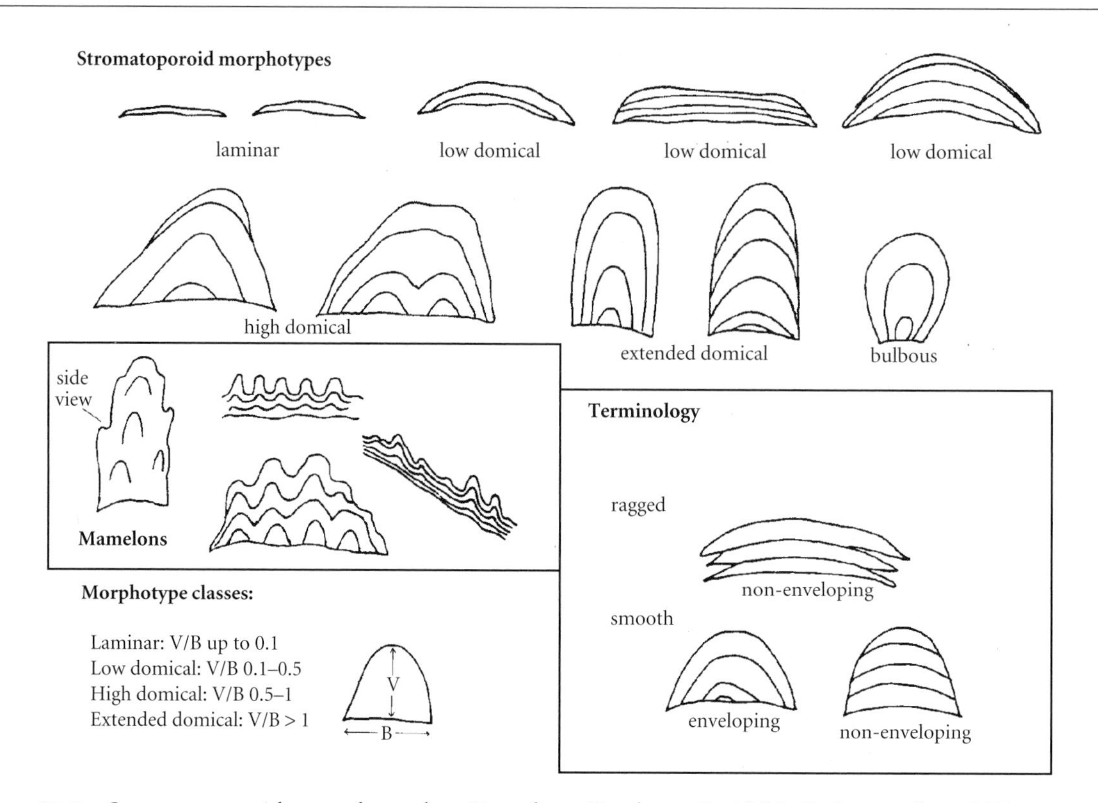

Figure 11.8 Stromatoporoid growth modes. (Based on Kershaw, S. 1984. *Palaeontology* 27.)

poroid skeleton; both bivalves and brachiopods have been seen in borings within stromatoporoids that may have provided some of the first cryptic habitats for Phanerozoic biotas.

Animals with a stromatoporoid grade of organization have been identified from rocks of Botomian age; however these forms were apparently short lived. *Pseudostylodictyon* from the Middle Ordovician of New York and Vermont may be the oldest true stromatoporoid, derived from a soft-bodied, sponge-like ancestor in the Early Ordovician. Stromatoporoids formed the basis for reef ecosystems during the Silurian and Devonian, becoming largely extinct during the end-Frasnian (Late Devonian) event. The group revived in the Mid and Late Jurassic when stromatoporoids again participated in reef frameworks. Nevertheless, most groups disappeared at the end-Cretaceous mass extinction. However, some living sponges have a stromatoporoid grade of organization; *Astrosclera* and *Calcifimbrospongia* are both calcified demosponges with a stromatoporoid architecture.

Archaeocyatha

The Archaeocyatha or "ancient cups" are one of only a few major animal groups that are entirely extinct. They appear to have been an evolutionary dead end. The group exploited calcium carbonate during the early part of the Cambrian radiation to construct porous cup- or cone-like skeletons, usually growing together in clumps and often living with stromatolites to form reefs. The Archaeocyatha dominated shallow-water marine environments, usually in tropical paleolatitudes. From an Early Cambrian origin on the Siberian Platform, the group spread throughout the tropics, forming the first Paleozoic reefs. However, by the end of the Early Cambrian and the start of the Middle Cambrian, archaeocyaths are known only from Australia, the Urals and Siberia. They disappeared at the end of the Cambrian.

Current studies suggest that the Archaeocyatha have a grade of organization similar to poriferans; in fact most authorities would place the group firmly within the sponges as a separate class. Because no living

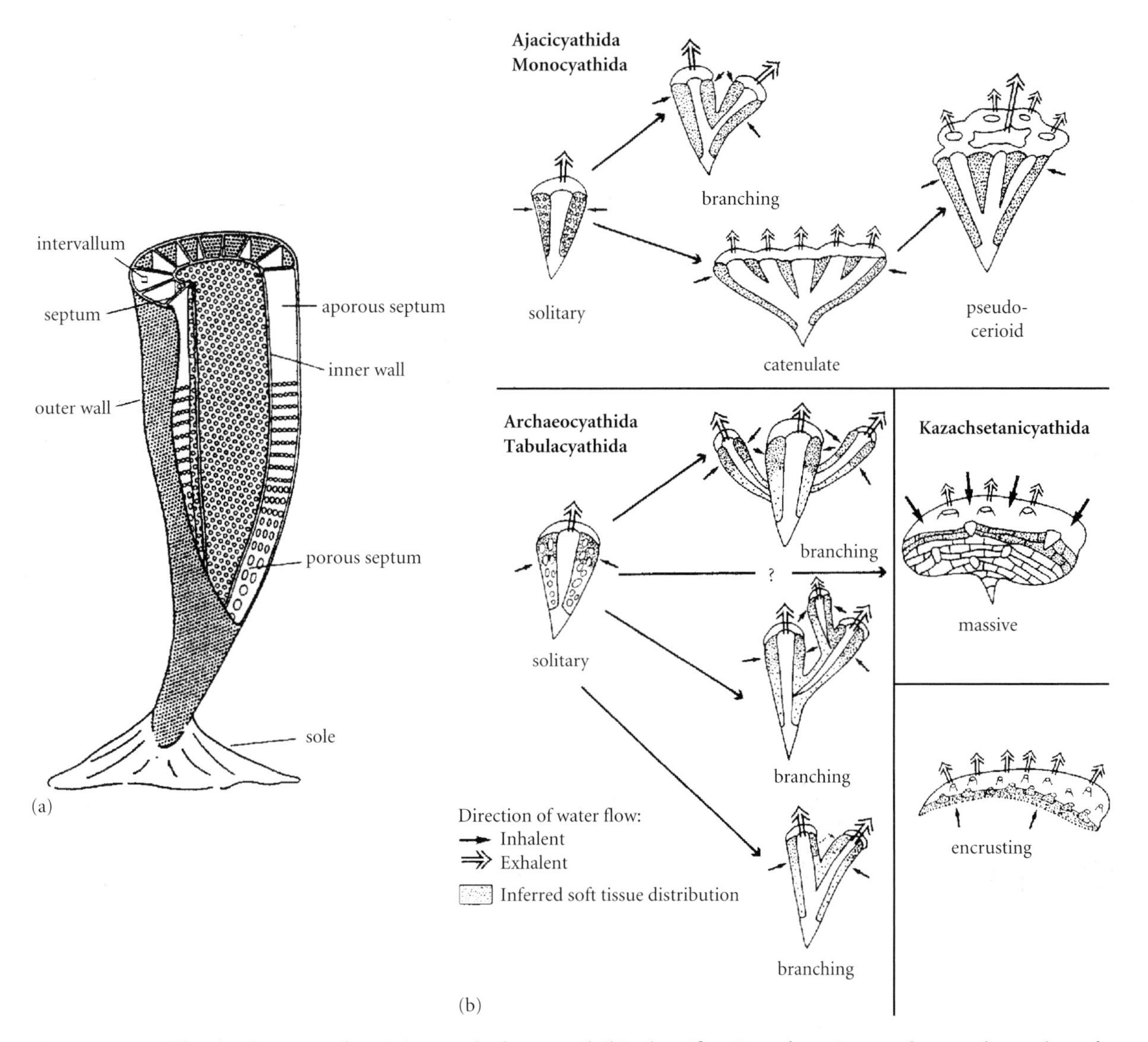

Figure 11.9 The Archaeocyatha: (a) morphology and (b) classification, function and growth modes of the main groups. (Based on Wood et al. 1992.)

representatives of the group exist there has been, in the past, considerable speculation about the taxonomic affinities of the archaeocyaths: they have been classified with algae, calcified protozoans, poriferan-grade metazoans, animals with a grade of organization intermediate between protozoans and metazoans, and cnidarians – none of which now seems likely.

Morphology and classification: archaeocyath individuals and modules

Archaeocyaths are most commonly found in carbonates, and details of their morphology are usually reconstructed from thin sections. Unfortunately many Cambrian carbonates have been recrystalized, often destroying the details of skeletal morphology. The exoskeleton of the archaeocyathan animal is **aspiculate** and usually composed of a very porous, inverted cone composed of two nested concentric walls separated from each other by radially arranged, vertical septa (Fig. 11.9). Both the inner and outer walls are densely perforated and together define the **intervallum**, or central cavity, partitioned into a number of segments (**loculi**) by the radial septa, which are often less porous than the walls or sometimes aporous. The inner wall circumscribes the central cavity, open at the top and closed at its base to form a tip. The apex of the

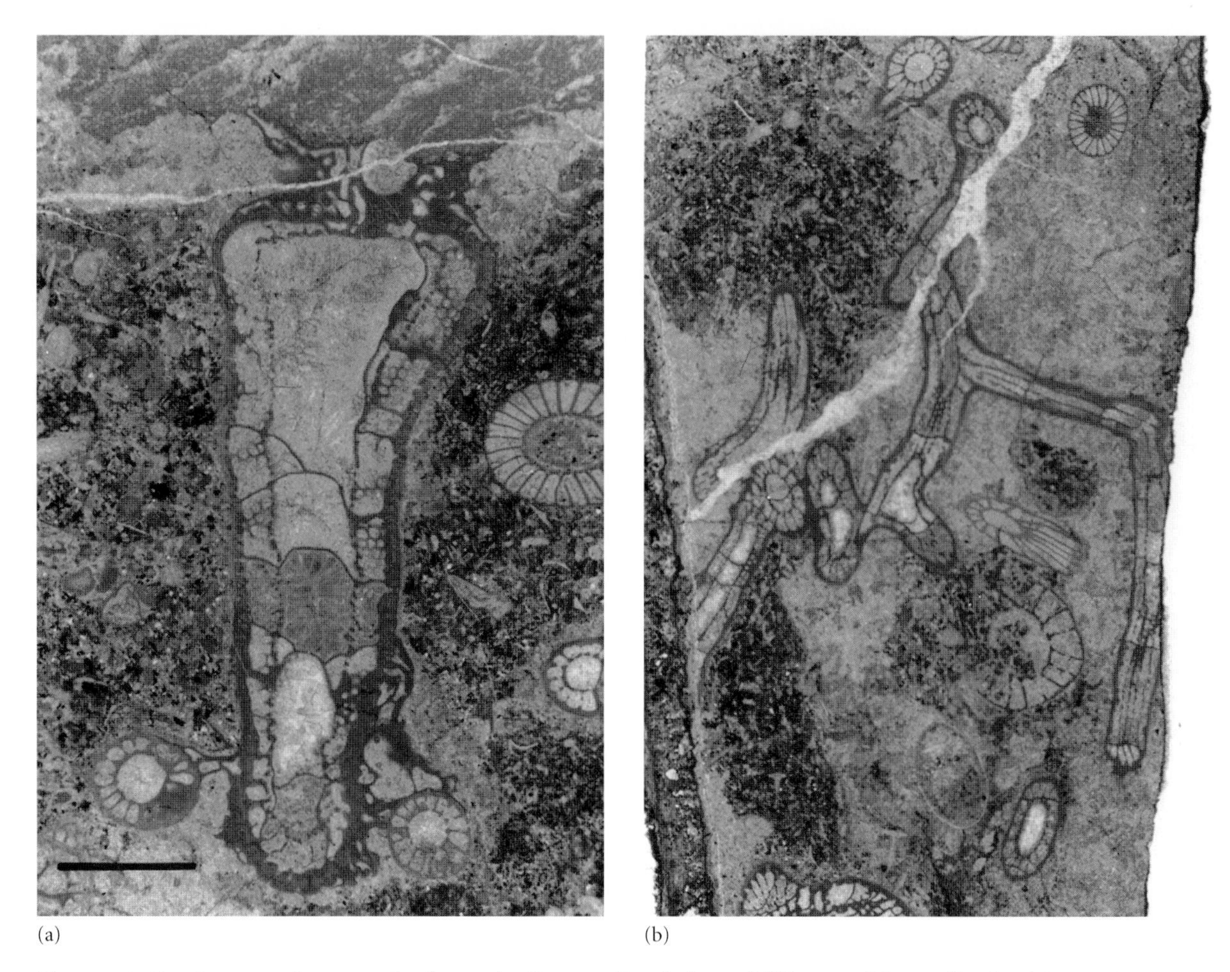

Figure 11.10 Some archaeocyaths from the Lower Cambrian of Western Mongolia, in thin section: (a) cryptic, solitary individual of *Cambrocyathellus* showing holdfast structures (×7.5), and (b) branching *Cambrocyathellus tuberculatus* with skeletal thickening between individuals associated with transverse sections of *Rotundocyathus lavigatus* (×5). (Courtesy of Rachel Wood.)

skeleton is usually buried in the sediment with a basal flange and roots or holdfasts adding anchorage and stability. In some taxa, the intervallum is partitioned horizontally firstly by porous shelves or **tabulae** or secondly with aporous, convex **dissepiments**, often extending into the adjacent central cavity.

Two main subdivisions have been defined within the group: the "Regulares" and the "Irregulares". The regular forms have an initial aporous, single-walled stage lacking dissepiments; soft tissue filled the entire body. The inner and outer walls are punctuated by septa and tabulae developed either singly or together. The irregular forms have initial aporous, single-walled stages with dissepiments. The twin walls have irregular pore structures, always dissepiments, and the skeleton is asymmetric; soft tissue was restricted by the development of secondary skeletal material. These groupings have now been shown to have little taxonomic value, reflecting rather ecological preferences (Debrenne 2007). Most archaeocyaths are "Regulares", including the orders Ajacicyathida and Coscinocyathida; however the apparent abundance of regular genera may be due to excessive taxonomic splitting. There are fewer "Irregulares" but this ecogroup includes the orders Archaeocyathida and Kazachsetanicyathida.

Synecology: archaeocyath reefs

The archaeocyaths were exclusively marine, probably living at depths of 20–30 m on carbonate substrates. The phylum developed an innovative style of growth based on modular organization (Fig. 11.10). Such modularity permitted encrusting abilities and the possi-

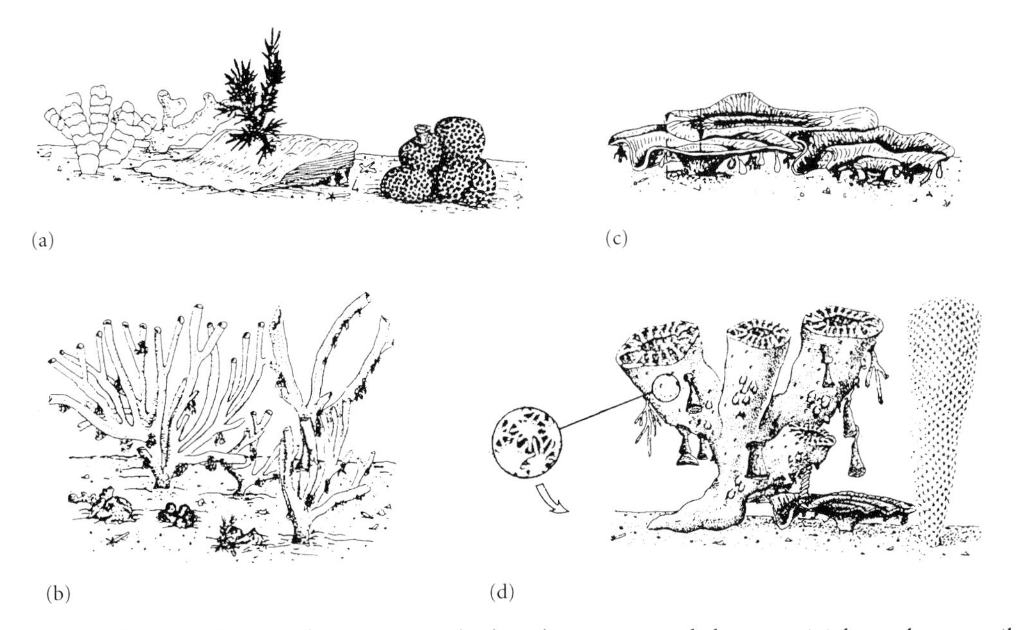

Figure 11.11 Archaeocyathan reef structures which, when preserved, become (a) boundstones, (b) bafflestones, (c) bindstones or (d) bioherms. (Based on Wood et al. 1992.)

bilities of secure attachment on a soft substrate; moreover growth to large size was enabled, together with a greater facility for regeneration (Wood et al. 1992). The archaeocyaths were thus key elements of the first reef-type structures of the Early Cambrian (Fig. 11.11), in intervals of high turbulence and rates of sedimentation. However, although archaeocyathan reefs were probably not particularly impressive, usually up to 3 m thick and between 10 and 30 m in diameter, they were nevertheless amongst the first animals to establish complex biological frameworks, processing large amounts of seawater through their bodies (Box 11.2). Archaeocyathan reefs were always associated with calcimicrobes that may have been the main frame builders. There are also some examples of cryptic organisms living within the reef cavities, including other sponges.

Distribution: Cambrian world of the archaeocyaths

The first archaeocyaths are known from the lowest Cambrian (Tommotian) rocks of the Siberian Platform and are represented by mainly solitary regulars. During the Early Cambrian, the phylum diversified, migrating into areas of North Africa, the Altai Mountains of the former Soviet Union, North America and South Australia (Fig. 11.13). Archaeocyaths were most common in the Mid to Early Cambrian (Botomian) when a number of distinct biogeographic provinces can be defined, but by the Lenian Stage the group was very much in decline. Few genera have been recorded from the Middle Cambrian and only one is known from Upper Cambrian strata. Archaeocyath history demonstrates a progressive move towards a more modular architecture in response to conditions of high turbulence. In general, solitary taxa dominated the Early Cambrian; but following the late Botomian, modular morphotypes continued after the extinction of most solitary forms (Fig. 11.14; Box 11.3). One advantage is that the abundance and diversity of the group in some parts of the world, particularly in Lower Cambrian rocks, has allowed its effective use in biostratigraphic correlation when there were few other organisms around that could act as zone fossils (see p. 28).

CNIDARIA

The bottom was absolutely hidden by a continuous series of corals, sponges, actiniæ [sea anemones] and other marine productions, of magnificent dimensions,

Box 11.2 Plugging the leaks: experimental morphology of archaeocyaths

It is often extremely difficult to reconstruct the life modes of long-extinct organisms that apparently lack modern analogs (see p. 150), particularly when the entire phylum is extinct. In an innovative experimental biomechanical study Michael Savarese (then at Indiana University) constructed models of the three main archaeocyathan morphotypes (aseptate, porous septate and aporous septate), and subjected each to currents of colored liquid in a flume (Fig. 11.12). The first morphotype, a theoretical reconstruction, performed badly with fluid escaping through the intervallum while also leaking through the outer wall. The porous septate form, however, suffered some slight leakage through the outer wall but no fluid passed through the intervallum. The aporous septate form was most efficient with no leakage through the outer walls and no flow through the intervallum. Significantly, ontogenetic series of the fossils show that an initially porous septate morphotype become aporous in later life, perhaps to avoid leakage through the outer wall (Savarese 1992). This was clearly a great advantage to an organism that survived by pumping huge volumes of seawater through its system!

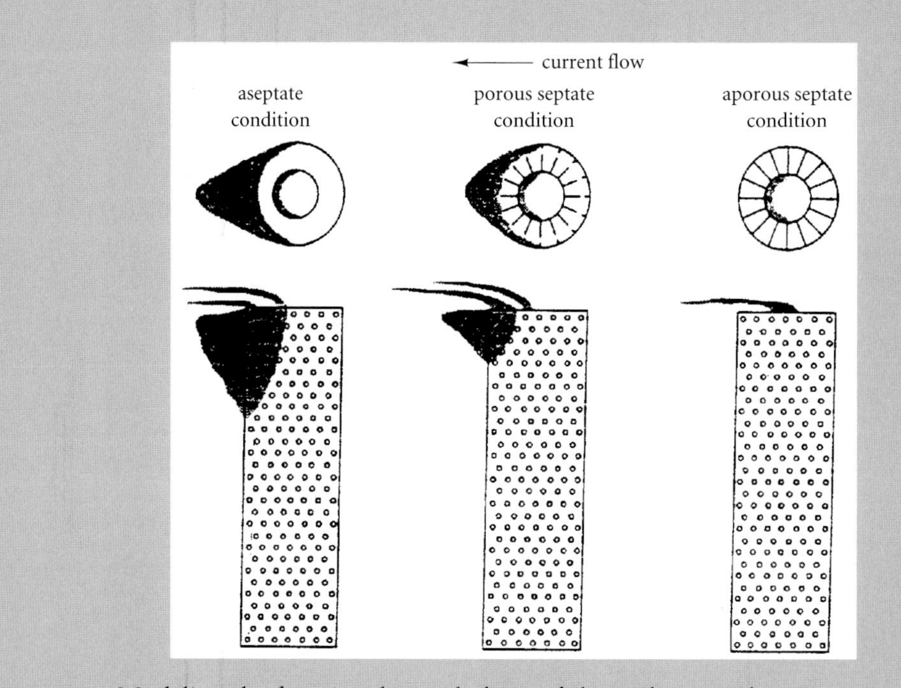

Figure 11.12 Modeling the functional morphology of the archaeocyaths. (From Savarese 1992.)

varied forms, and brilliant colours. . . . In and out among [the rocks and living corals] moved numbers of blue and red and yellow fishes, spotted and banded and striped in the most striking manner, while great orange or rosy transparent medusæ [jellyfish] floated along near the surface. It was a sight to gaze at for hours, and no description can do justice to its surpassing beauty and interest. For once, the reality exceeded the most glowing accounts I had ever read of the wonders of a coral sea.

Alfred R. Wallace (1869)
The Malay Archipelago

The cnidarians (or "nettle-bearers") include the sea anemones, jellyfish and corals and are the least complex of the true metazoans (eumetazoans), having cells organized into a

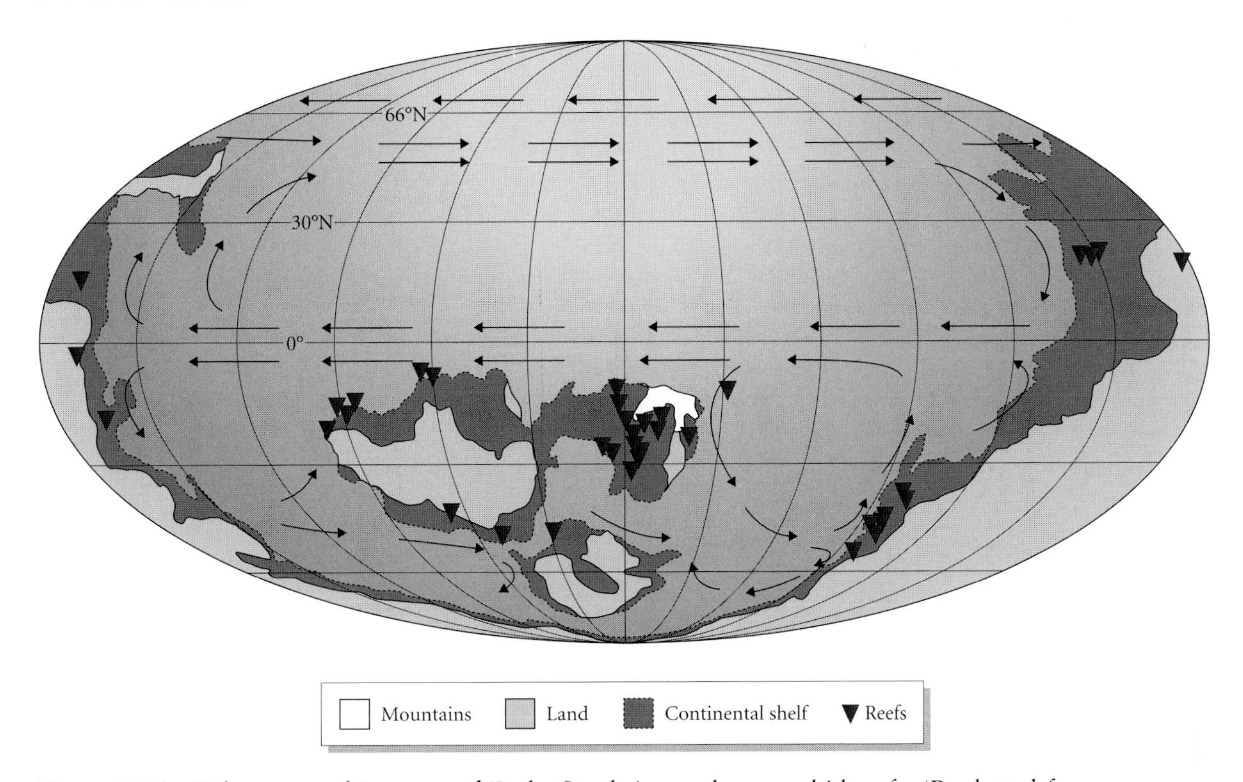

Figure 11.13 Paleogeographic range of Early Cambrian archaeocyathid reefs. (Replotted from Debrenne 2007.)

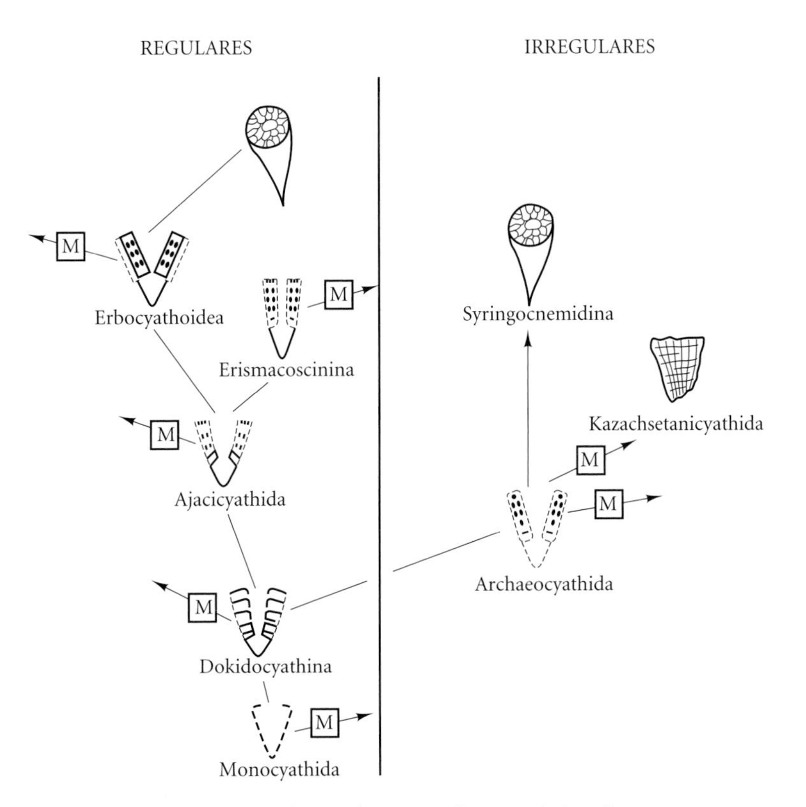

Figure 11.14 Evolutionary trends within the archaeocyaths; modular forms, appearing iteratively, are indicated by M. (Based on Wood et al. 1992.)

Box 11.3 Neoproterozoic colonies

When was the transition complete from the isolated protist way of life to the loosely integrated colonies of cells in the earliest poriferans? The Neoproterozoic rocks of Namibia yield some clues. Rachel Wood and her colleagues (2002) have described a giant, fully-mineralized, complex colonial skeleton, *Namapoikea*, from the Northern Nama Group, dated at about 550 Ma (Fig. 11.15). This postdates some of the earliest putative cnidarians and sponges in the Ediacara biota, but predates currently known metazoan reef-type ecosystems. *Namapoikea* is huge (up to 1 m in diameter), robust, with an irregular structure in transverse section but apparently lacking any internal features. It is uncertain whether this is a sponge or a coral but clearly large, modular, skeletal metazoans were already around in the Late Neoproterozoic, providing a hitherto unexpected complexity to terminal Proterozoic reefs and with the potential to provide both open surface and cryptic habitats. Perhaps these encrusting sheets provided shelter for some of the first micromorphic skeletal metazoans?

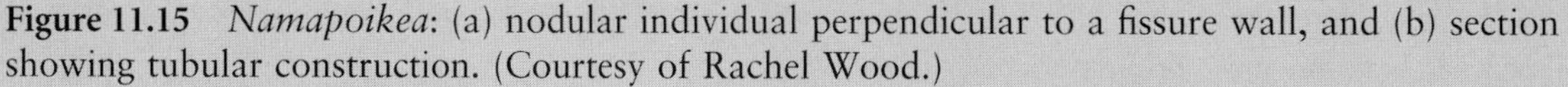

Figure 11.15 *Namapoikea*: (a) nodular individual perpendicular to a fissure wall, and (b) section showing tubular construction. (Courtesy of Rachel Wood.)

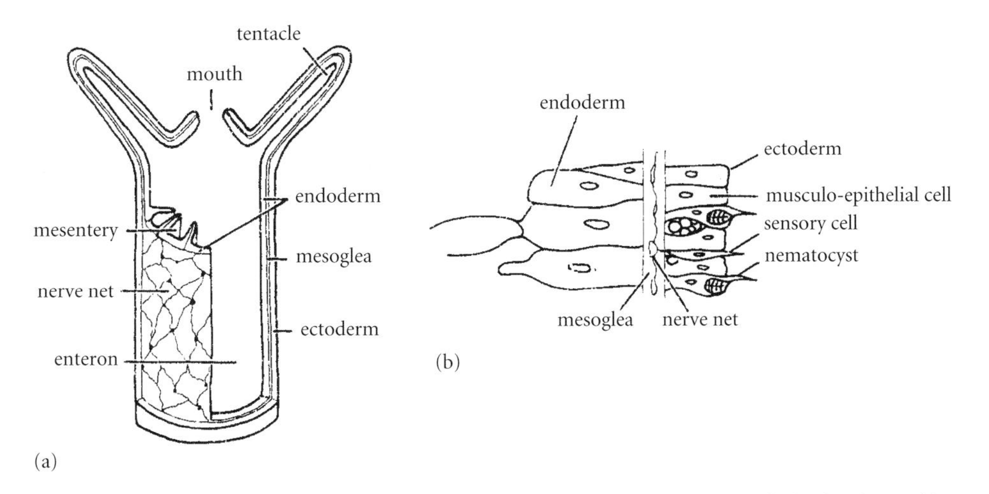

Figure 11.16 Morphology of Hydra: (a) general body plan, and (b) detail of the body wall.

relatively few different tissue types in a radial plan. They are typified by the well-known hydra (Fig. 11.16). Although there are no specialized organs and only a few tissue types, they are more complex than the parazoans. The group was, in the past, referred to as the Coelenterata, but because that phylum also included the sponges and the gelatinous ctenophores or comb-jellies, the more restricted term Cnidaria is now generally preferred. Two basic life strategies occur (Fig. 11.17): **polyps** are usually sessile or attached, although some can jump and somersault, while **medusae** swim, trailing their tentacles like the deadly and vicious snakes that adorned the head of the mythical Medusa. Although medusoids and polyps appear different, they are essentially the same structures but inverted. Many cnidarians exhibit both forms through their life cycles, others only one. The Portuguese man-of-war, for example, is a spectacular and scary colonial form with a medusoid module for floatation and various types of polyps that help feeding, locomotion and reproduction. As a whole the group is carnivorous, attacking crustaceans, fishes, worms and even microscopic diatoms, with their poisonous stinging cells (**cnidoblasts**) – the reason they are called "nettle-bearers".

Morphology: the basic cnidarian

The cnidarians are multicellular, having a single body cavity or **enteron**; the opening at the top (or bottom in most medusae), sur-

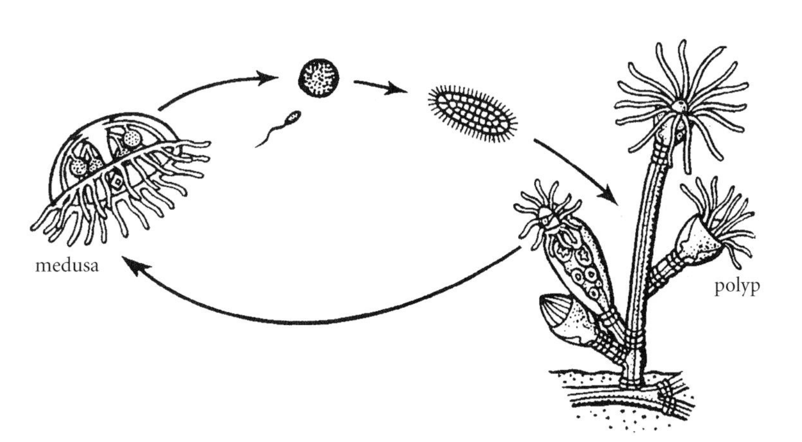

Figure 11.17 Cnidarian life cycles: generalized view of the life of the hydrozoan *Obelia*, alternating between the conspicuous polyp and medusa stages.

rounded by tentacles with stinging cells or **nematocysts**, functions both as a mouth and an anus. There is thus no head or tail, and nutrients and waste pass through the same opening. The body itself, although **diploblastic**, is, in fact, composed of three layers; the inner endoderm and the outer ectoderm both consist of living cells while the intervening **mesoglea** is a gelatinous, acellular substance containing rare cells. The outer layer of the body wall contains cnidoblast cells that contain the primed stings or nematocysts that are usually confined to the tentacles. A primitive nerve net is embedded in the mesoglea. Fingers of endoderm commonly poke into the enteron, forming radial partitions that increase the area of absorption of nutrients. These mesenteries can, in the case of the corals, secrete calcium carbonate to form solid, calcified partitions or septa. Most species are found in marine environments although hydrozoans can be very abundant in freshwater habitats.

Classification: design and relationships of the main groups

The phylum Cnidaria is usually split into three classes: hydrozoans, scyphozoans and anthozoans (Box 11.4). The hydrozoans

Box 11.4 Classification of Cnidaria

The phylum is characterized by radial symmetry, with the ectoderm and endoderm separated by the mesoglea; the enteron has a mouth surrounded by tentacles with stinging cells. The phylum ranges from Upper Precambrian to Recent. The putative medusoid *Brooksella*, which predates the Ediacara fauna may, in fact, be a trace fossil. The group has a wide range of body plans (Fig. 11.18).

Class HYDROZOA

- This includes six main orders of small, usually polymorphic forms. Each has an undivided enteron and solid tentacles, and may form colonies. There are six main orders; the Chondrophora contains some of the oldest cnidarians
- Ediacaran to Recent

Class SCYPHOZOA

- Mainly jellyfish, contained in the Scyphomedusae, which are only preserved in Lagerstätten. The extinct Conulata is often included here since the group has a tetrameral symmetry and apparently has tentacles. Their long conical shells, for example *Conularia*, are composed of chitinophosphate; conulates appeared in the Cambrian and were extinct by the Mid Triassic
- Ediacaran to Recent

Class ANTHOZOA

- These are exclusively marine, and most are sessile, colonial forms (though they have mobile planula larvae). The three subclasses, Ceriantipatharia, Octocorallia and Zoantharia (including the orders Rugosa, Tabulata and Scleractinia), all lack medusoid stages, possess hollow tentacles and have the enteron divided, longitudinally, by vertical septa. Both solitary and colonial forms occur. The class includes corals, sea anemones and sea pens. Octocorals often produce spicules that occur as microfossils
- Ediacaran to Recent

Class CUBOZOA

- The sea wasps and box jellyfish have both medusae and polyps and are mainly restricted to tropical and subtropical latitudes
- Carboniferous to Recent

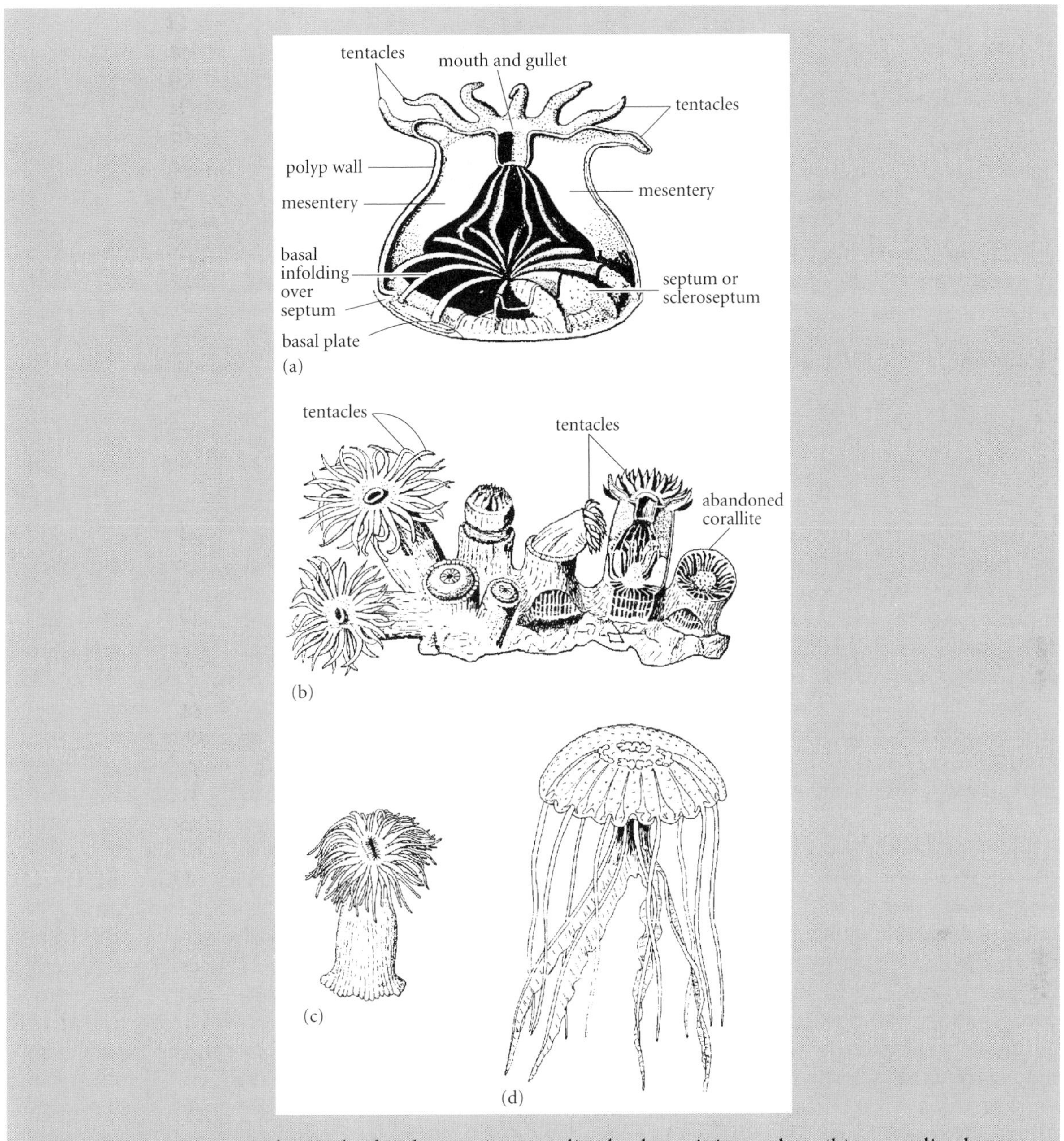

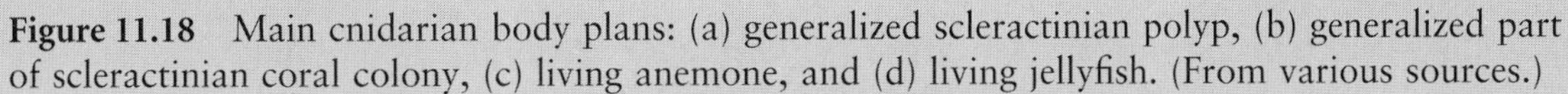
Figure 11.18 Main cnidarian body plans: (a) generalized scleractinian polyp, (b) generalized part of scleractinian coral colony, (c) living anemone, and (d) living jellyfish. (From various sources.)

include freshwater and colonial forms together with the fire corals and most kinds of "jellyfish". There are over 3000 living species inhabiting water depths up to 8000 m, mainly in marine environments. Supposed hydrozoans have been recorded from the Late Precambrian Ediacara fauna (see p. 242), where genera such as *Eoporpita* and *Ovatoscutum* may be the oldest sessile members of the phylum. Hydrozoans reproduce either sexually or by asexual budding; the polyp stage is asexual and the medusoid normally sexual. The scyphozoans are mainly free-swimming medusae or jellyfish often inhabiting open-ocean environments. Some elements of the Ediacara fauna may be scyphozoans, for example *Conomedusites* and *Corumbella*; however many of the best-preserved fossil

Table 11.1 Features of the main coral groups.

Feature	Rugosa	Tabulata	Scleractinia
Growth mode	Colonial and solitary	Colonial	Colonial and solitary
Septa	6 prosepta; later septa in only 4 spaces	Septa weak or absent	6 prosepta; later septa in all 6 spaces
Tabulae	Usual	Well developed	Absent
Skeletal material	Calcite	Calcite	Aragonite
Stability	Poor	Poor	Good with basal plate
Range	Ordovician to Permian	Ordovician to Permian	Triassic to Recent

forms have been collected from the Late Jurassic Solnhofen Limestone of Bavaria. Living members of the group include *Aurelia*, the moon jellyfish, and the compass jellyfish, *Chrysaora*. Although the anthozoans include the sea anemones, sea fans, sea pens and sea pansies, the class also includes the soft and stony corals. Following a short, mobile, planula larval phase, all members of the group pursue a sessile life strategy as polyps.

Corals

Corals are probably best known for their place in one of the planet's most diverse but most threatened ecosystems, the coral reef. Shallow-water coral reefs form only in a zone extending 30° degrees north and south of the equator and reef-forming corals generally do not grow at depths over 30 m or where the water temperature falls below 18°C, although certain groups of corals can also form structures in deep-water environments. Corals are not the only reef-forming organisms but throughout geological time they have constructed three main types of reefs: fringing reefs, barrier reefs and atolls. These structures formed the basis for Charles Darwin's then cutting-edge analysis *Coral Reefs* published in 1842. Unfortunately, such structures are under current threat, including damage from increased bleaching, coastal development, temperature change of seawater, tourism, runoff containing agricultural chemicals, abrasion by ships' hulls and anchors, smothering by sediment, poisoning or dynamiting during fishing, overfishing of important herbivores and predators, and even harvesting for jewelry. There seems little hope for this spectacular habitat unless more attention is paid to conservation.

The anthozoans are the most abundant fossil cnidarians, pursuing a polypoid lifestyle. The class Anthozoa contains two subclasses with calcareous skeletons. Whereas the Octocorallia have calcified spicules and axes, the Zoantharia include the more familiar fossil coral groups, the orders Rugosa, Tabulata and Scleractinia (Table 11.1). The Octocorallia, including the Alcyonaria, have eight complete mesenteries and a ring of eight hollow tentacles; the skeleton lacks calcified septa but calcareous or gorgonin spicules and axes comprise solid structures in the skeleton. Although the group is only sporadically represented in Silurian, Permian, Cretaceous and Tertiary rocks, the octocorals are important reef dwellers today. Some familiar genera include *Alcyonium* (dead men's fingers), *Gorgonia* (sea pen) and *Tubipora* (organ-pipe coral).

Morphology: general architecture

There are four main elements to the zoantharian coral skeleton: radial and longitudinal structures, together with horizontal and axial elements. Corals have **planula larvae**. Following the planula larval stage the coral polyp initially rests on a basal plate or disk termed the **holotheca** and begins the secretion of a series of vertical partitions or **septa** in a radial

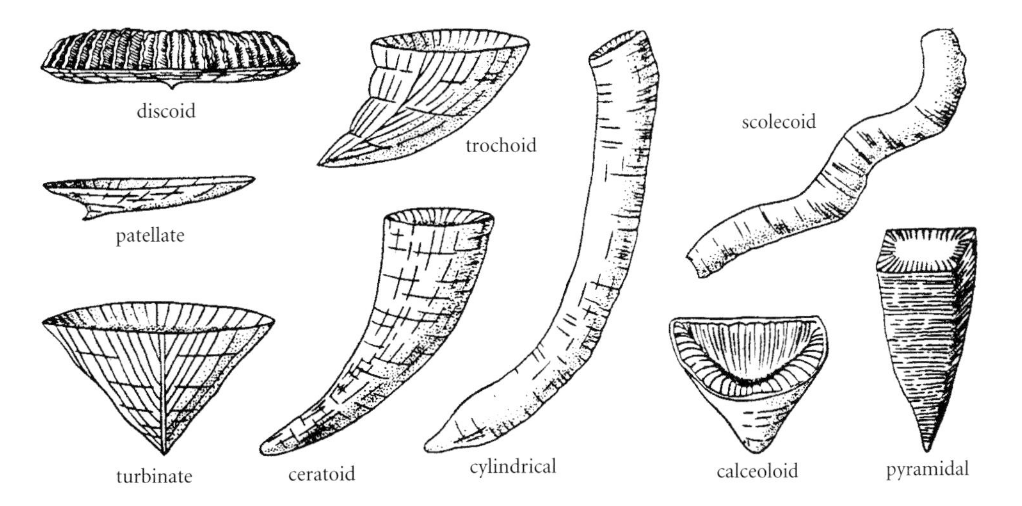

Figure 11.19 Terminology for the main modes of solitary growth in corals. (From *Treatise on Invertebrate Paleontology*, Part F. Geol. Soc. Am. and Univ. Kansas Press.)

arrangement. At the circumference, the septa are joined to the theca or skeletal wall, which extends longitudinally from the apex of the **corallum** to the **calice** where the polyp is attached. During growth the polyp may secrete a series of horizontal sheets, or tabulae, together with smaller curved or angled plates or dissepiments. The **columella**, usually arising from the fusion of the axial edges of the septa, occupies the core region of the corallum. The vertical walls or septa radiate outwards from the columella and divide the corallite. Despite the apparent simplicity of the coral skeleton, there is a great deal of variation in both solitary and colonial growth programs and the end result is a remarkable array of shapes and sizes of corals.

The three main subclasses of stony corals have colonial or compound growth modes whereas only the Scleractinia and Rugosa have solitary skeletons. The solitary growth forms include conical, ceratoid or horn-shaped, calceoloid, cylindrical, discoid, patellate, scolecoid, trochoid and turbinate skeletons (Fig. 11.19). Colonial corals with corallites have adopted either fasciculate or massive growth modes. Fasciculate styles exhibit either dendroid or phaceloid strategies with either no or poor integration. The halysitid or cateniform chain-like growth strategy is a further variation on this pattern. The massive colonies are much more varied, with cerioid, astraeoid, aphroid, thamnasteroid, meandroid and hydnophoroid together with coenenchymal or coenostoid growth programs (Fig. 11.20). Moreover colonies with imperforate walls may exhibit phaceloid, cateniform, cerioid and meandroid forms, whereas those with perforate walls have only phaceloid and cerioid growth modes together with coenenchymal structures in some taxa, such as the sarcinulids. These growth modes are variably developed across the rugosans, tabulates and scleractinians – but meandroid and hydnophoroid modes were developed during the Mesozoic and are thus restricted to the scleractinians.

Colonial integration usually involves a loss of individuality. Many organisms display a transition from solitary growth modes, through morphologies with asexually budded modules, to a fully integrated colony with the growth or **astogeny** of the compound structure showing little variation across the individual corallites. The degree of integration of a colony is usually measured by the amount of cohesion between the individual skeletal parts and soft tissues and by the range of form observed between individual components. Clearly there is a spectrum from phaceloid modes with little or no integration to thamnasteroid and meandroid (and coenenchymal) modes with high levels of integration. Individual polyps are no longer separated by corallite walls and may share a common enteron and nervous system. This suggests a high degree of integration where the colony approaches the body plan of a typical metazoan. These modes have varied through time (Fig. 11.21).

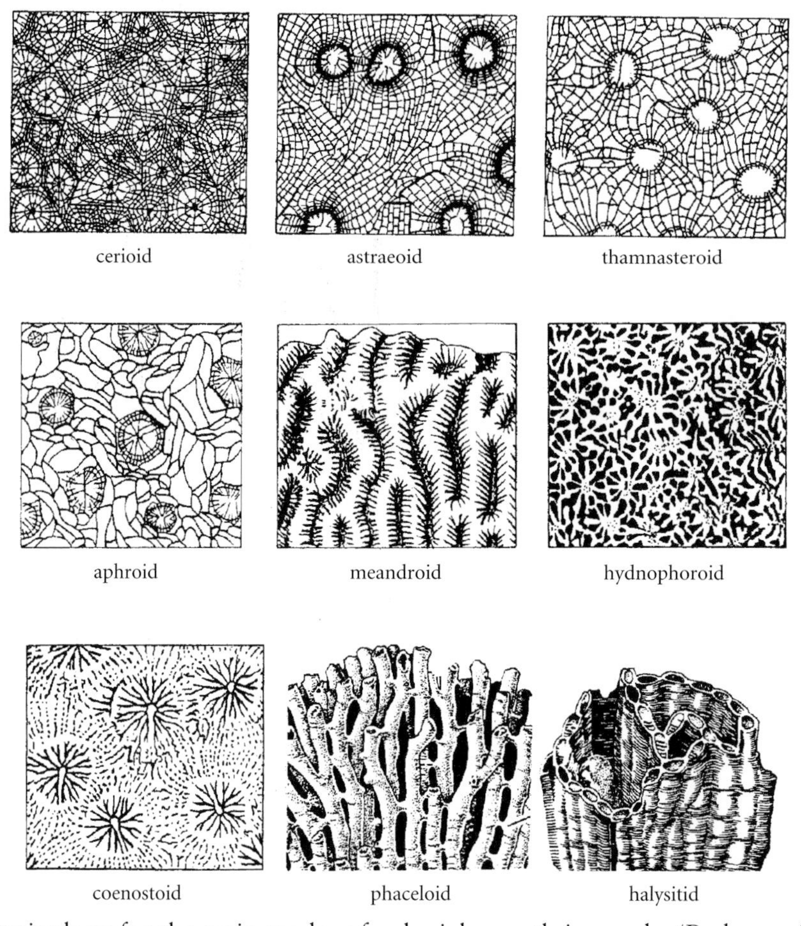

Figure 11.20 Terminology for the main modes of colonial growth in corals. (Redrawn from various sources.)

Scleractinian corals may be highly integrated because they have symbiotic zooxanthellae (see p. 285). The relatively low levels of integration seen in the Rugosa and some Tabulata colonies perhaps suggests a lack of algal symbionts. There has been a great deal of argument about this. Some rugosans are in fact quite highly integrated, and it is questionable whether high integration should only be associated with the presence of zooxanthellae.

Coral experts also use quantitative approaches in describing colony shapes. Key measurements are made on the colony and these are plotted on a ternary diagram. A series of fields can be mapped out within the triangle – for example, bulbous, columnar, domal, tabular and branching colonies are discriminated (Fig. 11.22). These different growth strategies may be ecophenotypic (see p. 123), commonly reflecting ambient environmental conditions.

Rugose corals

Rugose corals are generally robust, calcitic forms with both colonial and solitary life modes, more varied than those of tabulates. Rugosans have well-organized septal arrangements with six cardinal or primary septa. Secondary septa are inserted in four spaces around the corallum – between the **cardinal** septa and the two **alar** septa and also between the two counterlateral septa and lateral septa (Fig. 11.23a). Horizontal structures such as the tabulae, dissepiments and **dissepimentaria** are also well developed across the order. Undoubted rugosans, such as *Streptelasma*, with short secondary septa and lacking a dissepimentarium, are not recorded until the Mid Ordovician. By the Late Ordovician, rugose faunas were well established with the development of a wide variety of morphologies (Fig. 11.23b; Box 11.5). For example, the

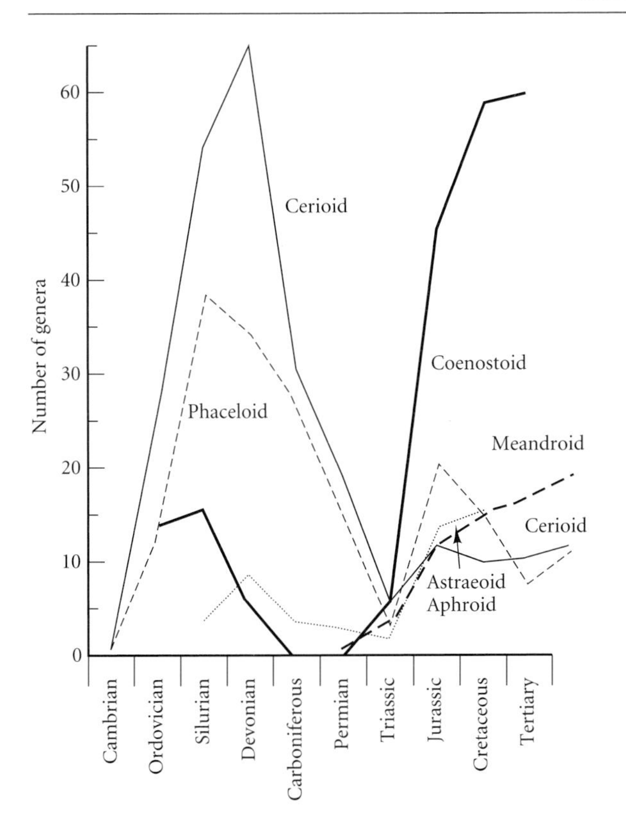

Figure 11.21 Schematic graph of the distribution of colonial growth modes through the Phanerozoic. (Based on data in Coates, A.G. & Oliver, W.A. Jr. 1973. *In Animal Colonies: Development and function through time.* Dowden, Hutchinson and Ross.)

Silurian *Goniophyllum* was pyramidal with a deep calyx, whereas the Devonian *Calceola* was a slipper-shaped form with a semicircular lid and the compound *Phillipsastrea* had a massive, astraeoid growth mode (Fig. 11.25).

Diverse rugosan faunas occurred during the Carboniferous Period. Solitary forms such as the large horn-shaped to cylindrical *Caninia*, the cylindrical *Dibunophyllum* with a marked dissepimentarium, the long cylindrical *Palaeosmilia,* and the smaller horn-shaped *Zaphrentis* are often conspicuous members of Carboniferous coral assemblages. The fasciculate, phaceloid *Lonsdaleia* and *Lithostrotion* with usually massive, cerioid growth modes are locally common. The order declined during the Permian until there were only 10 families left, and these disappeared by the end-Permian mass extinction (see p. 170).

Tabulate corals

As the name suggests, tabulate corals have well-developed tabulae (Fig. 11.26). The septa are usually very much reduced to short spines or are absent, and dissepiments are variably developed (Fig. 11.27). The group is varied, with erect, massive, sheet-like and chain-like colonies and branching forms; some authors have suggested that some tabulates, such as

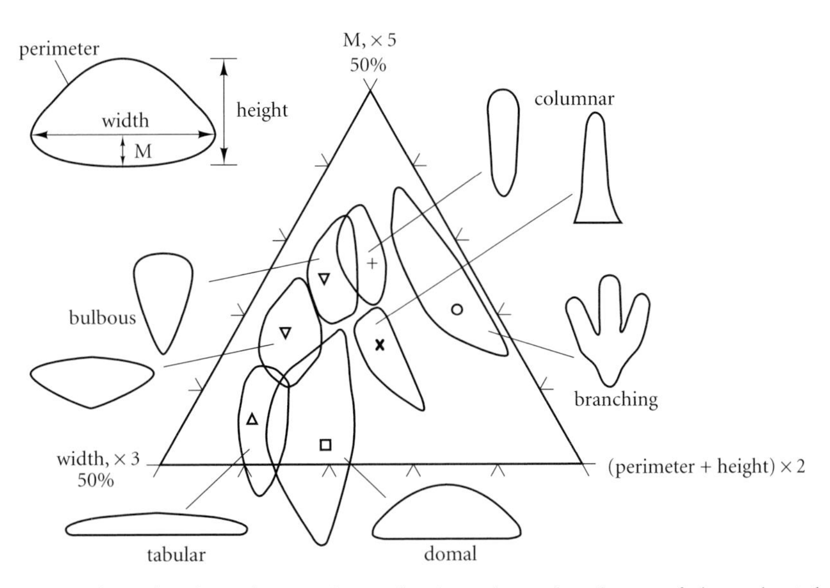

Figure 11.22 Ternary plot of colonial growth modes based on the shape of the colonial coral. (Based on data in Scrutton, C.T. 1993. *Cour. Forsch. Inst. Senckenberg* **164**.)

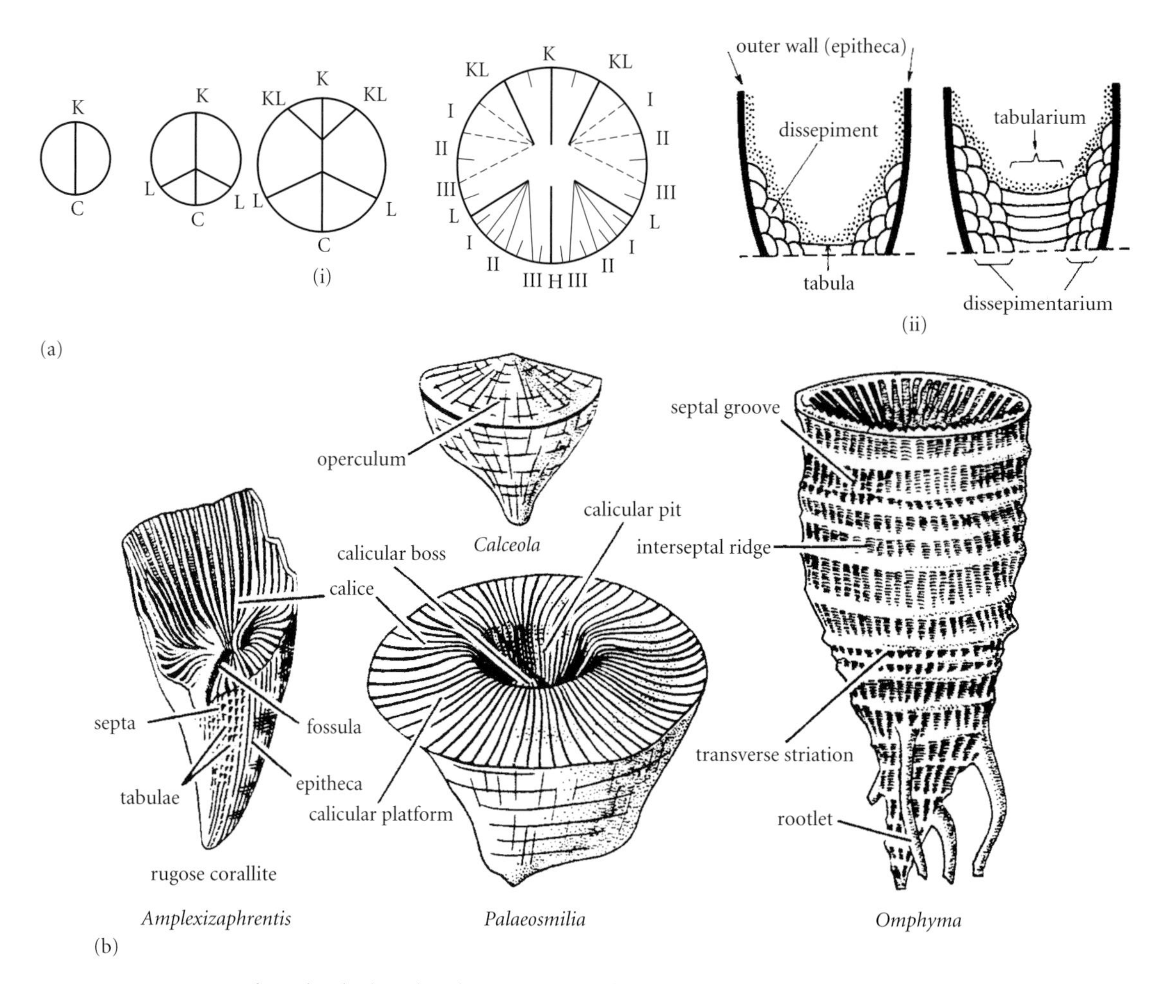

Figure 11.23 (a) Septal and tabular development in solitary rugose corals with (i) details of vertical partitions, and (ii) details of horizontal structures. C, cardinal septa; K, counter-cardinal septa; KL, counterlateral septa; L, alar septa. (b) Rugose coral morphology: external morphology of a variety of solitary rugose corals. (Based on various sources.)

the heliolitids, may not even be cnidarians. The occurrence of fossilized polyps in Silurian tabulates clearly demonstrates that at least some of them were corals. Only colonial or compound growth forms evolved in this order, usually with small, elongate corallites ranging from 0.5 to 5 mm in diameter. Commonly, the corallite walls are perforated by minute holes or mural pores. Tabulate corals first appeared in the Early Ordovician, probably predating the first rugosans. Forms such as *Lichenaria* have been recorded from Tremadocian rocks in the United States, although more definitive reports of the same genus are from the Darriwilian. Tabulates such as *Catenipora*, *Paleofavosites* and *Propora* became widespread during the later Ordovician.

Silurian tabulate coral faunas were dominated by massive to domal *Favosites* with cerioid corallites, *Halysites*, the chain coral with a series of linked, long cylindrical corallites of elliptical cross section, and *Heliolites*, the sun coral, with short, stubby septa. Similarly distinctive tabulates were characteristic of the Devonian. *Aulopora* usually comprised branching, encrusting colonies (Box 11.6), similar to the bryozoan *Stomatopora*; the extraordinary *Pleurodictyum* with large mural pores and thorn-like septa was virtually always associated with the commensal worm *Hicetes*.

Carboniferous tabulates such as *Michelinia*, with small colonies possessing large, massive, thick-walled corallites, and the long-ranging

Box 11.5 Rugosan life strategies

Despite the apparent simplicity of rugosan architecture, these corals may have pursued a number of different life strategies (Fig. 11.24). A number of corals, for example *Dokophyllum*, probably sat upright in the sediment rooted by fine holdfasts extending from the epitheca. Other taxa, such as *Holophragma*, were initially attached to a patch of hard substrate but subsequently toppled over to rest on the seabed. *Grewingkia* was cemented to areas of hard substrate. The small discoidal *Palaeocyclus*, however, may have been mobile, creeping over the substrate on its tentacles. A number of strongly curved rugosans, for example *Aulophyllum*, probably lay within the substrate, concave upwards. Successive increments of growth were directed more or less vertically giving the coral exterior a stepped appearance. Many other solitary corals exhibit a similar terraced theca, which may be due to changes in growth direction associated with adjustments following toppling of the corallum during slight turbulence or storms.

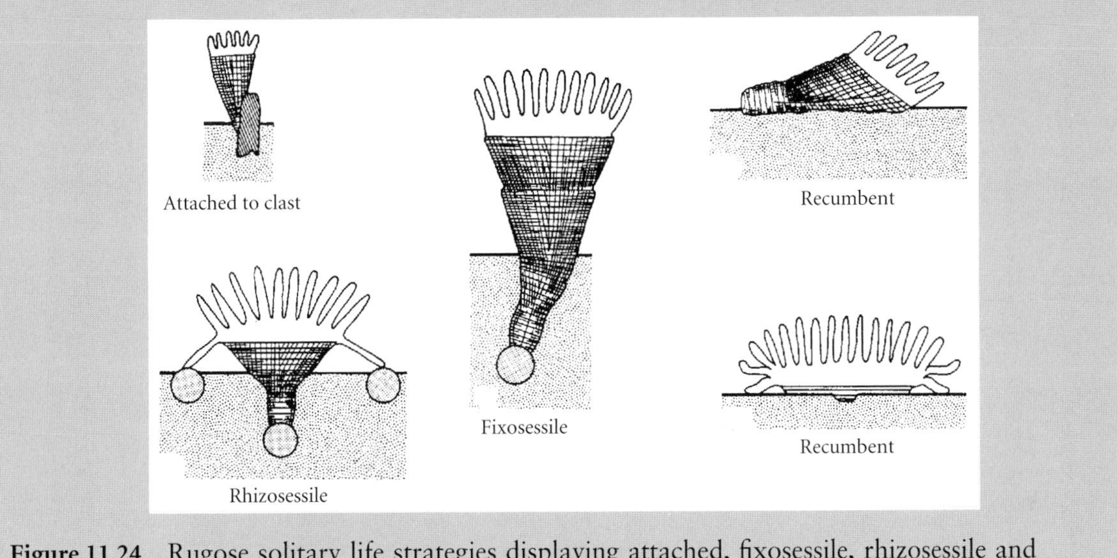

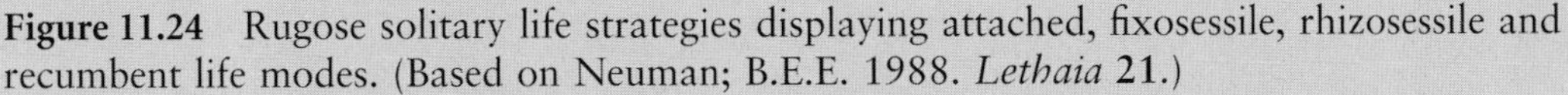

Figure 11.24 Rugose solitary life strategies displaying attached, fixosessile, rhizosessile and recumbent life modes. (Based on Neuman; B.E.E. 1988. *Lethaia* 21.)

phaceloid *Syringopora*, with long, thin, cylindrical corallites, characterize the coral faunas of the period. By the Late Permian the group was very much in decline following a long period of deterioration after the Frasnian extinctions; only five families survived to the end of the period.

Scleractinian corals

The scleractinians are elegant zoantharian corals with relatively light, porous skeletons composed of aragonite (Fig. 11.29). Both solitary and colonial modes exist with even more varied architectures than those of the rugosans. Secondary septa are inserted in all six spaces between the primary or cardinal septa. In further contrast to septal insertion in the rugosans, each cycle of six is fully completed before the next cycle of insertion commences. Tabulae are absent, although dissepiments and dissepimentaria are developed. Moreover the scleractinian skeleton, although relatively light and porous, has the stability of a basal plate which aids anchorage in the substrate. Additionally, the scleractinian polyp can secrete aragonite on the exterior of the corallite, often in the form of attachment structures. Both adaptations provided a much greater potential for reef building than the less

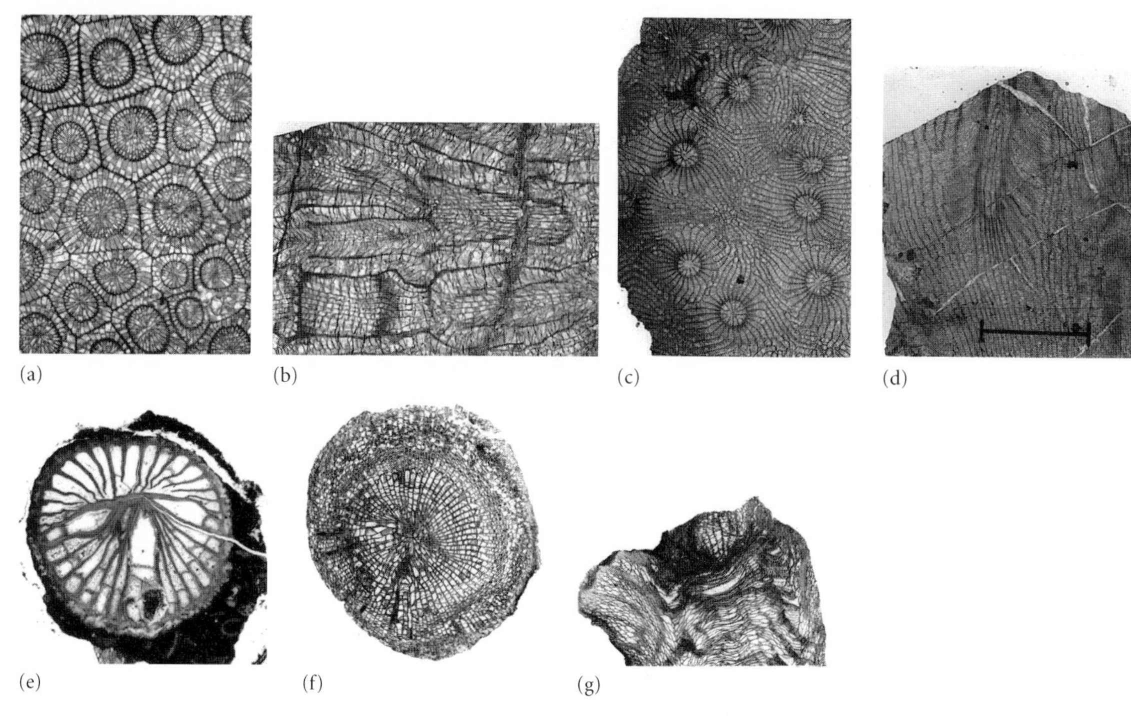

Figure 11.25 Some rugose corals: (a, b) cross and longitudinal sections of *Acervularia* (Silurian); (c, d) cross and longitudinal sections of *Phillipsastrea* (Devonian); (e) *Amplexizaphrentis* (Carboniferous); and (f, g) cross and longitudinal sections of *Palaeosmilia* (Carboniferous). Magnification approximately ×2 (a–d), ×3 (e), ×1 (f, g). Note that here and elsewhere, age assignments refer to the specimen figured and not to the entire stratigraphic range of the taxon. (Courtesy of Colin Scrutton.)

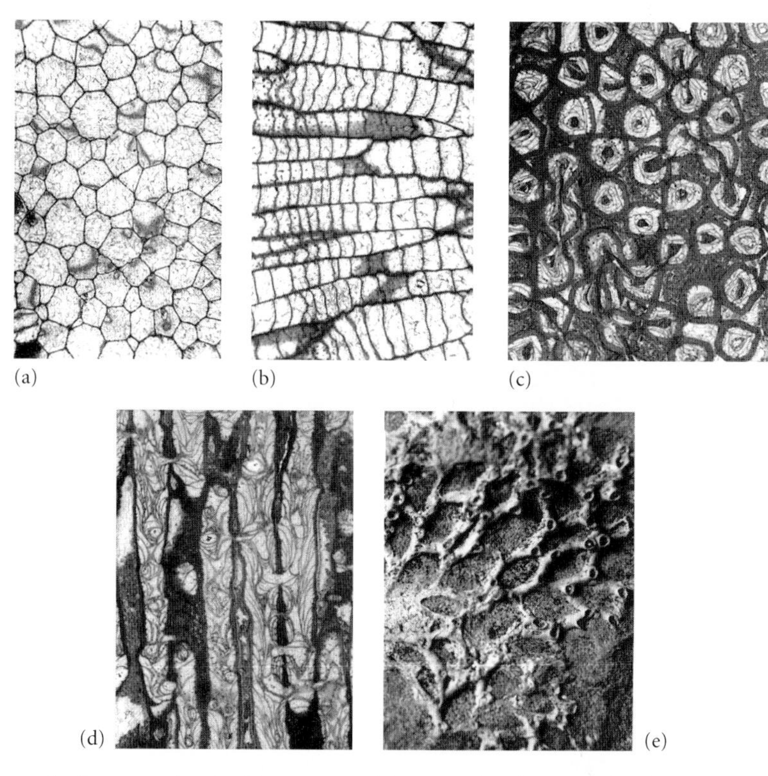

Figure 11.26 Some tabulate corals: (a, b) cross and longitudinal sections of *Favosites* (Silurian); (c, d) cross and longitudinal sections of *Syringopora* (Carboniferous); and (e) *Aulopora* (Silurian). Magnification approximately ×2. (Courtesy of Colin Scrutton.)

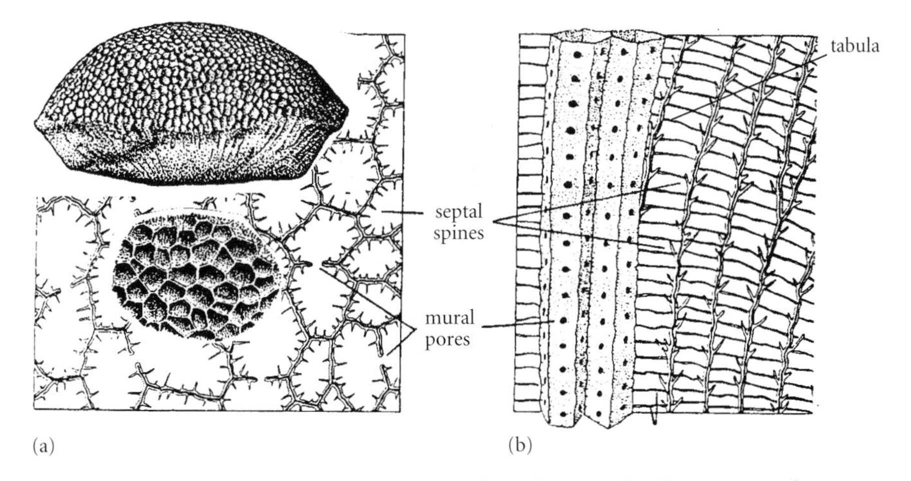

Figure 11.27 Tabulate morphology: (a) transverse and (b) longitudinal sections of *Favosites*. The insets on (a) show the lateral and upper surfaces of the entire *Favosites* colony.

stable rugose and tabulate corals of the Paleozoic. Finally, scleractinians have a distinctive ultrastructure composed of aragonite and a widespread development of coenosarc. Although scleractiniomorph corals are now known from both the Cambrian and Ordovician (Box 11.7), the scleractinians first appeared in the Mid Triassic with forms such as *Thamnasteria* becoming quickly widespread throughout Europe.

The scleractinians developed a wide range of morphologies (Fig. 11.31). For example, *Montlivaltia* is a small, cup-shaped coral common from the Early Jurassic to the Cretaceous. *Thecosmilia* is a small, dendroid to phaceloid colonial form with similar corallites that ranges from the Middle Jurassic to the Cretaceous; the massive cerioid *Isastraea* has a similar range. Scleractinians are now the dominant reef-building animals in modern seas and oceans where they form reef structures in a variety of settings, usually in the tropics.

Synecology: corals and reefs

Virtually all fossil corals were benthic. Two ecological groups have been recognized among Recent scleractinians. **Hermatypic corals** are associated with **zooxanthellae** (dinoflagellates) and are restricted to the photic zone to maintain this symbiosis. Symbiosis between the dinoflagellates and cnidarians is widespread across the living representatives of the phylum, with algae associating not only with corals but also anemones and gorgonians. The zooxanthellae are endosymbionts living in the tentacles and mouth of the cnidarian where they recycle nutrients, accelerate the rate of skeletal deposition and convey organic carbon and nitrogen to the cnidarian in return for support and protection from grazers. Hermatypic corals are commonly multiserial forms, with small corallites displaying a high degree of integration. **Ahermatypic corals**, lacking algal symbionts, are commonly solitary or uniserial compounds with large, poorly integrated corallites.

Some have suggested that coral morphology may help predict the presence of symbionts in fossil coral communities. It is probable that many tabulates were zooxanthellate whereas the rugosans were not. In broad terms, there may be parallels between the platform and basin associations of rugose and tabulate corals of the Paleozoic and reef-building and non-reef-building scleractinian corals of the Mesozoic and Cenozoic.

Reefs are biological frameworks with significant topography (Box 11.8). Three main types of structure occur in tropical shallow water: (i) **fringing reefs** develop directly adjacent to land areas; (ii) **barrier reefs** have an intervening lagoon; and (iii) **atolls** completely surround lagoons and are usually of volcanic origin. The last will continue to grow as the volcanic island subsides until eventually only a barrier reef, enclosing a lagoon, remains.

Paleozoic corals were not particularly successful reef builders; many preferred firm substrates and lacked structures that allowed anchorage and aided stability; calcareous

Box 11.6 Computer reconstruction of colonies

The colonial tabulate *Aulopora* had a long geological history and mainly occupied an encrusting niche, coating brachiopods, stromatoporoids and other, larger corals. *Aulopora* grew by dichotomous branching, pursuing a creeping or reptant life mode, efficiently siting its corallites adjacent to potential sources of food at, for example, the inhalant currents through brachiopod commissures. Colin Scrutton (University of Durham) has reconstructed colonies of the free-living animals in three dimensions using a computer-based technique (Fig. 11.28). Serial sections of the colony were digitized and assembled on a micro-VAX mainframe with software routinely used for building up three-dimensional views of diseased kidneys. Both the ontogeny of the procorallites and the astogeny of the colony as a whole were established in considerable detail by these techniques. With the development of desk and laptop microcomputers such modeling is now, more or less, routine.

Halysitids were tabulate corals that dominated some Ordovician and Silurian assemblages. As each colony grew, budding chains were able to find their way back to the colony, instead of heading off in random directions. Perhaps they could sense the gradient of a diffusive field of "pheromones", their waste products or the depletion of nutrients set up by the colony. In a simulation by Hammer (1998), new protocorallites are introduced into random positions, simulating "polyplanulate" astogenesis and the diffusive zones are established by numerically solving the differential equation for diffusion and decay.

Other fossil simulations are available at http://www.blackwellpublishing.com/paleobiology/.

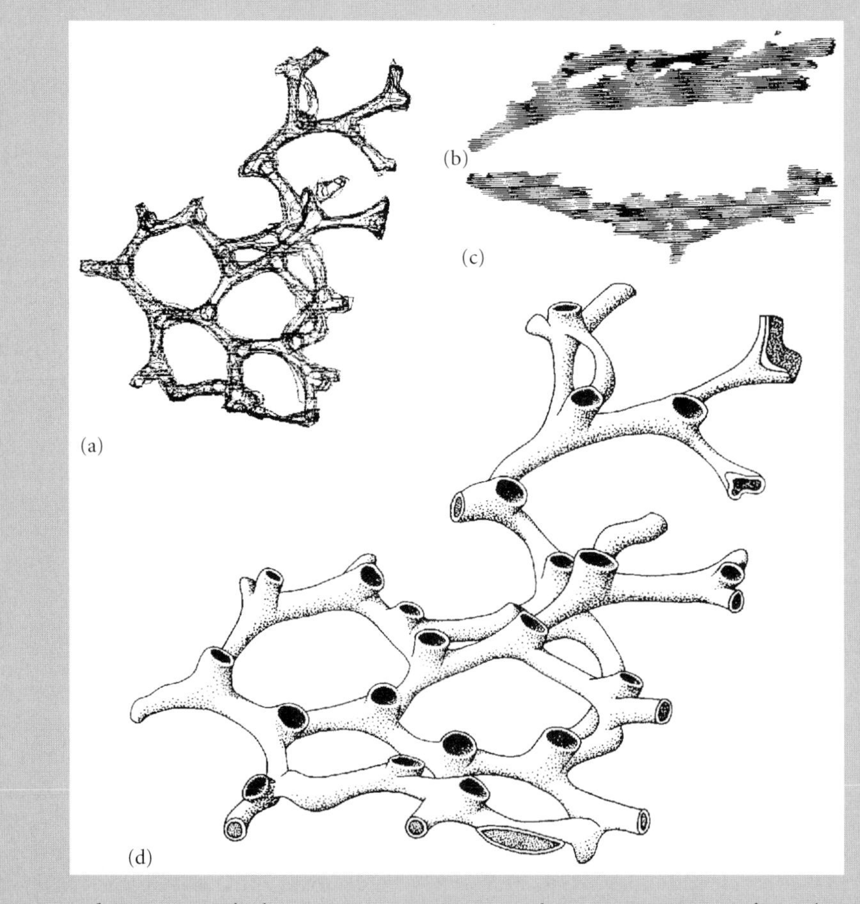

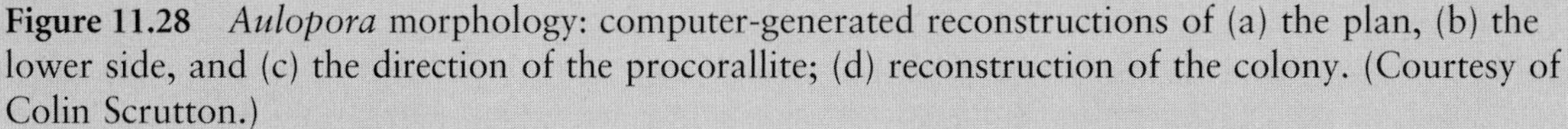

Figure 11.28 *Aulopora* morphology: computer-generated reconstructions of (a) the plan, (b) the lower side, and (c) the direction of the procorallite; (d) reconstruction of the colony. (Courtesy of Colin Scrutton.)

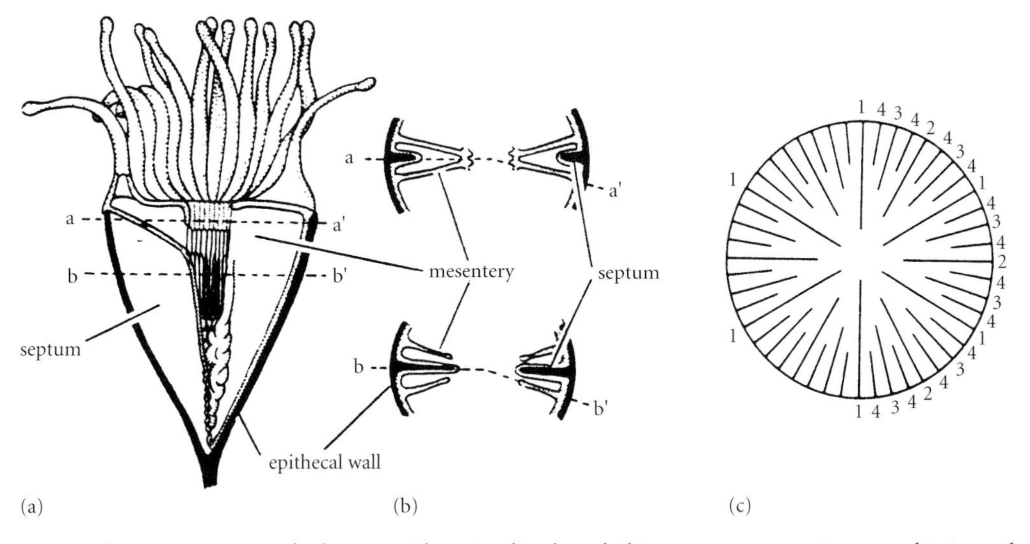

Figure 11.29 Scleractinian morphology: (a) longitudinal and (b) transverse sections, and (c) mode of septal insertion.

algae and stromatoporoids were usually more important. Nevertheless, frameworks dominated by colonial tabulates, and to a lesser extent rugosans, do occur, particularly during the Mid Paleozoic. Growth bands on the latter have provided us with a Paleozoic calendar (Box 11.9).

Pioneer and climax communities have been described from a number of Silurian and Devonian successions (Fig. 11.34). The scleractinians gradually became the dominant reef builders during the Mesozoic and Cenozoic. Modern coral reef associations have been documented in detail from eastern Australia, the eastern Pacific and the Caribbean.

The Great Barrier Reef on the continental shelf of eastern Australia is the largest coral structure on Earth, approaching 3000 km

Box 11.7 Kilbuchophyllida and iterative skeletalization

Did the scleractinian corals have a long cryptic history through the Paleozoic? When the coral *Kilbuchophyllum* (Fig. 11.30a) was described from the Middle Ordovician rocks of southern Scotland, it caused a sensation, at least amongst coral workers. *Kilbuchophyllum* seemed to have patterns of septal insertion and a microstructure identical to those of modern scleractinians, and quite unlike the contemporary rugosans and tabulates. At first, some paleontologists said this was an aberrant local form, but specimens have been found in the Silurian too. It is unlikely that *Kilbuchophyllum* was the stem group for the scleractinians; however, clearly other groups of soft-bodied anemones with the potential of skeletalization were around early in the history of the group. Following the end-Permian mass extinction, when the rugose and tabulate corals finally disappeared, calcification of other scleractinian-type morphs during the Triassic marked a new start of another highly successful calcified coral group. Similarly calcified, scleractinian-type polyps are known from the Permian, implying that this skeletal type re-evolved iteratively, that is time and time again. But what did the naked scleractinian-type polyps look like? Hou Xian-guang (Yunnan University) and his colleagues (2005) have described the sea anemone-like *Archisaccophyllia* from the Early Cambrian Chengjiang fauna (Fig. 11.30b; see p. 386). This organism may well have been one of a group of naked polyps that generated various scleractiniomorph corals during the Paleozoic and probably were responsible for seeding the Mesozoic radiation of the most successful reef builder in the oceans today.

Continued

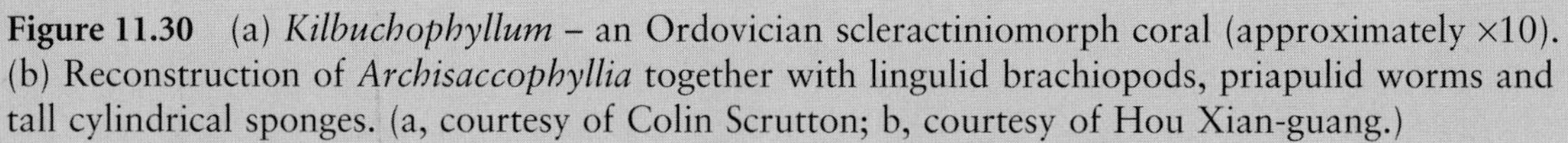

Figure 11.30 (a) *Kilbuchophyllum* – an Ordovician scleractiniomorph coral (approximately ×10). (b) Reconstruction of *Archisaccophyllia* together with lingulid brachiopods, priapulid worms and tall cylindrical sponges. (a, courtesy of Colin Scrutton; b, courtesy of Hou Xian-guang.)

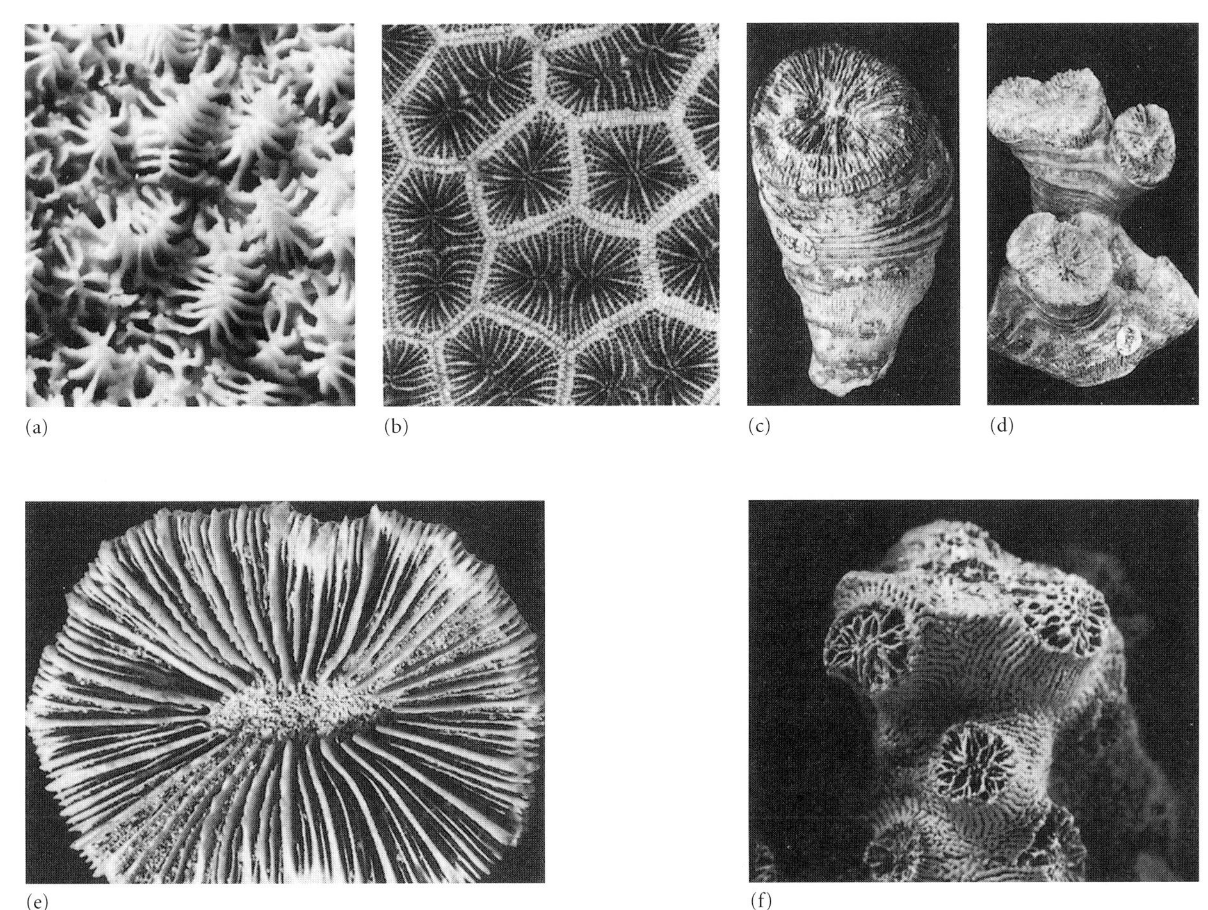

Figure 11.31 Some typical scleractinian corals: (a) *Hydnophora* (Recent); (b) *Gablonzeria* (Triassic); (c) *Montlivaltia* (Jurassic); (d) *Thecosmilia* (Jurassic); (e) *Scolymia* (Miocene); and (f) *Dendrophyllia* (Eocene). All natural size. (From Scrutton & Rosen 1985.)

Box 11.8 Reef building through time

Reefs were not just corals! Throughout geological time, a whole range of mainly modular organisms have contributed to these calcareous structures (Wood 2001), providing, too, significant carbonate factories often spalling off the continental shelves into deeper waters. While Early Paleozoic shallow-marine environments were dominated by various microbes, bryozoans, corals and sponges (including archaeocyaths and stromatoporoids), the Mesozoic and Cenozoic were characterized by scleractinian corals (Fig. 11.32). Through time, the more restricted environments were home to stromatolites during the Paleozoic and early Mesozoic; these too popped up after some extinction events, in for example the Early Silurian and Early Triassic, as disaster species. The later Mesozoic and Cenozoic saw the arrival of serpulid and oyster reefs in these more stressed brackish or hypersaline habitats. Deeper-water environments were the domain of the spicular sponges together with occasional symbiotic (ahermatypic) scleractinian corals.

Continued

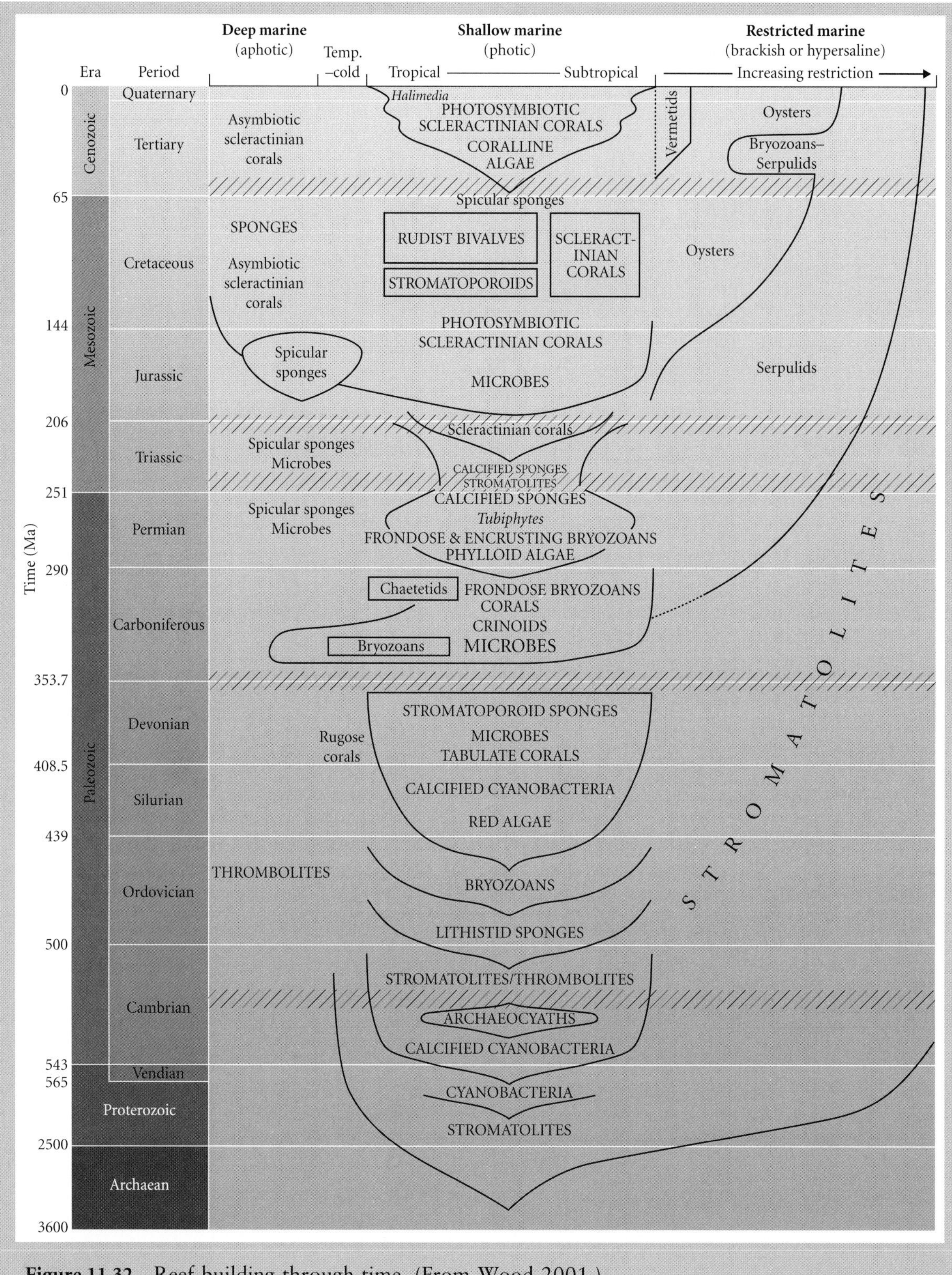

Figure 11.32 Reef building through time. (From Wood 2001.)

Box 11.9 Corals and the Earth's rotation

Through time, the Earth has changed its rate of rotation, and days have become longer. This extraordinary discovery has come from detailed analysis of the growth bands on coral epithecae. Well-preserved corals often display fine growth lines, grouped together into thicker bands; the former are thought to reflect daily growth while the latter bands are monthly growth cycles, controlled by the lunar orbit. A set of more widely spaced bands may represent yearly growth. In a classic study, John Wells of Cornell University counted the growth lines on a variety of Devonian corals (Fig. 11.33) and suggested that the Devonian year had about 400 days. The implication that Devonian days were shorter suggests the Earth's rate of rotation is decreasing due to the gravitation pull of the moon.

Ivan Gill of the University of New Orleans and his colleagues (2006) have taken the story much further. Using a range of more sophisticated techniques, including the scanning electron microscope and backscattered electron imaging, it is now possible to identify with much more precision microscale banding in some coral species that could ultimately act as proxies for daily changes in our environment, highlighting short-term climatic and other events. Moreover, this style of banding can help decipher a great deal more about the detailed mechanisms and timing of skeletalization within the corals as a whole.

Figure 11.33 Devonian banded coral, *Heliophyllum halli* (×3). (Courtesy of Colin Scrutton.)

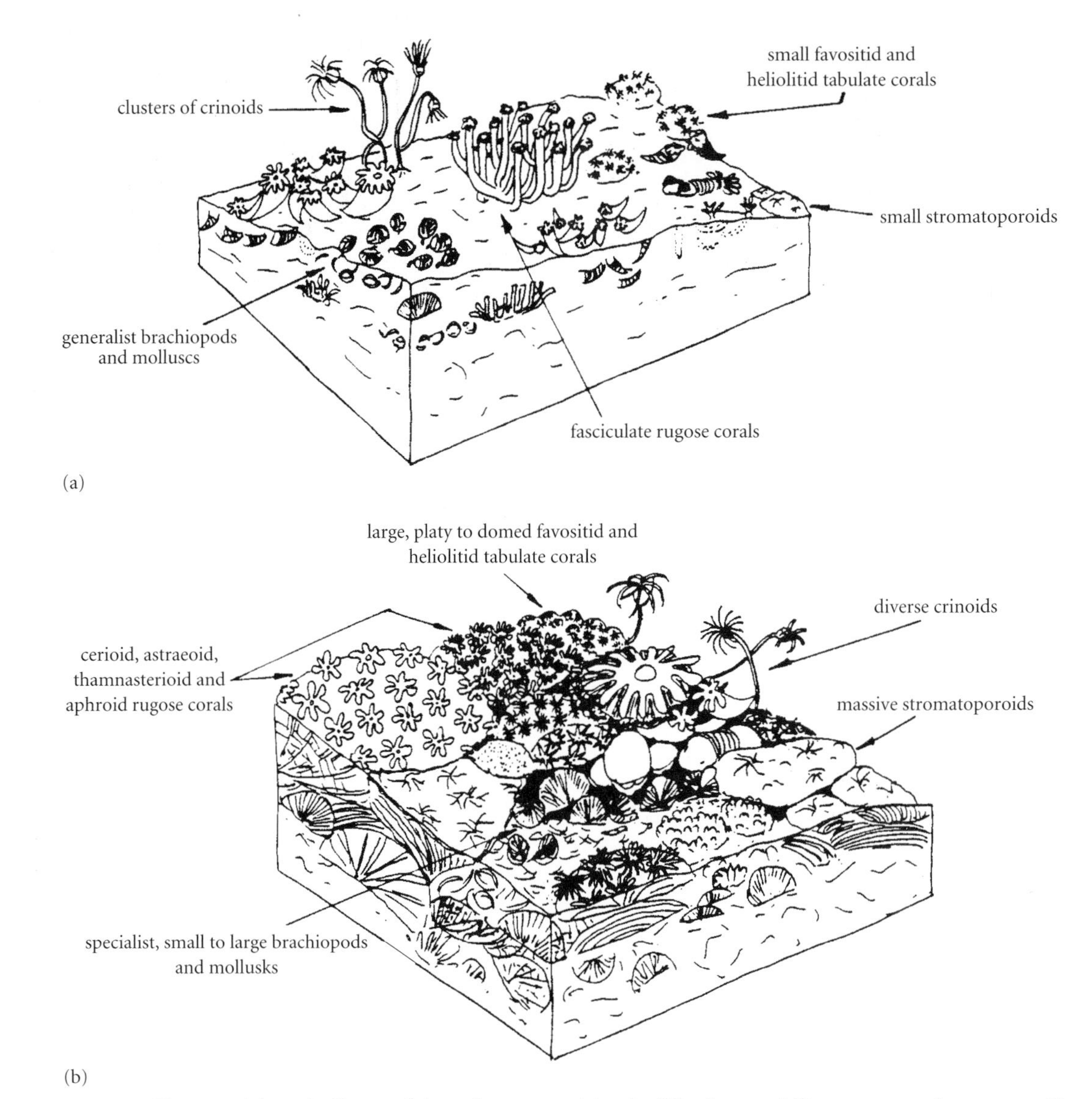

Figure 11.34 Pioneer (a) and climax (b) reef communities in Silurian and Devonan reef systems. (From Copper, P. 1988. *Palaios* 3.)

long and up to 300 km wide and visible from space. It is a long-lived structure dating back to the Miocene. The reef extends from 9° to 25° south and comprises many multicolored scleractinian corals together with many other invertebrates and calcareous algae. The fore-reef deposits tumble eastwards into the western Pacific; landward back-reef lagoons are developed against eastern Australia. Can such reef constellations really be recognized in the fossil record? On the adjacent continent the Upper Devonian rocks of the Canning Basin contain fossil barrier reefs dominated by calcareous algae and tabulate and rugose corals together with stromatoporoids and microbiolites. The reef and its associated facies can be mapped in considerable detail, as the Windjana Gorge dissects the near-horizontal strata of the northern margin of the Canning Basin (Fig. 11.35). An unbedded core of calcareous algae, corals and stromatoporoids sheltered a back-reef and lagoonal environment packed with calcareous algae, corals, stromatoporoids and crinoids together with brachiopods, bivalves, cephalopods and gastropods. In front the fore-reef was steep and littered by reef talus. However, during the Late Devonian extinction event, at the end of the Frasnian, associations dominated by rugose and tabulate corals together with stromatoporoids disappeared; this type of reef ecosystem never recovered.

(a) (b)

Figure 11.35 Devonian reefs of the Canning Basin, Australia: (a) main face, and (b) Windjana Gorge. The fore-reef slope in the foreground has large blocks of unbedded reef material in the background; the reef is prograding over the fore-reef toward the viewer. (Courtesy of Rachel Wood.)

Distribution: corals through time

Although some coral-like forms have been described from the Cambrian, most lack typical zooantharian structures. *Cothonion*, for example, with poorly integrated corallite-like clusters and opercula was probably a Cambrian experiment with coralization (Fig. 11.36). The first tabulates appeared during the Early Ordovician with cerioid growth modes; tabulae were rare and septa and mural pores were absent. Nevertheless, by the Mid and Late Ordovician the more typical characters of the Tabulata had evolved when they dominated coral faunas. Some workers have removed the heliolitids, with individual corallites mutually separated by extensive coenosteum, from the Tabulata, as a distinct order. The group was common until the Early Silurian when the more open structures of the favositids with massive cerioid colonies began to dominate, although they were already abundant in the Ordovician.

The Rugosa appear during the Mid Ordovician. Many of the evolutionary trends across the order have been repeated many times in different families. In general terms the group evolved more complex, heavier skeletons prior to extinction at the end of the Permian.

The first scleractinians were established by the Mid Triassic, derived from multiple ancestors among the sea anemones. The Triassic taxa were probably photosymbiotic, forming patch reefs in parts of the Tethyan belt. The group, however, expanded significantly during the Jurassic with the radiation of both reef-building and non-reef-building groups in shallow- and deep-water environments, respectively. Scleractinian evolution was marked by a number of morphological trends: solitary life strategies were eventually superseded by a dominance of colonial forms that display transitions from low levels of integration in phaceloid growth modes to higher levels in meandroid styles, common in modern reefs.

Corals have been used effectively for the correlation of Silurian (tabulates) and Devonian (rugose) strata but they have been proved most useful for Carboniferous biostratigraphy. During the early 1900s, Arthur Vaughan studied in detail the distribution of Lower Carboniferous corals in Belgium and Britain and he argued they would be of great value in Carboniferous biostratigraphy. Corals are very common, often widespread, usually distinctive, and well preserved in the Lower Carboniferous rocks of Europe. However, more modern studies on Carboniferous biostratigraphy using microfossils such as conodonts and foraminiferans, together with sequence stratigraphy, have shown that the occurrences of corals are controlled as much by rock facies as by time, and so they cannot be used for global correlation. Nevertheless, many corals are still useful for local correlations, and Lower Carboniferous stratigraphy (Fig. 11.37) has been refined on the basis of Vaughan's pioneer work and more modern techniques (Riley 1993).

But was coloniality amongst the bilaterians a derived condition or, more controversially, a primitive state (Box 11.10)?

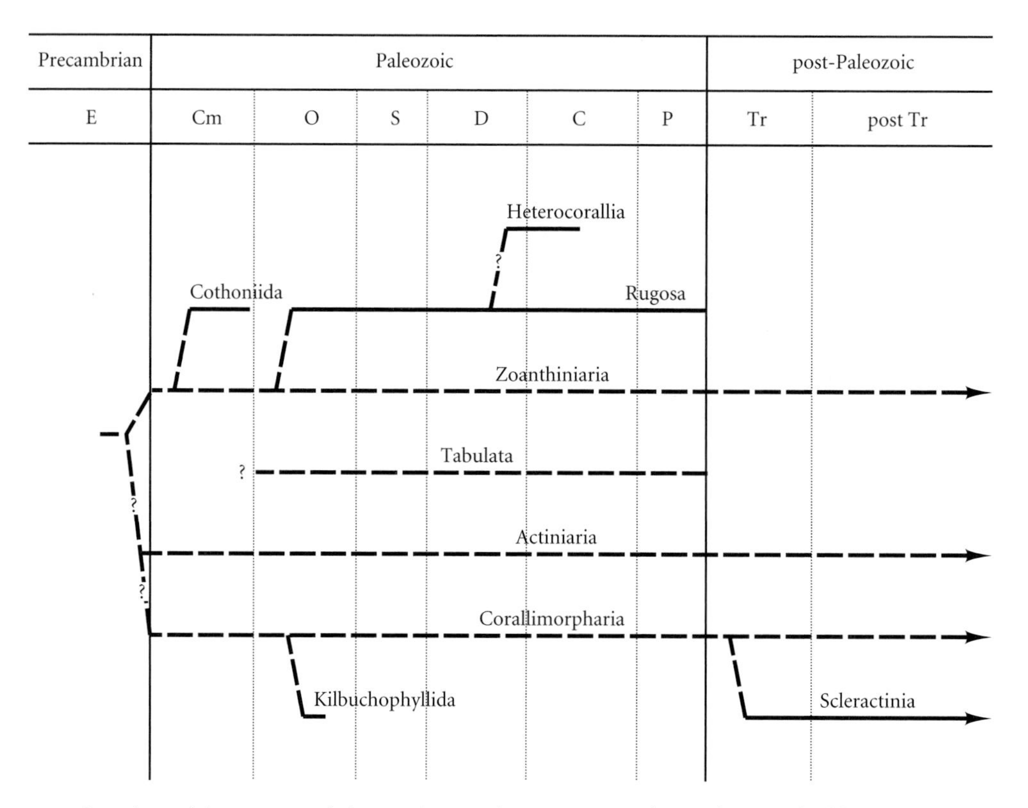

Figure 11.36 Stratigraphic ranges of the main coral groups. Geological period abbreviations are standard, running from Ediacarian (E) to Triassic (Tr). (Replotted from Clarkson 1998.)

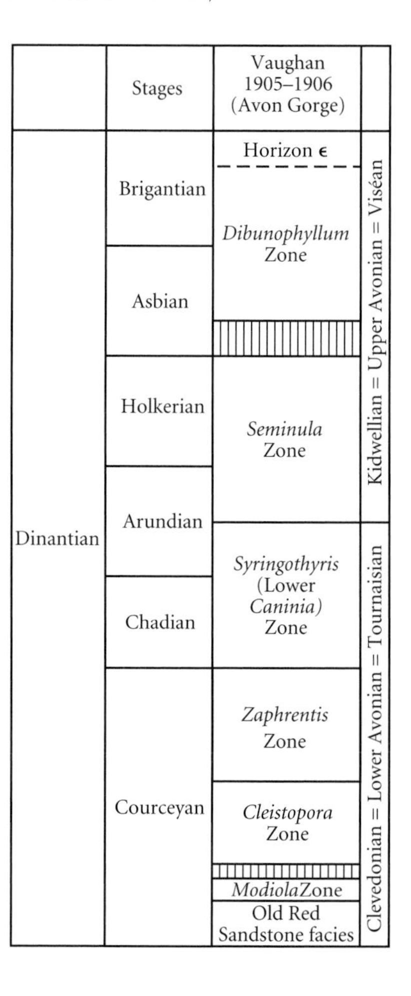

Figure 11.37 Coral biostratigraphy for the Dinantian. (Redrawn from various sources.)

Box 11.10 Colonies: the source of the first bilaterians?

Perhaps colonial organisms in the Late Precambrian had a deep significance for animal evolution. Is it possible that the complex bilateralians we see today originated within a colonial structure prior to the Cambrian explosion? Ruth Dewel (Appalachian State University, Boone) has developed a model involving the individuation of colony modules. Colonial organisms tend to develop greater degrees of integration and internal specialization through time as they begin to function as superorganisms. In this model an organism with bilaterian features, i.e. bilateral symmetry, with three body regions and epithelium-lined body compartments, can apparently break away from a complex, integrated cnidarian colony to form something like a pennatulacean octocoral that may have formed the stem group to both the cnidarians and bilateralians (Dewel 2000). A pathway from sponge to cnidarian to bilateralian body plans in her model is plausible (Fig. 11.38). Pure fantasy? Why then are outgroups to the early bilaterians large and simple whereas the bilaterians, themselves, are small and complex? It is an interesting hypothesis; but such hypotheses are there to be rigorously tested and falsified.

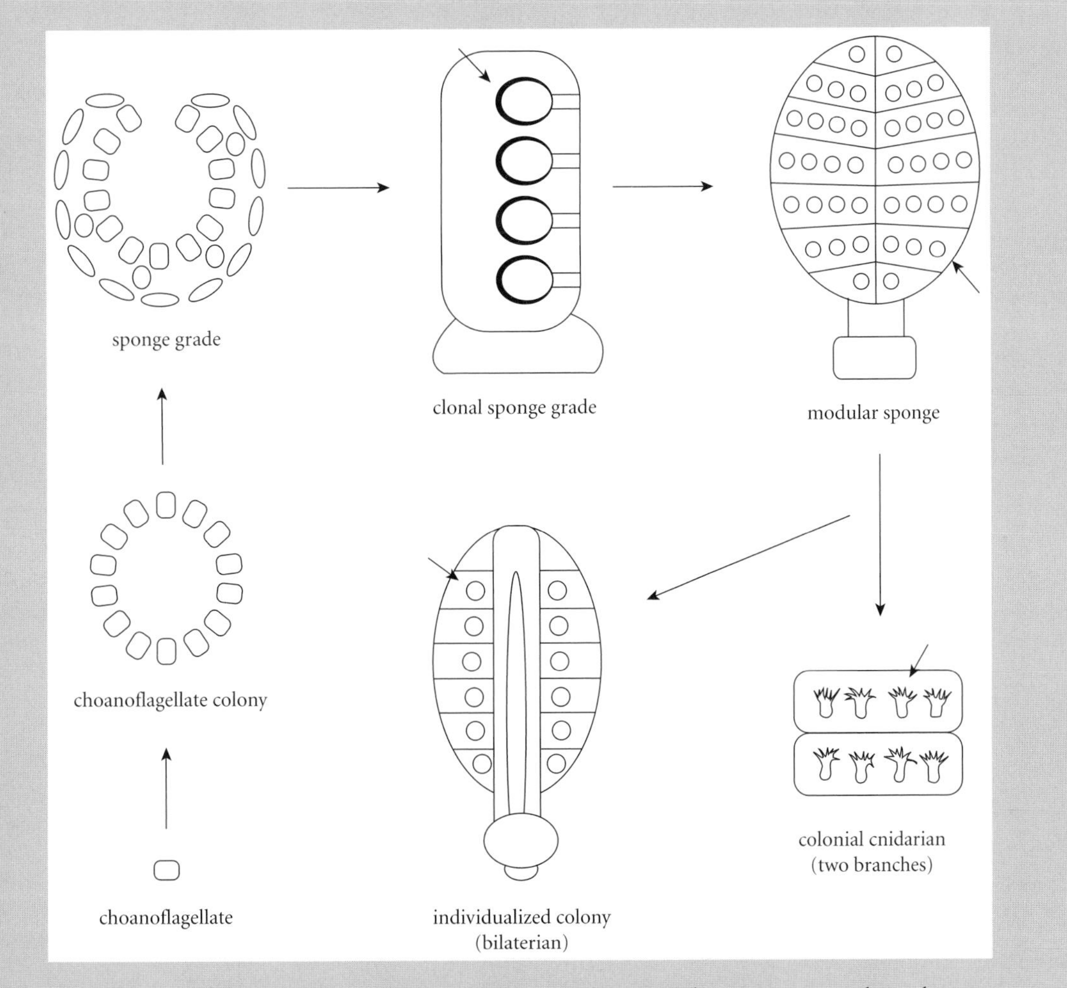

Figure 11.38 A possible origin for bilaterians in the colonies? The process involves the development of multicellularity, followed by multifunctional modules (short arrows) and finally a shift in their functional morphology within the cnidarians and the bilaterians. (From Dewel 2000.)

Review questions

1 Superficially sponges seem to be a compact morphological group but modern molecular data indicate that they are not monophyletic. Are there in fact morphological differences between the main sponge groups that back this up?
2 The archaeocyaths were some of the first metazoan reef builders, dominating the Early to Mid Cambrian tropics. How did their reef communities differ from the previous buildups of the Late Proterozoic *Namapoikea* and those later dominated by the corals and the stromatoporoids?
3 Tabulate corals were important frame-building organisms during intervals in the Paleozoic. Is there any evidence to suggest that they were associated with zooanthellae?
4 What do aberrant cnidarian taxa such as *Archisaccophyllia* and *Kilbuchophyllum* tell us about the possible track of coral evolution?
5 Metazoan reefs have been an important part of the marine ecosystem since the Early Cambrian. But during intervals of extreme stress, for example just after severe extinction events, such reefs disappear and the planet momentarily returns to a "stromatolite world". How can such an ecosystem, most characteristic of the Proterozoic, re-establish itself?

Further reading

Clarkson, E.N.K. 1998. *Invertebrate Palaeontology and Evolution*, 4th edn. Chapman and Hall, London. (An excellent, more advanced text, clearly written and well illustrated.)

Rigby, J.K. 1987. Phylum Porifera. *In* Boardman, R.S., Cheetham, A.H. & Rowell, A.J. (eds) *Fossil Invertebrates*. Blackwell Scientific Publications, Oxford, UK, pp. 116–39. (A comprehensive, more advanced text with emphasis on taxonomy; extravagantly illustrated.)

Rigby, J.K. & Gangloff, R.A. 1987. Phylum Archaeocyatha. *In* Boardman, R.S., Cheetham, A.H. and Rowell, A.J. (eds) *Fossil Invertebrates*. Blackwell Scientific Publications, Oxford, UK, pp. 107–15. (A comprehensive, more advanced text with emphasis on taxonomy; extravagantly illustrated.)

Rigby, J.K. & Scrutton, C.T. 1985. Sponges, chaetetids and stromatoporoids. *In* Murray, J.W. (ed.) Atlas of Invertebrate Macrofossils. Longman, London, pp. 3–10. (A useful, mainly photographic review of the group.)

Scrutton, C.T. 1997. The Palaeozoic corals, I: origins and relationships. *Proceedings of the Yorkshire Geological Society* **51**, 177–208. (First of two useful review papers.)

Scrutton, C.T. 1998. The Palaeozoic corals, II: structure, variation and palaeoecology. *Proceedings of the Yorkshire Geological Society* **52**, 1–57. (Second of two useful review papers.)

Scrutton, C.T. & Rosen, B.R. 1985. Cnidaria. *In* Murray, J.W. (ed.) *Atlas of Invertebrate Macrofossils*. Longman, London, pp. 11–46. (A useful, mainly photographic, review of the group.)

Wood, R. 1999. *Reef Evolution*. Oxford University Press, Oxford, UK. (Comprehensive overview of reefs through time.)

References

Debrenne, F. 2007. Lower Cambrian archeocyathan bioconstructions. *Comptes Rendus Palevol* **6**, 5–19.

Dewel, R.A. 2000. Colonial origin for Eumetazoa: major morphological transitions and the origin of bilateralian complexity. *Journal of Morphology* **243**, 35–74.

Gill, I.P., Dickson, J.A.D. & Hubbard, D.K. 2006. Daily banding in corals: implications for paleoclimatic reconstruction and skeletalization. *Journal of Sedimentary Research* **76**, 683–8.

Hammer, Ø. 1998. Regulation of astogeny in halysitid tabulates. *Acta Palaeontologica Polonica* **43**, 635–51.

Hou Xian-guang, Stanley, G.D. Jr., Zhao Jie & Ma Xiao-ya 2005. Cambrian anemones with preserved soft tissue from the Chengjiang biota, China. *Lethaia* **38**, 193–203.

Kershaw, S. 1990. Stromatoporoid palaeobiology and taphonomy in a Silurian biostrome in Gotland, Sweden. *Palaeontology* **33**, 681–706.

Riley, N.J. 1993. Dinantian (Lower Carboniferous) biostratigraphy and chronostratigraphy in the British Isles. *Journal of the Geological Society, London* **150**, 427–46.

Savarese, M. 1992. Functional analysis of archaeocyathan skeletal morphology and its paleobiological implications. *Paleobiology* **18**, 464–80.

Sperling, E.A., Pisani, D. & Peterson, K.J. 2007. Poriferan paraphyly and its implications for Precambrian paleobiology. *Special Paper Geological Society, London* **286**, 355–68.

Wood, R. 1990. Reef-building sponges. *American Scientist* **78**, 224–35.

Wood, R. 2001. Biodiversity and the history of reefs. *Geological Journal* **36**, 251–63.

Wood, R., Grotzinger, J.P. & Dickson, J.A.D. 2002. Proterozoic modular biomineralized metazoan from the Nama Group, Namibia. *Science* **296**, 2383–6.

Wood, R., Zhuravlev, A.Yu., Debrenne, F. 1992. Functional biology and ecology of Archaeocyatha. *Palaios* **7**, 131–56

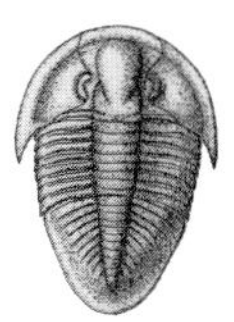

Chapter 12

Spiralians 1: lophophorates

Key points

- Three spiralian invertebrate groups have lophophores, a filamentous feeding organ: brachiopods, bryozoans and phoronids.
- Brachiopods are twin-valved shellfish, with a lophophore and usually a pedicle, adapted to a wide range of life strategies on the seafloor.
- The phylum Brachiopoda is currently divided into the linguliformeans, with organophosphatic shells, and the craniiformeans and rhynchonelliformeans, both with calcareous shells.
- Paleozoic communities were dominated by orthides and strophomenides, together with a variety of spire-bearing forms; rhynchonellides and terebratulides are typical of the lower-diversity post-Paleozoic brachiopod assemblages.
- Brachiopods dominated the filter-feeding benthos of the Paleozoic but never fully recovered in abundance or diversity from losses during the end-Permian mass extinction.
- Living brachiopods are relatively rare, occupying mostly cryptic and deep-water habitats.
- Bryozoans are colonial invertebrates with lophophores, commonly displaying marked non-genetic variation across a wide range of environments.
- The Stenolaemata dominated Paleozoic bryozoan faunas, with only the cyclostomes surviving the combined effects of the end-Permian and end-Triassic mass extinctions; as the cyclostomes continued to decline after the end-Cretaceous extinction event, the cheilostomes radiated to dominate Cenozoic assemblages.

We may consider here under the name Molluscoidea, the two groups of animals which are known respectively as the Polyzoa [Bryozoa] and the Brachiopoda. These two groups, in many respects closely allied to one another, present affinities on the one hand to the Worms and on the other hand to the Mollusca . . .

R.A. Nicholson and R. Lydekker (1890) *Manual of Palaeontology*, 3rd edn

What do lampshells, moss animals and the rare tube-dwelling phoronids, or horseshoe worms, have in common? They may look very different, but these three phyla, the Brachiopoda, Bryozoa and Phoronida, all possess a complex feeding organ, the **lophophore**, and have similar body cavities or celoms. Nevertheless the relationships among the three are not yet fully resolved, although the phoronids probably lie close to or may even be part of the group, the bryozoans are more distantly related. Our understanding has not changed much since 1890, but new molecular studies may help resolve these uncertainties in the next 10 years.

The phoronids are tube-dwelling, worm-like lophophorates, with the 10 or so described species divided between two genera, *Phoronis* and *Phoronopsis*. These animals lack a mineralized skeleton and pursue burrowing or boring life strategies with near-cosmopolitan distributions. The phylum has a long though questionable geological history, as some authors suggest that Precambrian and Lower Paleozoic records of the vertical burrow *Skolithos* (see p. 523) may possibly be the work of phoronids. The ichnogenus *Talpina*, present as borings in both Cretaceous belemnite rostra and Tertiary mollusk shells, may also have been constructed by phoronids.

BRACHIOPODA

> *It is no valid objection to this conclusion, that certain brachiopods have been but slightly modified from an extremely remote geological epoch; and that certain land and fresh-water shells have remained nearly the same, from the time when, as far as is known, they first appeared.*
>
> Charles Darwin (1859)
> *On the Origin of Species*

The brachiopods are one of the most successful invertebrate phyla in terms of abundance and diversity. They appeared first in the Early Cambrian and diversified throughout the Paleozoic to dominate the low-level, suspension-feeding benthos; a wide range of shell morphologies and sizes characterize the phylum, from the tiny acrotretides (microns in length) to the massive gigantoproductids (nearly 0.5 m wide). Although only about 120 genera of brachiopods, also known as lampshells, survive today, they occupy a wide range of habitats from the intertidal zone to the abyssal depths. The brachiopods are entirely marine, bilaterally symmetric animals with a ciliated feeding organ, or lophophore, contained within a pair of shells or valves. Internal structures such as **teeth** and **sockets, cardinal processes** and various muscle scars are all associated with the opening and closing of the two valves during feeding cycles. Brachiopods have featured in many paleoecological studies of Paleozoic faunas, when they dominated life on the seabed in terms of numbers of both individuals and species. Their use in paleobiogeographic analysis is well documented (see Chapter 4). Nevertheless brachiopods have also been widely used in regional biostratigraphy and, during the Silurian, a number of orthide, pentameride and rhynchonellide lineages show good prospects for international correlation.

Despite their relative low diversity today, living brachiopods are actually quite widespread, represented mainly by forms attached by pedicles to a variety of substrates across a spectrum of water depths. At high latitudes brachiopods range from intertidal to basinal environments at depths of over 6000 m. They are most common in fjord settings in Canada, Norway and Scotland and in the seas around Antarctica and New Zealand. The association of the brachiopod *Terebratulina retusa* growing on the horse mussel, *Modiolus modiolus*, a bivalve, is widespread in the northern hemisphere. In the tropics, however, many species are minute, exploiting cryptic habitats, hiding in reef crevices or in the shade of corals and sponges. Larger forms live in deeper-water environments, out of the range of predators, like sea urchins, that graze on the sumptuous meadows of newly attached larvae.

Morphology: brachiopod animal

The brachiopod soft parts are enclosed by two morphologically different shells or valves that are opened and closed by a variety of muscles; this arrangement is modified differently across the three subphyla – the linguliformeans, craniiformeans and rhynchonelliformeans (Fig. 12.1a–f; Box 12.1). In contrast to the bivalves, where the right valve is a mirror image of the left, the plane of symmetry in brachiopods bisects both valves perpendicular to the plane along which the valves open, or the **commissure**. The larger of the two valves is generally the **ventral** or **pedicle** valve; in many brachiopods the fleshy stalk or pedicle pokes through the apex of this valve and attaches the animal to the substrate. The pedicle can vary from a thick, fleshy stalk to a bunch of delicate, thread-like strands, which can anchor the brachiopod in fine mud. Some extinct brachiopods lost their pedicles during ontogeny and adopted a free-living mode of life, lying recumbent on or partially in the sediments on the seafloor. The **dorsal** or **brachial** valve contains the extendable food-gathering organ or lophophore together with its supports. A number of types of lophophore have evolved (Fig. 12.1 g). The earliest growth stage, the **trocholophe**, is an incomplete ring of filaments, still retained by the pedomorphic (see p. 146) microbrachiopod *Gwynia*. By the **schizolophe** stage a bilobed outline has developed, which probably characterized many of the smaller Paleozoic taxa. The more complex **plectolophe**, **ptycholophe** and **spirolophe** styles are characteristic of the articulated brachiopods.

The linguliformeans (see Fig. 12.1a, b) have **organophosphatic** shells with pedicles that either emerge between both valves or through an opening called the foramen. The shells develop from a **planktotrophic**, or plankton-feeding, larval stage, and linguliformeans are characterized by an alimentary tract ending in an anus. In the lingulates, the opening and closing of the valves is achieved by a complex system of muscles and the pedicle emerges between both valves. Withdrawal of the soft parts posteriorly causes a space problem that can force the valves apart; relaxation allows the animal to expand again forwards allowing the valves to close. The paterinates are the oldest group of brachiopods, appearing in the lowest Cambrian Tommotian Stage. Although linked to the other linguliformeans on the basis of an organophosphatic shell substance, the shell structure of the group is quite different and the shells have true **interareas, delthyria** and **notothyria** and apparently had a functional **diductor** muscle system.

The craniiformeans (see Fig. 12.1c) include a diverse, yet probably monophyletic, group of morphologies centered on *Crania* but including *Craniops* and the bizarre trimerellids. The shells consist of **organocarbonate** and the animal developed separate dorsal and ventral mantle lobes after the settlement of the larvae on the seabed during a nektobenthonic stage.

The rhynchonelliformeans (see Fig. 12.1d–f) have a pair of calcitic valves that contain a fibrous secondary layer, with variable convexity, hinged posteriorly and opening anteriorly along the commissure. The mantle lobes are fused posteriorly, where the interareas are secreted; their margins form the hinge between the ventral and dorsal valves. Articulation was achieved by a pair of ventral teeth and dorsal sockets, and the valves were opened and closed by opposing **diductor** and **adductor** muscle scars. In the majority of rhynchonelliformeans, the valves were attached to the substrate by a pedicle, emerging through a foramen in the delthyrial region. The subphylum contains five classes, the Chileata, the Obolellata, the Kutorginata, the Strophomenata and the Rhynchonellata. Already by the Early Cambrian, representatives of four of the five classes were present. However the two latter classes, containing respectively over 1500 and 2700 genera, dominated Phanerozoic brachiopod faunas.

Brachiopods possess both planktotrophic and lecitotrophic larvae. The planktotrophic stage may have been the most primitive, spending some time in the plankton, whereas **lecitotrophic** larvae lurking in the benthos may have developed at least twice. This obviously has important consequences for brachiopod dispersion. Since many linguliformeans are widespread it is assumed they had planktotrophic larvae in contrast to the more endemic rhynchonelliformeans with possible lecitotrophic larvae (Fig. 12.4).

Brachiopod shells can be very variable in shape. A single species can even mimic the outlines of a range of different orders. For example specimens of *Terebratalia transversa* from around the San Juan islands, western USA, show *Spirifer-*, *Atrypa-* and *Terebratula*-type morphs with increasing strengths of currents (Fig. 12.5). Moreover a number of brachiopods, such as the strophomenides, especially the productoids, may markedly

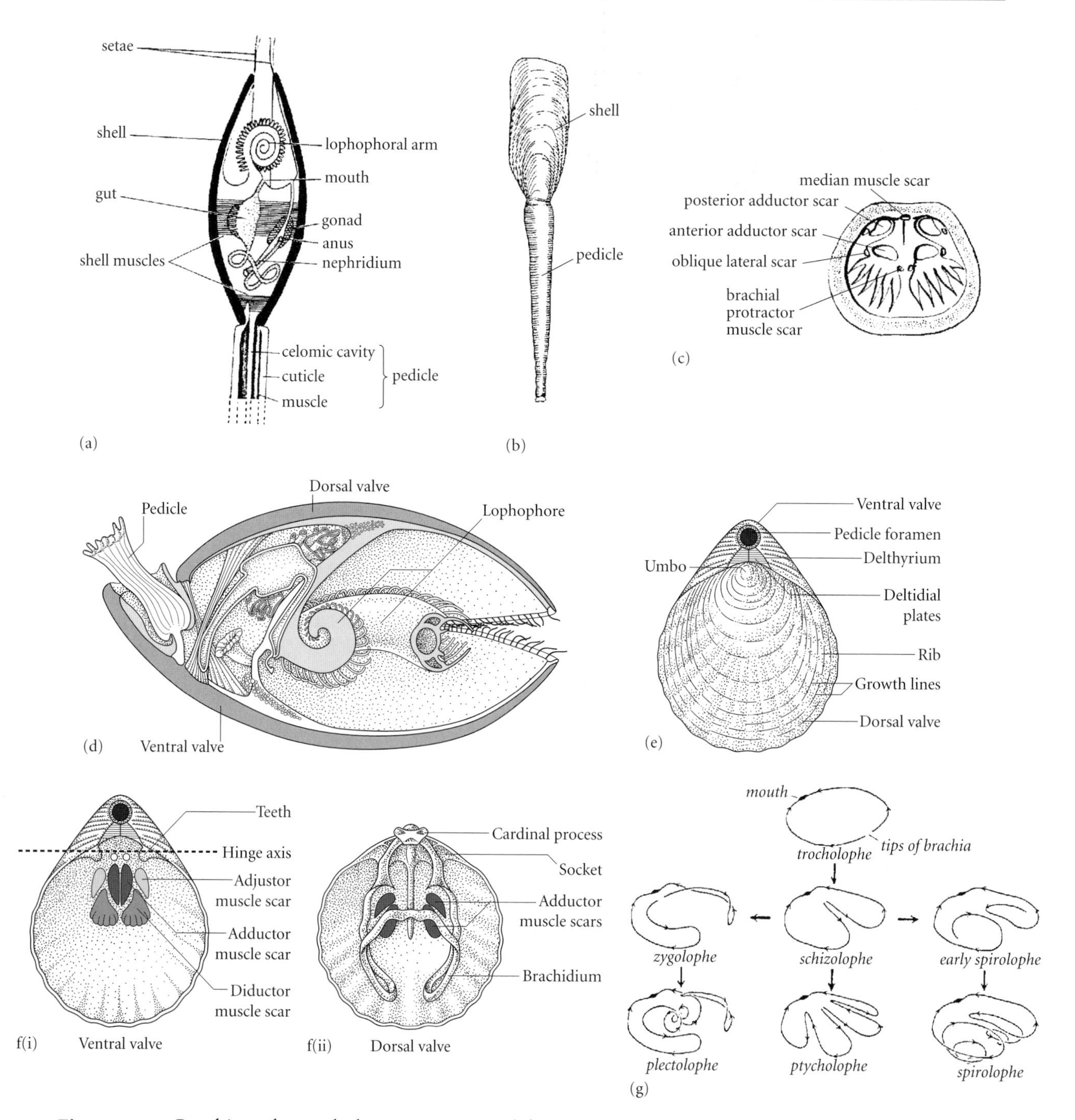

Figure 12.1 Brachiopod morphologies: (a) internal features of a lingulate, (b) exterior of a burrowing lingulate, (c) internal terminology of a craniform calciate, (d) internal features of a terebratulide, (e) external terminology of a typical articulate, (f) internal terminology of both valves of a terebratulide, and (g) main types of brachiopod lophophore.

change their shape and life mode during ontogeny from being attached to the seabed to lying untethered in the mud.

Ultramorphology: brachiopod shell

The brachiopod shell is a multilayered complex of both organic and inorganic material that has proved of fundamental importance in classification. The shells of most rhynchonelliformean brachiopods consist of three layers (Fig. 12.6). The outer layer (**periostracum**) is organic, and underneath are the mineralized primary and secondary layers. These layers are sequentially secreted by cells within the generative zone of the mantle, forming first a gelatinous sheath followed by the organic periostracum, and then the granular calcite of the primary layer. The subsequent secondary layer is thicker and composed of calcite fibers, and in some brachiopods a third prismatic layer is secreted. There are a number of variations of this basic template. The linguliformeans, for example, have phosphatic material as part of their shell fabric. The shells of rhynchonelliformean brachiopods are composed of low-magnesian calcite; these shells may have fibrous, laminar or cross-bladed laminar shell fabrics in their secondary layers. The mineral fabrics themselves, when investigated at the nanoscale, may be of particular ecological importance. Those with calcite seminacre, rather like mother-of-pearl, can cement directly to the seafloor whereas those with fibrous shells can not (Pérez-Huerta et al. 2007).

Many shells are perforated by small holes or **punctae**, in life holding finger-like extensions of the mantle or **ceca**. Their function is uncertain but they increased the amount of the brachiopod's soft tissue. Some strophomenates have **pseudopunctae**, with fine inclined calcite rods or **taleolae** embedded in the shell fabric.

The relatively stable brachiopod shell substance can tell much about the secretion of the shell but also about environmental conditions

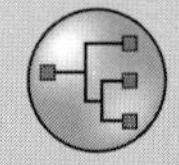

Box 12.1 Brachiopod classification

Recent cladistic and molecular phylogenetic analyses have shown that the traditional split of the phylum Brachiopoda into the Inarticulata and Articulata is incorrect, and instead there are three subphyla, the Linguliformea, Craniiformea and Rhynchonelliformea. All three have quite different body plans and shell fabrics (Fig. 12.2). The linguliformeans contain five orders united by organophosphatic shells; the inclusion of the paterinides is the most problematic since the group shares some morphological characters with the rhynchonelliforms. The craniiformeans include three rather disparate groups with quite different morphologies but which together possess an organocarbonate shell. Most scientists now accept 14 articulated orders in the rhynchonelliformeans, not counting the chileides, dictyonellides, obolellides and kutorginides, mainly based on the nature of the cardinalia and the morphology of the other internal structures associated with the attachment of muscles and the support of the lophophore. Recently the more deviant chileides, obolellides and kutorginides have been added to the subphylum. In addition, the articulated taxa have been split into those with deltidiodont (simple) and cyrtomatodont (complex) dentitions; the former group includes the orthides and strophomenides whereas the latter include the spire bearers.

Cladistic-based investigations have developed a phylogenetic framework for the phylum (Williams et al. 1996), supporting the three subphyla (Fig.12.2); their defining characters are based on shell structure and substance. The mutual relationships among these groups are still unclear as are the relationships between the many primitive articulated and non-articulated groups that appeared during the Cambrian explosion together with the origin of the phylum as a whole (Box 12.2).

A data matrix containing all the data from Williams et al. (1996) is available at http://www.blackwellpublishing.com/paleobiology/.

Continued

Subphylum	Order	Key characteristics	Stratigraphic range
Linguliformea	Lingulida	Spatulate valves with pedicle usually emerging between both shells	Cambrian to Recent
	Acrotretida	Micromorphic forms with conical ventral valve; dorsal valve with platforms	Cambrian to Devonian
	Discinida	Subcircular shells with conical ventral valve and distinctive pedicle foramen	Ordovician to Recent
	Siphonotretida	Subcircular, biconvex valves with spines and elongate pedicle foramen	Cambrian to Ordovician
	Paterinida	Strophic shells with variably developed interareas	Cambrian to Ordovician
Craniiformea	Craniida	Usually attached by ventral valve; dorsal valve with quadripartite muscle scars	Ordovician to Recent
	Craniopsida	Small oval valves with internal platforms and marked concentric growth lines	Ordovician to Carboniferous
	Trimerellida	Commonly gigantic, aragonitic shells, with platforms and umbonal cavities	Ordovician to Silurian
Rhynchonelliformea	Chileida	Strophic shells lacking articulatory structures but with umbonal perforation	Cambrian
	Dictyonellida	Biconvex valves with large umbonal opening commonly covered by a colleplax	Ordovician to Permian
	Naukatida	Biconvex shells with articulatory structures and apical foramen	Cambrian
	Obolellida	Oval valves with primitive articulatory structures	Cambrian
	Kutorginida	Strophic valves with interareas but lacking articulatory structures	Cambrian
	Orthotetida	Biconvex shells, commonly cemented, with bilobed cardinal process	Ordovician to Permian
	Billingsellida	Usually biconvex with transverse teeth and simple cardinal process	Cambrian to Ordovician
	Strophomenida	Concavoconvex, usually with a bilobed cardinal process; recumbent life mode; cross-laminar shell structure with pseudopunctae	Ordovician to Permian
	Productida	Concavoconvex valves with complex cardinalia; recumbent or cemented life mode; often with external spines	Ordovician to Triassic
	Protorthida	Well-developed interareas, primitive articulation and ventral free spondylium	Cambrian to Devonian
	Orthida	Biconvex, usually simple cardinal process; pedunculate; delthyria and notothyria open	Cambrian to Permian
	Pentamerida	Biconvex, rostrate valves with cruralia and spondylia variably developed	Cambrian to Devonian
	Rhynchonellida	Usually biconvex, rostrate valves with variably developed crurae	Ordovician to Recent
	Atrypida	Biconvex valves with dorsally-directed spiralia and variably developed jugum	Ordovician to Devonian
	Athyridida	Usually biconvex valves with short hinge line and posterolaterally-directed spiralia	Ordovician to Jurassic
	Spiriferida	Wide strophic valves with laterally-directed spiralia; both punctate and impunctate taxa	Ordovician to Jurassic
	Thecideida	Small, strophic shells with complex spiralia including brachial ridges and median septum	Triassic to Recent
	Terebratulida	Biconvex valves with variably developed long or short loops	Devonian to Recent

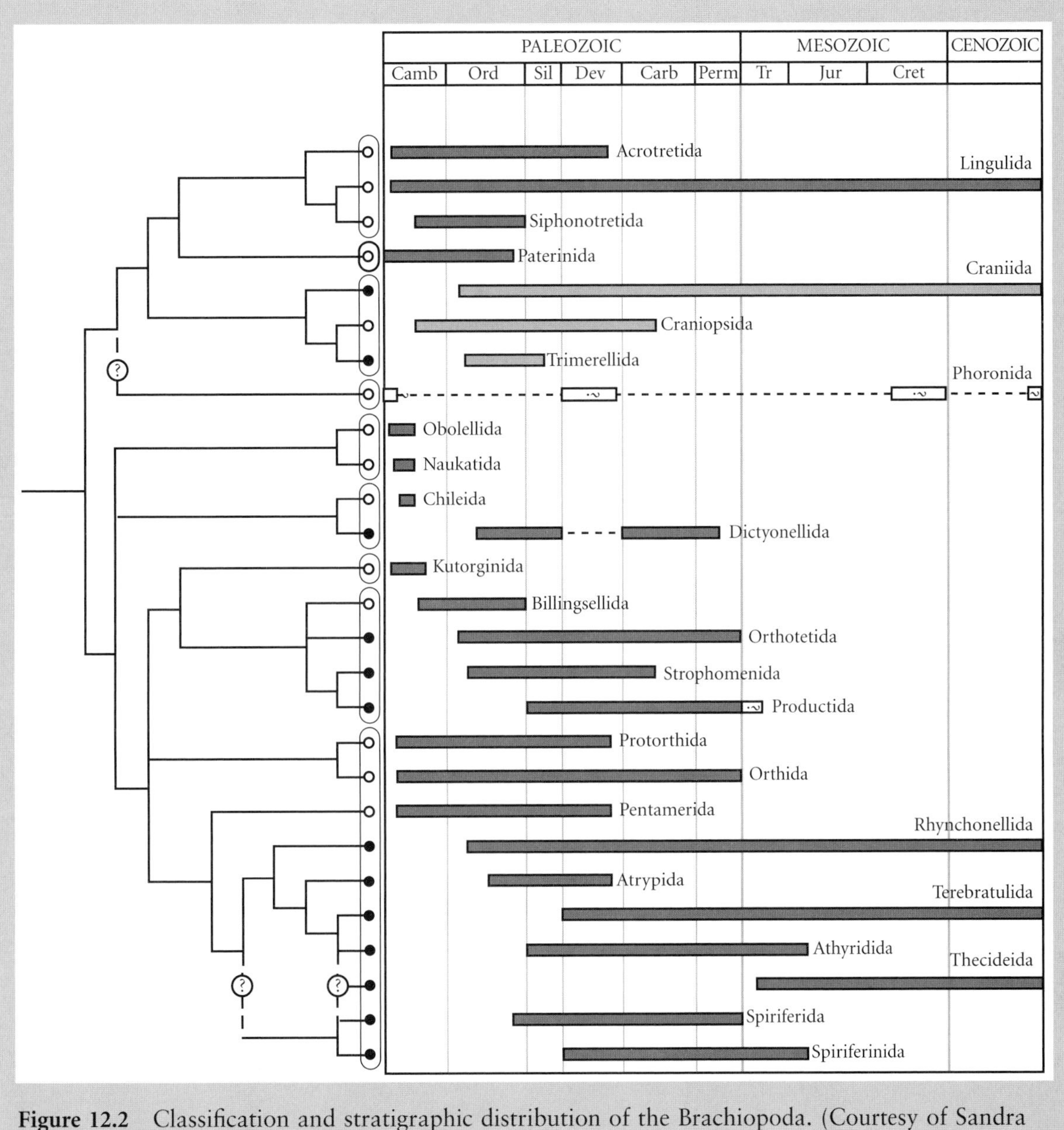

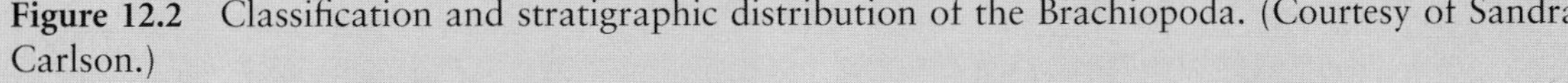
Figure 12.2 Classification and stratigraphic distribution of the Brachiopoda. (Courtesy of Sandra Carlson.)

at the time of deposition. The ratio of isotopes within the crystal lattice of the brachiopod shell was often controlled by the provenance of the chemical elements (marine or terrestrial) and temperature and salinity of the seawater. Carbon, oxygen and strontium isotopes are particularly useful. Devonian brachiopod shells from North America, Spain, Morocco, Siberia, China and Germany analyzed for stable isotopes ($\delta^{13}C$, $\delta^{18}O$ and $^{87}Sr/^{86}Sr$) have provided many new data on the termination of the Caledonian Orogeny (decrease in the $^{87}Sr/^{86}Sr$ ratio due to limited influx of freshwater), uplift during the Variscan Orogeny (increase in $^{87}Sr/^{86}Sr$ ratio due to increased influx of freshwater) and Devonian climate warming (negative $\delta^{18}O$ excursions) together with increased rates of carbon burial signaled by positive $\delta^{13}C$ excursions (van Geldern et al. 2006).

Box 12.2 The brachiopod fold hypothesis and the search for stem-group brachiopods

Already by the Early Cambrian a range of diverse brachiopods populated nearshore environments. But where can we find their ancestors and what sort of animals are we looking for? Many have assumed that a prototype brachiopod probably arose in the Late Precambrian with a phosphatic shell substance and an apparently simple *Lingula*-like morphology. But did it evolve from a burrow-dwelling sessile organism or from a mobile, slug-like ancestor? A careful study of the early development of the non-articulated brachiopod *Neocrania* by Claus Nielsen (University of Copenhagen) has yielded a few, exciting clues. During ontogeny the embryo actually curls over at both ends (Fig. 12.3). The resulting embryo has the posterior end of the animal forming the dorsal surface (or valve) and the anterior end, the ventral surface. This process, subsequently called the **brachiopod fold hypothesis** (Cohen et al. 2003), provides an elegant model for how a brachiopod could have evolved from a flat, possibly worm-like, animal with shells at its anterior and posterior ends. Care must be taken in locating such possible ancestors. *Halkieria*, for example, has shells at its anterior and posterior end but is a mollusk (see p. 331); however shells such as *Micrina* and *Mickwitzia* may have belonged to a slug-like stem-group brachiopod. The mystery may be solved only when some exceptionally well-preserved fossil is found.

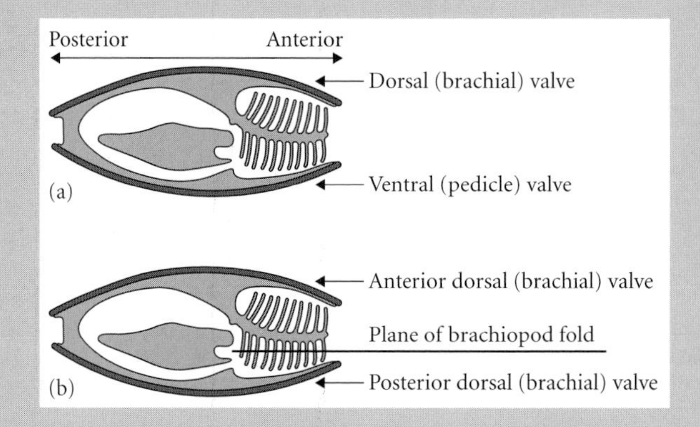

Figure 12.3 (a) The traditional body plan with an upper dorsal and a lower ventral shell. (b) The brachiopod fold hypothesis plan implies that the brachial valve is the anterior one and the pedicle posterior – both were previously on the dorsal surface of the animal. (From Cohen et al. 2003.)

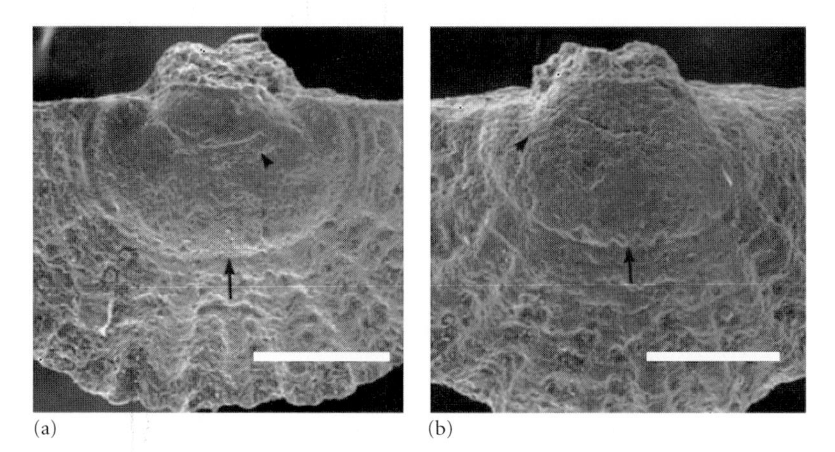

Figure 12.4 Brachiopod larvae. (a) Ventral and (b) dorsal valves of the brachiopod *Onniella*. Black arrows indicate the anterior extent of the larval shell. Scale bars, 200 μm. (From Freeman & Lundelius 2005.)

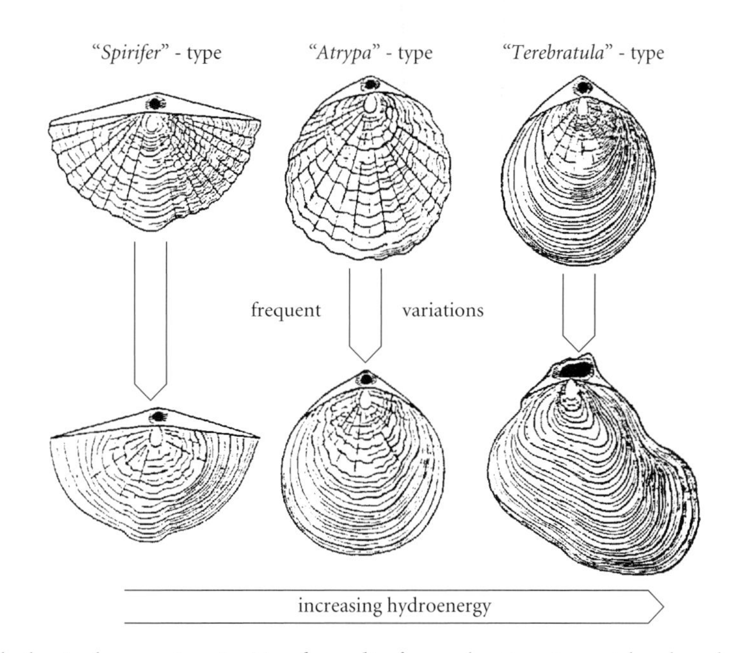

Figure 12.5 Morphological variation in *Terebratalia* from the San Juan islands related to changing hydrodynamic conditions. (From Schumann 1991.)

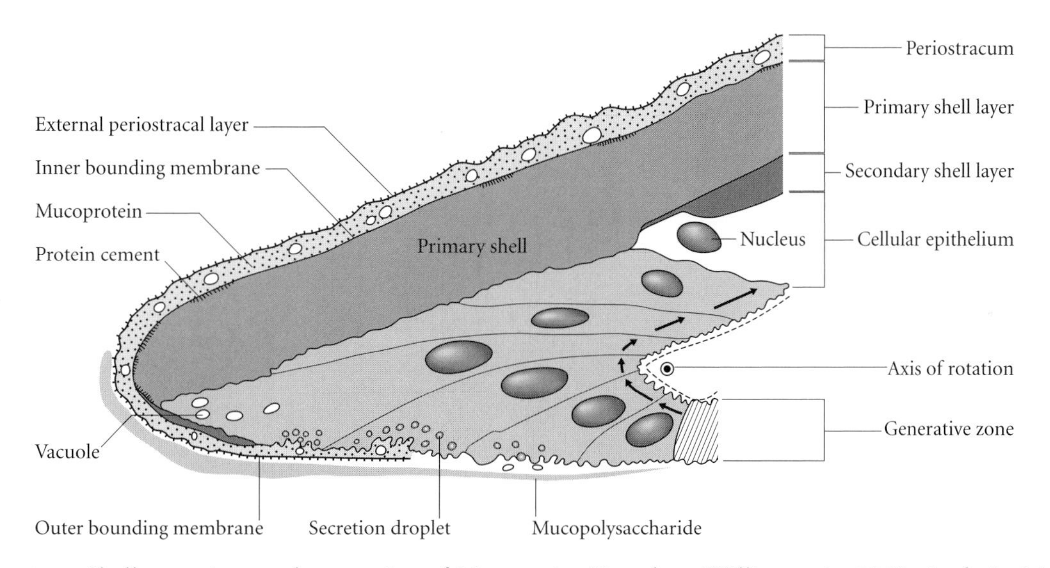

Figure 12.6 Shell secretion at the margins of *Notosaria*. (Based on Williams, A. 1968. *Lethaia* **1**.)

Distribution in time: extinctions and radiations

The make up of the Cambrian, Paleozoic and Modern brachiopod faunas are fundamentally different, represented by a dominance of different orders; some key representatives are illustrated in Fig. 12.7. Cambrian faunas were dominated by a range of non-articulated groups together with groups of disparate articulated taxa such as the chileides, naukatides, obolellides, kutorginides, billingsellides, protorthides, orthides and pentamerides. These brachiopods were members of a variety of loosely-structured, nearshore paleocommunities.

During the Ordovician radiation, the deltidiodont orthides and strophomenides dominated faunas. These first evolved around Early Ordovician island complexes and came to dominate the shelf benthos, where they began to move offshore and diversify around carbonate mounds. These communities formed the basis of the Paleozoic brachiopod fauna.

Figure 12.7 Representatives of the main orders of non-articulates and articulates. Non-articulates: (a) *Pseudolingula* (Ordovician lingulide), (b) *Nushibella* (Ordovician siphonotretide), (c) *Numericoma* (Ordovician acrotretide), (d) *Dinobolus* (Silurian trimerellide) and (e) *Crania* (Paleogene craniide). Articulates: (f) *Sulevorthis* (Ordovician orthide), (g) *Rafinesquina* (Ordovician strophomenide), (h) *Grandaurispina* (Permian productide), (i) *Marginifera* (Permian productide), (j) *Cyclacantharia* (Permian richthofeniid), (k) *Neospirifera* (Permian spiriferide), (l, m) *Rostricelulla* (Ordovician rhynchonellide) and (n, o) *Tichosina* (Pleistocene terebratulide). Magnification approximately ×2 (a, e–g, l, m), ×8 (b), ×60 (c), ×1 (d, h–k, n, o). (Courtesy of Lars Holmer (a), Michael Bassett (g), Robin Cocks (j) and Richard Grant (h, i, k, l).)

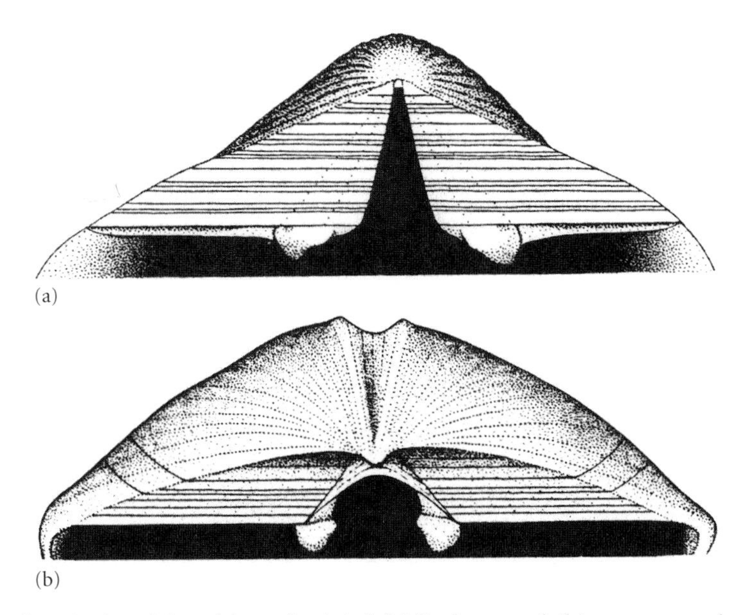

Figure 12.8 Teeth of articulated brachiopods: (a) deltidiodont and (b) cyrtomatodont dentition.

The brachiopods experienced five main extinction events followed by recoveries and radiations of varying magnitudes. The end-Ordovician event occurred in two phases against a background of glaciation and accounted for the loss of almost 80% of brachiopod families. The recovery and subsequent radiation is marked by the decline of deltidiodont groups such as the orthides and strophomenides, whereas the spire-bearing atrypides, athyridides and the spiriferides with cyrtomatodont dentition (Fig. 12.8), together with the pentamerides, achieved greater dominance, particularly in carbonate environments. Late Devonian events, at the Frasnian–Famennian Stage boundary, were also associated with climate change and removed the atrypides and pentamerides and severely affected the orthides and strophomenides, whereas the spiriferides and rhynchonellides survived in deeper-water environments and staged an impressive recovery. A particular feature of the post-Frasnian fauna was the diversity of recumbent brachiopod megaguilds (see p. 91), dominated by the productides. The Carboniferous and particularly the Permian were intervals of spectacular experimentation: some brachiopods mimicked corals or developed extravagant clusters of spines while a number of groups reduced their shells, thus presenting soft tissues to the outside environment.

Not unexpectedly, the end-Permian mass extinction saw the demise of over 90% of brachiopod species, including some of the most ecologically and taxonomically diverse groups. The post-extinction fauna was first dominated by a variety of disaster taxa (see p. 179), including lingulids; nevertheless the brachiopod fauna later diversified within a relatively few clades dominated by the rhynchonellides and terebratulides. The end-Triassic event removed the majority of the remaining spiriferides and the last strophomenides. The agenda set by the end-Permian event, involving the subsequent dominance of rhynchonellide and terebratulide groups, was continued after the end-Triassic event. The end-Cretaceous event may have been responsible for the loss of about 70% of chalk brachiopod faunas in northwest Europe; nevertheless, many genera survived to diversify again in the Danian limestones. Despite the post-Permian decline of the phylum, Modern brachiopods exhibit a remarkable range of adaptations based on a simple body plan and a well-defined role in the fixed, low-level benthos.

Ecology: life on the seabed

Living and fossil brachiopods have developed a wide range of lifestyles (Fig. 12.9). Most were attached by a pedicle cemented to a hard substrate or rooted into soft sediment. A

LIFESTYLE	BRACHIOPOD TAXA	ADAPTATIONS
Attached by pedicle		
Epifaunal – hard substrate (1) (plenipedunculate)	Orthides, rhynchonellides, spiriferides and terabratulides	
Epifaunal – soft substrate (2) (rhizopedunculate)	*Chlidonophora* and *Cryptopora*	
Cryptic	*Argyrotheca* and *Terebratulina*	
Interstitial	*Acrotretides* and *Gwynia*	
Cemented	*Craniops* and *Schuchertella*	
Encrusting (3)	Craniids and disciniids	
Clasping spines (4)	*Linoproductus* and *Tenaspinus*	
Mantle fibers	Orthotetoids	
Unattached		
Cosupportive (5)	Pentamerids and trimerellids	
Coral-like (6)	Gemmellaroids and richthofeniids	
Recumbent	Strophomenides	
Pseudofaunal (7) and inverted (8)	*Waagenoconcha* and *Marginifera*	
Free-living (9, 10)	*Cyrtia, Chonetes, Neothyris* and *Terebratella*	
Mobile		
Infaunal (11)	Linguloids	
Semi-infaunal (12)	*Camerisma* and *Magadina*	

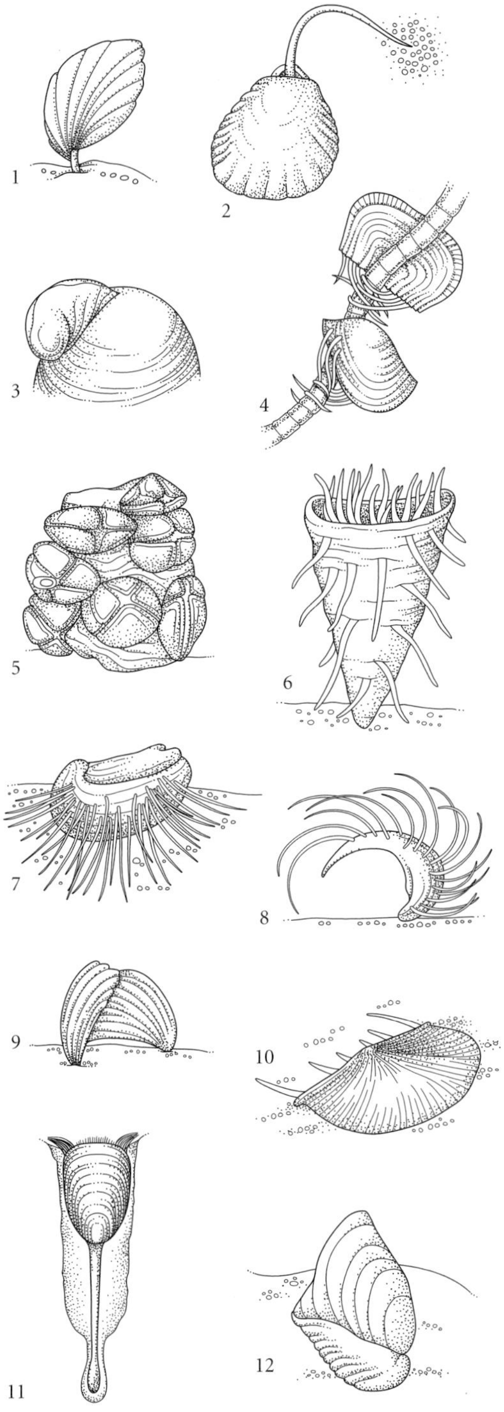

Figure 12.9 Brachiopod lifestyles. (Courtesy of David Harper and Roisin Moran.)

number of quite different non-articulated and articulated taxa were cemented to the substrate, whereas some groups evolved clasping spines to help stabilize their shells. In a number of groups the pedicle atrophied during ontogeny. Many taxa thus developed strategies involving inverted, **pseudoinfaunal** and **recumbent** life modes; a number lived in **cosupportive** clusters and others mimicked corals. Not all brachiopods were sessile; a few, such as *Lingula*, adopted an **infaunal** lifestyle (Box 12.3), whereas the articulated forms *Camerisma* and *Magadina* were semi-infaunal.

Throughout the Phanerozoic the brachiopods have participated in a spectrum of level-bottom, benthic paleocommunities. Pioneer studies on Silurian brachiopods suggested that their paleocommunities were depth related, and a predictable succession of faunas, each characterized by one or more key brachiopods, has been identified (Fig. 12.11). The onshore–offshore assemblages of the *Lingula*, *Eocoelia*, *Pentamerus*, *Stricklandia* (or its close relative *Costistricklandia*) and *Clorinda* paleocommunities, first identified in the Silurian of Wales, form the basis of benthic assemblage (BA) zones 1–5, ranging from intertidal environments to the edge of the continental slope; more basinal environments are included in BA6. Parallel studies on Mesozoic brachiopods have, on the other hand, suggested that brachiopod-dominated paleocommunities were controlled by substrate rather than depth (Fig. 12.12). Clearly, in reality, a combination of these and other factors controlled the distributions of the Brachiopoda in a complex system of suspension-feeding guilds.

Brachiopods have also acted as substrates for a variety of small epifaunal animals (see p. 97). The progressive and sequential colonization of Devonian spiriferids, by *Spirorbis*, itself a possible lophophorate (Taylor & Vinn 2006), *Hederella*, *Paleschara* and *Aulopora* marked the development of an eventual climax paleocommunity on the actual brachiopod shell itself. Were they feeding on incoming brachiopod food or just waste? The general view is that these animals congregated beside the inhalant currents on the median parts of the anterior commissure, and benefited from the indrawn particles of food. An alternative view, and it is hard to prove or disprove, is that they were taking advantage of waste being ejected from the brachiopod.

Brachiopods not only acted as suitable substrates for an epifauna, they were also prone to attack (Box 12.4) and drill holes suggest predation and in some cases attachment of other brachiopods themselves (Robinson & Lee 2008).

Brachiopods, functional morphology and paradigms

Martin Rudwick, an English brachiopod expert just beginning his career in the 1960s (he is now a distinguished historian of geology), proposed the **paradigm approach** in functional interpretation of fossils. His idea was to create an engineering model for a function, such as water flow in feeding. For example, does the **costation**, the zig-zag pattern of ridges and furrows, of the anterior commissure of the brachiopod have a real functional significance? In numerical terms it can be shown that costation increases the length of commissure and hence the intake area that may be held open without increasing

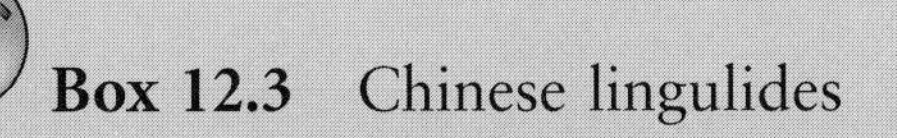

Box 12.3 Chinese lingulides

Did early lingulides live in burrows like many of their descendants? *Xianshanella haikouensis* from the Lower Cambrian Chengjiang fauna, South China was a subcircular animal with horny setae and a massive pedicle. Zhang Zhifei and his colleagues (2006) have shown that these earliest brachiopods did not live in burrows, but actually attached themselves to the shells of other invertebrates – an epibenthonic rather than infaunal mode of life (Fig. 12.10). Moreover the Chengjiang lingulide has a lophophore, a U-shaped digestive tract and an anteriorly-located anus; these advanced features were already present in the lingulate brachiopod lineage right from the start it seems.

Continued

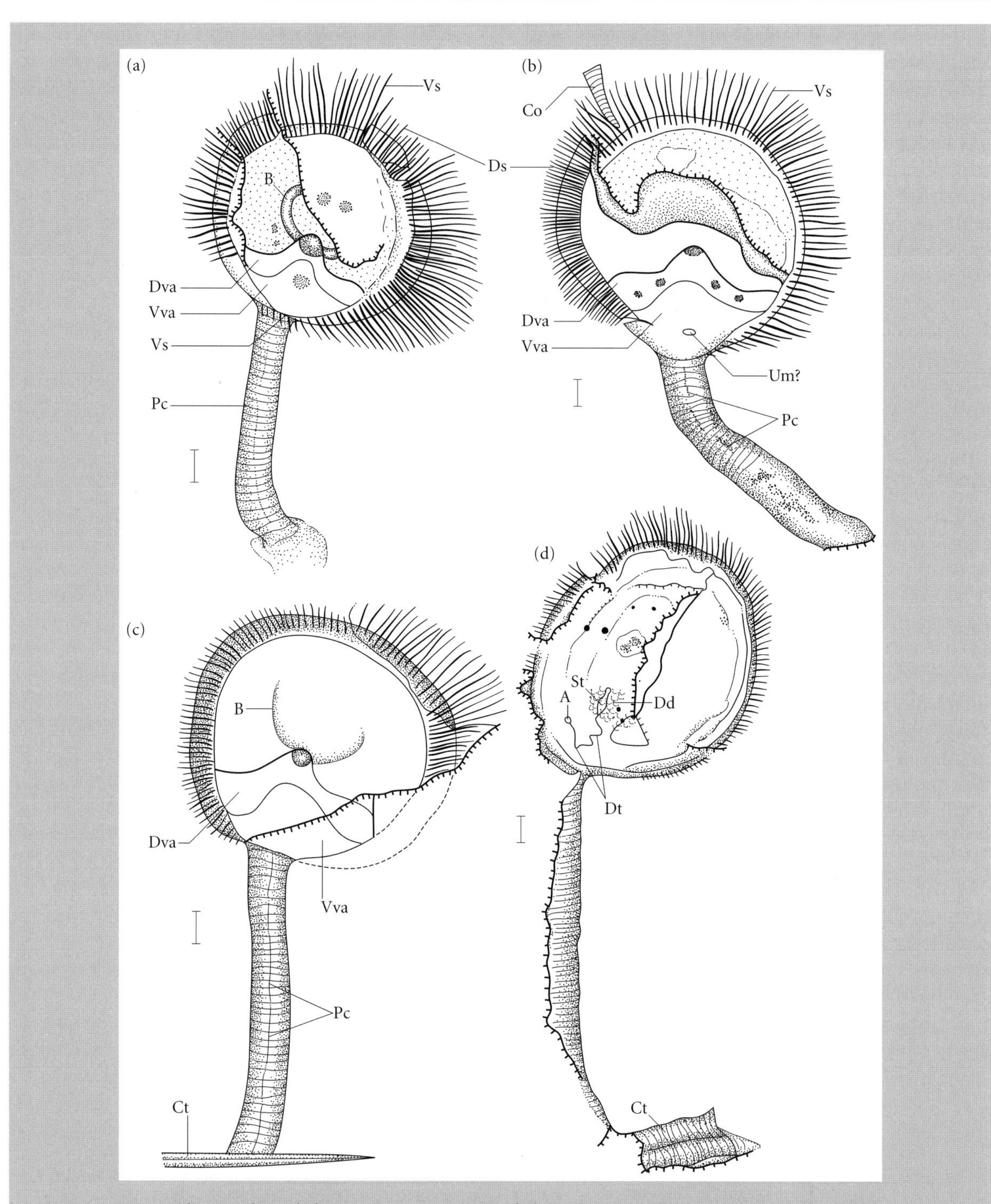

Figure 12.10 Chinese lingulides: Reconstruction of the Chengjiang lingulid *Xianshanella*. A, anal opening; B, brachial arm; Co, cone-like organisms; Ct, cheek of trilobite; Dd, digestive tract; Dva, dorsal visceral area; Pc, pedicle cavity; St, stomach; Um?, possible umbonal muscle; Vs, setae fringing ventral valve; Vva, ventral visceral area. Scale bars, 2 mm. (From Zhang et al. 2006.)

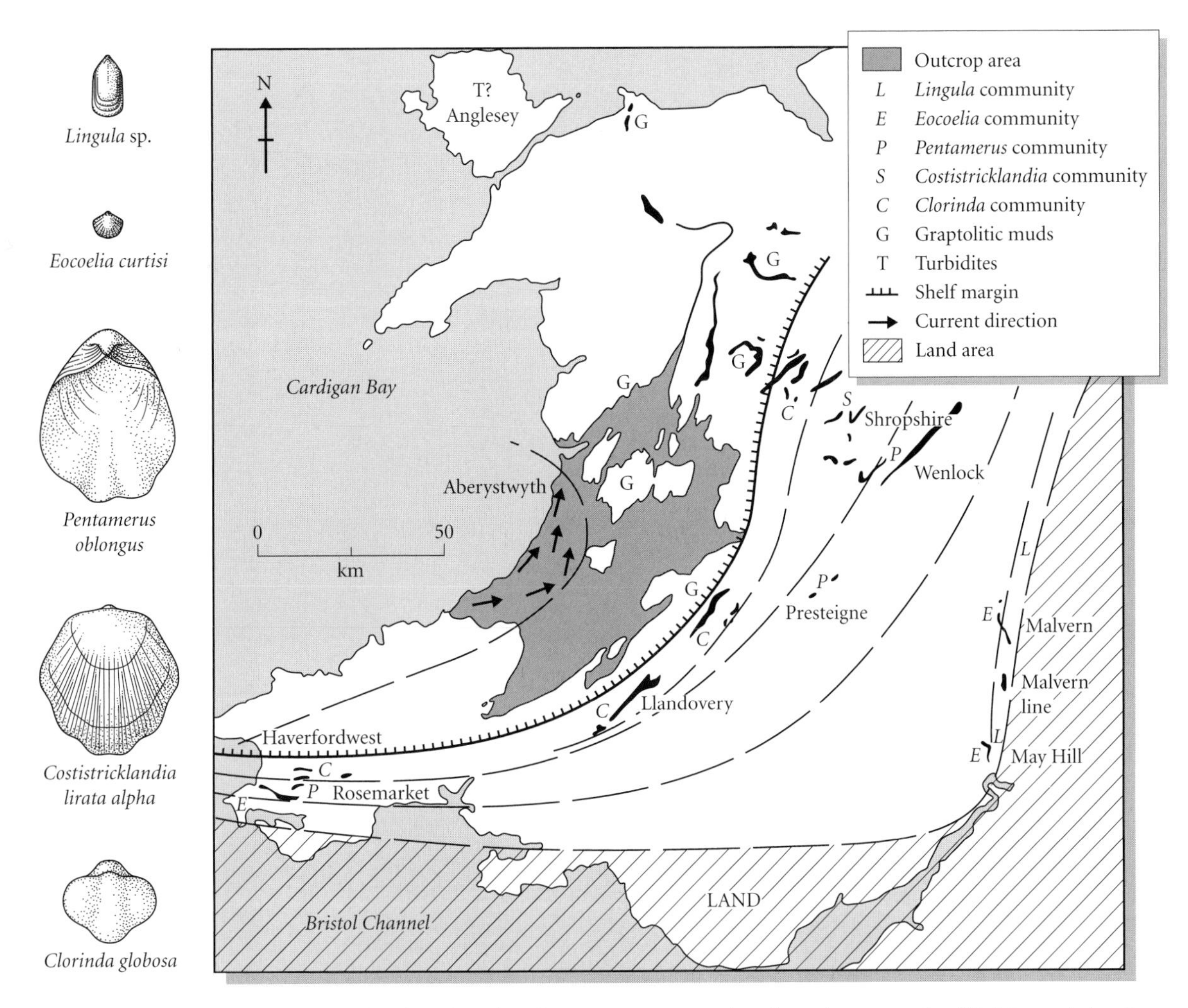

Figure 12.11 Lower Silurian depth-related paleocommunities developed across the Welsh and Anglo-Welsh region. (Based on Clarkson 1998.)

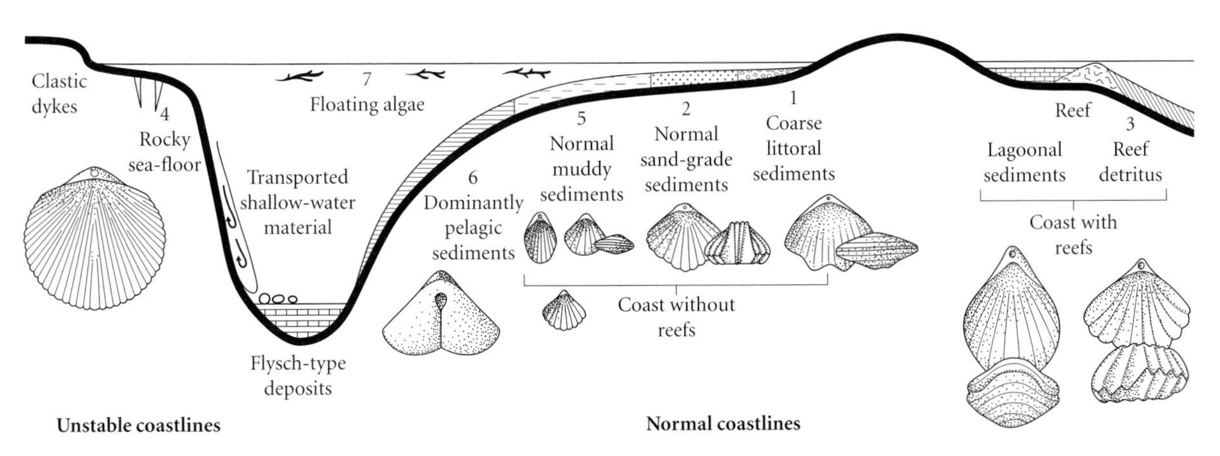

Figure 12.12 Mesozoic palaeocommunities developed across Alpine Europe. Numbers 1 to 7 refer to the seven different biotypes described on the figure. (Based on Ager, D.V. 1965. *Palaeogeogr. Palaeoclimatol. Palaeoecol.* **1**.)

Box 12.4 Brachiopod predation

Brachiopods were eaten by gastropods, arthropods and other predators, and the best evidence is found in Paleozoic examples, especially in the Devonian. Many predatory gastropods feed by drilling into the shells of their prey, and two types of drill hole are commonly present in Paleozoic brachiopods: small, cylindrical holes made by *Oichnus simplex* and larger, often beveled holes made by *O. paraboloides*; these are, of course, ichnogenera (see p. 525) and not actual brachiopods (Fig. 12.13). After the Devonian peak in drilling diversity, there was apparently a marked drop in the frequency of drilled shells, particularly after the Mid Carboniferous. Many Carboniferous and Permian groups such as the productides have thickened shells with an armor of frills, lamellae and spines, all perhaps acting as defense against marauders. Maybe the prey had won this early arms race or perhaps the introduction of mollusks into these communities provided fresh and preferable seafood for the predators. Nevertheless, if we use the Recent Antarctic benthos as a model for the Paleozoic fauna, there is a lack of fast-moving **durophagous** predators (Harper 2006). Some authors have speculated that the toxins within the flesh of some modern groups, such as the rhynchonellids, may have protected them from attack.

Figure 12.13 Brachiopod predation: boring of *Oichnus paraboloides* in the conjoined valves of *Terebratulina* from the Pleistocene rocks of Barbados. Scale bar is in millimeters. (Courtesy of Stephen Donovan.)

the gape of the two shells. Thus an increased volume of nutrient-laden fluid may flow into the mantle cavity while grains of sediment with diameters exceeding the shell gape will still be excluded. So far so good.

During the Permian, a group of aberrant productoids, called the richthofeniids, mimicked corals and built biological frameworks that may be found as fossils in the Salt Ranges of Pakistan and the Glass Mountains of Texas. These brachiopods have a cylindrical pedicle valve attached to the substrate and a small, cap-like brachial valve. It is difficult to understand how these animals fed. A possible scenario involves the flapping of the upper, brachial valve to generate currents through the brachiopod's mantle cavity. Rudwick filmed the flow of water through the cylindrical, lower, pedicle valve as the upper valve was moved up and down. Fluid did in fact move efficiently through the animal, bringing in nutrients and flushing out waste. The paradigm, however, failed the test of field-based evidence. Specimens of the athyride *Composita* apparently in life position occur attached to the upper valve of the richthofeniid. Vigorous flapping of the valve was thus unlikely and it would not have been an ideal attachment site for an epifauna. Rather, these aberrant animals may have developed lophophores with a ciliary pump action to move currents through the valves. One hypothesis has been rejected, and another stands as a possibility – we cannot prove how the richthofeniid brachiopods functioned, but the paradigm approach offers a reasonably objective way for paleontologists to approach these problems.

Distribution in space: biogeography

The biogeographic patterns of the linguliformean brachiopods were quite different from those of the craniiformeans and rhynchonelliformeans. The former had planktotrophic larval phases (see p. 241) with a facility for wide dispersal; in contrast the lecithotrophic larvae of the latter were short-lived and thus individual species were less widely distributed. Cambrian brachiopods were organized into tropical and polar realms. Linguliformeans developed widespread distributions in shelf and slope settings; rhynchonelliformeans were more diverse in the tropics, preferring shallow-water carbonate and mixed carbonate-siliciclastic environments. In the Ordovician, brachiopod provincialism generally decreased during the period. Provinciality was most marked during the Early Ordovician, when a range of platform provinces associated with the continents of Baltica, Gondwana, Laurentia and Siberia (see Appendix 2) were supplemented by centers of endemism associated with a range of microcontinents and volcanic arcs and island complexes.

Provincialism was reduced during the Silurian with the close proximity of many major continents. By the Wenlock, however, two broad provinces, the cool-water *Clarkeia* and the mid-latitudinal *Tuvaella* faunas, emphasized an increasing endemism, climaxing during the Ludlow and Prídolí epochs. Provinciality was particularly marked during the Mid Devonian coincident with peak diversities in the phylum. Clear biogeographic patterns continued into the Carboniferous, but the Permian was characterized by higher degrees of provinciality probably associated with steep climatic gradients.

During the Triassic, brachiopod faunas, following an interval of cosmopolitan disaster taxa, became organized into Boreal (high-latitude) and Tethyan (low-latitude) realms (Box 12.5). This pattern continued throughout the Mesozoic, but with centers of endemism and occasional modifications due to ecological factors such as the circulation of ocean currents and the local development of chemosynthetic environments. Biogeographic patterns among living forms reflect their Cenozoic roots: a southern area, the northern Pacific, and a northern area (Atlantic, Mediterranean, North Sea and the circumpolar northern oceans) are based on a variety of articulated brachiopod associations. The linguliformeans have more widespread, near-cosmopolitan distributions.

BRYOZOA

> *Besides these, there were the Bryozoa, a small kind of Mollusk allied to the Clams, and very busy then in the ancient Coral work. They grew in communities, and the separate individuals are so minute that a Bryozoan stock looks like some delicate moss. They still have their place among the Reef-Building Corals, but play an insignificant part in comparison with that of their predecessors.*
>
> *Atlantic Monthly* (April, 1863)

Box 12.5 Tethyan brachiopods in Greenland: a Cretaceous Gulf Stream current?

Brachiopods can give clues about ancient ocean currents. Today, the Gulf Stream runs out from the Caribbean, sweeps up the eastern seaboard of North America, and then detaches from the coast just north of New York and heads across the Atlantic to wrap the shores of Britain and western Europe in warmer-than-expected waters. Has the Gulf Stream always flowed the same way? Some Cretaceous brachiopods give us a clue. David Harper and colleagues (2005) showed how some Early Cretaceous brachiopod faunas from East Greenland were a mix of animals from two ocean provinces, Tethyan (low latitude) and Boreal (high latitude). The Boreal, shallow-water assemblage is dominated by large terebratulids and ribbed rhynchonellids, and occurs adjacent to a fauna containing Tethyan elements, more typical of deeper water, including *Pygope* (see p. 311). How did these exotic, tropical visitors travel so far north? Harper and colleagues suggested that an Early Cretaceous out-of-Tethys migration was helped by the early and persistent northward track of a proto-Gulf Stream current (Fig. 12.14). These kinds of studies of changing patterns of paleobiogeography through time are critical for understanding modern climate and ocean patterns.

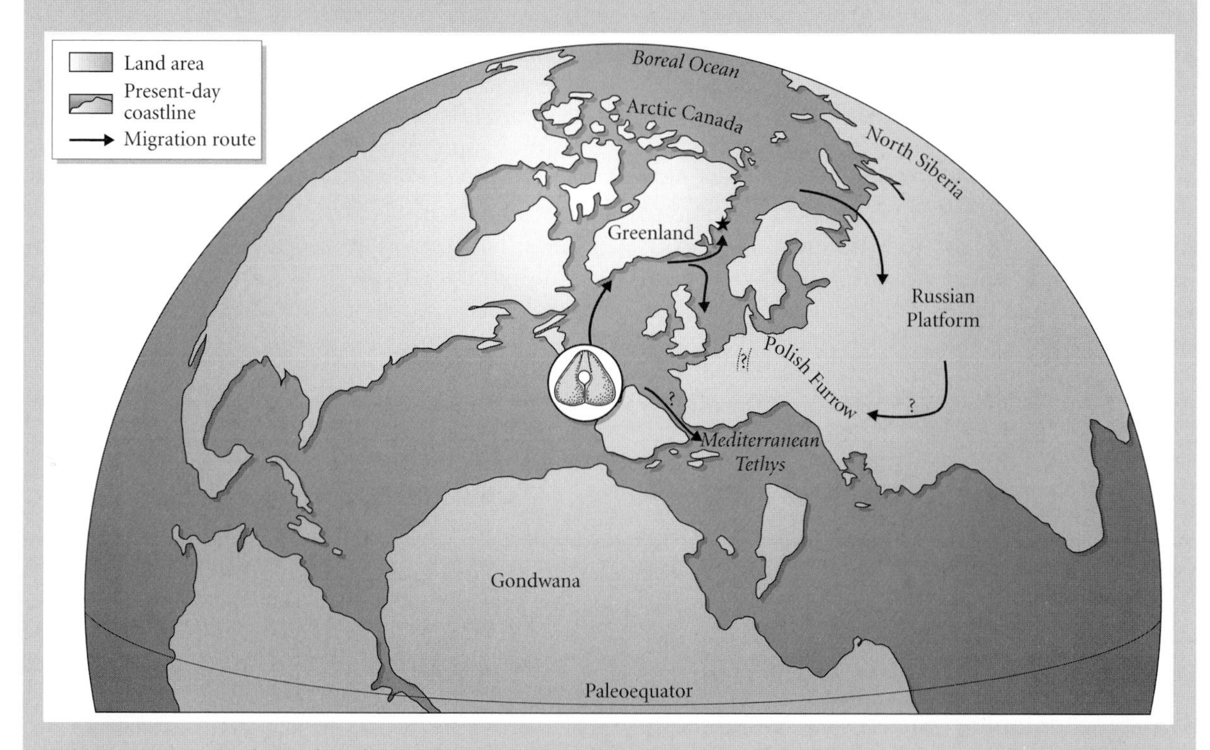

Figure 12.14 Tethyan brachiopods in East Greenland: *Pygope* and the proto-North Atlantic current (arrows), one of its possible migration routes. The star indicates the Lower Cretaceous, East Greenland locality.

Bryozoans are the only phylum in which all species are colonial. Many skeletons are exquisitely designed, but fragment very easily after death. Although relatively common, bryozoans are among the least well-known invertebrates. There are about 6000 living and 16,000 fossil species, and most are marine (Box 12.6). Superficially resembling the corals and hydroids, the bryozoans ("moss animals") are like minute colonial phoronids (see p. 298) with tiny individuals or **zooids**, commonly less than 1 mm in diameter. Each zooid is **celomate** with a separate mouth and anus together and a circular or horseshoe-shaped lophophore equipped with a ring of 8–100 tentacles – a major organizational jump from the cnidarians. The bryozoan lophophore is constructed differently from those of the brachiopods and phoronids and it may be a mistake to think that all three groups are closely related just because they possess ciliated feeding organs. Individual zooids are enclosed by a gelatinous, leathery or calcareous exoskeleton, usually in the form of slender tubes or box-like chambers called **zooecia**. The primary function of most zooids is the capture of food, but some are specialists in defense, reproduction or sediment removal; the bryozoan colony thus functions as a well-organized unit.

Morphology: *Bowerbankia*

The genus *Bowerbankia* is a relatively simple bryozoan useful for illustrating the general anatomy of bryozoan zooids (Fig. 12.15). Each living zooid is enclosed by a body wall or **cystid**. The lophophore, with its beating cilia, extends outwards from the zooid and comprises a ring of 10 tentacles, directing food to a central mouth leading into a U-shaped gut; the feces finally exit out through an anus. A **funiculus** extends along the **stolon** connecting all the zooids. This is thought to be a homolog of the blood vessels found in other animals. The individual zooids are hermaphrodites, developing eggs and sperm at different times; the eggs are usually fertilized in the tentacle sheath, developing later into **trochophore** larva.

Ecology: feeding and colonial morphology

Feeding strategies of bryozoans have had a major influence on the style of colony growth. Feeding behavior patterns are correlated with the shape of the colony and the size of the zooids. Bryozoan colonies can grow in a variety of modes from encrusting runners, uniserial or multiserial branches that split, and sheets where growth occurs around the entire margin, to more erect type forms that have complex three-dimensional morphologies (Box 12.7). Many elegant forms have evolved such as the bush- and tree-like trepostomes of the Paleozoic, the spiral *Archimedes* and vase-shaped *Fenestrella*, in both of which the entire colony may have acted like a sponge. But bryozoan colonies can also move. For example, colonies of *Selenaria* can scuttle across the seafloor. Stilt-like appendages or setae project downwards from specialized zooids and as the setae move in waves, the colony is transported across the seabed. Such a lifestyle can be traced back to the Late Cretaceous when free-living colonies, the so-called **lunulitiforms**, evolved their regular shape, without interference from adjacent objects on the seafloor.

Zooid size can give important clues about environment and particularly water temperature. Increased ranges of seasonal variation in temperature seem to be correlated with an increased amount of variation in the size of zooids in the colony (O'Dea 2003). It is not clear why there is this relationship, but nevertheless zooid size may also be a useful environmental proxy.

Evolution: main fossil bryozoan groups

The oldest bryozoans in the fossil record occur in the Tremadocian Stage of the Lower Ordovician, but it is very likely that primitive, soft-bodied bryozoans existed during the Cambrian but have not been fossilized; numerous families of bryozoans are found in the succeeding Floian Stage. The Stenolaemata dominated Paleozoic bryozoan faunas (Fig. 12.17). The trepostomes or stony bryozoans commonly had bush-like colonies with prismatic zooecia having polygonal apertures. The group diversified during the Ordovician to infiltrate the low-level benthos. Genera such as *Monticulipora*, *Prasopora* and *Parvohallopora* are typical of Ordovician assemblages.

The cryptostomes, although originating during the Early Ordovician, were more abundant during the Mid and Late Paleozoic as the trepostomes declined; in some respects the group forms a link with the net-like fenestrates that were particularly common in the Carboniferous (Fig. 12.18). *Fenestella*, itself, may be in the form of a planar mesh, cone or funnel. The branches of the colony are connected by dissepiments; rectangular spaces or fenestrules separate the branches that contain the biserially-arranged zooids. *Archimedes*, however, has a meshwork wound around a screw-shaped central axis. Richard Cowen and his colleagues (University of California)

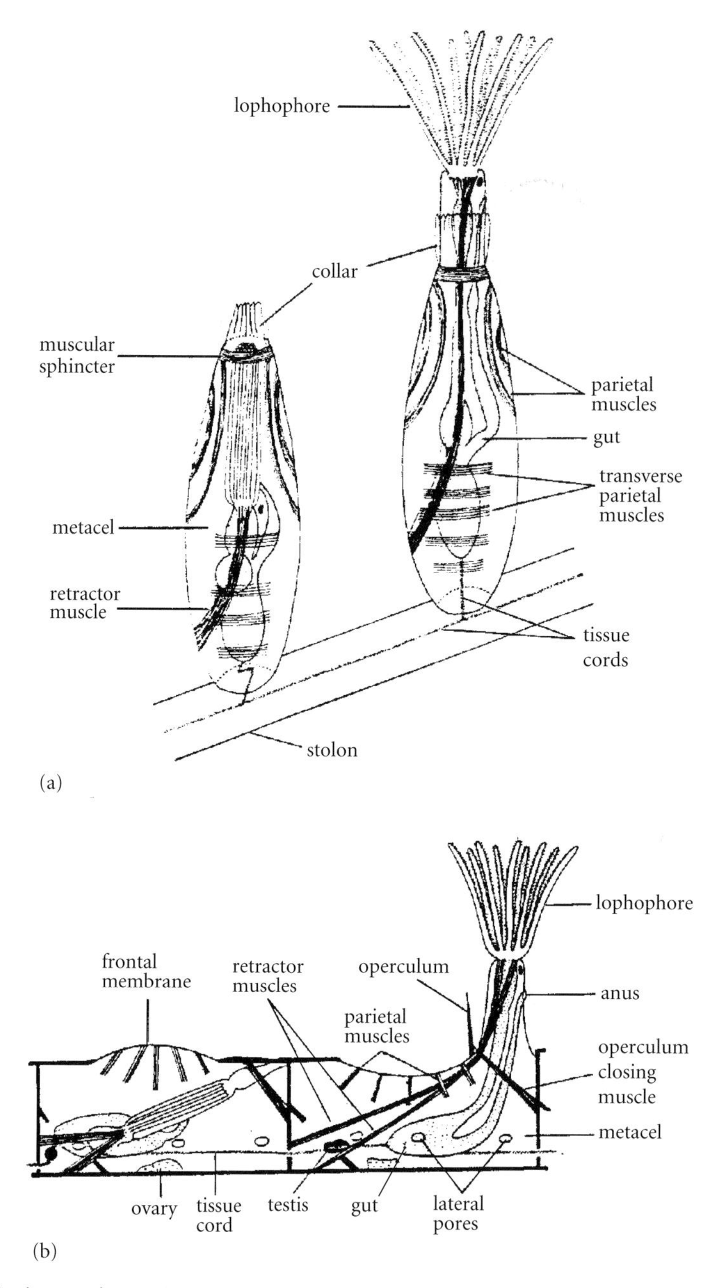

Figure 12.15 Morphology of two living bryozoans: (a) a stenolaemate and (b) a gymnolaemate. (Based on various sources.)

have modeled the feeding strategies of these screw-shaped colonies and other fenestrates. Carboniferous fenestrate colonies usually had inward-facing zooids and probably drew water in through the top of the colony and flushed it out through the fenestrules at the sides. On the other hand, Silurian colonies had outward-facing zooids and sucked in water through the fenestrules, expelling it out of the open top of the colony.

In general both the cryptostomes and fenestrates outstripped the trepostomes during the Late Paleozoic, many of the fenestrates populating reef environments. Although both groups disappeared at the end of the Permian or soon after, they were still conspicuous members of the Late Permian benthos; both *Fenestella* and *Synocladia* form large, vase-shaped colonies in the communities of the Zechstein reef complex in the north of England

Box 12.6 Bryozoan classification

Class PHYLACTOLAEMATA

- Cylindrical zooids with horseshoe-shaped lophophore. Statoblasts arise as dormant buds. Freshwater with non-calcified skeletons. Over 12 genera
- Triassic, possibly Permian to Recent

Class STENOLAEMATA

- Cylindrical zooids with calcareous skeleton. Membraneous sac surrounds each polypide; lophophore protrudes through an opening at the end of the skeletal tube. Marine, with an extensive fossil record. Contains the following orders: trepostomes (Ordovician–Triassic), cystoporates (Ordovician–Triassic), cryptostomes (Ordovician–Triassic), cyclostomes (Ordovician–Recent) and fenestrates (Ordovician–Permian). About 550 genera
- Ordovician (Tremadoc) to Recent

Class GYMNOLAEMATA

- Cylindrical or squat zooids of fixed size with circular lophophore, usually with a calcareous skeleton. The majority are marine but some are found in brackish and freshwater environments. Includes the cheilostomes (Jurassic–Recent). Over 650 genera
- Ordovician (Arenig) to Recent

Box 12.7 Module iteration: building a Lego bryozoan

Bryozoan colonies grow by iteration, repeating the same units again and again until the colony is built. But is this process just a simple addition of individual units (zooids) within the colony? If so, the opportunity for evolution and morphological complexity would be very limited. There may be a whole hierarchy of types of modules that are in fact iterated (repeatedly re-evolved). For example, much more variability will be generated if a branch rather than a zooid is duplicated and attached to various parts of the colony in various different orientations. Steven Hageman of Appalachian State University suggested just this in a paper published in 2003: there is a hierarchy of such modules and those second-order blocks will have a much greater effect on morphological change and evolution of the colony than simply duplicating the zooids. This can be easily demonstrated by an analogy with a Lego model. The individual blocks, if iterated, will form only fairly simple patterns, but build a structure and iterate that and suddenly considerable morphological complexity can be generated from relatively simple building blocks (Fig. 12.16).

Continued

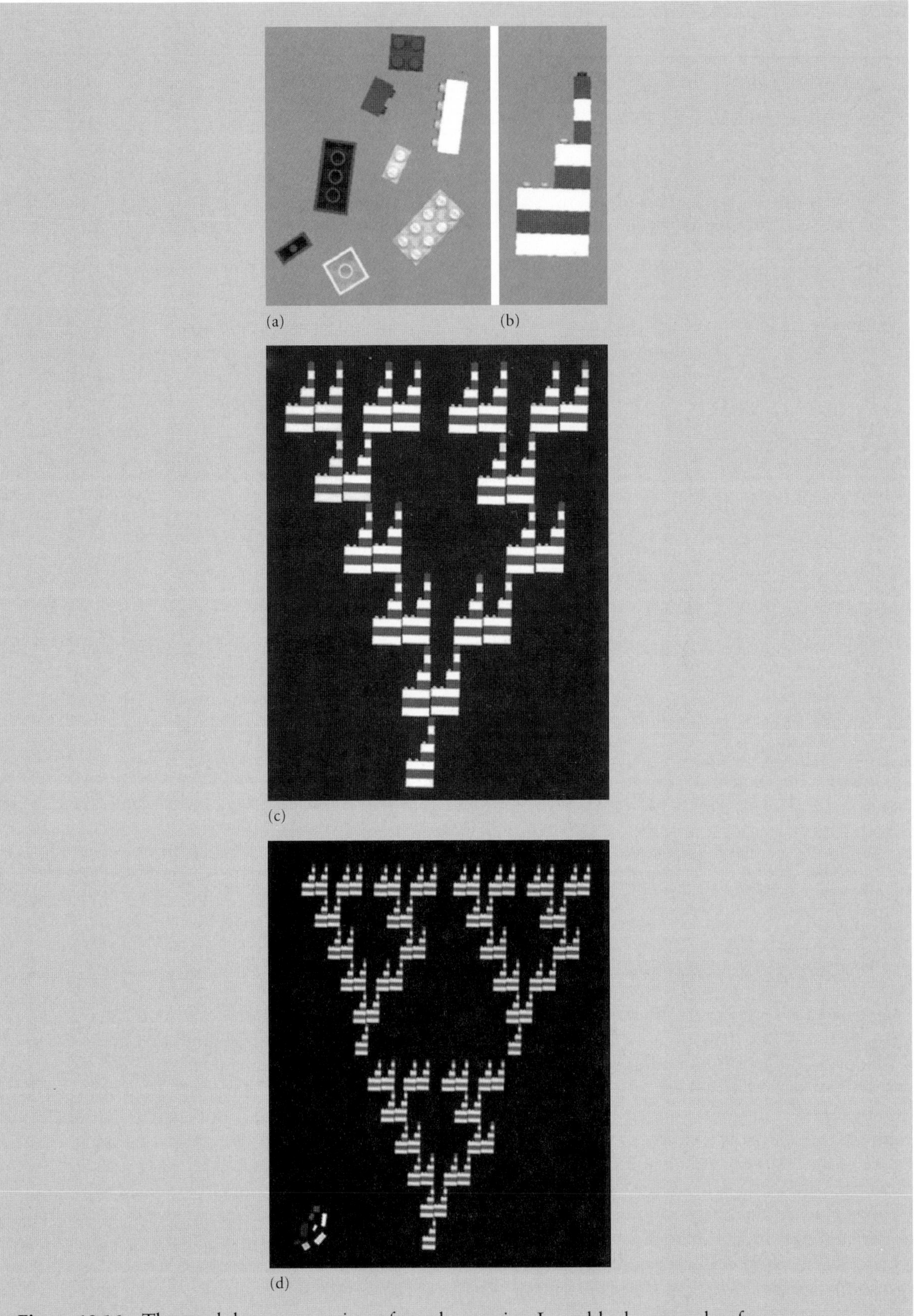

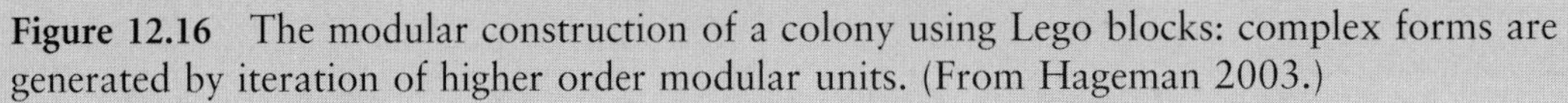

Figure 12.16 The modular construction of a colony using Lego blocks: complex forms are generated by iteration of higher order modular units. (From Hageman 2003.)

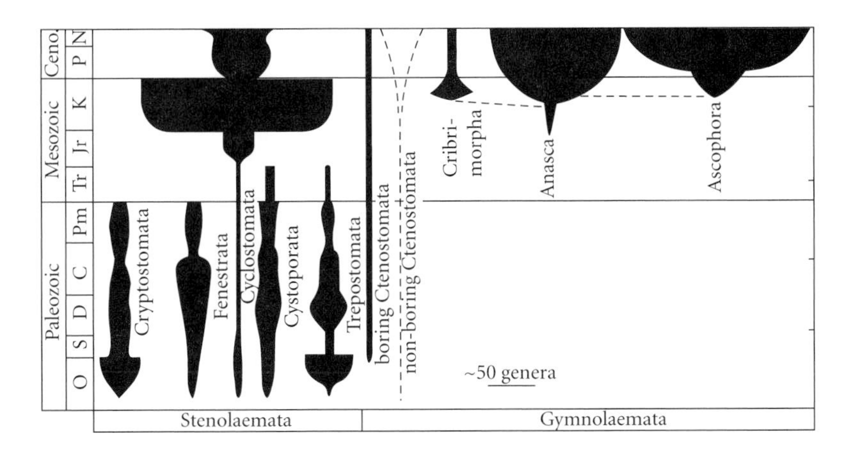

Figure 12.17 Stratigraphic ranges and absolute abundances of the main bryozoan groups. Geological period abbreviations are standard, running from Ordovician (O) to Neogene (N). (From Taylor, 1985.)

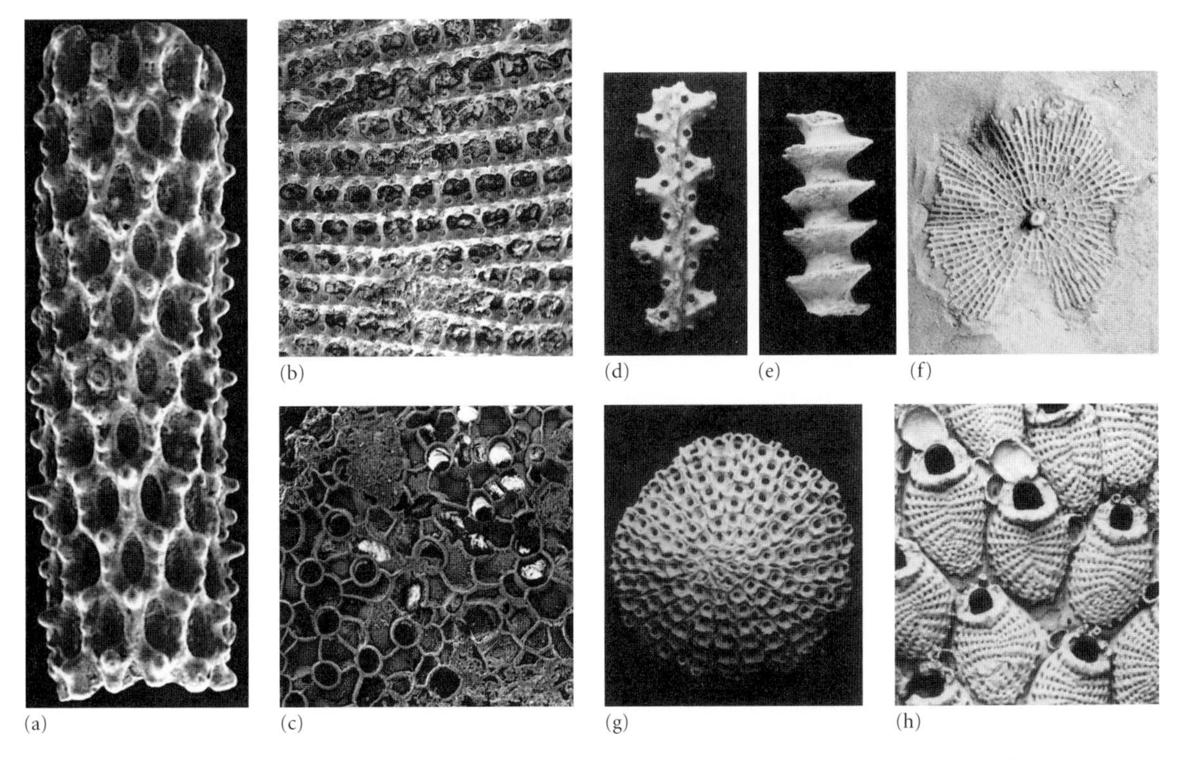

Figure 12.18 Some bryozoan genera: (a) *Rhabdomeson* (Carboniferous cryptostome), (b) *Rectifenestella* (Carboniferous fenestrate), (c) *Fistulipora* (Carboniferous cystopore), (d) *Penniretepora* (Carboniferous fenestrate), (e) *Archimedes* (Carboniferous fenestrate), (f) *Archaeofenestella* (Silurian fenestrate), (g) *Lunulites* (Cretaceous cheilostome), (h) *Castanapora* (Cretaceous cheilostome). Magnification approximately ×30 (a), ×15 (b, c), ×1 (d–f), ×5 (g), ×20 (h). (a–c, courtesy of Patrick Wyse Jackson; d–h, from Taylor 1985.)

and elsewhere. The trepostomes, however, lingered on until the Late Triassic.

The cyclostomes have tube-shaped zooecia and often grew as branching tree-like colonies or alternatively encrusting sheets or ribbons. The first representatives of the order are known in Lower Ordovician rocks, but the group peaked during the mid-Cretaceous in spectacular style, with a diversity of over 70 genera. Many genera such as *Stomatopora*, consisting of a series of bifurcating, encrusting branches, have very long stratigraphic ranges; moreover *Stomatopora* may have pursued an opportunist life strategy, rapidly spreading their zooids over hard surfaces.

The Gymnolaemata are represented in the fossil record by two orders, the ctenostomes and the cheilostomes. The ctenostomes first appeared in the Early Ordovician and many genera have since pursued boring and encrusting life strategies. *Penetrantia* and *Terebripora* are borers whereas the modern genus *Bowerbankia* has an erect colony with semi-spirally arranged zooecia clustered around a central branch. The cheilostomes, however, dominate the class and are most diverse of all the bryozoan groups (Box 12.8). Cheilostomes typically have polymorphic zooids, adapted for different functions, which are usually linked within the highly integrated colony. This advanced group appeared during the Late Jurassic; they are particularly common in shallow-water environments of the Late Cretaceous and Paleogene of the Baltic and Denmark. *Lunulites*, for example, is discoidal and free-living, whereas *Aechmella* is an encrusting form often associated with sea urchins.

Ecology and life modes

Virtually all bryozoans are part of the sessile benthos, mainly occurring from the sublitto-

Box 12.8 Competition and replacement in cyclostome and cheilostome clades: what really happened at the KT boundary?

Perhaps one of the most obvious changes in bryozoan faunas through time involves the relative decline of the cyclostomes and the diversification of the cheilostomes leading up to the Cretaceous–Tertiary (KT) boundary. Since both groups occupied similar ecological niches and are comparable morphologically, many workers have assumed that the cyclostomes, originating during the Ordovician and diversifying in the Cretaceous, were outcompeted by the cheilostomes at the end of the Cretaceous. However Scott Lidgard (Field Museum of Natural History, Chicago) and his colleagues have analyzed this transition in detail and the results are far from conclusive (Lidgard et al. 1993). Both groups continued to participate together in bryozoan communities during the Cenozoic and much of the apparent decline in the cyclostome numbers may be due to the greater diversification or expansion of the cheilostomes that began to dominate these assemblages in the Cenozoic. Perhaps this expansion had already been seeded in the Jurassic, when the poor and sporadic bryozoan fauna provided the ecological space for the expansion of the cheilostomes. A detailed statistical study based on generic-range data from Sepkoski's database (McKinney & Taylor 2001) has confirmed that origination within the cheilostome clades was the driving force behind the apparent takeover by this group (Fig. 12.19).

See http://www.blackwellpublishing.com/paleobiology/.

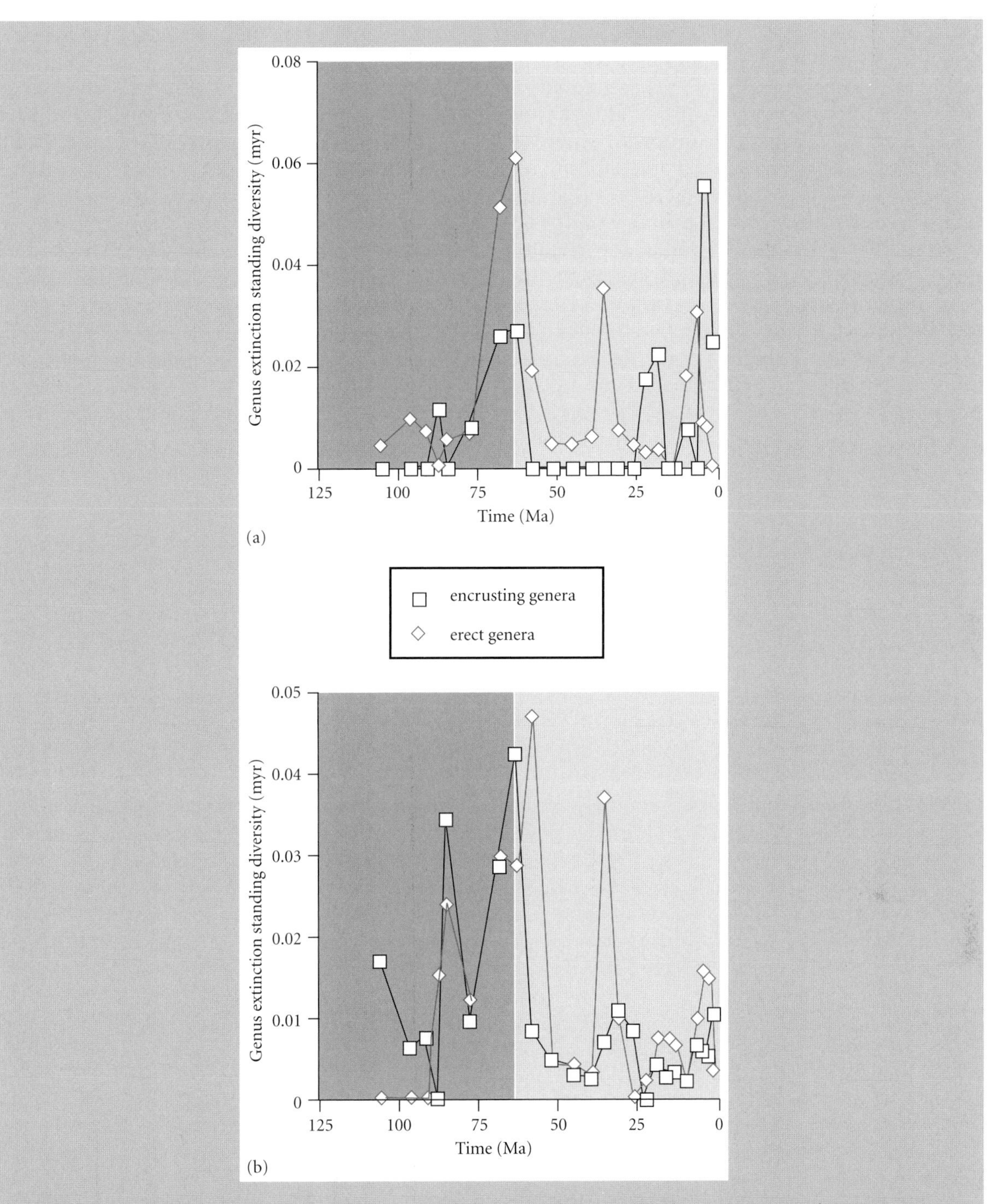

Figure 12.19 Distribution of (a) cyclostome and (b) cheilostome bryozoans across the Mesozoic–Cenozoic boundary: the cheilostomes suffered the heaviest losses while the erect genera of both groups suffered more than the encrusters. (Replotted from McKinney & Taylor 2001.)

ral zone to the edge of the continental shelf at depths of about 200 m. Nevertheless a few intertidal forms are known, while some bryozoans have been dredged from depths of over 8 km in oceanic trenches; moreover numerous species have been recorded from the hulls of ships. Most species are sensitive to substrate types, turbulence, water depth and temperature together with salinity. The shape of colonies can be very plastic, adapting to environmental conditions, with erect, tree-like colonies varying their branch thickness according to depth. In addition spines may be induced by high current velocities or by the presence of predators (Taylor 2005). Bryozoans are thus typical facies fossils exhibiting marked ecophenotypic variation (Box 12.9).

Bryozoans have successfully pursued several different life modes. Encrusting, erect, unattached or rooted phenotypes all reflect adaptive strategies in response to ambient environmental conditions. Shallow-water colonies, particularly in the subtidal zone, are and were dominated by encrusting, erect, rooted and free-living forms. But deeper-water environments, over 1 km deep, are characterized by mainly attached and rooted forms. Nevertheless bryozoan colonies have occasionally formed reefs or bryoherms, particularly during the mid-Silurian and Carboniferous.

Box 12.9 Bryozoans and environments

The majority of bryozoans grow as mounds, sheets or runners parallel to the substrate, many grow erect colonies perpendicular to the seabed and some colonies are actually mobile. There have been a number of growth–mode type classifications, some associated with particular genera, constructional geometry or based on autecology. A more comprehensive way at looking at these complex colonies is to combine attachment modes, construction orientation and the geometry of the individual zooids (Hageman et al. 1997). Such a hierarchical growth–mode classification can be used to describe regional biotas and predict paleoenvironments on limited datasets. However, as in many ecological studies, the most common species or growth forms can swamp the overall ecological signal; some form of scaling is needed. We can ask a couple of questions: How important is D at locality 1 relative to other occurrences of D and how important is D relative to all the other localities? Firstly a simple data table is set up with growth forms along the y-axis and localities along the x-axis (see below). One method of standardizing the data is to: (i) divide the number of growth type D at locality 1 by the product of all the different growth types and the total at this one locality [10/(45 * 22)]; and (ii) this is then multipled by 100^2 to scale values to roughly between 0 and 100. This equals 101; this growth is clearly important at this locality. The relative importance of each growth form at each locality can be plotted in a histogram.

	Form A	Form B	Form C	Form D	Sum
Locality 1	20	10	5	10	45
Locality 2	20	40	10	5	75
Locality 3	20	40	40	5	105
Locality 4	20	120	60	2	202
Sum	80	210	115	22	

This type of study has been expanded to an analysis of the distribution of growth forms across the shelf-slope transition on the Lacepede Platform, southern Australia. A distinct pattern emerged with free-living forms most important on the inner shelf and rigid cone-disk forms most important on the deep slope (Fig. 12.20).

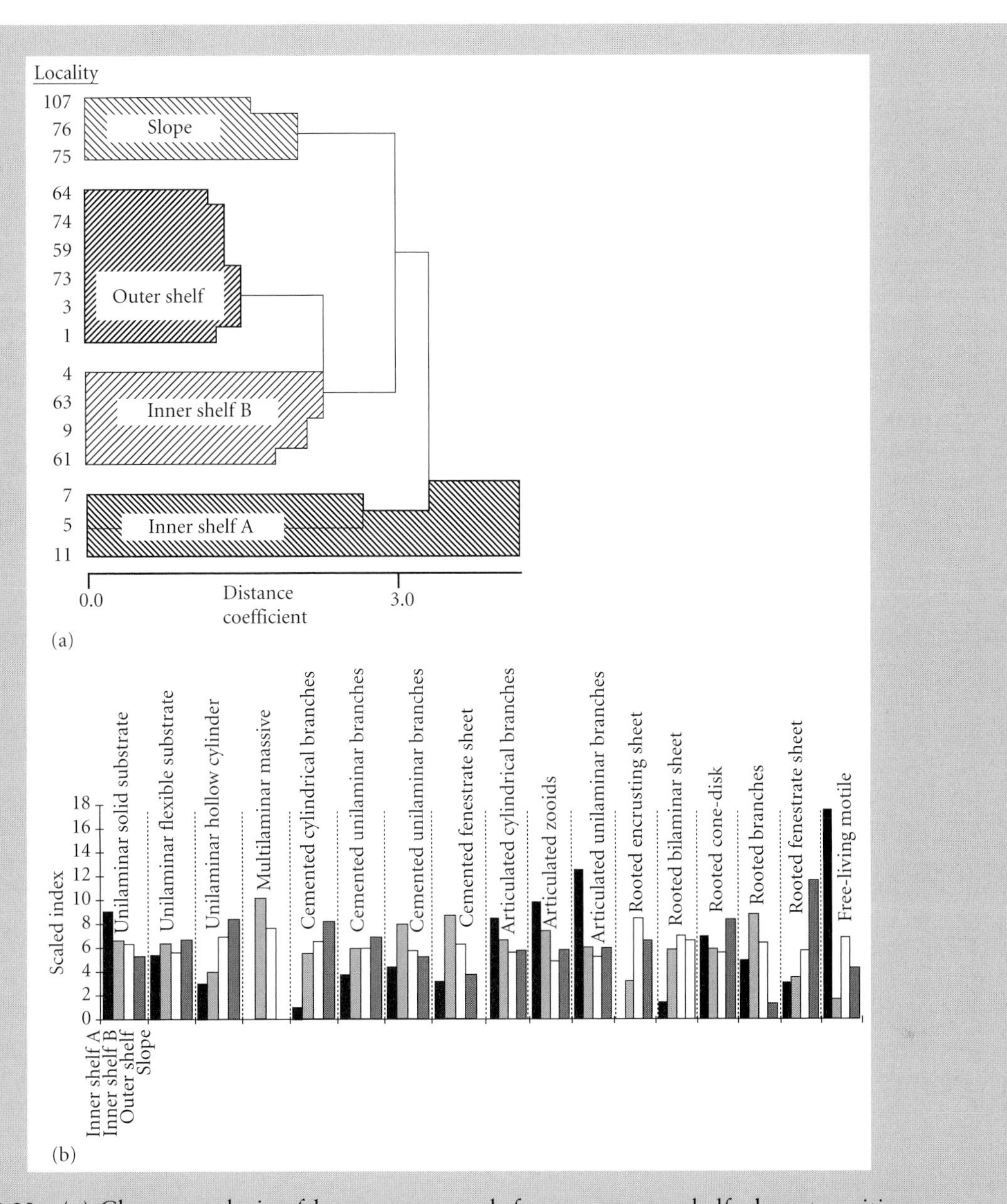

Figure 12.20 (a) Cluster analysis of bryozoan growth forms across a shelf–slope transition, showing an inner shelf A (clastic dominated), inner shelf B (carbonate dominated), outer shelf and slope. The cluster analysis, using a distance coefficient (x-axis) and average group linkage, indicates the presence of four distinctive assemblages. (b) Distribution of growth forms across the onshore–offshore gradient within the assemblages identified by cluster analysis. (Based on Hageman et al. 1997.)

Review questions

1 Current brachiopod research suggests that the phylum Brachiopoda can be split into three subphyla: Linguliformea, Craniiformea and Rhynchonelliformea. What sort of criteria can we use to discover how each subphylum was related to each other and the stem-group brachiopod?
2 Brachiopod shells store a huge amount of data, not only about the secretion of the shell, but also about its surrounding environment. How have brachiopod shells, particularly their stable isotopes, contributed to our understanding of climate change?
3 Although the thick-shelled and ornate productid brachiopods of the Late Paleozoic were resistant to attack, why did brachiopods apparently not feature much in the Mesozoic marine revolution or Mesozoic arms race?
4 The "dawn of the Danian" witnessed a marked change in bryozoan faunas with the dominance of the cheilostomes over the cyclostomes. Both are ecologically similar so why were the cheilostomes relatively more successful after the KT extinction event?
5 Brachiopods and bryozoans were both conspicuous members of the filter-feeding Paleozoic evolutionary fauna. Why then are brachiopods a relatively minor part of the Recent marine fauna but bryozoans continue to flourish?

Further reading

Boardman, R.S. & Cheetham, A.H. 1987. Phylum Bryozoa. *In* Boardman, R.S., Cheetham, A.H. & Rowell, A.J. (eds) *Fossil Invertebrates*. Blackwell Scientific Publications, Oxford, UK, pp. 497–549. (A comprehensive, more advanced text with emphasis on taxonomy; extravagantly illustrated.)

Carlson, S.J. & Sandy, M.R. (eds) 2001. *Brachiopods Ancient and Modern. A tribute to G. Arthur Cooper*. Paleontological Society Papers No. 7. University of Yale, New Haven, CT. (Diverse aspects of contemporary brachiopod research.)

Clarkson, E.N.K. 1998. *Invertebrate Palaeontology and Evolution*, 4th edn. Chapman and Hall, London. (An excellent, more advanced text; clearly written and well illustrated.)

Cocks, L.R.M. 1985. Brachiopoda. *In* Murray, J.W. (ed.) *Atlas of Invertebrate Macrofossils*. Longman, London, pp. 53–78. (A useful, mainly photographic review of the group.)

Harper, D.A.T., Long, S.L. & Nielsen, C. (eds) 2008. Brachiopoda: Fossil and Recent. *Fossils and Strata* **54**, 1–331. (Most recent proceedings from an international brachiopod congress.)

Kaesler, R.L. (ed.) 2000–2007. *Treatise on Invertebrate Paleontology. Part H, Brachiopoda* (revised), vols 1–6. Geological Society of America and University of Kansas, Boulder, CO/Lawrence, KS. (Up-to-date compendium of most aspects of the phylum.)

McKinney, F.K. & Jackson, J.B.C. 1989. *Bryozoan Evolution*. Unwin Hyman, London. (Evolutionary studies of the phylum.)

Rowell, A.J. & Grant, R.E. 1987. Phylum Brachiopoda. *In* Boardman, R.S., Cheetham, A.H. & Rowell, A.J. (eds) *Fossil Invertebrates*. Blackwell Scientific Publications, Oxford, UK, pp. 445–96. (A comprehensive, more advanced text with emphasis on taxonomy; extravagantly illustrated.)

Rudwick, M.J.S. 1970. *Living and Fossil Brachiopods*. Hutchinson, London. (Landmark text.)

Ryland, J.S. 1970. *Bryozoans*. Hutchinson, London. (Fundamental text.)

Taylor, P.D. 1985. Bryozoa. *In* Murray, J.W. (ed.) *Atlas of Invertebrate Macrofossils*. Longman, London, pp. 47–52. (A useful, mainly photographic review of the group.)

Taylor, P.D. 1999. Bryozoa. *In* Savazzi, E. (ed.) *Functional Morphology of the Invertebrate Skeleton*. Wiley, Chichester, UK, pp. 623–46. (Comprehensive review of the functional morphology of the group.)

References

Clarkson, E.N.K. 1998. *Invertebrate Palaeontology and Evolution*, 4th edn. Chapman and Hall, London.

Cohen, B.L., Holmer, L.E. & Luter, C. 2003. The brachiopod fold: a neglected body plan hypothesis. *Palaeontology* **46**, 59–65.

Freeman, G. & Lundelius, J.W. 2005. The transition from planktotrophy to lecithotrophy in larvae of lower Palaeozoic Rynchoneliiform brachiopods. *Lethaia* **38**, 219–54.

Geldern, van, R., Joachimski, M.M., Day, J., Jansen, U., Alvarez, F., Yolkin, E.A. & Ma, X.-P. 2006. Carbon, oxygen and strontium isotope records of Devonian brachiopod shell calcite. *Palaeogeography, Palaeoclimatology, Palaeoecology* **240**, 47–67.

Hageman, S.J. 2003. Complexity generated by iteration of hierarchical modules in Bryozoa. *Integrated Comparative Biology* **43**, 87–98.

Hageman, S.J., Bone, Y., McGowran, B. & James, N.P. 1997. Bryozoan colonial growth-forms as palaeoenvironmental indicators: evaluation of methodology. *Palaios* **12**, 405–19.

Harper, D.A.T., Alsen, P., Owen, E.F. & Sandy, M.R. 2005. Early Cretaceous brachiopods from North-

East Greenland: biofacies and biogeography. *Bulletin of the Geological Society of Denmark* **52**, 213–25.

Harper, E.M. 2006. Dissecting arms races. *Palaeogeography, Palaeoclimatology, Palaeoecology* **232**, 322–43.

Lidgard, S., McKinney, F.K. & Taylor, P.D. 1993. Competition, clade replacement, and a history of cyclostome and cheilostome bryozoan diversity. *Paleobiology* **19**, 352–71.

McKinney, F.K. & Taylor, P.D. 2001. Bryozoan genetric extinctions and originations during the last 100 million years. *Palaeontologia Electronica* **4**, 26 pp.

O'Dea, A. 2003. Seasonality and zooid size variation in Panamanian encrusting bryozoans. *Journal of the Marine Biological Association* **83**, 1107–8.

Perez-Huerta, A., Cusack, M., Zhu, W.-Z., England, J. & Hughes, J. 2007. Material properties of the brachiopod ultrastructure by nanoindentation. *Interface* **4**, 33–9.

Robinson, J.H. & Lee, D.E. 2008. Brachiopod pedicle traces: recognition of three separate types of trace and redefinition of *Podichnus centrifugalis* Bromley & Surlyk, 1973. *Fossils and Strata* **54**, 219–25.

Schumann, D. 1991. Hydrodynamic influences in brachiopod shell morphology of Terebratalia transversa (Sowerby) from the San Juan Islands. *In* MacKinnon, D.I., Lee, D.E. & Campbell, J.D. (eds) *Brachiopods through Time*. A.A. Balkema, Rotterdam.

Taylor, P.D. 1985. Bryozoa. *In* Murray, J.W. (ed.) *Atlas of Invertebrate Macrofossils*. Longman, London, pp. 47–52.

Taylor, P.D. 2005. Bryozoans and palaeoenvironmental interpretation. *Journal of the Palaeontological Society of India* **50**, 1–11.

Taylor, P.D. & Vinn, O. 2006. Convergent morphology in the small spiral worm tubes ("*Spirobis*") and its palaeoenvironmental implications. *Journal of the Geological Society, London* **163**, 225–8.

Williams, A., Carlson, S.J., Brunton, C.H.C., Holmer, L.E. & Popov, L. 1996. A supra-ordinal classification of the Brachiopoda. *Philosophical Transactions of the Royal Society B* **351**, 1171–93.

Zhang Zhifei, Shu Degan, Han Jian & Liu Jianni. 2006. New data on the rare Chengjiang (Lower Cambrian, South China) linguloid brachiopod *Xianshanella haikouensis*. *Journal of Paleontology* **80**, 203–11.

Chapter 13

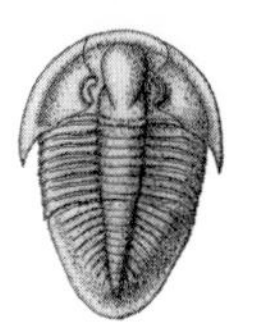

Spiralians 2: mollusks

Key points

- The Phylum Mollusca can be traced back to at least the Late Precambrian, when *Kimberella* probably fed on algae in Ediacaran communities.
- Early mollusks were characterized by some short-lived, unusual forms but with the molluskan features of a mantle, mineralized shell and radula; these were members of the small shelly fauna.
- Mollusk shell shape and even ornament can be modeled by a variety of microcomputer-based software packages; only a small percentage of theoretical morphospace is occupied by living and fossil mollusks.
- Bivalves are characterized by a huge variety of shell shapes, dentitions and muscle scars, adapted for a wide range of life strategies in marine and some freshwater environments.
- Most gastropods undergo torsion in early life; they have a single shell, often coiled. The group adapted to a wide range of environments from marine to terrestrial.
- Cephalopods are the most advanced mollusks, with a well-developed head, senses and a nervous system; they include the nautiloids, ammonoids and the coleoids. The group is carnivorous.
- During the Mesozoic many mollusks developed a number of protective strategies such as robust armor or deep infaunal life modes. The group may also have relied on multiformity of shape and color to confuse predator search images.
- Annelid worms were a sister group to the mollusks; their jaws, the scolecodonts, are relatively common in Paleozoic faunas.

She sells seashells on the seashore;
The shells that she sells are seashells I'm sure.
So if she sells seashells on the seashore,
I'm sure that the shells are seashore shells.

Old nursery rhyme

This famous tongue twister was first recited over 200 years ago in England, and it is very likely based on the exploits of Mary Anning, the most famous fossil collector of her time. She is best known for her spectacular discoveries of marine reptiles in the Lower Jurassic rocks of her native Lyme Regis in southern England; but she made most of her regular income from selling fossil ammonites and other mollusk fossils to visitors. Most of us have found seashells while playing or walking on the beach and have been amazed by their colors, shapes and ornaments. Not only are clams, oysters and scallops good to eat, but their shells, throughout historic times, have featured as ornaments, tools and even currency. The Mollusca is the second largest animal phylum after the Arthropoda, with records of over 130,000 living species and a history extending back into the Precambrian.

MOLLUSKS: INTRODUCTION

The Phylum Mollusca includes the slugs, snails, squids, cuttlefish and octopuses in addition to all manner of marine shellfish such as clams, mussels and oysters (Box 13.1). Although some mollusks are the size of sand grains, the giant squid *Architeuthis* can grow to over 20 m in length, the largest and possibly the most frightening genus of all living invertebrates. Mollusks are probably the most common marine animals today, occupying a very wide range of habitats, from the abyssal depths of the oceans across the continental shelves and intertidal mudflats to forests, lakes and rivers. Mollusks are usually unsegmented, soft-bodied animals with a body plan based on four features:

1 The head contains the sensory organs, and a rasping feeding organ, the **radula**, composed of chitin and designed to scrape and in some cases drill.
2 The **foot** is primitively a sole-like structure on which the animal crawls, but is considerably modified in many mollusks.
3 The **visceral mass** of the digestive, excretory, reproductive and circulatory organs is enclosed in the celomic cavity.
4 The **mantle** is a sheet of tissue lying dorsally over the visceral mass that is responsible for secreting the shell.

Molluskan shells are secreted as calcium carbonate, mainly aragonite, with an organic matrix and an outer organic layer. In the case of the bivalves, a range of shell fabrics have evolved from simple prismatic structures, through nacreous and prismatic, to crossed-lamellar aragonitic and prismatic and foliated calcite fabrics. Shell structure has been used in the higher classification of the group (as in the brachiopods). Crossed-lamellar structures evolved independently in some gastropods. Beneath the mantle, the mantle cavity lies behind the visceral mass and is the respiratory chamber that houses the molluskan gills (**ctenidia**); the openings of the excretory and reproductive ducts and the anus open into the mantle cavity and their products are carried out on the exhalant current.

From simple beginnings as a limpet-like crawler back in the Precambrian, mollusks have evolved a spectacular range of shapes and sizes, and their hard, calcareous shells are readily fossilized.

A simple method of visualizing molluskan evolution is to consider the hypothetical ancestor, or **archemollusk**, with a minimal molluskan morphology; this approach has been modified and merged with a recent cladogram for the phylum (Fig. 13.1). There is still a great deal of uncertainty about the identity of the first mollusks, and new finds constantly change the picture (Boxes 13.2, 13.3). The most recent common ancestor of the mollusks probably had seven- to eight-fold serial repetition, the presence of valves and a foot, and had a crawling mode of life (Sigwart & Sutton 2007).

EARLY MOLLUSKS

The Early Cambrian was a time of experimentation, with a variety of short-lived, often bizarre, molluskan groups, such as the helcionelloids, dominating many faunas (Peel 1991). Most workers now agree that the first mollusks were descended from forms like living flatworms – probably spiculate animals with radula and gills situated posteriorly. These mollusks were similar to modern soft-bodied aplacophorans, a group of shell-less mollusks. The aplacophorans and the shelled mollusks shared a common ancestor probably during the Late Precambrian. Significantly, the articulated remains of a halkieriid mollusk from

Box 13.1 Classification of Mollusca

Class CAUDOFOVEATA

- Worm-like, shell-less mollusks living inverted in burrows in the seabed
- Recent

Class APLACOPHORA

- Worm-like, spiculate mollusks
- Possibly Carboniferous (or older) to Recent

Class MONOPLACOPHORA

- Limpet-like, cap-shaped shells with segmented soft parts
- Cambrian (Lower) to Recent

Class DIPLACOPHORA

- Anterior and posterior shell separated by elongate zone of scale-like sclerites
- Cambrian (Lower)

Class POLYPLACOPHORA

- Segmented shell usually with eight plates, large muscular foot and a series of gill pairs
- Cambrian (Upper) to Recent

Class TERGOMYA

- Exogastrically coiled, univalved, bilaterally symmetric, often planispirally coiled or cap-shaped mollusks
- Cambrian (Middle) to Recent

Class HELCIONELLOIDA

- Endogastrically coiled, univalved, untorted mollusks
- Cambrian (Lower) to Devonian (Pragian)

Class GASTROPODA

- Univalved, shell usually coiled, having head with eyes and other sense organs, muscular foot for locomotion. Internal organs rotated through 180° during torsion early in ontogeny
- Cambrian (Upper) to Recent

Class BIVALVIA

- Twin-valved, joined along dorsal hinge line commonly with teeth and ligament; lacking head but with well-developed muscular foot and often elaborate gill systems
- Cambrian (Lower) to Recent

Class ROSTROCONCHIA

- Superficially similar to bivalves but with shells fused along dorsal midline
- Cambrian (Lower) to Permian (Kazanian)

Class SCAPHOPODA

- Long, cylindrical shell, open at both ends
- Devonian to Recent

Class CEPHALOPODA

- Most advanced mollusks with head and well-developed sensory organs together with tentacles
- Cambrian (Upper) to Recent

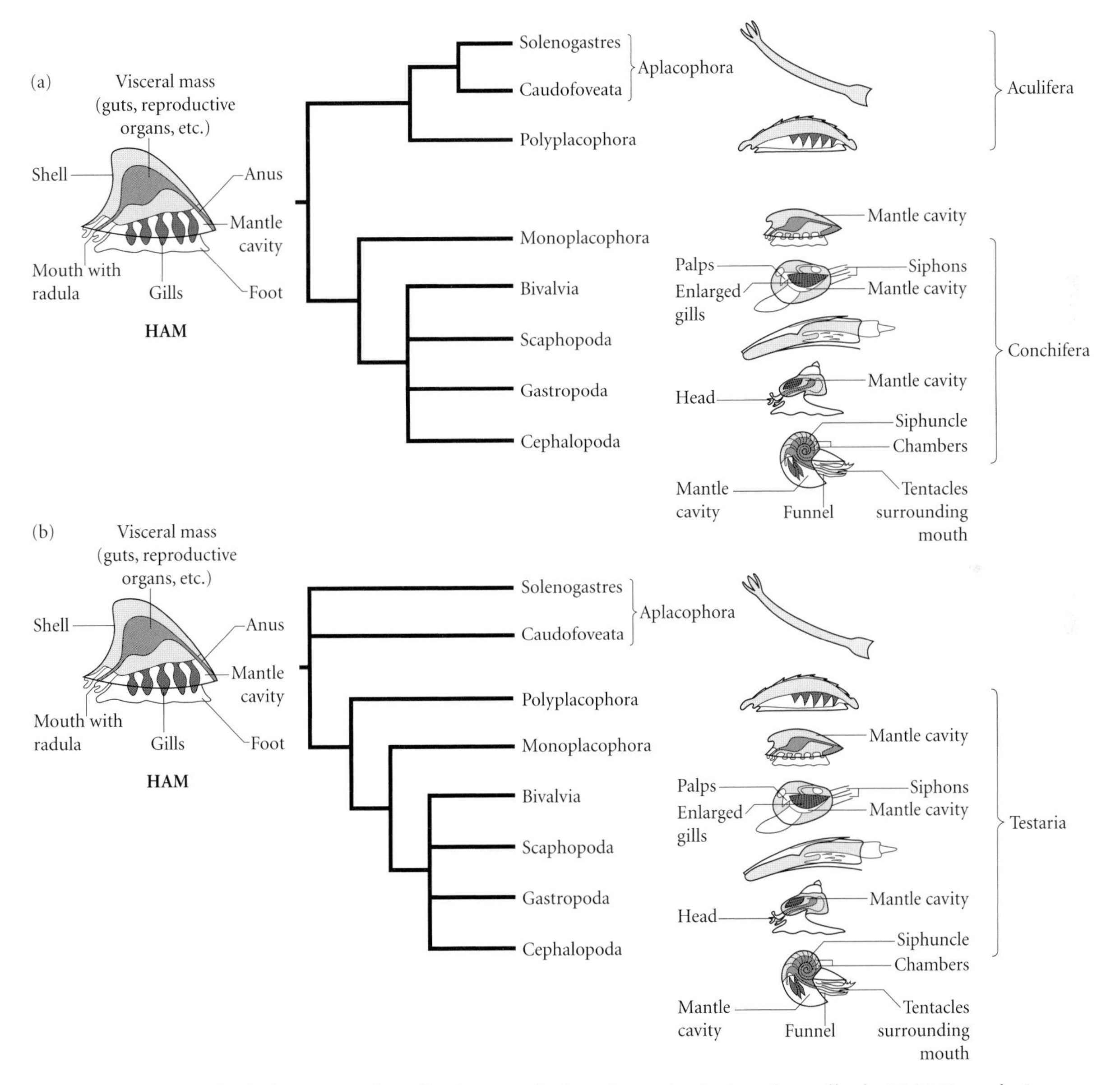

Figure 13.1 Pseudocladograms of molluskan evolution: hypothetical archemollusk (HAM) evolution integrated with a cladistic-type framework. Model (a) demonstrates a split into the Aculifera and Conchifera, whereas (b) indicates a division into the Aplacophora and Testaria. (Based on Sigwart & Sutton 2007.)

Box 13.2 *Kimberella* and *Odontogriphus* join the mollusks

A modest-sized, disk-shaped fossil from the Late Precambrian, named *Kimberella* in 1959, has suffered mixed fortunes. First described from the Ediacaran rocks of Australia as a jellyfish and later a cubozoan, Mikhail Fedonkin and Ben Waggoner (1997) then reconstructed *Kimberella* as a bilaterally symmetric, benthic crawler with a non-mineralized, single shell, on the basis of new material from the White Sea, Russia. *Kimberella* is linked with a variety of trace fossils suggesting mobility and a feeding strategy that must have involved a radula. The body fossils and trace fossils place *Kimberella* near the base of the molluskan clade and suggest a deep origin for the phylum (Fig. 13.2), and for the bilateralians, significantly earlier than the Cambrian explosion. But who were its closest relatives? A new investigation by Jean-Bernard Caron and his colleagues (2006) offers some clues. They studied another enigmatic animal, *Odontogriphus* from the Burgess Shale. *Odontogriphus* had previously been allied with the brachiopods, bryozoans, phoronids and even early vertebrates. The new study shows that *Odontogriphus* possesses a radula, a broad foot and a stiffened dorsum, so placing it firmly within the mollusks, close to *Kimberella*, together with *Wiwaxia* (another enigmatic soft-bodied organism covered with possible scierites), which also possesses a radula, and another enigma, *Halkieria* (Box 13.3).

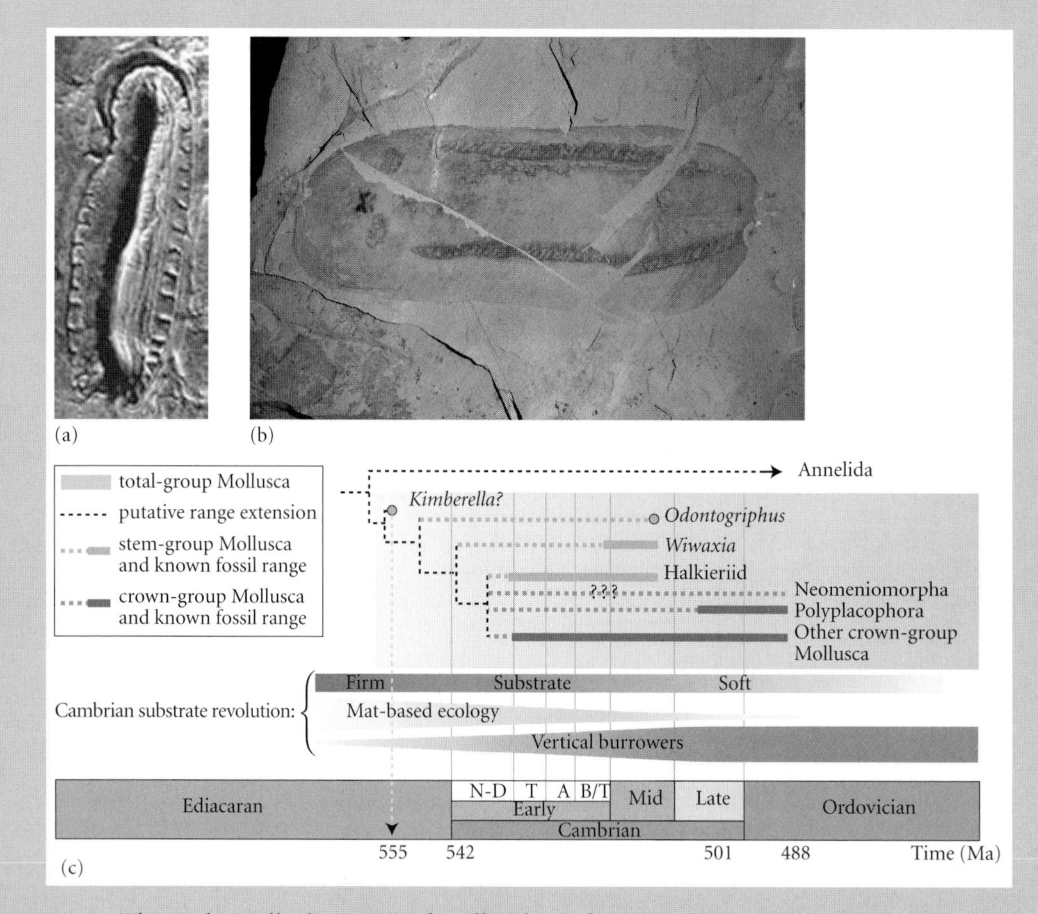

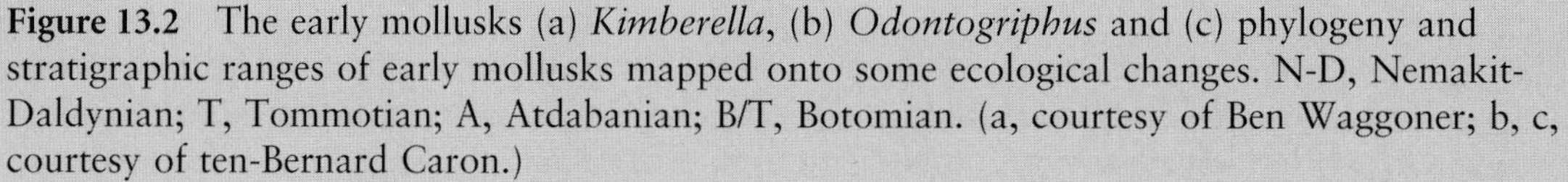

Figure 13.2 The early mollusks (a) *Kimberella*, (b) *Odontogriphus* and (c) phylogeny and stratigraphic ranges of early mollusks mapped onto some ecological changes. N-D, Nemakit-Daldynian; T, Tommotian; A, Atdabanian; B/T, Botomian. (a, courtesy of Ben Waggoner; b, c, courtesy of ten-Bernard Caron.)

Box 13.3 *Halkieria*: from stem-group brachiopod to new class of mollusk

Halkieria was first described on the basis of disarticulated shells from the Cambrian rocks of the Danish island of Bornholm. But the discovery in the 1980s of articulated specimens from the Early Cambrian Sirius Passet fauna from North Greenland (see p. 386) generated huge excitement. The animal was in fact an elongate, worm-like creature with two mollusk-like shells at the front and the back separated by an armor of sclerites between (Fig. 13.3), quite bizarre and quite different from previous interpretations of the animal. Initial attempts to place it together with the mollusks were superseded by its placement as a stem-group brachiopod; reasonable enough because both shells are very similar to the dorsal and ventral valves of some non-articulated brachiopods. However, to become a brachiopod, *Halkieria* would have had to lose its foot, develop a lophophore as a feeding organ and convert its sclerites to chaetae. Jakob Vinther and Claus Nielsen (University of Copenhagen) in 2004 dissected the fossil in detail and compared it with a range of living mollusks. There was a simpler solution. *Halkieria* is in fact a mollusk, possessing most of the features that define the phylum, but a number of characters (such as the shells at the anterior and posterior of the animal) have formed the basis for a new class of mollusk, the Diplacophora.

Figure 13.3 The mollusk *Halkieria* from Sirius Passet (natural size).

the Lower Cambrian rocks of north Greenland has promoted new discussion on the identity of the earliest mollusks (Box 13.3). The halkieriid not only displays the articulation of a series of **sclerites**, or plates, commonly described in the past as discrete organisms, but also two large mollusk-like shells at the front and back of the worm-like animal. The many, often bizarre but distinctive, early mollusks formed the basis for subsequent radiation of the phylum particularly during the Late Cambrian and Early Ordovician. The shapes of these and other mollusk shells have formed the basis numerical modeling, demonstrating that fossil and living shell shapes, and indeed many unknown in nature, can be generated by computers (Box 13.4).

The hyoliths – long, conical, calcareous shells with an operculum-covered aperture – have often been called mollusks. The group ranges from the Cambrian to Permian with some of the 40 known genera reaching lengths of 200 mm. Current studies assign the group to its own phylum, related to the mollusks and the peanut worms, the Sipunculida.

CLASS BIVALVIA

Bivalves are among the commonest shelly components of beach sands throughout the world. Many taxa are farmed and harvested for human consumption, and pearls are a valuable by-product of bivalve growth. The bivalves developed a spectacular variety of shell shapes and life strategies, during a history spanning the entire Phanerozoic, and all are based on a simple bilaterally symmetric exoskeleton. The first bivalves were marine shallow burrowers; epifaunal, deep

Box 13.4 Computer-simulated growth of mollusks

Most valves of any shelled organism can be modeled as a coil and, in fact, the ontogeny of living *Nautilus* was known to approximate to a logarithmic spiral in the 18th century. David Raup (University of Chicago), in an influential study, defined and computer-simulated the ontogeny of shells on the basis of a few parameters: (i) the shape of the generating curve or axial ratio of the ellipse; (ii) the rate of whorl expansion after one revolution (W); (iii) the position of the generating curve with respect to the axis (D); and (iv) the whorl translation rate (T). Shells are generated by translating a revolving generating curve along a fixed axis (Fig. 13.4). For example, when $T = 0$, shells lacking a vertical component such as bivalves and brachiopods, are simulated, whereas those with a large value of T are typical of high-spired gastropods. Only a small variety of possible shell shapes occur in nature. Raup's (1966) original simulations were executed on a mainframe system. Andrew Swan (1990) adapted the software for microcomputers and has simulated a wide variety of shell shapes. More recent work has applied more complex techniques to simulate ammonite heteromorphs. Nevertheless only a relatively small percentage of the theoretically available morphospace has actually been exploited by fossil and living mollusks. Clearly some fields map out functionally and mechanically improbable morphologies – perhaps the aperture is too small for the living animal to feed from within the shell, or the shape would not allow the animal to move; other fields have yet to be tested in evolution. Raup's morphospace is, however, non-orthogonal and it has been argued that the mosaic of morphospace occupation is merely an artifact of presentation. Theoretical morphospace has been explored for a range of other groups including bryozoans, echinoids, graptolites, some fishes and some plants (Erwin 2007).

There have been many modifications of Raup's original algorithm and a number of web interfaces that can generate shell shapes; one of the simplest may be accessed via http://www.blackwellpublishing.com/paleobiology/.

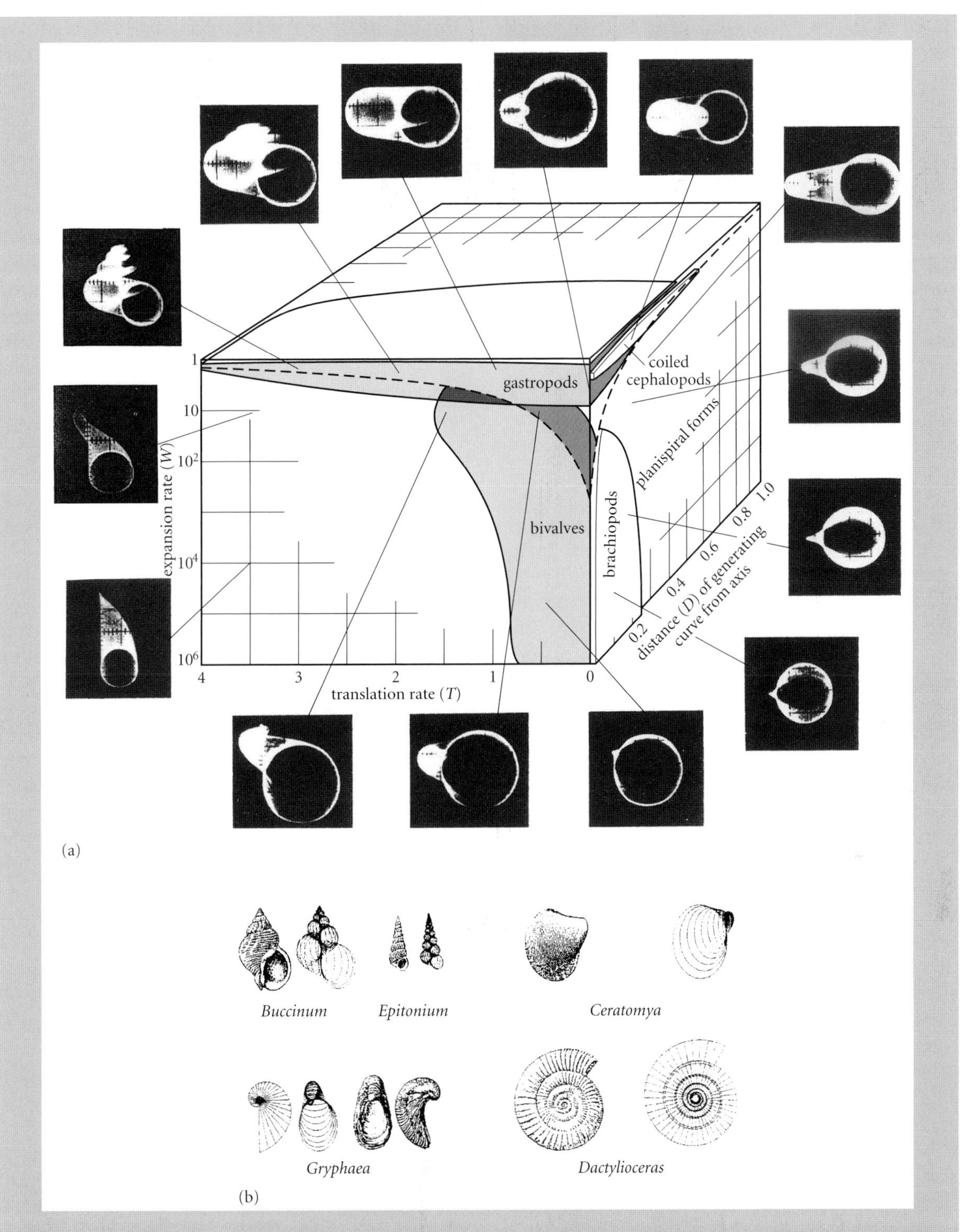

Figure 13.4 Theoretical morphospace created by the computer simulation of shell growth (a) and some computer simulations matched with reality (b). (a, based on Raup 1966; b, from Swan 1990.)

burrowing and boring strategies together with migrations to freshwater habitats were secondary innovations. There are over 4500 genera of living bivalves, with fewer than half of that number described from the fossil record. In view of the wide range of life strategies and their relationships to particular sediments, the bivalves are good facies fossils. Although non-marine bivalves have been used extensively, in the absence of other groups, to zone parts of the Upper Carboniferous and by Charles Lyell in his classic work in the 1820s and 1830s to subdivide the Tertiary (the increasing proportion of living forms in fossil faunas through the Tertiary was used to subdivide the system; see p. 29), their biostratigraphic precision is limited.

Basic morphology

Bivalves are twin-valved shellfish superficially resembling the brachiopods and common in modern seas (Fig. 13.5). In contrast to the Brachiopoda, bivalve shells are always composed of calcium carbonate, usually aragonite, and many have a plane of symmetry parallel to the commissure separating the left and right valves from each other, i.e. the two valves are virtually mirror images of each other. Bivalves have sometimes been termed lamellibranchs or pelecypods, but they were first named Bivalvia by Linnaeus in 1758.

In the bivalves the molluskan head is lost, only the anterior mouth indicates its position. Sensory organs are concentrated instead on the mantle margins and include eye-spots, **chemoreceptors** and **statocysts**. The bivalve exoskeleton has two lateral valves, left and right, essentially mirror images of each other, united dorsally along the hinge line by an elastic **ligament** and usually interlocking teeth and sockets; the valves open ventrally. The valves are secreted by mantle lobes. The attachment of the mantle is marked by the pallial line, which may be indented posteriorly with the extension of the siphons. The earliest-formed parts of each shell, the beaks or **umbones**, may be separated by the cardinal area supporting the dorsal ligament. When the valves are closed, a pair of adductor muscles, situated anteriorly and posteriorly, is in contraction. While the shells are closed, the hinge ligament is constrained between the dorsal parts of the shells; when the adductors relax, the ligament expands and the shells spring open. The scars of these shell-closing muscles may be seen usually as clear roughened and depressed areas inside both valves.

Classification of the bivalves is based primarily on gill structure (Fig. 13.6a). Dentition is of secondary importance (Fig. 13.6b). Teeth may be all along the hinge line or separated into discrete cardinal (subumbonal) and lateral (both anterior and posterior of the hinge line) teeth. The three most important tooth arrangements are: (i) **taxodont** – numerous subequal teeth arranged in a subparallel pattern; (ii) **actinodont** – teeth radiating out from beneath the umbo; or (iii) **heterodont** – a mixture of cardinal (beneath umbo) and lateral teeth. Various other terms have been employed in the past when the teeth are thickened, modified or reduced, but are now less commonly used.

In most cases the umbones of the valves point or face obliquely anteriorly, the **pallial sinus** (if present) is situated posteriorly and the posterior adductor is usually the larger of the two scars. In some forms the anterior adductor is lost, together with the foot. When the valves are held with the commissure between the two valves vertically, the anterior end pointing away from the observer and the umbones at the top, then the right and left valves are in the correct orientation.

Main bivalve groups

The Bivalvia are classified by zoologists mainly on the basis of soft-part morphology such as features of the digestive system and the gills; paleontologists have usually attempted to use details of the hinge structures. There are seven basic features that are of use for classification at various levels within the Bivalvia: gill structure (subclass and infrasubclass levels), dentition (all levels), ligament insertion (infrasubclass down to ordinal levels), adductor muscle scars (superfamily to orders), pallial line (family level and below), shell shape (all levels), and shell fabric (infrasubclass down to superfamily level). Two subclasses are recognized: (i) the Protobranchia with simple **pro-**

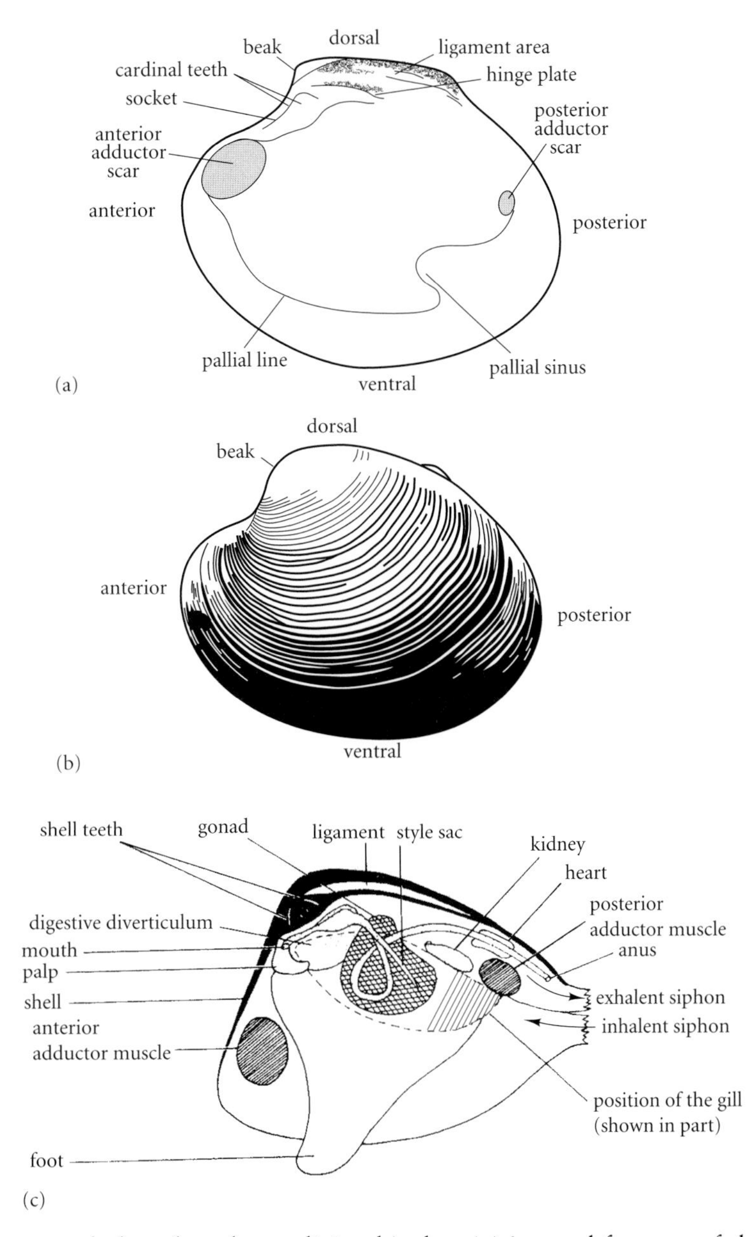

Figure 13.5 Bivalve morphology based on a living bivalve: (a) internal features of the right valve, (b) external features of the left valve, and (c) reconstruction of the internal structures attached to the right valve. (Based on *Treatise on Invertebrate Paleontology*, Part N. Geol. Soc. Am. and Univ. Kansas.)

tobranch gills very like those of the archetype mollusk, that are deposit feeders; and (ii) the Autolamellibranchiata that mostly have large **leaf-like gills** modified for food gathering as well as for respiration (**filibranch** and **eulamellibranch** types), but some have lost their gills altogether and use the mantle cavity for respiration (**septibranch**) (Fig. 13.6a).

A number of taxa from the two subclasses, the protobranchs and autolamellibranchs, are illustrated in Fig. 13.7.

Protobranchs

The Nuculoida is the oldest and most primitive infrasubclass, characterized by

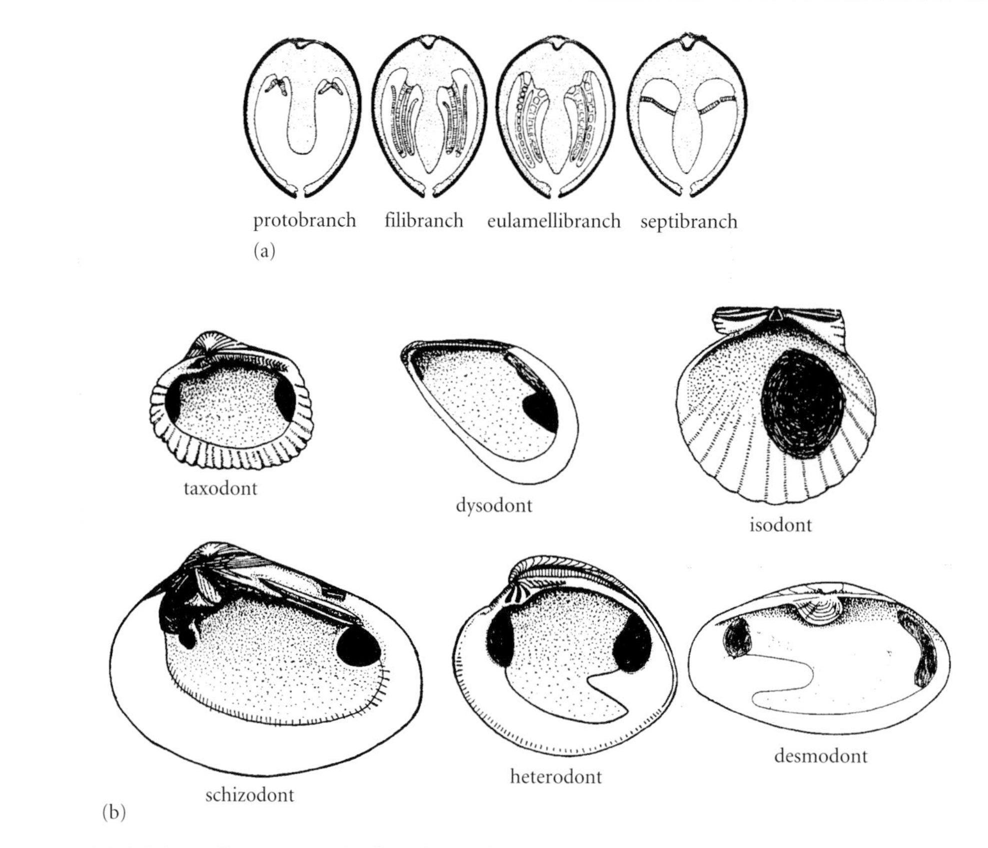

Figure 13.6 (a) Main gill types in the bivalves. (b) Main types of bivalve dentition.

Figure 13.7 Some bivalve genera: (a) *Glycimeras* (Miocene), (b) *Trigonia* (Jurassic), (c) *Gryphaea* (Jurassic), (d) *Chlamys* (Jurassic), (e) *Mya* (Recent), (f) *Pholas* (Recent), and (g) *Spondylus* (Cretaceous). Magnification ×0.75 for all.

prismato-nacreous shells, taxodont dentition, equivalved shells and protobranch gills. Most are detritus-feeding infaunal marine animals, such as *Nucula*, and most abundant today in deeper-water environments. *Ctenodonta* has a typical taxodont dentition, an elliptical shell and an external ligament; it is principally Ordovician in age.

Members of the infrasubclass Solemyoida are specialized, infaunal burrowers with an anteriorly elongate shell. Most have symbiotic autochemotrophic bacteria allowing them to live in fetid muds, ranging in age from Early Ordovician to Recent.

Autolamellibranchs

Autolamellibranchs were derived from the protobranchs by the earliest Ordovician, possibly via a group of nuculoids that developed hinge-teeth allowing greater opening of the valves. This is necessary to avoid sediment inadvertently trapped by the gills during the food-gathering process.

The pteriomorphs are mainly marine, fixed benthos, attached by a **byssus**, or pad of sticky threads, modified from the foot, or they may be cemented. They are an important part of bivalve faunas from the earliest Ordovician and most had an outer mineralized shell layer of calcite; the gills are of filibranch grade. The group includes the mussels *Modiolus* and *Mytilus* and the ark shells *Arca* and *Anadara*, the scallops *Chlamys* and *Pecten*, and the oysters *Crassostrea* and *Ostrea*.

The heteroconchs are a mixed bag of mainly suspension feeders, important in bivalve faunas from the earliest Ordovician and radiating during the Mesozoic when mantle fusion and the development of long siphons promoted a deep-infaunal life mode. They are and were very successful burrowers. Gill grades are mainly eulamellibranch and many have crossed-lamellar or complex crossed-lamellar shell microstructures. This group includes the typical clams such as the giant clam *Tridacna*, the horse-hoof clam *Hippopus* and the surf-clam *Donax*, together with the razor shells *Ensis* and *Tagelus*, the ship-worm *Teredo* and the cockle *Cerastoderma*.

The anomalodesmatans are predominantly suspension-feeding marine forms with prismato-nacreous shells and reduced dentitions, such as *Pholadomya*. They have eulamellibranch or septibranch gill grades. These too are found from the earliest Ordovician but only form a minor part of bivalve faunas.

Lifestyles and morphology

There are seven main bivalve forms that relate to their modes of life (Stanley 1970): infaunal shallow burrowing, infaunal deep burrowing, epifaunal attached by a byssus, epifaunal with cementation, free lying, swimming, and borers and cavity dwellers. Specific assemblages of morphological features are associated with each life mode; these are summarized in Fig. 13.8. Steven Stanley's studies have been adapted by a number of authors for similar bivalve-dominated communities throughout the Phanerozoic (Fig. 13.9). Most bizarre were the rudists that built extensive reefs in the Cretaceous (Box 13.5).

Bivalve evolution

The earliest known bivalves have been reported from the basal Cambrian. Two Early Cambrian genera are the praenuculid *Pojetaia* from Australia and China and *Fordilla* from Denmark, North America and Siberia. Both genera have two valves separated by a working hinge with a ligament, together with muscles and teeth. These probably came about 10 myr after the oldest rostroconch, *Heraultipegma*, and so the bivalves might just have evolved from rostroconchs (see p. 357) or something like them. The class evolved rapidly in the Early Ordovician to include basal forms of all bivalve infrasubclasses. Not only were taxodont, actinodont and heterodont dentitions established, but a variety of feeding types had also developed following the Tremadocian and Floian radiation.

Following this major diversification, the group stabilized during the remaining part of the Paleozoic, although some groups evolved extensive siphons that aided deep-burrowing life modes. This adaptation, together with the mobility provided by the bivalve foot, were important advantages over most brachiopods, which simultaneously pursued a fixed epifaunal existence. The earliest **autolamellibranchiate** forms are known from the Early Tremadocian. The early Mesozoic radiation of the group featured siphonate forms with desmodont and heterodont dentitions,

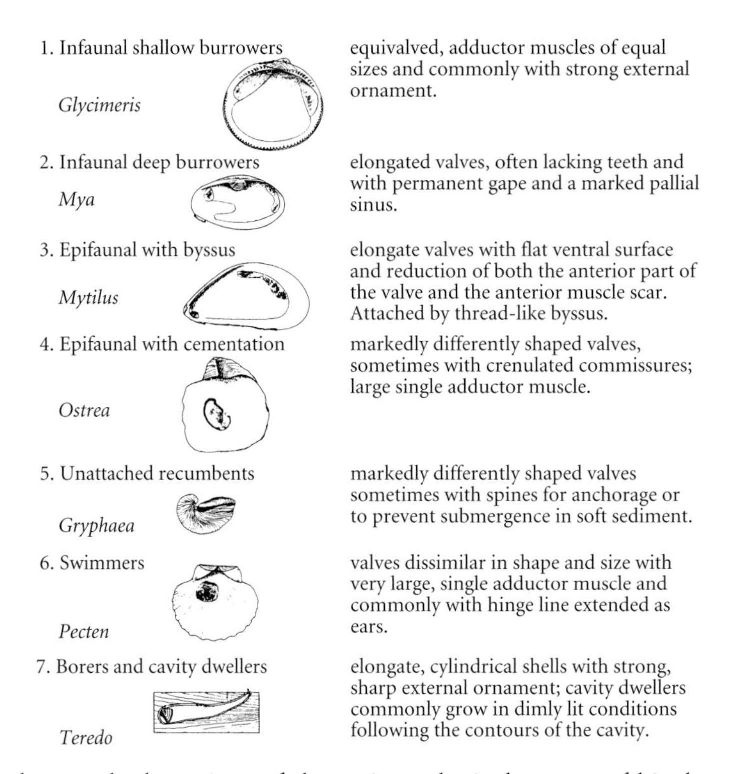

Figure 13.8 Morphology and adaptations of the main ecological groups of bivalve mollusk.

equipped to handle life deep in the sediments of nearshore and intertidal zones where they diversified.

CLASS GASTROPODA

The gastropods, the "belly-footed" mollusks, are the most varied and abundant of the molluskan classes today. The group includes the snails and slugs, forms both with and without a calcareous shell. During a history spanning the entire Phanerozoic, gastropods evolved creeping, floating and swimming strategies together with grazing, predatory and parasitic trophic styles.

Most gastropods are characterized by **torsion** in which the mantle cavity containing the gills and anus, excretory and reproductive openings comes to lie above the head (Fig. 13.11). The advantages of this arrangement are unclear. In fact, torsion seems to be distinctly disadvantageous because it involves the loss of one of the gills and/or development of a **peristomal slit** allowing separation of inhalant and exhalant currents. The first larval stage, the trochophore, is usually fixed. However, the second, veliger, phase is free-swimming and unique to the mollusks. During development, the head and foot remain fixed but all the visceral mass, the mantle and the larval shell are, in effect, rotated through 180°. The process of torsion is characteristic of the Gastropoda, although in some groups there may be secondary reversal. The coiling of the gastropod shell is unrelated to the rotation of the soft parts. Following torsion, the mantle cavity and anus are open anteriorly and the shell is coiled posteriorly in an endogastric position, in contrast to the exogastric style of the "monoplacophoran" grade shell.

The gastropod shell is usually aragonitic, usually conical with closure posteriorly at the pointed apex, and open ventrally at the aperture. Each revolution of the shell or whorl meets adjacent whorls along a suture, and the whorls together comprise the spire. Tight coiling about the vertical axis generates a central pillar or **columella**. The aperture is commonly oval or subcircular and is circumscribed by an outer and inner lip. The head emerges at the anterior margin of the aperture, where the aperture may be notched or extended as a siphonal canal supporting inhalant flow through the siphons. Material is ejected

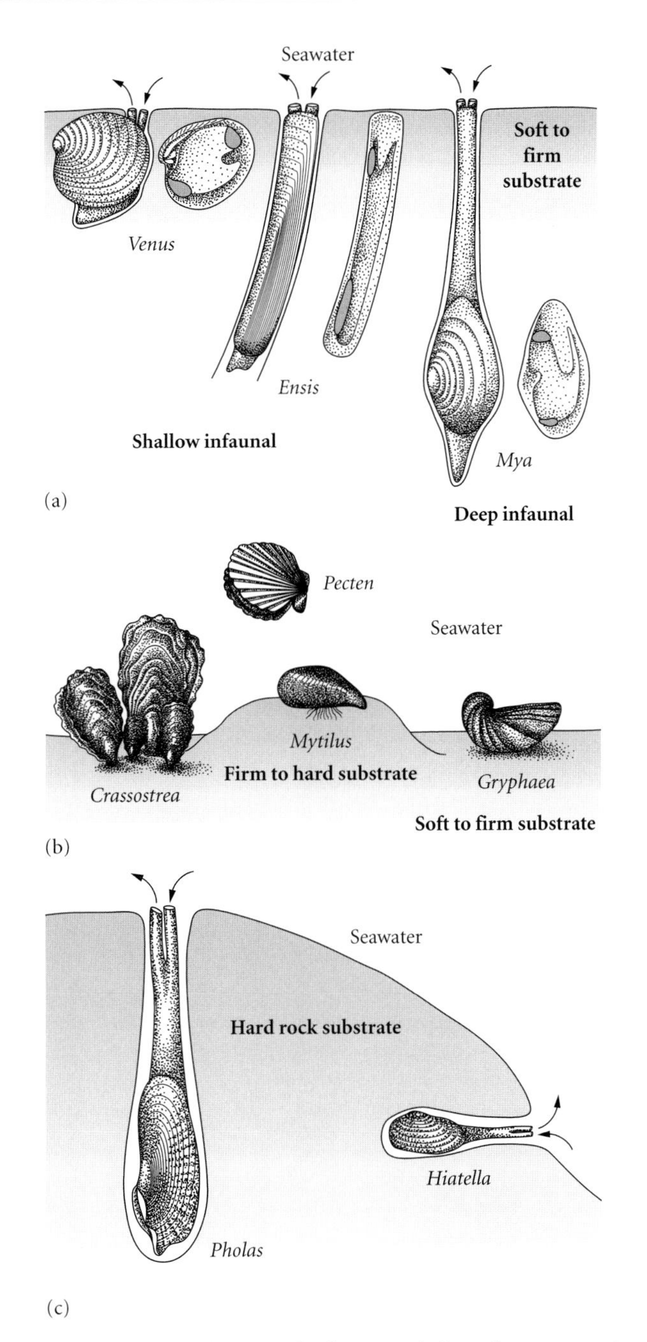

Figure 13.9 Life modes of bivalve mollusks: (a) shallow and deep burrowers into soft to firm substrates, (b) epifaunal swimming, attached or resting on soft to firm substrates, and (c) boring into hard substrates. (From Milsom & Rigby 2004.)

Box 13.5 Rudists: bivalves disguised as corals

The rudists were aberrant heteroconch bivalves that range in age from the Late Jurassic to the Late Cretaceous and occupied the Tethyan region. During a relatively short interval they developed a bizarre range of morphologies, and although many groups apparently mimicked corals, the rudists were probably not reef-building organisms. The rudists were inequivalved with a large attached valve, usually the right valve of conventional terminology, and a small cap-like free valve. Virtually all rudists had a single tooth flanked by two sockets in the attached valve, and two corresponding teeth and a socket in the free valve. The valves functioned with an external ligament and pairs of adductors attached to internal plates or myophores. Three growth strategies have been identified (Fig. 13.10). Elevators had tall conical shells with a commissure raised above the sediment–water interface to free the animal from the risk of ingesting sediment. The elevators were thus similar to solitary corals, suggesting a possible reef-building strategy. Clingers or encrusters were flat, bun-shaped forms that usually adhered to hard substrates. The recumbents had large shells, extending laterally extravagantly over the seafloor like large calcified bananas. The rudists occupied carbonate shelves throughout the Tethys region, with their larvae island hopping around the tropics, often growing together in a gregarious habit; clusters or clumps probably trapped mud in molluskan-rich structures. As noted above it now seems likely that the rudists were never true reef-building organisms although they came close to fulfilling that mode of life.

Thomas Steuber (University of Bochum) has developed a comprehensive database on rudist bivalves together with spectacular pictures of rudist accumulations. Study of this comprehensive database, and a smaller dataset that can be used to reconstruct ancient paleogeographic associations at http://www.blackwellpublishing/paleobiology/, can be used for a variety of exercises. The small dataset investigates the biogeography of Campanian rudists, emphasizing their relationship to the paleotropics (Tethyan province) on http://www.blackwellpublishing.com/paleobiology/.

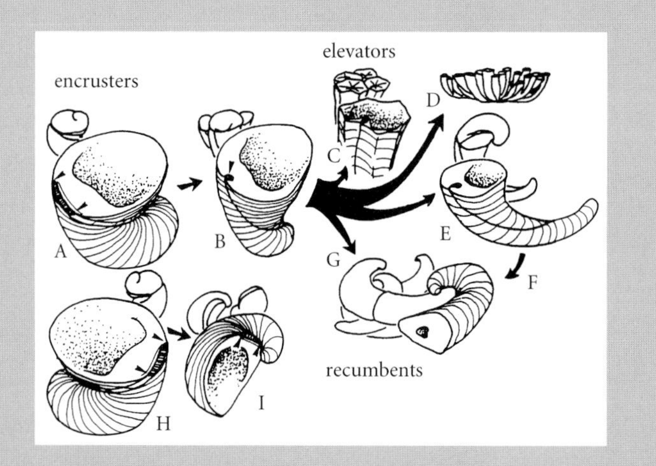

Figure 13.10 Rudist growth strategies: encrusters (A, B, H and I), elevators (C, D and E) and recumbents (F, G). (From Skelton, P.W. 1985. *Spec. Pap. Palaeont.* 33.)

through the exhalant slit in the outer lip. During ontogeny the inactive track of the slit is successively overgrown with shell material to form the **selenizone**, the calcified track of the slit band separating the siphons from the mouth.

The gastropod shell is normally oriented with the aperture facing forward and the apex facing upwards. If the aperture is on the right-hand side, the shell is coiled clockwise in a so-called **dextral** mode; **sinistral** shells have

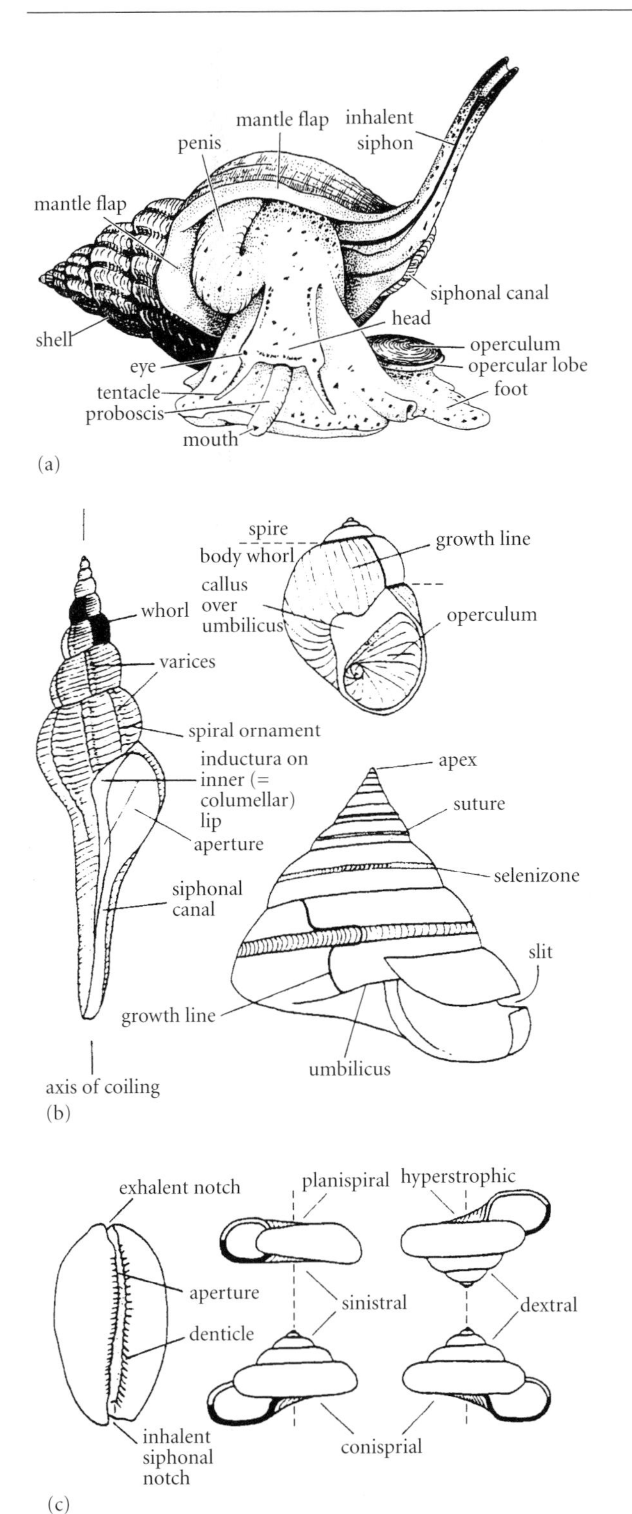

Figure 13.11 Gastropod morphology: (a) annotated reconstruction of a living gastropod, (b) annotated shell morphology of three gastropod shell morphotypes, and (c) main types of gastropod coiling strategy.

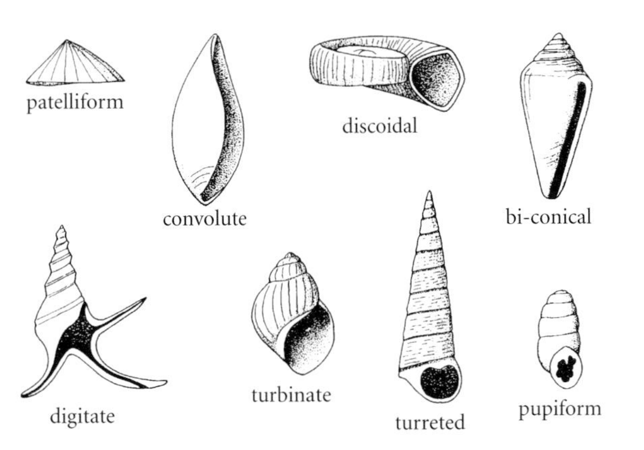

Figure 13.12 Gastropod shell shapes.

the opposite sense of coiling. The shell surface is commonly modified by strong growth lines, ribs, tubercles and projections. Many gastropods have an operculum covering the aperture.

Gastropods developed a variety of shell shapes. Eight different morphologies ranging from the simple patelliform to the complex digitate shell are illustrated in Fig. 13.12 as a sample of the large amount of exoskeletal variation in the group.

Main gastropod groups and their ecology

Gastropods have been divided into three classes largely based on information from their soft parts. Three subclasses are traditionally defined on the basis of the radula and their respiratory and nervous systems, although some of the groups may not be truly monophyletic: (i) the Prosobranchia are fully torted with one or two gills, an anterior mantle cavity and cap-shaped or conispiral shells; (ii) the Opisthobranchia are untorted (having gone through torsion followed by detorsion) with the shell reduced or absent, and the mantle cavity posterior or absent; and (iii) the Pulmonata are untorted with the mantle cavity modified as a lung, and the shells are usually conispiral. Fossil taxa are usually assigned to these categories on the basis of similarities in shell morphology with their living representatives.

The prosobranchs are mainly part of the marine benthos with a few freshwater and terrestrial taxa. The primitive members of the group, the Eogastropoda, are marine, mainly grazing herbivores with cap-shaped or low-spired forms and include a diverse set of superfamilies including the following groups. Macluritines have large, thick shells lacking a slit-band; for example *Maclurites* is planispirally coiled, hyperstrophic with a robust operculum and ranged from the Ordovician to the Devonian. The pleurotomariines have variably shaped shells, usually conispiral. They dominated shallow-water Paleozoic environments, although today the group is restricted to deeper-water settings. *Pleurotomaria* had a trochiform shell with a broad selenizone; the older Ordovician-Silurian *Lophospira* had a turbinate shell. The trochines are typical of rocky coasts, grazing on algae; Paleozoic taxa, for example the Ordovician-Silurian *Cyclonema*, were probably scavengers, whereas some, such as the Devonian *Platyceras*, are commonly attached to the anal tubes of crinoids and were parasites. The patellines, such as the limpets like *Patella*, have cap-like shells and they graze on algae on rocks in the intertidal zone. The euomphalines were mainly discoidal, such as *Euomphalus*, which ranged from the Silurian to the Permian.

The murchisoniines were a more advanced group that ranged from the Ordovician to the Triassic, possessing high-spired shells with a siphonal notch. *Murchisonia* is a long-ranging genus (Silurian-Permian).

Finally, the precise systematic position of the bellerophontines is still unresolved; they were planispirally-coiled shells with a well-developed slit, ranging in age from the Cambrian to the Triassic. The long-ranging *Bellerophon* was very common in the Early Carboniferous.

The order Mesogastropoda consists of prosobranchs that have lost the right gill and usually have conispiral shells with siphonal notches. These taxa have diversified in marine, freshwater and terrestrial environments. *Turritella* is a high-spired, multiwhorled shell with strong ribs and a simple aperture, whereas *Cypraea* is involute with the earlier whorls completely enclosed by the final whorl.

The order Neogastropoda contains conispiral, commonly fusiform, shells with a siphonal notch; most of the order is carnivorous and members dominated marine environments from the Tertiary onwards. *Neptunea* has a large body whorl and a short siphonal canal whereas *Conus* is biconical with a narrow aperture and a siphonal notch.

The subclass Opisthobranchia includes marine gastropods with reversed torsion and commonly lacking shells. Pteropods and sea slugs are typical opisthobranchs.

The subclass Pulmonata contains detorted gastropods, with the mantle cavity modified as an air-breathing lung. The group probably ranges in age from the Jurassic to the present, and is characteristic of terrestrial environments. *Planorbis* has a smooth, planispiral shell with a wide umbilicus whereas *Helix* is smooth and conispiral and *Pupilla* has a smooth pupiform shell.

The gastropods show a considerable diversity of form across the entire class (Fig. 13.13). It is difficult to relate given morphotypes to particular life modes although the overall morphology of the shell can reflect its trophic function (Wagner 1995). In general terms, however, gastropods occupying high-energy environments have thick shells and are commonly cap-shaped or low-spired, whereas shells with marked siphonal canals are adapted to creeping across soft substrates. Carnivores are usually siphonal whereas herbivores have complete apertural margins and commonly grazed on hard substrates. Thin-shelled taxa are typical of freshwater and terrestrial environments.

Gastropod evolution

There is no general agreement on the origin of the gastropods. Currently the group is thought to have been derived from a monoplacophoran-type ancestor by torsion and development of an exogastric condition, where the shell is coiled away from the animal's head. An origin from among coiled forms such as *Pelagiella* may link the monoplacophoran grade through the Tommotian *Aldanella* to the gastropods.

The monophyly of the gastropods has been questioned. It is possible that many of the traditional groups, for example the archaeogastropods, mesogastropods, opisthobranchs and pulmonates may be grades of gastropod organization, forming a series of parallel-evolving clades. In particular the archaeogastropods have been shown to be polyphyletic and they are no longer considered to be a natural grouping. Nevertheless, the neogas-

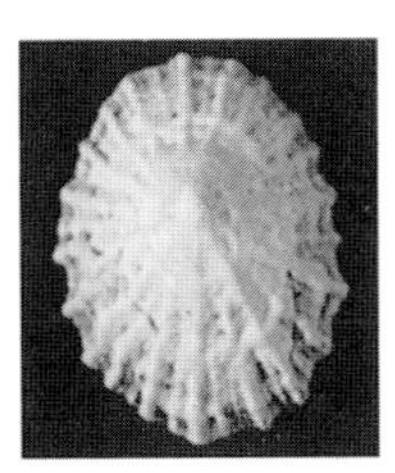

(a) (b) (c) (d)

(e) (f) (g) (h)

Figure 13.13 Some gastropod genera: (a) *Murchisonia* (Devonian) (×1.25), (b) *Euomphalus* (Carboniferous) (×0.5), (c) *Lophospira* (Silurian) (×0.5), (d) *Patella* (Recent) (×1), (e) *Platyceras* (Silurian) (×1), (f) *Neptunea* (Plio-Pleistocene) (×0.6), (g) *Viviparus* (Oligocene) (×0.8), and (h) *Turritella* (Oligocene) (×1). (Courtesy of John Peel.)

tropods appear to comprise a unified group derived from either advanced eogastropods or primitive mesogastropods during the Late Mesozoic.

Most Paleozoic gastropods were probably herbivores or detritus feeders. Drill holes in brachiopod shells, however, suggest that a few genera were carnivores and some, such as *Platyceras*, were parasites. The class became more important during the Late Paleozoic and the Mesozoic when many more predatory groups evolved. However, during the Cenozoic, gastropods reached their acme with the neogastropods in particular dominating molluskan nektobenthos.

Gastropods are not particularly good zone fossils, although nerineid gastropods are stratigraphically useful in parts of the English Middle Jurassic in the absence of ammonites. Gastropods are generally associated with particular facies and few rapidly evolving lineages are known in detail. Nevertheless, microevolutionary sequences in the genus *Poecilizontes* from the Pleistocene of Bermuda, described in detail by Stephen Jay Gould, suggest that new subspecies evolving by allopatric speciation arose suddenly by pedomorphosis (see p. 145). These rapid speciation events, separated by intervals of stasis, are strong supportive evidence of the punctuated equilibrium model of microevolutionary change. Moreover, in a classic study of Late Tertiary snails from Lake Turkana, Kenya, Peter Williamson (1981) suggested there had been punctuated changes in 14 separate lineages (see also p. 123).

CLASS CEPHALOPODA

The cephalopods are the most highly organized of the mollusks, with the greatest complexity of any of the spiralian groups. The close association of a well-defined head with the foot modified into tentacles is the source of their name, meaning "head-footed". High metabolic and mobility rates, a well-developed nervous system, and sharp eyesight associated with an advanced brain, are ideal adaptations for a carnivorous predatory life mode. The funnel or **hyponome** is also modified from the foot, and squirts out water from the mantle cavity providing the animal with a form of jet propulsion.

Modern cephalopods belong to two groups. Firstly, living *Nautilus* has an external coiled shell with a thin internal mantle and nearly 100 tentacles. Only five species of this genus are extant although it was once used as an analog for the behavior of all extinct externally-shelled cephalopods such as the ammonoids. Secondly, the coleoids; these have internal shells and thick external mantles. They include the 10-tentacled extinct belemnites, the squids and cuttlefish; the octopods have eight tentacles and have lost their skeleton. These living forms are most common in shallow-water belts around the ocean margins.

A tripartite division of the cephalopods into three subclasses includes: (i) Nautiloidea, with straight or coiled external shells with simple sutures (Late Cambrian to Recent); (ii) Ammonoidea, with coiled, commonly ribbed external shells with complex sutures (Early Devonian to latest Cretaceous, possibly earliest Paleogene); and (iii) Coleoidea, with straight or coiled internal skeletons (Carboniferous to Recent).

The origin of the cephalopods remains controversial, although most agree the group was derived from a monoplacophoran-like ancestor. John Peel (1991) suggested that the group is derived from within the class Helcionelloida; both groups are characterized by endogastric coiling and, moreover, the helcionelloids predate the appearance of the cephalopods by some 10 million years. Another group of gastropod-like shells, the tergomyans, with apical septa, might also have been ancestral, only they lack perforate septa.

Nautiloidea

Most information about nautiloids comes from studies of the behavior and morphology of the living *Nautilus* that occurs mainly in the southwest Pacific, normally at depths of 5–550 m (Box 13.6). It pursues a nocturnal, nektobenthonic life mode as both a carnivore and scavenger; however it is prey to animals with powerful jaws such as the perch, marine turtles and sperm whales.

Living *Nautilus* has its head, tentacles, foot and hyponome concentrated near the aperture of the body chamber; the visceral mass containing other vital organs is situated to the rear of the body chamber (Fig. 13.14). The surrounding mantle extends posteriorly as the siphuncular cord connecting all the previous, now empty, chambers that together constitute the phragmocone. Each chamber is partitioned from those adjacent by a sheet of calcareous material, the septum; the suture is formed where each septum is cemented to the outer shell. The form of the suture, or the suture pattern, is used in the classification of externally-shelled cephalopods. The conch is usually oriented as follows: anterior at the aperture, posterior at the point furthest from the aperture, the venter on the side with the hyponome, usually the outside, and the dorsum opposite. Despite the simplicity of this arrangement, fossil nautiloids developed a wide range of shell morphologies (Fig. 13.15).

Ammonoidea

The ammonite usually had a planispirally coiled shell comprising the protoconch, phragmocone and body chamber (Fig. 13.16). The protoconch or larval shell records the earliest

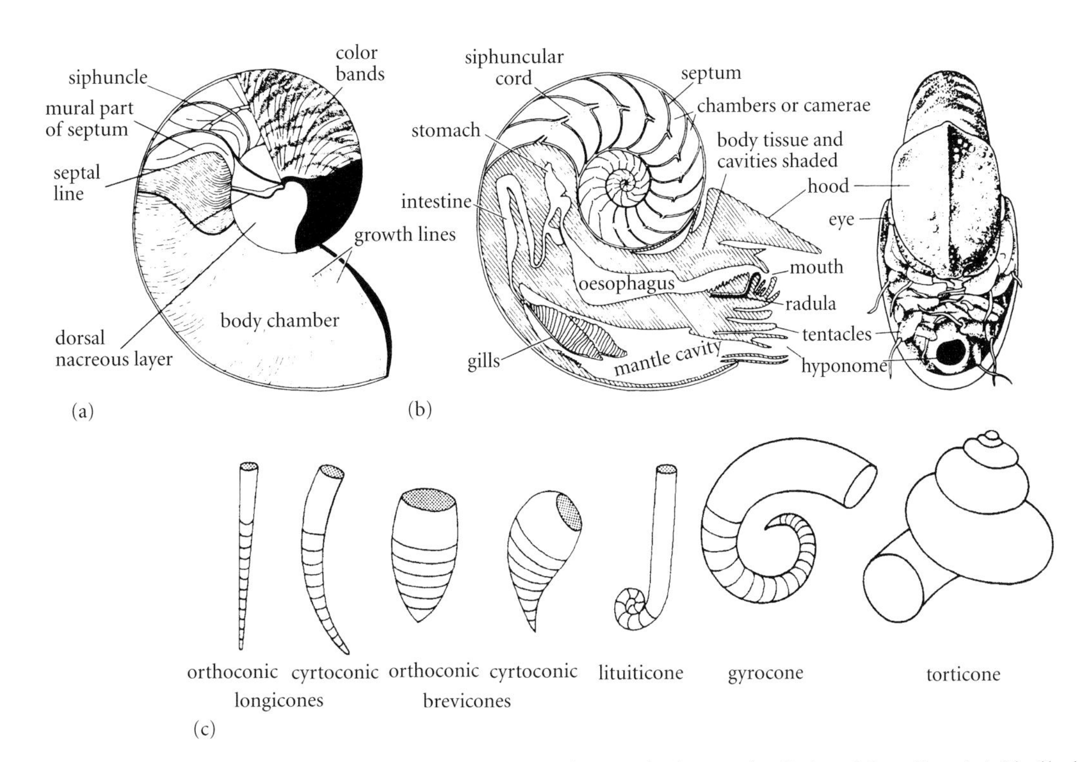

Figure 13.14 (a) Features of the shell and (b) internal morphology of a living *Nautilus*. (c) Shell shapes of the nautiloids.

Box 13.6 Living *Nautilus*

Living *Nautilus* has allowed biologists and paleontologists to model the functions and life modes of the ancient ammonites by using a modern analog. But despite the similarity of their respective shells, coleoids are, in fact, more closely related to ammonites than modern nautiloids, and thus better behavioral analogs may be found within the coleoids (Jacobs & Landman 1993). It is probable that coleoid-type swimming mechanisms probably evolved prior to the loss of the body chamber in the coleoids. Ammonoids thus probably had a coleoid-like mantle and thus may have operated quite differently from living *Nautilus*. But how far could an empty ammonite shell travel? Ryoji Wani and his colleagues (2005) have demonstrated that the phragmocone of living *Nautilus pompilius* becomes waterlogged only after the mantle tissue decomposes. Water is then sucked into the shell because of its lower internal gas pressures. This is actually more common for smaller shells, generally with diameters less than 200 mm, and these fill up with water more quickly. Only larger shells had the ability to drift long distances. Since the ratio of volumes of the body chamber to the phragmocone in nautiloids is similar to that of the ammonoids, they probably behaved similarly. The small shells sank and the large shells drifted.

ontogeny of the animal. The phragmocone is chambered, with each chamber marking successive occupation by the animal, and sealed off from previous chambers by a septum, complex in structure at its margins, like a sheet of corrugated iron. Where the septum is welded to the shell, a suture is developed, commonly with a complex pattern of frilled lobes and saddles.

Five main sutural types are recognized among cephalopods (Fig. 13.17). The **orthoceratitic** pattern, with broad undulations or

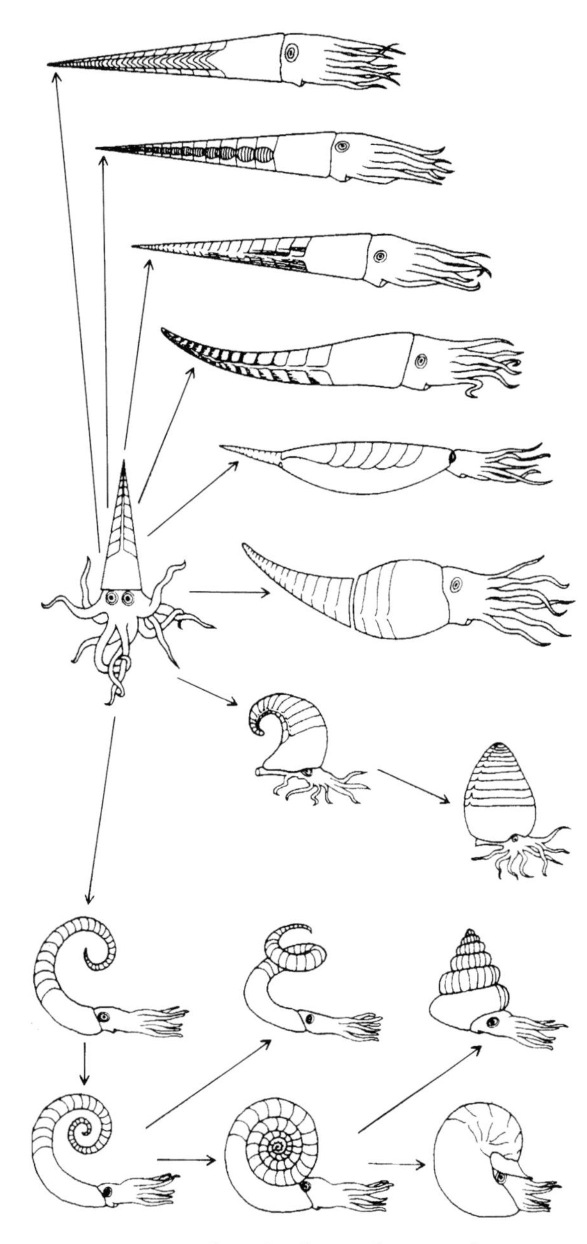

Figure 13.15 Life attitudes and external morphologies of the nautiloids. (From Peel et al. 1985.)

rounded lobes and saddles, characterizes mainly nautiloids ranging in age from Late Cambrian to Late Triassic. **Anarcestid** and **agoniatitic** patterns, however, have a narrow mid-ventral lobe and a broad lateral lobe with additional lobes and saddles, and range in age from the Early to Mid Devonian. **Goniatitic** sutures are characterized by sharp lobes and rounded saddles, and are found in Late Devonian-Permian ammonoids. **Ceratitic** sutures show frilled lobes and undivided saddles, and **ammonitic** patterns have both the lobes and saddles fluted and frilled. Based on these sutural patterns, three groups among the ammonoids can be recognized in a general way: the goniatites are typical of the Devonian-Permian, the ceratites of the Triassic, and the ammonites dominated the Jurassic and Cretaceous. Nevertheless, these sutural patterns may be cross-stratigraphic, with Cretaceous taxa having both goniatitic and ceratitic grades of suture in homeomorphs of more typical Devonian and Triassic forms.

The siphuncle connects the outer body chamber with the phragmocone that includes all the empty, previous chambers. Septal necks act like washers, guiding the passage of the siphuncle through each septum. Excepting the clymeniids, the siphuncle is situated along the outer ventral margin of the shell. Seawater may be pumped in or out of the chambers through the siphuncle in order to alter the buoyancy of the ammonite, similar to mechanisms in the nautiloids and in submarines.

The body chamber contains the soft parts of the ammonite. The aperture may be modified laterally with lappets and ventrally with the rostrum. In many taxa, **aptychi** sealed the aperture externally, although these plates may also have been part of the jaw apparatus.

Main ammonoid groups

The subclass Ammonoidea is currently split into nine orders. The first three, the Anarcestida, the Clymeniida and the Goniatitida, have goniatitic sutures and are included in the order Goniatitida. The anarcestides characterize Early to Mid Devonian faunas when forms such as *Anarcestes* and *Prolobites* displayed tightly coiled shells together with a ventral siphuncle. The clymeniids were the only ammonoids with a dorsal siphuncle; they radiated in Late Devonian faunas in Europe and North Africa, where the group is important for biostratigraphic correlation. The order developed a variety of shell shapes: *Progonioclymenia* is evolute with simple ribs, *Soliclymenia* evolved triangular whorls, and *Parawocklumeria* is a globular involute form with a trilobed appearance. As a whole, the goniatitides ranged in age from the Mid Devonian to Late Permian, with typical goniatitic sutures consisting of eight lobes and ventral siphuncles. *Goniatites*, for example, was a spherical inflated form with spiral striations,

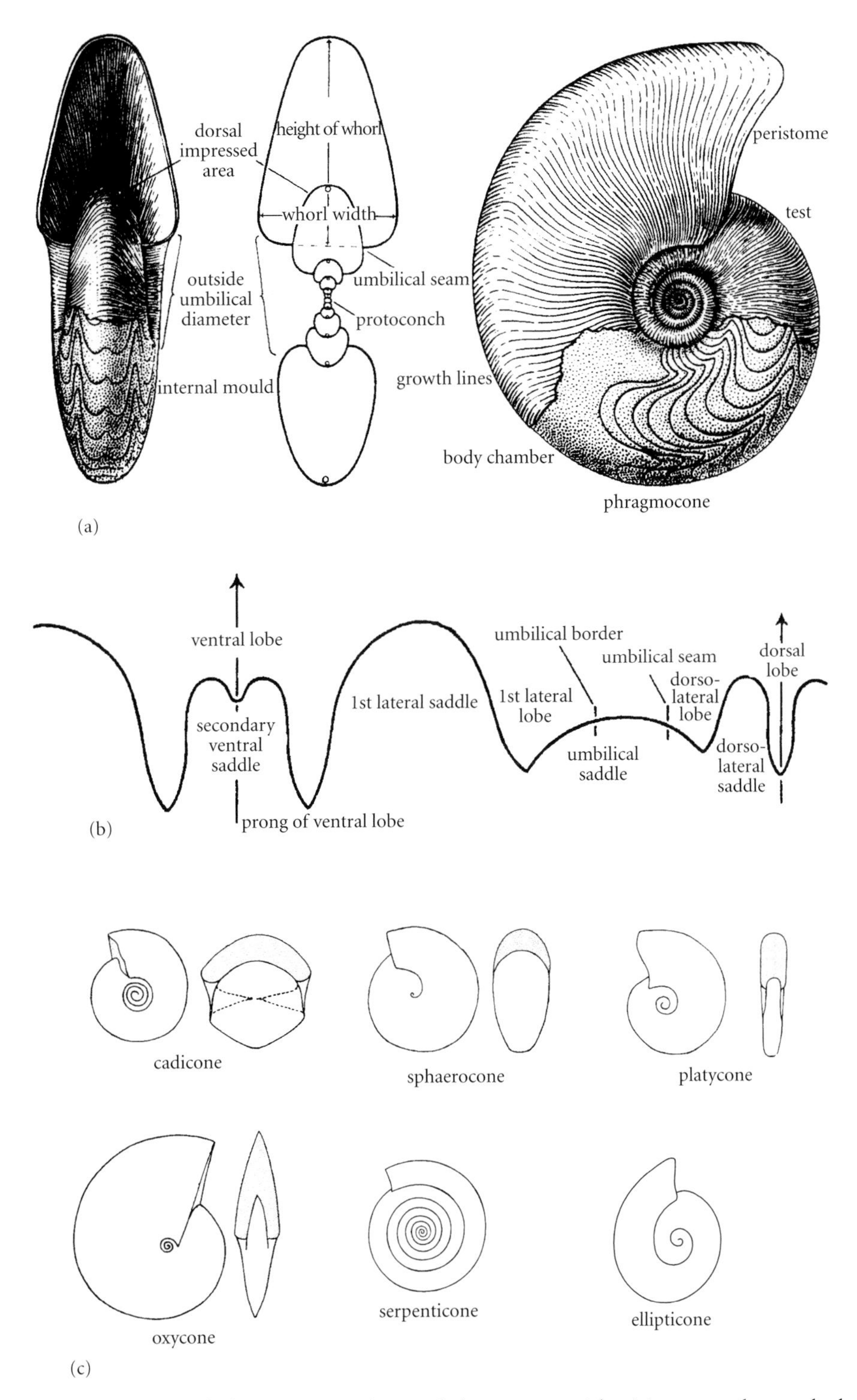

Figure 13.16 Morphology and shape terminology of the ammonoids: (a) external morphology, (b) suture pattern, and (c) shell shapes.

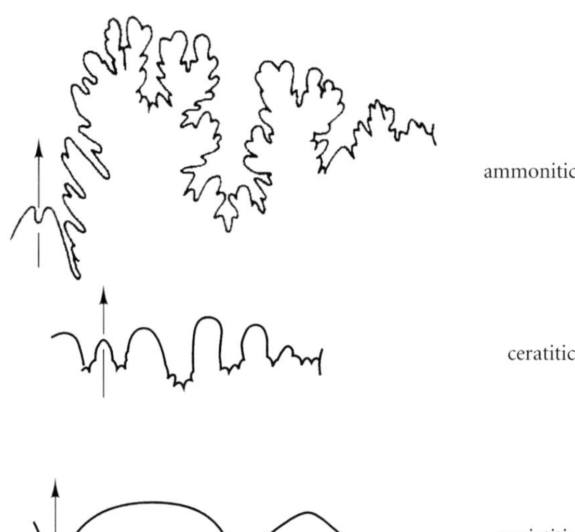

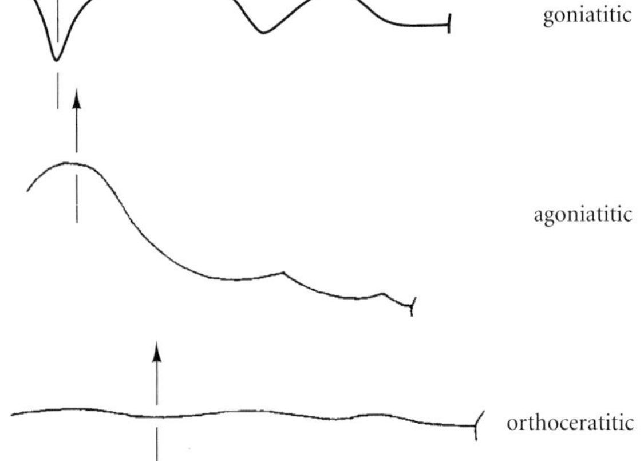

Figure 13.17 Evolution of suture patterns: the five main types; arrows point towards the frontal aperture.

whereas *Gastrioceras* was a depressed, tuberculate form.

The order Ceratitida includes the suborders Prolecantida and Ceratitida. The prolecantidines (Early Carboniferous to Late Permian) had large, smooth shells with wide umbilici, and sutures grading from goniatitic to ceratitic. *Prolecanites*, for example, was evolute with a wide umbilicus. The ceratitides include most of the Triassic ammonoids with ceratitic suture patterns and commonly elaborate ornamented shells. Nevertheless, some taxa developed ammonitic-grade sutures and a number of lineages evolved heteromorphs (Box 13.7).

The ammonites proper (Fig. 13.18) comprise four orders, the Phylloceratida, the Lytoceratida, the Ammonitida and the Ancyloceratida. The ammonitides appeared first in the Early Triassic with ammonitic sutures, commonly ornamented shells and ventral siphuncles. The first members of the order Phylloceratida, such as *Leiophyllites*, appear in Lower Triassic faunas and, according to some, this stem group probably gave rise to the entire ammonite fauna of the Jurassic and Cretaceous (Fig. 13.19). The morphologically conservative *Phylloceras* survived from the Early Jurassic to near the end of the Cretaceous with virtually no change, after having generated many of the major post-Triassic lineages. The phylloceratides were smooth, involute (with the last whorl covering all the previous ones), compressed forms; the suture had a marked leaf-like or phylloid saddle and a crook-shaped or lituid internal lobe. Although the group had a near-cosmopolitan distribution, its members were most common in the Tethyan province, but were characteristic of open-water environments.

The lytoceratides originated near the base of the Jurassic, with evolute (all previous whorls visible), loosely coiled shells, as seen in *Lytoceras* itself, which had a near-cosmopolitan distribution particularly during high stands of sea level. Like the phylloceratides, the order remained conservative; however, it too generated many other groups of Jurassic and Cretaceous ammonites.

The ammonitides included the true ammonites and ranged from the Lower Jurassic to the Upper Cretaceous, whereas the ancyloceratides included most of the bizarre heteromorph ammonites, ranging from the Upper Jurassic to the Upper Cretaceous.

Figure 13.18 (*opposite*) Ammonite taxa: (a) *Ludwigia murchisonae* (macroconch) from the Jurassic of Skye, (b) cluster of *Ludwigia murchisonae* (microconchs) from the Jurassic of Skye, (c) *Quenstedtoceras henrici* from the Jurassic of Wiltshire, (d) *Quenstedtoceras henrici* (showing a characteristic suture pattern) from the Jurassic of Wiltshire, and (e) *Peltomorphites subtense* from the Jurassic of Wiltshire, (f) *Placenticeras* (Cretaceous), (g) *Lytoceras* (Jurassic), (h) *Hildoceras* (Jurassic) and (i) *Cadoceras* (Cretaceous). Magnification ×1 (a–e), ×0.5 (f–i). (a–e, courtesy of Neville Hollingworth.)

(a)

(b)

(c)

(d)

(e)

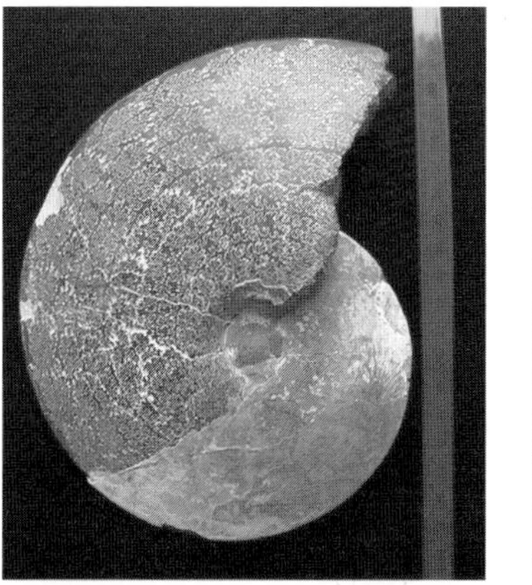
(f)

(g)

(h)

(i)

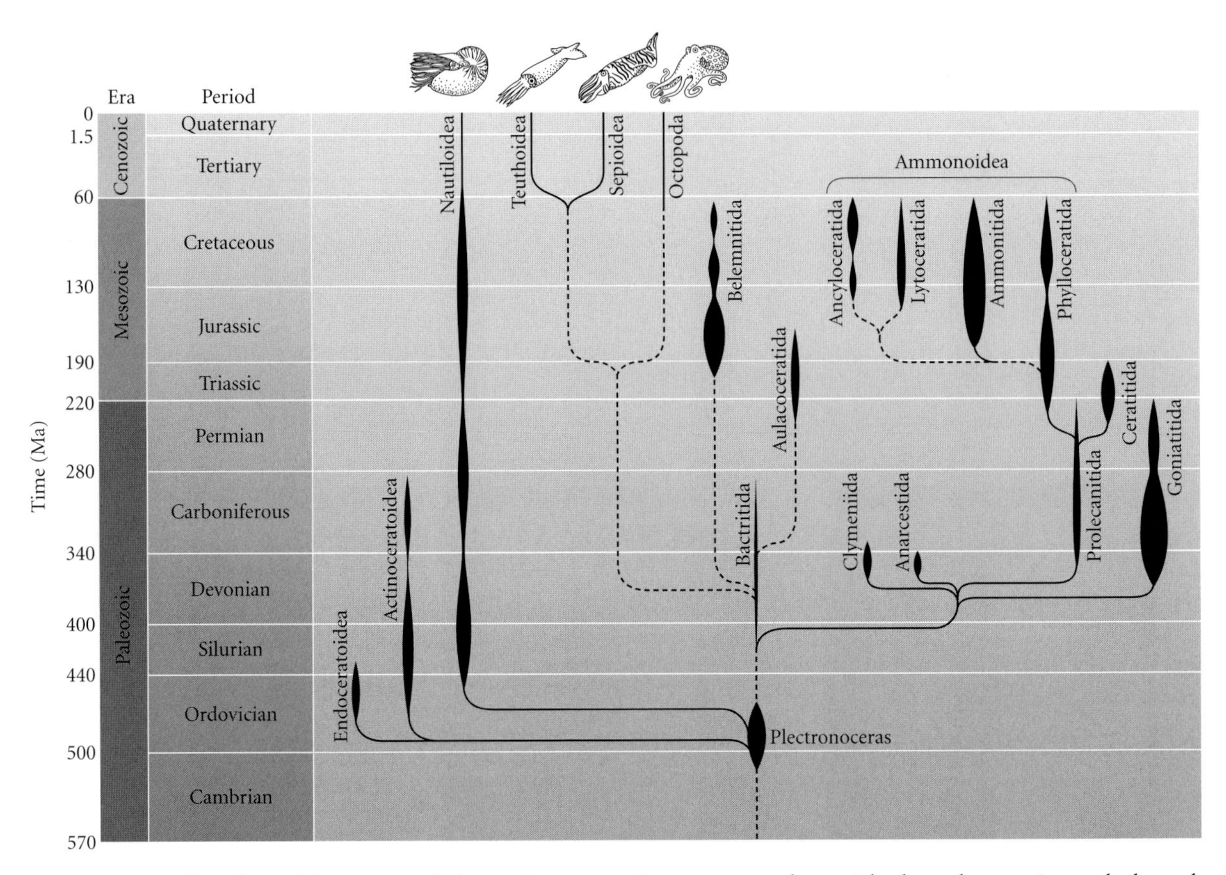

Figure 13.19 Stratigraphic ranges of the main ammonite taxa together with the other main cephalopod groups. (Based on Ward, P. 1987. *Natural History of Nautilus*. Allen & Unwin, Boston.)

Ammonoid ecology and evolution

The pioneer work by Arthur Trueman (University of Glasgow) on the buoyancy and orientation of the ammonite shell established the probable life attitudes for even the most bizarre heteromorph forms (Fig. 13.21a). Theoretically, at least, virtually all ammonoids could favorably adjust their attitude and buoyancy in the water column. Most ammonoids were probably part of the mobile benthos, although after death their gas-filled shells could be widely distributed by oceanic currents. Many groups of ammonoids were endemic, and the shovel-like jaws of some groups were most efficient at the sediment–water interface. Richard Batt's studies (1993) on Cretaceous ammonite morphotypes from the United States have established a series of shell types related to life modes and environments (Fig. 13.21b). For example, evolute heavily ornamented forms were probably nektobenthonic, as were spiny cadicones and spherocones, nodose spherocones and platycones, together with broad cadicones. Evolute planulates and serpenticones, together with small planulates, were probably pelagic in the upper parts of the water column. However, most oxycones were restricted to shallow-shelf depths. Some heteromorphs were nektobenthonic, whereas a few floated in the surface waters.

In many ammonite faunas the consistent co-occurrence of large and small similarly ornamented mature shells at specific horizons suggests that the macroconch and microconch may be related sexual dimorphs (the male and female of the species). The macroconch was probably the female, though this may not always have been the case. The ammonoids probably originated from the bactritid orthocone nautiloids, with protoconchs and large body chambers, during the earliest Devonian. The anarcestide goniatites, with simple sutural patterns, were relatively scarce during the Mid Devonian. However, by the Famennian, other groups such as the clymeniids, with a dorsally situated siphuncle, were common. The goniatitides expanded during the

Box 13.7 Ammonite heteromorphs

One of the more spectacular aspects of ammonite evolution was the appearance of bizarre **heteromorphic** ("different shape") shells in many lineages at a number of different times (Fig. 13.20). Heteromorphs first appeared during the Devonian, but were particularly significant in Late Triassic and Late Cretaceous faunas. Some such as *Choristoceras*, *Leptoceras* and *Spiroceras* appeared merely to uncoil; *Hamites*, *Macroscaphites* and *Scaphites* partly uncoiled and developed U-bends; whereas *Noestlingoceras*, *Notoceras* and *Turrilites* mimicked gastropods and *Nipponites* adopted shapes based on a series of connected U-bends. Initially, the heteromorph was considered as a decadent degenerate animal anticipating the extinction of a lineage. Nevertheless, some heteromorphs apparently gave rise to more normally coiled descendants and their association with extinction events only is far from true. Additionally, functional modeling suggests many were perfectly adapted to both nektobenthonic and pelagic life modes. Moreover Stephane Reboulet and her colleagues (2005) have shown that among the ammonites in the Albian rocks of the Vocontian Basin, southern France, heteromorphs probably were better adapted to compete in meso- and oligostrophic conditions than many other groups.

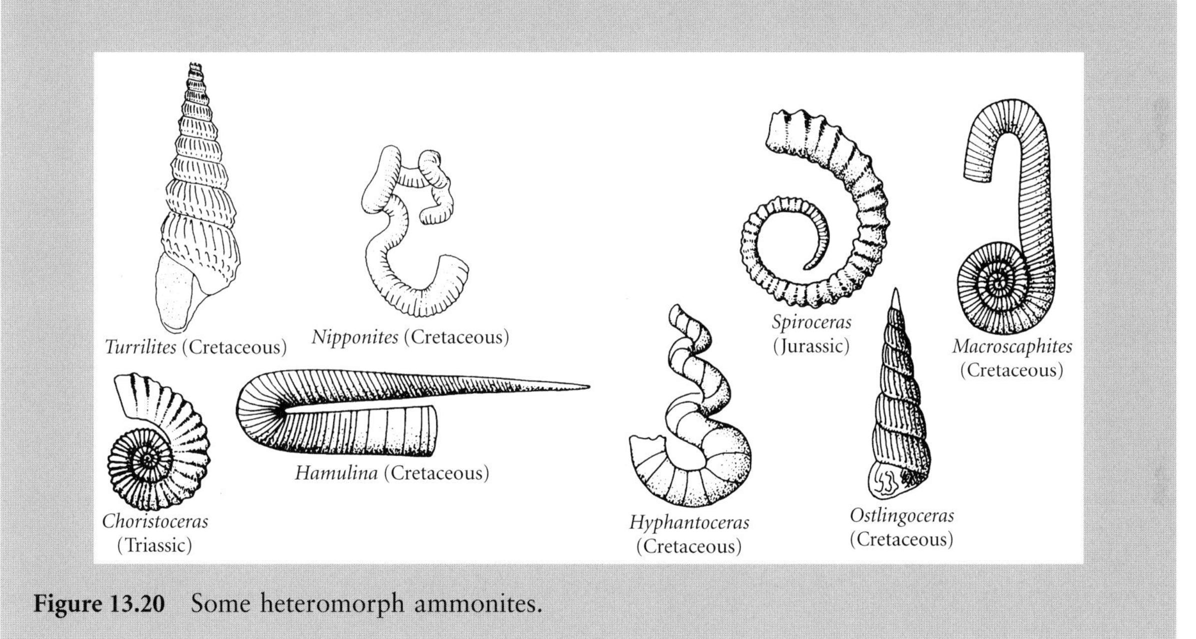

Figure 13.20 Some heteromorph ammonites.

Carboniferous, together with the prolecantides, where all the subsequent ammonoids probably originated. During the Triassic, the ceratitides diversified, peaking in the Late Triassic; but by the Jurassic the smooth involute phylloceratides, the lytoceratides and the ammonitides were all well established. Complex septa and sutures may have increased the strengths of the ammonoid phragmocone, protecting the shell against possible implosion at deeper levels in the water column. More intricate septa also provided a larger surface area for the attachment of the soft parts of the living animal, perhaps aiding more vigorous movement of the animal and its shell.

Coleoidea

The subclass Coleoidea contains cuttlefish, squids and octopuses, the latter including the paper nautilus, *Argonauta*. Coleoids show the dibranchiate condition, with a single pair of gills within the mantle cavity. Although argonauts can be traced back to the Mid Tertiary, the living coleoid orders generally have a poor fossil record, but preservation of arms, ink

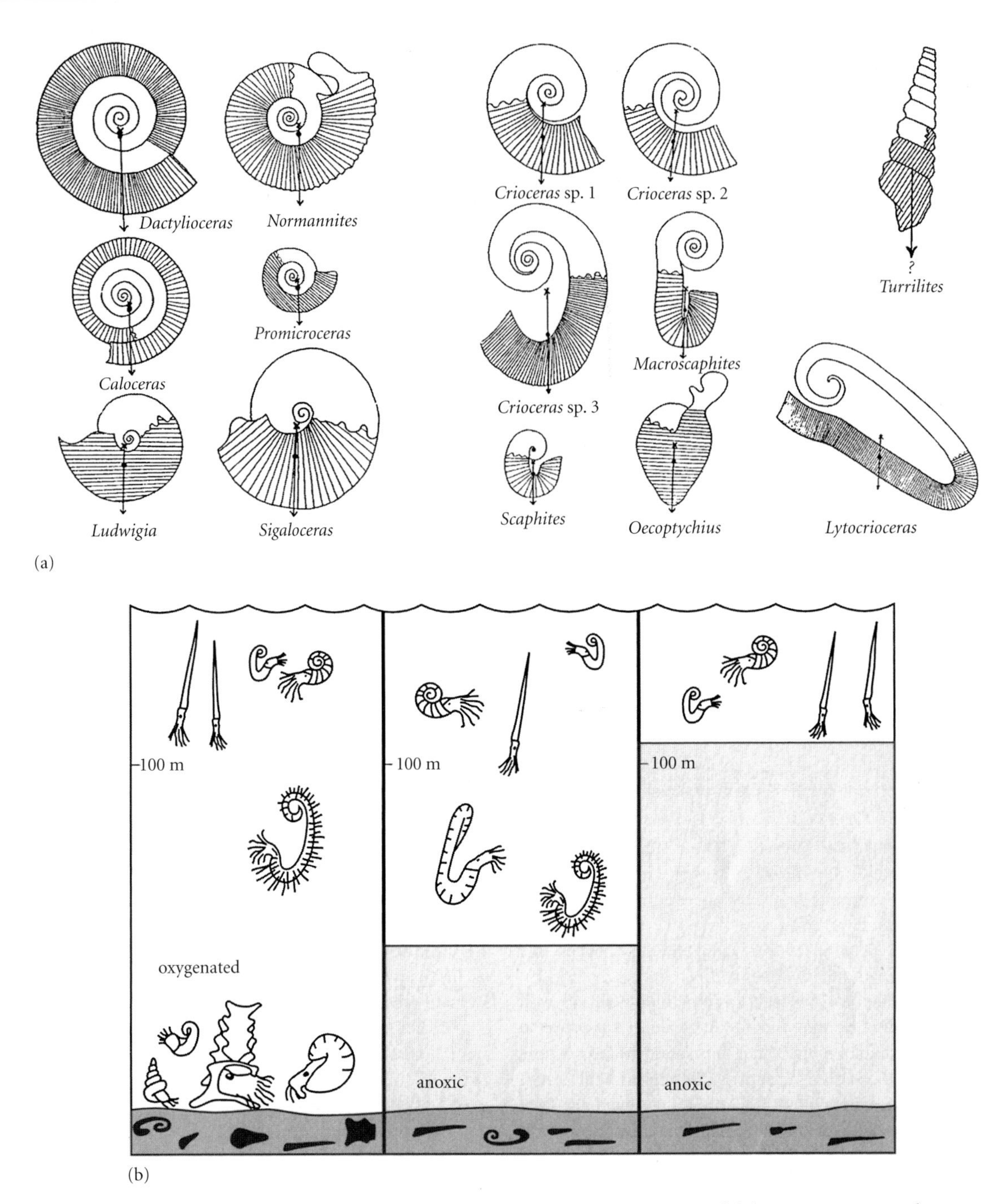

Figure 13.21 Life attitudes and buoyancy of the ammonites. (a) Supposed life orientations of a selection of ammonite genera, with the center of gravity marked ×; the center of buoyancy is marked with a dot and the extent of the body chamber is indicated with subparallel lines. (b) Relationship of some ammonite morphotypes to water depth and the development of anoxia. (a, from Trueman, A.E. 1940. *Q. J. Geol. Soc. Lond.* **96**; b, from Batt 1993.)

sacs and body outline is well known from several localities in the Jurassic. In contrast, the skeletons of extinct belemnites are locally abundant in Jurassic and Cretaceous rocks. Belemnites had an internal skeleton, contrasting with the exoskeletons of the shelled cephalopods such as the nautiloids and ammonoids. The belemnite skeleton is relatively simple, consisting of three main parts: the bullet-shaped **guard** is solid and composed of radially arranged needles of calcite with, at its anterior end, a conical depression or **alveolus** that houses the apical portion of the conical phragmocone, consisting of concave septa and a ventral siphuncle, and the spatulate **pro-ostracum** (corresponding to the dorsal wall of the body chamber of ectocochliate forms) that extends anteriorly (Fig. 13.22). This assemblage, situated on the dorsal side of the animal, is analogous to the chambered shells of nautiloids and ammonoids. Soft parts of belemnites, including the contents of ink sacs and tentacle hooks, are also occasionally preserved.

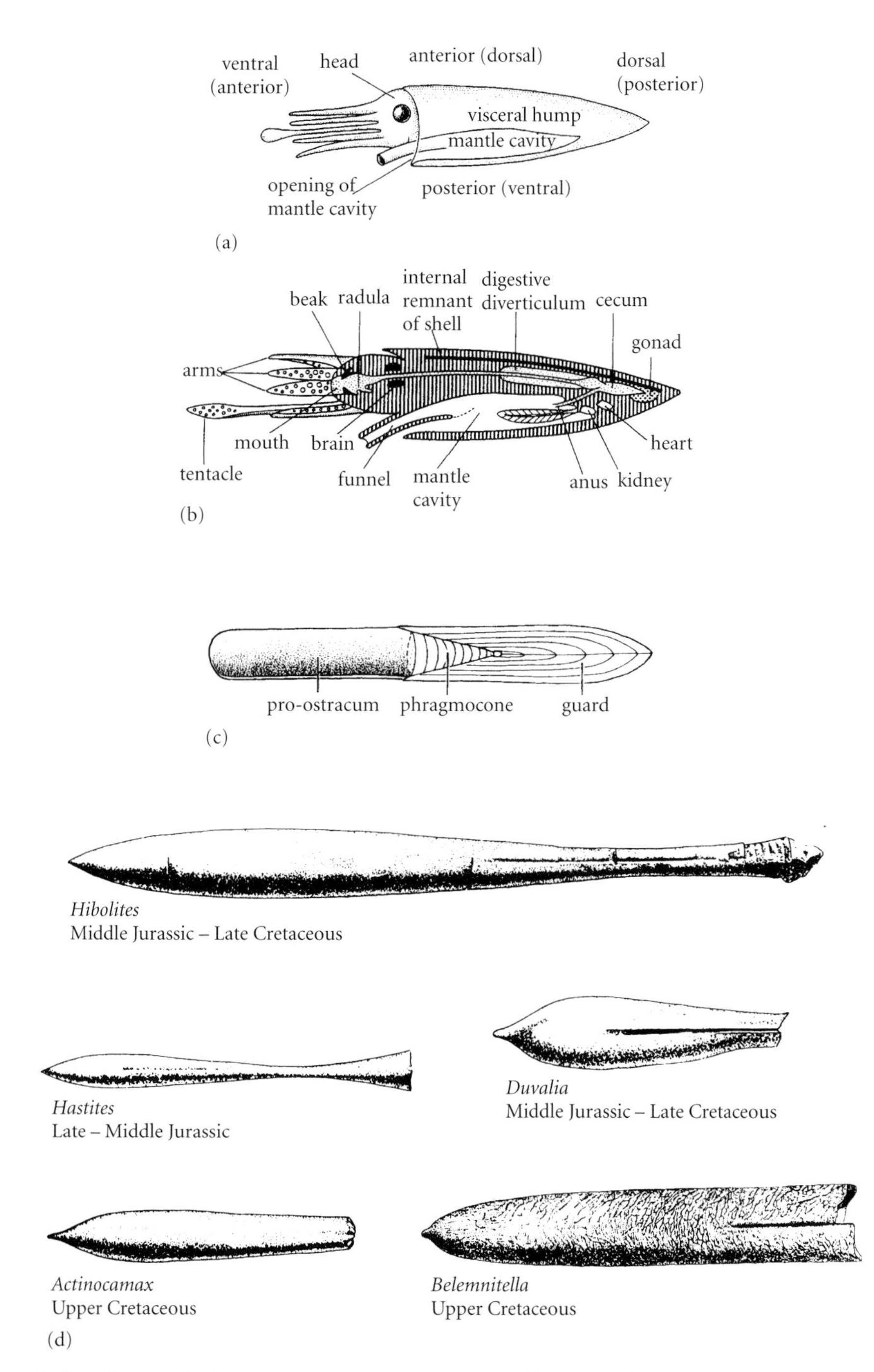

Figure 13.22 Coleoid morphology: (a) reconstruction of a living belemnite, (b) soft-part morphology of the belemnites, (c) internal skeleton of the belemnites, and (d) some belemnite genera. (From Peel et al. 1985.)

By analogy with modern squids, the belemnites were probably rapidly-moving predators living in shoals with their body level regulated by the guard. The animal thus probably maintained a horizontal attitude within the water column, preferring the open ocean. Data from the stomach contents of ichthyosaurs confirm that these mollusks formed part of their diet.

Some of the oldest records of belemnites, for example *Jeletzkya* from the mid-Carboniferous of Illinois, are tentative. The first unequivocal belemnites are from the Middle Triassic rocks of Sichuan Province, China where several species of *Sinobelemnites* occur. Belemnites became extinct at the end of the Cretaceous; later records are reworked or based on misinterpretations.

Some of the first supposed belemnites, like the Carboniferous *Paleoconus*, were relatively short stubby forms. In the Early Jurassic, *Megateuthis* was a long, slender form, whereas *Dactyloteuthis* was laterally flattened; the later Jurassic *Hibolites* is spear-shaped. The Cretaceous *Belemnitella* has a large bullet-shaped guard, whereas that of *Duvalia* has a flattened spatulate shape (Fig. 13.22d). However, despite differences in the detailed morphology of the endoskeltons across genera, many authorities consider that most of the Mesozoic belemnites probably looked very similar, but there are still enough features to measure on their skeletons and discriminate taxa (Box 13.8).

The compact calcareous guards of the belemnites have proved ideal for the analysis of oxygen isotope ratios ($O^{16}:O^{18}$) relating to paleotemperature conditions in the Jurassic and Cretaceous seas. These data have indicated warm peaks during the Albian and the Coniacian-Santonian (mid-Cretaceous) with a gradual cooling from the Campanian (Late Cretaceous) onwards. And as with many other Mesozoic groups, belemnite distributions show separate low-latitude Tethyan and high-latitude Boreal assemblages.

Spectacular mass accumulations of belemnite rostra are relatively common in Mesozoic sediments and, although some authors have used these assemblages in paleocurrent studies, few have addressed their mode of accumulation. Dense accumulations of bullet-shaped belemnite rostra have promoted the term "belemnite battlefields" for such distinctive shell beds (Fig. 13.23). These accumulations conform to five genetic types (Doyle & MacDonald 1993): (1) post-spawning mortalities (Fig. 13.23a); (2) catastrophic mass mortalities; (3) predation concentrates, either in situ or regurgitated (Fig. 13.23b); (4) condensation deposits perhaps aided by winnowing and sediment by-pass; and (5) resedimented deposits derived from usually condensed accumulations. Many of these so-called belemnite battlefields are then partly natural occurrences, reflecting the biology of the animals (numbers 1–3), but it is important to distinguish these from sedimentary accumulations (numbers 4 and 5) that say nothing about belemnite behavior.

CLASS SCAPHOPODA

Scaphopods are generally rare as fossils. The Scaphopoda, or elephant-tusk shells, have a single, slightly curved high conical shell, open at both ends (Fig. 13.25a). They lack gills and eyes, but have a mouth equipped with a radula and surrounded by tentacles; they also possess a foot, similar to that of the bivalves, adapted for burrowing. Scaphopods are mainly carnivorous, feeding on small organisms such as foraminiferans and spending much of their life in quasi-infaunal positions within soft sediment in deeper-water environments. The first scaphopods appeared during the Devonian and apparently had similar lifestyles to living forms such as *Dentalium*.

CLASS ROSTROCONCHA

Relatively recently a small class of mollusks, superficially resembling bivalves but lacking a functional hinge, has been documented from the Paleozoic. Over 35 genera have been described; most were originally described as bivalved arthropods. The rostroconchs probably had a foot that emerged through the anterior gape between the shells. However, the two shells are in fact fused along the mid-dorsal line, and posteriorly the shells are extended as a platform or rostrum (Fig. 13.25b). Ontogeny occurs from an initial **dissoconch** with the bilobed form developing from the disproportionate growth of shell from the lateral lobes of the mantle. The group appeared first during the Early Cambrian when, for example, *Heraultipegma* and

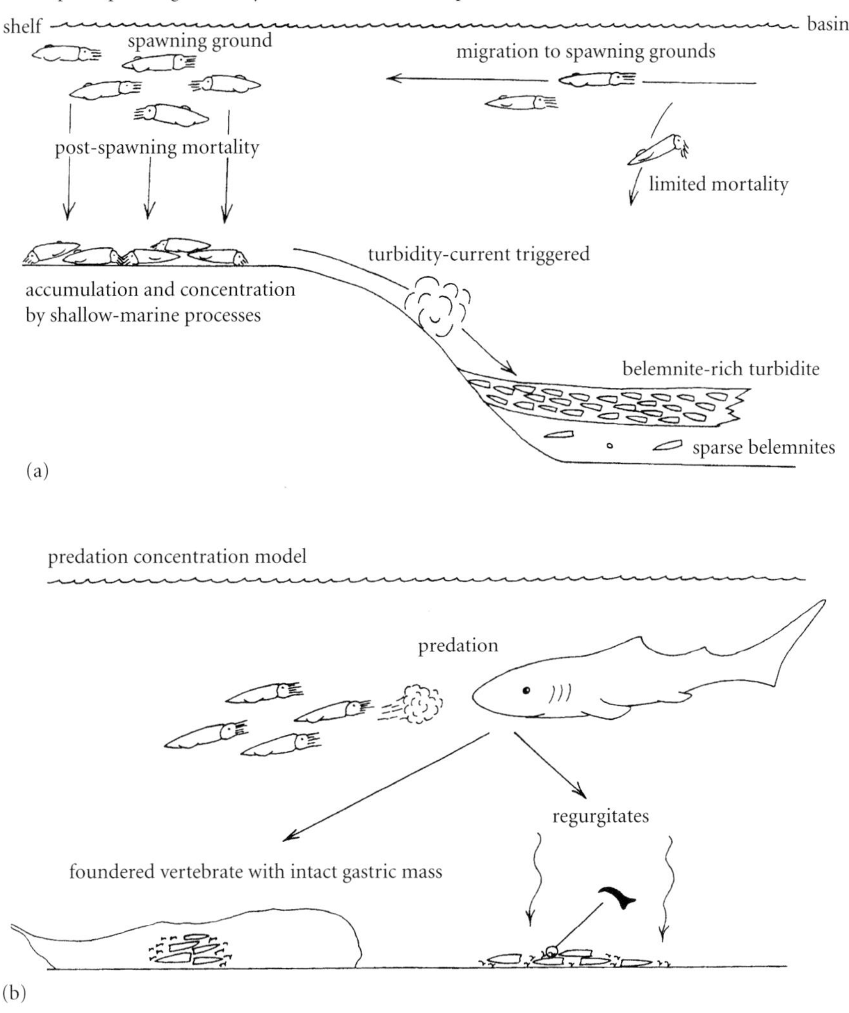

Figure 13.23 Belemnite battlefields and their possible origin: (a) post-spawning mortality model and (b) predation concentration model. (From Doyle & MacDonald 1993.)

Watsonella dominated rostroconch faunas of the Tommotian. The rostroconchs diversified during the Ordovician to reach an acme in the Katian when all seven families were represented. They probably occupied similar ecological niches to those of the bivalves. However, there followed a decline in abundance and diversity until final extinction at the end of the Permian when only conocardiodes such as *Arceodomus* were still extant.

The rostroconchs occupy a pivotal position in molluskan evolution (Runnegar & Pojeta 1974). The group developed from within the monoplacophoran plexus with a loss of segmentation; the rostroconchs themselves generated both the bivalves and the scaphopods whereas the gastropods and cephalopods were probably derived independently from a separate monoplacophoran ancestor. In some respects, the rostroconchs may represent a missing link between the univalved and bivalved molluskan lineages, while their unlikely morphology may have contributed towards their late discovery.

EVOLUTIONARY TRENDS WITHIN THE MOLLUSCA

A spectacular variety of mollusk morphotypes and life modes evolved during the Phanerozoic, from the simple body plan of the archemollusk. Despite the diversity of early mollusks in the Cambrian, the phylum was not notably conspicuous in the tiered suspension-feeding

Box 13.8 Gradualistic evolution of belemnites

There are relatively few long fossil lineages that can be used to demonstrate either phyletic gradualism or punctuated equilibria. Most of the best case studies (see p. 124) are based on mobile or sessile benthic organisms. The Cretaceous (Campanian) belemnite faunas, particularly the genus *Belemnitella*, of North Germany are abundant, well preserved and known in great detail and provide an unequalled opportunity to test these models using a pelagic group of organisms (Christensen 2000). There is not much variation in lithology throughout the succession in Lower Saxony – they are mainly rather boring, monotonous marly limestones. There are thus limited opportunities for facies shifts to influence the morphological record of *Belemnitella* by the migration of more exotic morphotypes in and out of the basin. Although samples of *Belemnitella* through the section are superficially similar, several measurements show gradualistic trends when treated quantitatively (Fig. 13.24). Not all changes are unidirectional, some exhibit reversals. It seems most likely that the Campanian belemnites of northern Germany conformed to continuous, gradual phyletic evolution in narrowly fluctuating, slowly changing environments.

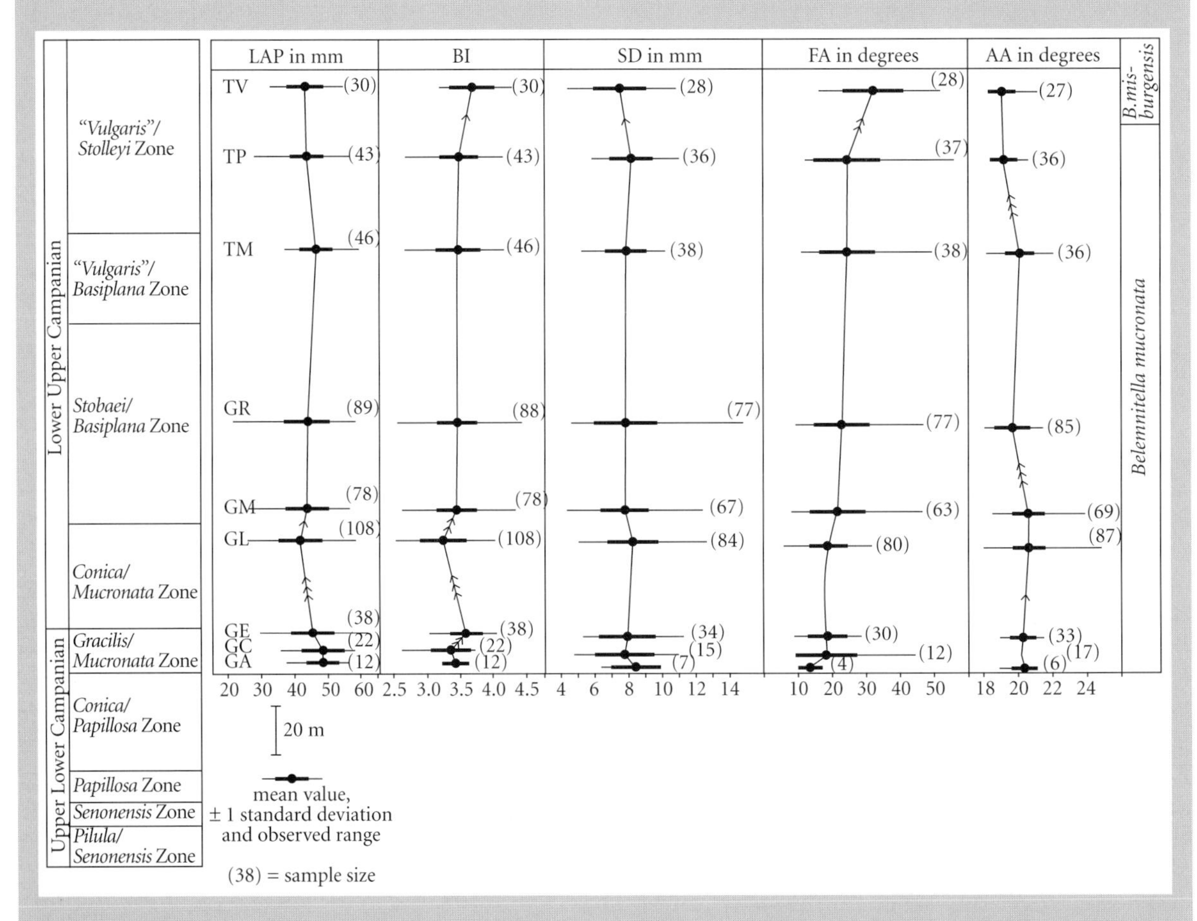

Figure 13.24 Gradualistic evolution of Cretaceous belemnites from North Germany. Summary of changes of the length from the apex to the protoconch (LAP), Birkelund index (BI), Schatzky distance (SD), fissure angle (FA) and alveolar angle (AA) of nine samples of *Belemnitella*. Successive mean values are different at the 5% level (one arrow), 1% level (two arrows) and 0.1% level (three arrows). (Courtesy of the late Walter Kegel Christensen.)

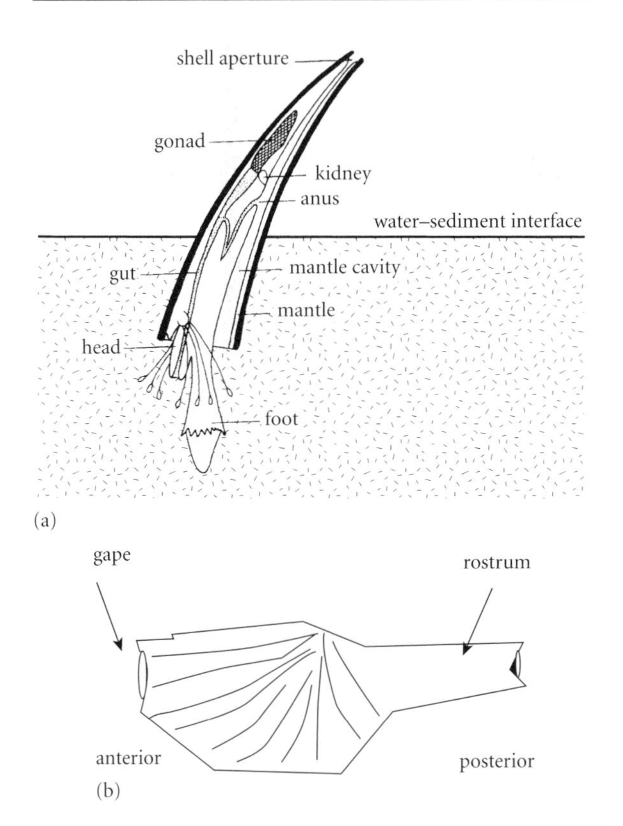

Figure 13.25 (a) Scaphopod morphology and (b) rostroconch morphology.

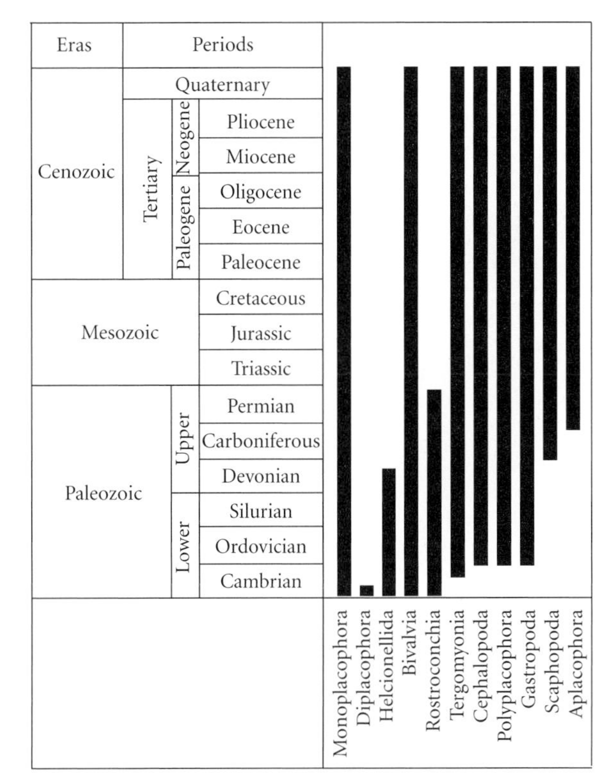

Figure 13.26 Stratigraphic range of the main mollusk groups.

benthos of the Paleozoic, although many more localized, often nearshore, assemblages were dominated by mollusks.

During the Paleozoic, bivalves were common in nearshore environments, often associated with lingulide brachiopods, although the class also inhabited a range of deeper-water clastic environments; and by the Late Paleozoic bivalves had invaded a variety of carbonate environments. However, at the end of the Paleozoic, the appearance of more typical bivalves in shallow-water belts may have displaced the Paleozoic associations seaward. During the Late Mesozoic and Cenozoic the significant radiation of infaunal taxa may have been a response to increased predation.

The majority of Paleozoic gastropods were Eogastropoda that commonly dominated shallow-water marine environments and some carbonate reef settings. The Mesozoic, however, was dominated by the Mesogastropoda, which grazed on algal-coated hard substrates. The Cenozoic, marks the acme of the group with the radiation of the siphonal carnivorous neogastropods, and with a further diversification of mesogastropods (Fig. 13.26).

The cephalopods evolved through the development of a chambered shell with a siphuncle, which gave them considerable control over attitude and buoyancy; this system was refined in the nautiloid groups. The evolution of complex folded sutures in the ammonoids, the exploitation of a pelagic larval stage and a marginal position for the siphuncle apparently set the agenda for the further radiation of the group during the Mesozoic.

Throughout the Phanerozoic, the fleshy mollusks provided a source of nutrition for many groups of predators. The evolution of the phylum was probably in part influenced by the development of predator–prey relationships and minimization of predator success. Thick armored shells were developed in some groups while the evolution of deep-infaunal life modes was also part of a defensive strategy. Predation and the development of avoidance strategies, together with the so-called arms race, had an important influence on molluskan evolution. Predators develop a particular search image when seeking their favored prey. Living terrestrial snails show a wide range of color patterns and the purpose of this variability may be to confuse predators like the song thrush by presenting a wide range of images. If a predator targets as prey one particular variant in the population, then other variants would be free

Box 13.9 Mesozoic marine revolution

The post-Paleozoic seas and oceans were probably different in many ways from those before. One key difference is the more intense predator–prey relationships, signaled by the Mesozoic marine revolution (MMR). During this interval, shell predation by, for example, crushing and drilling, became commoner. A Mesozoic arms race, with predators evolving more highly developed weapons of attack, was balanced by prey evolving better defensive mechanisms and structures. Thus whereas crustaceans developed the efficiency of their claws, jaws and pincers, mollusks grew thicker, more highly-ornamented shells and perhaps burrowed deeper and faster into the sediment. This form of escalation is somewhat different from the mechanism of coevolution; organisms adapt to each other rather than merely change together. In this system, predators will always be one step ahead of their prey. Liz Harper (2006) has reviewed the evidence for post-Paleozoic escalation, plotting the ranges of durophagous body and trace fossils that may have been predatory together with evidence for crushing and drilling of shells (Fig. 13.27). The MMR may have been a complex series of events: (i) a Triassic radiation of decapods, sharks and bony fishes; (ii) Jurassic-Cretaceous radiations of malacostracans and marine reptiles; (iii) a Paleogene explosion of neogastropods, teleosts and sharks; and (iv) the Neogene appearance of mammals and birds.

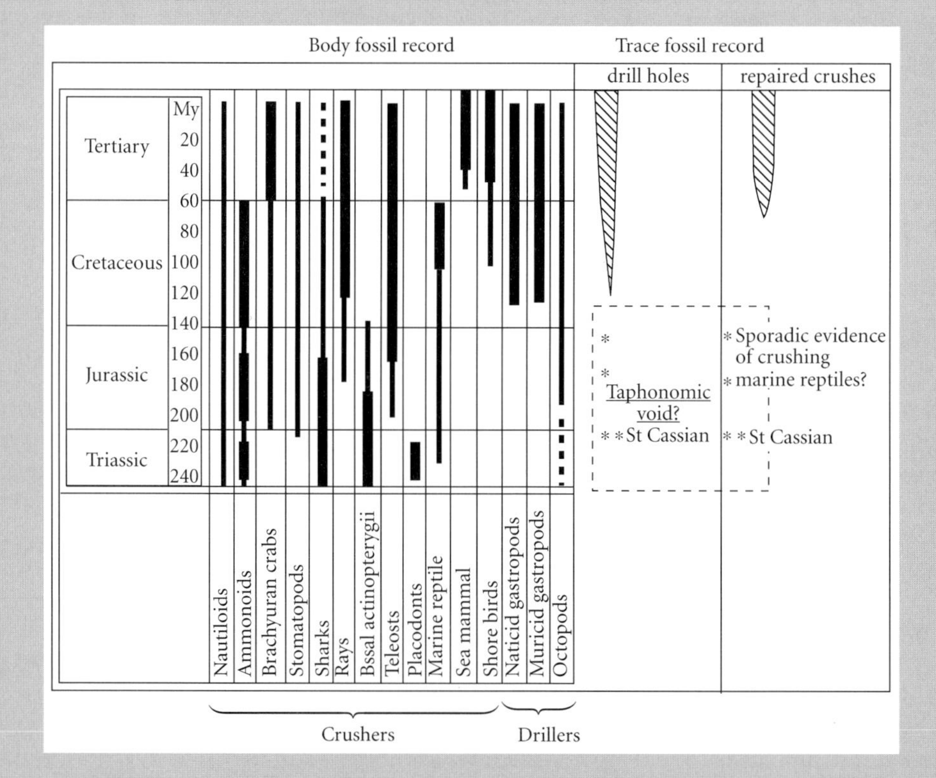

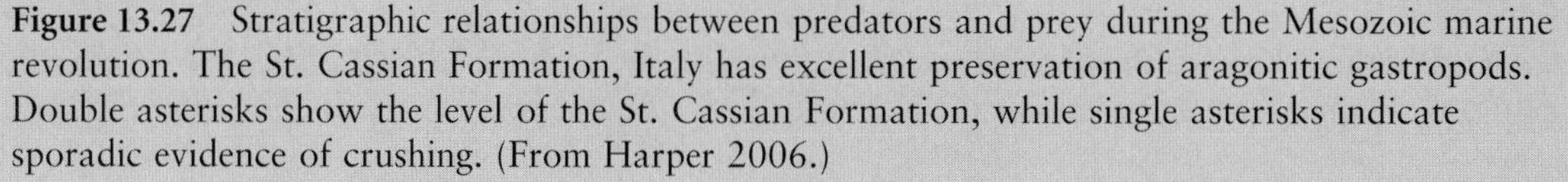

Figure 13.27 Stratigraphic relationships between predators and prey during the Mesozoic marine revolution. The St. Cassian Formation, Italy has excellent preservation of aragonitic gastropods. Double asterisks show the level of the St. Cassian Formation, while single asterisks indicate sporadic evidence of crushing. (From Harper 2006.)

to recover until a switch in images was produced. Although such relationships are documented for some Mesozoic (Box 13.9) and Cenozoic faunas, data are sparse for the Paleozoic. On the other hand, close relatives of the mollusks, the annelids, may have been important predators equipped with an efficient jaw apparatus (Box 13.10).

Box 13.10 Fossil annelids and their jaws

The annelids are segmented protostomes that are represented today by animals such as the earthworms and leaches. Recent species are important, widely distributed, benthic predators and occur from intertidal to abyssal depths. Modern molecular studies suggest they form a sister group to the mollusks and, in fact, share a number of morphological characters such as the possession of chaetae. In general the group has a fairly sparse fossil record, appearing fleetingly in Lagerstätte deposits such as the Burgess Shale and Mazon Creek fauna. However many residues of acid-etched Paleozoic limestones contain scolecodonts (Fig. 13.28). These were the jaws of ancient annelids and are abundant and diverse at many horizons. They were similar to conodonts (see p. 429), forming multielement apparatuses with similar functions but were composed of collagen fibers and various minerals such as zinc. The group first appeared in the Lower Ordovician and diversified rapidly to become common in Upper Ordovician-Devonian carbonate facies. Scolecodonts were relatively rare after the Permian, but nevertheless have proved useful in biostratigraphic and thermal maturation studies.

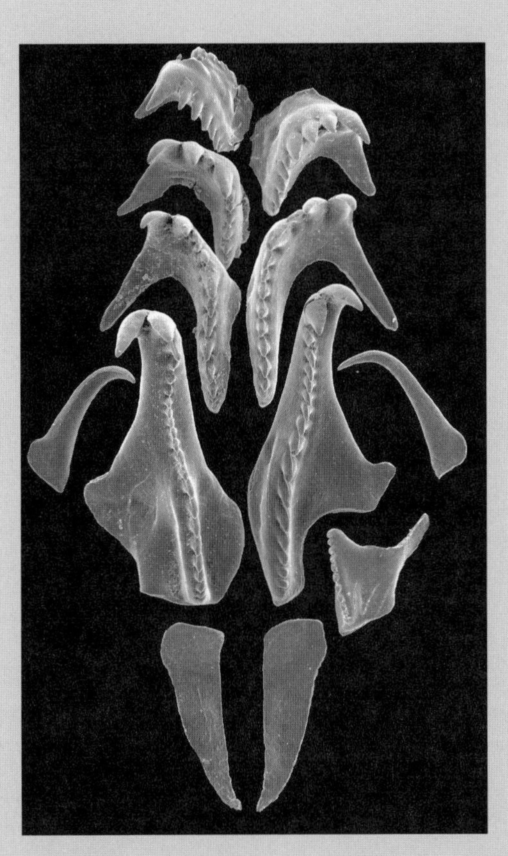

Figure 13.28 Scolecodont morphology. Reconstruction of the polychaete jaw apparatus of the Ordovician *Ramp hoprion* Kielan-Jaworowska. (Courtesy of Olle Hints.)

Review questions

1 There has been some difficulty identifying the first mollusk. What are the key features of the phylum and how would they be recognized in the first mollusk?
2 Many taxa that form part of the Early Cambrian biota are undoubtedly mollusks. Which mollusk groups are already present in the small shelly fauna?
3 Theoretical morphospace is a useful tool to investigate shell morphology. Some groups are more constrained in their developmental opportunities than others. What advantages should univalved mollusks have over bivalved mollusks in a quest to generate extreme morphotypes?
4 Belemnites seem an unlikely group to test models for microevolution. What conditions should be met in such tests of microevolutionary hypotheses?
5 The Mesozoic marine revolution (or arms race) was a complex ecological event that set the agenda for marine life in the Modern evolutionary fauna. How did mollusks react to predation pressures?

Further reading

Clarkson, E.N.K. 1998. *Invertebrate Palaeontology and Evolution*, 4th edn. Chapman and Hall, London. (An excellent, more advanced text; clearly written and well illustrated.)

Lehmann, U. 1981. *The Ammonites – their Life and their World*. Cambridge University Press, Cambridge, UK.

Morton, J.E. 1967. *Molluscs*. Hutchinson, London.

Peel, J.S., Skelton, P.W. & House, M.R. 1985. Mollusca. *In* Murray, J.W. (ed.) *Atlas of Invertebrate Macrofossils*. Longman, London. (A useful, mainly photographic review of the group.)

Pojeta, J. Jr., Runnegar, B., Peel, J.S. & Gordon, M. Jr. 1987. Phylum Mollusca. *In* Boardman, R.S., Cheetham, A.H. & Rowell, A.J. (eds) *Fossil Invertebrates*. Blackwell Scientific Publications, Oxford, UK, pp. 270–435. (A comprehensive, more advanced text with emphasis on taxonomy; extravagantly illustrated.)

Vermeij, G.J. 1987. *Evolution and Escalation. An Ecological History of Life*. Princeton University Press, Princeton, NJ. (Visionary text.)

References

Batt, R. 1993. Ammonite morphotypes as indicators of oxygenation in a Cretaceous epicontinental sea. *Lethaia* **26**, 49–63.

Caron, J.-B., Acheltema, A., Schander, A. & Rudkin, D. 2006. A soft-bodied mollusc with radula from the Middle Cambrian Burgess Shale. *Nature* **442**, 159–163.

Christensen, W.K. 2000. Gradualistic evolution in *Belemnitella* from the middle Campanian of Lower Saxony, NW Germany. *Bulletin of the Geological Society of Denmark* **47**, 135–63.

Doyle, P. & MacDonald, D.I.M. 1993. Belemnite battlefields. *Lethaia* **26**, 65–80.

Erwin, D.H. 2007. Disparity: morphological pattern and developmental context. *Palaeontology* **50**, 57–73.

Fedonkin, M. & Waggoner, B.M. 1997. The Late Precambrian fossil *Kimberella* is a mollusk-like bilaterian organism. *Nature* **388**, 868–71.

Harper, E.M. 2006. Disecting post-Palaeozoic arms races. *Palaeogeography, Palaeoclimatology, Palaeoecology* **232**, 322–43.

Jacobs, D.K. & Landman, N.H. 1993. *Nautilus* – a poor model for the function and behavior of ammonoids. *Lethaia* **26**, 101–11.

Milsom, C. & Rigby, S. 2004. *Fossils at a Glance*. Blackwell Publishing, Oxford.

Peel, J.S. 1991. Functional morphology, evolution and systematics of early Palaeozoic univalved molluscs. *Grønlands Geologiske Undersøgelse* **161**, 116 pp.

Raup, D.M. 1966. Geometric analysis of shell coiling: general problems. *Journal of Paleontology* **40**, 1178–90.

Reboulet, S., Giraud, F. & Proux, O. 2005. Ammonoid abundance variations related to changes in trophic conditions across the Oceanic Anoxic Event 1d (Latest Albian, SE France). *Palaios* **20**, 121–41.

Runnegar, B. & Pojeta, J. 1974. Molluscan phylogeny: the palaeontological viewpoint. *Science* **186**, 311–17.

Sigwart, J.W. & Sutton, M.D. 2007. Deep molluscan phylogeny: synthesis of palaeontological and neontological data. *Proceedings of the Royal Society B* **274**, 2413–19.

Stanley, S.M. 1970. Relation of shell form to life habits of Bivalvia. *Geological Society of America Memoir* **125**, 296 pp.

Swan, A.R.H. 1990. Computer simulations of invertebrate morphology. *In* Bruton, D.L. & Harper, D.A.T. (eds) *Microcomputers in Palaeontology*. Contributions from the Palaeontological Museum, University of Oslo, Vol. 370. pp. 32–45. University of Oslo, Oslo.

Vinther, J. & Nielsen, C. 2004. The Early Cambrian *Halkieria* is a mollusc. *Zoologica Scripta* **34**, 81–9.

Wagner, P.J. 1995. Diversity patterns among early gastropods: contrasting taxonomic and phylogenetic descriptions. *Paleobiology* **21**, 410–39.

Wani, R., Kase, T., Shigeta, Y. & De Ocampo, R. 2005. New look at ammonoid taphonomy, based on field experiments with modern chambered nautilus. *Geology* **33**, 849–52.

Williamson, P.G. 1981. Palaeontological documentation of speciation in Cenozoic molluscs from Turkana basin. *Nature* **293**, 140–2.

Chapter 14

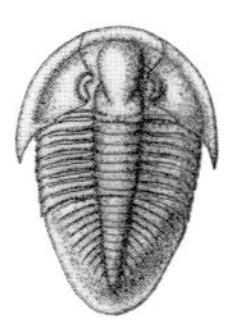

Ecdysozoa: arthropods

Key points

- Arthropods – such as lobsters, spiders, beetles and trilobites – have legs, a segmented body plan with jointed appendages and the ability to molt.
- The first major arthropod faunas of the Early Cambrian appear bizarre by modern standards but probably were no more morphologically different to each other than are living faunas.
- A number of arthropod-like animals in the Ediacara biota suggest an ancient origin for the phylum.
- Trilobites appeared in the Early Cambrian and during the Paleozoic evolved advanced visual systems and enrolment structures while pursuing a variety of benthic and pelagic life styles.
- The largest arthropods were the chelicerates and included the giant eurypterids that patrolled marine marginal environments during the Silurian and Devonian.
- Myriapods represent the earliest terrestrial body fossils in the Mid Ordovician, but trackways indicate euthycarcinoids (i.e. stem-group mandibulates) moved onto land even earlier, in the Late Cambrian.
- Insects first appeared during the Early Devonian and diversified rapidly; there are probably 10 million species of living insects.
- Insects had probably already evolved flight before the Mid Carboniferous, when giant dragonflies patrolled the forests.
- The crustaceans include many familiar groups such as crabs, lobsters and shrimps, together with the barnacles and ostracodes.
- Much of our knowledge of the early history of the phylum has come from exceptionally preserved fossils from the Cambrian Burgess Shale, Chengjiang and Sirius Passet faunas.

Whence we see spiders, flies, or ants entombed and preserved forever in amber, a more than royal tomb.

Francis Bacon, English philosopher (1561–1626)

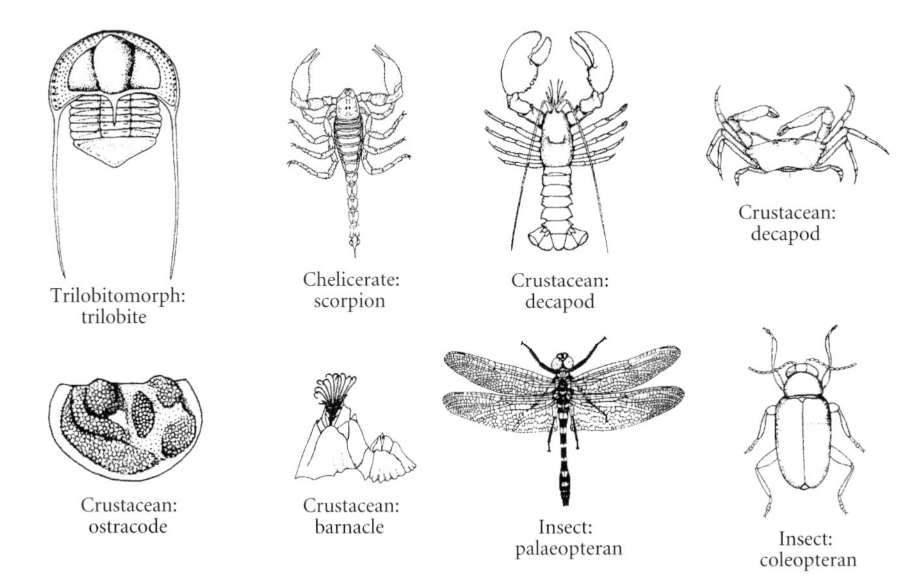

Figure 14.1 Some of the main arthropod groups: a variety of forms based on a simple body plan of a tough exoskeleton and jointed limbs.

ARTHROPODS: INTRODUCTION

Arthropods are a very common and spectacularly diverse group of legged invertebrates accounting for about three-quarters of all species living on the planet today, largely because of the phenomenal abundance of the insects. The basic body plan – conspicuously segmented, with jointed appendages adapted for feeding, locomotion and respiration – together with a tough **exoskeleton**, first appeared during the Early Cambrian and has since been exploited by a huge variety of living and fossil arthropods that pursue many lifestyles. All members of this phylum have both segmented bodies and appendages (Fig. 14.1); moreover the animal is differentiated into a head, thorax and abdomen, with often the head and thorax fused to form the **cephalothorax**. The possession of **mandibles**, or hard mouthparts, equipped many arthropods with the ability to process a wide variety of foods.

The arthropod exoskeleton is constructed mainly from the organic substance chitin. This is often hardened or sclerotized by calcium carbonate or calcium phosphate, so the potential for preservation is excellent across the group. The exoskeleton acts as a base for the attachment of locomotory muscles, permitting rapid movement, and is not usually mineralized. Although many arthropods undergo metamorphosis, virtually all the main groups grow by molting or **ecdysis**; first the endoskeleton is dissolved and second the old exoskeleton is detached along sutures while the new exoskeleton is generated. **Exuviae**, or cast-off coverings, are all that remain of the previous skeleton or cuticle of the animal: one arthropod can thus produce many potential skeletal fossils in its lifetime.

During a geological history of at least 540 million years, the five subphyla of arthropods (Box 14.1) have adapted to life in marine, freshwater and terrestrial environments. For a long time their closest living relatives were thought to be the segmented annelid worms, but new studies show that the closest sister group of arthropods is a clade of unsegmented worms that includes the priapulids and the nematodes or round worms. Their segmentation may thus either have arisen independently to that of the annelids, or may have been inherited from a very deep ancestor to both groups.

EARLY ARTHROPOD FAUNAS

A huge variety of bizarre arthropod types formed much of the basis for the Cambrian explosion (see p. 249). Over 20 groups of arthropod have been described from the Mid

Box 14.1 Classification of arthropods

There are currently differences in the status given to the main arthropod groups. If the Arthropoda is in fact a superphylum, the following groupings are phyla. However, some authorities have assigned the following superclass status within the phylum Arthropoda. The classification here is a compromise. Basal to the phylum are a number of minor but evolutionarily important groups, such as the tardigrades (water bears), that are now known from the Cambrian.

Subphylum TRILOBITOMORPHA

- Trilobites and their relatives; animals with a cephalon, thorax and pygidium; the body, lengthwise, has an axial lobe and two lateral pleural lobes
- Cambrian to Permian

Subphylum CHELICERATA

- Large group with a body divided into two tagmata; the prosoma (which bears six pairs of appendages, the first being the chelicerae or pincer-like appendages, giving the group its name), and the opisthosoma with an extended tail or telson
- ?Cambrian to Recent

Subphylum MYRIAPODA

- Includes the flexible centipedes together with the millipedes
- ?Ordovician, Silurian to Recent

Subphylum HEXAPODA

- Highly-diverse group, with a head, thorax and abdomen and six legs; includes the ants, beetles, dragonflies, flies and wasps
- Devonian to Recent

Subphylum CRUSTACEA

- Includes the bivalved phyllocarids; Early Paleozoic taxa were ancestral to the crabs, shrimps and lobsters
- Cambrian to Recent

Cambrian Burgess Shale and related deposits (see Box 14.8); some have even been assigned to new phyla, emphasizing the expansive nature of the explosion, truly evolution's "big bang". Stephen Jay Gould, in his bestseller *Wonderful Life* argued that morphological disparity during the Cambrian was greater than at any time since. Nevertheless, cladistic, and phenetic analyses of both morphological and taxonomic criteria suggested otherwise (Briggs et al. 1993). Rather, the morphological disparity among the Cambrian arthropods is not markedly different from that seen across living taxa, they just look stranger to us. But it is nonetheless remarkable that very early in their history arthropods attained high levels of morphological disparity not really exceeded during the next 500 million years of evolution. Moreover, our knowledge of the Cambrian arthropod record, particularly that of soft-bodied organisms, is probably not nearly as complete as that of the modern fauna and we should expect further surprises as more Cambrian Lagerstätten are investigated (Box 14.2).

SUBPHYLUM TRILOBITOMORPHA

The trilobitomorphs are highly derived arthropods lacking specialized mouthparts, and with **tagmata** comprising the **cephalon, thorax** and **pygidium**, together with trilobitomorph appendages that have lateral branches developed from the walking limbs. The trilobitomorphs include mainly the trilobites and over 15,000 species are known. Trilobites were a unique and very successful arthropod group, common throughout the Paleozoic until their extinction at the end of the Permian. There is no doubt that the trilobites are one of the most attractive fossils groups, much prized by both amateur and professional collectors alike. Some of the earliest arthropods were trilobites, and marine Cambrian strata (as well as many later deposits) are usually correlated on the basis of trilobite assemblages. The group formed an important part of the mobile benthos, although a few groups were adapted to pelagic life modes.

Trilobite morphology

The trilobite exoskeleton (Fig. 14.3), as the name trilobite ("three-lobed") suggests, is divided longitudinally into three lobes; the **axial lobe** protects the digestive system, whereas the two **pleural lobes** cover appendages. In virtually all trilobites a well-defined cephalon, thorax and pygidium are developed; the trilobite exoskeleton is composed almost entirely of calcite.

The cephalon has a raised axial area, the **glabella**, with a series of glabellar furrows. Eyes are commonly developed laterally (Box 14.3), with the **facial** or **cephalic suture** separating the inner fixed and the outer free cheeks. Although many trilobites lacked eyes, this may be a secondary condition; despite loss of vision the cephalic sutures remained. The sutures themselves are very important for understanding the functional morphology and classification of the group. There are four main types of suture (Fig. 14.5): the **proparian** mode extends posteriorly in front of the genal angle, whereas the **opisthoparian** mode cuts the posterior margin of the cephalon behind the genal angle, and the **gonatoparian** suture bisects the genal angle and lateral sutures follow the margin of the cephalon. In the rarer **metaparian** condition the suture extends from near the genal angle on the posterior margin, around the eye to finish farther along the same margin.

On the ventral surface, underneath the cephalon, three plates were associated with the anterior soft parts including the mouth. The **rostral plate** is situated at the anterior margin. Posterior to the rostral plate, the **hypostome**, a plate of variable shape and size, is usually sited under the glabella. The shape and position of the hypostome is of great help in classifying the group. The small **metastoma** is known from only a few taxa and apparently lay behind the mouth. The dorsal margin was protected by a ventral flange or **doublure**.

Box 14.2 Ediacaran arthropods?

Are they or aren't they? Some paleontologists believe they can identify some of the Ediacaran animals as arthropods or proto-arthropods; others dispute this. *Parvancorina* (Fig. 14.2), for example, is a possible candidate, with its shield-shaped outline, strong axial ridge and arched anterior lobes, together with a convex profile. It really looks like a juvenile trilobite molt stage, but did not have a mineralized skeleton. Not convinced? Beautifully preserved fossils from a new Cambrian Lagerstätte, the Chinese Kaili fauna seem to confirm it. Specimens of the genus *Skania*, first described from the Burgess Shale, have many similarities to *Parvancorina* but this genus has an exoskeleton and a better-defined cephalon and dorsal trunk (Lin et al. 2006). *Skania* together with *Parvancorina* and *Primicaris* may have formed a sister group to the Arachnomorpha (which includes the spiders). Moreover this relationship establishes a Proterozoic root for the Arthropoda and is the first arthropod crown group that demonstrably ranges through the Precambrian–Cambrian boundary. A Proterozoic origin for the arthropods may help pin down more precisely their time of divergence from the last common ancestor of the arthropods and priapulid worms.

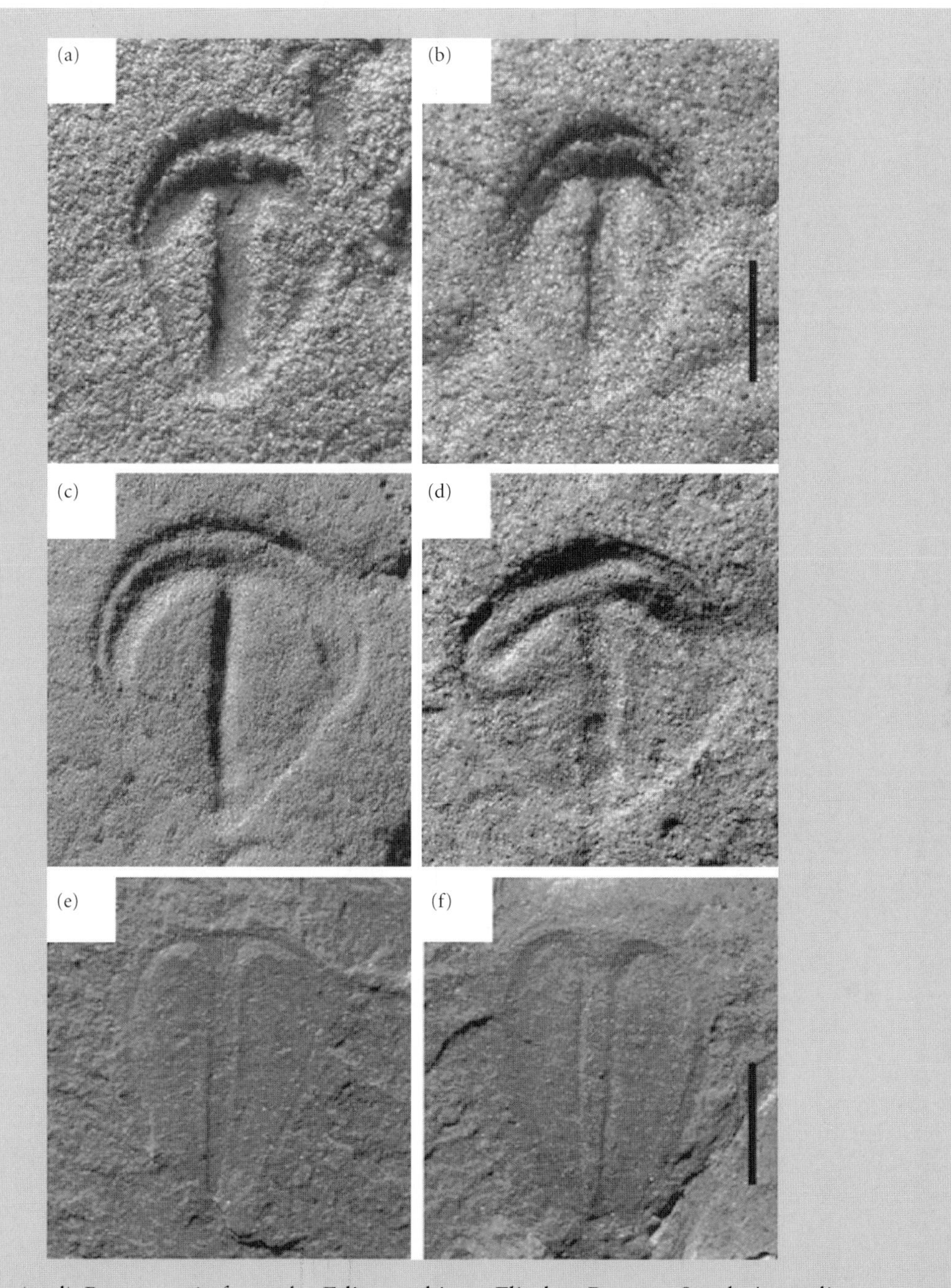

Figure 14.2 (a–d) *Parvancoria* from the Ediacara biota, Flinders Ranges, South Australia; (e, f) *Skania* from the Middle Cambrian of Guizhou Province, South China. Scale bar: 3.5 mm (a), 4 mm (b), 10 mm (c, d), 2 mm (e, f). (Courtesy of Jih-Pai (Alex) Lin.)

With the exception of the agnostids and eodiscids, which have two and two to three thoracic segments, respectively, trilobites are **polymeric**, that is usually having up to about 40 thoracic segments. The trilobite pygidium is usually a plate of between one and 30 fused segments. Most Cambrian trilobites have small, **micropygous** pygidia, whereas later forms are either **heteropygous**, where the pygidium is smaller than the cephalon,

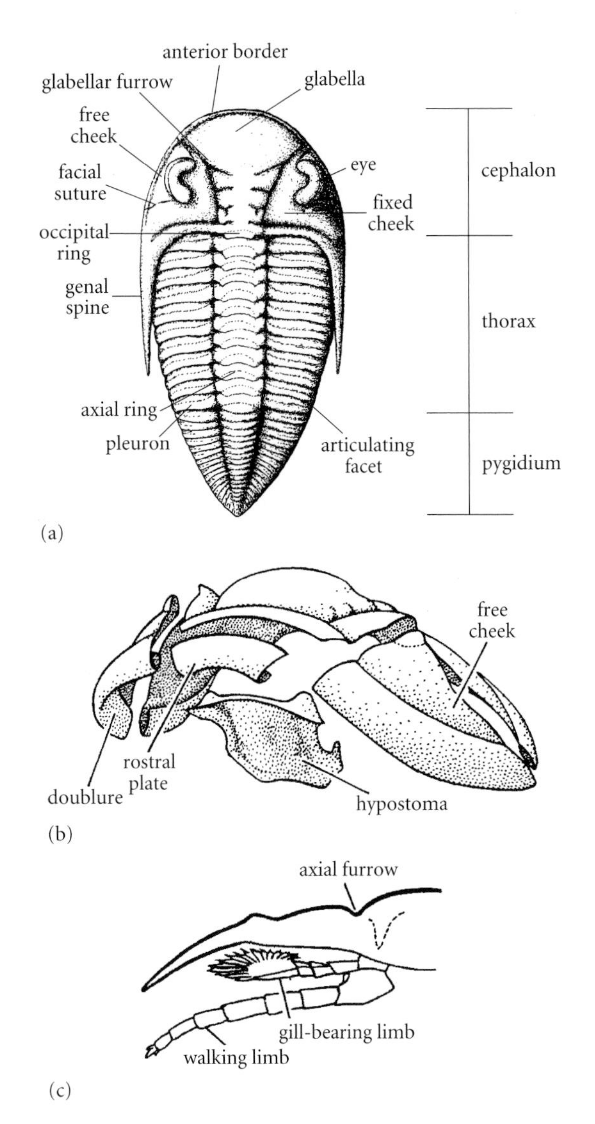

Figure 14.3 Trilobite morphology: (a) external morphology of the Ordovician trilobite *Hemiarges*; (b) generalized view of the anterior of the Silurian trilobite *Calymene* revealing details of the underside of the exoskeleton; and (c) details of the limb pair associated with a segment of the exoskeleton.

or **macropygous**, where the pygidium is larger.

Like virtually all arthropods the trilobites grew by **ecdysis** or molting (Fig. 14.6). Ontogeny involved the periodic discarding of spent exoskeletons or exuviae. Initial molt stages were quite different from those of adults. After a **phaselus** larval stage that swam freely in the plankton, the **protaspis** stage is a minute disk with a segmented median lobe destined to become the glabella. The next, **meraspis**, stage has a discrete, transitory pygidium where thoracic segments form at its anterior margin and are released at successive molts to form the thorax. The **holaspis** stage has a full complement of thoracic segments for the species but growth continues through further molts and maturity may not be reached until some time after the holaspis stage was reached. Clearly in many trilobite-dominated faunas, counts of skeletal remains will significantly over-represent the relative numbers of living animals in the community. Many researchers divide the number of exuviae by about six to eight to obtain a more realistic census of the trilobite population in a typical community (see p. 93).

During times of stress, to avoid unpleasant environmental conditions or perhaps an attentive predator, most trilobites could roll up like a carpet. During the Paleozoic, a number of groups, including asaphids, calymenids, phacopids and trinucleids (see p. 374), evolved a variety of sophisticated structures to enhance this behavior, although Cambrian taxa probably had a limited ability to curl up. **Spheroidal** enrolment involved articulation of all the thoracic segments to form a ball, whereas in the less common **discoidal** mode of enrolment the thorax and pygidium were merely folded over the cephalon. Cambrian trilobites could certainly enrol, but it was not until the Ordovician that true **coaptative** structures, locking parts of the skeleton against each other, first appeared. For example, in the phacopids, tooth and socket pairs were developed on the cephalic and pygidial doublure, respectively; these opposing structures clicked together to hold the trilobite in a tight ball, presenting only the exoskeleton to the world outside (Bruton & Haas 2003).

Main trilobite groups and lifestyles

Although some workers have split the trilobites into two orders, the Agnostida and the Polymerida, most currently recognize about nine orders of trilobite based on a spectrum of characters, including the anatomy of their ontogenetic stages and more recently the location and morphology of the hypostome. In the most primitive **conterminant** condition, the hypostome is similar in shape to the glabella and is attached to the anterior part of the doublure. **Natant** hypostomes were not attached to the skeleton, whereas the **impendent** hypostome was attached to the doublure, but its shape was quite different from the glabella above.

Some authorities have excluded the distinctive agnostids from the Trilobitomorpha and there is now strong evidence to suggest they were crustaceans. Agnostids were small to minute, usually blind animals with subequal cephala and pygidia and only two thoracic segments; they were probably planktonic, which may account for their very wide distribution.

The redlichiids include *Olenellus*, with 18–44 spiny thoracic segments and typical of the Atlantic province, *Redlichia* itself, more typical of the Pacific province, and the large, spiny, micropygous *Paradoxides*, common in the high latitudes of the Mid Cambrian.

Corynexochid trilobites were a mixed bag of taxa; the order includes genera with conterminant hypostomes such as *Olenoides* and large smooth forms such as *Bumastus* and *Illaenus*, having impendent hypostomes.

The lichids contain mainly spiny forms with conterminant hypostomes. Apart from *Lichas* itself the order also includes the spiny odontopleurids such as *Leonaspis*.

Phacopids were mainly proparian trilobites with schizochroal eyes (Box 14.3) and lacking rostral plates that ranged from the Lower Ordovician to Upper Devonian. The order includes the large tuberculate *Cheirurus*, *Calymene* with a marked gonatoparian suture, and *Dalmanites* with long genal spines, kidney-shaped eyes, spinose thoracic segments and the pygidium extended as a long spine.

The ptychopariids are all characterized by natant hypostomes and include some specialized groups. For example *Triarthrus* was modified for burrowing, *Conocoryphe* was blind and *Harpes* had a sensory fringe round the cephalon.

Asaphids had either conterminant or impendent hypostomes and include *Asaphus*

Box 14.3 Vision in trilobites: from corrective lenses to sunshades

Trilobites have the oldest known visual system based on eyes: paleontologists can even look through the ancient lenses and see the world as trilobites saw it! Trilobite eyes are **compound**, consisting of many lenses, just like those of the crustaceans and insects. Euan Clarkson's classic studies (1979) emphasized the functions of the two main types of lens arrangement found in trilobites (Fig. 14.4). The trilobite eye generally consists of many lenses of calcite with the c-axis (the main optical axis) perpendicular to the surface of the eye. The more primitive and widespread **holochroal** eye has many close-packed lenses, all about the same size, covered by a single membrane. The more advanced and complex **schizochroal** condition has no modern analog and has larger, discrete lenses arranged in rows or files. It is uncertain how this system operated in detail; presumably it offered higher-quality images than those of the holochroal systems. Moreover both mature holochroal and schizochroal configurations apparently developed from immature schizochroal conditions. Thus early growth stages of holochroal eyes in quite different groups such as those of the Cambrian eodiscid *Shizhudiscus* with the oldest visual system in the world and *Phacops* from the Devonian with a schizochroal system have broadly similar arrangements, suggesting that the latter system developed by pedomorphosis (see p. 145). A third, less well known optical system, **abathochroal**, is confined to a short-lived Cambrian group, the eodiscids (most of which were blind). Less is known about them than other visual systems and their origins remain obscure.

But could such visual systems cope with bright sunlight? Probably not, and this suggests that many groups were nocturnal. But not all. A remarkable Devonian phacopid trilobite, *Erbenochile*, from Morocco, actually has a type of sunshade covering the top of a column of lenses (Fortey & Chatterton 2003). The animal could scan the seafloor for potential prey without the distraction of direct sunlight.

Continued

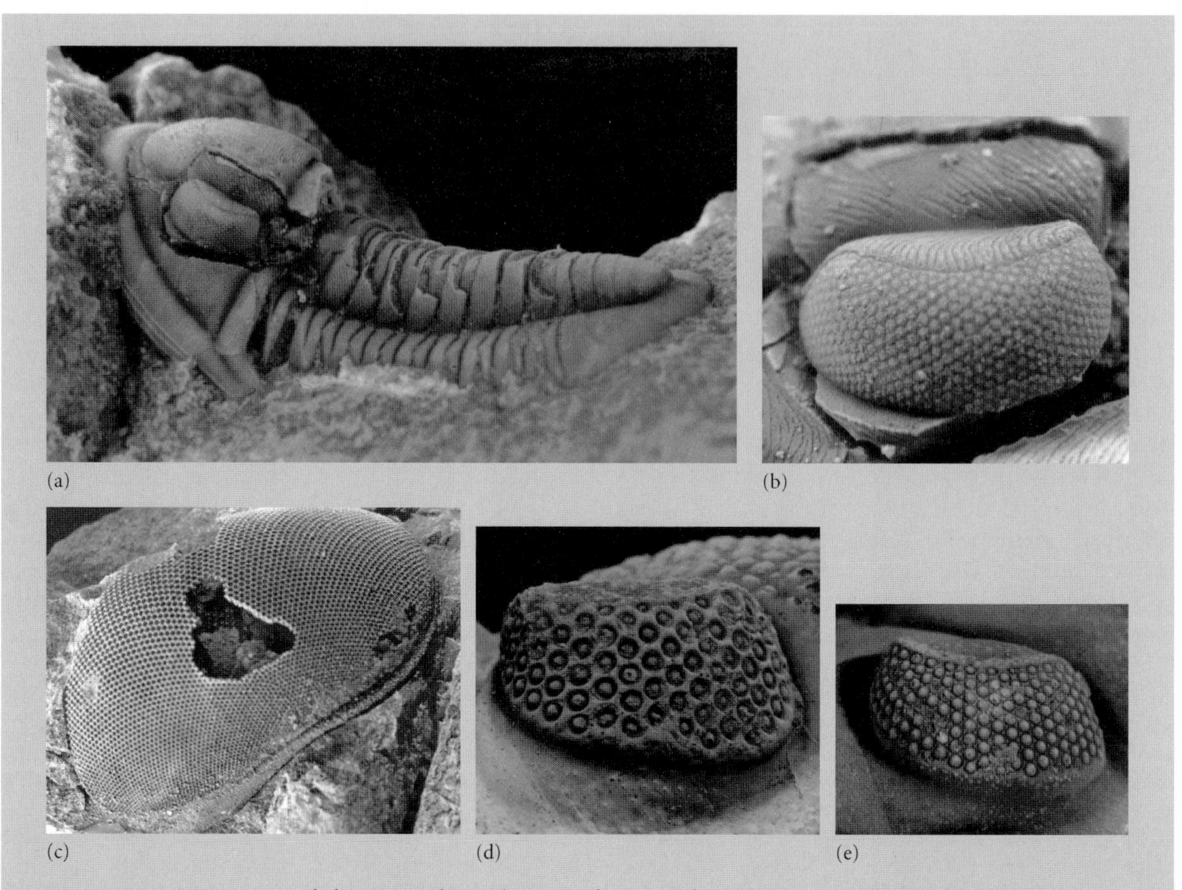

Figure 14.4 Vision in trilobites: (a) lateral view of a complete specimen of *Cornuproetus*, Silurian, Bohemia (×4); (b) detail of the compound eye of *Cornuproetus* (×20); (c) holochroal compound eye of *Pricyclopyge*, Ordovician, Bohemia (×6); (d) schizochroal compound eye of *Phacops*, Devonian, Ohio (×4); and (e) schizochroal compound eye of *Reedops*, Devonian, Bohemia (×5). (Courtesy of Euan Clarkson.)

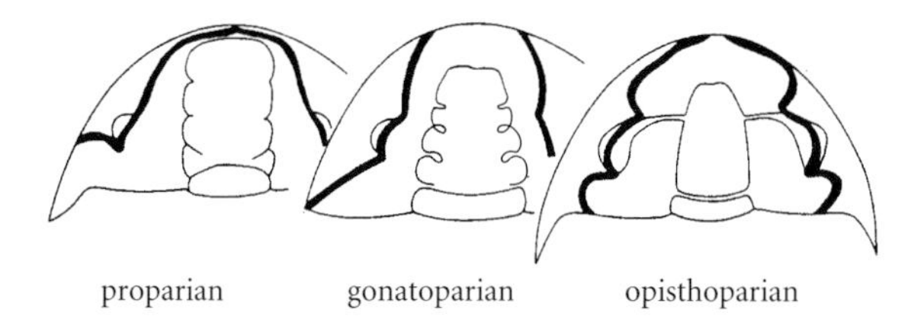

Figure 14.5 Facial sutures: the tracks of the proparian, gonatoparian and opisthoparian sutures. The lateral suture (not illustrated) follows the lateral margin of the cephalon.

and *Ceratopyge* together with pelagic forms such as *Cyclopyge* and *Remopleurides*, and the stratigraphically important trinucleids such as *Onnia*, *Cryptolithus* and *Tretaspis*.

The proetids were isopygous forms with large glabellae and long hypostomes having genal spines and large holochroal eyes. The group ranged from the Lower Ordovician to the Upper Permian. *Proetus* was a small form with a relatively large, inflated and often granular glabella, known from the Ordovician to the Devonian. *Phillipsia*, one of the youngest members of the order, was a small isopygous genus with large crescent-shaped eyes and an opisthoparian suture.

The naraoids, including *Naraoia* itself and *Tegopelte*, have often been included with the trilobites. They were not calcified and lacked thoracic segments. The group was restricted to the Middle Cambrian. *Naraoia* was first described from the Burgess Shale as a branchiopod crustacean, but it has only more

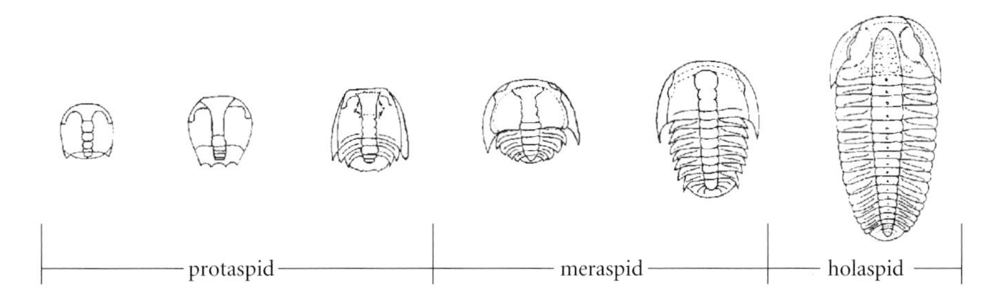

Figure 14.6 Molt phases of the Bohemian trilobite *Sao hirsuta* Barrande. Magnifications: protaspid stages approximately ×9, meraspid stages approximately ×7.5 and the holaspid stages approximately ×0.5. (Based on Barrande 1852.)

recently been reclassified as a soft-bodied trilobite. It is now known from other Cambrian Lagerstätten together with a number of related taxa. The group is probably a sister group to the trilobites + agnostids (Edgecombe & Ramsköld 1999) and recent cladistic analyses confirm this phylogenetic position, basal to Trilobitomorpha, and within the larger clade Arachnomorpha (Cotton & Braddy 2004).

Trilobite morphology is hugely variable, presumably reflecting their broad range of adaptations (Fig. 14.7). Most trilobites were almost certainly benthic or nektobenthic, leaving a variety of tracks and trails in the marine sediments of the Paleozoic seas (see Chapter 19). With the exception of the phacopids that may have hunted, the simple mouthparts of the trilobites suggest a diet of microscopic organisms and a detritus-feeding strategy.

Many trilobites developed spinose exoskeletons. The spines reduce their weight:area ratio and this suggested that these trilobites adopted a floating, planktonic life strategy, supposedly backed up by the fact they occasionally had inflated glabellae. More recently, however, the suggestion that their glabella was filled with gas has been shown to be a little fanciful, and it seems more likely that these forms used their long spines to spread the weight on a soft muddy substrate. Downward-directed spines probably held the thorax and pygidium well above the sediment–water interface. In some forms, the spines probably aided shallow burrowing when the body flexed. Spines are most extravagantly developed in the odontopleurids.

Some trilobites such as *Cybeloides* and *Encrinurus* evolved eyes on stalks or others, for example *Trinucleus*, lost them altogether in favor of possible sensory **setae** (stiff hair-like structures). These specialized forms may have periodically concealed themselves in the sediment. *Trimerus* had a cephalon and pygidium fashioned in the shape of a shovel that might have helped it plow through the sediment. The cyclopygid *Opipeuter*, from the Lower Ordovician of Spitzbergen, Ireland and Utah, on the other hand, seems to have been an active pelagic swimmer; it had a long, slender body with a flexible exoskeleton and large eyes, just like a modern shrimp-like amphipod, together with a widespread distribution.

Trilobites show extensive convergence: the same broad morphotypes appear repeatedly in different lineages, presumably reflecting repeats of the same life strategies. Richard Fortey and Robert Owens documented seven ecomorphic groups ranging from the turberculate, mobile phacomorphs to the smooth, infaunal illaenimorphs (Fig. 14.8) and these were related to their wide variety of lifestyles (Fig. 14.9).

Distribution and evolution: trilobites in space and time

Trilobite faunas have formed the basis for many paleogeographic reconstructions of the Cambrian and Ordovician world. During the Cambrian, biogeographic patterns were complex, but some provinces have been defined, such as the high-latitude Atlantic region (with redlichiids) and the low-latitude Pacific region (with olenellids). Statistical analysis of Ordovician trilobite faunas in the early 1970s established a low-latitude bathyurid province (Laurentia), an intermediate to high-latitude asaphid province (Baltica) and a

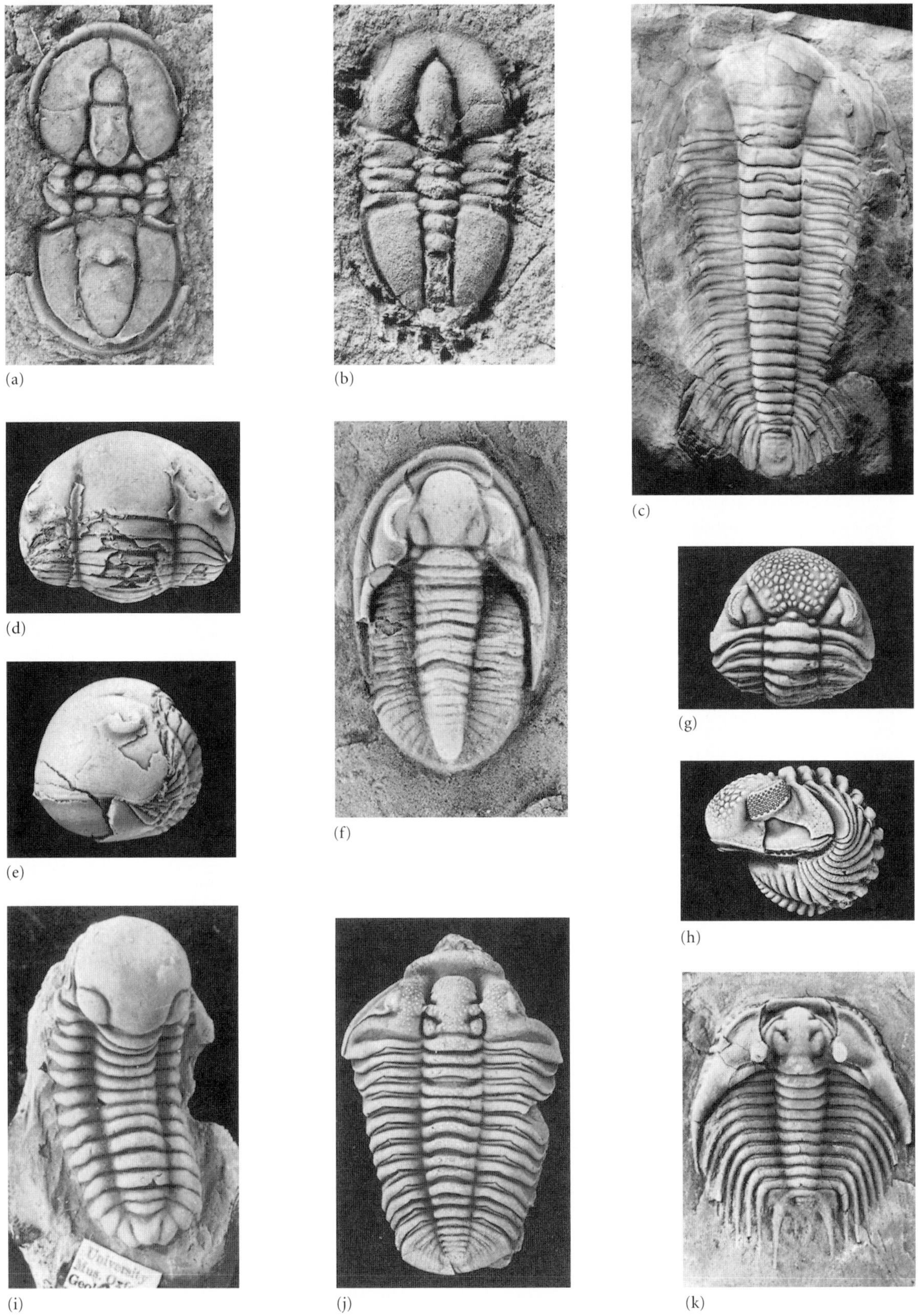

(a)

(b)

(c)

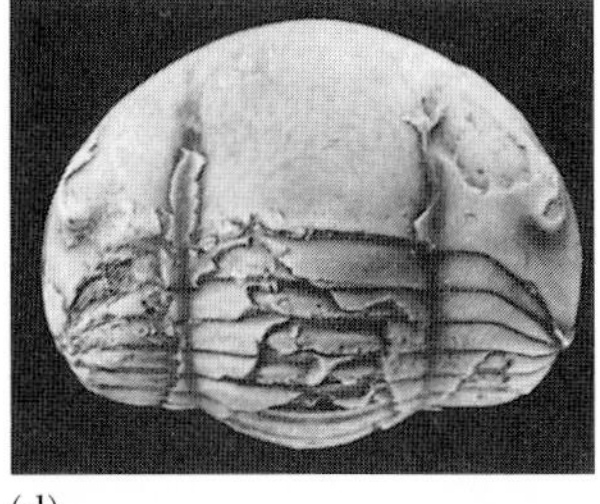

(d)

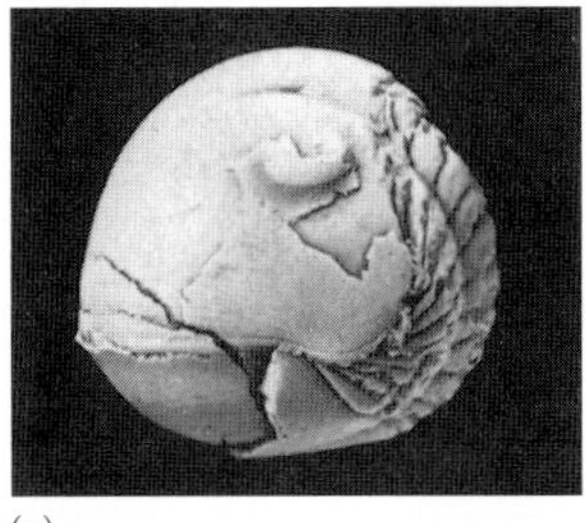

(e)

(f)

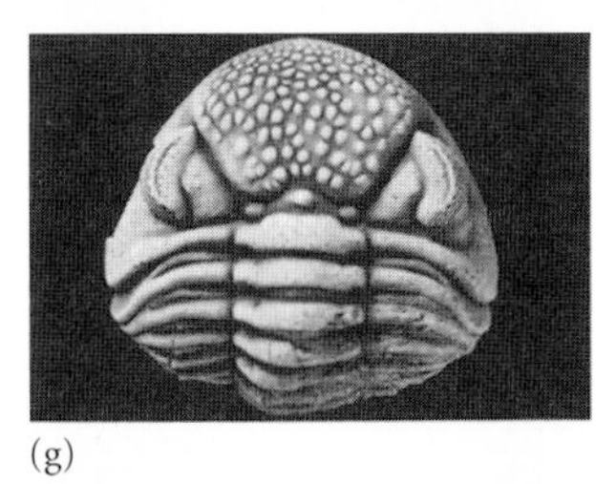

(g)

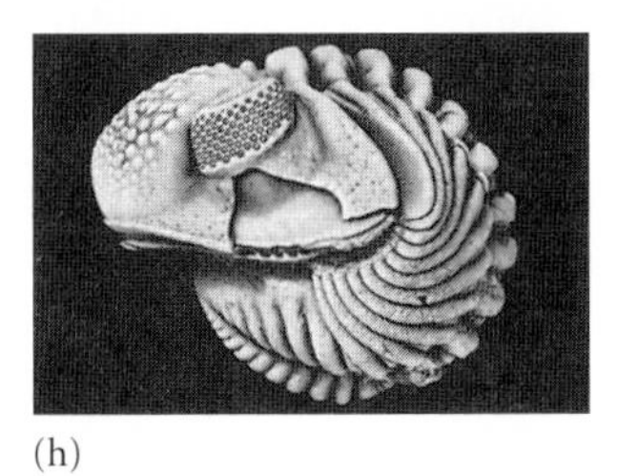

(h)

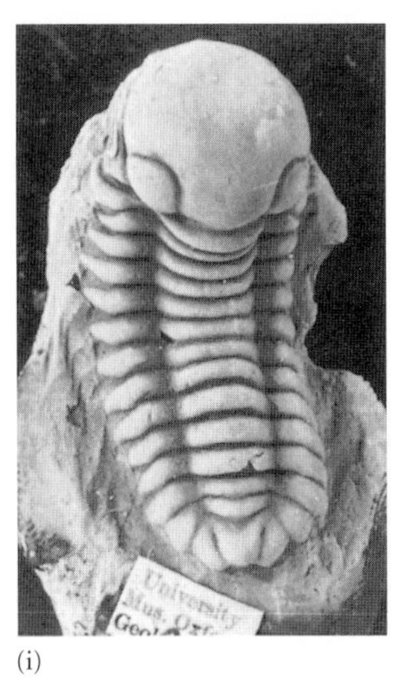

(i)

(j)

(k)

Figure 14.7 Some common trilobite taxa: (a) *Agnostus* (×10), (b) *Pagetia* (×5), (c) *Paradoxides* (×0.5), (d, e) *Illaenus* (×1), (f) *Warburgella* (×3), (g, h) *Phacops* (×0.75), (i) *Spherexochus* (×0.75), (j) *Calymene* (×0.75), (k) *Leonaspis* (×2). Magnifications are approximate.

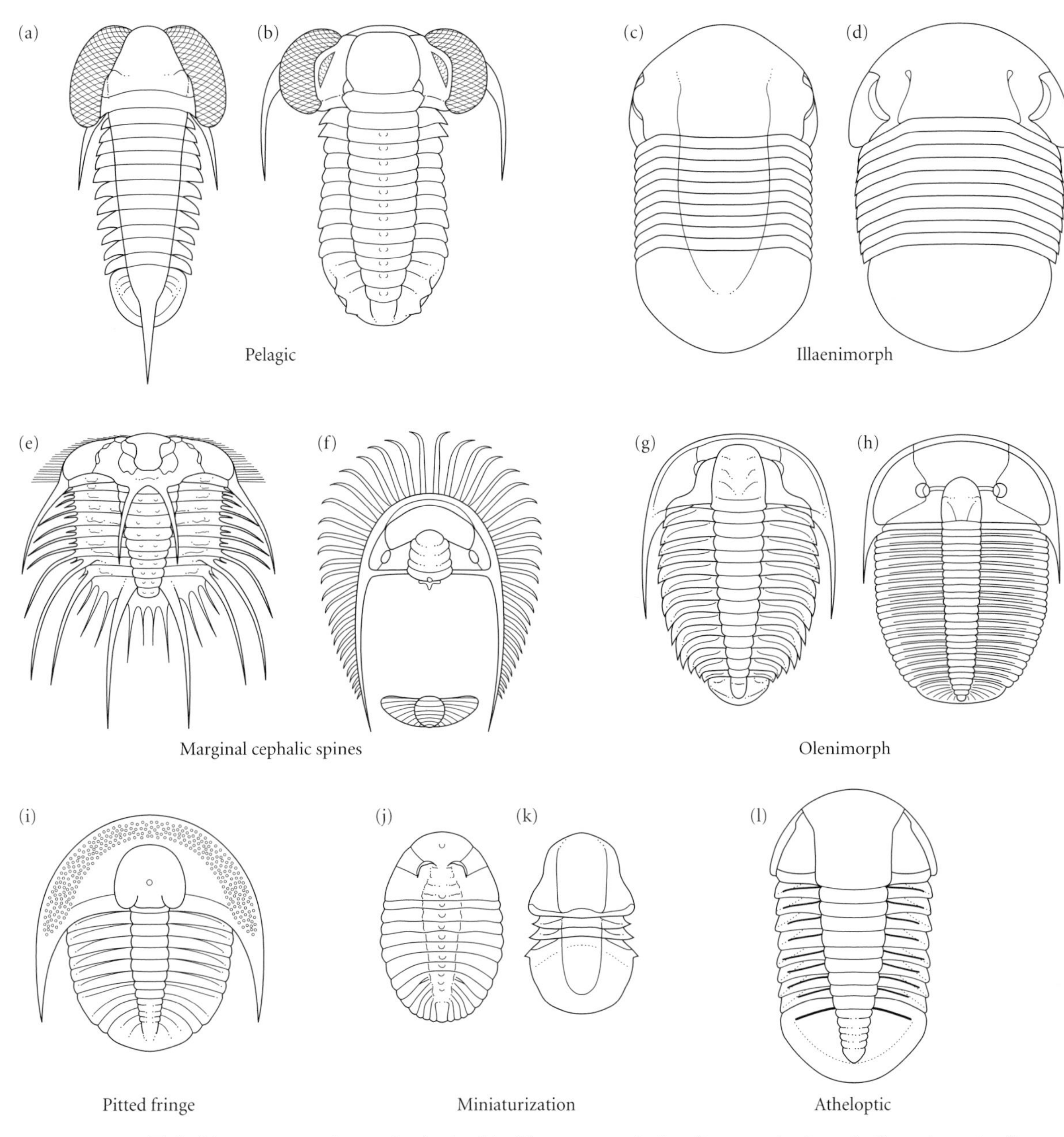

Figure 14.8 Trilobite ecomorphs: pelagic (a, b), illaenomorph (c, d), marginal cephalic spines (e, f), olenimorph (g, h), pitted fringe (i), miniature (j, k) and atheloptic (blind) (l) morphotypes. (Based on Fortey & Owens 1990.)

high-latitude *Selenopeltis* province (Gondwana). Despite a number of modifications, this basic pattern is generally accepted (see also Chapter 2).

Some Early Paleozoic trilobite communities may also be interpreted as showing an onshore–offshore spectrum, from shallow-water illaenid–cheirurid associations to deep-water olenid communities (Fig. 14.10). In general terms, the shallow-water, pure carbonate, illaenid–cheirurid communities apparently lasted the longest.

Trilobites (such as *Choubertella* and *Schmidtiellus*) first appeared in the Early Cambrian and the group survived until the end of the Permian, when the last genera, such as *Pseudophillipsia*, disappeared (Fig. 14.11). In a history of 350 million years, the basic body plan was essentially unchanged, but many modifications promoted trilobite

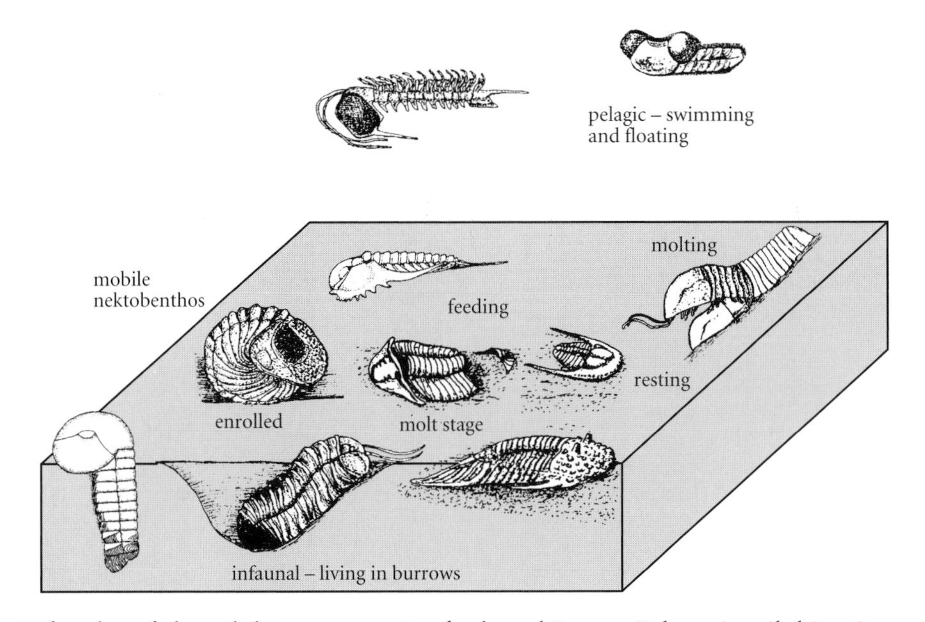

Figure 14.9 Lifestyles of the trilobites: a mosaic of selected Lower Paleozoic trilobites in various life attitudes.

abundance and diversity. Not surprisingly the Trilobitomorpha has been a major source of evolutionary data and there have been many studies on the functional morphology of the group (e.g. Bruton & Haas 2003).

Trilobites have provided key evidence in studies of macroevolution, especially in the controversy over punctuated equilibrium and phyletic gradualism (see Chapter 5). Trilobites have complex morphologies that can be easily measured and analyzed statistically (Box 14.4). The studies of Niles Eldredge and Stephen Jay Gould on the number of lens files of the Devonian trilobite *Phacops rana* formed the basis for their punctuated equilibrium model. On the other hand Peter Sheldon's investigation of over 15,000 specimens of Mid Ordovician trilobites demonstrated gradual changes in the number of pygidial ribs, possibly a slower, adaptive, fine-tuning to more stable environments. Euan Clarkson's survey of microevolutionary change in Upper Cambrian olenid trilobites from the Alum Shales of Sweden provided evidence of similar gradual change (Fig. 14.13). Macroevolutionary change in trilobites was effected by heterochrony (see p. 145). Pedomorphosis during ontogeny of the animal as a whole or applied to particular organs such as the eyes generated new species and new biological structures.

Trilobites show a number of evolutionary trends. Through time, for example, those trilobites that adopted enrolment as a defensive strategy became better at it: the spines and sockets around their exoskeletons came to fit and lock better and better. Early trilobites probably rolled up into a rough ball, but could be prized apart by a persistent predator; later enrolling trilobites were impenetrable. There was a reduction in the size of the rostral plate and in some groups there was an increase in spinosity and a trend from micropygy to isopygy. The evolution of schizochroal visual systems appeared, by pedomorphosis, during the Early Ordovician in the phacopids.

Trilobite abnormalities and injuries

Trilobites have left a rich record of abnormalities and injuries, some evidence that they faced problems during ecdysis and that they were attacked by predators (Fig. 14.14). There are three main types of abnormality (Owen 1985):

1. Injuries sustained during molting.
2. Pathological conditions resulting from disease and parasitic infestations.
3. Teratological effects arising through some embryological or genetic malfunctions.

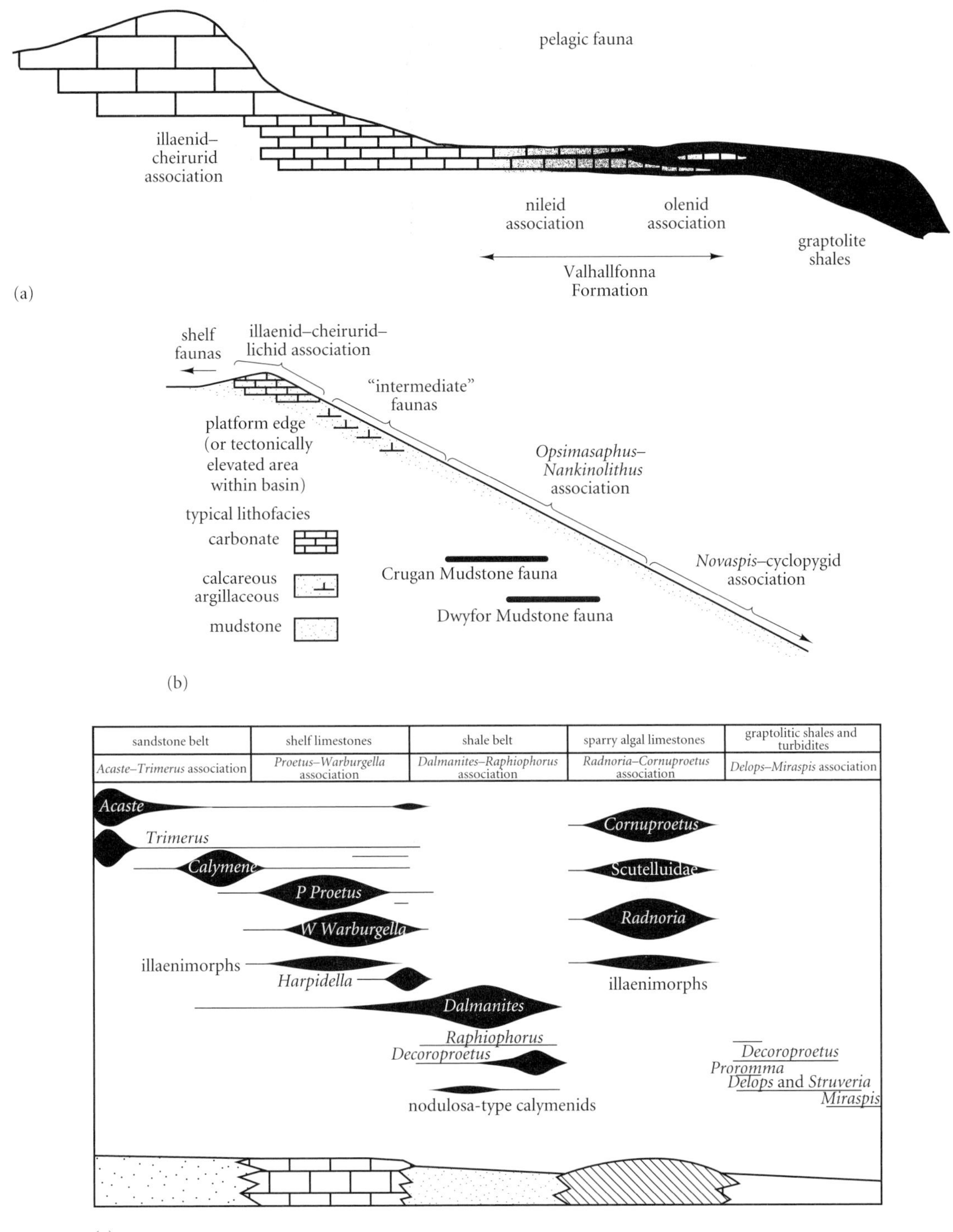

Figure 14.10 Trilobite communities: overview of (a) Early Ordovician (Arenig), (b) Late Ordovician (Ashgill) and (c) Mid Silurian (Wenlock) trilobite associations in relation to water depth and sedimentary facies. (a, from Fortey, R.A. 1975. *Fossils and Strata* 4; b, from Price, D. 1979. *Geol. J.* 16; c, from Thomas, A.T. 1979. *Spec. Publ. Geol. Soc. Lond.* 8.)

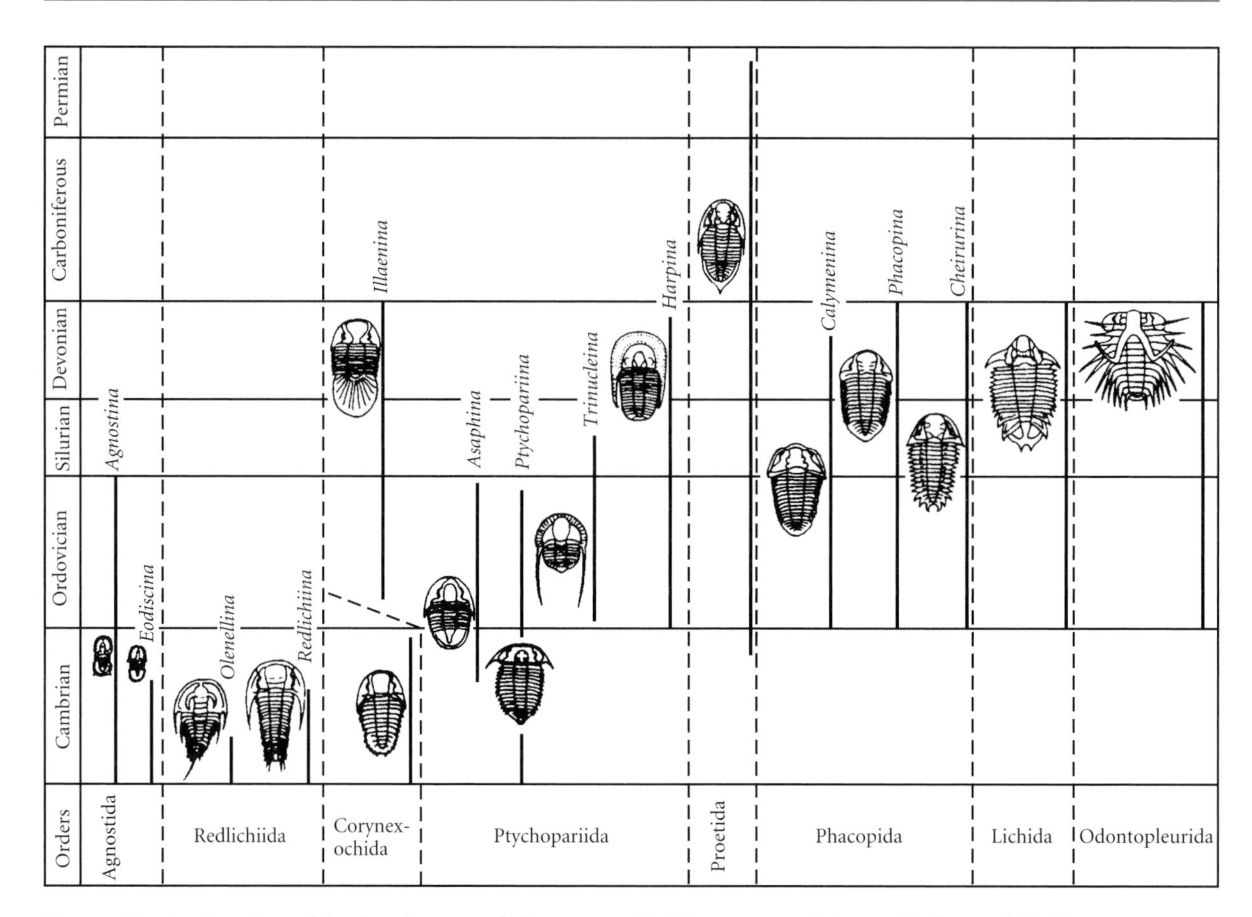

Figure 14.11 Stratigraphic distributon of the main trilobite groups. (From Clarkson 1998.)

Box 14.4 Landmarks: the Silurian trilobite *Aulacopleura*

Landmarks, as the name suggests, are recognizable geographic features. Such features can also be defined on fossil organisms and they form the basis for geometric **morphometrics**. The aim of these statistical techniques is to define precisely how shapes differ from each other, and the landmarks are the fixed points of comparison. Each landmark can be recorded as a set of coordinates or the distances between points, and they can be recorded from digital photographs or image analysis systems and stored in spreadsheets. For example, 22 landmarks were necessary to define shape variations in the exoskeletons of well-preserved *Aulacopleura* from the Silurian rocks of Bohemia (Fig. 14.12). The data can be used in a variety of ways. For example it is relatively easy to see, visually, how the trilobite actually grew; the most substantial growth took place in the thoracic region during ontogeny. In some studies it is necessary to translate this into quantitative terms, and landmark analysis is the key.

A large dataset is available at http://www.blackwellpublishing.com/paleobiology/. These data may be analyzed and manipulated using a range of morphometric techniques such as principal component analysis (see also Hammer & Harper 2005).

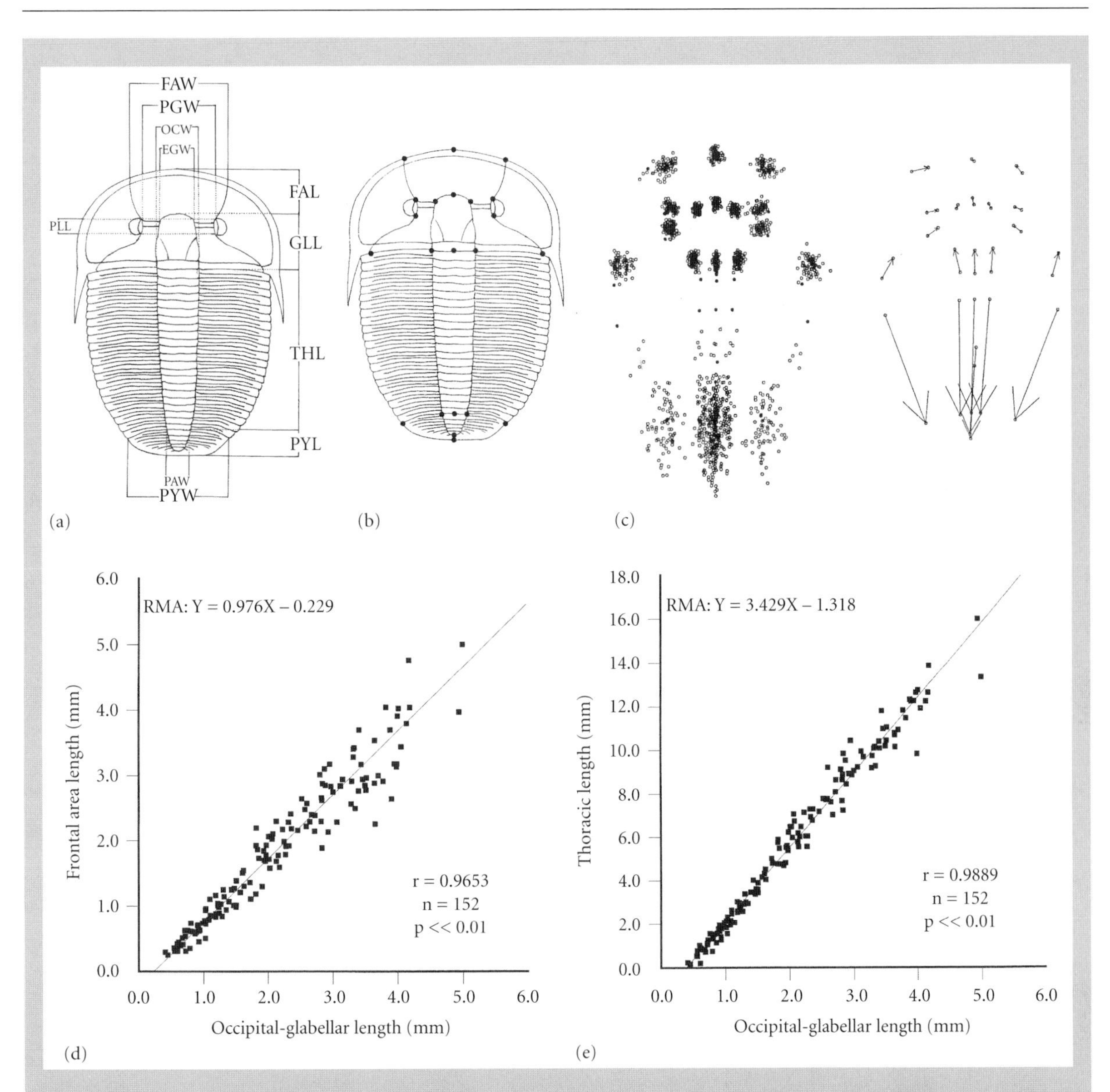

Figure 14.12 Landmark analysis of *Aulacopleura*. (a) Measurements, (b) landmarks, (c) plot of landmarks, (d) bivariate plot of occipital–glabellar length versus frontal area length, and (e) bivariate plot of occipital–glabellar length versus thoracic length. FAW, width of frontal area; PGW, OCW, width of occipital glabella; EGW, FAL, length of frontal area; PLL, GLL, THL, length of thorax; PYL, length of pygidium; PAW, PYW, width of pygidium; RMA, reduced major axis. (Courtesy of Nigel Hughes.)

In addition there are signs of predation that often show an asymmetric distribution. Predation scars are three times as likely to be present on the right-hand side of the exoskeleton as the left. If predators preferred to attack from the right then perhaps there was already a lateralization of their nervous system and other organs (Babcock 1993). However, it can be argued that these specimens were the survivors and that predators preferred to attack on the left-hand side of the trilobite, and we never see those victims.

SUBPHYLUM CHELICERATA

The chelicerates are a diverse and heterogeneous group including the mites, scorpions and spiders. The familiar horseshoe crab,

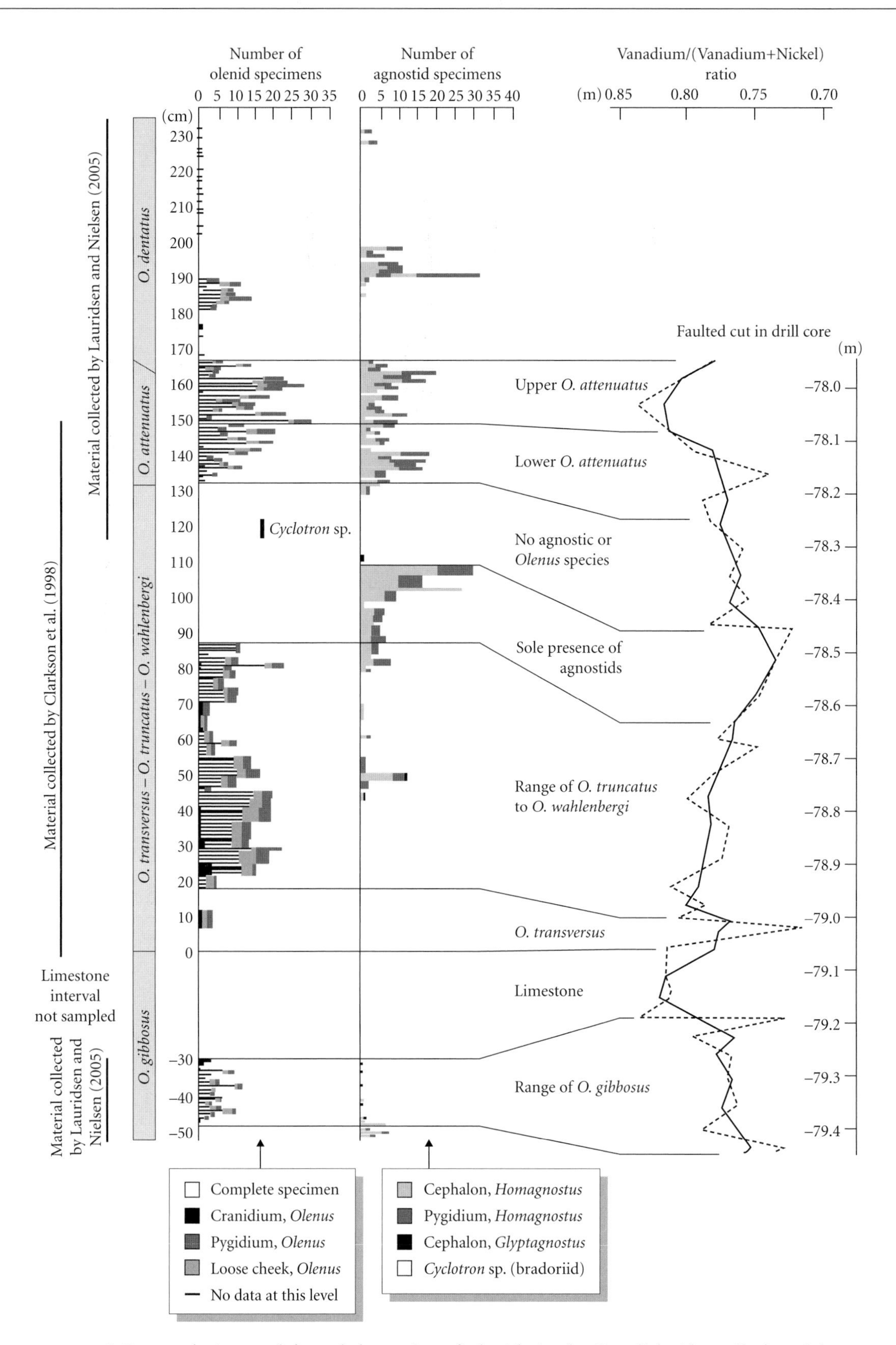

Figure 14.13 Microevolution and faunal dynamics of olenids in the Swedish Alum Shales. *Olenus* species evolve gradually up through the section. (Based on Clarkson et al. 1998.)

Figure 14.14 Pathological trilobites: (a) *Onnia superba* – the fringe in the lower part of the photograph has an indentation and a smooth area, probably regeneration following an injury during molting (×4); (b) *Autoloxolichas* – the deformed segments on the left-hand side may be either genetic or the result of repair following injury (×3); and (c) *Sphaerexochus* – only two ribs are developed on the right-hand side, probably a genetic abnormality (×25). (Courtesy of Alan Owen.)

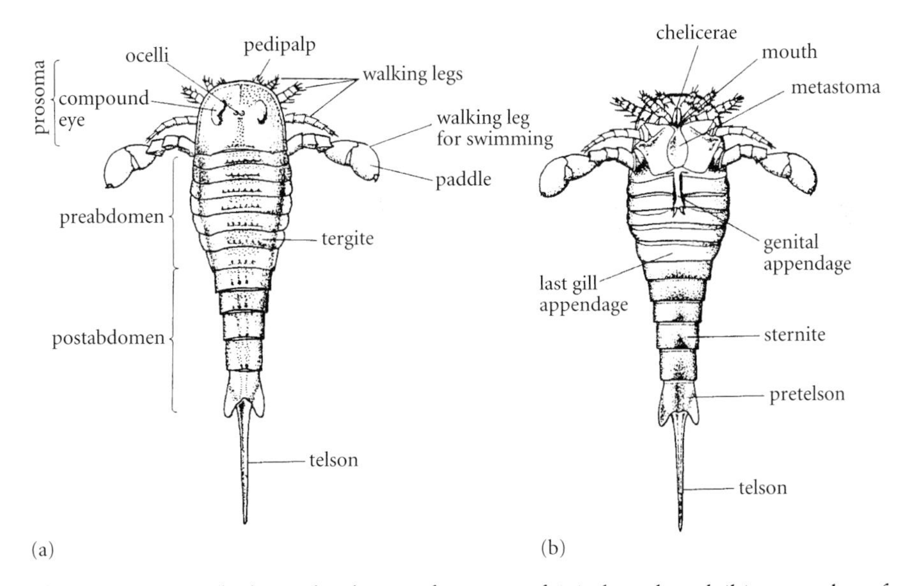

Figure 14.15 Chelicerate morphology displaying features of (a) dorsal and (b) ventral surfaces. (Based on McKinney 1991.)

Limulus, together with the extinct sea scorpions, the eurypterids, are also members of the subphylum, defined in terms of a prosoma (head and thorax) with six segments (bearing appendages), an opisthosoma (abdomen) with at most 12 segments, and a pair of chelicerae (pincers) attached to the first segment of the prosoma (Fig. 14.15).

There have traditionally been two main chelicerate groups: (i) the merostomes, including the aquatic horseshoe crab *Limulus* and the giant sea scorpions or eurypterids; and (ii) the arachnids that mainly comprise the terrestrial spiders and scorpions. But this traditional split of the subphylum into marine meristomes and non-marine arachnids has been challenged; both groups probably had marine and non-marine representatives. The bizarre *Sanctacaris* from the Middle Cambrian Burgess Shale may be a basal outgroup to the clade Chelicerata, whereas the so-called "great appendage" arthropods such as *Emeraldella* and *Sidneyia* together with the aglaspids belong within the clade (Cotton & Braddy 2004).

The xiphosures, or horseshoe crabs, have a relatively large, convex prosoma, approximately equal in length to the opisthosoma, which usually contains less than 10 segments. The **telson**, or tail spine, is commonly long

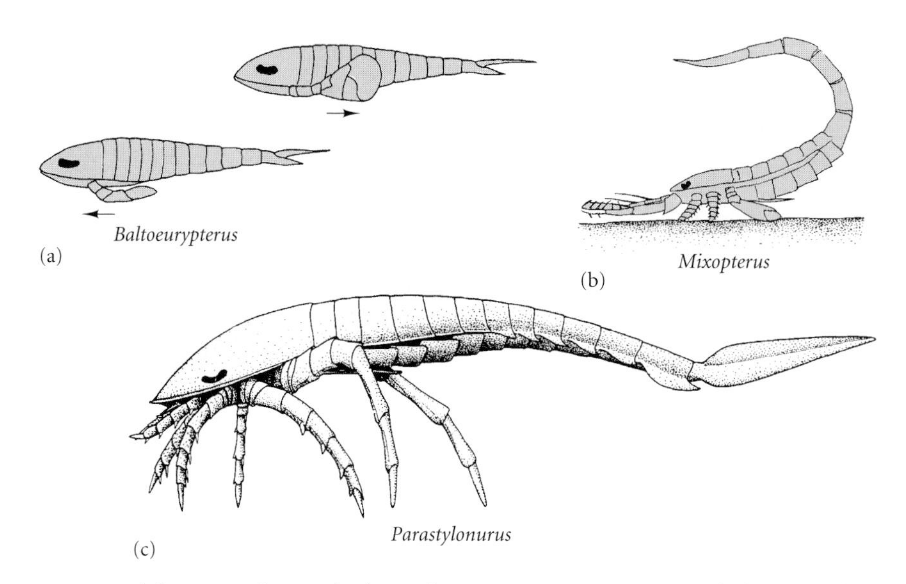

Figure 14.16 Eurypterid functional morphology showing (a) swimming and (b, c) walking life modes. (From Clarkson 1998.)

and spiny, and the ophthalmic ridges and cardiac lobe are usually well preserved. Although some enigmatic xiphosure-like taxa, such as *Eolimulus*, have been described from the Lower Cambrian, the first are probably of Early Ordovician age. A trend towards larger size and a shorter fused abdomen is seen in most groups. Carboniferous taxa, for example *Belinurus* and *Euproops*, have well-developed cardiac lobes and ophthalmic ridges together with fused abdomen. *Mesolimulus* from the Upper Jurassic Solnhofen Limestone, however, is smaller than living taxa but it too was marine and left clear evidence of its appendages in a trackway in the lagoonal muds of Bavaria.

The eurypterids include the largest known and most scary of the arthropods, some approaching 2 m in length, that range in age from Ordovician to Permian. They occupied a variety of environments from marine to freshwater and some may have been amphibious. Much of our knowledge of eurypterid morphology has been derived from superbly preserved specimens from Silurian dolomites on the island of Saaremaa that were acid-etched from the rock by the Swedish paleontologist Gerhard Holm at the end of the 1800s. The exoskeleton is long and relatively narrow. The subrectangular prosoma bears a variety of appendages; the first pair of appendages were chelicerae adapted for grasping while others were modified for movement, with the last pair of large, paddle-like appendages probably adapted for swimming. The opisthosoma, comprising the pre- and postabdomen, consists of 12 visible segments. The telson was variably developed as a long spine or a flattened paddle.

With the exception of generalist feeders such as *Baltoeurypterus*, most eurypterids were predators, attacking fishes and other arthropods. Moreover, where relatively common, a number of eurypterid-dominated communities have been described, emphasizing the range of habitats occupied by these large, versatile animals. In the Silurian rocks of the Anglo-Welsh area, *Pterygotus* and its allies are associated with normal marine faunas, whereas *Eurypterus* itself preferred inshore environments. *Hughmilleria* and related forms dominated brackish to freshwater communities.

Many more than 50 genera of eurypterids have been described. The group was most abundant during the Silurian and Devonian, but only two families, the adeloopthalmids and the hibbertopterids, survived into the Permian. The varied styles of locomotion of the group suggest a diversity in lifestyles (Fig. 14.16).

The arachnids are a huge group of terrestrial carnivores containing mites, scorpions, spiders (Box 14.5) and ticks. There are probably over 100,000 known species of arachnids. The prosoma consists of six segments with a pair of chelicerae, a pair of sensory or feeding pedipalps and four pairs of walking

Box 14.5 Fossil spider webs

Beautifully intricate but lethal spider webs are an integral part of terrestrial arthropod ecosystems. But how old is this specialized and unique mode of predation? Arthropods trapped in spider silk webs have, in fact, been recovered from Early Cretaceous amber (Fig. 14.17). Enrique Peñalver and his colleagues (2006) have illustrated a remarkable mosaic of silk strands, some with sticky droplets, which ambushed a fly and a mite, whereas another piece of amber shows a trapped wasp. The webs were elastic and armed with glue droplets – no match for stray insects. Preservation of spider silk in Lower Cretaceous amber from the Lebanon and from Spain suggests this form of predation was already well established by the Cretaceous. The diversification of the insects during the later Mesozoic was tracked by a similar diversification in spiders; perhaps the evolution of aerial webs and winged insects was linked, evidence of an arms race in the airways of the Cretaceous.

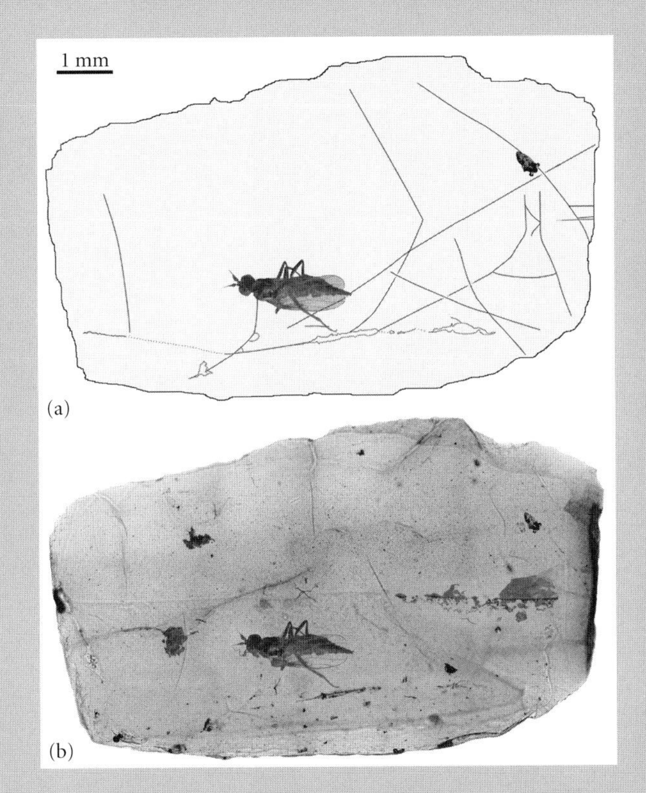

Figure 14.17 Insects trapped in a Cretaceous spider's web: (a) reconstruction and (b) actual specimen. Strands of the web have been emphasized on the reconstruction together with droplets; a fly (center left) and mite (top right) were both caught in the web. (Courtesy of Enrique Peñalver.)

legs. The opisthosoma usually has 12 segments. Arachnids breathe mainly through so-called book lungs or tracheae or both.

SUBPHYLUM MYRIAPODA

The myriapods are a varied group comprising the millipedes, centipedes, symphylans and pauropods. They first appeared during the Mid Silurian, when *Kampecaris*-like forms were responsible for a variety of terrestrial trails (Box 14.6). Some of the largest forms, for example the giant *Arthropleura*, nearly 2 m long, hoovered their way through the lush, green vegetation of the Late Carboniferous forests.

Box 14.6 Invasion of the land

The myriapods were the first animals to colonize the land. Heather Wilson and Lyall Anderson (2004) have described the few Silurian and Devonian taxa from Scotland (Fig. 14.18). One of the oldest genera from the Middle Silurian, *Cowiedesmus*, is named after Cowie Harbour near Stonehaven, near Aberdeen and occurs together with *Pneumodesmus*, which shows clear evidence of a respiratory system. *Cowiedesmus* is so distinctive and different from other millipedes that it forms the basis for a new order, the Cowiedesmeda. These animals suggest that terrestrialization amongst the arthropods had already begun by the Mid Silurian and millipedes were breathing and scuttling across the emerging new landscapes of the Caledonian mountain belt in Scotland. But there may be older indirect evidence. Trackways from Middle Ordovician rocks in the English Lake District (Johnson et al. 1994) suggests that arthropods were on land about 50 myr earlier, while trace fossils from Cambrian rocks in Ontario push arthropod life on land back even further into the Cambrian (MacNaughton et al. 2002), suggesting the presence then of large, amphibious euthycarcinoids. Euthycarcinoids are an enigmatic group of arthropods with antennae and mandibles, that have been placed in phylogeny somewhere near the origin of myriapods and insects.

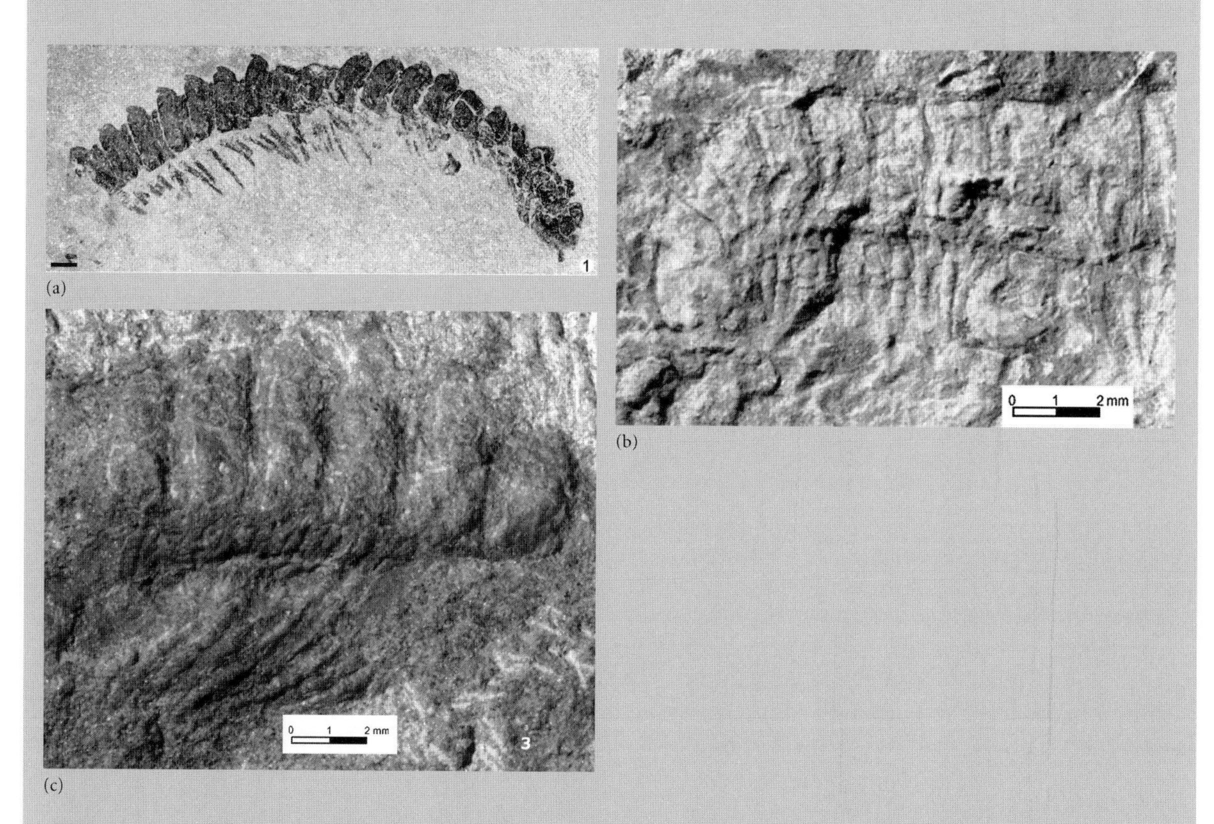

Figure 14.18 The millipedes: (a) *Archidesmus* (Lower Devonian), (b) *Cowiedesmus* (Middle Silurian) and (c) *Pneumodesmus* (Middle Silurian), from Scotland. Scale bars, 2 mm. (Courtesy of Lyall Anderson.)

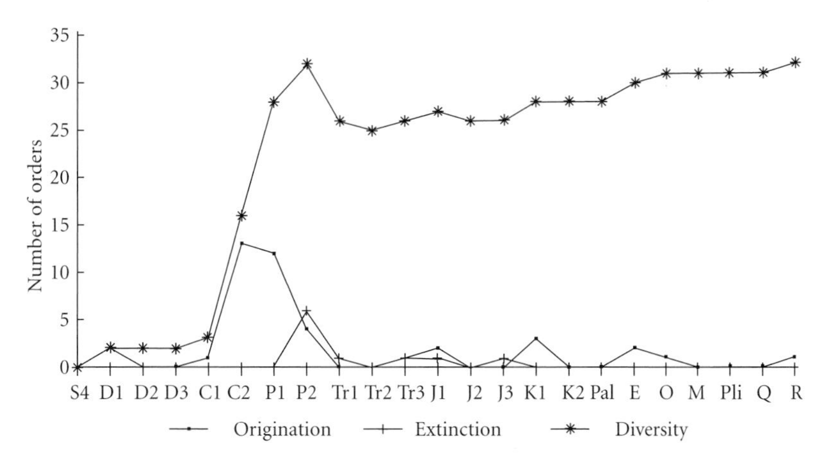

Figure 14.19 Ranges of selected insect orders. Geological period abbreviations are standard, running from Silurian (S) to Recent (R). (Based on Jarzembowski, E.A. & Ross, A.J. 1996. *Geol. Soc. Spec. Publ.* **102**.)

SUBPHYLUM HEXAPODA

The Hexapoda, essentially the insects, can be divided into **pterygotes** (with wings) and **apterygotes** (without wings) and include the springtails, dragonflies, cockroaches and locusts. The group may prove to have as many as 10 million living species when the rich faunas of the tropics have been completely described. The subphylum also includes the onychophorans, with flexible segmented bodies and unjointed limbs propelled by changes in blood pressure analogous to the water vascular system of the sea urchins. The hexapods have unbranched or uniramous appendages, a simple gut, a single pair of antennae and a pair of mandibles, together with a toughened head capsule. Insects have six limbs.

The oldest insect is probably the springtail *Rhyniella praecursor* from the Lower Devonian Rhynie Chert of the Orcadian Basin of northeast Scotland (Fig. 14.19). Conrad Labandeira (Smithsonian Institution, Washington) and his colleagues have shown that insects diversified earlier than had been thought (Labandeira 2006), and the group probably originated in freshwater during the Late Silurian, which may account for the poor fossil record of the group before the Devonian (Glenner et al. 2006). Early and Mid Devonian faunas are now well known from Rhynie, Gaspé, Québec and Gilboa, New York State and these probably coincided with the diversification of land plants. And by the Late Carboniferous a very diverse insect fauna had evolved, with forms such as the dragonflies and mayflies capable of powered flight (Box 14.7). By the end of the Permian, most of the familiar insect orders had appeared. During the later Mesozoic and Cenozoic, significant coevolutionary relationships were established between plants and insects, particularly between flowering plants and insect pollinators, and possibly even between spiders and flies (see Box 14.6). Moreover, by the Miocene, fossil hair trapped in amber together with the sand fly *Lutzomyia* suggests that these blood suckers were already feeding on mammals in arboreal nests during the Mid Tertiary (Peñalver & Grimaldi 2006).

SUBPHYLUM CRUSTACEA

As the name suggests, the crustaceans have a hard, crusty carapace. The group, which first appeared in the Cambrian, is aquatic, mainly marine, with gills, mandibles, two pairs of antennae and stalked compound eyes. The heavily armored crabs and lobsters typify this diverse subphylum; but the barnacles and ostracodes are also crustaceans with a notable geological record.

There are at least eight main classes of crustacean, but with the exception of the ostracodes, which are usually considered part of the microfauna, only two groups, the Cirripedia and the Malacostraca have significant geological records.

Box 14.7 Insects take to the airways

Fossil insects with functional wings are first reported from Mid Carboniferous strata. These insects were extraordinary (Fig. 14.20); the dragonfly *Meganeura* had an incredible wingspan of 70 cm. Intense aerobic activity such as powered flight suggests that atmospheric oxygen levels at the time were unusually high. But how effective was *Meganeura* as a flyer? Robin Wootton and his colleagues (1998) have identified so-called smart features that capitalized on both upstroke and downstroke movements of the animal's large wings. This form of smart engineering helps depress the trailing edges of the wings, rather like an aircraft's flaps during takeoff and landing, and helps wing twisting. It is unlikely that the giant Mid Carboniferous dragonflies could actually hover like modern forms but they had good maneuvrability. These winged giants had already developed a predatory lifestyle and, being about the size of a seagull, would have made a highly visible addition to terrestrial life in the forests of the Carboniferous (see p. 488).

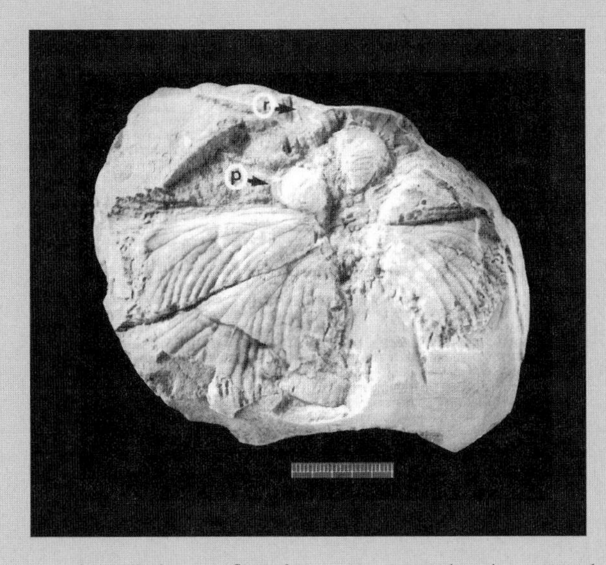

Figure 14.20 Giant Carboniferous dragonflies from Ayr, Scotland. p, prothoracic lobe; r, rostrum. Scale bar is in millimeters. (Courtesy of Ed Jarzembowski.)

The cirripedes or barnacles have shells, or **capitula** (singular, capitulum), consisting of several plates and these animals are adapted to an encrusting lifestyle. Two groups, the acorn barnacles and goose barnacles, have contrasting life strategies. The acorn barnacles, such as *Balanus*, have capitula consisting of overlapping plates and they are attached to rocks and other shells. The group rapidly diversified from an origin during the Late Cretaceous and are locally common. The goose barnacles are pseudoplankton, living attached to floating debris, that have a relatively poor fossil record.

The malacostracans include two subclasses, the phyllocarids and the eumalacostracans. The phyllocarids have large bivalved carapaces, seven abdominal somites and a telson with a pair of **furcae** (forked extensions; singular, furca), extending posteriorly. *Canadaspis* from the Burgess Shale may be one of the first crustaceans. Living phyllocarids are usually minute, in contrast to their larger Paleozoic ancestors. The eumalacostracans include decapods – shrimps, lobsters and crabs – together with the less common branchiopods. Some of the most spectacular Carboniferous eumalacostracans have been

Figure 14.21 Carboniferous shrimps: (a) *Tealliocaris woodwardi* from the Gullane Shrimp Bed, near Edinburgh (×4); (b) *Waterstonella grantonensis* from the Granton Shrimp Bed, near Edinburgh (×2); (c) *Crangopsis socialis* and *Waterstonella grantonensis* from the Granton Shrimp Bed (×2). (Courtesy of Euan Clarkson.)

described from the Granton Shrimp Bed by Euan Clarkson (University of Edinburgh) and his colleagues (Fig. 14.21).

Ostracodes

Ostracodes are crustacean arthropods, abundant and widespread in aquatic environments. They have small bivalved carapaces, hinged along the dorsal margin (Fig. 14.22a). The carapace is perforate and completely covers the entire animal when closed. Most ostracodes are benthic, swimming, crawling or burrowing at the sediment–water interface in muds or silts with abundant organic material. A few, such as the myodocopids, are planktonic and some are commensal or parasitic. They are very useful for environmental reconstructions and, at some levels in the stratigraphic record, have been used for correlation.

Ostracodes have weak segmentation with a poorly defined head, thorax and abdomen; the animal is contained within the two shells, with the carapace united dorsally by an elastic ligament and a variably developed hinge. Growth is by periodic ecdysis or molting. Following each molt phase the carapace initially develops as a pair of chitinous valves enclosing the animal; most of the carapace is then calcified, except the dorsal margin that remains as a chitinous ligament forcing the valves apart when the internal adductor muscles relax. The central muscle scars vary across the class (Fig. 14.22b), from complex patterns in the Leperditicopida to a single scar in some members of the Palaeocopida.

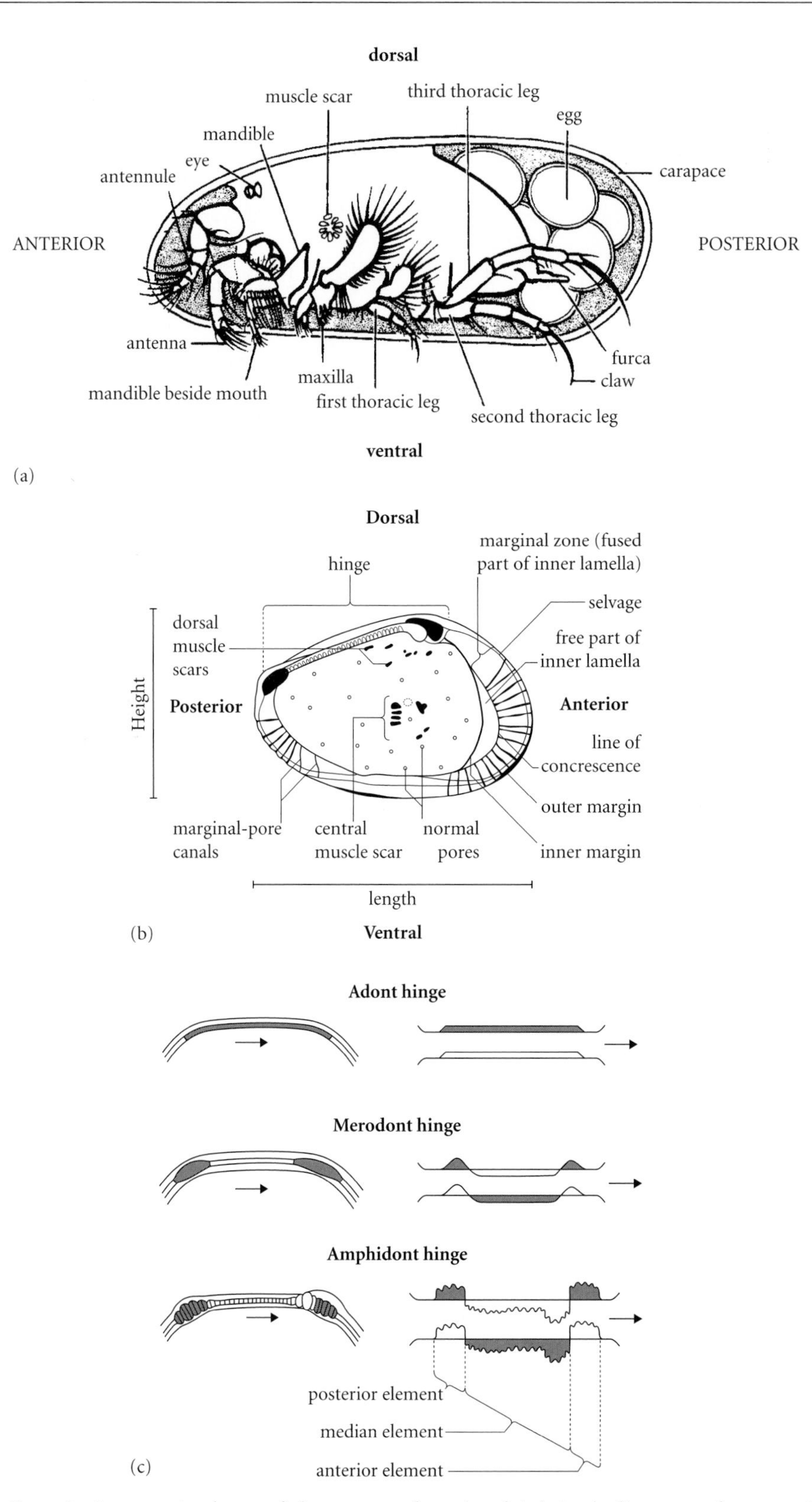

Figure 14.22 Descriptive terminology of the ostracode animal (a), including muscle scars (b) and hinge structures (c). (Based on Armstrong & Brasier 2005.)

Articulatory structures are variably developed along the hinge line. Three main types of hinge are known (Fig. 14.22c). Adont hinges lack teeth but have a long median element on the right valve that fits a socket on the left valve. The merodont hinge has long striated terminal elements on the right valve fitting respective sockets on the left valve. Amphidont hinges have short terminal elements with well-developed teeth on the right valve.

The carapace is perforated by canals holding setae that communicate with the exterior. The body is suspended within the carapace, attached by muscles. It is equipped with seven pairs of appendages, three in front of the mouth and four behind. The appendages are specialized, acting as sensory organs, limbs for the capture and processing of food; moreover they allow locomotion and general cleaning and housekeeping within the carapace. The animal has a digestive system, sophisticated genitalia and a nervous system; commonly a median eye is located behind a tubercle.

Sexual dimorphism is common and often reflected in the ostracode carapace (Fig. 14.23). Males commonly have a greater length : height ratio than the females, whereas in some benthic Paleozoic ostracodes the female had a brood pouch in the carapace wall. Females are often called heteromorphs while the males, lacking the brood pouch, are the tecnomorphs.

Ostracodes appeared first during the Early Cambrian. The archaeocopids were a bizarre group of large taxa with distinctive appendages quite different from more typical ostracodes. The group was short lived, disappearing during the latest Cambrian to earliest Ordovician. The later history of the group shows a number of clear trends: evolution of small size, simpler muscle systems and shorter hinge lines; the functional significance of these changes is not immediately obvious.

Large Leperditicopida and Palaeocopida appeared during the Ordovician, dominating ostracode faunas until the Devonian, when deep-water limestones were locally characterized by the small spiny myocopids. Many new groups appeared near the end of the Paleozoic, but hitherto important groups such as the palaeocopids eventually disappeared in the Triassic after a decline during the Permian.

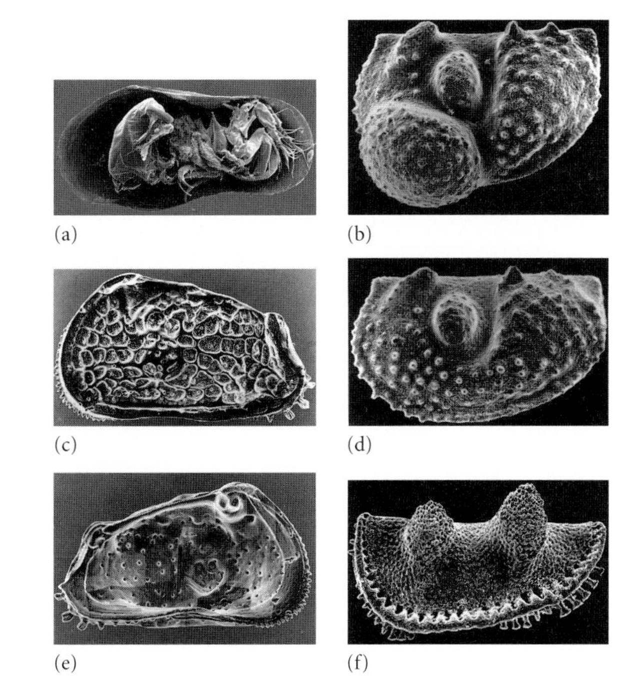

Figure 14.23 Some ostracode genera: (a) left valve of a male living *Limnocythene* showing details of appendages (×30); (b, d) left valves of female and male heteromorphs of *Beyrichia* (Silurian) (×18); (c, e) external and internal views of the left valve of living *Patagonacythene* (×30); (f) palaeocopid *Kelletina* (Carboniferous) (×30). (Courtesy of David Siveter.)

Although Early Jurassic ostracode assemblages are of low diversity the platycopines, cypridaceans and cytheraceans radiated steadily during the Jurassic. By the Cenozoic, the cypridaceans dominated lake environments whereas the cytheraceans were established in marine settings.

Any doubts that real ostracodes did not actually exist in the Paleozoic have been dispelled by some remarkable soft-part preservation, digitally reconstructed from material from the Silurian Lagerstätten at Hereford, England (Siveter et al. 2003). The precise details of the animal's morphology, down to the enormous male copulatory organ, confirm the ostracode identity of the specimen; it seems even very similar to living myodocopids. Lagerstätten, such as the Hereford biota, have provided a remarkable series of windows on arthropod evolution through time, right back to the Cambrian (Box 14.8).

Box 14.8 Exceptional arthropod-dominated faunas

Arthropods are common in a number of Lagerstätten deposits, suggesting that they were much more diverse in the past than the regular fossil record suggests. More than 40% of the animals described from the Mid Cambrian Burgess Shale are arthropods. Apart from typical trilobites such as *Olenoides* there are also soft-bodied taxa, for example *Naraoia* and the larger *Tegopelte*. However the commonest and first discovered Burgess arthropod is the elegant, trilobitomorph *Marrella*. The fauna contains many other arthropods such as *Canadaspis*, probably the first phyllocariid crustacean. There are many unique arthropods in the fauna that are difficult to classify: *Anomalocaris*, *Emeraldella*, *Leanchoilia*, *Odaraia*, *Sidneyia* and *Yohoia* are not easily aligned with established groups. The small and bizarre *Hallucigenia* was probably an onychophoran, while *Sanctacaris* was a stem-group chelicerate. The slightly older faunas at Chengjiang, South China, and Sirius Passet, North Greenland, have also yielded a spectacular array of enigmatic arthropod faunas, further contributing to our knowledge of the Cambrian explosion.

Calcareous concretions (or orsten) from the Upper Cambrian of the Baltic area have yielded a phosphatized fauna dominated by stem- and crown-group crustaceans and ostracodes together with agnostid trilobites. Many of these diverse forms were minute, living in microhabitats within or on the muds of the Cambrian seas (Fig. 14.24). These faunas are quite distinct from the earlier Burgess Shale-type faunas and provide a window on a habitat occupied by a wide range of body plans on a microscopic scale, possibly adapted to life below the sediment–water interface. Recent work by Dieter Walossek (Ulm Universität) on, for example, remarkably preserved complete ontogenetic series of *Rehbachiella* from orsten has helped elucidate the life cycle, habits and functional morphology of these animals. Moreover some of the most remarkable of all the arthropods, the pycnogonids, or sea spiders, are now known from the Cambrian orsten banks, the Silurian Herefordshire fauna and the Devonian Hunsrückschiefer (Budd & Telford 2005).

The Early Devonian faunas of the Hunsrückschiefer of the German Rhineland contain beautifully preserved phyllocariid crustaceans such as *Nahecaris*, together with a number of other arthropods apparently lacking living counterparts such as *Cheloniellon* (a large, ovoid creature with a pair of antennae, nine segments and conical telson) or *Mimetaster*, which is similar to *Marrella* from the Burgess Shale.

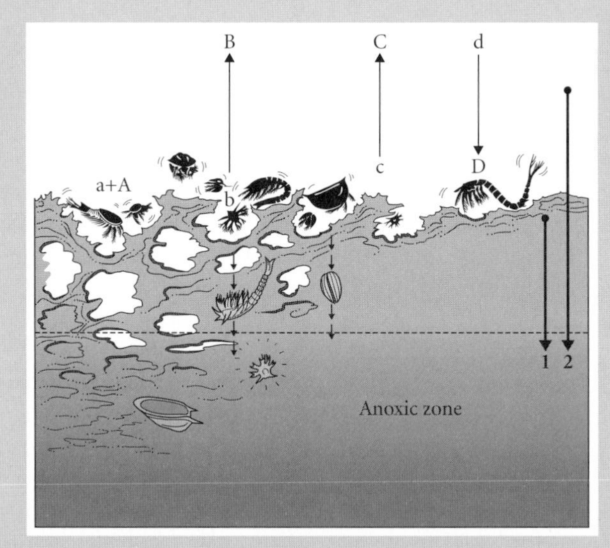

Figure 14.24 Composite of Mid Cambrian and Late Cambrian forms and reconstructions. Lower case letters (a–d), larvae; upper case letters (A–D), adult stages. Distance of sinking into the zone of preservation: 1, short distance; 2, long distance. (Redrawn from Walossek, D. 1993. *Fossils and Strata* 32.)

The Late Carboniferous Mazon Creek fauna of Illinois occurs across two facies associated with a deltaic system. The marine, Essex fauna developed on the delta front and is dominated by fishes, including coelacanths and some of the earliest lampreys. However, huge crustaceans are found together with the weird *Tullimonstrum* whose affinities are uncertain but might be a heteropod gastropod. The non-marine Braidwood assemblage is a diverse array of arthropods including 140 species of insects together with centipedes, millipedes, scorpions and spider-like arachnids. The fauna, together with a flora of over 300 species of land plant, occupied a lowland swamp milieu between the sea and coal forests. Shrimps and ostracodes apparently inhabited ponds within the swamps.

More recent terrestrial assemblages such as the Montsech fauna from the Lower Cretaceous of northeast Spain have yielded new information on the evolution of spiders. Paul Selden (University of Kansas) has described three web-weaving species equipped to attack an abundant insect life inhabiting settings around coastal lagoons.

It is clear from these extraordinarily well-preserved faunas that numerous ancient communities, marine and non-marine, were dominated by arthropods, just as today.

Review questions

1 The spectacular arthropod faunas of Burgess, Chengjiang and Sirius Passet suggest an early diversification of these ecdysozoan taxa. Was this really evolution's "big bang" or are these arthropods just too weird to comprehend when compared to modern faunas?
2 Trilobites were an integral part of the Paleozoic fauna for over 200 million years yet they finally became extinct at the end of the Permian. What sorts of animals filled their niches in the Modern evolutionary fauna?
3 Trilobites have featured in a number of evolutionary schemes, some showing gradualistic trends and others showing punctuated trends. Are these different patterns correlated with different groups of trilobite or perhaps to their life environments?
4 Insects are and probably were the most numerically dominant life of Earth. Why have they a relatively poor fossil record?
5 Exceptionally-preserved biotas occur sporadically throughout the Phanerozoic. Arthropods are usually well represented. Why?

Further reading

Briggs, D.E.G., Thomas, A.T. & Fortey, R.A. 1985. Arthropoda. *In* Murray, J.W. (ed.) *Atlas of Invertebrate Macrofossils*. Longman, London, pp. 199–229. (A useful, mainly photographic review of the group.)

Clarkson, E.N.K. 1998. *Invertebrate Palaeontology and Evolution*, 4th edn. Chapman and Hall, London. (An excellent, more advanced text; clearly written and well illustrated.)

Fortey, R. 2000. *Trilobite: Eyewitness to Evolution*. HarperCollins Publishers, London. (Fascinating personal voyage of discovery.)

Gould, S.J. 1989. *Wonderful Life. The Burgess Shale and the Nature of History*. Hutchinson Radius, London. (Inspirational analysis of evolution's "big bang".)

Robison, R.A. & Kaesler, R.L. 1987. Phylum Arthropoda. *In* Boardman, R.S., Cheetham, A.H. & Rowell, A.J. (eds) Fossil Invertebrates. Blackwell Scientific Publications, Oxford, UK, pp. 205–69. (A comprehensive, more advanced text with emphasis on taxonomy; extravagantly illustrated.)

Whittington, H.B. 1985. *The Burgess Shale*. Yale University Press, New Haven, NJ. (Classic description of the Burgess Shale and its fauna.)

References

Armstrong, H.A. & Brasier, M.D. 2005. *Microfossils*, 2nd edn. Blackwell Publishing, Oxford, UK.

Babcock, L.E. 1993. Trilobite malformations and the fossil record of behavioral symmetry. *Journal of Paleontology* 67, 217–29.

Barrande, J. 1852. Systèm Silurien du Centre de la Bohème. Recherches Paléontologiques, Vol. 1, *Planches, Crustacés, Trilobites*. Prague and Paris.

Briggs, D.E.G., Fortey, R.A. & Wills, M.A. 1993. How big was the Cambrian evolutionary explosion? A taxonomic and morphological comparison of Cambrian and Recent arthropods. *In* Lees, D.R. & Edwards, D. (eds) *Evolutionary Patterns and Processes*. Linnean Society of London, London, pp. 33–44.

Bruton, D.L. & Haas, W. 2003. Making *Phacops* come alive. *Special Papers in Palaeontology* 70, 331–47.

Budd, G.E. & Telford M.J. 2005. Along came a sea spider. *Nature* **437**, 1099–102.

Clarkson, E.N.K. 1979. The visual system of trilobites. *Palaeontology* **22**, 1–22.

Clarkson, E.N.K. 1998. *Invertebrate Palaeontology and Evolution*, 4th edn. Chapman and Hall, London.

Clarkson, E.N.K., Ahlberg, A. & Taylor, C.M. 1998. Faunal dynamics and microevolutionary investigations in the Cambrian *Olenus* Zone at Andrarum, Skåne, Sweden. *GFF* **120**, 257–67.

Cotton, T.J. & Braddy, S.J. 2004. The phylogeny of arachnomorph arthropods and the origin of the Chelicerata. *Transactions of the Royal Society of Edinburgh: Earth Sciences* **94**, 169–93.

Edgecombe, G.D. & Ramsköld, L. 1999. Relationships of Cambrian Arachnata and the systematic position of Trilobita. *Journal of Paleontology* **73**, 263–87.

Fortey, R.A. & Owens, R.M. 1990. Trilobites. *In* McNamara, K.J. (ed.) *Evolutionary Trends*. University of Arizona Press, Tucson, pp. 121–42.

Glenner, H., Thomsen, P.F., Hebsgaard, M.B., Sørensen, M.V. & Willerslev, E. 2006. The origin of insects. *Science* **314**, 1883–4.

Gould, S.J. 1989. *Wonderful Life. The Burgess Shale and the nature of history*. W.W. Norton & Co., New York.

Fortey, R.A. & Chatterton, B. 2003. A Devonian trilobite with an eyeshade. *Science* **301**, 1689.

Johnson, E.W., Briggs, D.E.G., Suthren, R.J., Wright, J.L. & Tunnicliff, S.P. 1994. Non-marine arthropod traces from the subaerial Ordovician Borrowdale Volcanic Group, English Lake District. *Geological Magazine* **131**, 395–406.

Labandeira, C.C. 2006. The four phases of plant-arthropod associations in deep time. *Geologica Acta* **4**, 409–38.

Lauridsen, B.W. & Nielsen, A.T. 2005. The Upper Cambrian trilobite *Olenus* at Andrarum, Sweden: a case of interative evolution? *Palaeontology* **48**, 1041–56.

Lin Jih-Pai, Gon III, S.M., Gehling, J.G. et al. 2006. A *Parvancorina*-like arthropod from the Cambrian of South China. *Historical Biology* **18**, 33–45.

MacNaughton, R.B., Cole, J.M., Dalrymple, R.W., Braddy, S.J., Briggs, D.E.G. & Lukie, T.D. 2002. First steps on land: Arthropod trackways in Cambrian-Ordovician eolian sandstone, southeastern Ontario. *Geology* **30**, 391–4.

McKinney, F.K. 1991. *Exercises in Invertebrate Paleontology*. Blackwell Scientific Publications, Oxford, UK.

Owen, A.W. 1985. Trilobite abnormalities. *Transactions of the Royal Society of Edinburgh: Earth Sciences* **76**, 255–72.

Peñalver, E., Grimaldi, D.A. & Declòs, X. 2006. Early Cretaceous spider web with its prey. *Science* **312**, 1761.

Peñalver, E. & Grimaldi, D. 2006. Assemblages of mammalian hair and blood-feeding midges (Insecta: Diptera: Psychodidae: Phlebotominae) in Miocene amber. *Transactions of the Royal Society of Edinburgh: Earth Sciences* **96**, 177–95.

Siveter, D.J., Sutton, M.D., Briggs, D.E.G. & Siveter, D.J. 2003. An ostracode crustacean with soft parts from the Lower Silurian. *Science* **302**, 1749–51.

Wilson, H.M. & Anderson, L.I. 2004. Morphology and taxonomy of Paleozoic millipedes (Diplopoda: Chilognatha: Archipolypoda) from Scotland. *Journal of Paleontology* **78**, 169–84.

Wootton, R.J., Kukalová-Peck, J., Newman, D.J.S. & Muzón, J. 1998. Smart engineering in the mid-Carboniferous: how well could Palaeozoic dragonflies fly? *Science* **282**, 749–51.

Chapter 15

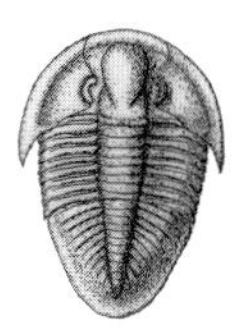

Deuterostomes: echinoderms and hemichordates

Key points

- Echinoderms today include sea urchins, starfish and sea cucumbers. They are all equipped with a water vascular system, a mesodermal skeleton of calcitic plates with a stereom structure, pentameral symmetry and tube feet.
- During the Cambrian radiation many bizarre forms evolved. The spindle-shaped *Helicoplacus* may be part of the stem group for the entire phylum but did not survive the Cambrian substrate revolution.
- Pelmatozoans were mainly fixed echinoderms and include the blastoids, crinoids and cystoids; the crinoids include four classes: the Inadunata, Flexibilia, Camerata and Articulata.
- The echinoids were part of the mobile benthos. During the Mesozoic irregular groups, adapted for burrowing, evolved from the more regular forms that characterized the Paleozoic.
- Asteroids (starfish) were more important in post-Paleozoic rocks; their Triassic radiation may have inhibited the re-radiation of some key groups of brachiopod.
- Carpoids are traditionally classed with the echinoderms, although some have argued they were ancestral to chordates; they were probably stem-group echinoderms.
- Graptolites are hemichordates closely related to the living rhabdopleurids with similarly constructed rhabdosomes and ultrastructure.
- Dendroids, with autothecae and bithecae together with many stipes, and graptoloids, with generally fewer stipes and only one type of theca, are the two most common graptolite orders.
- Graptolites probably pursued benthic (dendroids), planktic (dendroids and graptoloids) and automobile (graptoloids) lifestyles.
- Graptoloids evolved rapidly and were widespread, the ideal zone fossils in rocks of Ordovician-Silurian and Early Devonian age.

The echinoderms and hemichordates appear to be two very different groups of animals, one characterized by five-fold symmetry and a water vascular system, the other a group of odd stick-like colonial organisms. Surprisingly, both are closely related to each other and, moreover, are not so distant from ourselves, the chordates. Both groups are deuterostomes; the first opening to develop in the embryo is the anus and a second forms the mouth. The group has a dipleurula larva and a body cavity that developed from an extensions of the embryonic gut (see p. 240). Modern morphological and molecular analyses indicate that the echinoderms and hemichordates are in fact sister groups (Smith 2005). A small, extinct group – the Vetulicolia – so far known only from the Cambrian, has also been related to the deuterostomes because of similar gill structures and the absence of limbs. But although recent finds from Utah have suggested that this group has more in common with the arthropods and probably belongs to the ecdysozoans (see p. 361), the group remains an enigma (see Box 15.10).

ECHINODERMS

> *Clearly we stood among the ruins of some latter-day South Kensington! Here, apparently, was the Palæontological Section, and a very splendid array of fossils it must have been. . . . The place was very silent. The thick dust deadened our footsteps. Weena, who had been rolling a sea urchin down the sloping glass of a case, presently came, as I stared about me, and very quietly took my hand and stood beside me. And at first I was so much surprised by this ancient monument of an intellectual age, that I gave no thought to the possibilities it presented. Even my preoccupation about the Time Machine receded a little from my mind.*
>
> H. G. Wells (1898) *The Time Machine*

Echinoderms today are one of the most abundant marine animal groups, and as fossils they can sometimes be rather robust, as Weena from *The Time Machine* found. Sea urchins are common in many intertidal environments, and out in the deep sea the ocean floors are covered by brittle stars and sea cucumbers. The phylum Echinodermata has an unusual five-fold symmetry and is uniquely equipped with a water vascular system in which water is forced around the plumbing by muscular action, while tube feet, extending from the system, are often modified for food processing, locomotion and respiration. The 6000 or so living echinoderm species include familiar forms such as sea lilies, sea urchins, sand dollars, starfish and sea cucumbers (Fig. 15.1). Although many species today live in the intertidal or subtidal zones, the group is most diverse in the deep sea. Echinoderms also occupied a wide range of marine environments and pursued a variety of life strategies in the geological past. Fossil echinoderms are relatively common, and because many echinoderm skeletons disintegrate rapidly after death, many limestones are packed with the distinctive skeletal debris of calcitic plates.

Apart from the water vascular system, echinoderms have a number of other distinctive features. All members of the phylum have a **mesodermal** skeleton constructed from porous plates of calcite; each plate is usually a single crystal of calcite and easy to recognize in thin sections. In addition, the plates have a unique ultrastructure of rods linked to form a three-dimensional lattice. This network, or **stereom**, is permeated by finger-like pieces of soft tissue that occupy the spaces, or stroma, in the lattice. Finally, five-rayed or pentameral symmetry, occasionally modified by a secondary bilateral symmetry, is typical of the echinoderms. The phylum is generally split into the mobile, non-stalked **eleutherozoans** and the mainly fixed, stalked **pelmatozoans** (Box 15.1), but the earliest forms are hard to classify (Box 15.2).

The multiplated echinoderm skeleton disintegrates very rapidly after death; although individual plates or ossicles have high preservation potential, the complete skeletons do not. Nevertheless, occasionally rapid burial or transportation into anoxic conditions may result in the preservation of complete echinoderm skeletons. Starfish beds, usually characterized by accumulations of complete echinoderms, occur sporadically throughout the fossil record. The Leintwardine Starfish Bed of the England–Wales border area con-

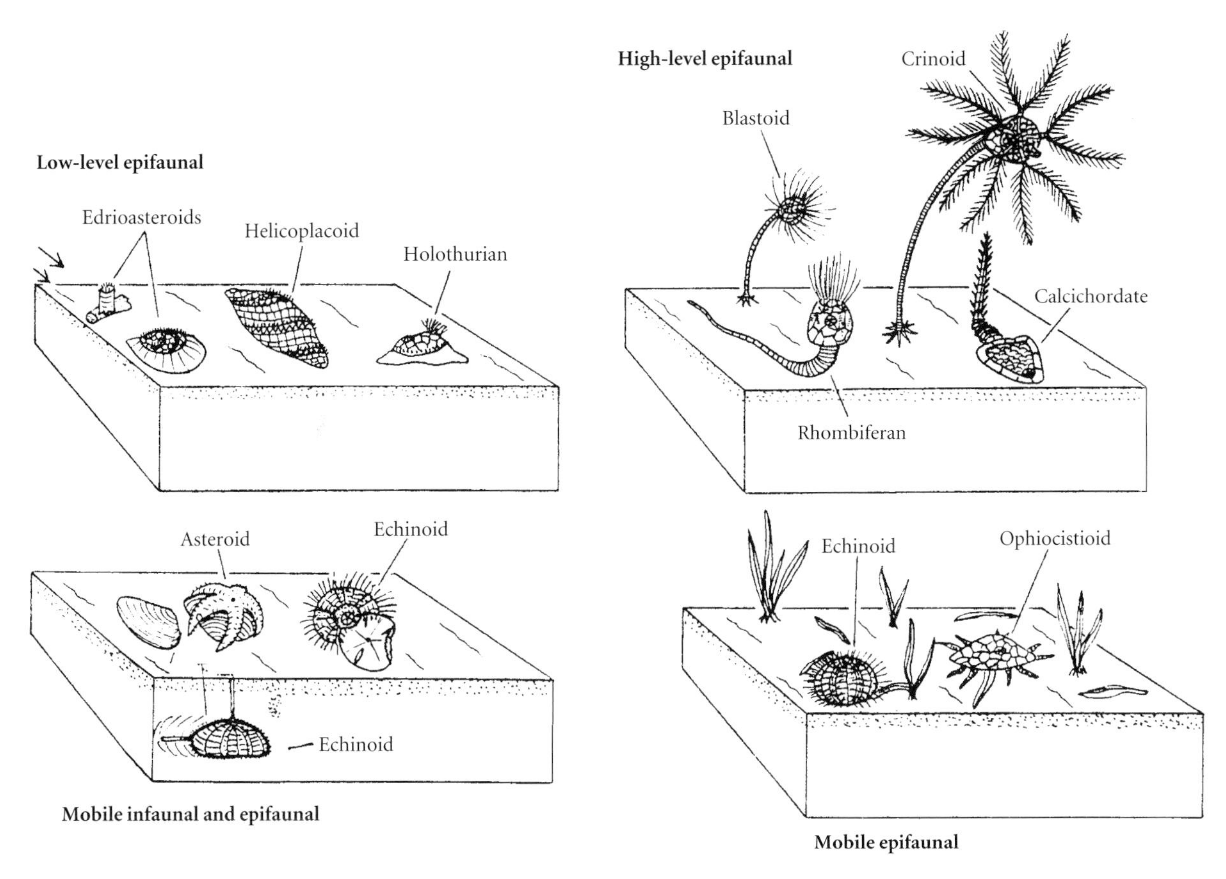

Figure 15.1 Life modes of the main echinoderm body plans. (Based on Sprinkle 1980.)

tains Late Silurian echinoderms within fine-grained turbidites, and the Lower Jurassic Starfish Bed of South Dorset, England is dominated by ophiuroids suddenly buried by a thick layer of sandstone. However, one of the most remarkable echinoderm Lagerstätte occurs in the Upper Ordovician succession of the Craighead inlier, north of the Girvan valley, southwest Scotland. Here, the Lady Burn Starfish Bed is one of several sandstone

Box 15.1 Echinoderm classification

In broad terms, the Echinodermata may be divided into two main sister groups – the stalked pelmatozoans and the mobile eleutherozoans. But there are a number of more bizarre Lower Paleozoic forms, known from only a few specimens at single localities, which are difficult to classify at present. The classification is based on a number of key features: the main body of the animal, enclosed by plates (the theca or test), areas bearing tube feet (ambulacra) with perforations or holes (brachioles) and, in the case of the pelmatozoans, the possession of a cup (calyx) and arms (brachia).

Subphylum PELMATOZOA
Class EOCRINOIDEA

- Globular or flat theca with 2–5 ambulacra bearing brachioles
- Cambrian (Lower) to Silurian

Continued

Class PARACRINOIDEA

- Irregularly arranged plates comprising globular to lenticular theca; 2–5 ambulacra commonly with pinnules. Hydropore adjacent to mouth. Stem attached to three basal plates
- Ordovician (Darriwilian to Hirnantian)

Class BLASTOIDEA

- Flask-shaped theca commonly with three basal plates; ambulacra with elongate lancet plate and rows of side plates
- Ordovician (Katian) to Permian (Tatarian)

Class DIPLOPORITA

- Globular theca with many plates in an irregular to regular pattern; 3–5 food grooves with brachioles. Diplopores perforate thecal plates
- Cambrian (Middle) to Devonian (Eifelian)

Class RHOMBIFERA

- Globular theca with 2–5 ambulacra extending from the mouth to, commonly, the edge of the upper surface. Pore structures cross adjacent thecal plates arranged in a rhomboid pattern, and comprise the respiratory system
- Cambrian (Upper) to Devonian (Frasnian)

Class CRINOIDEA

- Calyx with lower cup and upper tegmen. Sea lilies and feather stars
- Ordovician (Tremadocian) to Recent.

Subphylum ELEUTHEROZOA
Class EDRIOASTEROIDEA

- Disk-shaped thecae with straight or curved ambulacra with the mouth situated centrally and the anus sited on the interambulacra
- ?Precambrian (Ediacaran), Cambrian (Lower) to Carboniferous (Gzelian)

Class ASTEROIDEA

- Between 5 and 25 arms with large tube feet extend from a central disk. Starfishes or sea stars
- Ordovician (Floian) to Recent

Class OPHIUROIDEA

- Five long, thin, flexible arms, consisting of vertebrae and with small tube feet, extend from large, circular central disk. Brittle stars or basket stars
- Ordovician (Floian) to Recent

Class ECHINOIDEA

- Test is usually globular with plates differentiated into ambulacral and interambulacral areas. Mouth on underside, anus on upperside or sited posteriorly. Sea urchins, heart urchins and sand dollars
- Ordovician (Katian) to Recent

Class HOLOTHUROIDEA

- Body is cucumber-shaped with leathery skin with muscular mesoderm and spicules. A ring of modified tube feet surround the mouth. Sea cucumbers
- Ordovician (Floian) to Recent

Box 15.2 Origin of the echinoderms and the status of the helicoplacoids

During the major Early Cambrian radiation of echinoderms, many rather bizarre forms appeared suddenly with very different morphologies. At least nine genera were present, of which about half had pentameral symmetry, but the others were not pentameral at all. One such non-pentameral group, the helicoplacoids (Fig. 15.2), is unique in having only three ambulacral areas with tube feet wrapped around their spindle-shaped bodies. Moreover, the group lacked appendages and individuals probably lived with their shorter ends anchored to the sediment. However, helicoplacoids have many plates with the distinctive stereom structure, ambulacra and a mouth sited laterally together with an apical anus. The helicoplacoids have thus been interpreted as primitive echinoderms, surviving by suspension feeding in the sessile benthos. *Helicoplacus* may be very close to the stem group of all subsequent Echinodermata, and something like this animal might have given rise to the pelmatozoan and eleutherozoan body plans. Other groups of echinoderms were already diverse and widespread during the Early Cambrian, but the helicoplacoids were apparently restricted to western North America, where they were very abundant during only the Early Cambrian. Their extinction may have been a very important ecological signal. Such groups of unattached "sediment stickers" were well adapted to the algal mat substrates of the Neoproterozoic. Perhaps they could not cope with the increased bioerosion and bioturbation of soft substrates that were part of the move away from seafloors covered by microbial mats that prompted the Cambrian substrate revolution (Bottjer et al. 2000).

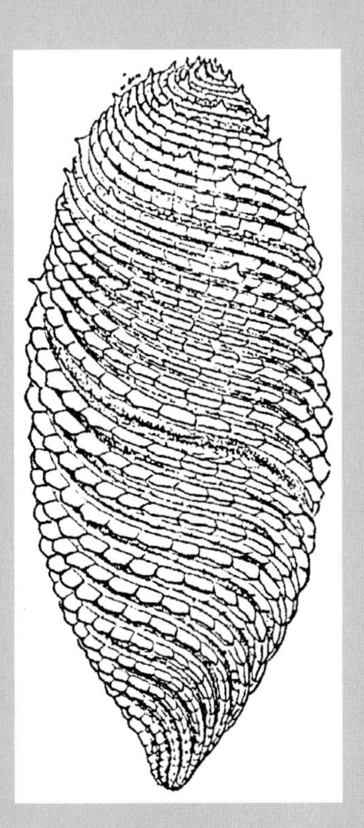

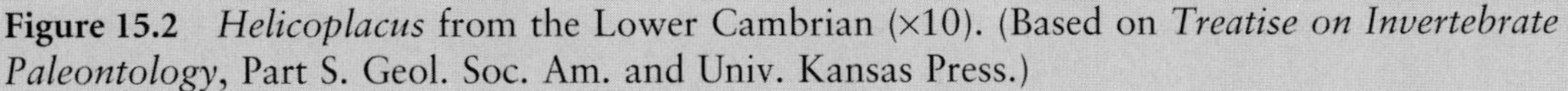

Figure 15.2 *Helicoplacus* from the Lower Cambrian (×10). (Based on *Treatise on Invertebrate Paleontology*, Part S. Geol. Soc. Am. and Univ. Kansas Press.)

Box 15.3 Columnal classification

The majority of crinoid assemblages are represented by disarticulated ossicles. Conventional taxonomy based on a description of complete, articulated specimens is thus not possible. Nevertheless, ossicles have many distinctive features, arguably with more well-defined characteristics than many groups of macrofossils (Fig. 15.3). Single stems consist of many ossicles with a central canal or lumen usually carrying nerve fibers. Both the ossicles and lumens have distinctive shapes that are the basis of a **form taxonomy** of the group. Form taxonomy helps us classify the shapes of fossils, in the same way that we can classify nuts and bolts. It is a useful method of organizing our data, but since it is not biologically meaningful, cannot be used in phylogenetic analyses. Stems may be either homeomorphic, composed of similarly shaped ossicles, or heteromorphic with a variety of different-shaped ossicles. Moreover stems may be subdivided into zones that may be internally homeomorphic or heteromorphic. Columnal taxonomy has proved useful in describing taxa (so-called col. taxa) of pelmatozoan, particularly crinoid, ossicles of stratigraphic significance.

Figure 15.3 Some crinoid ossicle types. (a) Articular facet of a columnal of the bourgueticrinid *Democrinus* (?) sp., with a fulcral ridge of the synarthrial articulation; the lumen opens at the bottom of the "8"-shaped depression (×15). (b) Cirral scar on a nodal of the isocrinoid *Neocrinus* with well-preserved stereom microstructure and knob-like synarthrial fulcrum (×18). (c) Articular facet of a columnal of the isocrinoid *Neocrinus* with symplectial articulation around the five petal-like areola areas (×9). (Courtesy of Stephen Donovan.)

units within a deep-water mudstone sequence. Entire crinoids, cystoids, echinoids and calcichordates were carried downslope and rapidly buried on the unstable slopes of a submarine fan system.

Crinoidea

Although the crinoids, famously called "sea lilies", look more like plants than animals, there is no doubt they are animals, and echinoderms at that. They are usually sessile, with characteristic echinoderm pentameral symmetry, rooted by a stalk, for at least part of their life cycle, to the seabed; but some forms after a short fixed stage are entirely free living. Modern forms live in dense clusters or "forests" ranging from the warm waters of the tropics to the icy conditions of polar latitudes. The "feather stars" prefer the clear-water conditions of the continental shelf, living in nooks and crevices, and emerging at night to perch on ridges. The fixed sea lilies occupy the deep-water environments of the continental slope. The majority of fossil forms were almost certainly part of the shallow-water sessile benthos. The success of the crinoids may be measured by the fact we know over 6000 fossil species and an age range from the Early Ordovician to the present day.

Morphology and life modes

The crinoids consist of a segmented stalk or stem composed of columnals or ossicles (Box 15.3) fixed to the seabed by root-like structures or **holdfasts**. Attached to the top of the stalk is the case containing the main functional part of the animal called alternatively the calyx, aboral cup or theca. The **calyx** is built of two rings of calcitic plates – the basals and the overlying radials in a monocyclic configuration. In a number of taxa, the dicyclic forms, a second circle of smaller plates, the **infrabasals**, interface between the basals and the stem, providing further articulation. The upper, oral surface of the calyx is covered by a flexible membrane or tegmen and houses a number of important structures. These are the mouth, which is usually situated centrally at the convergence of five radially arranged feeding grooves; the anus, which is sited posteriorly with the outlet often modified by an anal tube enhancing the efficiency of waste disposal; and the arms or **brachials**, which extend upwards from the calyx and together form the crown.

As already noted, two main life strategies were pursued by the crinoids (Fig. 15.4). The majority of fossil crinoids and about 25 Recent

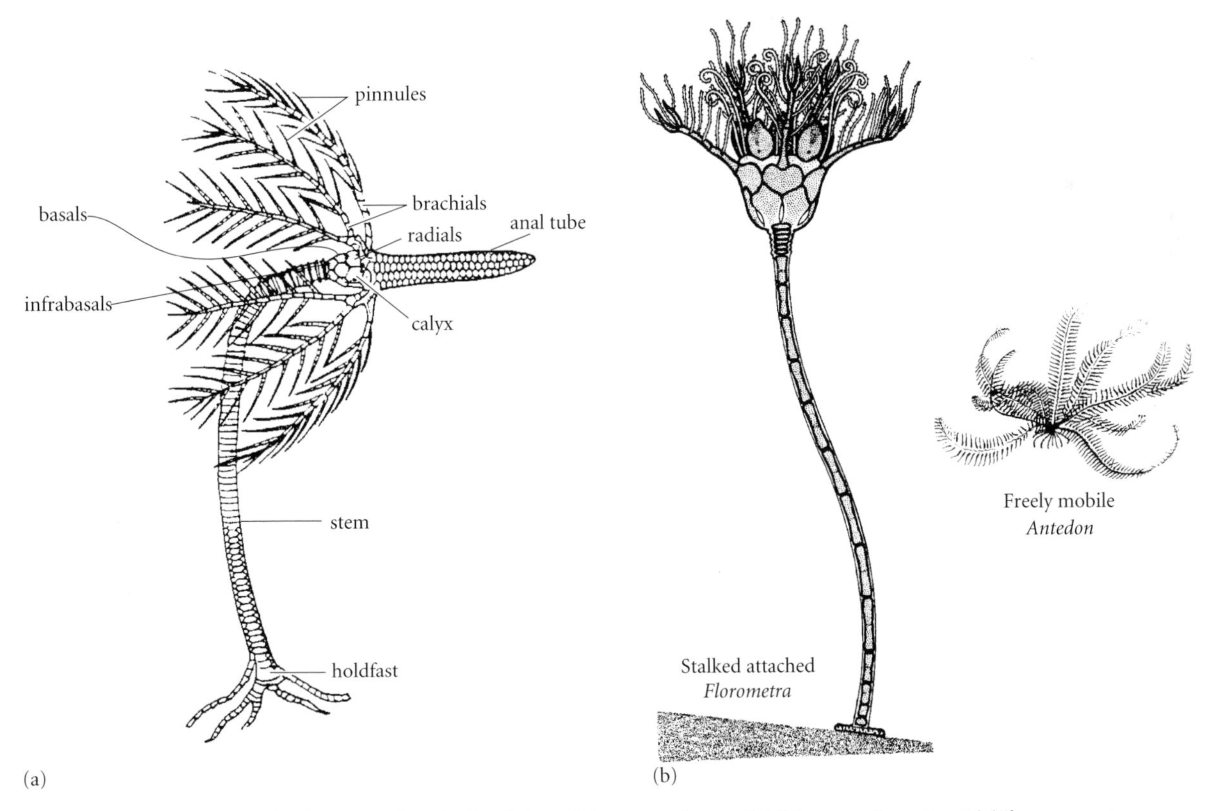

Figure 15.4 (a) Morphology of the Ordovician *Dictenocrinus*. (b) Two main crinoid life strategies, fixed and mobile. (Redrawn from various sources.)

genera are stalked forms, attached to the seabed. Modern oceans, however, are dominated by mobile comatulids that move about like pneumatic umbrellas, pumping their long arms in unison. *Antedon* is one of nearly 100 non-stalked genera that, after a short fixed stage, are free to crawl and swim with the aid of flexible arms and cirri.

Classification and evolution

The oldest reported crinoid, *Echmatocrinus brachiatus* from the Middle Cambrian Burgess Shale, has uniserial or single brachials, in contrast to the biserial arms of contemporary eocrinoids. *Echmatocrinus* has few other unequivocal crinoid characters and there is wide agreement that it is actually an octocoral. More recognizable crinoids with more typically constructed cups and columnal-bearing stems, such as *Dendrocrinus*, appear some time later during the Tremadocian. A major expansion in the Early Ordovician tropics marked a period of intense morphological experimentation and many adaptive radiations.

Virtually all Paleozoic crinoids were stalked, and traditionally have been grouped into three subclasses, the Inadunata, Flexibilia and Camerata (Fig. 15.5). Inadunate crinoids comprise a large and varied group, originating in the Early Ordovician and continuing until the Triassic. They have a rigid calyx with either free or loosely attached brachials, and monocyclic or dicyclic calyx bases. Camerate crinoids are characterized by large cups with both monocyclic and dicyclic plate configurations. The uniserial or biserial brachials, decorated with pinnules, are firmly attached to the cup and the tegmen is heavily plated, obscuring the food grooves and mouth, but developed laterally with an anal tube. The Flexibilia, comprising some 60 genera, have a dicyclic plate configuration comprising three infrabasals. The brachials are uniserial and lacking pinnules, and the tegmen is flexible with a mosaic of small plates. Their stems have circular cross-sections and lack cirri. These groups are especially well known in the Carboniferous (Box 15.4).

The fourth subclass of crinoids, the Articulata, with the exception of some Triassic inadunates, includes all post-Paleozoic crinoids. A few Paleozoic forms with articulate similarities such as *Ampelocrinus* and *Cymbiocrinus* may be stem-group articulates. Over 250 genera are recognized with almost two-thirds of known genera extant. Microcrinoids are a highly specialized crinoid morphotype developed within both the Inadunata, during the Paleozoic, and the Articulata, during the Mesozoic. Microcrinoids are minute, never more than 2 mm in size; they may be pedomorphic forms living together with more typical crinoid communities.

Blastozoans

Blastozoans are an informal grouping that includes three of the more minor, yet nevertheless important, echinoderm groups that are all extinct: the cystoids, blastoids and eocrinoids. These pelmatozoans were usually equipped with a short stem but often lacked brachia or arms. Blastomorphs were probably high-level filter feeders, particularly characterized by pores or brachioles punctuating the thecal plates. Eocrinoids are included by some authors in the cystoids, appearing near the base of the Cambrian and becoming extinct during the Silurian. The eocrinoids, however, probably included ancestors to both the cystoids and the crinoids.

Cystoids

Mid Paleozoic blastozoans with respiratory pore structures modifying the thecal plates have been traditionally placed within the Cystoidea. This mixed bag includes two classes, the Diploporita and the Rhombifera, that became very widespread during the Mid Paleozoic. They had spherical or sac-like thecae, commonly with 1000 or more irregularly arranged plates. Moreover the group has brachioles lacking pinnules and characteristically the plates are usually equipped with distinctive pore structures. A variety of such pore structures have been recognized in the cystoids (Fig. 15.7), and they are fundamental in the higher-level classification of the group.

Diploporita The diploporites had thecal plates punctuated by pairs of pores either covered with soft tissue (diplopores) or a layer of stereom with the pore pairs joined by a network of minor canals (humatipores). These pores probably held a bulbous respiratory bag and allowed for the efficient entry and exit of celomic fluid. Both stalked and non-stalked forms are present in this group, suggesting a

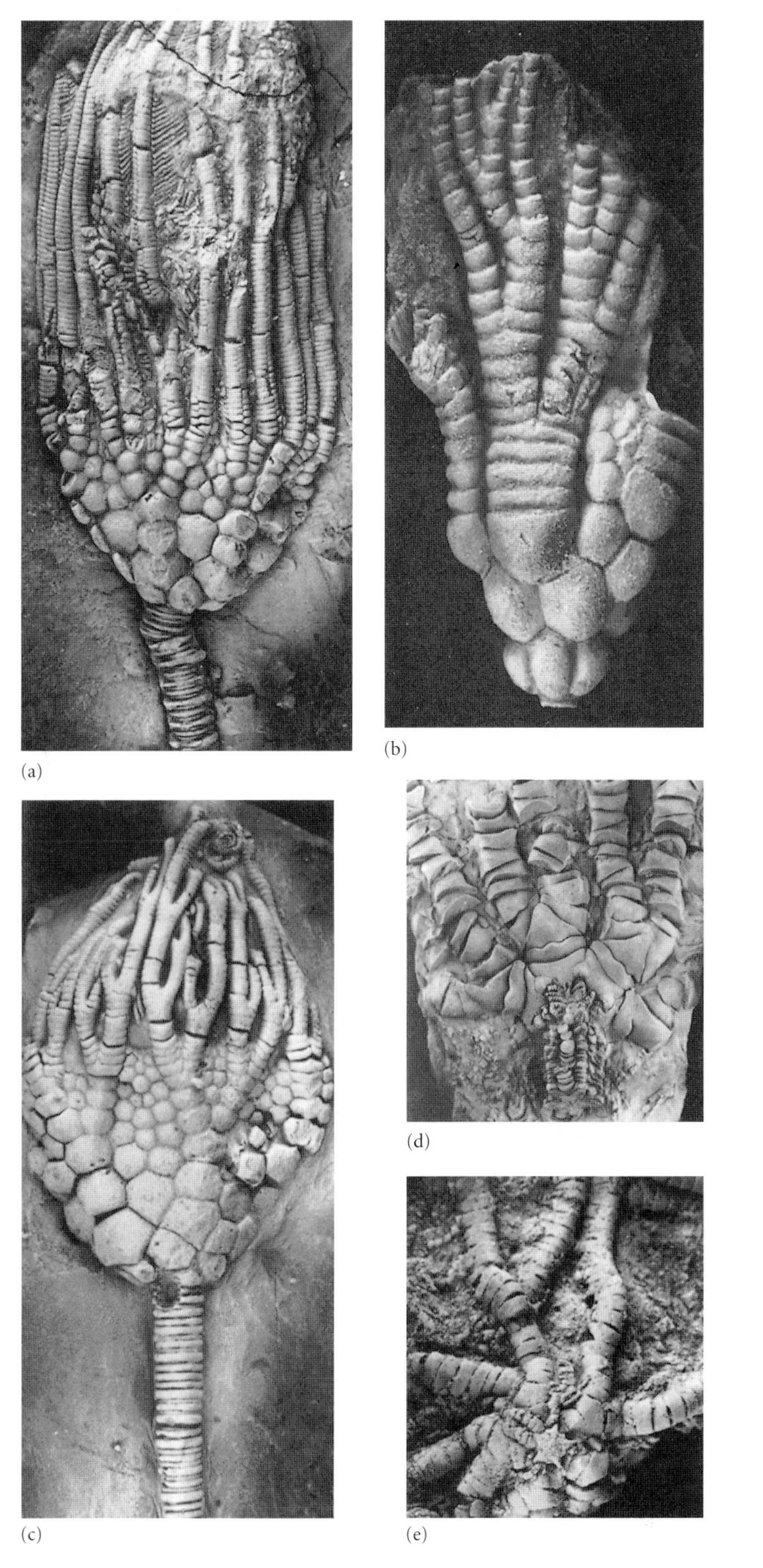

Figure 15.5 Some crinoid genera: (a) *Dimerocrinites* (Silurian; Camarata), (b) *Cupalocrinus* (Ordovician; Indunata), (c) *Sagenocrinites* (Silurian; Flexibilia), (d) *Chladocrinus* (Jurassic; Articulata) and (e) *Paracomatula* (Jurassic; Articulata comatulide). Magnification approximately ×1 (a, c), ×2 (b, d, e). (From Smith & Murray 1985.)

Box 15.4 The age of crinoids: an Early Carboniferous diversity spike

Early Carboniferous (Mississippian) crinoids were abundant and diverse, so much so that this interval is often called the "Age of crinoids". Limestones of this age often consist of over 50% pelmatozoan debris, and are known as encrinites. Why then were crinoids so abundant at this time? Two factors seem to have contributed to these extensive shoals of crinoids (Kammer & Ausich 2006). Firstly, five major groups were in various states of recovery after the Frasnian-Famennian extinction event, particularly the advanced cladids (Fig. 15.6). Secondly, with the disappearance of the shelf-edge coral–stromatoporoid buildups at the end of the Devonian, platform geometries were quite different. There was improved and unimpeded water circulation, which promoted stenohaline conditions that encouraged the growth of crinoid communities. With new ecospace and a lack of predation pressures, crinoid diversity exploded. Sadly, the good times came to an end with regression and the cooler-water conditions associated with the Late Carboniferous glaciation. Crinoids were never again so diverse.

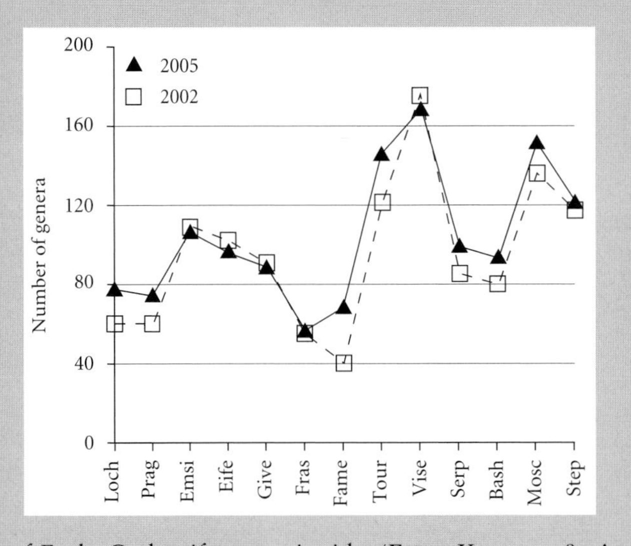

Figure 15.6 Diversity of Early Carboniferous crinoids. (From Kammer & Ausich 2006.)

wide range of strategies from a fixed sessile mode to free-living recumbent styles. The diploporites were very widespread from the Early Ordovician to the Early Devonian and probably evolved from a Late Cambrian blastozoan ancestor.

Rhombifera The rhombiferans appeared during the Late Cambrian equipped with brachioles and distinctive rhombic patterns of respiratory pores crossing thecal plate sutures (Fig. 15.7). They are classified according to the pattern and shape of their pores; these separate the order Dichoporita from the Fistulipora. The rhombiferans became common during the Early Ordovician and continued with a near cosmopolitan distribution until the Late Devonian, and were probably replaced by the better adapted blastoids during the Silurian and Devonian.

Blastoids

The extinct blastoids were small, pentamerally symmetric animals with short stems and hydrospires adapted for respiration (Fig. 15.8). They are represented by over 80 genera in rocks of Silurian to Permian age. The blastoid cup or theca is usually globular and composed of a ring of three basal plates, surmounted by a circle of five larger radial plates. The mouth is often surrounded by five large openings or **spiracles** associated with the respiratory system. Although relatively rare,

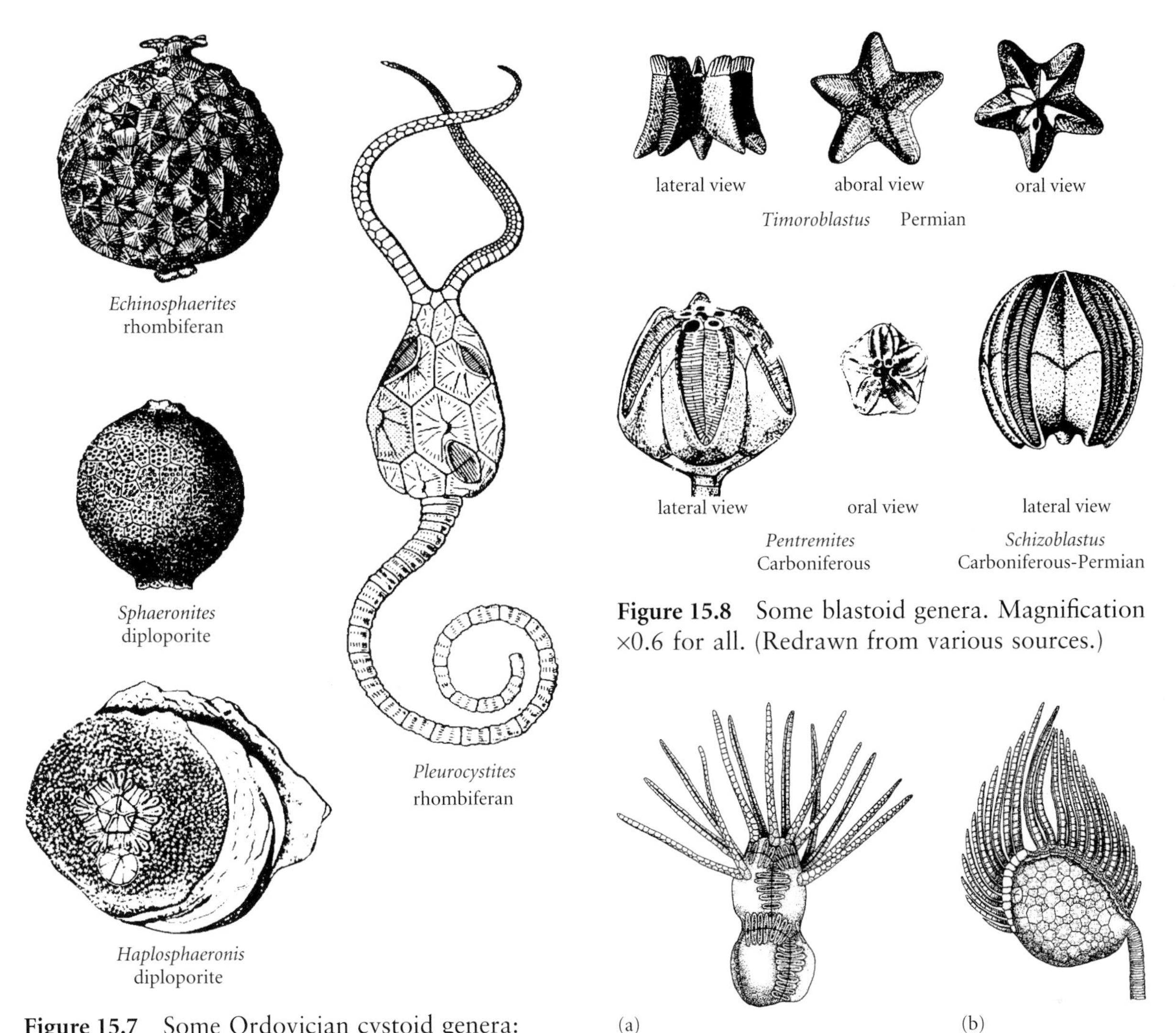

Figure 15.7 Some Ordovician cystoid genera: *Echinosphaerites* and *Sphaeronites*, (×0.75), *Haplosphaeronis* and *Pleurocystites* (×1.5). (Based on *Treatise on Invertebrate Paleontology*, Part S. Geol. Soc. Am. and Univ. Kansas Press.)

Figure 15.8 Some blastoid genera. Magnification ×0.6 for all. (Redrawn from various sources.)

Figure 15.9 (a) An eocrinoid, and (b) a paracrinoid. (Based on *Treatise on Invertebrate Paleontology*, Part S. Geol. Soc. Am. and Univ. Kansas Press.)

a few horizons are packed with blastoids, particularly when the diversity of the group peaked in the Early Carboniferous. Viséan reefal facies in northern England yield abundant blastoids, as do the Permian limestones on the island of Timor where, for example, *Timoroblastus* and *Schizoblastus* occur.

The blastoids first appeared during the Silurian, probably evolving from an Ordovician ancestor with brachioles and a reduced number of plates. They initially competed, ecologically, with the rhombiferan cystoids. The evolutionary history of the group was marked by changes in the shape of the theca and variations in the length of the ambulacra. Two main groups are recognized: the more basal Fissiculata characterized by hydrospire folds, and the Spiraculata with, as the name suggests, well-developed spiracles.

Eocrinoids

The eocrinoids were the earliest of the brachiole-bearing echinoderms. They had a huge range of thecal shapes with primitive holdfasts and an irregular to regular arrangement of plates (Fig. 15.9a). Sutural pores rather than thecal pores, along the joins between the plates, were characteristic of the earliest eocrinoids; in others there is a total lack of respiratory structures. Eocrinoids differ from the crinoid groups in having biserial brachial appendages. Over 30 genera have been described from rocks of Early Cambrian to Late Silurian age. The origins of the other blastozoan classes are probably to be found within this heterogenous group; for example the aberrant Late Cambrian eocrinoid *Cambrocrinus* has been cited as an ancestor for the rhombiferan cystoids. Whereas many eocrinoids were high-level suspension feeders with the first columnal-constructed stems, some lay reclined or recumbent on the seabed. The Ordovician *Cryptocrinus*, for example, has a globular theca with a more irregular arrangement of plates.

Paracrinoids

The paracrinoids (Fig. 15.9b) are a small, odd group of arm-bearing echinoderms that have globular thecae and numerous irregularly-arranged plates together with two to five arm-like, food-gathering structures. They are so different that some scientists have suggested that they represent a separate subphylum. The group is restricted to North America, where they are common in the Middle Ordovician.

Echinoidea

Echinoids, the well-known sea urchins and sand dollars, have robust, rigid endoskeletons, or **tests**, composed of plates of calcite coated by an outer skin covered by spines. The tests are usually either globular or discoidal to heart-shaped (Smith 1984). Echinoids are most common in shallow-water marine environments where they congregate in groups as part of the nektobenthos. Their classification (Box 15.5) is based on the arrangement of plates and their mouth structures.

Echinoids have a long history from their first radiation in the Ordovician (Paul & Smith 1984). Two of the most significant evolutionary events in the history of the subphylum were marked by sudden divergences from the regular morphology to generate irregular burrowing echinoids. The first, in the Jurassic, led subsequently to a range of irregular burrowers, and the second, during the Paleocene, to the quasi-infaunal sand dollars. Both events were probably rapid and permitted major adaptive radiations of parts of the group into new ecological niches.

Basic morphology

The exoskeleton or test of most regular echinoids, for example the common sea urchin *Echinus esculentus*, is hemispherical and displays all the main features of the group (Fig. 15.11). The lower, adapical or oral, surface is perforated by the mouth whereas the upper, apical or aboral, surface has the anal opening. The sea urchin is part of the active mobile benthos, in contrast to the sand dollars which were quasi-infaunal.

The test is built of a network of many hundreds of interlocking calcite plates organized into 10 segments, radiating from the oral surface and converging on the aboral surface. Five narrower segments or **ambulacral areas** (ambs) carry the animal's tube feet and are in contact with the ocular plates. The ambs alternate with the wider **interambulacral areas** (interambs), are armed with spines and abut against the genital plates. Together the ambs and interambs comprise in total 10 areas and 20 columns, which make up the corona – the majority of the test.

The central part of the aboral surface has a ring of five genital plates, each perforated by a hole to allow the release of gametes; the **madreporite** is commonly larger than the other genital plates and has numerous minute pores interfacing, beneath, with the water vascular system. These alternate with the ocular plates, terminating the ambulacral areas, and each houses further outlet holes for the water vascular system. This part of the apical system surrounds the **periproct**, or anal opening, which is partially covered by a number of smaller plates attached to a membrane. On the underside of the test, the **peristome**, containing the mouthparts, is also covered by a membrane coated with small plates. The mouth holds a relatively sophisticated jaw apparatus comprising five individual jaws each with a single, curved, saber-like tooth, operating like a mechanical grab and forcing particles into the animal's digestive system. The great ancient Greek naturalist Aristotle, who described the structure first, compared it to a "horn lantern with the panes of horn left out", and the echinoid jaw is often called **Aristotle's lantern**. In crown-group forms, muscles attached to the lantern are anchored to the perignathic girdle, developed around the edge of the peristome.

Box 15.5 Echinoid classification

The traditional split of the class into regular and irregular forms is no longer considered to reflect the true phylogeny of the echinoids. Whereas the irregular echinoids are probably monophyletic, arising only once, the regular echinoids do not form a clade. The group was traditionally subdivided into three subclasses (Fig. 15.10) – the Perischoechinoidea, Cidaroidea and Euechinoidea – the first, however, has been shown to be polyphyletic and the term stem-group echinoids is preferred.

Stem-group ECHINOIDEA

- Regulars with ambulacra in more than two columns, interambulacra with many columns; in total the test is composed of over 20 columns. Lantern with simple grooved teeth and lacking a perignathic girdle
- Upper Ordovician to Permian

Crown-group ECHINOIDEA
Subclass CIDAROIDEA

- Regulars with test consisting of 20 columns of plates; two columns in each ambulacra and interambulacral areas. Interambulacral plates have large tubercle. Teeth are crescentic to U-shaped and the perignathic girdle includes only interambulacral elements
- Lower Permian to Recent

Subclass EUECHINOIDEA

- Post-Paleozoic taxa, both regular and irregular. Both ambulacra and interambulacra with twin columns. Perignathic girdle composed on ambulacral projections
- Middle Triassic to Recent

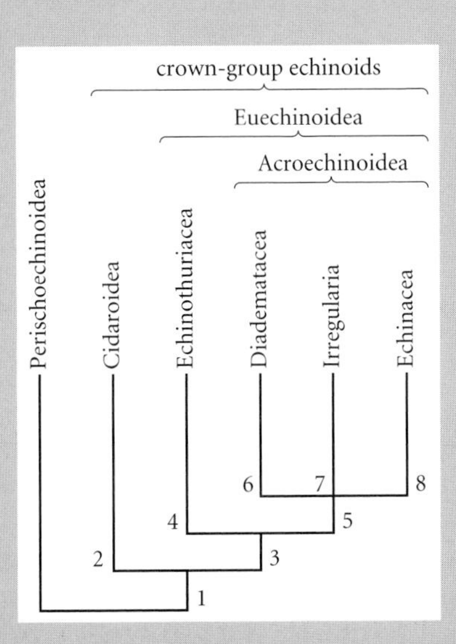

Figure 15.10 Echinoid classification based mainly on cladistic analysis: 1, 10 ambulacral and 10 interambulacral areas; 2, upright lantern without foramen magnum; 3, distinctive perignathic girdle; 4, distinctive ambulacral areas; 5, upright lantern with deep foramen magnum; 6, grooved teeth; 7, stout teeth; 8, keeled teeth.

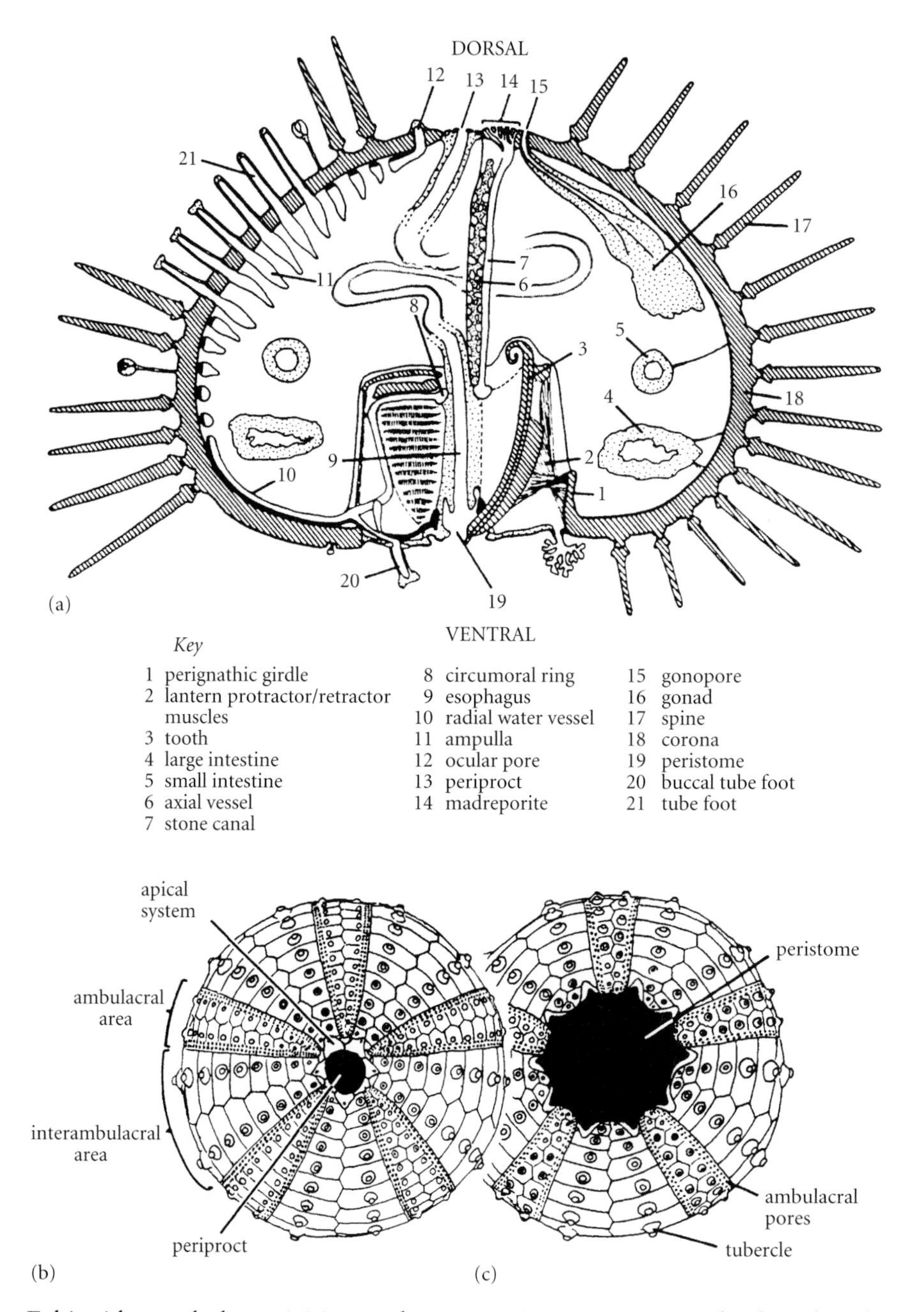

Figure 15.11 Echinoid morphology: (a) internal anatomy in cross-section; (b) dorsal and (c) ventral views of *Echinus*. (Based on Smith 1984.)

The echinoid's various organs are suspended within the test and supported by fluid. The water vascular system copes with a number of functions. The stone canal rises vertically, from the central ring around the esophagus, to unite with the madreporite. Five radial water vessels depart from the central ring to service the ambulacral areas; smaller vessels are attached to each tube foot and its ampulla where variations in water pressure drive the animal's locomotory system.

The echinoid digestive system lacks a stomach and operates through the esophagus together with a large and small intestine; waste material is expelled through the rectum into the anus and externally by way of the periproct. An unsophisticated nervous system comprising a nerve ring and five radial nerves connects with the ambulacral pores where the nerve ends divide externally to form a sensory net.

The regular to irregular transition

Regular echinoids, the sea urchins, with a compact, symmetric morphology are most basal. Their shape contrasts with the irregulars, with marked bilateral symmetries and specialized for forward movement. The irregular echinoid morphotype evolved rapidly and apparently involved some large architectural changes to adapt the animal to burrowing. *Plesiechinus hawkinsi* is one of the first irregular echinoids, appearing early in the Early Jurassic (Sinemurian) with an asymmetric test, short numerous spines, large adapical pores and a posteriorly placed periproct together with presumed keeled teeth. Ten million years later, by the Toarcian, much of the "toolkit" of adaptations had evolved for a burrowing life mode. Secondary bilateral symmetry was superimposed on the existing pentameral symmetry to form a heart-shaped or flattened ellipsoidal test. The periproct migrated from a position on the apical surface to the posterior side of the test to eject waste laterally. By the Early Cretaceous, one of the ambulacral areas had become modified to form a food groove and a series of tube feet were extendable with flattened ends to assist respiration.

One of the earliest sand dollars, the clypeasteroid *Togocyamus*, appeared during the Paleocene, and some 20 million years later in the Eocene more typical sand dollars had evolved to command a cosmopolitan distribution. The flattened test was adapted for burrowing, whereas the accessory tube feet could encourage food along the food grooves and draped the test with sand. The highly accentuated petals helped respiration by providing an increased surface area for the tube feet, and the development of a low lantern with horizontal teeth signaled changes in feeding patterns.

Ecology: modes of life

The regular *Echinus* and the irregular *Echinocardium* probably mark the ends of a spectrum of life modes from epifaunal mobile behavior to a number of infaunal burrowing strategies (Fig. 15.12). Mobile regular forms such as *Echinus* grazed on both hard and soft substrates and in caves and crevices on the seafloor; these sea urchins may have been omnivores, carnivores or herbivores. Irregular forms display a range of adaptations appropriate to an infaunal mode of life where burrows were carefully constructed in low-energy environments. Extreme morphologies were developed in the sand dollars or Clypeasteroidea, permitting rapid burial just below the sediment–water interface in shifting sands. Echinoids generally lived in shallow seawaters, but some went deeper; the timing of this move offshore has been controversial (Box 15.6).

Life modes and evolution: microevolution of Micraster One of the classic case studies of evolutionary patterns in fossils is seen in *Micraster*, an infaunal, irregular echinoid. Paleobiologists have repeatedly used this example to test phyletic gradualist and punctuated equilibria models (see p. 121) and as the raw material for the rigorous statistical analysis of both ontogenetic and phylogenetic change. In the best-known lineage, *M. leskei–M. decipiens–M. coranguinum*, the following morphological changes occurred (Fig. 15.14):

1 The development of a higher, broader (heart-shaped) form associated with an increase in size and thickness of the test.
2 The peristome (mouth) moved anteriorly and the posteriorly situated periproct (anus) had a lower position on the side of the test with a broader subanal fasciole.
3 The madreporite increased in size at the expense of the adjacent specialized plates.
4 More tuberculate and deeper anterior ambulacra evolved.
5 More granulated periplastronal areas developed.

These morphological changes are associated with life in progressively deeper burrows. But there is a lack of associated trace fossils that might prove this. On the other hand, the adaptations may have been geared to greater burrowing efficiency, probably in shallow depths in the chalk where such traces were destroyed by reworking of the sediments.

Microevolutionary trends have been tested in other echinoid lineages. The irregular *Discoides* occurs abundantly through an Upper Cretaceous section at Wilmington, south Devon, England. The height and diameter of

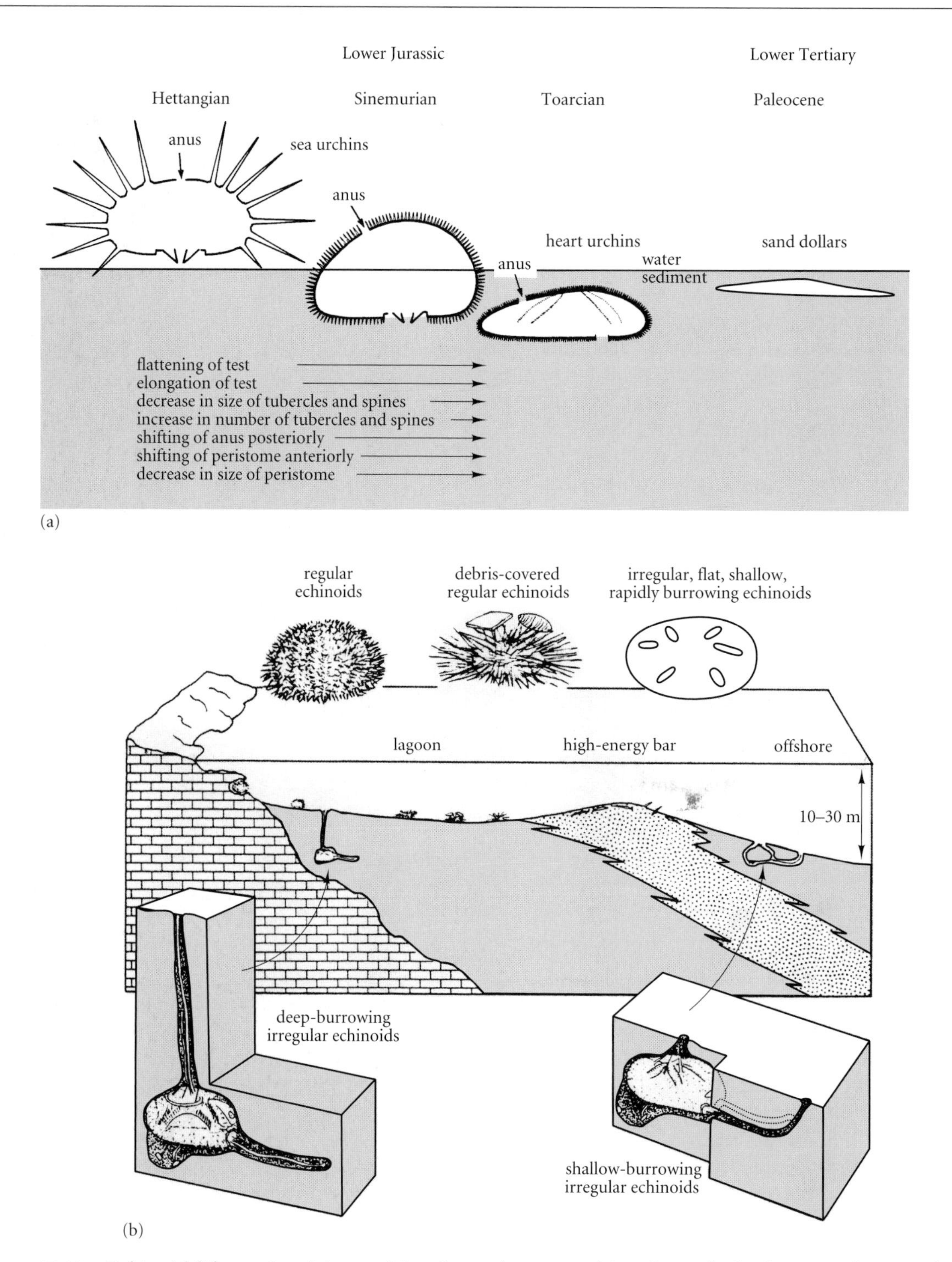

Figure 15.12 Echinoid life modes: (a) transition from the sea urchins through the heart urchins to the sand dollars; (b) habits and modes of life of echinoids. (a, based on Kier, P. 1982. *Palaeontology* **25**; b, based on Kier, P. 1982. *Smithson. Contr. Paleobiol.* **13**.)

the echinoids change through the section, but this seems to be related to the grain size of the sediment, where high, narrow forms favor fine sediment. The case history is available at http://www.blackwellpublishing.com/paleobiology/.

Evolution

The first echinoids had appeared by the Mid Ordovician but it is only in Lower Carboniferous rocks that echinoids become relatively

Box 15.6 Into the deep: but not until the Late Cretaceous?

A number of animal groups common in the Paleozoic evolutionary fauna, such as the brachiopods and crinoids, are common in deep-water environments. This reinforces the view that the deep sea is some sort of refuge for archaic taxa that have been forced down the continental slope by predation or unsuccessful competition on the shelf. Andrew Smith and Bruce Stockley (2005) have developed a quite different model, however, based on molecular clock estimates and the phylogeny of echinoid taxa. Results show that the modern deep-sea omnivore fauna appeared gradually over the last 150–200 myr; detritivores, however, were in place during a much shorter time span between 75 and 50 Ma (Fig. 15.13). This 25 myr window of seaward migration appears to be associated with marked increases in seasonality, continental sediment discharge and surface productivity. The increased availability of organic carbon and nutrients in the deep sea provided the means for habitat expansion rather than an escape from competition and predation on the shelf.

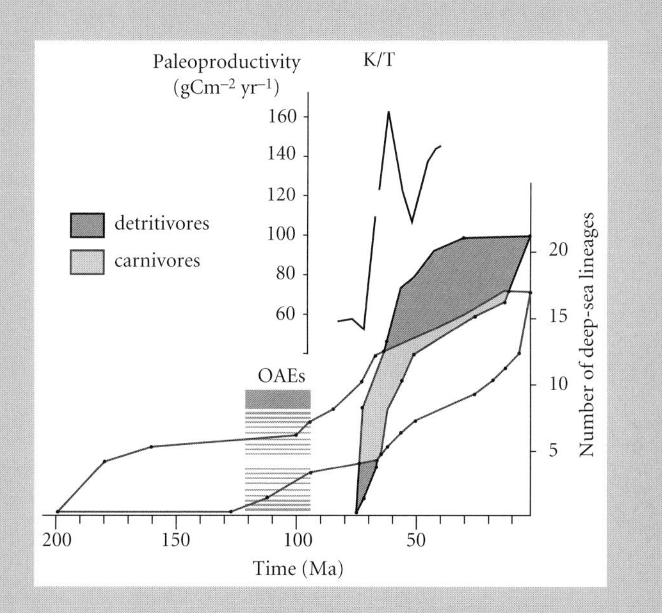

Figure 15.13 Events in the deep sea: cumulative frequency polygons for maximum and minimum times of origin of 38 clades of extant, carnivore and detritivore deep-sea echinoids (Smith & Stockley 2005). K/T, Cretaceous–Tertiary boundary; OAEs, oceanic anoxic events.

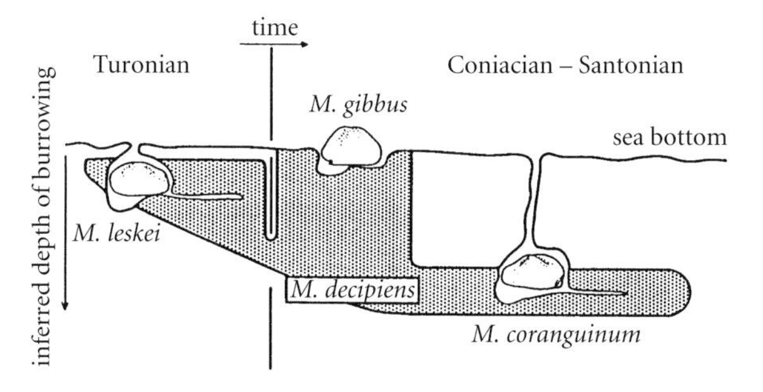

Figure 15.14 Evolution of the Late Cretaceous heart urchin, *Micraster*. (Based on Rose, E.P.F. & Cross, N.E. 1994. *Geol. Today* 9.)

abundant. The sparse early record of the group might reflect a relatively fragile skeleton that quickly disintegrated after death; on the other hand the echinoids were probably not a common element of the Paleozoic benthos. The enigmatic *Bothriocidaris*, described from the Ordovician of Estonia and from southwest Scotland, has been variously classified as an echinoid, cystoid or holothurian. Some authorities consider that *Bothriocidaris* and *Eothuria* might be unclassifiable echinoids, hopeful monsters that arose during the rapid Ordovician radiation of the group.

Aulechinus from the Upper Ordovician of southwest Scotland is one of the most primitive echinoids and the first with only two plate columns in the ambulacral areas. During the Paleozoic there was generally an increase in the number and size of ambulacral areas and the sophistication of Aristotle's lantern, although most genera remained relatively small (Fig. 15.15).

There was a significant decline in echinoid diversity during the Late Carboniferous. By the Permian only half a dozen species are known, and they belonged to two primary groups: detritus feeders and opportunists. Large proterocidarids were highly specialized detritus feeders, and the small omnivorous *Miocidaris* and *Xenechinus* were opportunists. Two lineages, including *Miocidaris*, survived the end-Permian extinction event to radiate extravagantly during the early Mesozoic, thus ensuring the survival of the echinoids. Following the end-Permian extinctions the regular echinoids diversified during the Late Triassic and Early Jurassic with more advanced regulars dominating the early Mesozoic record. The irregulars appeared during the Early Jurassic and substantially increased in numbers during the period. Diversity was severely reduced by the Cretaceous-Tertiary extinction event but both the regulars and irregulars recovered rapidly during the early Cenozoic.

Asteroidea

Starfish are common on beaches today, and their biology has made them hugely successful. Some feed by preying on shellfish and other slow-moving shore and shallow-marine animals. Their feeding mode is unusual but deadly: they simply sit on top of their chosen snack, turn their stomachs inside out and absorb the flesh of their victim. The majority are benthic deposit feeders that ingest prey or filter feed. Starfish are also unusual in that they have eyes at the ends of their arms – these are actually light-detecting cells, not true eyes, but the adaptation is novel nonetheless.

Asteroids appeared first during the Early Ordovician. The subphylum contains two main groups: the asteroids or starfish and the ophiuroids or brittle stars. These animals have a star-shaped outline with usually five arms radiating outwards from the central body or disk. The water vascular system is open. The mouth is situated centrally on the underside of the animal on the oral or dorsal surface whereas the anus, if present, opens ventrally on the adoral surface. The asterozoans are characterized by a mobile lifestyle within the benthos, where many are carnivores. Asterozoan skeletons disintegrate rapidly after death due to feeble cohesion between the skeletal plates. Thus, recognizable fossils are relatively rare. Nevertheless there are a number of starfish Lagerstätten deposits where asterozoans are extremely abundant and well preserved.

Distribution and ecology of the main groups

Three classes of asterozoans have been recognized: the basal Somasteroidea, the Asteroidea or starfish and the Ophiuroidea or brittle stars. The Somasteroidea include some of the earliest starfish-like animals, described from the Tremadocian of Gondwana. These echinoderms have pentagonal-shaped bodies with the arms initially differentiated from around the oral surface. In some respects this short-lived group, which probably disappeared during the Mid Ordovician, displays primitive starfish characters intermediate between a pelmatozoan ancestor and a typical asterozoan descendant. Typical asteroids have five arms radiating from the disk, which is coated by loosely fitting plates permitting considerable flexibility of movement (Fig. 15.16). Additional respiratory structures, called **papulae**, project from the celom through the plates of the upper surface. This backup system aids the high metabolic rates of these active starfish.

The first true starfish were probably derived from the somasteroids during the Early Ordovician and were relatively immobile, infaunal sediment shovelers. Some of the first starfish, for example *Hudsonaster*, from the Middle Ordovician, have similar plate configurations to the young growth stages of living forms such as *Asterias*. Although relatively uncommon in Paleozoic rocks, the group was important during the Mesozoic and Cenozoic and is now one of the most common echinoderm classes.

The ophiuroids first appeared during the Early Ordovician (Arenig), but the group, as presently defined, may be paraphyletic. Classification is based on arm structure and disk plating. The ophiuroid body plan is distinctive, with a subcircular central disk and five

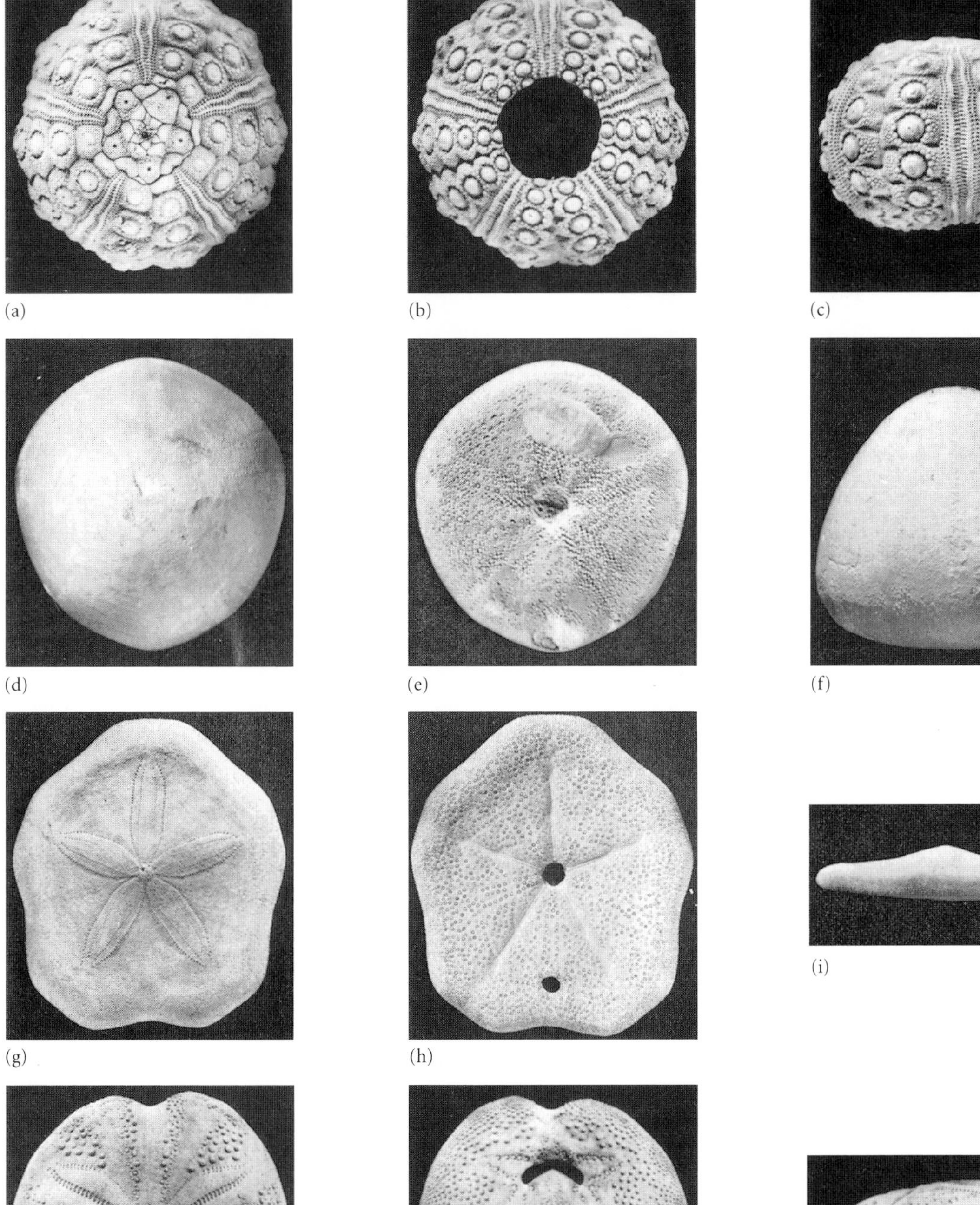

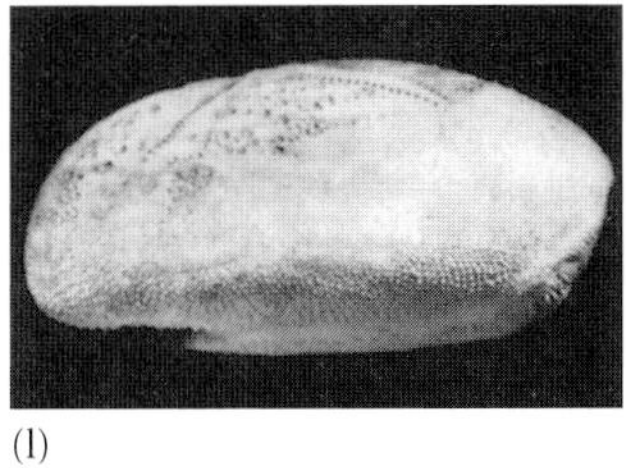

Figure 15.15 Aboral, oral and lateral views of some echinoid genera: (a–c) *Cidaris* (Recent; regular), (d–f) *Conulus* (Cretaceous; irregular), (g–i) *Laganum* (Recent; sand dollar) and (j–l) *Spatangus* (Recent; heart urchin). All approximately natural size. (From Smith & Murray 1985.)

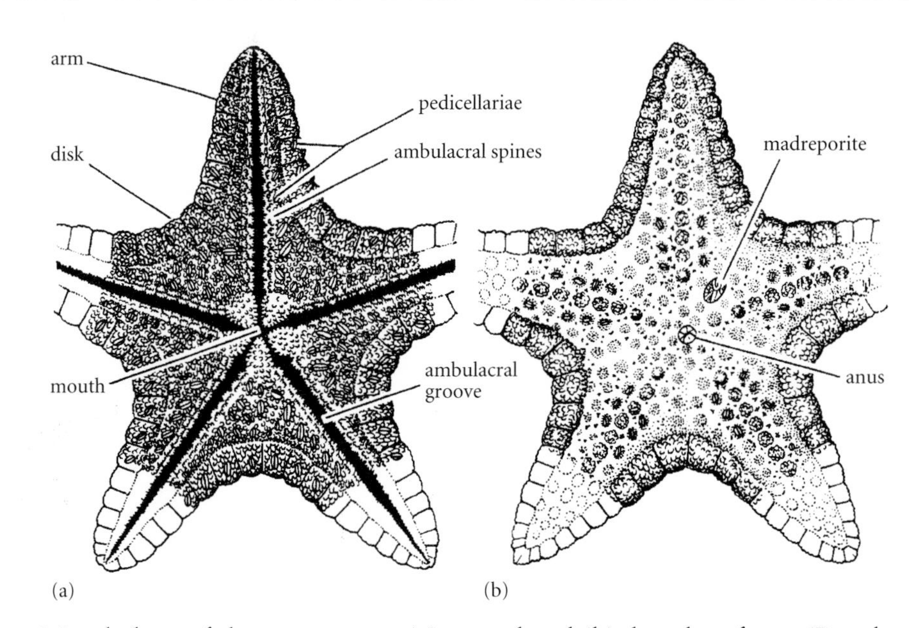

Figure 15.16 Morphology of the asterozoans: (a) ventral and (b) dorsal surfaces. (Based on *Treatise on Invertebrate Paleontology*, Part U. Geol. Soc. Am. and Univ. Kansas Press.)

long, thin, flexible arms. The mouth is situated centrally on the lower surface of the disk. Most of the disk is filled by the stomach and, in the absence of an anus, waste products are regurgitated through the mouth. The arms consist of highly specialized ossicles or vertebrae. Ophiuroids are common in modern seas and oceans, preferring deeper-water environments below 500 m. Their basic architecture differs little from some of the first members of the group, for example *Taeniaster* from the Middle Ordovician of the United States.

A few modern starfish are vicious and voracious predators enjoying a diet of shellfish. Asteroids can prize apart the shells of bivalves with their sucker-armored tube feet far enough to evert their stomachs through their mouths and into the mantle cavity of the animal, where digestion of the soft parts takes place. Stephen Donovan and Andrew Gale (1990) suggested that this predatory life mode significantly inhibited the post-Permian diversification of some brachiopod groups. The strophomenides, the most diverse Permian brachiopods, largely pursued a reclined, quasi-infaunal life strategy and they may have presented an easy kill for the predatory asteroids.

Carpoidea

The carpoids include some of the most bizarre and controversial fossil animals ever described. Variably described as carpoids, homalozoans or calcichordates, depending on preference, most authorities consider the group to be very different to the radiate Echinodermata; indeed, carpoids show some puzzling similarities to the chordates.

The carpoids were marine animals ranging in age from Mid Cambrian to possibly Late Carboniferous, with a calcitic, echinoderm-type skeleton lacking radial symmetry (Fig. 15.17). Two main types of carpoid are recognized: the cornutes and the mitrates. The cornutes were often boot-shaped and appear to have a series of gill slits on the left side of the roof of the head, whereas the mitrates, derived from a cornute ancestor, were more bilaterally symmetric with covered gill slits on both sides.

It might seem unexpected, but the carpoids have featured at the center of a long-running and heated debate that has hit the headlines over the past 50 years. After much careful study, Richard Jefferies (1986) presented detailed evidence that carpoids and chordates share many characters, the so-called "calcichordate hypothesis". He based his conclusion on painstaking studies of their anatomy and the anatomy of embryos of modern echinoderms and chordates. A chordate-implied reconstruction of carpoids suggests that the body consists of a head and a tail used for locomotion (Sutcliffe et al. 2000) (Fig. 15.18).

Moreover, Jefferies (1986) described structures indicating a fish-like brain, cranial nerves, gill slits and a filter-feeding pharynx similar to that in tunicates (often known as the sea squirts). In the calcichordate hypothesis, Hemichordata is identified as a sister group to Echinodermata + Chordata, a clade that Jefferies called Dexiothetica. Reappraisals of the anatomy of carpoids have shown, however, that they may be interpreted rather more convincingly as echinoderms, and that the calcichordate hypothesis fails (Box 15.7). Further, when these redescriptions of the fossil material are combined with new molecular evidence on phylogeny, the case is lost (Ruta 1999). Molecular phylogenetic analyses (Winchell et al. 2002; Delsuc et al. 2006) show that Hemichordata is the sister group of Echinodermata, forming together the Ambulacraria, and that Ambulacraria is the sister group of Chordata. Dexiothetica does not exist.

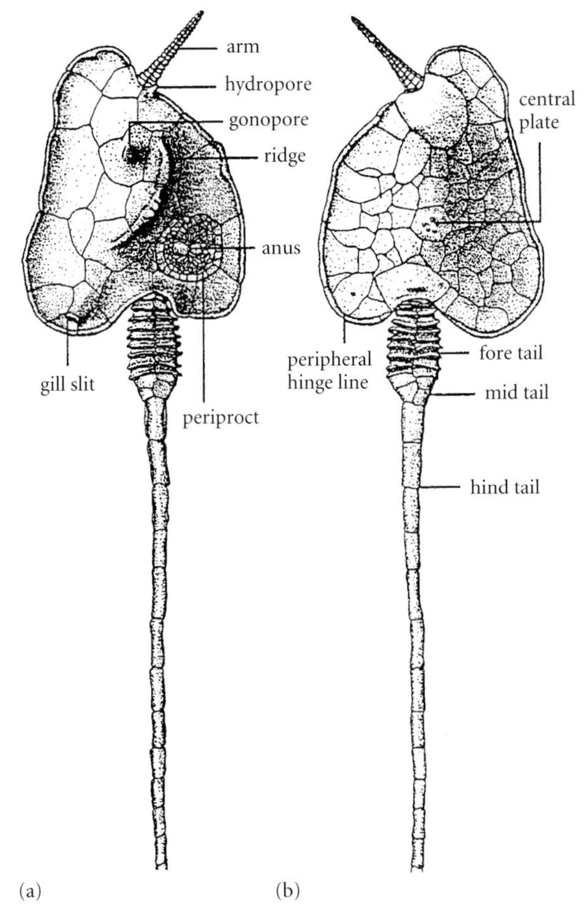

Figure 15.17 Morphology of the carpoids: (a) dorsal and (b) ventral surfaces. (From Jefferies & Daley 1996.)

HEMICHORDATES

What was the character of the vegetation that clothed this earliest prototype of Europe is a question to which at present no definite answer is possible. We know, however, that the shallow sea which spread from the Atlantic southward and eastward over most of Europe was tenanted by an abundant and characteristic series of invertebrate animals – trilobites, graptolites, cystideans, brachiopods, and cephalopods, strangely unlike, on the whole, to anything living in our waters now, but which then migrated freely along the shores of the arctic land between what are now America and Europe.

Sir Archibald Geikie from a lecture delivered to the Royal Geographical Society (1897)

The hemichordates form a small phylum of only a few hundred species and are unfamiliar to most people, but their importance for the study of vertebrate evolution cannot be underestimated. Their most common fossil representatives, the graptolites, were abundant in

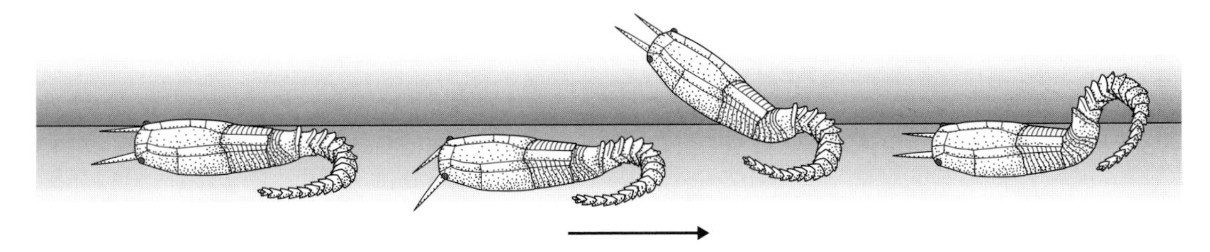

Figure 15.18 Reconstruction of a living carpoid: the Devonian *Rhenocystis* moving across and through the sediment from left to right. (From Sutcliffe et al. 2000.)

Box 15.7 The oldest stylophorans: echinoderms with a locomotory organ

How much further can we go with the debate on the affinities of the carpoids? New material and new investigative techniques will always help. The oldest stylophoran carpoid is from the Middle Cambrian rocks of Morocco. Sebastien Clausen and Andrew Smith (2005) have analyzed the morphology of the animal in great detail, particularly its microstructure with the scanning electron microscope. *Ceratocystis* in fact has a stereom microstructure typical of most echinoderms, but its appendage was covered by articulating plates and filled with muscle tissue and ligaments (Fig. 15.19). It seems that this bizarre asymmetric animal has an echinoderm skeletal structure but also possessed a muscular locomotory appendage rather similar to its sister taxon, the pterobranchs. This group has all the features of a stem-group echinoderm prior to acquiring five-fold symmetry and presumably a water vascular system, and it rather conclusively disproves a close phylogenetic association between carpoids and chordates.

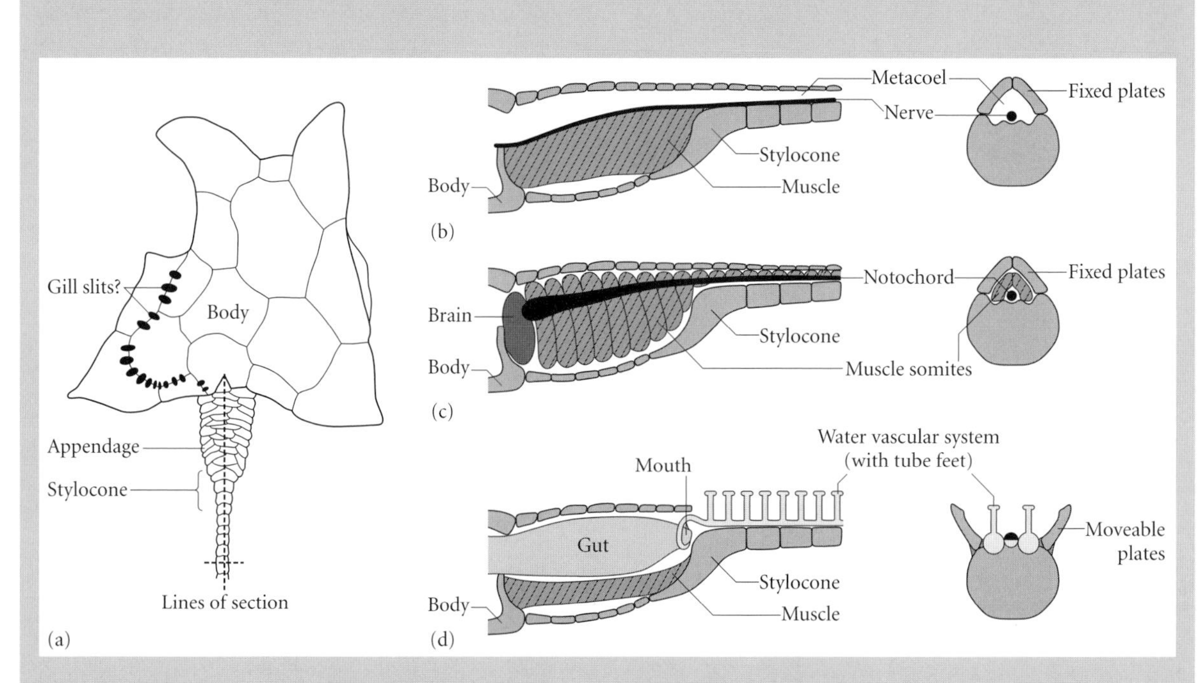

Figure 15.19 *Ceratocystis* from North Africa. (a) Basic anatomic features. (b–d) Three current interpretations of the soft-tissue anatomy of the stylophoran appendage in proximal longitudinal (left) and distal transverse (right) section: (b) primitive echinoderm model, (c) calcichordate model and (d) crinozoan model. (Based on Clausen & Smith 2005.)

the ancient seas of the Early Paleozoic, in communities and environments quite different from those of today. Graptolites are widely used for correlation because of their abundance, widespread distribution and rapid evolution. Although graptolites are extinct, and their life styles are difficult to interpret, they were hemichordates – a phylum containing about 100 living species characterized by a rod-like structure, the **notochord**. They were small, soft-bodied animals with bilateral symmetry and a lack of segmentation. The phylum contains two very different classes: first, the tiny, mainly colonial, pterobranchs that lived in the sessile benthos and, second, the larger infaunal acorn or tongue worms, the enteropneusts that lived in burrows mainly in subtidal environments.

The hemichordates have a mixture of characters suggesting links with the lophophorates, the echinoderms and the chordates. They have been closely related to the cephalochordates and urochordates or tunicates, but molecular and other data suggest that the latter two groups are more closely related to the chordates than the hemichordates. Although the notochord is now known to be unrelated to a true backbone, the hemichordates have, nevertheless, gill slits and a nerve cord.

Modern hemichordate analogs

Pterobranchs superficially resemble the bryozoans – both are colonial animals and the individual **zooids** feed with tentaculate, ciliated arms (Fig. 15.20). The group has a long geological history with early records such as *Rhabdotubus* from the Middle Cambrian and *Graptovermis* from the Tremadocian. The living genera *Cephalodiscus* and *Rhabdopleura* (Fig. 15.20) have been used as analogs for many aspects of graptolite morphology, ontogeny and paleoecology. These living genera and the graptolites both have a **periderm**, or skin, with fusellar tissue, while the dendroid **stolon**, a tube that connects the thecae to each other, may be related to the pterobranch **pectocaulus**. *Rhabdopleura* is known first from the Middle Cambrian and occurs in oceans today mainly at depths of a few thousand meters. The genus is minute with a creeping colony hosting a series of exoskeletal tubes, each containing a zooid with its own lophophore-like feeding organ comprising a pair of arms. The zooids are budded from a stolon and interconnected by a contractile stalk, the pectocaulus.

Cephalodiscus, however, is rather different, being constructed from clusters of stalked

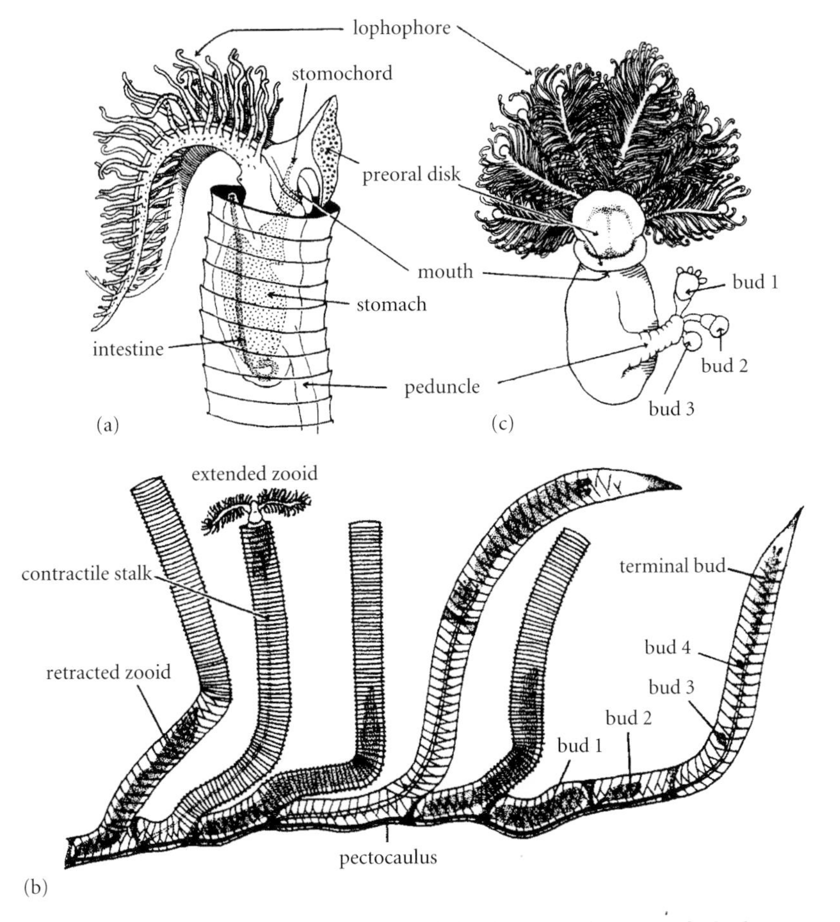

Figure 15.20 Rhabdopleurid morphology: (a, b) *Rhabdopleura* and (c) *Cephalodiscus*. (Based on *Treatise on Invertebrate Paleontology*, Part V. Geol. Soc. Am. and Univ. Kansas Press.)

tubes budded from a basal disk. Moreover, in further contrast to *Rhabdopleura*, species of *Cephalodiscus* usually have five pairs of ciliated feeding arms. Individual zooids in the *Cephalodiscus* colony can actually crawl outside the colony along its exterior and often farther afield onto adjacent surfaces. The zooids of living *Cephalodiscus*, with their considerable freedom of mobility, can even construct external spines from outside the skeleton.

Graptolites

The graptolites, or Graptolithina, are generally stick-like fossils, very common in many Lower Paleozoic black shales. In fact the group is so prevalent that it has proved to be of key importance in correlating Lower Paleozoic strata. The majority of Ordovician and Silurian biozones are based on graptolite species or assemblages. Graptolites, from the Greek "stone writing", usually occur in black shales as flattened carbonized films resembling hieroglyphics. Graptolite fossils often show evidence of having been transported by currents, although fortunately complete, unflattened specimens have been extracted from cherts and limestone by acid-etching techniques. The affinities of the group were largely unknown until the 1940s, when the Polish paleontologist Roman Kozłowski identified a notochord in three-dimensional material isolated from limestones. There are several groups of graptolites and graptolite-like animals (Box 15.8).

Morphology: the graptolite colony

The basic graptolite architecture consists of a probably collagenous skeleton characterized by a growth pattern of half rings of periderm interfaced by zigzag sutures, similar to the construction of the pterobranchs (Fig. 15.21). Each colony or **rhabdosome** grew from a small cone, the **sicula**, as one or a series of branches or **stipes**. The stipes may be isolated or linked together by lateral struts to resemble a reticulate lattice. A series of variably cylindrical tubes are developed along the stipes; these thecae house the individual zooids of the colony. Aggregates of rhabdosomes, **synrhabdosomes**, have been documented for some species. These complex structures have generally been explained by asexual budding or common attachment to a single float or patch of substrate. A more recent taphonomic explanation, however, suggests they formed by entrapment of clusters of rhabdosomes using

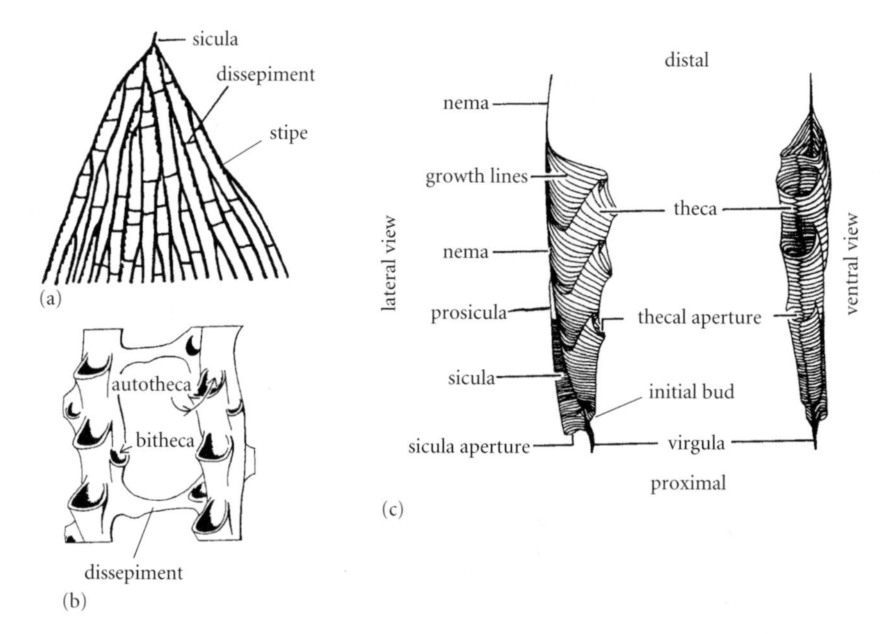

Figure 15.21 Graptolite morphology: (a) dendroid morphology with a detail of the thecae (b), and (c) graptoloid morphology.

Box 15.8 Graptolite classification

A complete classification of the group is presented here although in practice it is only the dendroids and graptoloids that have good fossil records. (The stolonoids, an encrusting or sessile group, restricted to Poland, may be Pterobranchia.)

Class GRAPTOLITHINA
Order DENDROIDEA

- Multibranched colonies; stipes, commonly supported by dissepiments, have autothecae, bithecae and a stolotheca. Anisograptids, intermediate between the dendroids and graptoloids, are retained here
- Cambrian (Middle) to Carboniferous (Namurian)

Order TUBOIDEA

- Similar to dendroids but characterized by irregular branching and reduced stolothecae. Autothecae and bithecae commonly form clusters
- Ordovician (Tremadocian) to Silurian (Wenlock)

Order CAMAROIDEA

- Encrusting life mode; endemic to Poland. Autothecae with expanded, sack-like bases. Bithecae small and irregularly spaced. Stolotheca black and hard
- Ordovician (Tremadocian-Darriwilian)

Order CRUSTOIDEA

- Encrusting life mode; endemic to Poland. Autothecae with complex apertures
- Ordovician (Floian-Darriwilian)

Order DITHECOIDEA

- Sister group to the dendroids and graptoloids; central axis with a holdfast
- Cambrian (Middle) to Silurian (Lower)

Order GRAPTOLOIDEA

- The jury is still out on the detailed classification of this group. The position of the anisograptids (included here with the dendroids) is uncertain, as is the status of the dichograptids; the retiolitids are aberrant diplograptids. Colonies with few stipes (one to eight), a nema and sicula and a single type of theca
- Ordovician (Tremadocian) to Devonian (Pragian)

Suborder DICHOGRAPTINA

- Basal graptoloids lacking both bithecae and virgellae
- Ordovician (Tremadocian-Katian)

Suborder VIRGELLINA

- Virgella always present
- Ordovician (Floian) to Devonian (Pragian)

marine snow, a bonding material composed of organic debris and mucus; this seems less likely because the synrhabdosomes are remarkably symmetric, which suggests they grew that way. About six orders are now recognized in the class Graptolithina, but only two, the Dendroidea and Graptoloidea have important geological records. The patterns of evolution linking these groups are uncertain (Box 15.9).

Box 15.9 The first graptolites: a cryptic Cambrian dimension?

By the Ordovician, the graptolites were represented by a number of well-defined groups including the familiar dendroids and graptoloids and the less well-known camaroids, crustoids, dithecoids and tuboids. It has long been a mystery where these diverse groups came from because the Cambrian record was virtually non-existent. Barrie Rickards and Peter Durman (2006) have reassessed all the possible ancestors, Cambrian specimens that have been variably assigned to graptolites, hydroids or algae from the Middle and Upper Cambrian. They reassigned some of these cryptic Cambrian specimens to the rhabdopleurids and excluded a number of them from the graptolites. The graptolites and rhabdopleurids therefore probably shared a common ancestor in the Early Cambrian (Fig. 15.22). The rhabdopleurids are remarkable animals; Cambrian forms are virtually identical to modern rhabdopleurids, making them true living fossils. The common ancestor to the graptolites and rhabdopleurids was probably a solitary, worm-like animal, equipped with a lophophore, and living in pseudocolonial filter-feeding clumps on the seafloor. Thus the graptolites, which dominated the Early Paleozoic water column, started out as rather anonymous benthic filter feeders in the shadow of the more obvious early arthropods, crinoids and mollusks of the Cambrian evolutionary fauna.

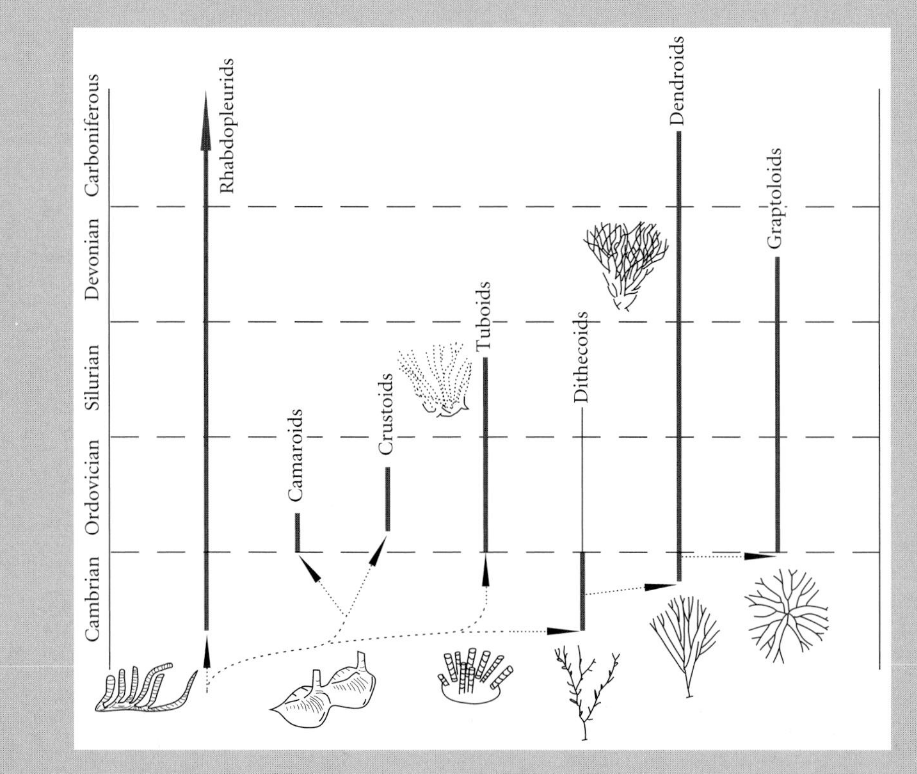

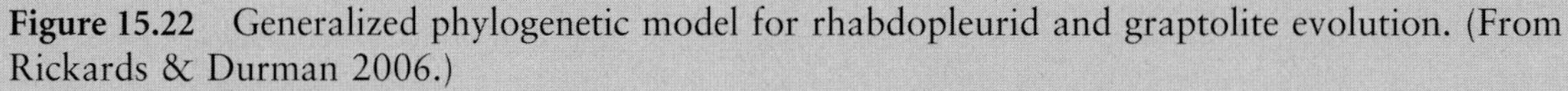

Figure 15.22 Generalized phylogenetic model for rhabdopleurid and graptolite evolution. (From Rickards & Durman 2006.)

Dendroidea

The Dendroidea is the older of the two main groups with important geological records, first appearing in the Middle Cambrian and disappearing during the Late Carboniferous. The dendroid rhabdosome was multibranched, like a bush, with its many stipes connected laterally by struts or **dissepiments**. Two types of theca, of different sizes, the **autotheca** and **bitheca**, grew along the stipes. The earlier genera were benthic, attached to the seafloor by a short stalk and basal disk. Probably during the latest Cambrian a few genera, including *Rhabdinopora*, detached themselves to evolve a new lifestyle in the plankton; together with minute brachiopods and the occasional trilobite, they probably formed a major part of the preserved Early Paleozoic plankton.

Dendroid taxa *Dendrograptus* was a benthic genus, bush-like, erect and attached to the seafloor by a rooting structure or holdfast. *Dictyonema* was also benthic and ranged in age from the Late Cambrian to the Late Carboniferous. The rhabdosome was conical to cylindrical in shape. Planktonic dendroids similar to *Dictyonema* are placed in *Rhabdinopora*.

The following anisograptid genera are in some ways intermediate between the typical dendroids and graptoloids and may be classified with either group. Here they are included with the dendroids. *Radiograptus*, for example, developed large spreading colonies. Both *Kiaerograptus* and some early species of *Bryograptus* had both auto- and bithecae, and the latter had triradiate rhabdosomes with, initially, three primary stipes. *Clonograptus* had a horizontal, biserially symmetric rhabdosome with stipes generated by dichotomous branching from an initial biradiate configuration.

Graptoloidea

Compared with the dendroids, the graptoloid rhabdosome is superficially simpler and consists of an initial sicula, divided into an upper **prosicula** and a lower **metasicula**, with at its apex, distally, a long thin, spine, the **nema**. The metasicula, like the rest of the rhabdosome, was composed of **fusellar tissue**, bundles of short, branching fibrils. The **virgella** projected below the secular aperture, proximally and is characteristic of the suborder Virgellina. The thecae grew out from the sicula and subsequent thecae grew in sequence as the rhabdosome developed.

Graptoloid taxa The architecture of the graptoloid skeleton depended on three sets of structures: the number of stipes or branches, their mutual attitudes and the shape of the thecae. Morphology in this order is thus based on permutations of these structures; the following genera illustrate this variation (Fig. 15.23).

Tetragraptus, common during the Floian (later Early Ordovician), typically had four stipes arranged in horizontal, pendent or reclined attitudes with simple, overlapping thecae. *Didymograptus* was twin-stiped or **biramous**, commonly with the branches in horizontal, pendent or reclined orientations; thecae were simple. *Isograptus*, however, had two relatively wide stipes, reclined with a long, thread-like sicula. *Nemagraptus* had a very distinctive rhabdosome consisting of two sigmoidal stipes, initially diverging from the sicula at about 180°, with additional stipes, curved, and arising at intervals along the main branches. Thecae were long, thin and diverged at small angles from the stipes. *Dicellograptus* had a pair of stipes that adopted reclined attitudes but often the branches were curved or even coiled; the thecae were characterized by extravagant sigmoidal shapes and incurved apertures. *Monograptus* was a uniserial scandent form with a straight or curved rhabdosome and a nema embedded in the dorsal wall that projected distally. *Rastrites* possessed long, straight, widely separated thecae, often with hooked ends. *Cyrtograptus* had a spirally coiled rhabdosome with secondary branches or cladia oriented like the arms of a spiral galaxy. *Corynoides* was minute, consisting of a sicula and three to four thecae.

Retiolitids

The retiolitids are a spectacular group of apparently scandent, diplograptid biserials with a reduced, minimalist periderm consisting of a network of bars or lists probably surrounded by a net-like structure, termed the **ancora sleeve** in Silurian forms (Fig. 15.24). The group appeared in the Mid Ordovician and continued successfully, for almost 50 myr,

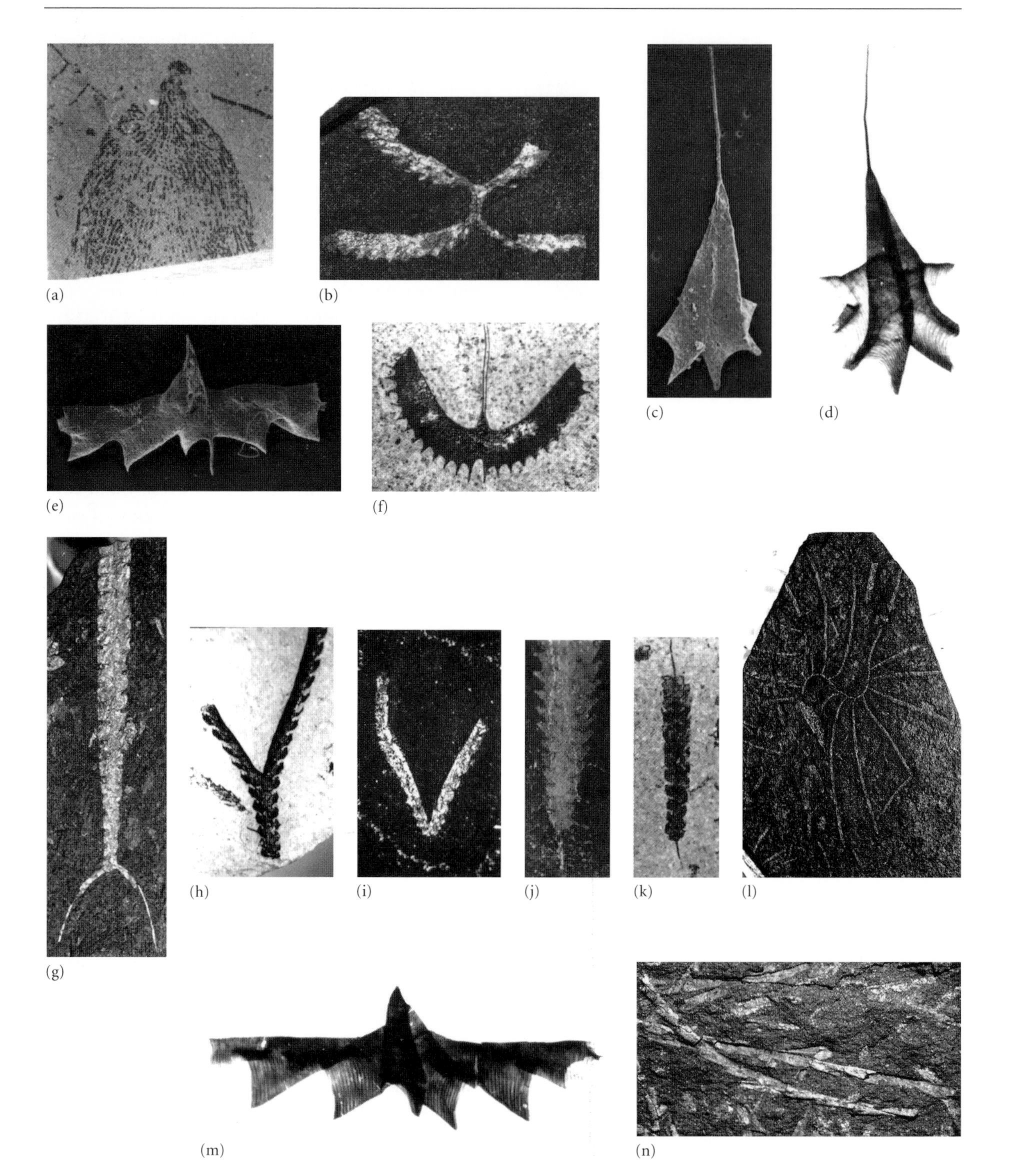

Figure 15.23 Some graptolite genera: (a) *Rhabdinopora* (×2), (b) *Tetragraptus* (×2), (c) *Tetragraptus*, proximal end (×20), (d) *Isograptus*, proximal end (×20), (e) *Xiphograptus* (×20), (f) *Isograptus* (×10), (g) *Appendispinograptus* (×2), (h) *Dicranograptus* (×2), (i) *Dicellograptus* (×2), (j) *Orthograptus* (×2), (k) *Undulograptus* (2), (l) *Nemagraptus* (×2), (m) *Didymograptus* (*Expansograptus*) (×20) and (n) *Atavograptus* (×2). (a) An Early Ordovician dendroid, (b–f, k, m) Early Ordovician graptoloids; (g–j, l) Late Ordovician graptoloids; and (n) a Silurian monograptid. (Courtesy of Henry Williams.)

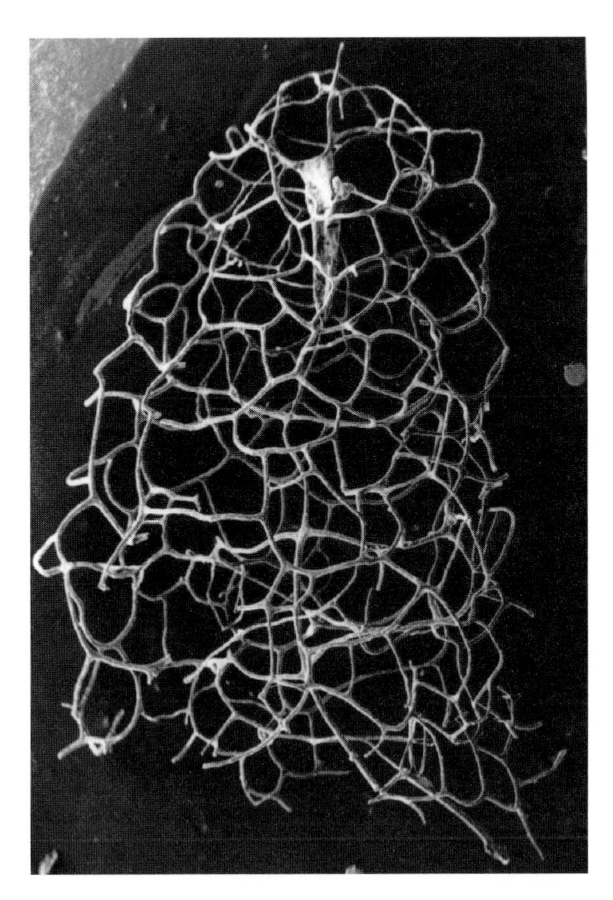

Figure 15.24 Retiolitid *Phorograptus* (Middle Ordovician) (×30). (Courtesy of Denis Bates.)

until the latest Silurian (Kozłowski-Dawidziuk 2004). The retiolitids probably represent a polyphyletic grade of organization where the rhabdosome of various groups may have functioned like a sponge, drawing in fluid and nutrients through the periderm and expelling waste upwards.

Growth and ultrastructure of the graptolites

Detailed studies on the ultrastructure of the graptolite using both scanning and transmission electron microscopes has identified two types of skeletal tissue. Fusellar tissue occurs together with **cortical tissue** in the form of longer parallel fibers. Fusellar material was secreted as a series of half rings with the cortical tissue overlapping the fusellar layer both inside and outside the rhabdosome (Fig. 15.25). The cortical tissue itself was secreted as "bandages", looking rather like multiple overlapping band-aids. Secretion may have been by mobile zooids, free to patrol the exterior of the colony while still attached by a flexible cord to the rest of the colony, rather like an astronaut maintaining a space station, or the entire rhabdosome may have been surrounded by soft tissue.

Although graptolites are abundant and important fossils in many Early Paleozoic assemblages, it is notoriously difficult to discover what they were actually made of. Most assemblages occur in black shales that have been compacted, diagenetically altered and often metamorphosed within or around orogenic belts. Moreover, graptolite periderm, when actually preserved, consists mainly of an aliphatic polymer, immune to base hydrolysis. It lacks protein even though both the structure, as well as chemical analyses, of the periderm of living *Rhabdopleura* suggest that it was originally composed of collagen. Previous studies suggested that the collagen had been replaced by macromolecular material from the surrounding sediment. New analyses suggest that the aliphatic composition of graptolite periderm reflects direct incorporation of lipids from the organism itself by in situ polymerization (Gupta et al. 2006). A similar process may account for the preservation of many other groups of organic fossils (see p. 60).

Colonial growth of the graptoloids The growth of a colony lends itself to graphic and mathematical simulations. A few authors have devised computer models based on a set of simple rules that dictate such growth modes. These models are usually deterministic and static. For example, Andrew Swan (1990) generated a series of theoretical morphotypes based on a model of dichotomous branching at given stipe lengths; the orientation of the bifurcation together with the stipe length and width was varied. Additionally, soft tissue could be added to the computer reconstructions. Swan showed that the shapes of most graptolite colonies could be simulated using variations in only a few parameters, and he was able to test the efficiency of each colony for particular functions. Swan targeted the efficiency of the graptolite feeding strategy and tested the efficacy of nutrient capture for a sequence of computer-generated colonies, and he showed that known graptolite colonies pass the test as being the most efficient shapes

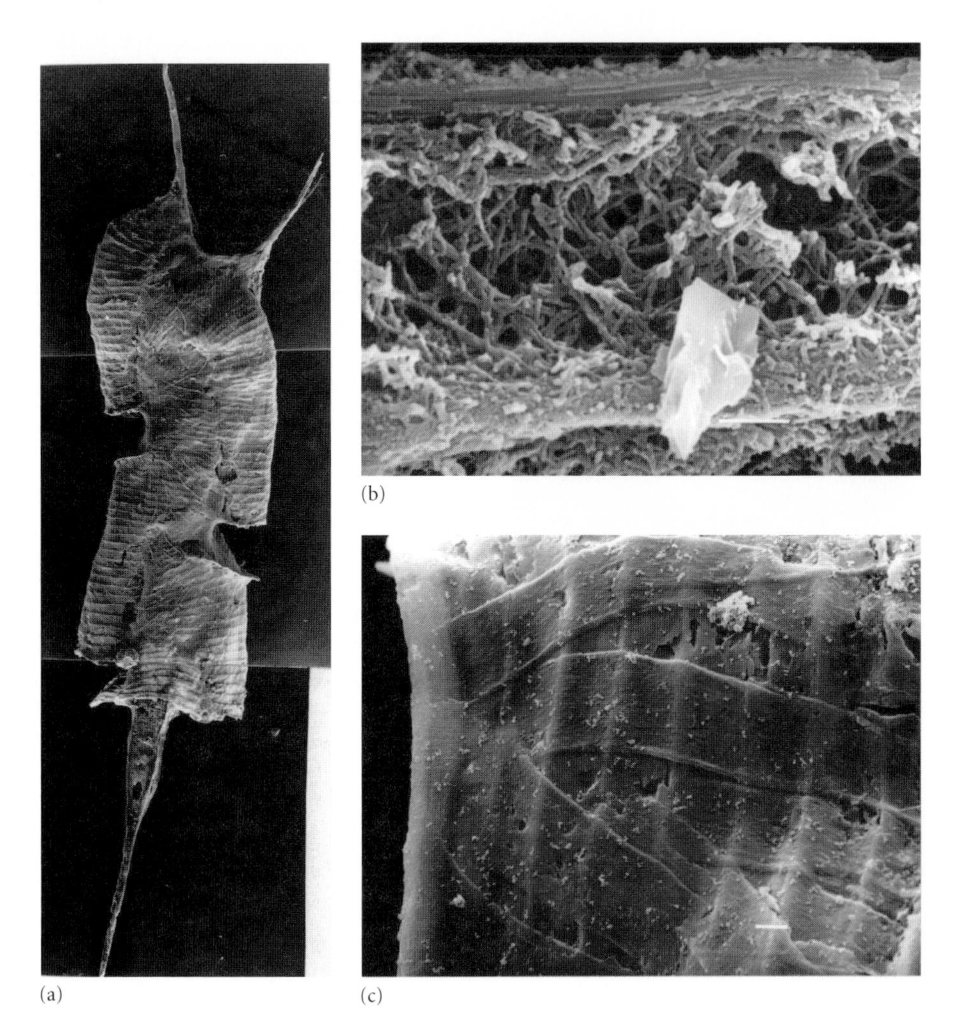

Figure 15.25 Graptolite ultrastructure: (a) collage of *Geniculograptus* rhabdosome showing banded fusellar tissue (×50); (b) detailed section through part of a rhabdosome showing relatively thin, parallel sheet fabric (top) and criss-cross fusellar fabric (below) (×1000); and (c) detail of aperture exterior of *Geniculograptus* showing the development of bandages (×500). (Courtesy of Denis Bates.)

for capturing most food in the shortest time from a given water volume.

Ecology: modes of life and feeding strategy

There is little doubt that the earliest bush-like dendroids were attached to the seabed and functioned as part of the sessile benthos. Detachment in various benthic genera occurred at the beginning of the Ordovician with genera such as *Rhabdinopora* entering the plankton. More controversial is the mode of life of the various graptoloid groups (Fig. 15.26). Conventionally the graptoloids were considered to be passive drifters, their flotation being aided by fat and gas bubbles in their tissues or even by vane-like extensions to the nema. But they clearly occupied different levels in the water column (Underwood 1994).

The suggestion by Nancy Kirk (1969) that far from being passive members of the plankton, the graptolites were automobile, moving up and down in the water column, has stimulated considerable and continued interest and research on the life habits of these extinct organisms. During intervals of intense feeding, a reactive upward movement of the colony in the water column would have occurred. At night the colony could move vertically into the nutrient-rich photic zone and later, when replete, the rhabdosomes would sink to positions in the water column where the specific

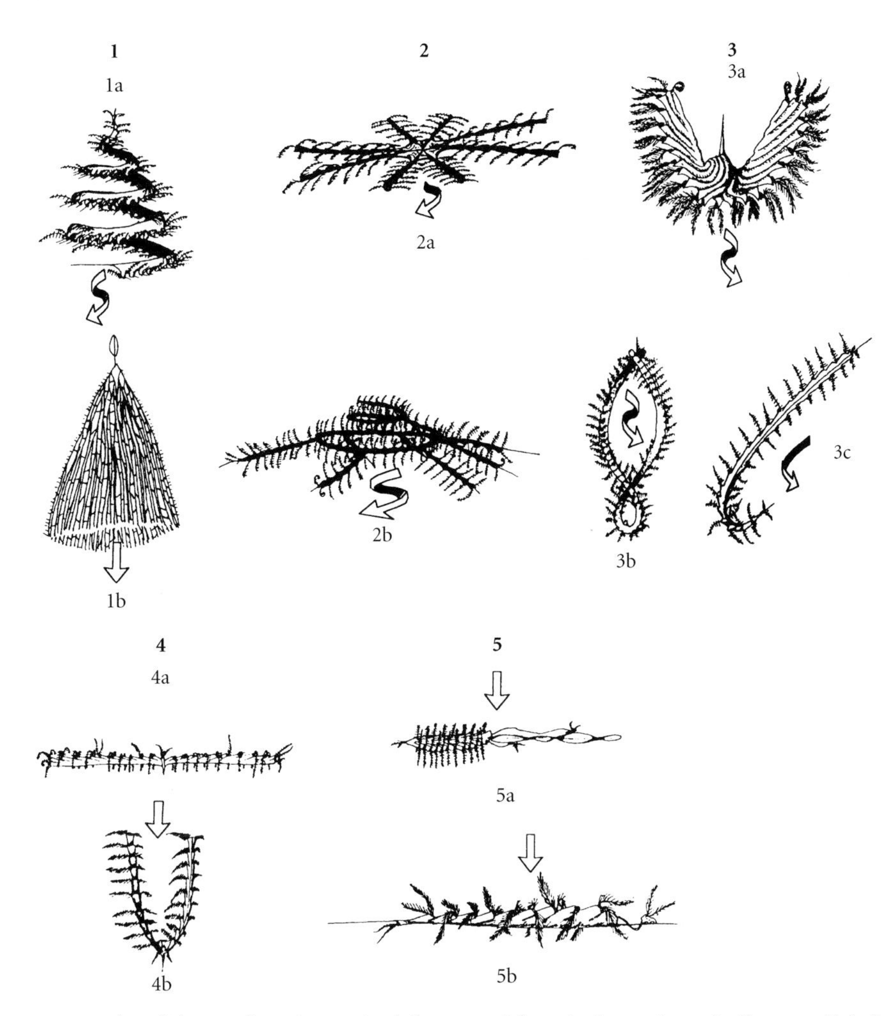

Figure 15.26 Graptolite life modes: 1, conical forms with spiral motion; 2, flat or slightly conical forms with slow, slightly spiral velocities; 3, mono- or biramous forms with spiral movement due to asymmetry; 4, forms with high angles between stipes having linear movement; 5, straight forms with mainly linear descent. (Based on Underwood 1994.)

gravity of the colony matched that of the surrounding seawater.

Computer models and physical models, including exposure to wind-tunnel conditions that mimic the effects of water currents, have emphasized the importance of harvesting strategies for the success of the colony. These probably exerted an important influence on the evolutionary pathways that the graptoloids followed.

Evolution: graptolite stipes and thecae

Graptolite evolution has been described in terms of four main stages of morphological development:

1 The transition from sessile to planktonic strategies in the dendroids during the Late Cambrian and Early Ordovician.
2 At the end of the Tremadocian (early Early Ordovician), the appearance of the single-type thecae of the graptoloids.
3 The development of the biserial rhabdosome in the Floian (late Early Ordoviaian).
4 Finally, the origin of the uniserial monograptids.

The small, stick-like benthic organisms reported from Middle Cambrian rocks on the Siberian platform and ascribed to the graptolites may be better assigned to the

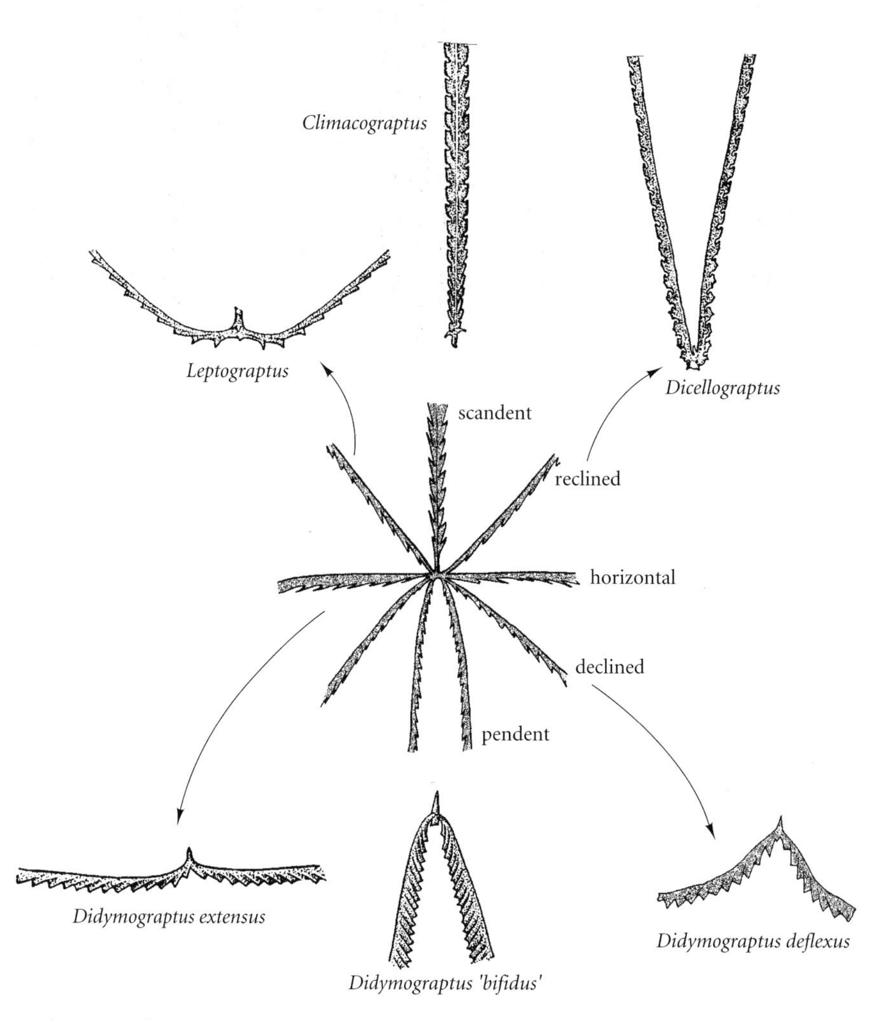

Figure 15.27 Evolution of stipes.

rhabdopleurids. The first undoubted graptolites include the dendroids *Callograptus*, *Dendrograptus* and *Dictyonema* occurring in Middle Cambrian rocks of North America. But by the Late Cambrian, the diversity of the dendroid fauna had markedly increased. The fauna included genera such as *Aspidograptus* and *Dictyonema*, which resembled small shrubs and were attached to the substrate by holdfasts or more complex root-like structures. During the Late Cambrian and Early Ordovician, some dendroids made the jump from the sessile benthos to the plankton; attachment disks continuous with the nema suggest these genera may have hung suspended in the surface waters and pursued an epiplanktonic life strategy. Both *Radiograptus* and *Dictyonema* have been cited as possible ancestors for the planktonic graptolites, and perhaps *Staurograptus* was in fact the first planktonic graptolite. The Tremadocian seas witnessed the radiation of the anisograptids.

The explosion of dichograptid genera during the Floian introduced a variety of symmetric graptolites with from about eight to two stipes oriented in declined, pendent and scandent attitudes (Fig. 15.27). A twin-stiped dichograptid was probably ancestral to the next wave of graptolites, the diplograptids, which radiated in the Mid Darriwilian (Middle Ordovician).

The single-stiped monograptids dominated Silurian graptolite faunas and, despite their apparent simplicity, the group developed a huge variety of forms (Fig. 15.28). The last graptolites, species of *Monograptus*, disap-

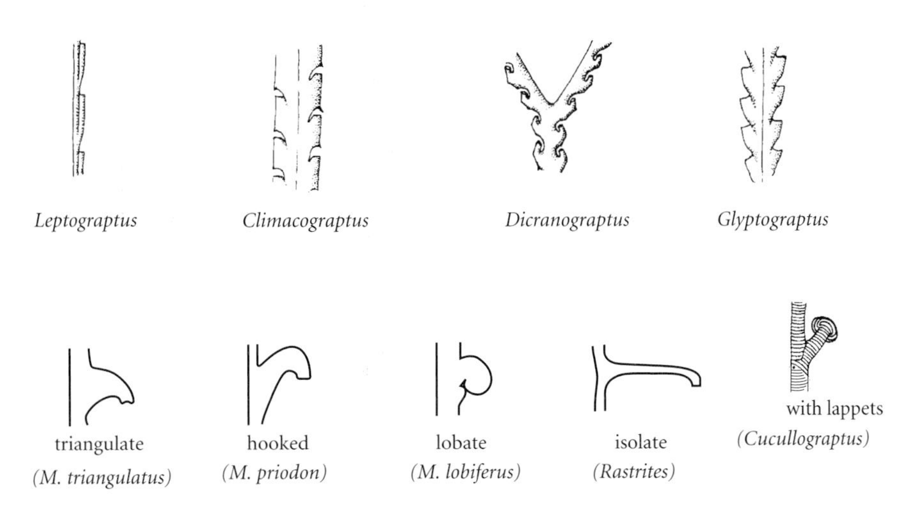

Figure 15.28 Evolution of thecae. *M*, *Monograptus*.

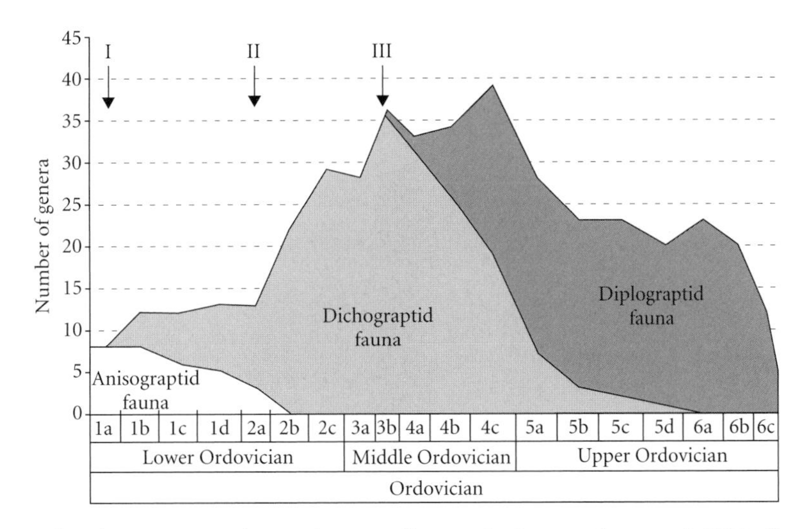

Figure 15.29 Graptolite biostratigraphy and graptolite evolutionary faunas. I–III indicate the three main radiations: anisograptid, dichograptid and diplograptid; 1a–6c represent 19 time slices through the Ordovician Period. (Based on Chen et al. 2006.)

peared during the Early Devonian (Pragian) in China, Eurasia and North America. Nevertheless this uniserial morphology had survived for over 30 million years and may have continued after the Early Devonian in lineages that did not secrete a preservable skeleton. Why should a trend towards a reduction in stipes be such an advantage? Perhaps the simpler stipe configuration was hydrodynamically more stable, better adapted to turbulence and aided the motion of the graptolites through the water column on feeding forays. It may also have prevented the interference between thecae on adjacent stipes, providing a simpler, more efficient colony structure.

Graptolite morphology and stratigraphy have formed the basis for the definition of evolutionary faunas within Ordovician assemblages that include anisograptid, dichograptid and diplograptid evolutionary faunas (Chen et al. 2006). These faunas not only help us understand better the Ordovician radiation of the graptolites but add an additional dimension to the zonal framework of the Ordovician System.

Biostratigraphy: graptolites and time

Graptolites are among the best zone fossils (see p. 28) and are excellent for biostratigraphic correlation. Traditionally, four sequential graptolite faunas have been recognized through the Early Ordovician to Early Devonian interval (Fig. 15.29). The **anisograptid fauna**, with *Rhabdinopora* and allied genera, characterizes the Tremadocian; although Upper Tremadocian graptolite faunas are rare, the genera *Bryograptus*, *Kiaerograptus* and *Aorograptus* are important and have been described in detail from western Newfoundland (Williams & Stevens 1991). The appearance of the Floian **dichograptid fauna** is signaled by *Tetragraptus*, associated with didymograptids and some relict anisograptids. The later **diplograptid** fauna contains four smaller units: the *Glyptograptus-Amplexograptus* (Darriwilian), *Nemagraptus-Dicellograptus* (Sandbian), *Orthograptus-Dicellograptus* (Katian) and *Orthograptus-Climacograptus* (Hirnantian) subfaunas. The **monograptid** fauna contains a variety of evolving single-stiped forms. The last graptoloids disappeared in the Pragian (Lower Devonian).

In some parts of the world graptolites have provided the basis for some high-resolution stratigraphy. The Upper Ordovician–Lower Silurian clastic succession in the Barrandian basin in the Czech Republic was deposited on an outer shelf, influenced by the end-Ordovician Gondwanan glaciation and an aftermath that included a persistent post-glacial anoxia related to upwelling systems. High-resolution graptolite stratigraphy based on some 19 biozones has provided a framework to link sedimentary environments with graptoloid faunal dynamics and fluctuations in organic content (Štorch 2006). The resulting analyses have provided an accurate time line through four major transgressive cycles. On these are superimposed glacial and interglacial events, intervals of upwelling and oceanic perturbations (Fig. 15.30). Moreover these new data suggest that, far from being a quiet period, the Silurian was punctuated by a number of significant extinction events associated with large climatic and environmental fluctuations. Critical is the accurate correlation between sections in the Barrandian basin with sections elsewhere, otherwise we cannot show these changes were indeed global. Higher in the Silurian, Lennart Jeppsson and Mikel Calner (2003) have reported that the Mulde Secundo-Secundo Event (Wenlock), first identified in the Silurian platform carbonates of the Swedish island of Gotland, includes three extinctions, widespread deposition of carbon-rich sediments, and wild sea-level fluctuations together with a glaciation event. These extinctions are related to a severe reduction in primary plankton productivity. Amazingly, such precision and the recognition of these important events was only made possible by the accurate graphic correlations of sections based on the rapid evolution of these small, beautiful creatures.

Biogeography: graptolites in space

Since the majority of graptolites lived either in the water column or within the plankton, quite different factors influenced their distribution in contrast to, say, that of the coeval benthos. Provinciality was most marked in the earlier Ordovician (Darriwilian) when two main provinces, the Atlantic and Pacific, were recognized. The Atlantic province, including the then high-latitude regions of Avalonia and Gondwana, was characterized by pendent *Didymograptus* species. The Pacific province, including low latitude, tropical regions such as the Laurentian margins, was more diverse with isograptids, cardiograptids and oncograptids. The isograptid biofacies, itself, was more pandemic, occupying deeper water and associated with the world's continental margins. During the end-Ordovician extinction events (see p. 169), the Pacific province graptolites suffered particularly badly, and although graptolites again diversified during the Silurian their provinciality developed a different if less obvious pattern. During the Early Silurian, the Gondwanan province was characterized by endemic taxa while later, in the Mid Silurian, the equatorial region hosted taxa not known from elsewhere.

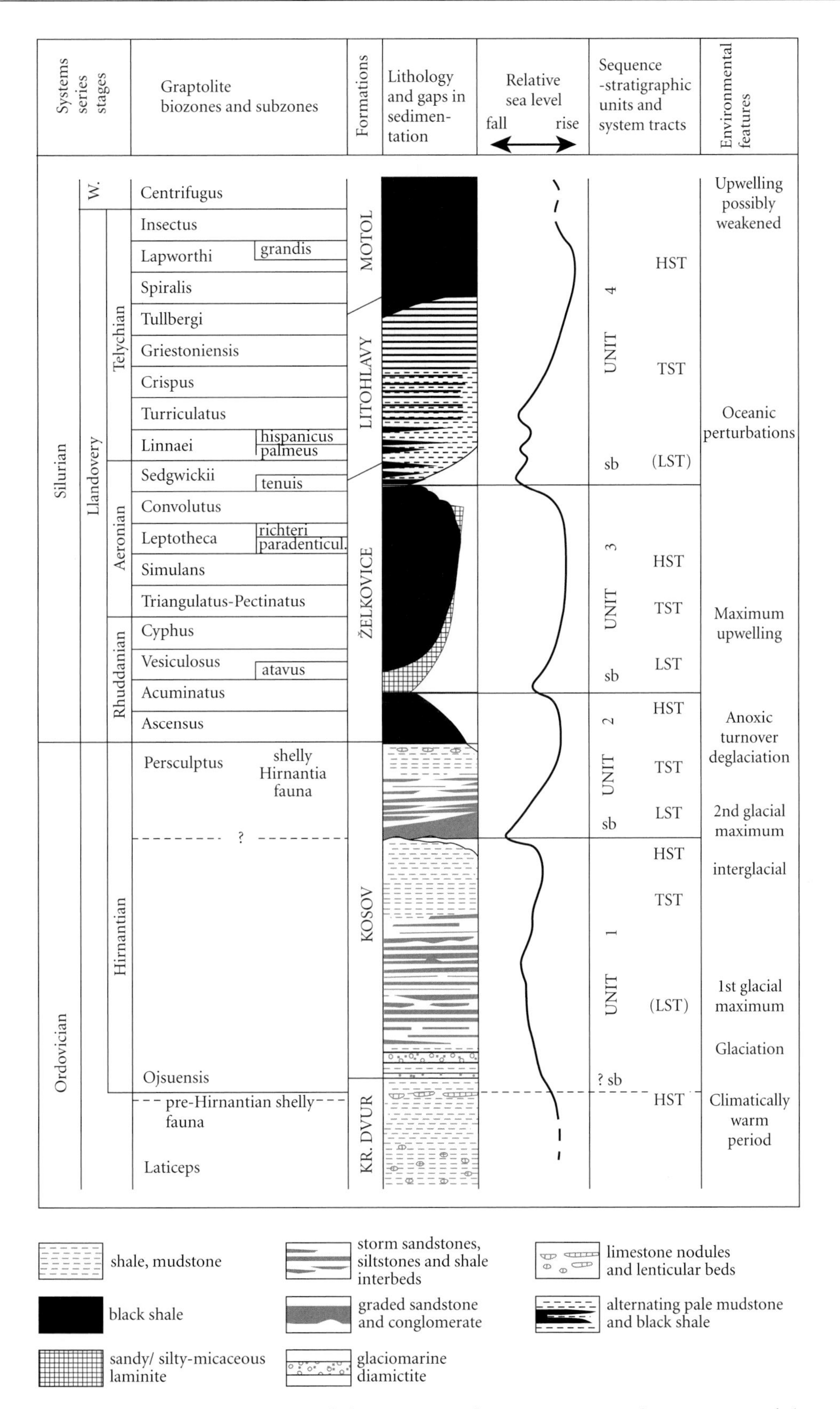

Figure 15.30 Graptolite biostratigraphy of the Upper Ordovician–Lower Silurian strata of the Barrandian basin. HST, highstand systems tract; TST, transgressive systems tract; LST, lowstand systems tract. (Based on Štorch 2006.)

Box 15.10 The Vetulicolians: protostomes, deuterostomes or phylogenetic orphans?

Sometimes the fossil record throws out a weird animal that it is just impossible to classify. The material may be common, distinctive and well preserved but there are simply not enough key characters to link it with other groups. The vetulicolians have been characterized as unusual arthropods, stem-group deuterostomes and even tunicates (Aldridge et al. 2007). They have been reported from a number of Cambrian Lagerstätten and two classes have been recognized, the Vetulicolida and Banffozoa. They were probably active, nektobenthic animals with the facility to both deposit and filter feed. But what were they? In simple terms they lack limbs, making assignment to the arthropods difficult, whereas they have gills similar to those of the deuterostomes (Fig. 15.31). If they were, in fact, deuterostomes they probably lay close to the tunicates as stem vertebrates. But despite well-preserved material from the Chengjiang fauna and careful phylogenetic analyses, it remains impossible to classify them. Their unique combination of characters is thus still an enigma awaiting the discovery of new animals that could link the vetulicolians to a crown group.

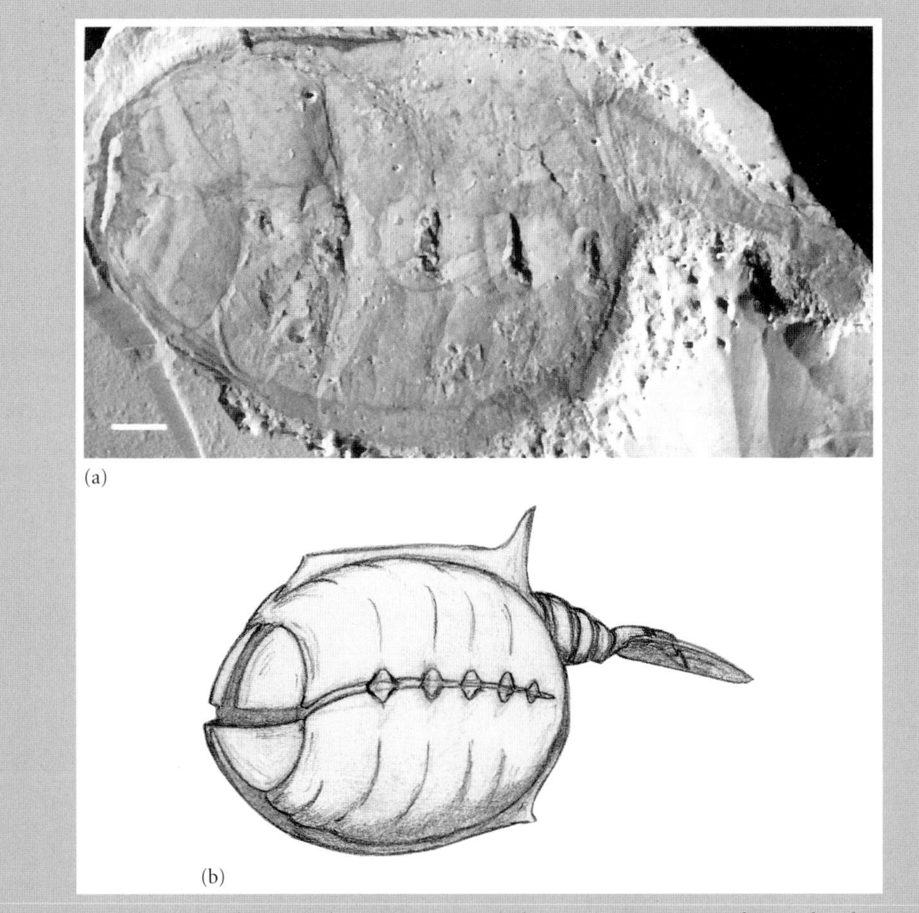

Figure 15.31 (a) Photograph (scale bar, 5 mm) and (b) reconstruction of *Vetulicola*. (Courtesy of Dick Aldridge.)

Review questions

1 The deuterostomes include two apparently morphologically different groups, the echinoderms and hemichordates. What sort of characters could be used to unite them?
2 Crinoids are most common in deep-water environments but probably exploited much shallower-water environments during the Paleozoic. When and why did they move to deeper water?
3 Echinoids have a long history. Why did it take over 250 myr to develop the buried (sand dollars) and burrowing (sea urchin) life strategies?
4 Graptolites evolved through time by reducing their numbers of stipes and developing more complex thecae. What were the ecological advantages of this more streamlined body plan with more elaborate zooid openings?
5 The vetulicolians highlight one of the difficulties of the fossil record, identifying definitive characters of phylogentic significance in bizarre taxa. Should new higher taxa, for example classes of phyla, be established to accommodate such material or should it be shoehorned into existing taxa?

Further reading

Berry, W.B.N. 1987. Phylum Hemichordata (including Graptolithina). *In* Boardman, R.S., Cheetham, A.H. & Rowell, A.J. (eds) *Fossil Invertebrates*. Blackwell Scientific Publications, Oxford, pp. 612–35. (A comprehensive, more advanced text with emphasis on taxonomy; well illustrated.)

Clarkson, E.N.K. 1998. *Invertebrate Palaeontology and Evolution*, 4th edn. Chapman and Hall, London. (An excellent, more advanced text; clearly written and well illustrated.)

Rickards, R.B. 1985. Graptolithina. *In* Murray, J.W. (ed.) *Atlas of Invertebrate Macrofossils*. Longman, Harlow, Essex, pp. 191–8. (A useful, mainly photographic review of the group.)

Smith, A.B. & Murray, J.W. 1985. Echinodermata. *In* Murray, J.W. (ed.) *Atlas of Invertebrate Macrofossils*. Longman, Harlow, Essex, pp. 153–90. (A useful, mainly photographic review of the group.)

Sprinkle, J. & Kier, P.M. 1987. Phylum Echinodermata. *In* Boardman, R.S., Cheetham, A.H. & Rowell, A.J. (eds) *Fossil Invertebrates*. Blackwell Scientific Publications, Oxford, pp. 550–611. (A comprehensive, more advanced text with emphasis on taxonomy; extravagantly illustrated.)

References

Aldridge, R.J., Hou Xian-Guang, Siveter, D.J., Siveter, D.J. & Gabbott, S.E. 2007. The systematics and phylogenetic relationships of vetulicolians. *Palaeontology* **50**, 131–68.

Bottjer, D.J., Hagadorn, J.W. & Dornbos, S.Q. 2000. The Cambrian substrate revolution. *GSA Today* **10**, 1–7.

Chen Xu, Zhang Yuan-Dong & Fan Jun-Xuan. 2006. Ordovician graptolite evolutionary radiation: a review. *Geological Journal* **41**, 289–301.

Clausen, S. & Smith, A.B. 2005. Palaeoanatomy and biological affinities of a Cambrian deuterostome (Stylophora). *Nature* **438**, 351–4.

Delsuc, F., Brinkmann, H., Chourrout, D. & Philippe, H. 2006. Tunicates and not cephalochordates are the closest living relatives of vertebrates. *Nature* **444**, 85–8.

Donovan, S.K. & Gale, A.S. 1990. Predatory asteroids and the decline of the articulate brachiopod. *Lethaia* **23**, 77–86.

Gupta, N.S., Briggs, D.E.G. & Pancost, R.D. 2006. Molecular taphonomy of graptolites. *Journal of the Geological Society, London* **163**, 897–900.

Jefferies, R.P.S. 1986. *The Ancestry of the Vertebrates*. British Museum (Natural History), London.

Jefferies, R.P.S. & Daley, P. 1996. *In* Harper, D.A.T. & Owen, A.W. (eds) *Fossils of the Upper Ordovician*. Field Guide to Fossils No. 7. Palaeontological Association, London.

Jeppsson, L. & Calner, M. 2003. The Silurian Mulde event and a scenario for secundo-secundo events. *Transactions of the Royal Society of Edinburgh: Earth Sciences* **93**, 135–54.

Kammer, T.W. & Ausich, W.I. 2006. The "Age of crinoids": a Mississippian biodiversity spike coincident with widespread carbonate ramps. *Palaios* **21**, 238–48.

Kirk, N. 1969. Some thoughts on the ecology, mode of life, and evolution of the Graptolithina. *Proceedings of the Geological Society of London* **1659**, 273–93.

Kozłowska-Dawidziuk, A. 2004. Evolution of retiolitid graptolites – a synopsis. *Acta Palaeontologica Polonica* **49**, 505–18.

Paul, C.R.C. & Smith, A.B. 1984. The early radiation and phylogeny of echinoderms. *Biological Reviews* **59**, 443–81.

Rickards, R.B. & Durman, P.N. 2006. Evolution of the earliest graptolites and other hemichordates. *In* Bassett, M.G. & Diesler, V.K. (eds) *Studies in Palaeozoic Palaeontology*. Geological Series No. 25. National Museum of Wales, Cardiff, pp. 5–92.

Ruta, M. 1999. Brief review of the stylophoran debate. *Evolution and Development* **1**, 123–35.

Smith, A.B. 1984. *Echinoid Palaeobiology*. Special Topics in Palaeontology No. 1. George Allen and Unwin, London.

Smith, A.B. 2005. The pre-radial history of the echinoderms. *Geological Journal* **40**, 255–80.

Smith, A.B. & Murray, J.W. 1985. Echinodermata. *In* Murray, J.W. (ed.) *Atlas of Invertebrate Macrofossils*. Longman, Harlow, Essex, pp. 153–90.

Smith, A.B. & Stockley, B. 2005. The geological history of deep-sea colonization by echinoids: roles of surface productivity and deep-water ventilation. *Proceedings of the Royal Society B* **272**, 865–9.

Sprinkle, J. 1980. *Echinoderms: Notes for a short course*. Studies in Geology No. 3. University of Tennessee.

Štorch, P. 2006. Facies development, depositional settings and sequence stratigraphy across the Ordovician-Silurian boundary: a new perspective from the Barrandian area of the Czech Republic. *Geological Journal* **41**, 163–92.

Sutcliffe, O.E., Südkamp, W.H. & Jefferies, R.P.S. 2000. Ichnological evidence on the behaviour of mitrates: two trails associated with the Devonian mitrate *Rhenocystis*. *Lethaia* **33**, 1–12.

Swan, A.H.R. 1990. A computer simulation of evolution by natural selection. *Journal of the Geological Society, London* **147**, 223–8.

Underwood, C.J. 1994. The position of graptolites within Lower Paleozoic planktic ecosystems. *Lethaia* **26**, 198–202.

Williams, S.H. & Stevens, R.K. 1991. Late Tremadoc graptolites from western Newfoundland. *Palaeontology* **34**, 1–47.

Winchell, C.J., Sullivan, J., Cameron, C.B., Swalla, B.J. & Mallatt, J. 2002 Evaluating hypotheses of deuterostome phylogeny and chordate evolution with new LSU and SSU ribosomal DNA data. *Molecular Biology and Evolution* **19**, 762–76.

Chapter 16

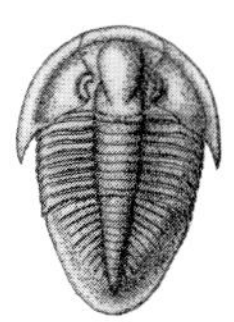

Fishes and basal tetrapods

Key points

- Vertebrates are characterized by a skeleton made from bone (apatite).
- The oldest vertebrates are small fish-like creatures from the Early Cambrian of China.
- Armored fishes were abundant in Devonian seas and lakes.
- After the Devonian, the cartilaginous and bony fishes radiated in several phases.
- Conodonts commonly occur as tooth-like elements that are useful in biostratigraphy, as are some other fish teeth and scales (ichthyoliths).
- Tetrapods arose during the Devonian from lobe-finned fish ancestors, and fish-eating amphibians diversified in the Carboniferous.
- The first reptiles were small insect eaters.
- Synapsids dominated ecosystems on land during the Permian and Triassic.
- These groups were heavily hit by the end-Permian mass extinction event, and diapsid reptiles, most notably the dinosaurs, were key forms through the Mesozoic.

Most species do their own evolving, making it up as they go along, which is the way Nature intended. And this is all very natural and organic and in tune with mysterious cycles of the cosmos, which believes that there's nothing like millions of years of really frustrating trial and error to give a species moral fiber and, in some cases, backbone.

Terry Pratchett (1991) *Reaper Man*

The backbone is the key. Human beings are vertebrates, and so are horses, sparrows, alligators, turtles, frogs and trout. What they all share is their bony internal skeleton, and, in particular, **vertebrae** – the individual elements of the backbone. The skeleton consists of a backbone, a skull enclosing the brain and sense organs, and bones supporting the fins or limbs. Vertebrates are important today because humans are such a successful species, and also because of the huge diversity and abundance of species of bony fishes, birds and mammals. Other groups, such as insects and microbes, are even more abundant and diverse, but vertebrates include the largest animals on land, in the sea and in the air.

Vertebrates are a subgroup of the Phylum Chordata, a major deuterostome clade. Current views and debates about the nearest relatives of vertebrates are considered in Chapter 14. In this chapter, we look at the origin of vertebrates, the evolution of fishes from the Cambrian to the present day, and the Paleozoic tetrapods. The end-Permian mass extinction reset the clock for vertebrates on land, so we save the dinosaurs and their allies and the mammals for Chapter 17. If the vertebrate skeleton is so significant, what is so special about it?

ORIGIN OF THE VERTEBRATES

The skeleton

The skeleton of vertebrates is made from bone and cartilage. Bone consists of a network of **collagen** fibers on which needle-like crystals of **hydroxyapatite** (a form of apatite, calcium phosphate, $CaPO_4$) accumulate. Hence bone has a flexible component and a hard component, which explains why bones may undergo a great deal of strain before they break, and also why bones do not break along simple brittle faces. **Cartilage** is a flexible, gristly tissue, usually unmineralized, and containing collagen and elastic tissues. In humans, most of the bones are laid down in the early embryo in the form of cartilage, and this progressively mineralizes by deposition of apatite. In adult humans, cartilage can be found in flexible parts like the ears and the nose, as well as at the ends of the ribs and some limb bones.

The first vertebrates probably had a cartilaginous skeleton. Some of the oldest fish fossils, such as *Sacabambaspis* from the Ordovician of Brazil (Fig. 16.1a), had the beginnings of a bony skeleton, but only on the outside of the body, and there is no trace preserved of an internal mineralized skeleton. The rigid armor is made up of lots of little tooth-like structures, each equivalent to an individual shark scale, but united by continuous sheets of bone arranged like plywood.

This shows how adaptable the vertebrate skeleton can be, and this is perhaps why vertebrates became such a diverse and abundant group. The internal skeleton of vertebrates has a unique property – it allows them to

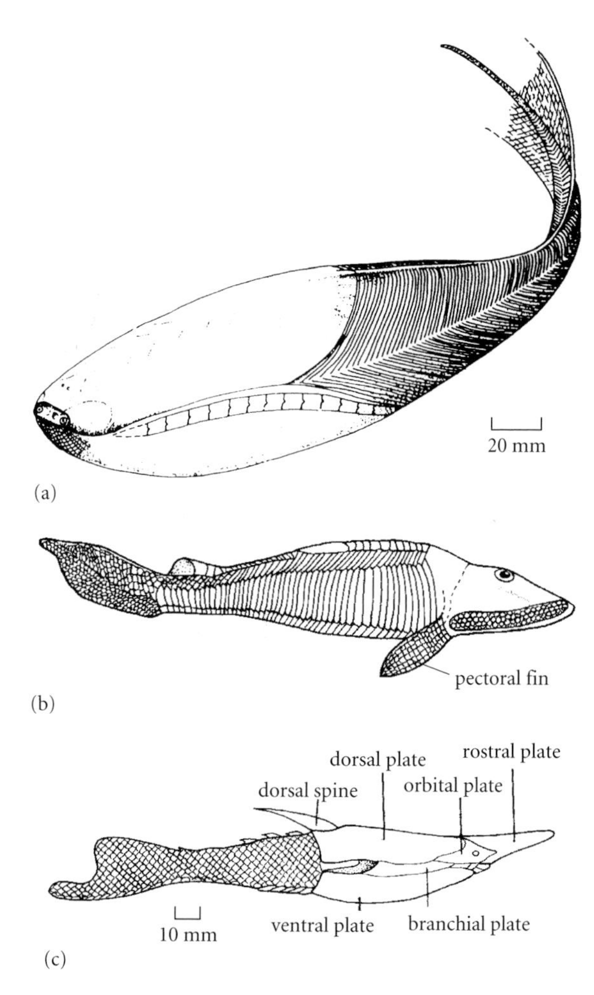

Figure 16.1 Early jawless fishes: (a) *Sacabambaspis* from the Mid Ordovician of Brazil, the oldest well-preserved fish; (b) the osteostracan *Hemicyclaspis* from the Devonian; and (c) the heterostracan *Pteraspis*, also from the Devonian. (a, b, based on Gagnier 1993; c, based on Moy-Thomas & Miles 1971.)

grow very large because the skeleton can grow with the animal. An external skeleton cannot grow so fast, and is less adaptable in supporting a large volume of soft tissues. Further, the external skeleton is vulnerable to damage and either has to be repaired by extending fleshy parts outside the shell (mollusks, graptolites) or by molting the skeleton (arthropods), a wasteful process that uses up energy and leaves the animal vulnerable until the new exoskeleton hardens. By contrast, the vertebrate skeleton is maintained and remodeled constantly within the body, and can act as a support for small, medium, large and massive organisms.

Jawless fishes: slurping rather than biting

Two key defining characters of vertebrates are the head and neural crest tissues. Our head is so essential that we rarely stop to think that actually only vertebrates have heads – indeed vertebrates are sometimes called craniates, meaning "with a skull". Mollusks, worms, brachiopods and echinoderms do not have heads – we might call the front end of a worm its "head", but it really is not any more than its front end. The vertebrate head is unique in providing an organized structure that contains the brain, the major sense organs and the mouth.

The vertebrate head is formed from cells derived from the **neural crest**, a second key apomorphy of vertebrates. The neural crest appears in the early embryo as a strip of cells lying just below the outer skin, the ectoderm, of the embryo, above the line of where the backbone will develop. As tissues begin to differentiate in the early embryo, cells derived from the neural crest spread through the embryo and stimulate the development of muscles, nerves and blood vessels along the trunk and around the heart and gut, but a major target is the head region. The cranial neural crest cells give rise to bones, cartilage, nerves and connective tissue in the head and neck region, forming the face, teeth, eyes, inner ear, the thymus, thyroid and parathyroid glands, and the gills and gill arches of fishes.

The first vertebrates had no jaws (Fig. 16.1). Until recently, these first fishes were said to be Ordovician in age, but controversial new specimens from the remarkable fossil sites at Chengjiang in China (Box 16.1) have pushed the range back to the Early Cambrian. In the Late Cambrian and Ordovician, the commonest vertebrates were the conodont animals. Fishes became common and diverse during the Late Silurian and Devonian.

The jawless fishes are sometimes referred to as **ostracoderms** (Box 16.2). Ostracoderms were jawless, they were generally armored, although some were not, and they had their heyday in the Devonian. Osteostracans like *Hemicyclaspis* (Fig. 16.1b) have a semicircular head shield bearing openings on top for the eyes and nostrils, as well as porous regions round the sides that may have served for the passage of electrical sense organs, perhaps used in detecting other animals by their movements in the water. Heterostracans like *Pteraspis* (Fig. 16.1c), are more streamlined in shape, and were perhaps more active swimmers. Both forms have their mouths underneath the head shield, and they probably fed by sieving organic matter from the sediment. These armored jawless fishes died out at the end of the Devonian, and their place was taken over by fishes with jaws.

Jawless fishes still exist today, the 50 or so species of lampreys and hagfishes, eel-shaped animals. Hagfishes scavenge on dead flesh, while lampreys are often parasitic. Although they have no jaws, their mouths are filled with tooth-bearing bones, and these are used to grip prey animals and to rasp off lumps of flesh. Salmon and trout are commonly caught in the American Great Lakes with huge circular craters in the sides of their bodies, where flesh has been torn out by a sea lamprey.

Conodonts: animals of mystery

The commonest early vertebrates were the conodont animals (Sweet & Donoghue 2001). For over 150 years conodonts had been a mystery, known only from their jaw elements – no one knew which animal had produced them.

Conodonts were first identified by the Latvian embryologist and paleontologist Christian Pander in 1856. They occur as phosphatic tooth-like microfossils, termed **elements**. Three main conodont groups have been established (Fig. 16.3): (i) protoconodonts such as *Hertzina* are simple cones with deep basal cavities; (ii) paraconodonts like

Box 16.1 The world's oldest vertebrates

There was a sensation in 1999 when Shu Degan and colleagues (Shu et al. 1999) announced a new fossil vertebrate, *Myllokunmingia*, from the Early Cambrian locality Chengjiang in China. This site has become celebrated for the exceptional preservation of all kinds of animal fossils, and it rivals the Burgess Shale (see p. 249) as a window into Cambrian life. Until 1999, the oldest vertebrates were much debated, with some tentative Middle and Late Cambrian candidates, but nothing really certain until the Ordovician.

Myllokunmingia (Fig. 16.2) is tiny, less than 30 mm long – you could hold a hundred or so of them, like a handful of wriggling whitebait. The head is poorly defined, but there seems to be a mouth at one end. Relatives seem to show detail in the head, possible eyes and a brain. If it has such differentiated head features, it is a vertebrate. Behind the "head" are six gill pouches, a possible heart cavity and a gut. Above these are the notochord, a key chordate character (see p. 410), and **myotomes** or V-shaped muscle blocks. There is a narrow dorsal fin along the back, and possibly a ventral fin below. *Myllokunmingia* presumably swam by flicking its body and fins from side to side and wriggling forward through the water. None of the Chinese specimens have mineralized bone – but this does not rule them out as vertebrates. Evidently, the vertebrate skeleton began as a cartilaginous structure in early forms, and became mineralized with apatite later in the Cambrian.

In the same paper, Shu et al. (1999) also named *Haikouichthys*, a similar early vertebrate from Chengjiang. A rival team, Hou et al. (2002), suggest that *Haikouichthys* was the same as *Myllokunmingia*, although Shu and colleagues disagree. The two groups, led by Shu and Hou, also disagree over the identification of different organs within these fossils, and this affects where they are placed in the vertebrate phylogeny. The Chengjiang fossils are preserved in grey or yellow sediment, and the fossils may be grey or reddish, with the internal organs picked out in grey, brown and black colors. Interpreting these multicolored blobs and squiggles would test the patience of a saint, and yet it is remarkable that such details have been preserved for 500 Myr. There are now more than 500 specimens of these early vertebrates, so further intensive study may clarify their anatomy further.

See http://www.blackwellpublishing.com/paleobiology/ for relevant web links.

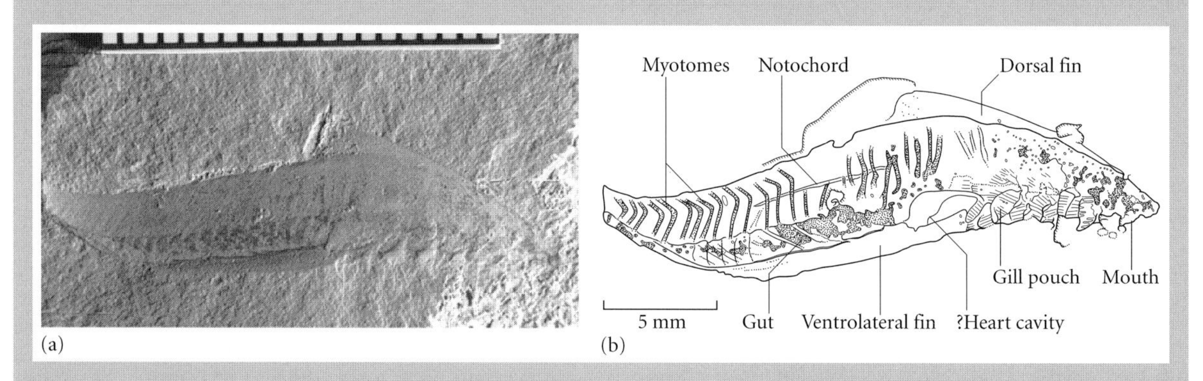

Figure 16.2 The basal vertebrate *Myllokunmingia* from the Early Cambrian of Chengjiang, China: (a) photograph of specimen, and (b) interpretive drawing showing possible identities of the internal organs. (Courtesy of Shu Degan.)

Box 16.2 Classification of fishes

"Fishes" form a paraphyletic grouping, consisting of several distinctive clades of swimming vertebrates. Ordovician and Silurian records of placoderms, acanthodians, chondrichthyans and osteichthyans are mainly isolated scales and teeth; these groups are best known from the Devonian onwards.

Subphylum VERTEBRATA
"Class AGNATHA"

- A paraphyletic group of jawless fishes, including armored and unarmored Paleozoic ostracoderms, and modern lampreys and hagfishes
- Late Cambrian to Recent

Class PLACODERMI

- Heavily armored fishes with jaws and a hinged head shield
- Mid Silurian to Late Devonian

Class CHONDRICHTHYES

- Cartilaginous fishes, including modern sharks and rays
- Late Ordovician to Recent

Class ACANTHODII

- Small fishes with many spines and large eyes
- Late Ordovician to Early Permian

Class OSTEICHTHYES

- Bony fishes, with ray fins (Subclass Actinopterygii) or lobe fins (Subclass Sarcopterygii), the latter including ancestors of the tetrapods
- Late Silurian to Recent

Furnishina are mainly simple cones; and (iii) euconodonts or true conodonts are more complex, with cones, bars and blades. The protoconodonts are almost certainly unrelated to true conodonts; they may be chaetognaths or arrow worms, a group of basal metazoans of uncertain affinities.

Euconodonts occur as three broad types of element, consisting of laminae of apatite. These are formed by outer accretion from an initial growth locus. White matter often occurs between, or crosscuts, lamellae; this material compares well with the composition and structure of vertebrate bone. The three main morphotypes of conodont element have been used in the past as the basis of a crude single element or form taxonomy (Fig. 16.4). The cones or **coniform elements** are the simplest, with the base surmounted by a cone-like cusp, tapering upward, and sometimes ornamented with ridges or costae (Fig. 16.4a, b). Bars or **ramiform elements** consist of an elongate blade-like ridge with up to four processes developed posteriorly, anteriorly or laterally to the cusp (Fig. 16.4c, d, g). Platforms or **pectiniform elements** have a wide range of shapes, with denticulate processes extending both anteriorly, posteriorly and/or laterally from the area of the basal cavity (Fig. 16.4e, f, h–j); some also have primary lateral processes. The cusp is attenuated, whereas the base may be expanded to form a platform with denticles on its upper surface. The basal cavity is filled by the basal body of the element in the form of a dentine-like material, although this is not always preserved.

Conodonts are common in certain marine facies from the Cambrian to the Triassic.

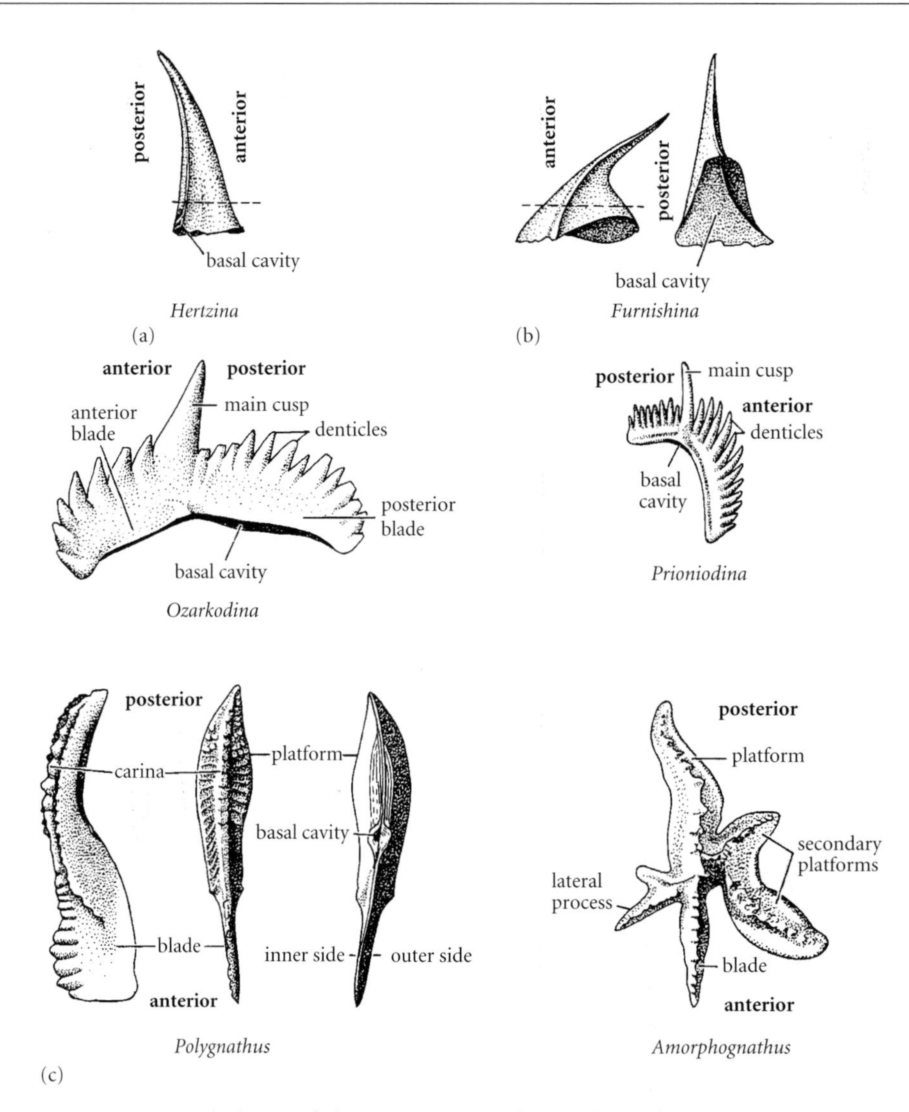

Figure 16.3 Descriptive morphology of the main types of conodont elements: (a) protoconodont *Herzina* (×40); (b) paraconodont *Furnishina* (×40); and (c) euconodonts *Ozarkodina* (×40), *Prionodina* (×20), *Polygnathus* (×40) and *Amorphognathus* (×40). (Based on Armstrong & Brasier 2004.)

Paraconodonts are reported from the Mid Cambrian; older records are doubtful. During the Late Cambrian, simple conical euconodonts appeared. In the Early Ordovician, apparatuses with coniforms, and some with coniform and ramiform element types, appeared. Conodont diversity peaked during the Mid Ordovician, with a global maximum of over 60 genera. During this interval of experimentation, there was a huge diversity of apparatus patterns never again matched; later apparatuses are relatively uniform, perhaps indicating stabilization of feeding modes. Pectiniform elements were common from the Early Ordovician, together with a wide variety of blades and platforms in the Mid to Late Ordovician. This great diversity of forms was wiped out by the Late Ordovician mass extinction (see p. 169). Silurian faunas are less variable, mainly apparatuses with ramiform and pectiniform elements. The conodonts again radiated during the Late Devonian, with specialized ramiform and pectiniform elements; over 1000 conodont taxa have been named from the Upper Devonian. Carboniferous conodonts (Fig. 16.5a) were characterized by a lack of coniform elements, together with pectiniform elements in the P apparatus position, whereas ramiform elements occupied the M and S positions (see

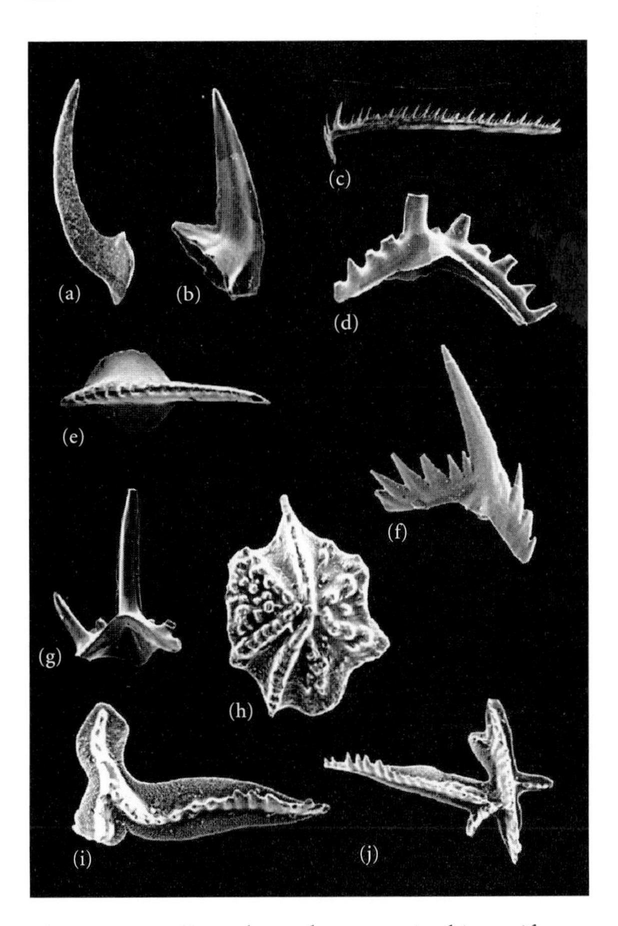

Figure 16.4 Conodont elements: (a, b) coniform, lateral view; (c, d) ramiform, lateral view; (e) straight blade, upper view; (f) arched blade, lateral view; (g) ramiform, posterior view; and (h-j) platform, upper view. Magnification ×20–35 for all. (Courtesy of Dick Aldridge.)

Figure 16.5 Homing in on the conodont animal: (a) natural assemblage of conodonts from the Carboniferous of Illinois (×24); and (b) the conodont animal from the Carboniferous Granton Shrimp Bed, Edinburgh, Scotland, with the head at left-hand end (×1.5). (Courtesy of Dick Aldridge.)

next paragraph). Conodonts became rarer during the Early Permian, and most Late Permian and Triassic species had small apparatuses. They became extinct at the end of the Triassic.

The first piece of evidence about the identity of the conodont animal is that the elements sometimes occur in a cluster or **apparatus** consisting typically of 15 elements, 14 of them arranged bilaterally and one symmetric element positioned on the midline. The elements are arranged in a particular way in the apparatus: pectiniform (P elements) at the back, makelliform (M elements) at the front, and symmetry transition series (S elements) in between. Generally bars and platforms occupy P positions, whereas bars and cones are found in M and S positions. The P, M and S positions may be defined more precisely with subscripts, for example P_a and P_b elements.

The first conodont apparatus was found in 1879, and this gave some idea about the function of conodonts, perhaps as some sort of teeth, and provided some clues about the whole animal. Several supposed conodont animals were identified in the 1960s, but most of these turned out to be predatory critters that had just eaten a conodont animal, and so had lots of conodont elements inside them!

Despite the mystery of their identity, conodonts became key tools in biostratigraphy (Box 16.3). In addition, because color changes of the elements can be related to changing temperature, conodonts are important indicators of thermal maturation. Now paleontologists believe they know what conodont animals looked like, but it took 150 years to work this out.

The solution came in 1983, when the first complete conodont animal was found in the Granton Shrimp Bed, a dark Carboniferous mudstone on the seacoast near Edinburgh, Scotland (Fig. 16.5b). This was an eel-like animal with a conodont apparatus at its front end. Detailed examination showed that the elements were in place and, this time, had not been merely eaten by the animal. Ten conodont animals have now been found, as well as examples from other localities (Aldridge et al. 1993a). The Scottish conodont animal is up to 55 mm long, and has a short, lobed head with large goggling eyes that are fossilized black, perhaps a stain produced by the visual pigments. Below and behind the eyes is the conodont apparatus, clearly located where the mouth should be, showing that conodont elements really did function as teeth. The

Box 16.3 Conodonts and biostratigraphy

Detailed biostratigraphic schemes based on conodonts have been established for many parts of the Paleozoic and Triassic. For example, over 20 conodont zones have been determined for the Ordovician System, while the Upper Devonian is the most congested interval, with over 30 biozones, each less than 500,000 years long. In northwest Europe the Carboniferous is routinely correlated on the basis of conodont zones.

Remarkable precision is now available in some zonal schemes. This has permitted the development of models for global environmental change during the Early Silurian (Fig. 16.6) tied to a tight conodont zonation (Aldridge et al. 1993b). Two oceanic states are recognized: those with oxygenated cool oceans that had a good vertical circulation and adequate supplies of nutrients (termed "primo"), and those with warm stratified oceans that had deep saline levels and poor nutrient supplies (termed "secundo"). Sudden changes between ocean states altered the vertical circulation and nutrient supply dramatically, perhaps causing extinction events.

These kinds of stratigraphic schemes may depend on geographic zonations. Cambrian conodont faunas were divided into equatorial (low latitude) warm-water associations and polar (high latitude) cool-water associations. During the Early Ordovician, these low- and high-latitutde assemblages further divided into six discrete provinces. Conodonts evolved independently at high latitudes, and there were only a few incursions from lower-latitude faunas. Towards the end of the Ordovician high-latitude, cold-water faunas migrated into lower latitudes. Thus Late Ordovician equatorial mid-continent assemblages originated in polar and subpolar regions and themselves formed the foundation for the Silurian fauna. During the Mid and Late Paleozoic, conodonts were mainly restricted to tropical latitudes. Devonian and Carboniferous faunas show some biogeographic differentiation among shelf associations. These differences among the geographic provinces can affect the stratigraphic schemes and the possibility of correlation from area to area.

Conodonts occur in a wide range of marine and marine-marginal environments, although the group is most common in nearshore carbonate facies, commonly in the tropics. Distinct environment-related conodont paleocommunities have been identified in many parts of the Paleozoic, and statistical analysis may discriminate, for example, deeper-water from shallow-water assemblages. It is important to be aware of the influence of depth and other factors on the distribution of communities before they are used in establishing biostratigraphic zones. It would clearly be a mistake to identify distinctive depth-determined conodont assemblages and then to interpret them as indicators of different time intervals.

Web links are available through http://www.blackwellpublishing.com/paleobiology/.

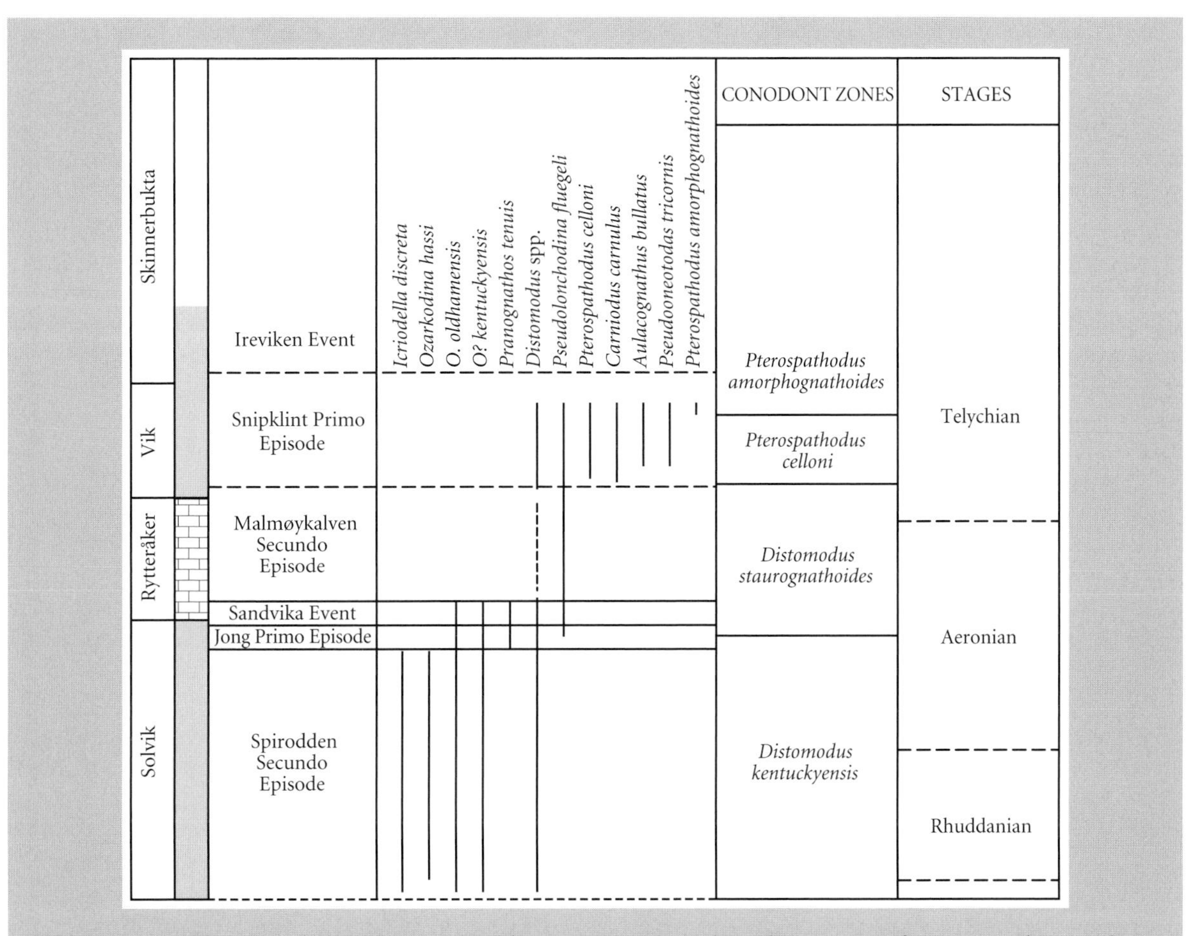

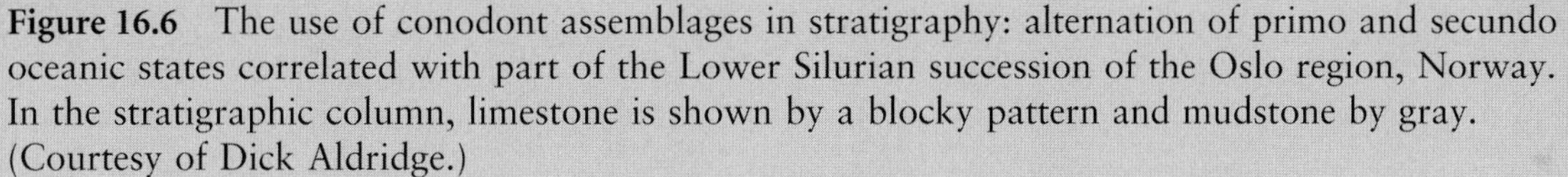

Figure 16.6 The use of conodont assemblages in stratigraphy: alternation of primo and secundo oceanic states correlated with part of the Lower Silurian succession of the Oslo region, Norway. In the stratigraphic column, limestone is shown by a blocky pattern and mudstone by gray. (Courtesy of Dick Aldridge.)

anterior comb-like ramiform elements probably grasped prey items that were sucked towards the mouth, and the posterior pectiniform elements may have chomped the food before swallowing. One the greatest mysteries in paleontology had been solved.

JAWS AND FISH EVOLUTION

The first jaws

The basal vertebrates, including conodonts, lacked jaws, and jaws probably evolved during the Ordovician. Study of the anatomy of modern vertebrates suggests that jaws may have evolved from the strengthening bars of cartilage or bone between the gill slits, each of which consists of several elements, all linked by tiny muscles. The transition cannot be followed in fossils because the gill skeleton of jawless fishes was not mineralized. Molecular biologists have even suggested that the origin of jaws was so profound that it must have been associated with a dramatic genome duplication event – but the fossils say no (Box 16.4).

Some of the oldest jaw-bearing fishes were the placoderms, such as *Coccosteus* (Fig. 16.8a), which had an armor of large bony plates over the head and shoulder region, as in the ostracoderms, and a more lightly armored posterior region. They swam by beating this tail region from side to side. The edges of the jaws did not carry teeth, but instead sharp bony plates that would have been just as effective in snapping at prey.

Placoderms were fearsome predators, some of them, like *Dunkleosteus* from the Late Devonian of North America, reaching the impressive length of 10 m. This was the largest animal that had lived until then, and its size and fearsome jaws may explain why so many Devonian fishes were armored.

Other Devonian fishes were more modern in appearance. The first shark-like chondrichthyans, or cartilaginous fishes, came on the scene during the Early Devonian. Acanthodians were small fishes, mostly in the range 50–200 mm in length, and they bore numerous spines at the front of each fin and in

Box 16.4 Genome duplications and vertebrate evolution

Vertebrates have larger genomes than other animal groups. The **genome** is the entire sequence of genes contained on all the **chromosomes** within the nuclei of cells. Various worms and insects have around 15,000 **genes** in their genomes, while the figure is 31,000 for humans, 30,000 for the mouse and 38,000 for the pufferfish. However, vertebrates do not just have more genes than invertebrates, they have two, four or even eight copies of many individual invertebrate genes. At one time, molecular biologists thought that humans had as many as 100,000 genes, but the reduced figure was established in 2004 after the intense gene sequencing efforts of the Human Genome Project. What does genome size mean?

Some have suggested that genome size maps on to the complexity of an organism. Surely, a single-celled bacterium does not need many genes because it does not do much, and vertebrates, as much more complex organisms, would need more genes. Humans ought to have the largest genomes since we are somehow very complex and important. In fact, genome size is only loosely related to bodily complexity: the largest genome reported so far comes from a lungfish! Much of the genome is so-called **junk DNA**, or at least duplicate genes and non-coding sections, so the functional genome size might be a better correlate of function or bodily complexity.

Whether functional or not, molecular biologists have proposed that there were at least three **genome duplication events** (GDEs) in the history of vertebrates – times when evolutionary change was dramatic and large sectors of the genome duplicated. GDEs are identified at the origin of vertebrates, the origin of gnathostomes and the origin of teleosts, the hugely diverse modern bony fishes (Furlong & Holland 2004). Could the evolutionary jump have caused the GDE, or perhaps the GDE stimulated rapid and fundamental reorganization of the fishes at these three points?

Donoghue and Purnell (2005) suggest that molecular biologists have been misled. By omitting fossils, they see artificial morphological jumps in their cladograms, and then link this to the postulated GDE. In fact, when fossils are inserted, the "jumps" seem less clear. For the origin of gnathostomes, biologists have compared lampreys with sharks, and there is a wide gulf between these two groups, so suggesting quite a leap in terms of anatomic change and in terms of genome duplication. However, when fossils are inserted (Fig. 16.7), seven major ostracoderm and placoderm clades fall between the living groups, and the evolutionary transition is stretched. Some of the fossil groups (especially pteraspidimorphs, conodonts and placoderms) were diverse, and it is not clear that the GDE drove, or permitted, a single dramatic burst of speciation, as had been proposed. Further, it is not clear that there was a single reorganization of anatomy associated with the origin of jaws and the GDE: the fossils show step-by-step character changes over a long interval.

This is a developing field of study. The claim that genome duplication can drive major bursts of evolution is dramatic, and perhaps overstated. Paleontologists can make profound contributions in new areas of science by working hand-in-hand with molecular and developmental biologists.

Read more through http://www.blackwellpublishing.com/paleobiology/.

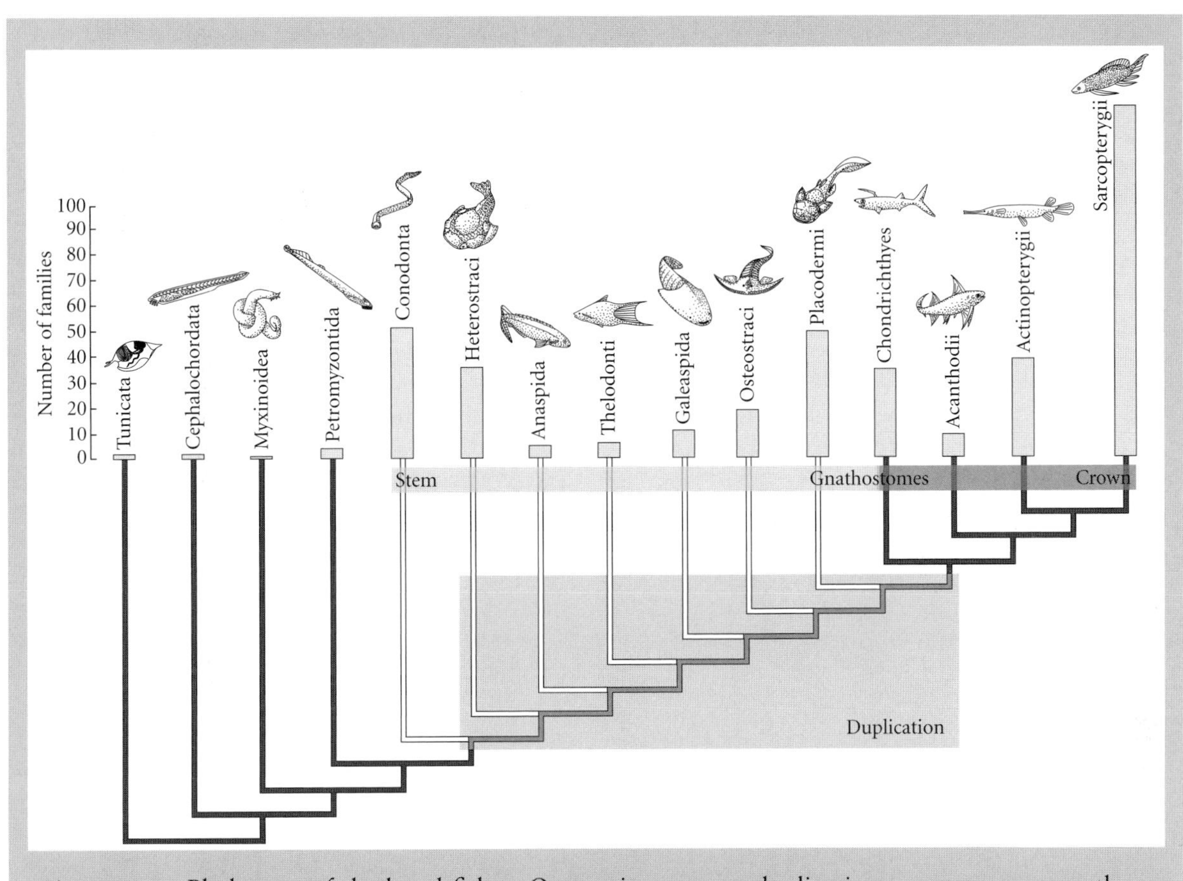

Figure 16.7 Phylogeny of the basal fishes. One major genome duplication event was apparently associated with the origin of jaws. When the fossil groups (open lines) are omitted, there is a large morphological and genomic leap from jawless lampreys and hagfishes; when the fossil groups are included, as here, the transition appear much more gradual. The timing of the genome duplication events is uncertain, and falls within the area of the gray box. The number of families within each living and fossil group is shown by the shaded vertical bars. (Courtesy of Phil Donoghue.)

spaced rows on their undersides (Fig. 16.8b). Acanthodians are often found preserved in vast numbers in the rock layers, so they probably swam in huge shoals in open water, perhaps feeding on small arthropods and plankton. They escaped predators by rapid darting from side to side in their shoals, and perhaps their exceptional spininess made them difficult to swallow.

Bony fishes: ray fins and lobefins

The osteichthyans, or bony fishes, also appeared in the Devonian. There are two groups: (i) those with ray-like fins, the actinopterygians, ancestors of most fishes today from carp to salmon, and seahorse to tuna; and (ii) the lobefins, the sarcopterygians, that had thick, muscular, limb-like fins. Today, the lobefins are rare, being represented by only three species of lungfishes and the rare coelacanth. The coelacanth *Latimeria* is a famous "living fossil". Until 1938, coelacanths were only known as Devonian to Cretaceous fossils, but in 1938 the world was astounded to hear that a living coelacanth had been fished out of deep waters off East Africa, and more have been caught since then.

The ray fins of the Devonian include *Cheirolepis* (Fig. 16.8c), which had a flexible body covered with small scales and a plated head. This was an active predator that may have fed on acanthodians. The Devonian lobefins include both lungfishes and "rhipidistians".

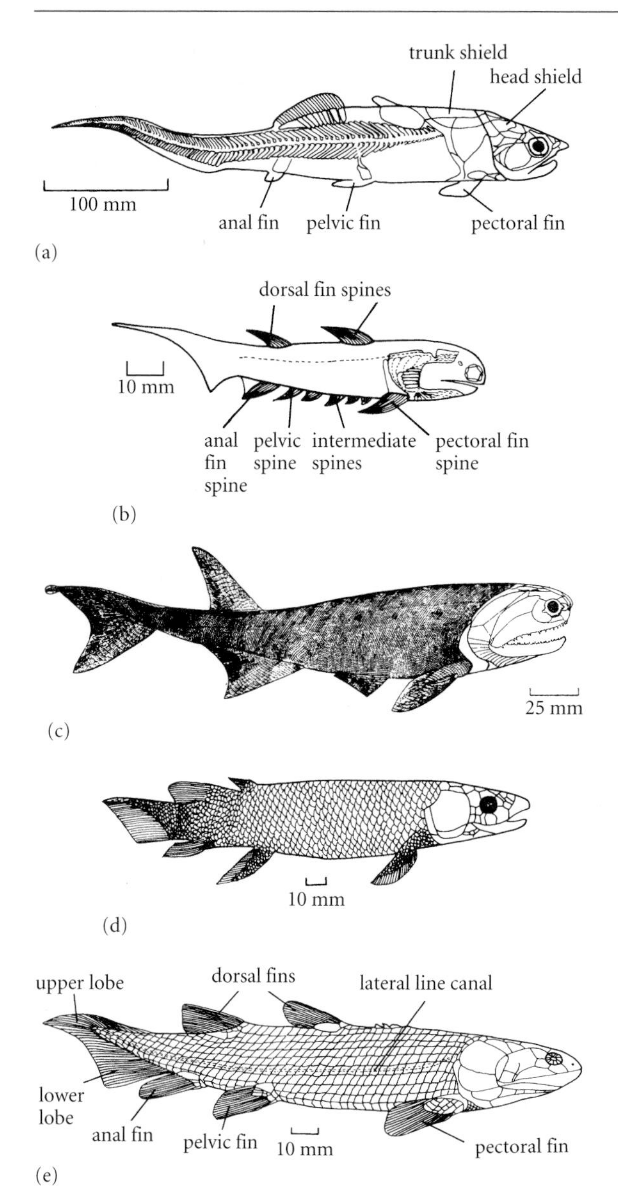

Figure 16.8 Jawed fishes of the Devonian: (a) the placoderm *Coccosteus*; (b) the acanthodian *Climatius*; (c) the actinopterygian bony fish *Cheirolepis*; (d) the lungfish *Dipterus*; and (e) the lobefin *Osteolepis*. (Based on Moy-Thomas & Miles 1971.)

The lungfish *Dipterus* (Fig. 16.8d) was a long, slender fish that hunted invertebrates and fishes, and crushed them with broad grinding tooth plates. The "rhipidistian" *Osteolepis* (Fig. 16.8e) was also long and slender, and was an active predator. These lobefins had muscular front fins, and could have used these to haul themselves over mud from pond to pond. Specimens of these fishes are known from the Devonian of many parts of the world (Box 16.5).

After the Devonian, the actinopterygians seem to have radiated three times. The first radiation (Devonian-Permian) consisted of the palaeonisciforms (Fig. 16.10a), a paraphyletic group of bony fishes with large bony scales and heavy skull bones. The second radiation of bony fishes, an assemblage termed the "holosteans", occurred in the Late Triassic and Jurassic. *Semionotus* (Fig. 16.10b), a small form that has been found in vast shoals, had more delicate scales than the palaeonisciforms, and a jaw apparatus that could be partly protruded, hence providing a wider gape.

The third and largest radiation of actinopterygian fishes, occurred in the Late Jurassic and Cretaceous (Fig. 16.10c), with the diversification of the teleosts. Teleosts are the most diverse and abundant fishes today, including 23,000 living species, such as eel, herring, salmon, carp, cod, anglerfish, flying fish, flatfish, seahorse and tuna. The huge success of this radiation may be the result of their remarkable jaws. Palaeonisciforms opened their jaws like a simple trapdoor, holosteans could enlarge their gape a little, but teleosts can project the whole jaw apparatus like an extendable tube (Fig. 16.10d). This came about because of great loosening of the elements of the skull: as the lower jaw drops, the tooth-bearing bones of the upper jaw (the **maxilla** and **premaxilla**) move up and forwards. Rapid projection of a tube-like mouth allows many teleosts to suck in their prey, while others use the system to vacuum up food particles from the seafloor, or to snip precisely at flesh or coral.

The evolution of sharks: an arms race with their prey?

During the Carboniferous, numerous extraordinary shark-like fishes arose, and these were clearly important marine predators. A second shark radiation took place in the Triassic and Jurassic. *Hybodus* (Fig. 16.11a) was a fast-swimming fish, capable of accurate steering using its large pectoral (front) fins. The hybodontiforms had a range of tooth types, from triangular pointed flesh-tearing teeth to broad button-shaped crushers, adapted for dealing with mollusks. It is rare to find whole

Box 16.5 The Scottish Old Red fishes

The Old Red Sandstone Continent of northern Europe and Canada lay close to the equator, in hot tropical conditions, during the Devonian. Fishes lived in the shallow seas and in landlocked lakes around this continent. One of the best collecting areas is in the north of Scotland, where the first specimens came to light nearly 200 years ago.

The fish beds were laid down in large deep lakes (Trewin 1986). Bulky armored ostracoderms and placoderms fed on the bottom in shallow waters, while shoals of silvery acanthodians darted and swirled near the surface. Bony fishes, such as the actinopterygian *Cheirolepis*, the "rhipidistian" *Osteolepis* and the lungfish *Dipterus*, moved rapidly through the plants near the water's edge seeking prey, and sometimes swam out through the deeper waters to new feeding grounds.

The fishes are usually found, beautifully preserved and nearly complete (Fig. 16.9a), in dark-colored siltstones and fine sandstones. These rocks were deposited in the deepest parts of the lake, probably in anoxic (low oxygen) conditions (Fig. 16.9b). There were repeated cycles of deposition, perhaps controlled by the climate. During times of high rainfall, great quantities of sand were washed into the lakes from the surrounding Scottish Highlands. Lake levels then fell during times of aridity, and in places the lakes dried out, leaving mud cracks and soils. Then flooding occurred, together with mass kills of fishes, perhaps as a result of **eutrophication** (oxygen starvation caused by decaying algae after an algal bloom) or following storms. In all, a pile of lake deposits some 2–4 km thick accumulated over the 40–50 myr of the Devonian, and there are dozens of fish beds throughout this thickness.

Read more through http://www.blackwellpublishing.com/paleobiology/.

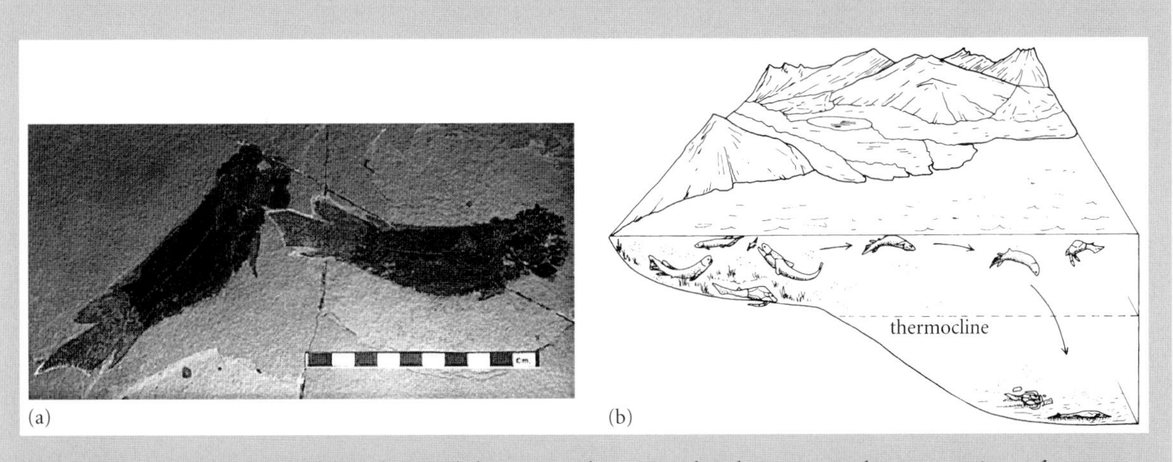

Figure 16.9 The Old Red Sandstone lake in northern Scotland: (a) typical preservation of two specimens of *Dipterus*; and (b) model of environmental cycles in the lake. Sediment is fed in from the surrounding uplands during times of heavy rainfall. Fishes inhabit shallow and surface waters, but carcasses may sink below the thermocline into cold, relatively anoxic waters, where they sink to the bottom and are preserved in undisturbed condition in dark grey laminated muds. (Courtesy of Nigel Trewin.)

shark fossils because the bulk of the skeleton is cartilaginous and rots away before fossilization. The apatite teeth and scales are more commonly found isolated, and these and other fish teeth and scales, sometimes called ichthyoliths, have proved useful in biostratigraphy (Box. 16.6).

Modern sharks, called collectively neoselachians, are faster swimmers and more ferocious flesh eaters than their precursors.

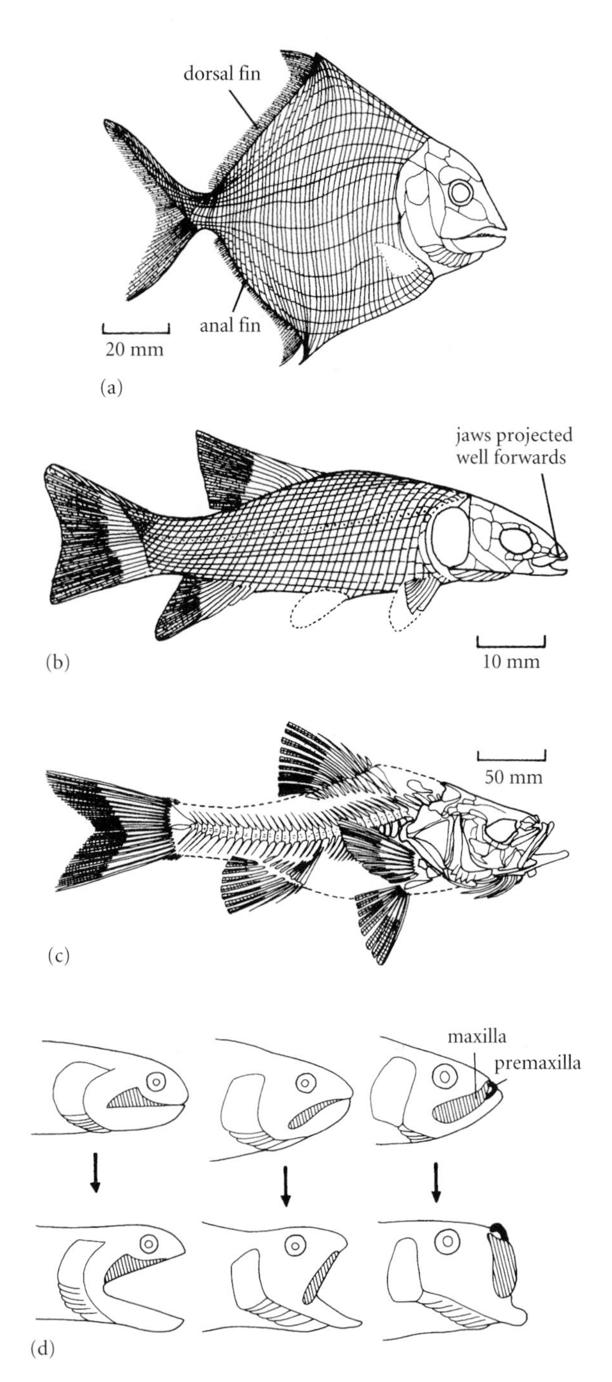

Figure 16.10 Evolution of the ray-finned bony fishes: (a) the Carboniferous palaeonisciform *Cheirodus*, a deep-bodied form; (b) the Triassic "holostean" *Semionotus*; (c) the Cretaceous teleost *Mcconichthys*; (d) evolution of actinopterygian jaws from the simple hinge of a palaeonisciform (left) to the more complex jaws of a holostean (middle) and the fully pouting jaws of a teleost (right). (a, b, based on Moy-Thomas & Miles 1971; c, based on Grande 1988; d, based on Alexander 1975.)

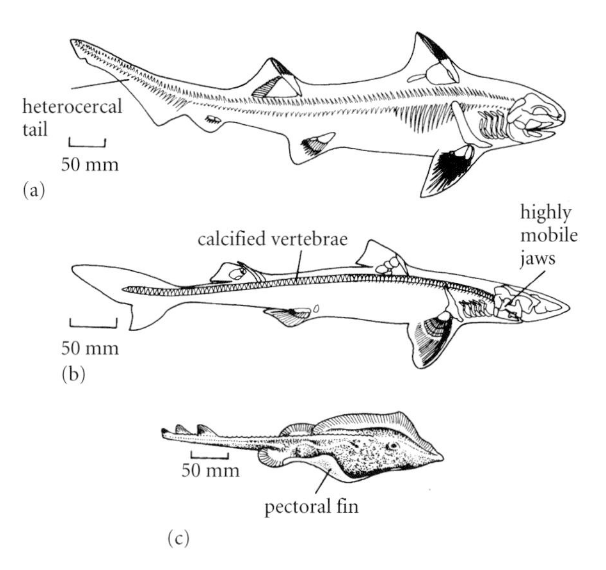

Figure 16.11 Sharks and rays, ancient and modern: (a) the Jurassic shark *Hybodus*; (b) the modern shark *Squalus*; and (c) the modern ray *Raja*. (Based on various sources.)

Neoselachians radiated dramatically during the Jurassic and Cretaceous to reach their modern diversity of 42 families. These sharks can open their mouths wider than their precursors, and they have adaptations for gouging masses of flesh from their prey. The body shape (Fig. 16.11b) is more bullet-like than in their ancestors, and the pectoral fins are wider and more flexible. Neoselachians range in size from common dogfishes (0.2–1 m long) to basking and whale sharks (16 m long), but the monster ones are not predators: they feed on krill, which they filter from the water. The skates and rays, unusual neoselachians, are specialized for life on the seafloor, having flattened bodies and broad pectoral fins for swimming by sending waves of up-and-down motion from front to back (Fig. 16.11c).

Sharks and their kin radiated three times during the Paleozoic, Mesozoic and Cenozoic, and this seems to match the three-phase radiation of bony fishes, palaeonisciforms, holosteans and teleosts. It is impossible to say which set of evolutionary radiations came first: the bony fishes had to swim faster to escape their sharky predators, and the sharks had to swim faster to catch their bony fish prey. This is a classic example of an **arms race**, where predator and prey keep upping the ante, but neither side wins.

Box 16.6 Fish teeth and scales

Isolated teeth and scale, commonly grouped as **ichthyoliths** ("fish stones"), can usually be assigned to their fishy originator (Fig. 16.12). However, there is great debate over the precise identity of many shark teeth, and especially over the amount of variation that may occur among the teeth from a single species – are they all the same, or do they vary in shape around the jaws? Also, many Cambrian and Ordovician ichthyoliths are a mystery – the original host animal is generally unknown. The thelodonts, an ostracoderm group, were known from a rich diversity of scales from the Ordovician to Devonian, but there are only a handful of partial or complete specimens of the whole fishes.

Ichthyoliths have been used to establish stratigraphic schemes in the Silurian, Carboniferous, Triassic, Cretaceous and Tertiary. In some Paleozoic strata, teeth and scales sometimes occur in association with conodonts, and sometimes in situations where there are no other biostratigraphically useful fossils. It is hard work to generate workable dating schemes using such ichthyoliths, not least because their morphology can be so intricate; but it is also frustrating that it is often impossible to relate the ichthyoliths to the complete fishes that might have produced them.

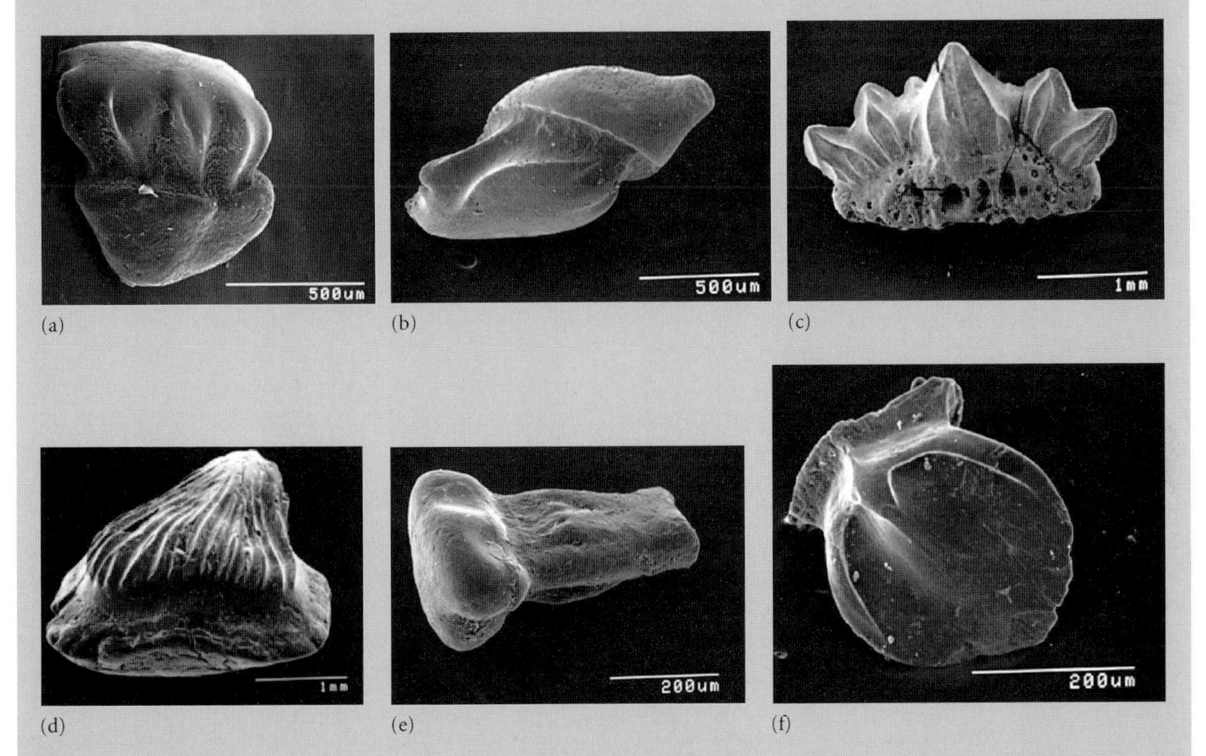

Figure 16.12 Some microvertebrate specimens: (a) thelodont scale (Devonian); (b) thelodont body scale (Devonian); (c) protacrodont shark tooth (Late Devonian to Early Carboniferous); (d) acanthodian scale (Devonian); (e) shark tooth-like scale (Triassic); and (f) shark scale (Triassic). (Courtesy of Sue Turner.)

Figure 16.13 Skull of the Late Devonian amphibian *Acanthostega*, showing the streamlined shape, deeply-sculpted bones and small teeth, all inherited from its fish ancestor. (Courtesy of Jenny Clack.)

TETRAPODS

The origin of tetrapods: fins to limbs

When a fish became a land animal, surely the key problem was breathing air? Not so. Early bony fishes almost certainly had *both* lungs and gills, and could already breathe air when necessary. The main problem for the first tetrapods ("four feet"), the four-legged land vertebrates, was support – in water, an animal "weighs" virtually nothing, but on land the body has to be held up from the ground, and the internal organs have to be supported in some way within a strong rib cage to prevent them from collapsing. In addition, reproductive, **osmotic** (water balance) and sensory systems had to adapt, but here the changes did not happen all at once.

Tetrapods arose from fishes during the Devonian. But what were the closest fishy relatives of the tetrapods? Most attention has focused on the lobefins, because of their complex bony and muscular **pectoral** (front) and **pelvic** (back) paired fins (Fig. 16.8d, e). Close study of lobefins such as *Osteolepis* and *Eusthenopteron* shows that they share many characters of the limbs and skull with the earliest tetrapods.

The first good evidence of tetrapods comes from the Late Devonian, and they have been studied intensively in recent years. The best-known forms are *Acanthostega* and *Ichthyostega*, which were 0.6 and 1 m long, respectively. The head is still very fish-like in shape and sculpturing (Fig. 16.13), and the limbs and tail are clearly still adapted for swimming. The limbs though are remarkable. We have always thought that our five fingers and toes are a fundamental feature of tetrapods, but new work shows that this was not the case (Box 16.7).

The amphibians: half-way land animals

Amphibians today are a minor group, consisting of 4000 species of mainly small animals that live in or close to water. They show many adaptations to life on land, but they still rely on the water for breeding and water balance.

Frogs and toads (anurans), known since the Triassic, have specialized in jumping: the hindlimbs are long and the hip bones reinforced to withstand the impact of landing. The head is broad, the jaws are lined with small teeth, and most feed by flicking a long sticky tongue out and trapping insects. The urodeles – salamanders and newts – date from the Jurassic, and consist of modest-sized, long-bodied swimming predators. The third living amphibian group, the caecilians, are small limbless animals that look rather like earthworms, and live largely in soil and leaf litter in tropical lands. The oldest fossil form, with reduced limbs, is Jurassic in age. All living amphibians appear to be closely related, forming a clade, the Lissamphibia (Box 16.8), characterized by the structure of their tiny teeth.

The lissamphibians form part of a larger clade, the batrachomorphs. The most important fossil batrachomorphs are the "temnospondyls", a paraphyletic group that was important in Carboniferous communities, and continued with reasonable success through the Permian and Triassic, finally dying out in the Early Cretaceous. Temnospondyls have a low round-snouted skull (Fig. 16.15a, b), and most of them appear to have operated like sluggish crocodiles, living in or near freshwaters and feeding on fishes. Some temnospondyls became fully terrestrial, and others evolved elongate gavial-like snouts for catching rapidly swimming fishes. Some Carboniferous temnospondyls had tadpole young, just as modern amphibians do. This proves that they had a similar developmental pattern to modern amphibians, with an aquatic **larval** stage, the tadpole, that metamorphoses into the adult land-living form. Relatives of the temnospondyls included small forms, the aquatic nectrideans and the aquatic and terrestrial microsaurs.

The second amphibian lineage, the reptiliomorphs (see Box 16.8), included important groups in the Carboniferous and Permian, as well as the ancestors of reptiles, birds and mammals. Anthracosaurus had a longer, narrower skull than the temnospondyls, but may have had similar lifestyles – hunting prey on land and in freshwaters. Some Permian reptiliomorphs, such as *Seymouria* (Fig. 16.15c) were seemingly adapted to a fully terrestrial life. *Seymouria* has long limbs and a relatively small skull, and probably hunted microsaurs and other small tetrapods.

REIGN OF THE REPTILES

Making the break: the origin of the reptiles

Amphibians only made it halfway on to land, and they still produce swimming tadpoles.

Box 16.7 The first tetrapods had seven or eight toes

New studies of *Ichthyostega* and *Acanthostega* from the Late Devonian of Greenland (Coates et al. 2002), and of other animals of the same age, including the "limbed fish" *Tiktaalik* from Arctic Canada (Daeschler et al. 2006; Shubin et al. 2006), show that the first tetrapods had more than five fingers and toes, indeed as many as seven or eight (Fig. 16.14b, c). It is possible to draw comparisons between the bones of the pectoral fin of a sarcopterygian (Fig. 16.4a) and those of the forelimb of an early amphibian (Fig. 16.14b). This caused a major rethink of the classic story of the evolution of vertebrate limbs: five digits must have become standard only *after* the origin of tetrapods. What if tetrapods had settled on seven, rather than five, fingers? Probably we would not use the decimal counting system, and imagine the changes to musical instruments and computer keyboards!

The implications are wider, because the new evidence suggests that particular features of an organism may not all be preprogrammed in the genetic code of the developing embryo. In other words, there is not a single gene that codes for each finger and toe. It seems that aspects of the developmental environment, rather than genetic programming, determine some details of adult structure: as a limb develops in the embryo, at first it has no fingers or toes, and then a pulse of information triggers the sprouting of digits at a particular time. In rare cases, humans may be born with a sixth finger, perhaps a genetic memory of our condition 400 Myr ago. Also, many tetrapods have only four (frogs), three (rhinos), two (cows) or one (horses) finger – perhaps losing digits is associated with the "switching" on or off of particular controlling genes.

Read more about the basal tetrapods and the fin to limb transition in Zimmer (1999), Clack (2002) and Shubin (2008) and at http://www.blackwellpublishing.com/paleobiology/.

Continued

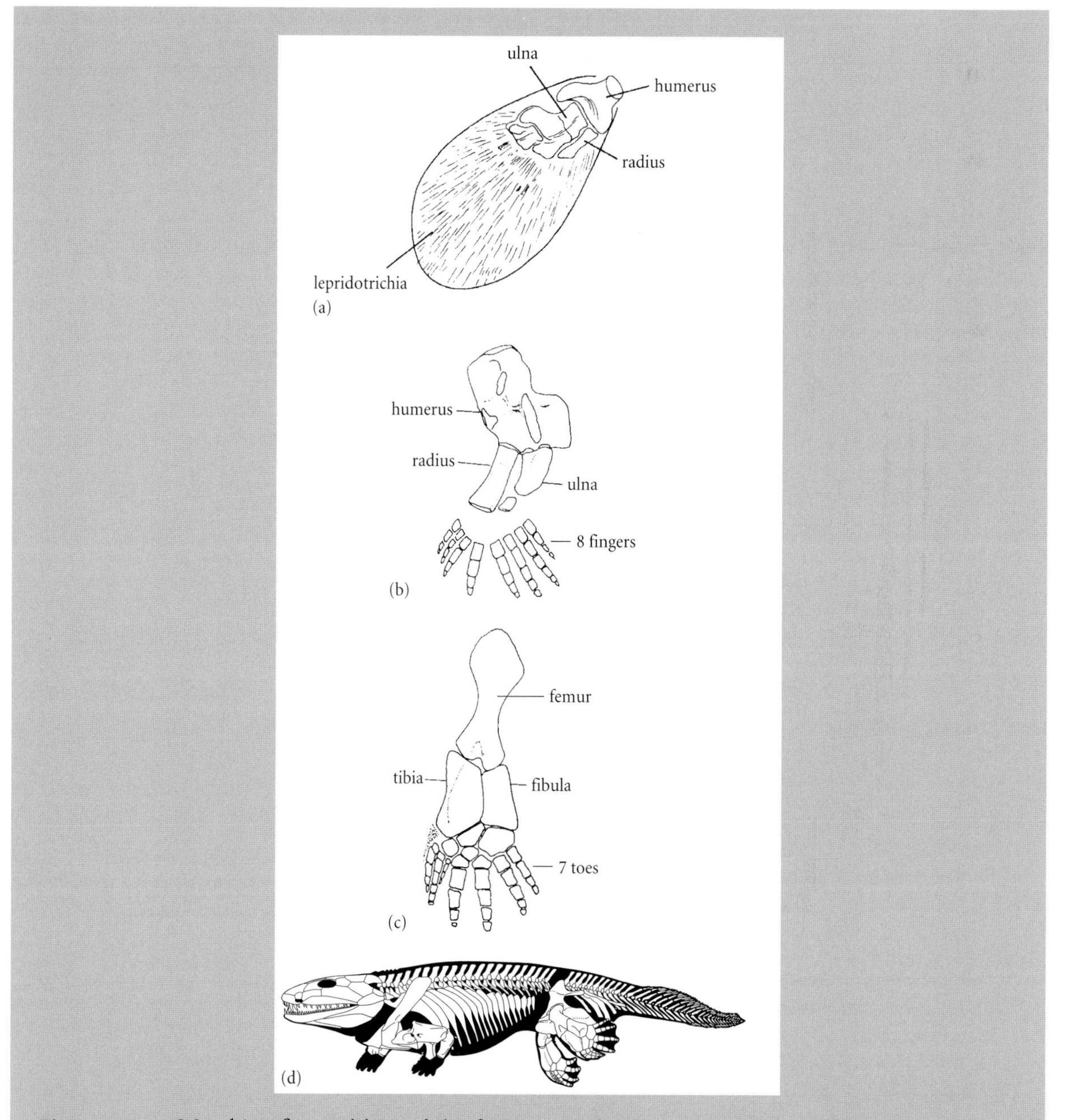

Figure 16.14 Matching fins and legs of the first tetrapods: the pectoral fin of the Devonian sarcopterygian fish *Eusthenopteron* (a) shows bones that are probable homologs of tetrapod arm bones, such as in the Devonian amphibian *Acanthostega* (b). *Acanthostega* had eight fingers and *Ichthyostega* had seven toes on its hindlimb (c). (d) The early tetrapod *Acanthostega*. (Courtesy of Mike Coates.)

Reptiles and their descendants made a clean break from the water by producing a kind of egg that did not have to be laid in water.

The **cleidoic** ("closed") egg, sometimes called the amniotic egg, is enclosed within a tough semipermeable shell, hence its name. This egg type is seen in all members of the clade Amniota (reptiles, birds, mammals): it is the familiar hen's egg that we eat for breakfast. Primitive mammals, such as the platypus, still lay eggs, but most mammals have suppressed the egg and it "hatches" inside the

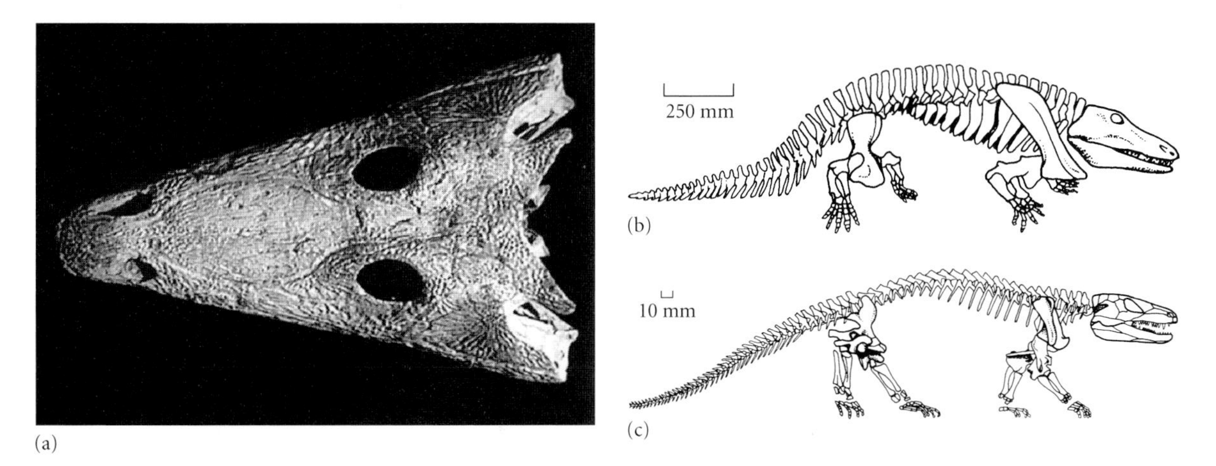

Figure 16.15 Fossil amphibians: (a) skull of the Early Triassic temnospondyl *Benthosuchus*; (b) skeleton of the Early Permian temnospondyl *Eryops*; and (c) skeleton of the Early Permian reptiliomorph *Seymouria*. (a, courtesy of Mikhail Shishkin; b, c, based on Gregory 1951/1957.)

Box 16.8 Classification of amphibians

The amphibians are a paraphyletic group as they exclude their descendants, the reptiles. Modern amphibians are clearly distinguishable from the diverse fossil groups.

Superclass TETRAPODA
Subclass BATRACHOMORPHA (Amphibia)
Order "TEMNOSPONDYLI"

- Broad-snouted, low-skulled amphibians, showing a range of sizes
- Early Carboniferous to Early Cretaceous

Infraclass LISSAMPHIBIA

- Frogs, salamanders (newts) and caecilians (gymnophionans)
- Early Triassic to Recent

Order NECTRIDEA

- Small slender aquatic forms
- Early Carboniferous to Late Permian

Order MICROSAURIA

- Terrestrial and aquatic long-bodied forms with deep skulls
- Early Carboniferous to Early Permian

Subclass REPTILIOMORPHA
Order ANTHRACOSAURIA

- Narrow-skulled, fish-eating amphibians
- Early Carboniferous to Late Permian

Order SEYMOURIAMORPHA

- High-skulled terrestrial amphibians
- Late Carboniferous to Late Permian

mother's womb. The amniotic eggshell is usually hard and made from calcite, but some lizards and snakes have leathery eggshells. The shell retains water, preventing evaporation, but allows the passage of gases, oxygen in and carbon dioxide out. The developing embryo is protected from the outside world, and there is no need to lay the eggs in water, nor is there a larval stage in development. Inside the eggshell is a set of membranes that enclose the embryo (the **amnion**), that collect waste (the **allantois**) and that line the eggshell (the **chorion**) (Fig. 16.16). The chorion is the thin papery tissue just inside the eggshell, which you peel off a hard-boiled egg. Food is in the form of **yolk**, a yellow material rich in protein.

The oldest-known amniote, *Hylonomus* from the mid-Carboniferous of Canada (Fig. 16.17a, b) is known only from its skeleton; no amniotic eggs are known from the Carboniferous. *Hylonomus* has been superbly well preserved inside ancient tree stumps, into which it crawled in pursuit of insects and worms, and then was overwhelmed by floodwaters. *Hylonomus* looks little different from some amphibians of the time, such as the microsaurs, but it shows several clearly amniote characters, a high skull, evidence for

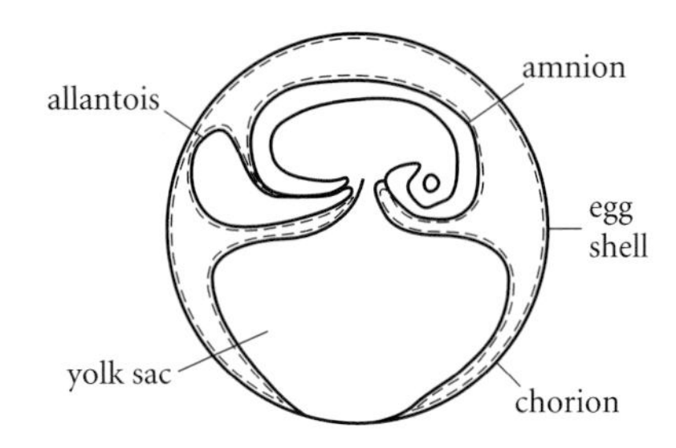

Figure 16.16 The cleidoic egg of amniotes in cross-section, showing the eggshell and extra-embryonic membranes.

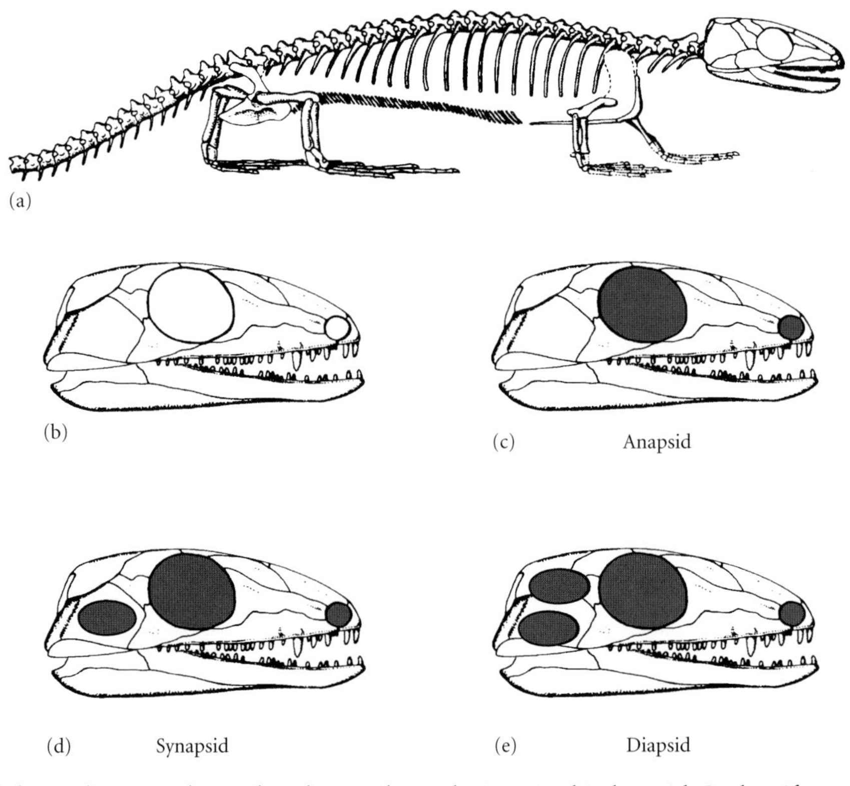

Figure 16.17 The earliest reptile, and early reptile evolution: (a, b) the mid-Carboniferous reptile *Hylonomus*, skeleton and skull; (c-e) the three major skull patterns seen in amniotes: anapsid, diapsid and synapsid. (Based on Carroll 1987.)

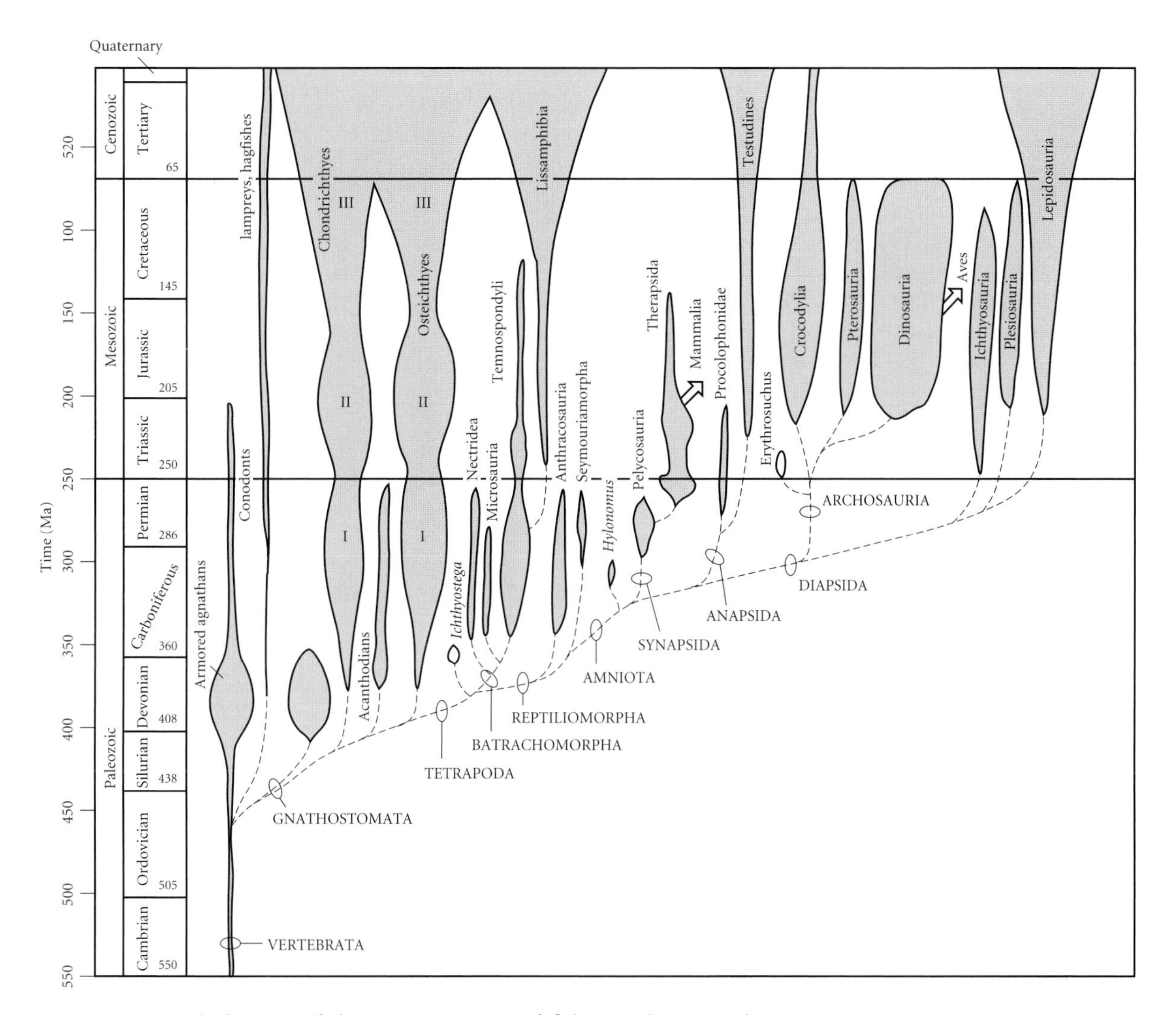

Figure 16.18 Phylogeny of the major groups of fishes and tetrapods.

additional jaw muscles and an **astragalus** bone in the ankle. We know *Hylonomus* laid cleidoic eggs because of the shape of the evolutionary tree: how can that be?

Amniotes radiated during the Late Carboniferous, giving rise to three main clades. These are distinguished by the pattern of openings in the side of the skull, especially the **temporal openings** behind the eye socket (Fig. 16.17c–e). The primitive state is termed the **anapsid** ("no arch") skull pattern, since there are no temporal openings. The two other skull patterns seen in amniotes are the **synapsid** ("same arch"), where there is a lower temporal opening, and the **diapsid** ("two arch") pattern, where there are two temporal openings. These temporal openings correspond to low-stress areas of the skull, and the edges serve as attachment sites for jaw muscles.

These three skull patterns diagnose the key clades among amniotes (Fig. 16.18; Box 16.9). The Anapsida include various early forms such as *Hylonomus*, as well as some Permian and Triassic reptiles, and the turtles. The Synapsida include the "mammal-like reptiles" and the mammals, and the Diapsida includes a number of early groups, as well as the lizards and snakes and the crocodiles, pterosaurs, dinosaurs and birds. All modern amniotes produce the cleidoic egg, and the structure is so similar that we can be sure this egg arose at the base of clade Amniota; so, *Hylonomus*, clearly located within the evolutionary tree of Amniota, must have shared the cleidoic egg.

Box 16.9 Classification of the reptiles

Reptiles are a paraphyletic group of Amniota that excludes the reptile descendants, birds and mammals. The basic classification is founded on skull pattern (Fig. 16.18).

Class REPTILIA

Subclass ANAPSIDA: no temporal openings
- Various basal groups, such as procolophonids
- Permian to Triassic

Order TESTUDINES (Chelonia)

- Turtles; bony carapace, retractable neck and limbs
- Late Triassic to Recent

Subclass SYNAPSIDA: one (lower) temporal opening
Order "PELYCOSAURIA"

- The sail-backed "mammal-like reptiles" and relatives
- Late Carboniferous to Early Permian

Order THERAPSIDA

- Synapsids with differentiated teeth, including the mammals
- Late Permian to Recent

Subclass DIAPSIDA: two temporal openings
Infraclass ARCHOSAURIA

- Thecodontians, crocodilians, pterosaurs, dinosaurs and birds; characterized by an antorbital fenestra
- Late Permian to Recent

Infraclass LEPIDOSAURIA

- Lizards, snakes and their ancestors
- Late Triassic to Recent

The Anapsida: turtles and relatives

The oldest anapsids were small insect eaters. During the Permian and Triassic, some unusual anapsids came on the scene. The most diverse of these were the procolophonids (Fig. 16.19a), small animals with triangular skulls and broad teeth adapted to a diet of tough plants and insects.

The turtles appeared first in the Late Triassic, being represented by *Proganochelys* (Fig. 16.19b). Modern turtles have no teeth, but *Proganochelys* still had some on its palate. The skull is solid, and the body is covered above and below by a bony shell. Turtles live on land, in ponds (Fig. 16.19c) and in the sea. Some marine turtles of the Cretaceous reached 3 m in length.

Our current understanding of amniote evolution (see Fig. 16.18) has been challenged by new molecular studies that suggest turtles might in fact be modified diapsid reptiles; if this is so, and it is still controversial (Lee et al. 2004), then the clade Anapsida might no longer have any meaning.

A world of synapsids

The first synapsids, known from the Late Carboniferous and Early Permian, are grouped loosely as "pelycosaurs". Most of these were small- to medium-sized insectivores and carnivores with powerful skulls and sharp, flesh-piercing teeth. Some later pelycosaurs, like *Dimetrodon* (Fig. 16.20a), had vast sails supported on vertical spines growing up from the

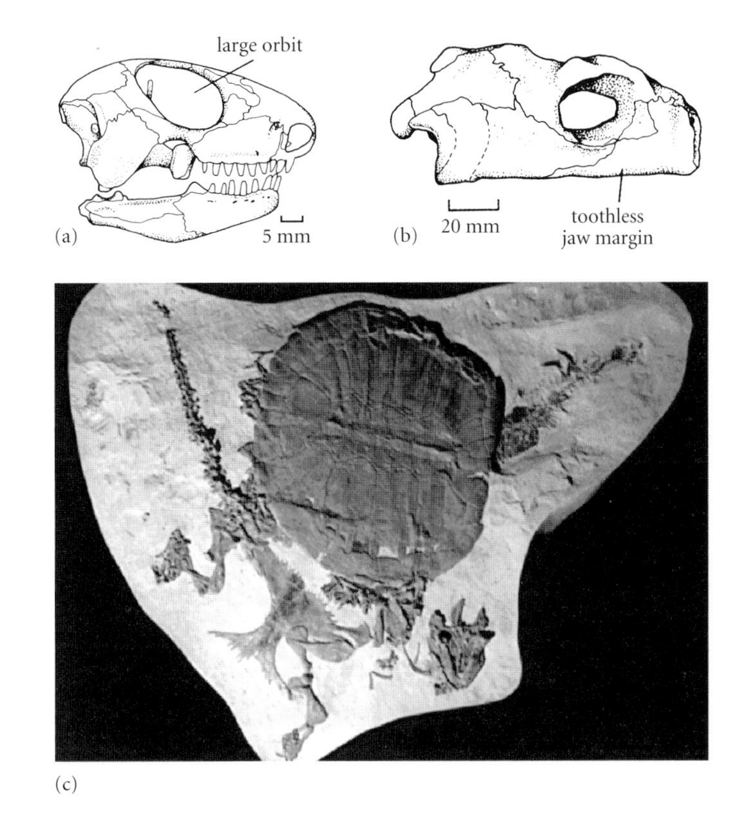

Figure 16.19 Fossil and recent anapsid reptiles: (a) skull of the Triassic procolophonid *Procolophon*; (b) skull of the Triassic turtle *Proganochelys*; (c) a fossilized snapping turtle, with the head (bottom right) and skeleton separated from the carapace, from pond sediments filling an impact crater at Steinheim, Germany. (a, based on Carroll & Lindsay 1985; b, based on Gaffney & Meeker 1983.)

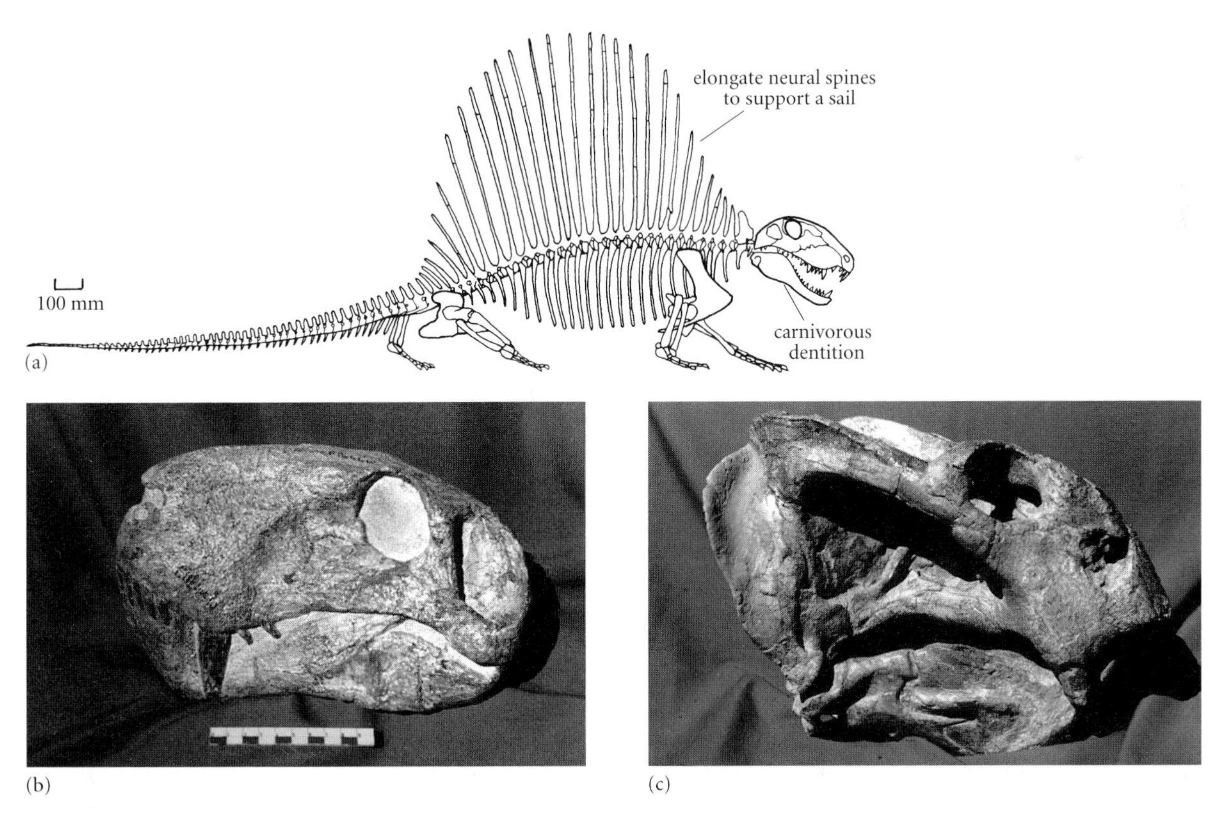

Figure 16.20 Synapsids of the Permian: (a) the carnivorous pelycosaur *Dimetrodon*; (b) the carnivorous gorgonopsian *Lycaenops*; and (c) the herbivorous dicynodont *Dicynodon*. (a, based on Gregory 1951/1957; b, c, courtesy of Gillian King.)

vertebrae, perhaps used in controlling body temperature. The pelycosaurs also include a number of groups that adapted to plant eating, among the first herbivorous land vertebrates.

Synapsids radiated dramatically in the Late Permian as a new clade, the Therapsida. The most astonishing carnivores were the gorgonopsians (Fig. 16.20b) with their large, wolf-like bodies and massive saber teeth that they probably used to attack the larger thick-skinned herbivores. The dicynodonts had bodies shaped like overstuffed sausages, and no teeth at all, or only two tusks (Fig. 16.20c). They were successful herbivores and some of the first animals to have a complex chewing cycle that allowed them to tackle a wide variety of plant foods. Late Permian therapsids are common in the continental sediments of the Karoo Basin in South Africa and the Urals in Russia. At the end of the Permian, at the time of a major mass extinction in the sea (see p. 170), most of these animals died out. The gorgonopsians disappeared and the dicynodonts were nearly wiped out – extinction on a huge scale.

The cynodonts were an important Triassic synapsid group. The Early Triassic form *Thrinaxodon* (Fig. 16.21a) looked dog-like. In the snout area of the skull, there are numerous small canals that indicate small nerves serving the roots of sensory whiskers. If *Thrinaxodon* had whiskers, it clearly also had the potential for hair on other parts of its body, and this implies insulation and temperature control. Cynodonts evolved along several lines during the Triassic, and gave rise to mammals, such as *Megazostrodon* (Fig. 16.21b), in the Late Triassic and Early Jurassic.

The transition from basal synapsid to mammal is marked by an extraordinary shift of the jaw joint into the middle ear. Reptiles typically have six bones in the lower jaw and the articular bone articulates with the quadrate in the skull (Fig. 16.21c). In mammals, on the other hand, there is a single bone in the lower jaw, the dentary, which articulates with the squamosal (Fig. 16.21d). The reptilian articular–quadrate jaw joint became reduced in Triassic cynodonts, and moved into the middle ear passage. That is why we have three tiny ear bones, the hammer, anvil and stirrup, which transmit sound from the ear drum to the brain, while reptiles have only one, the stirrup or stapes.

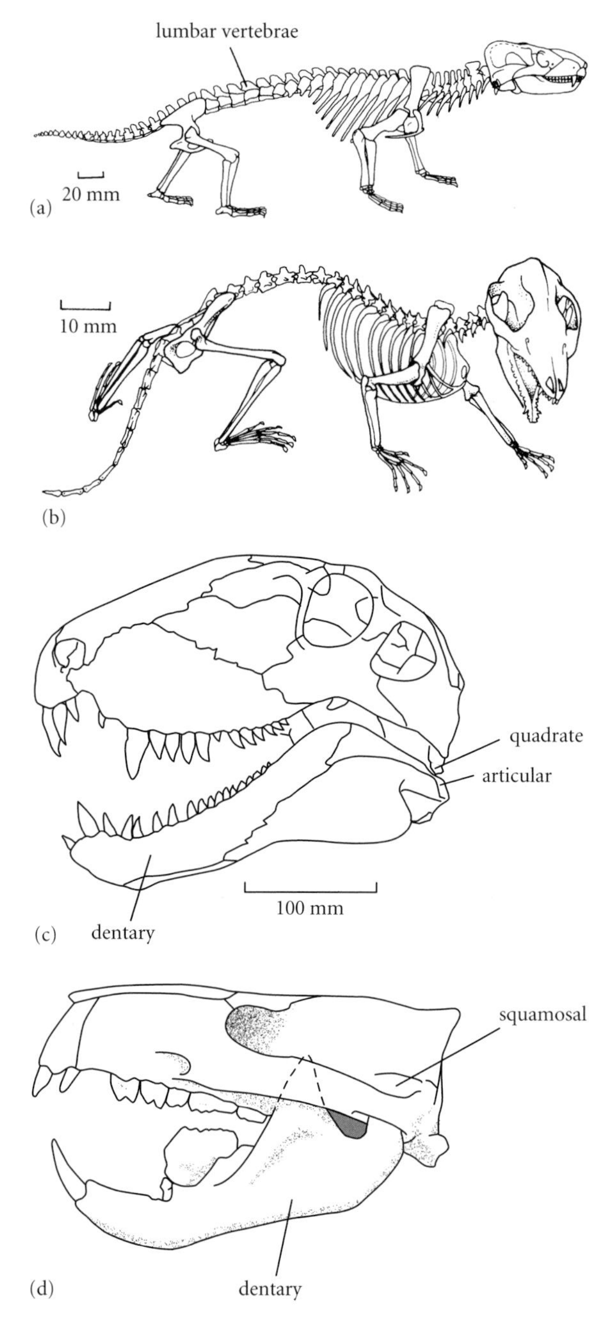

Figure 16.21 Transition to the mammals: (a) the Early Triassic cynodont *Thrinaxodon*; (b) the Early Jurassic mammal *Megazostrodon*; and (c, d) skulls of an early synapsid (c) and a mammal (d) to show the reduction in elements in the lower jaw and switch of the jaw joint. (a, based on Jenkins 1971; b, based on Jenkins & Parrington 1976; c, d, based on Gregory 1951/1957.)

Dinosaurs and mammals

People usually think of the dinosaurs as precursors of the mammals. Dinosaurs famously ruled the Earth for 160 Myr of the Mesozoic, and then were replaced by the mammals 65 Ma. However, as we have seen, the mammals arose in the Late Triassic, about the same time as the first dinosaurs. So, both groups evolved side by side through the Late Triassic, Jurassic and Cretaceous – the dinosaurs as large to very large beasts, and the mammals generally scuttling unobtrusively through the undergrowth. Our understanding of how both groups evolved has changed enormously in recent years, and this is discussed in Chapter 17.

Review questions

1 How has the application of cladistics (see p. 129) affected our ideas about basal vertebrate phylogeny? Look at older books and papers, and compile simple trees of Agnatha, Placodermi, Chondrichthyes, Acanthodii and Osteichthyes as accepted in 1960, 1970, 1980, 1990 and 2000.
2 Read about the typical Devonian fishes of either the Orcadian Basin in Scotland or Miguasha in Canada, and attempt to reconstruct a food web (see p. 88): what eats what?
3 How has the discovery of seven- and eight-fingered tetrapod fossils from the Late Devonian changed our views about the development of fingers and toes? Read about older and newer views on development (embryology) of limb buds and digits, and find out about the *Hox* genes (see p. 148).
4 How did global environments change through the Carboniferous and Permian, and how did this affect tetrapod evolution?
5 How did mammal-like characters appear in the synapsids of the Permian and Triassic? Draw up a simple cladogram of 10 key synapsid genera, and mark on the acquisition of key apomorphies.

Further reading

Armstrong, H.A. & Brasier, M. 2004. *Microfossils*, 2nd edn. Blackwell, Oxford. (Chapter on conodonts.)

Benton, M.J. 1991. *The Reign of the Reptiles*. Crescent, New York.

Benton, M.J. 2003. *When Life Nearly Died*. Norton, New York.

Benton, M.J. 2005. *Vertebrate Paleontology*, 3rd edn. Blackwell, Oxford.

Carroll, R.L. 1987. *Vertebrate Paleontology and Evolution*. Freeman, San Francisco.

Clack, J.A. 2002. *Gaining Ground: The Origin and Evolution of Tetrapods*. Indiana University Press, Bloomington, IN.

Cracraft, J. & Donoghue, M.J. 2004. *Assembling the Tree of Life*. Oxford University Press, New York.

Gould, S.J. (ed.) 2001. *The Book of Life*. Norton, New York.

Long, J. 1996. *The Rise of Fishes*. Johns Hopkins University Press, Baltimore.

Shubin, N.H. 2008. *Your Inner Fish*. Pantheon, New York.

Zimmer, C. 1999. *At the Water's Edge*. Touchstone, New York.

References

Aldridge, R.J., Briggs, D.E.G., Smith, M.P. et al. 1993a. The anatomy of conodonts. *Philosophical Transactions of the Royal Society of London B* **340**, 405–21.

Aldridge, R.J., Jeppsson, L. & Dorning, K.J. 1993b. Early Silurian oceanic episodes and events. *Journal of the Geological Society of London* **150**, 501–13.

Alexander, R.McN. 1975. *The Chordates*. Cambridge University Press, Cambridge.

Armstrong, H.A. & Brasier, M. 2004. *Microfossils*, 2nd edn. Blackwell, Oxford.

Carroll, R.L. 1987. *Vertebrate Paleontology and Evolution*. Freeman, San Francisco.

Carroll, R.L. & Lindsay, W. 1985. The cranial anatomy of the primitive reptile *Procolophon*. *Canadian Journal of Earth Sciences* **22**, 1571–87.

Coates, M.I., Jeffery, J.E. & Ruta, M. 2002. Fins to limbs: what the fossils say. *Evolution and Development* **4**, 390–401.

Daeschler, E.B., Shubin, N.H. & Jenkins Jr., F.A. 2006. A Devonian tetrapod-like fish and the evolution of the tetrapod body plan. *Nature* **440**, 757–63.

Donoghue, P.C.J. & Purnell, M.A. 2005. Genome duplication, extinction and vertebrate evolution. *Trends in Ecology and Evolution* **20**, 312–19.

Furlong, R.F. & Holland, P.W.H. 2004. Polyploidy in vertebrate ancestry: Ohno and beyond. *Biological Journal of the Linnean Society* **82**, 425–30.

Gaffney, E.S. & Meeker, L.J. 1983. Skull morphology of the oldest turtles: a preliminary description of *Proganochelys quenstedti*. *Journal of Vertebrate Paleontology* **3**, 25–8.

Gagnier, P.-Y. 1993. *Sacabambaspis janvieri*, vertébré Ordovicien de Bolivie: 1. analyse morphologique. *Annales de Paléontologie* **79**, 19–69.

Grande, L. 1988. A well preserved paracanthopterygian fish (Teleostei) from freshwater lower Paleocene deposits of Montana. *Journal of Vertebrate Paleontology* **8**, 117–30.

Gregory, W.K. 1951/1957. *Evolution Emerging*, Vols 1 and 2. Macmillan, New York.

Hou, X.-G., Aldridge, R.J., Siveter, D.J. et al. 2002. New evidence on the anatomy and phylogeny of the earliest vertebrates. *Proceedings of the Royal Society B* **269**, 1865–9.

Jenkins Jr., F.A., 1971. The postcranial skeleton of African cynodonts. *Bulletin of the Peabody Museum of Natural History* **36**, 1–216.

Jenkins Jr., F.A. & Parrington, F.R. 1976. The postcranial skeletons of the Triassic mammals *Eozostrodon*, *Megazostrodon* and *Erythrotherium*. *Philosophical Transactions of the Royal Society B* **173**, 387–431.

Lee, M.S.Y., Reeder, T.W., Slowinski, J.B. & Lawson, R. 2004. Resolving reptile relationships: molecular and morphological markers. *In* J. Cracraft & M.J. Donoghue (eds) *Assembling the Tree of Life*. Oxford University Press, Oxford, pp. 451–67.

Moy-Thomas, J.A. & Miles, R.S. 1971. *Palaeozoic Fishes*, 2nd edn. Chapman and Hall, London.

Shu, D.-G., Luo, H.L., Conway Morris, S. et al. 1999. Lower Cambrian vertebrates from South China. *Nature* **402**, 42–6.

Shubin, N.H., Daeschler, E.B. & Jenkins Jr., F.A. 2006. The pectoral fin of *Tiktaalik roseae* and the origin of the tetrapod limb. *Nature* **440**, 764–70.

Sweet, W.C. & Donoghue, P.C.J. 2001. Conodonts: past, present and future. *Journal of Paleontology* **75**, 1174–84.

Trewin, N.H. 1986. Palaeoecology and sedimentology of the Achanarras fish bed of the Middle Old Red Sandstone, Scotland. *Transactions of the Royal Society of Edinburgh: Earth Sciences* **77**, 21–46.

Chapter 17

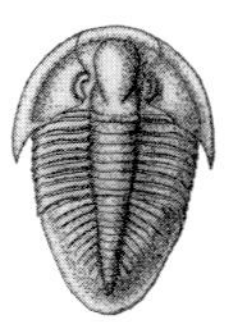

Dinosaurs and mammals

Key points

- After the end-Permian mass extinction event 251 Ma, diapsid reptiles diversified in the Triassic.
- Dinosaurs were a hugely successful group for 160 myr of the Mesozoic.
- Pterosaurs were key Mesozoic flyers, and the most important marine reptiles were the plesiosaurs and ichthyosaurs.
- Birds evolved from dinosaurs, and radiated particularly during the Tertiary.
- The first mammals were small insect eaters of the latest Triassic; the mammals achieved great diversity and abundance only after the extinction of the dinosaurs.
- Molecular and paleontological evidence show that modern mammals radiated in the Late Jurassic and Early Cretaceous, and that some basic splits were geographic – with major clades separated in South America, Africa, Australasia and the northern hemisphere.
- Humans arose 6–8 Ma, and fossil evidence points to repeated human migrations out of Africa.

The dinosaur's eloquent lesson is that if some bigness is good, an overabundance of bigness is not necessarily better.

Eric Johnston, President of the US Chamber of Commerce (1896–1963)

True; not true. Dinosaurs were big – and some were very big. To politicians, the word "dinosaur" is often a term of abuse, hurled at their enemies to characterize them as old-fashioned, over-blown, past it. Paleontologists (and 4-year-old kids) know better, since dinosaurs were of course one of the most successful animal groups of all time. Nonetheless, they were certainly big, and yet in their day large size was clearly a great advantage for them: after all, dinosaurs and mammals, the subject of this chapter, both arose at the same time, in the Late Triassic, and yet dinosaurs somehow dominated ecosystems in terms of diversity and size, and kept our hairy little ancestors on the fringes.

The tetrapods moved onto land some 380 Ma in the Devonian, and they diversified and occupied more and more ecospace through the Carboniferous and Permian (see pp. 442–52). The most successful tetrapod clade, the Amniota, became most diverse by the end of the Paleozoic, dominating many terrestrial habitats. Two amniote groups in particular rose to prominence. First were the synapsids, which dominated Permian lands but were then hit very hard by the end-Permian mass extinction (see p. 170). The synapsids recovered and gave rise to the mammals in the Late Triassic. The second major amniote group, the diapsids, were minor components of Permian ecosystems, but they diversified in the Triassic, giving rise to the dinosaurs in the Late Triassic. Perhaps the devastating end-Permian mass extinction gave diapsids, and especially the dinosaurs, their chance to diversify.

In this chapter, we take the evolution of vertebrates forward into the Mesozoic and Cenozoic. We explore first the rise of the diapsids, and especially the dinosaurs and their descendants, the birds. Then, we look at how the modest evolution of mammals through the Mesozoic set the scene for their explosive radiation at the beginning of the Cenozoic, after the dinosaurs had gone.

DINOSAURS AND THEIR KIN

The diapsids take over

The diapsids (see p. 447) were initially small to medium-sized carnivores that never matched the abundance of the synapsids in the Carboniferous or Permian. Things began to change during the Triassic, perhaps as a result of the end-Permian extinction event, which had such a devastating effect on therapsid communities. Small and large meat eaters such as *Erythrosuchus* (Fig. 17.1a) appeared, one of the first of the archosaurs, a group that was later to include the dinosaurs, pterosaurs, crocodilians and birds. Archosaurs are characterized by an additional skull opening between the orbit and the naris, termed the **antorbital fenestra**, whose function is unclear.

During the Triassic, diapsids diversified widely, some on land and some in the sea. Some archosaurs became large carnivores, others became specialized fish eaters, others adopted a specialized grubbing herbivorous lifestyle, yet others were small, two-limbed, fast-moving insectivores (crocodilians and dinosaurs), and some became proficient flyers (pterosaurs). It took another extinction event, near the beginning of the Late Triassic (about 220 Ma) to set the new age of diapsids fully in motion. Most of the synapsids died out then, as did various basal archosaur groups. Many new kinds of land tetrapods then radiated: the dinosaurs, pterosaurs, crocodilians and lizard ancestors, as well as the turtles, modern amphibians and true mammals.

The pterosaurs were proficient flapping flyers (Fig. 17.1b), with a lightweight body, narrow hatchet-shaped skull and a long narrow wing supported on a spectacularly elongated fourth finger of the hand. The bones of the arm and finger supported a tough flexible membrane that could fold away when the animal was at rest, and stretch out for flight. Pterosaurs were covered with hair, and were almost certainly **endothermic**. Some later pterosaurs were much larger than any known bird, such as *Pteranodon* with a wingspan of 5–8 m, and *Quetzalcoatlus* with a wingspan of 11–15 m. Most pterosaurs fed on fishes caught in coastal seas, but others were insectivorous.

Early crocodilians were largely terrestrial in habits, walked on all fours and had an extensive armor of bony plates (Fig. 17.1c). Crocodilians were more diverse and abundant during the Jurassic and Cretaceous than they are now. Some even became fully marine in adaptations, to the extent of having paddles instead of hands and feet, and a deep tail fin

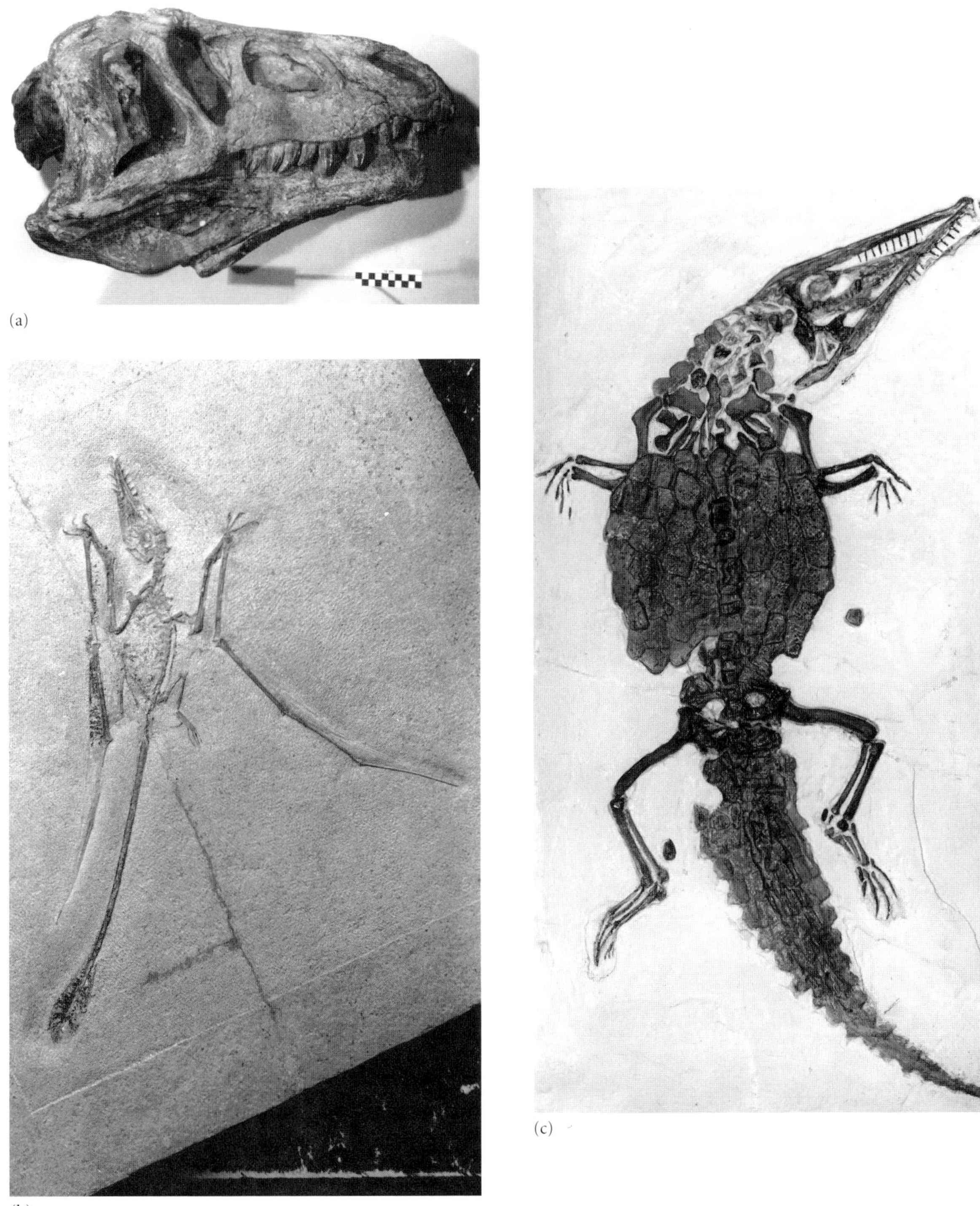

Figure 17.1 Archosaurs: (a) skull of the Early Triassic archosaur *Erythrosuchus* (×0.1); (b) the Late Jurassic pterosaur *Rhamphorhynchus*, showing the elongated wing finger on each side, and the long tail with its terminal "sail" made from skin (×0.3); and (c) the Late Jurassic crocodilian *Crocodilemus*, showing the skeleton and armor covering (×0.2). (Courtesy of David Unwin and Danny Grange.)

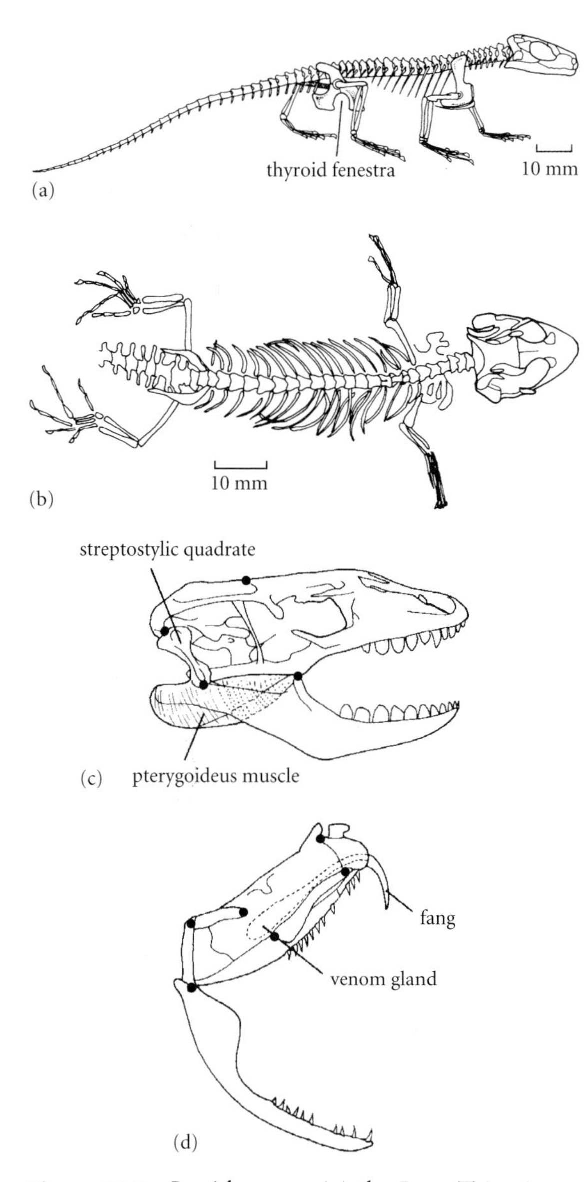

Figure 17.2 Lepidosaurs: (a) the Late Triassic sphenodontid *Planocephalosaurus*; (b) the Late Jurassic lizard *Ardeosaurus*; and (c, d) skulls of a modern lizard (c) and snake (d), showing the points of mobility that permit wide jaw opening. (a, based on Fraser & Walkden 1984; b, based on Estes 1983.)

to speed their swimming. The modern crocodilians – crocodiles, alligators and gavials – all arose in the Late Cretaceous.

The second major diapsid clade, the lepidosaurs, represented today by lizards and snakes, diversified in the Late Triassic. The key forms then were sphenodontids – snub-nosed, lizard-sized animals (Fig. 17.2a) that fed on plants and insects. The group dwindled after the Jurassic, except for a single living representative, *Sphenodon*, the tuatara of New Zealand, a famous "living fossil". The first true lizards are known from the Mid and Late Jurassic (Fig. 17.2b), and they show characteristic mobility of the skull: the bar beneath the lower temporal opening is broken, the quadrate is mobile, and the snout portion of the skull can tilt up and down (Fig. 17.2c). This process of loosening of the skull was taken even further in the snakes, a group known first in the Early Cretaceous. Snakes have such mobile skulls that they can open their jaws to swallow prey animals that are several times the diameter of the head (Fig. 17.2d).

The age of dinosaurs

Dinosaurs were the most important of the new diapsid groups of the Triassic, both in terms of their abundance and diversity, and in terms of the vast size reached by some of them. The first dinosaurs were modest-sized bipedal carnivores. After the Late Triassic extinction, a new group of herbivorous dinosaurs, the sauropodomorphs, radiated dramatically, some like *Plateosaurus* (Fig. 17.3a) reaching a length of 5–10 m during the Late Triassic. Later sauropodomorphs were mainly large and very large animals; some of them, such as *Brachiosaurus* (Fig. 17.3b) reaching lengths of 23 m or more and heights of 12 m. These giant dinosaurs pose fascinating biological problems (Box 17.1).

The theropods include all the carnivorous dinosaurs, and in the Jurassic and Cretaceous the group diversified to include many specialized small and large forms. *Deinonychus* (Fig. 17.5a) was human-sized, but immensely agile and intelligent (it had a bird-sized brain). Its key feature was a huge claw on its hindfoot, which it almost certainly used to slash at prey animals. *Tyrannosaurus* (Fig. 17.5b) is famous as probably the largest land predator of all time, reaching a body length of 14 m, and having a gape of nearly 1 m. The theropods and sauropodomorphs share the primitive reptilian hip pattern, in which the two lower elements point in opposite directions, the **pubis** forwards and the **ischium** backwards (Fig. 17.5b). They also share derived charac-

(a)

(b)

Figure 17.3 Sauropodomorph dinosaurs: (a) the Late Triassic prosauropod *Plateosaurus*; and (b) the Late Jurassic sauropod *Brachiosaurus*. (Courtesy of David Weishampel.)

ters of the skull and limbs that show they form a clade, the Saurischia.

All other dinosaurs share a unique hip pattern in which the pubis has swung back and runs parallel to the ischium (Fig. 17.6a), and these are termed the Ornithischia, all of which were herbivores. Two groups of armored ornithischians are the stegosaurs and the ankylosaurs. *Stegosaurus* (Fig. 17.6a) has a row of bony plates along the middle of its back that may have had a temperature control or display function. *Euoplocephalus* (Fig. 17.6b) is a massive, tank-like animal with a solid armor of small plates of bone set in the skin over its back, tail, neck and skull: it even had a bony eyelid. The tail club was a useful defensive weapon that it used to whack threatening predators such as *Tyrannosaurus*.

Most ornithischians were ornithopods, bipedal forms, initially small, but later often large. In the Late Cretaceous, the hadrosaurs were successful fast-moving plant eaters. Many of them have bizarre crests on top of their heads that may have been used for species-specific signaling, and their duck-billed jaws are lined by multiple rows of grinding teeth (Fig. 17.7). Close relatives of the ornithopods were the ceratopsians ("horn-faces"), like *Centrosaurus* (Fig. 17.6c), which had a single, long nose-horn and a great bony frill over the neck.

There has been a continuing debate about whether the dinosaurs were warm-blooded (endothermic) or not. Evidence for warm-bloodedness is strongest for the small active predators like *Deinonychus* that might have required the added stamina and speed. However, endothermy is costly in terms of the extra food required as fuel, and it is not clear whether the larger dinosaurs could have eaten fast enough. Indeed, larger dinosaurs would have maintained a fairly constant core body temperature simply because of their size, whether they were endothermic or not.

Dinosaur reproductive habits have also come under scrutiny recently. Discoveries of eggs and nests in North America and Mongolia have shown that many dinosaurs practiced parental care. They laid their eggs in earth nests scooped in the soil, and returned to feed the young when they hatched out. Some of the most spectacular finds are unhatched eggs with the tiny bones of the dinosaur embryos still inside (Box 17.2).

Dragons of the deeps

During the Mesozoic, several reptile groups became key marine predators. The

Box 17.1 Paleobiology of the largest animals ever

When the monster sauropods of the Late Jurassic were first discovered in the 19th century, many paleontologists thought that they were too big to have lived fully on land. It was assumed that the sauropods lived in lakes, supporting their bulk in the water, and feeding on waterside plants. New evidence shows, however, that life on land was quite possible, and sauropods, like elephants today, could move freely over vast plains, and in and out of the water at times as well.

Sauropods may have divided their feeding preferences by height. Modern herbivores show such **niche partitioning**, where each animal has its preferred food and feeding mode. Some sauropods such as *Brachiosaurus* (Fig. 17.3b) may have been like super giraffes, feeding on leaves from very tall trees. Most other sauropods, however, were designed for lower-level browsing, and were probably not able to raise their necks much above horizontal. This allowed several species of sauropods to live side by side, some feeding on low plants, others on mid-height shrubs, and yet others on the leaves of trees.

But how did sauropods get to be so big? *Brachiosaurus* and relatives reached lengths of 20 m or more, and some weighed as much as 50 tonnes. Did they take 70 or 100 years to reach sexual maturity, growing at the same rate as a modern crocodile, as some paleobiologists have suggested? Or did they grow fast, like modern mammals? An elephant reaches sexual maturity at about 15 years old, while a blue whale grows even faster, reaching sexual maturity at 5–10 years. Blue whales can put on as much as 90 kg per day (equivalent to 30 tonnes per year) during their fastest juvenile growth. How can you tell how fast a dinosaur grew?

The secret is in the bones. Modern reptiles grow in fits and starts – bursts of fast growth when food is plentiful, and very slow growth when they are starved. Typically, there is one good season and one poor season each year, and this is shown in growth rings in the bone. Greg Erickson, a paleontologist at Florida State University in Tallahassee, and Martin Sander, a paleontologist from the University of Bonn in Germany, have sectioned the ribs and leg bones of many dinosaurs, and they have counted the growth lines or "lines of arrested growth" (LAGs). The bones are cut through, and thin sections are glued on to large glass slides, and then ground down to a uniform thickness, so light can pass through. Erickson and Sander have examined these thin sections under a polarizing light microscope that highlights the crystalline apatite components of the fossil bone and allows them to count the LAGs easily (Fig. 17.4a).

Bone histologists, people who study the microscopic structure of bone, argue that each LAG represents a year because this is the case in living reptiles and other animals that have them. Paleontologists can count the LAGs to give the age when a dinosaur died, and then compare these ages with estimates of body masses assessed from the size of the bones. Erickson has looked at a growth series of bones, from tiny (well actually quite big) baby *Apatosaurus* through juveniles to fully adult specimens. These suggest that *Apatosaurus* juveniles stayed pretty small for the first 5 years of their life, and then there was a burst of growth from age 5 to 12, when they put on up to 5 tonnes a year, to reach a young adult body mass of 26 tonnes, presumably the age of sexual maturity (Fig. 17.4b). So the LAGs indicate stop–start seasonal patterns of growth, just like a modern reptile, but the rate of juvenile growth is much more like a bird or mammal. Sauropods did not have to wait until they were 100 years old before they could have sex.

Read more about sauropod bone histology in Erickson et al. (2001) and Sander and Klein (2005) and Sander et al. (2006) and find a selection of the best dinosaur web sites at http://www.blackwellpublishing.com/paleobiology/.

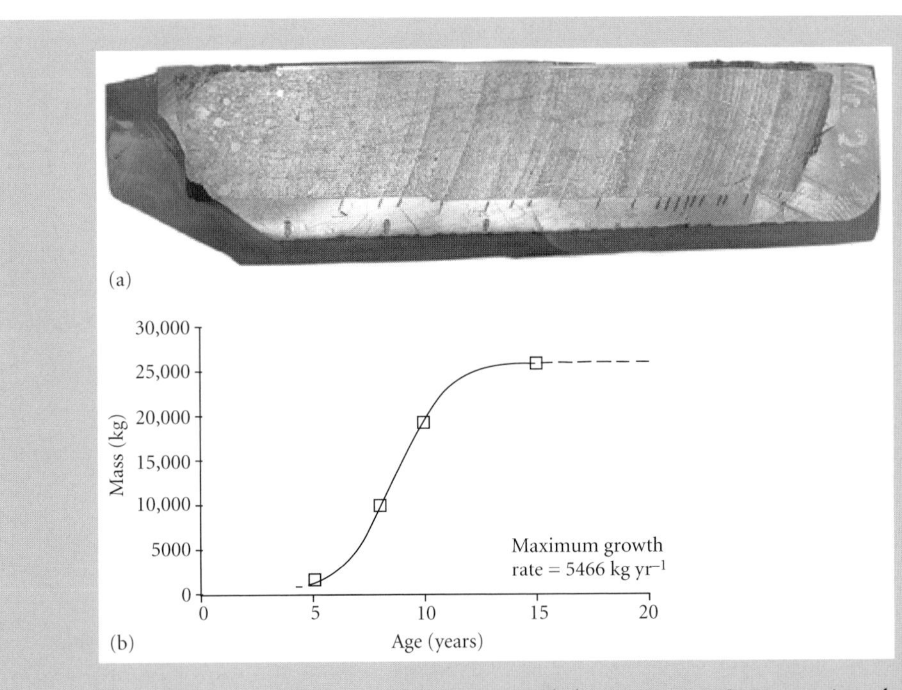

Figure 17.4 Measuring the growth rate of a sauropod dinosaur. (a) Cross-section through the bone wall of the femur of the sauropod *Janenschia* from the Late Jurassic of Tanzania; the animal was full grown and the femur was 1.27 m long. The section was made by drilling into the bone and extracting a core that was then cut through; the center of the bone is to the left, the outside to the right. Lines of arrested growth are the darker bands, where the bone structure is tighter, indicating a slow-down in growth. These are marked off with tick marks on the side of the slide. (b) Growth curve for the sauropod *Apatosaurus* based on sections from the limb bones and ribs of several individuals, juveniles and adults, showing how the animal reached adult size with a spurt of growth from years 5 to 12. (Courtesy of Martin Sander and Greg Erickson.)

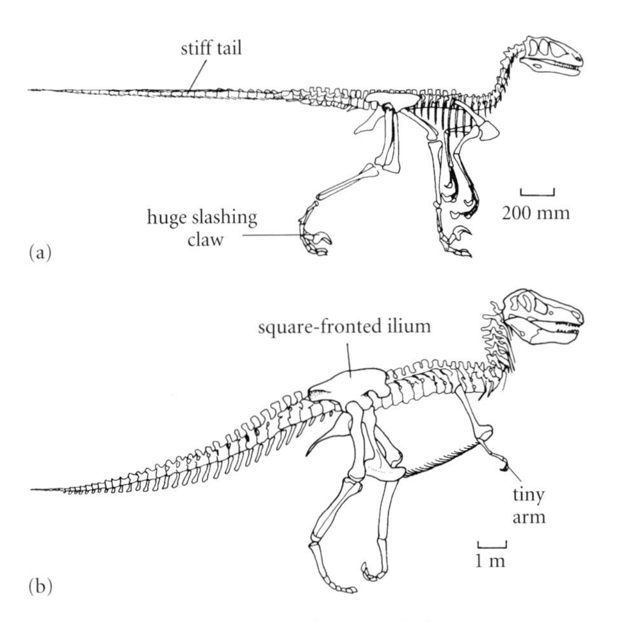

Figure 17.5 Cretaceous theropod dinosaurs: (a) *Deinonychus*, and (b) *Tyrannosaurus*. (a, based on Ostrom 1969; b, based on Newman 1970.)

ichthyosaurs (Fig. 17.9a) were fish-shaped animals, entirely adapted to life in the sea, and they evolved from land-living diapsids. Ichthyosaurs had long, thin, snouts lined with sharp teeth, and they fed on ammonites, belemnites and fishes. Exquisite preservation of many specimens shows the tail fin, dorsal fin and the paddle outlines. Ichthyosaurs swam by beating the body and tail from side to side, and they used the front paddles for steering. There are even some remarkable specimens of mothers with developing embryos inside their bellies: like whales and dolphins, ichthyosaurs could not flop up on to land to lay eggs, and they gave birth to live young while at sea.

The second major marine reptile group was the plesiosaurs. Most plesiosaurs had long necks and small heads (Fig. 17.9b), but the pliosaurs were larger and had short necks and large heads. Plesiosaurs fed mainly on fishes, using their long neck like a snake to dart after

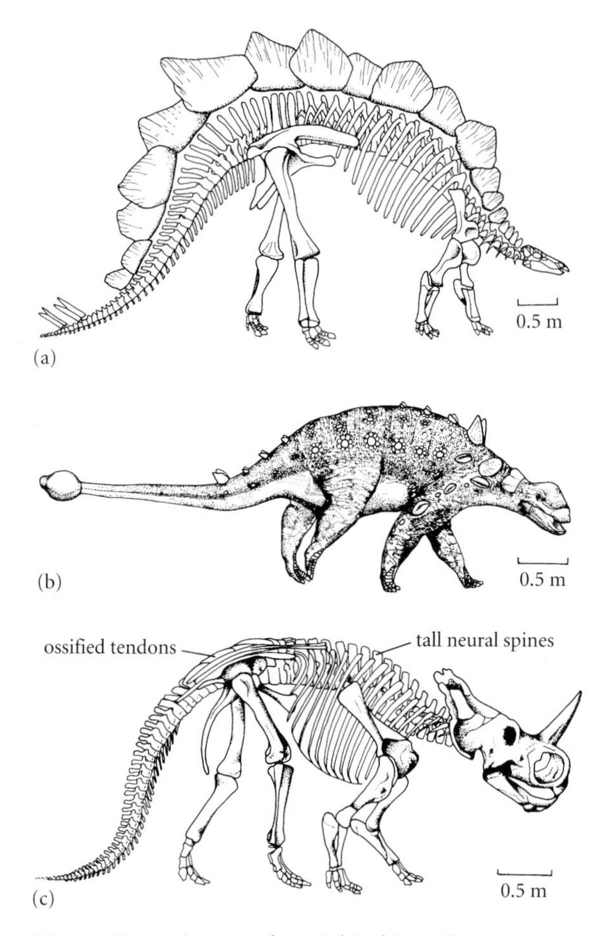

Figure 17.6 Armored ornithischian dinosaurs from the Jurassic (a) and Cretaceous (b, c): (a) *Stegosaurus*, (b) *Euoplocephalus*, and (c) *Centrosaurus*. (a, c, based on Gregory 1951; b, based on Carpenter 1982.)

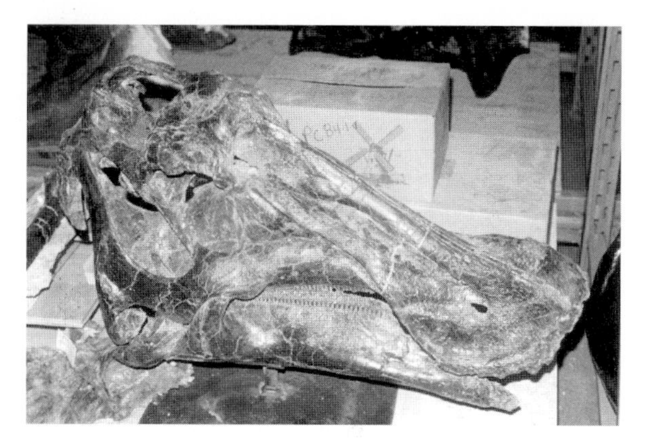

Figure 17.7 Skull of the Late Cretaceous hadrosaur *Edmontosaurus*.

fast-moving prey, and they swam by beating their paddles in a kind of "flying" motion. The extraordinary diversity of tetrapod predators in the sea came to an end 65 Ma during the great Cretaceous-Tertiary mass extinction (see pp. 174–7) that saw the end of the dinosaurs and pterosaurs too.

BIRD EVOLUTION

One of the most famous fossils is *Archaeopteryx*, the oldest known bird (Fig. 17.10). The first specimen was found in Upper Jurassic sediments in southern Germany in 1861, and was hailed as the ideal "missing link" or proof of evolution in action. Here was an animal with a beak, wings and feathers, so it was clearly a bird, but it still had a reptilian bony tail, claws on the hand, and teeth. Since 1861, nine more skeletons have come to light, the last two in 1992 and 2005.

Archaeopteryx was about the size of a magpie, and it fed on insects. The claws on its feet and hands suggest that *Archaeopteryx* could climb trees, and the wings are clearly those of an active flying animal. This bird could fly as well as most modern birds, and flying allowed it to catch prey that were not available to land-living relatives. The skeleton of *Archaeopteryx* is very like that of *Deinonychus* (see Fig. 17.5a), especially in the details of the arm and hindlimb, showing that birds are small flying theropod dinosaurs.

Until recently, birds remained rare until the Late Cretaceous, but now numerous spectacular fossils of birds and dinosaurs, with feathers preserved, have been reported from China, and astonishing new specimens are announced every month (Box 17.3). In the Late Cretaceous, new sea birds radiated. They still had teeth, but the bony tail was reduced to a short knob as in modern birds, and they had other modern features. Modern groups of birds appeared in the latest Cretaceous and Early Tertiary, including flightless ratites and ancestors of water birds, penguins and birds of prey. The perching birds or songbirds, consisting of 5000 species today, radiated in the Miocene.

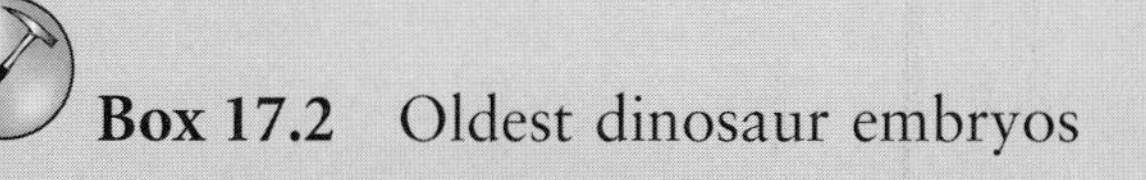

Box 17.2 Oldest dinosaur embryos

Dinosaur eggs and nests have been known since the 1860s when the first finds were made in France. The most famous finds were made by American expeditions to the Late Cretaceous of Mongolia in the 1920s, when whole nests with eggs were brought back to the American Museum of Natural History. Since then, hundreds of dinosaur nest sites have been found. There is no question that dinosaurs laid eggs, and they placed them in nests they kicked out of the dirt on the ground. Dinosaurs did not build nests in trees, for rather obvious reasons!

Maybe dinosaurs covered their nests with sand, and left them to develop, or maybe they covered them with plant debris, which formed a kind of compost to keep the babies warm, as some crocodilians do today. Some dinosaur mothers even incubated their eggs: a mother *Oviraptor* was found in the 1990s sitting on a nest in Mongolia.

Most spectacular of all are unhatched eggs containing embryos inside. The oldest examples were reported in 2005, when Robert Reisz and colleagues announced some eggs laid by the prosauropod *Massospondylus* in the Early Jurassic of South Africa. The researchers X-rayed the eggs, and saw the little bones inside, so Diane Scott, a skilled preparator, took a year of painstaking work to remove the eggshell on one side, and pick the mudstone off the bones with a fine needle, to reveal – a complete little baby, just about to hatch (Fig. 17.8). Reisz and colleagues believe the baby *Massospondylus* would have walked on all fours when it was born, but it had to be looked after. Its teeth were just too small for it to be able to tear up plants by itself. Is this the oldest evidence for parental care in the fossil record?

Read more about dinosaur eggs and nests at http://www.blackwellpublishing.com/paleobiology/. The full description is in Reisz et al. (2005).

Figure 17.8 Photograph of one of the *Massospondylus* eggs with a complete embryo skeleton inside, measuring some 15 cm in total length. It died just before hatching. As an adult, it would have grown to a length of 5 m. (Courtesy of Robert Reisz.)

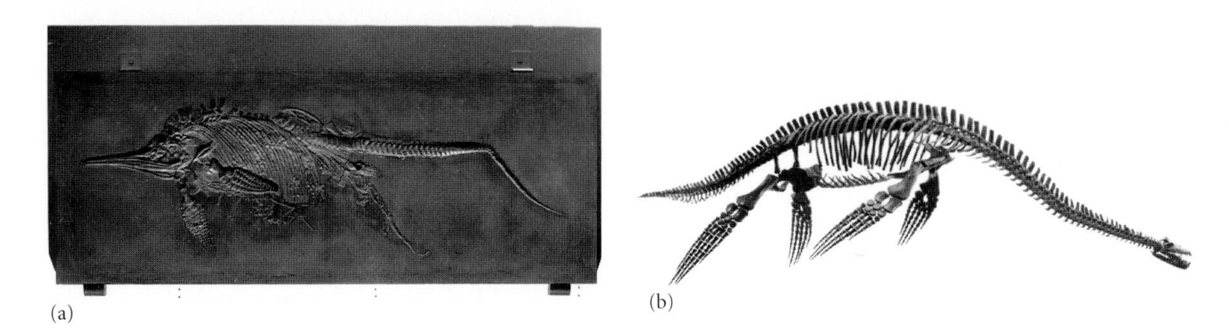

Figure 17.9 Jurassic marine reptiles: (a) the ichthyosaur *Stenopterygius* and (b) the plesiosaur *Cryptoclidus*. (Courtesy of Rupert Wild.)

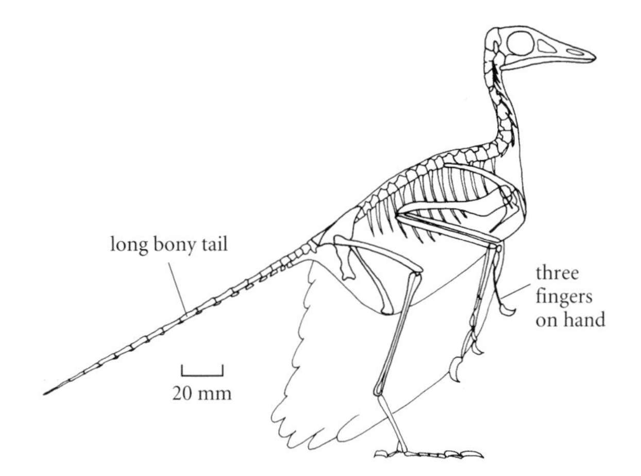

Figure 17.10 The oldest bird, *Archaeopteryx*, from the Late Jurassic. (Courtesy of Andrzej Elzanowski.)

RISE OF THE MAMMALS

Primitive forms, and then success

The first mammals, small insect eaters in the Late Triassic and Early Jurassic (see p. 450), probably hunted at night. Mammals remained small through most of the Mesozoic and they did not achieve high diversity, perhaps held in check in some way by the dinosaurs. Several lines of insectivorous, carnivorous and herbivorous forms appeared, some of them adapted to climbing trees. Most Mesozoic mammals were small, and a recent find from China proved to be an exception (Box 17.4). Nonetheless, most basal mammal groups did not outlive the Cretaceous-Tertiary mass extinction (see pp. 174–7). Three of the clades that did survive into the Tertiary were the monotremes, marsupials and placentals, the modern groups (Box 17.5).

Monotremes today are restricted to Australasia, being represented by the platypus and the echidnas. These mammals are unique in still laying eggs, as the cynodont ancestors of mammals presumably did. The young hatch out as tiny helpless creatures, and feed on their mother's milk until they are large enough to live independently.

It is often said that mammals owe everything to their teeth. The marsupials, and especially the placentals, radiated dramatically in the Tertiary, and this is often taken as a classic example of an adaptive radiation (see p. 544). It is notable that previously rather small, similar-looking animals had diversified within 10 myr to forms as disparate as bats and rats, monkeys and whales. Mammals uniquely have **differentiated** teeth, with incisors, canines and cheek teeth. Fishes, amphibians and reptiles have undifferentiated teeth – their teeth are pretty much identical from front to back. Differentiated teeth allow mammals to adopt a huge array of diets, and to become superefficient at biting, and especially chewing. High metabolic rates need lots of nutritious food, and **chewing**, moving the cheek teeth round and across the food, allows mammals to improve the efficiency of their digestive systems. Most dinosaurs (except ornithopods; see p. 457), like reptiles in general, could not chew, or at least not well – so they swallowed their food whole and probably failed to digest much of it. One thing is for sure: don't stand downwind of a dinosaur!

Marsupials: the pouched mammals

Marsupial young are also born tiny and helpless, and have to feed on maternal milk in a

Box 17.3 The spectacular birds of Liaoning

Back in 1984, local farmers in Liaoning Province, China, began to send fossils they had found in their fields to paleontologists in Beijing and Nanjing. The fossils were all fantastically well preserved: fishes with skin, insects with color patterns, birds and dinosaurs with feathers, and early mammals with hair. The Chinese paleontologists began to publish accounts of the specimens, and they mounted organized digs to recover many tonnes of fossiliferous sediments. It seems all these ancient creatures had swum, or fallen, into ponds where fine lime muds were accumulating and locking in the cadavers before they could decay.

So far, about 20 species of birds have been named from the Liaoning sites (Zhou et al. 2003), including *Confuciusornis*, perhaps the best-known new genus. The most amazing specimen of *Confuciusornis* shows a male and female bird of the same species sitting side by side on the slab (Fig. 17.11). The male has long, streamer-like tail feathers, almost certainly brightly colored in life and used in sexual displays. *Confuciusornis* was about the size of a rook, and it is advanced over *Archaeopteryx* in that it has no teeth in the jaws, and its bony tail has been reduced to a nubbin of bone, or **pygostyle**. But the Chinese bird is still primitive in having powerful fingers and claws on its wings. Nonetheless, its feathers are like those of any modern bird, and the confuciusornithids almost certainly flapped in and out of the trees in China 125 Ma, swooping after prey and landing gracefully on the branches, just like any modern bird.

Not only birds with feathers, but dinosaurs too! Since 1995, a string of reports of small theropod dinosaurs from Liaoning have shown that many flesh-eating dinosaurs also had feathers, even though they did not fly. The feathers were presumably initially for insulation, so the theropod dinosaurs at least must have been warm-blooded. So, in the evolution of birds, feathers came first, perhaps as early as the Early Jurassic and then wings and flight came in the Late Jurassic when *Archaeopteryx* evolved.

Read more about the Liaoning birds and dinosaurs in Zhou et al. (2003), and at web sites linked through http://www.blackwellpublishing.com/paleobiology/.

Figure 17.11 Two examples of the Early Cretaceous bird *Confuciusornis* from Liaoning, China, showing a male (below, with long tail streamers) and a female. (Courtsey of Zhou Zhonghe.)

Box 17.4 Mammal eats dinosaur shock!

A common view about the mammals of the Mesozoic is that they were all rather small, and that this was because the dinosaurs preyed on them and prevented any becoming large. Two new mammals from the mid-Cretaceous of China have turned this idea over: one was as big as a dog, the other as big as a cat, and one of them had just eaten a dinosaur – admittedly a baby dinosaur only 140 mm long.

The new fossils come from the classic localities around Liaoning, China, that have produced so many spectacular fossils of dinosaurs, birds (see Box 17.3), salamanders, fishes, insects and plants. Yaoming Hu, a graduate student at the American Museum of Natural History in New York and his colleagues from China described two new species, *Repenomamus giganticus* and *R. robustus*, both based on excellent skeletons (Hu et al. 2005), 1.0 and 0.4 m in length, respectively. *Repenomamus* is a triconodont, a group known otherwise mainly from isolated jawbones, and thought before to have specialized in eating insects. One specimen of *R. robustus* had the remains of a baby *Psittacosurus*, a ceratopsian dinosaur (see p. 457), torn into chunks inside its rib cage, in the region of the stomach (Fig. 17.12).

"This is a good man-bites-dog story", commented paleontologist Kevin Padian, when the discovery was announced. Read a review of Mesozoic mammals in Luo (2007), and find out more about *Repenomamus*, and see color images, at http://www.blackwellpublishing.com/paleobiology/.

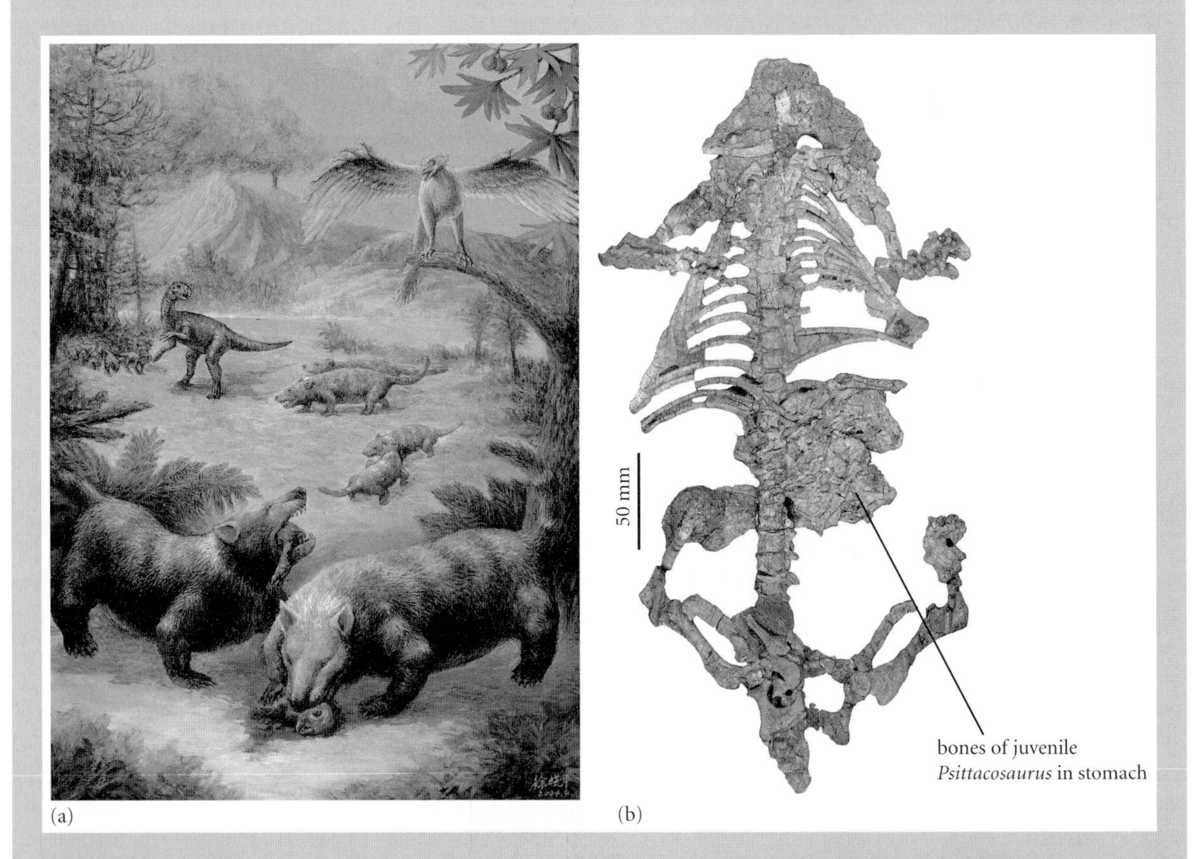

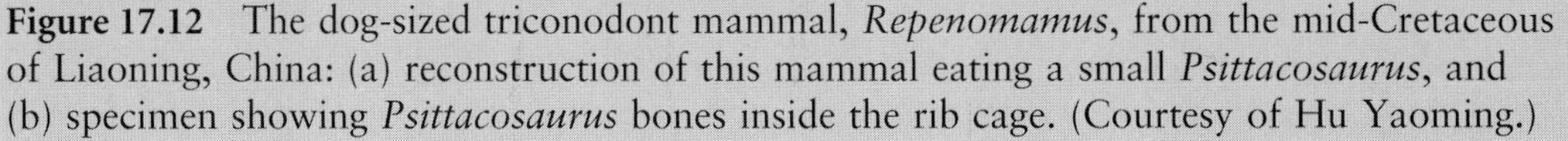

Figure 17.12 The dog-sized triconodont mammal, *Repenomamus*, from the mid-Cretaceous of Liaoning, China: (a) reconstruction of this mammal eating a small *Psittacosaurus*, and (b) specimen showing *Psittacosaurus* bones inside the rib cage. (Courtesy of Hu Yaoming.)

Box 17.5 Classification of mammals

Modern mammals fall into three groups – the monotremes, marsupials and placentals – characterized by their breeding modes. Various primitive groups are omitted.

Class MAMMALIA
Subclass MONOTREMATA
- Females lay eggs, and newborn young grow in pouch
- Early Cretaceous to Recent

Subclass METATHERIA (marsupials and extinct relatives)
- Young are born live, but continue development in pouch
- Late Cretaceous to Recent

Subclass EUTHERIA (placentals and extinct relatives)
- Young are born live at an advanced stage, having been nourished by a placenta while in the womb. Main orders only are listed
- Mid-Cretaceous to Recent

Infraclass AFROTHERIA
Order PROBOSCIDEA (elephants)
- Eocene to Recent

Infraclass XENARTHRA
Order EDENTATA (armadillos, tree sloths, anteaters)
- Paleocene to Recent

Infraclass LAURASIATHERIA
Superorder BOREOEUTHERIA
Order LIPOTYPHLA ("insectivores": hedgehogs, moles, shrews)
- Paleocene to Recent

Order CHIROPTERA (bats)
- Eocene to Recent

Order ARTIODACTYLA (pigs, hippos, camels, cattle, deer, giraffes, antelopes)
- Eocene to Recent

Order CETACEA (whales and dolphins)
- Eocene to Recent

Order PERISSODACTYLA (horses, rhinos, tapirs)
- Eocene to Recent.

Order CARNIVORA (dogs, bears, cats, hyaenas, seals)
- Paleocene to Recent

Superorder EUARCHONTOGLIRES
Order PRIMATES (monkeys, apes, humans)
- Paleocene to Recent

Order RODENTIA (mice, rats, squirrels, porcupines, beavers)
- Paleocene to Recent

Order LAGOMORPHA (rabbits and hares)
- Eocene to Recent

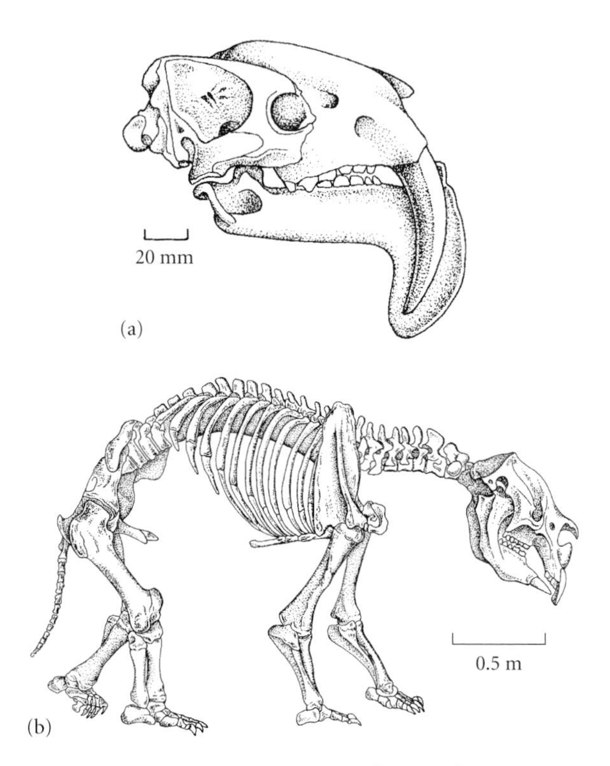

Figure 17.13 Extinct marsupials: (a) the sabretooth *Thylacosmilus* from South America, and (b) the giant herbivore *Diprotodon* from Australia. (Based on Gregory 1951.)

pouch for many months, but egg laying has been abandoned. The oldest marsupial fossils come from the mid-Cretaceous of North America. The group radiated successfully in South America during the Tertiary, and included several lines of insectivores, carnivores and herbivores, many of which were remarkably like unrelated placental mammals elsewhere. Some forms were dog-like, and *Thylacosmilus* (Fig. 17.13a) independently evolved all the characters of the placental saber-toothed cats of Europe and North America.

In Australia, the marsupials diversified even more, after reaching that continent in the Eocene by traveling across a much warmer Antarctica, which then linked the southern tip of South America with Australia. Once there, the Australian marsupials radiated to parallel placental mammals in functions and body forms, except of course for the unique kangaroos. In the Pleistocene, there were abundant and diverse faunas of large marsupials, including giant kangaroos and the hippopotamus-sized herbivorous *Diprotodon* (Fig. 17.13b).

Palaeogeography and diversification of the placentals

Placental mammals produce young that are retained in the mother's womb much longer than is the case in marsupials, and they are nourished by blood passed through the placenta. The oldest is the mid-Cretaceous *Eomaia* from Liaoning in China. Many fossil placental mammals have been reported from the Late Cretaceous, but most are rather incomplete, and sometimes their classification has been controversial.

Mammalogists have struggled for two or more centuries to understand the relationships of the major groups of living placentals – are cattle related to horses, bats to monkeys, whales to seals? Some morphological evidence was found to show that, for example, rabbits and rodents are sister groups, elephants are closely related to the enigmatic African hyraxes and the aquatic sirenians, but many other supposed relationships were hotly disputed. Now, however, everything seems to have been resolved (Box 17.6).

Some time early in the Late Cretaceous, the placental mammal clade split into four. First to split off were the Afrotheria, and that clade continued to evolve in Africa. Then the Xenarthra became isolated in South America. The Boreoeutheria remained in the northern hemisphere and split there into Laurasiatheria and Euarchontoglires. So the split into placentals in Africa, South America and Laurasia (North America–Europe–Asia) seems to have been central to the diversification of the group, and it ties perhaps with the split of major continents through the mid-Cretaceous, with the South Atlantic splitting South America from Africa, and with other oceans separating those southern continents from North America, Europe and Asia.

Placentals in southern continents

The Afrotheria ("African mammals") are known best by the elephants. The African and Indian elephants of today (Proboscidea) are a sorry remnant of a once-diverse group. Early elephant relatives such as *Moeritherium* (Fig. 17.15a) were small hippo-like animals that

Box 17.6 Mammals, morphology and molecules

The classification of living placental mammals has long been mysterious. We know what a whale or a bat or a primate is, but how do these major orders relate to each other? The techniques of molecular phylogeny estimation (see pp. 133–4) have revealed the answer.

The story began when Mark Springer and colleagues (1997) discovered the Afrotheria, a clade consisting of African animals, linking the elephants (Proboscidea), hyraxes and sirenians with the aardvarks (Tubulidentata), tenrecs and golden moles. The last three groups had all been assigned various positions in the classification of mammals, but their genes show they shared a common ancestor with the elephant – hyrax – sirenian group. After 1997, everything else fell into place (Fig. 17.14). The South American placentals, the edentates, formed a second major group, the Xenarthra. And the remaining mammalian orders formed a third major clade, the Boreoeutheria ("northern mammals"), split into Laurasiatheria (insectivores, bats, artiodactyls, whales, perissodactyls, carnivores) and Euarchontoglires (primates, rodents, rabbits). So, in the course of 2 or 3 years, several independent teams of molecular biologists solved one of the outstanding puzzles in the tree of life (e.g. Springer et al. 2003; Asher 2007).

But why had this proved to be such a phylogenetic puzzle? Some suggest that the major splits among placental mammals happened very rapidly, and there was no time for shared morphological characters to become fixed. But the morphologists are fired up to find such characters: if the Afrotheria is really a clade, then there must be some obscure anatomic feature shared among them all! The hunt goes on.

Read a review of Afrotheria in Tabuce et al. (2008), and find out more about the search for morphological characters of the clade at http://www.blackwellpublishing.com/paleobiology/.

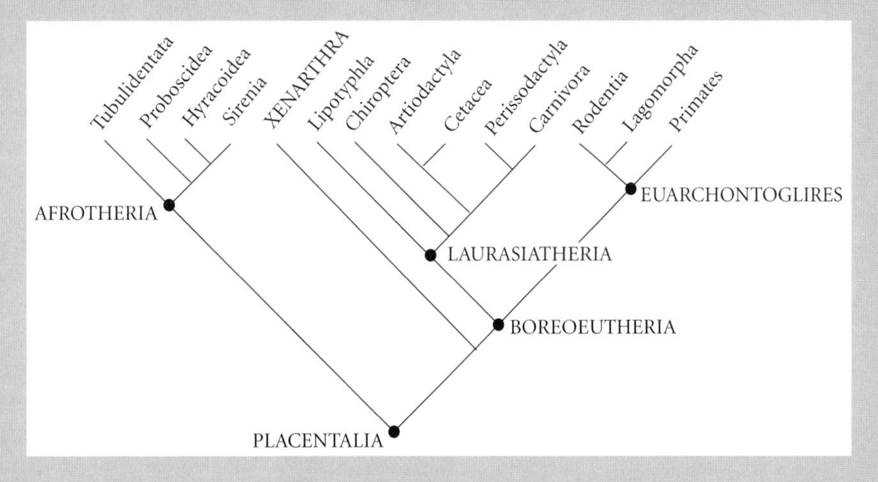

Figure 17.14 Cladogram of the major orders of placental mammals based on molecular evidence. The four deep splits among modern orders happened in the Late Cretaceous, but modern placentals did not become diverse until after the extinction of the dinosaurs.

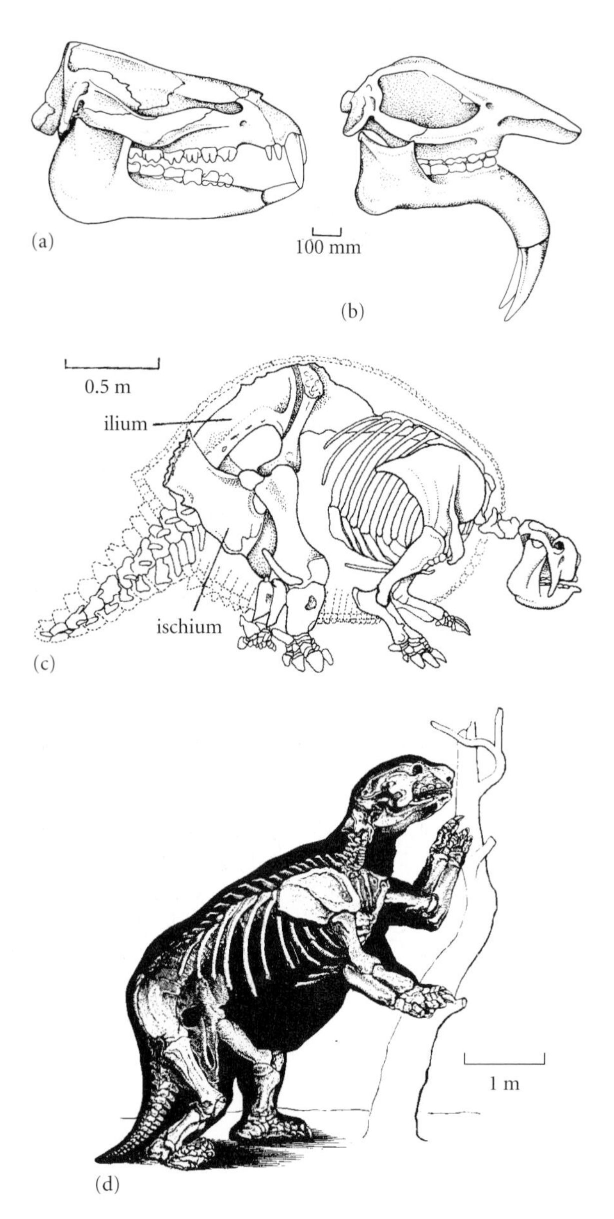

Figure 17.15 Afrotheres and xenarthrans: (a, b) skulls of the Eocene proboscidean *Moeritherium* (a) and the Miocene proboscidean *Deinotherium* (b); and (c, d) Pleistocene edentates from Argentina, *Glyptodon* (c) and *Mylodon* (d). (Based on Gregory 1951.)

probably fed on lush plants in the ponds and rivers of Africa. Later, many lines of proboscideans diverged, distinguished by an astonishing array of tusks, which are modified incisor teeth. Some had tusks in the upper jaw (as in modern elephants), others had tusks in the lower jaw, and others had tusks in both as in *Deinotherium* (Fig. 17.15b). The Pleistocene mammoths were abundant in cold northern ice age climates, but died out as the ice retreated 10,000 years ago.

Among the other afrotheres, the hyraxes and sirenians are close relatives of proboscideans. The aardvark (Tubulidentata), as well as golden moles and tenrecs, seem to form a second afrothere group.

The Xenarthra, or edentates ("no teeth"), are a peculiar South American group. They include modern armadillos, tree sloths and anteaters, as well as their remarkable ancestors, known especially from the Pliocene and Pleistocene. There were giant armadillos such as *Glyptodon* (Fig. 17.15c), and giant ground sloths such as *Mylodon* (Fig. 17.15d) that reached a length of 6 m and fed on coarse leaves from the treetops. The ground sloths survived in South America until 11,000 years ago, and their subfossil remains include clumps of reddish hair and caves full of unrotted dung that occasionally ignites spontaneously.

The "northern" placentals

The Laurasiatheria include about half of modern placental mammals, and they are as diverse as shrews, bats, cattle, whales and lions. The Lipotyphla (sometimes Insectivora) is composed of dozens of species of hedgehogs, moles and shrews, all small animals with long snouts that feed on insects. The oldest Lipotyphla are Paleocene in age, and for the most part the fossil forms probably looked like the modern ones. One exception is the giant spiny hedgehog *Deinogalerix*, which was 0.5 m long.

Next in the cladogram (see Fig. 17.14) are the Chiroptera, or bats, a diverse group today of some 1000 species. The earliest bats such as *Icaronycteris* from the Eocene (Fig. 17.16a) show the typical wing structure in which the flight membrane is supported on four fingers of the hand that spread out. The feet turn backwards, and *Icaronycteris* could have hung upside down. It also had large eyes and the ear region was modified for echolocation. *Icaronycteris*, like most modern bats, hunted insects at night, using its large eyes to pick up movements, and sending out high-pitched

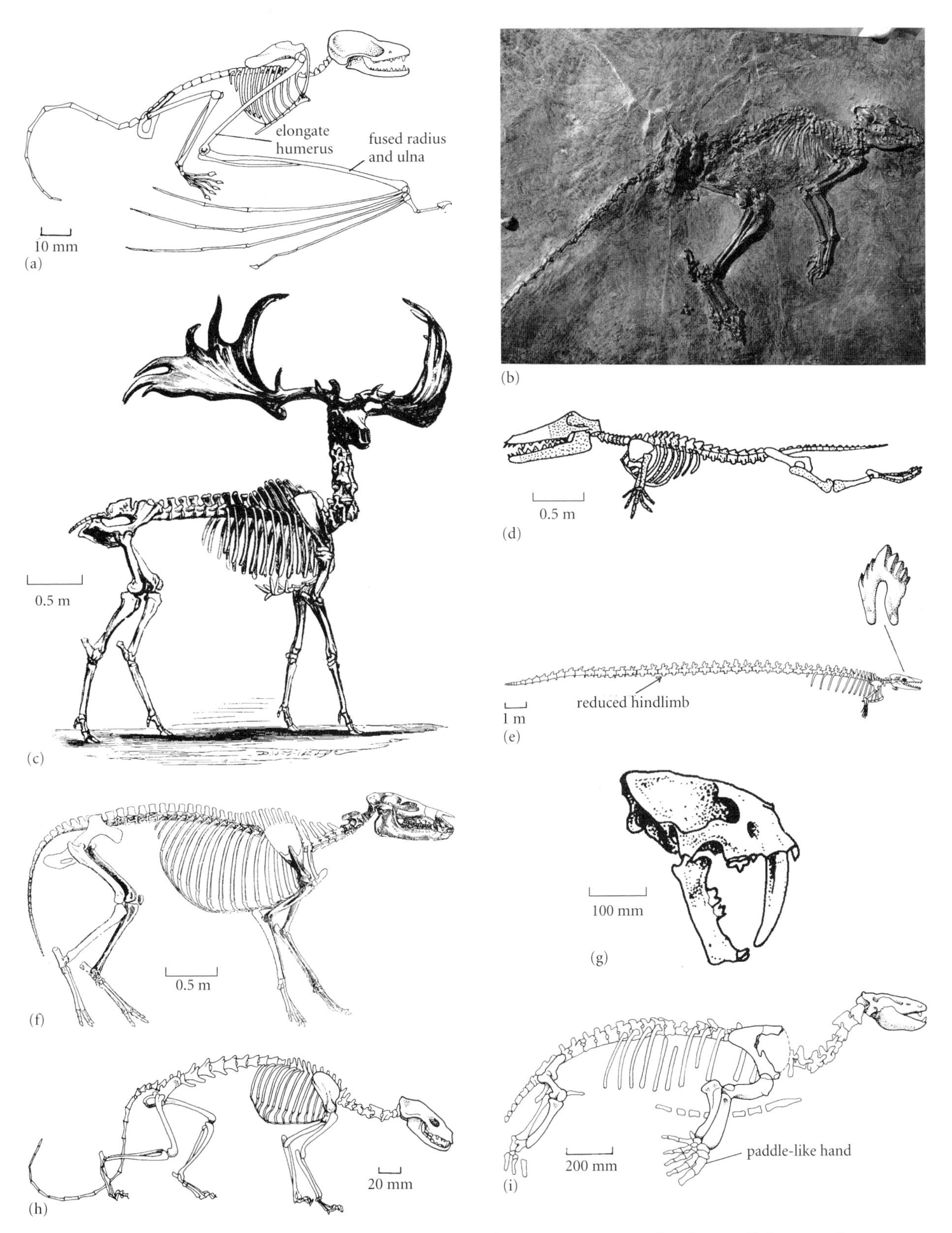

Figure 17.16 Diverse laurasiatherians: (a) the Eocene bat *Icaronycteris*; (b) the small four-toed artiodactyl *Messelobunodon*, showing the complete skeleton and a mass of chopped plant material in the stomach area, from the oil shale deposit of Messel, Germany; (c) the Pleistocene giant Irish deer *Megaloceros*; (d) the middle Eocene whale *Ambulocetus*; (e) the late Eocene whale *Basilosaurus*; (f) the Miocene horse *Neohipparion*; (g) the Pleistocene sabre-toothed cat *Smilodon*; (h) the Eocene dog *Hesperocyon*; and (i) the Miocene "seal" *Allodesmus*. (a, based on Jepsen 1970; b, courtesy of Jens Franzen; c, e–i, based on Gregory 1951; d, based on Thewissen et al. 1994.)

squeaks to detect its prey by the echoes they made.

The next clade is perhaps a little unexpected. Much evidence links whales and artiodactyls, the even-toed **ungulates** (larger plant eaters), as a clade called Cetartiodactyla. Whales and artiodactyls, for example, share a similar pulley-like ankle joint (do whales have ankles?) – not now, but the early forms did (see below). The artiodactyls arose in the Eocene (Fig. 17.16b), and the group includes pigs, hippos, camels, cattle, deer, giraffes and antelopes, all with an even number of toes (two or four). Pigs and hippos share ancestors in the Oligocene, at the same time as vast herds of oreodonts fed on the spreading grasslands of North America. Oreodonts are related to the camels and the ruminants. The first camels were long-limbed and lightly built North American animals: it was only later that camels moved to Africa and the Middle East and evolved adaptations for living in conditions of drought.

Most artiodactyls today are **ruminants**, animals that pass their food into a forestomach, regurgitate it (chew the cud) and swallow it again. The multiple digestive process allows ruminants to extract all the nourishment from their plant food, usually grass, and to pass limited waste material (compare the homogenous excrement of cattle with the fibrous undigested droppings of horses, which do not ruminate). Ruminants became successful after the mid-Miocene, when a great variety of deer, cattle and antelopes appeared. These animals usually have horns or antlers, seen in spectacular style in the Irish deer *Megaloceros* (Fig. 17.16c). The headgear is used in all cases for displays and fights between males seeking to establish territories and win mates.

The whales, Cetacea, evolved from a raccoon-sized artiodactyls that fed on aquatic plants along the edges of streams and ponds (Thewissen et al. 2007) – the oldest whale *Ambulocetus* (Fig. 17.16d) still has fully developed limbs, and these show the pulley-like ankle bone of artiodactyls. By the late Eocene, whales such as *Basilosaurus* (Fig. 17.16e) had become very large, at lengths of 20 m or more. *Basilosaurus* had a long thin body, like a mythical sea serpent, and a relatively small skull armed with sharp teeth. It was probably a fish eater, like the toothed whales today. The baleen whales, the biggest of all modern whales, arose later, and they owe their success to their ability to filter vast quantities of small crustaceans, krill, from polar seawaters.

The second major ungulate group, the Perissodactyla, consisting of horses, rhinos and tapirs, all have an odd number of toes – one or three. The horses provide a classic example of evolution (see p. 543). The first horse, *Hyracotherium*, was a small woodland-living animal that had four fingers and three toes, and low teeth used for browsing on leaves. During the Oligocene and Miocene, horses became adapted to the new grasslands that were replacing the forests, and they became larger, lost toes and evolved deep-rooted cheek teeth for grinding tough grass (Fig. 17.16f).

Tapirs and rhinoceroses are probably related. Eocene and Oligocene rhinos were modest-sized, running animals, not much different from some of the early horses. Tapirs later became a rare group, restricted to Central and South America and Southeast Asia. The rhinoceroses flourished for a while, producing monsters such as the Oligocene *Indricotherium*, the largest land mammal of all time: 5.5 m tall at the shoulder and weighing 15 tonnes.

The Carnivora – cats, dogs, hyenas, weasels and seals – are characterized by sharp cheek teeth (**carnassials**) used for tearing flesh. The cats have a long history during which dagger-toothed and saber-toothed forms evolved many times. The sabertooths such as *Smilodon* (Fig. 17.16g) preyed on large, thick-skinned herbivores by cutting chunks of flesh from their bodies. The saber-toothed adaptations evolved independently in some South American marsupials (cf. Fig. 17.13a). Early dogs such as *Hesperocyon* (Fig. 17.16h) were light, fast-moving animals, close to the ancestry of modern dogs and bears. Some carnivores related to raccoons and weasels entered the sea during the Oligocene, and gave rise to the seals, sealions and walruses. Early forms such as *Allodesmus* (Fig. 17.16i) had broad, paddle-like limbs and fed on fish.

The monkey-rabbits

The second boreoeutherian clade (see Fig. 17.14) is the Euarchontoglires, a mouthful that is made from a combination of the group names Archonta and Glires, and it means something like "true primate–treeshrew rodent–rabbits". The name tells the whole story!

Rats and rabbits, representing the Rodentia and Lagomorpha, belong to one group (Glires), and they shared an ancestor with buck teeth and a propensity for breeding. Rodents are the largest group, consisting of over 1700 species of mice, rats, squirrels, porcupines and beavers. They owe their success to their powerful gnawing teeth: the front incisors are deep-rooted and grow continuously, so that they can be used to grind wood, nuts and husks of fruit. The oldest rodents, such as *Paramys* (Fig. 17.17a) already had the front grinders, and this ability to chew materials ignored by other animals triggered several phases of rapid radiation. Beavers, porcupines and cavies radiated in the Miocene. The cavies include a giant Pliocene guinea pig that weighed 1 tonne, and was the size of a small car (Rinderknecht & Blanco 2008); why the South American rodents became so large is hard to understand. Rabbits and their relatives (Lagomorpha) have never been as diverse as the rodents. Fossil forms in the Oligocene have elongate hindlimbs used in jumping.

Primates, consisting of monkeys, apes and humans, are part of a larger clade, Archonta, which also includes the rare treeshrews and flying lemurs (dermopterans) – but it is the primates that have attracted most attention.

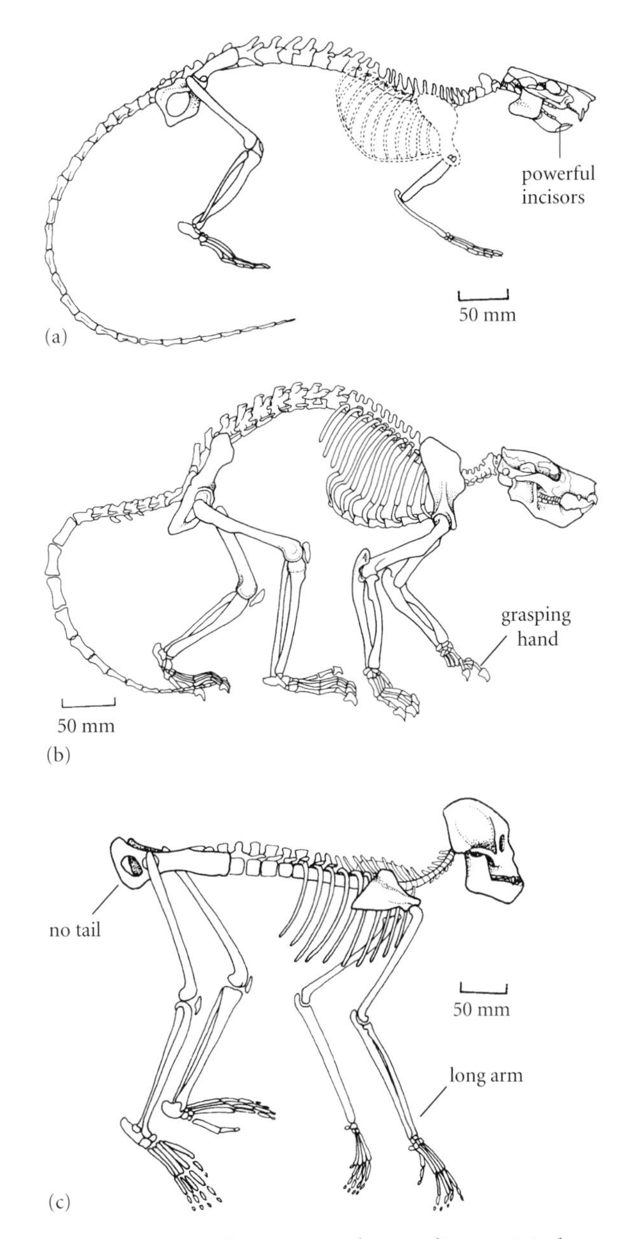

Figure 17.17 Diverse euarchontoglirans: (a) the Eocene rodent *Paramys*; (b) the Paleocene primate *Plesiadapis*; and (c) the Miocene ape *Proconsul*. (a, based on Wood 1962; b, based on Lewin 1999.)

THE LINE TO HUMANS

Early primates

The Primates is one of the oldest of the modern placental mammal groups. The name "primate" (from *primus*, "first") does not refer to this, but to the fact that humans are primates, and so "first" among animals: the namer, *Homo sapiens* ("wise person"), has the privilege of choosing the name! For most of their history, the primates were a rare and rather obscure group. All primates share a number of features that give them agility in the trees (mobile shoulder joint, grasping hands and feet, sensitive finger pads), a larger than average brain, good binocular vision and enhanced parental care (one baby at a time, long time in the womb, long period of parental care, delayed sexual maturity, long lifespan).

Early relatives of the primates such as *Plesiadapis* (Fig. 17.17b) were squirrel-like animals that may have climbed trees and fed on fruit, seeds and leaves. Various basal primates radiated in the Paleocene, Eocene and Oligocene, and gave rise to the modern lemurs, lorises and tarsiers.

True monkeys arose in the Eocene and they diverged into two groups, the New World monkeys of South America, and the Old World monkeys of Africa, Asia and Europe. The New World monkeys, such as marmosets and spider monkeys, have flat noses and **prehensile** tails that may be used as extra limbs in swinging through the trees. The Old World monkeys, such as macaques and baboons, have narrower projecting noses and non-prehensile tails, or no tails at all.

The apes arose from the Old World monkeys before the end of the Oligocene and the group radiated in Africa in the Miocene. Even early forms, such as *Proconsul* (Fig. 17.17c), have no tail, and a relatively large braincase, indicating high intelligence. These apes ran about on the ground and along low branches on all fours, and fed on fruit. The apes spread out from Africa into the Middle East, Asia and southern Europe by the mid-Miocene and gave rise to some of the modern ape groups at that time. Fossil and molecular evidence on phylogeny (see pp. 133–4) suggests that the gibbons of Southeast Asia are the most primitive living apes, having branched off 25–20 Ma, followed by the orangutan 20–15 Ma (Fig. 17.18). The focus of ape (and human) evolution remained in Africa.

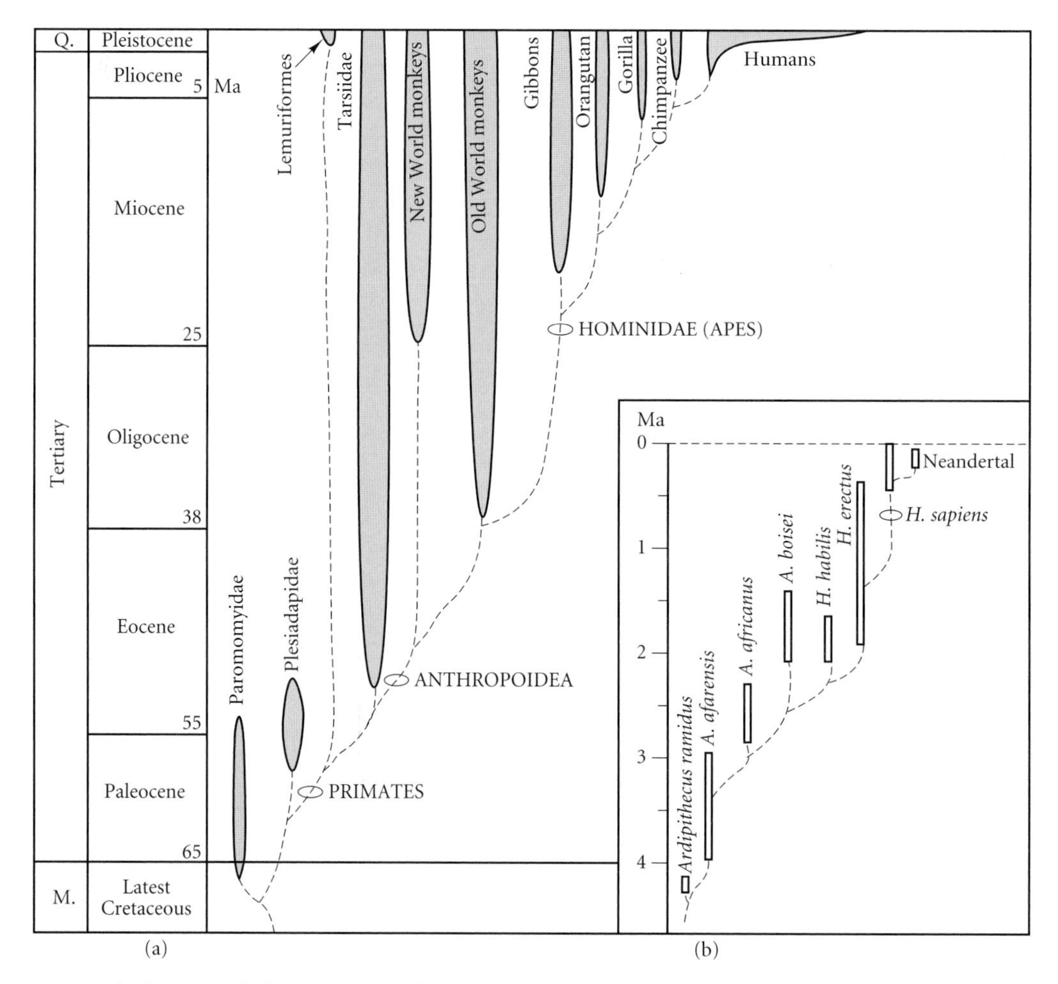

Figure 17.18 Phylogeny of the primates, showing some of the main fossil and living groups (a), and the detail of one view of human evolution (b). *A.*, *Australopithecus*; *H.*, *Homo*; M, Mesozoic; Q, Quaternary.

Gorillas, chimpanzees and humans appear to be very closely related: they share many anatomic characters and more than 94% of their DNA is identical. Gorillas seem to have diverged first, about 10 Ma, and the ancestors of humans and chimps separated about 8–6 Ma.

Human evolution

Humans are set apart from other primates by their large brain and their **bipedalism**, walking upright. We might like to think the large brain and human intelligence evolved first, but all the evidence shows that the move from four to two legs came first, and this may have been because of a climatic accident in Africa in the late Miocene. Much of Africa had been covered with lush forests in which the ancestral apes flourished, but climates then became arid and the East African rift valley began to open up, separating the forests in the west from the arid grasslands in the east. Tree-living apes (chimps and gorillas) retreated west and the remaining apes (our ancestors) remained in the eastern grasslands. They had to stand upright to look for enemies, to permit them to run long distances in search of food and to free their arms for carrying food. Humans evolved from bipedal ape-like ancestors that had no special high brainpower.

Until recently, the oldest human remains were about 4 Ma, which seemed to match the molecular estimate of 5 Ma for the split between humans and chimps. Then, in 2001 and 2002, there was a flurry of excitement when two French teams announced human fossils that were 6 Ma. There was immediate dispute about whether these fossils truly were human, and which was oldest, and the debate rumbles on (Box 17.7).

A number of incomplete human fossils were reported during the 1990s from Africa from rocks dated between 6 and 4 Ma. It may be that *Sahelanthropus* and *Orrorin* were already bipedal by 6 Ma, but the oldest clear evidence for bipedalism was the find of some human tracks in volcanic ash from Tanzania, dated at 3.75 Ma. The oldest substantial skeletons, of *Praeanthropus afarensis*, come from rocks dated at about 3.2 Ma and also show clear anatomic evidence for advanced bipedalism, but still an ape-sized brain. The famous skeleton of a female *P. afarensis* from Ethiopia, called Lucy by its discoverer Don Johanson in the 1970s (Fig. 17.20a), has a rather modern humanoid pelvis and hindlimb (Fig. 17.20b). The pelvis is short and horizontal, rather than long and vertical as in apes, the thighbone slopes in towards the knee, and the toes can no longer be used for grasping. Lucy's brain, however, is small, only 415 cm^3 for a height of 1–1.2 m – not much different from a chimpanzee.

The human genus *Australopithecus* continued to evolve in Africa from about 3 to 1.4 Ma, giving rise to further small species, and some large robust ones (Fig. 17.21a, b). The larger australopithecines reached heights of 1.75 m, but their brain capacities did not exceed 550 cm^3, a rather ape-like measure. The leap forward to modern human brain sizes only came with the origin of a new human genus, *Homo*. The first species, *H. habilis* (Fig. 17.21c), lived in Africa from 2.4 to 1.5 Ma, and had a brain capacity of 630–700 cm^3 in a body only 1.3 m tall. *H. habilis* may have used tools. It is a remarkable fact that, for over 1 myr, three or four different human species lived side by side in Africa.

So far, the focus of human evolution had been entirely in Africa, but a new species, *H. erectus*, which arose 1.9 Ma in Africa, spread to China, Java and central Europe. *H. erectus* had a brain size of 830–1100 cm^3 (Fig. 17.21d) in a body up to 1.6 m tall, and there is clear evidence that this early human species had semipermanent settlements, a basic tribal structure, knew about the use of fire for cooking, and made tools and weapons from stone and bone.

Modern peoples

Truly modern humans, *H. sapiens*, may have arisen as much as 400,000 years ago, and certainly by 150,000 years ago, in Africa, having evolved from *H. erectus*. It seems that all modern humans arose from a single African ancestor, and that the *H. erectus* stocks in Asia and Europe died out. *H. sapiens* spread to the Middle East and Europe by 90,000 years ago. The European story is particularly well known, and it includes a phase, from 90,000 to 30,000 years ago, when Neandertal man occupied much of Europe from Russia to Spain and from Turkey to southern England. Neandertals had large brains (on average,

Box 17.7 Vive la différence!

The world of paleoanthropology was rocked in 2001 when a team based in Paris, Senut et al. (2001), announced a new hominid, *Orrorin tugenensis* from Kenya, dated as 6 Ma. The remains were rather scrappy, as hominid remains often are: teeth, jaw fragments and limb bones. Brigitte Senut and her colleagues argued that the teeth were rather ape-like, and that the arm bones suggested *Orrorin* could brachiate – that is, swing through the trees arm-over-arm, like an ape. However, the femur (thigh bone) showed that *Orrorin* stood upright, and so this was a true early human. Other paleoanthropologists questioned these claims: they accept the great age of the fossils, but some have doubted whether the remains all belong to the same species, and others question whether *Orrorin* really could have walked upright.

Soon after, another French team, this time from Poitiers, announced a second ancient hominid, from Chad in North Africa: *Sahelanthropus* (Brunet et al. 2002). *Sahelanthropus* is based on a fairly complete skull (Fig. 17.19), some fragmentary lower jaws and teeth. The sediments in Chad are hard to date, but are about 7–6 Ma, perhaps the same as *Orrorin*, or maybe older. The *Sahelanthropus* skull indicates a brain volume of 320–380 cm^3, similar to a modern chimpanzee, but the teeth are more human-like, with small canines. The position of the **foramen magnum** is disputed: Michel Brunet and his colleagues claim the foramen magnum, the great opening in the skull through which the spinal cord passes, is located beneath the skull, so implying that *Sahelanthropus* stood upright. Critics have said the specimen is too shattered to be sure, and that the foramen magnum might lie more towards the back of the skull, so suggesting that the spinal cord exited horizontally and that *Sahelanthropus* stood on all fours. These critics have cheekily suggested *Sahelanthropus* ("Sahel man") should be renamed *Sahelpithecus* ("Sahel ape")!

Read more in the original papers by Senut et al. (2001) and Brunet et al. (2002), and at many web sites, linked from http://www.blackwellpublishing.com/paleobiology/.

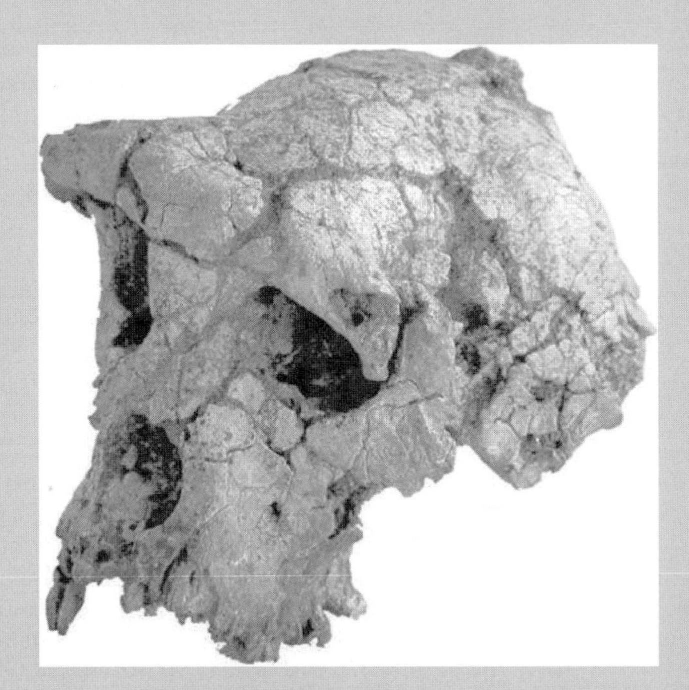

Figure 17.19 Our oldest ancestor? The spectacular skull of *Sahelanthropus* from the upper Miocene of Chad, over 6 Ma. (Courtesy of Michel Brunet.)

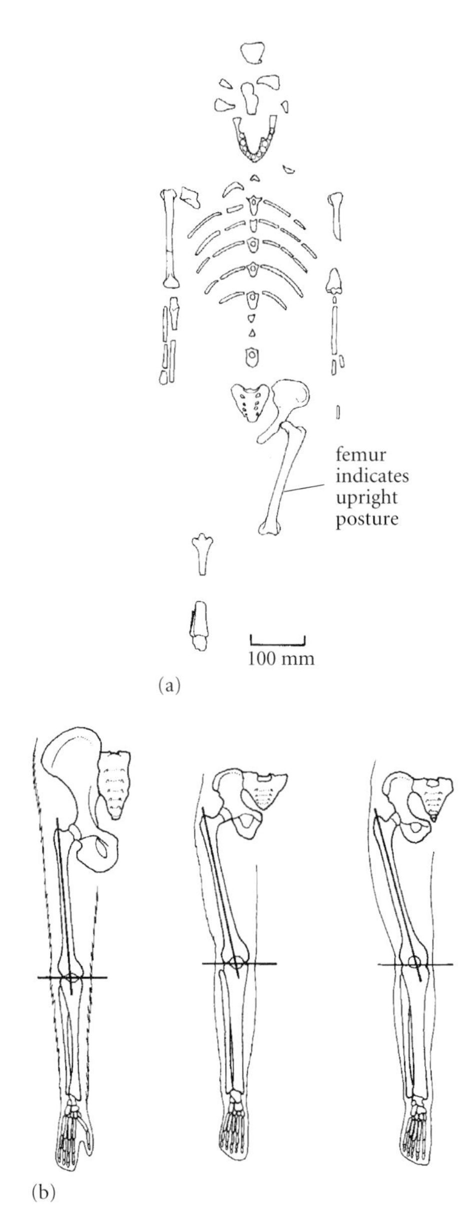

Figure 17.20 The origin of bipedalism in humans: (a) the Pliocene hominid *Praeanthropus afarensis*, known as "Lucy"; and (b) comparison of the hindlimb of an ape (left), Lucy (middle) and a modern human (right). (Based on Lewin 1999.)

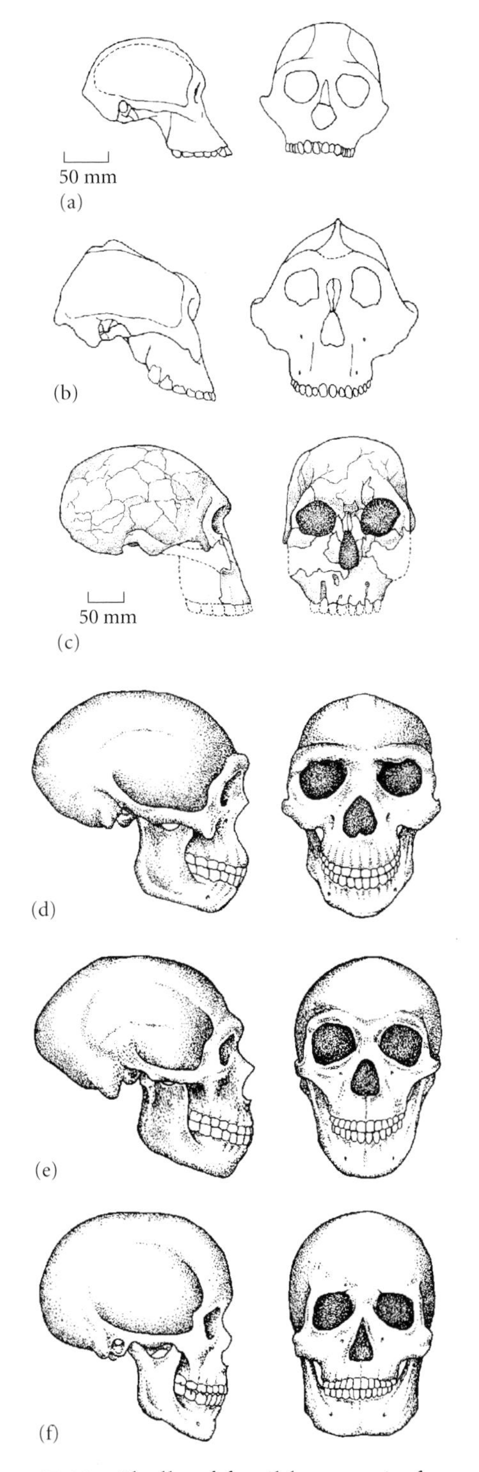

Figure 17.21 Skulls of fossil humans in front and side views: (a) *Australopithecus africanus*; (b) *A. boisei*; (c) *Homo habilis*; (d) *H. erectus*; (e) *H. sapiens*, Neandertal type; and (f) modern *H. sapiens*. (Based on Lewin 1999.)

1400 cm^3), heavy brow ridges (Fig. 17.21e) and stocky powerful bodies. They were a race of *H. sapiens* adapted to living in the continuous icy cold of the last ice ages, and had an advanced culture that included communal hunting, the preparation and wearing of sewn animal-skin clothes, and religious beliefs. Some paleoanthropologists see the

Box 17.8 The quest for the Hobbit

An extraordinary human fossil was unearthed in 2003 on the island of Flores, Indonesia, in Southeast Asia. The bones came from a woman who was fully grown, and yet barely 1 m tall, and she was named as the exemplar of a new human species, *Homo floresiensis*. This is an astonishing claim, that a distinct human species existed in Indonesia at the same time as *H. sapiens* was striding across northern Asia towards the Bering Strait, and Middle Stone Age peoples in Europe were painting wonderful hunting scenes in the caves of France.

The Flores remains consist of a more or less complete skull and skeleton, as well as the remains of seven other individuals. The skull (Fig. 17.22) shows this hominid had a brain size of only 380 cm^3. It was named as a clearcut new species, *H. floresiensis*, by its discoverers, Peter Brown of the University of New England, Australia, and colleagues (2004). They argued that the remains show evidence for a unique population of little people – soon dubbed the hobbits by the press – that, remarkably, hunted pygmy elephants and giant lizards on the island. The discoverers argued that this unique species had lived on the island for some time, unaware of its larger human relatives elsewhere, and that Flores man had become small as a result of **island dwarfing**. It is a relatively common observation that mammals isolated on islands may become smaller (or sometimes larger) because there are usually fewer species than on the neighboring mainland, and each species finds a new place in the ecosystem. The elephants on Flores doubtless became small too so they could survive in a smaller area.

As soon as the paper appeared, many paleoanthropologists were immediately critical. Surely this could not be a distinct species, merely a local variant of *H. sapiens*, either a pygmy race, or perhaps a **microcephalic** individual (Jacob et al. 2006), that is an individual with an unusually small head and brain? There have been accusations that researchers on both sides of the dispute have been less than willing to share the specimens, that others have borrowed specimens and then refused to hand them back, that specimens have been damaged, and other multifarious examples of skullduggery. The majority view still seems to be that *H. floresiensis* is truly a distinct species, but the inaccessibility of the location and legal problems may make further collecting difficult.

Read more about the hobbit in Morwood and van Oosterzee (2007) and Culotta (2007), as well as web presentations available through http://blackwellpublishing.com/paleobiology/.

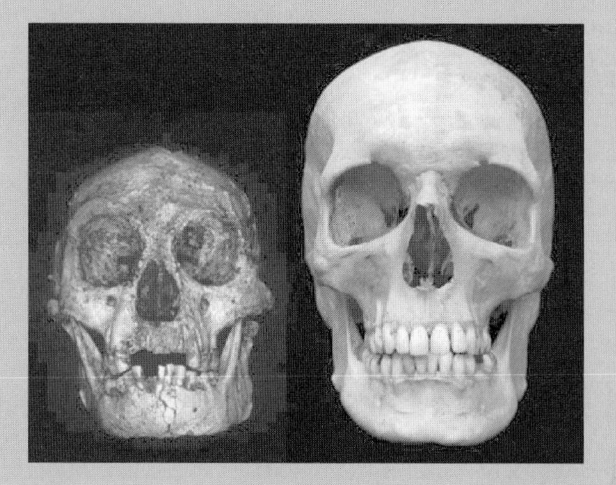

Figure 17.22 Skulls of Flores man, *Homo floresiensis* (left), and of a typical modern human, *H. sapiens*, to show the great difference in size. (Courtesy of Paul Morwood.)

Neandertals as distinct enough to be given their own species, *H. neanderthalensis*.

The Neandertals disappeared as the ice withdrew to the north, and more modern humans advanced across Europe from the Middle East. This new wave of colonization coincided with the spread of *H. sapiens* (Fig. 17.21f) over the rest of the world, crossing Asia to Australasia before 40,000 years ago. There is much debate about the dating of these migrations, and how the various human populations are related. New discoveries from Indonesia have stirred up real controversy over whether there was a unique small-sized human species, *H. floresiensis*, living there only 18,000 years ago (Box 17.8).

Equally controversial is the question of when modern humans reached the Americas. All agree that native Americans walked across from Siberia to Alaska, and colonized southwards over some hundreds or thousands of years. It is commonly said that humans reached North America 11,500 years ago, and yet apparently older remains are reported from time to time. These fully modern humans, found worldwide, with brain sizes averaging 1360 cm^3 (see Fig. 17.21f), brought more refined tools than those of the Neandertals, art in the form of cave paintings and carvings, and religion. The nomadic way of life began to give way to settlements and agriculture about 10,000 years ago.

Review questions

1 Why were dinosaurs so huge? Establish the typical size range of dinosaurs in comparison with modern mammals, and read about ideas past and present about why dinosaurs were an order of magnitude larger than mammals.
2 Were dinosaurs warm-blooded? Read around the topic, back to the debates in the 1970s and 1980s, and through to the present day. List the different lines of evidence used to suggest endothermy and evaluate the arguments for and against.
3 Did mammals radiate explosively after the Cretaceous-Tertiary event? Investigate the "classic" story of a massive diversification/adaptive radiation of mammals 65 Ma, and consider why molecular phylogenies and dates seem to indicate a much earlier diversification.
4 What happened to mammalian faunas 11,000 years ago, at the end of the ice ages? Read about the climate change and overkill hypotheses for Pleistocene extinctions, and decide which side of the debate has the best evidence.
5 Track the discoveries of new human fossils over the past 20 years, and focus on the question of dating human origins. What were the oldest human fossils in 1980, 1990 and 2000, and what is the view today?

Further reading

Benton, M.J. 2005. *Vertebrate Paleontology*, 3rd edn. Blackwell, Oxford.

Brunet, M., Guy, F., Pilbeam, D. et al. 2002. A new hominid from the upper Miocene of Chad, central Africa. *Nature* **418**, 145–51.

Carroll, R.L. 1987. *Vertebrate Paleontology and Evolution*. Freeman, San Francisco.

Chiappe, L.M. 2007. *Glorified Dinosaurs: The origin and early evolution of birds*. Wiley-Liss, New York.

Cracraft, J. & Donoghue, M.J. 2004. *Assembling the Tree of Life*. Oxford University Press, New York.

Culotta, E. 2007. The fellowship of the Hobbit. *Science* **317**, 740–2.

Delson, E., Tattersall, I., Van Couvering, J.A. & Brooks, A.S. 2002. *Encyclopedia of Human Evolution and Prehistory*, 2nd edn. Garland, New York.

Erickson, G.M., Curry-Rogers, K. & Yerby, S. 2001. Dinosaur growth patterns and rapid avian growth rates. *Nature* **412**, 429–33.

Farlow, J.O. & Brett-Surman, M.K. (eds) 1997. *The Complete Dinosaur*. Indiana University Press, Bloomington.

Fastovsky, D.E. & Weishampel, D.B. 2005. *The Evolution and Extinction of the Dinosaurs*, 2nd edn. Cambridge University Press, Cambridge.

Gould, S.J. (ed.) 2001. *The Book of Life*. Norton, New York.

Kemp, T.S. 2005. *The Origin and Evolution of Mammals*. Oxford University Press, Oxford.

Lewin, R. 2004. *Human Evolution*, 5th edn. Blackwell, Oxford.

Lewin, R. & Foley, R. 2003. *Principles of Human Evolution*, 2nd edn. Blackwell, Oxford.

Luo, Z.-X. 2007. Transformation and diversification in early mammal evolution. *Nature* **450**, 1011–19.

Morwood, M. & van Oosterzee, P. 2007. *A New Human*. Smithsonian Books, Washington, DC.

Reisz, R.R., Scott, D., Sues, H.-D. et al. 2005. Embryos of an Early Jurassic prosauropod dinosaur and their evolutionary significance. *Science* **309**, 761–4.

Rose, K.D. & Archibald, J.D. (eds) 2005. *Placental Mammals: Origin, Timing and Relationships of the*

Major Extant Clades. Johns Hopkins University Press, Baltimore.

Sander, P.M. & Klein, N. 2005. Developmental plasticity in the life history of a prosauropod dinosaur. *Science* **310**, 1800–2.

Sander, P.M., Mateus, O., Laven, T. & Knötschke, N. 2006. Bone histology indicates insular dwarfism in a new Late Jurassic sauropod dinosaur. *Nature* **441**, 739–41.

Savage, R.J.G. & Long, M.R. 1986. *Mammal Evolution*. British Museum (Natural History), London.

Senut, B., Pickford, M., Gommery, D. et al. 2001. First hominid from the Miocene (Lukeino Formation, Kenya). *Comptes Rendus de l'Académie de Sciences* **332**, 137–44.

Stringer, C.B. & Andrews, P. 2005. *The Complete World of Human Evolution*. Thames & Hudson, London.

Tabuce, R., Asher, R.J. & Lehmann, T. 2008. Afrotherian mammals: a review of current data. *Mammalia* **72**, 2–14.

Weishampel, D.B., Dodson, P. & Osmólska, H. (eds) 2004. *The Dinosauria*, 2nd edn. University of California Press, Berkeley.

Wood, B. & Richmond, B.G. 2007. *Human Evolution: A guide to fossil evidence*. Westview Press, Boulder, CO.

Zhou, Z.-H., Barrett, P.M. & Hilton, J. 2003. An exceptionally preserved Lower Cretaceous ecosystem. *Nature* **421**, 807–14.

References

Asher, R.J. 2007. A web-database of mammalian morphology and a reanalysis of placental phylogeny. *BMC Evolutionary Biology* **7**, 108. doi 10.1186/1471-2148-7-108.

Brown, P., Sutikna, T., Morwood, M.J. et al. 2004. A new small-bodied hominin from the Late Pleistocene of Flores, Indonesia. *Nature* **431**, 1055–61.

Brunet, M., Guy, F., Pilbeam, D. et al. 2002. A new hominid from the upper Miocene of Chad, central Africa. *Nature* **418**, 145–51.

Carpenter, K. 1982. Skeletal and dermal armor reconstruction of *Euoplocephalus tutus* (Ornithischia: Ankylosauridae) from the Late Cretaceous Oldman Formation of Alberta. *Canadian Journal of Earth Sciences* **121**, 689–97.

Estes, R. 1983. Sauria terrestria, Amphisbaenia. *Handbuch der Paläoherpetologie* **10A**, 1–249. Gustav Fischer, Stuttgart.

Fraser, N.C. & Walkden, G.M. 1984. The postcranial skeleton of the Upper Triassic sphenodontid *Planocephalosaurus robinsonae*. *Palaeontology* **27**, 575–95.

Gregory, W.K. 1951/1957. *Evolution Emerging*, Vols 1, 2. Macmillan, New York.

Hu, Y., Meng, J., Wang, Y. & Li, C. 2005. Large Mesozoic mammals fed on young dinosaurs. *Nature* **433**, 149–52.

Jacob, T., Indriati, E., Soejono, R.P. et al. 2006. Pygmoid Australomelanesian *Homo sapiens* skeletal remains from Liang Bua, Flores: population affinities and pathological abnormalities. *Proceedings of the National Academy of Sciences, USA* **103**, 13421–6.

Jepsen, G.L. 1970. *Biology of Bats. Vol. 1. Bat Origins and Evolution*. Academic, New York.

Lewin, R. 1999. *Human Evolution*, 4th edn. Blackwell, Oxford.

Newman, B.H. 1970. Stance and gait in the flesh-eating dinosaur *Tyrannosaurus*. *Biological Journal of the Linnean Society* **2**, 119–23.

Ostrom, J.H. 1969. Osteology of *Deinonychus antirrhopus*, an unusual theropod from the Lower Cretaceous of Montana. *Bulletin of the Peabody Museum of Natural History* **30**, 1–165.

Rinderknecht, A. & Blanco, R.E. 2008. The largest fossil rodent. *Proceedings of the Royal Society B* **275**, 923–8.

Senut, B., Pickford, M., Gommery, D. et al. 2001. First hominid from the Miocene (Lukeino Formation, Kenya). *Comptes Rendus de l'Académie de Sciences* **332**, 137–44.

Springer, M.S., Cleven, G.C., Madsen O. et al. 1997. Endemic African mammals shake the phylogenetic tree. *Nature* **388**, 61–4.

Springer, M.S., Murphy, W.J., Eizirik, E. & O'Brien, S. J. 2003. Placental mammal diversification and the Cretaceous–Tertiary boundary. *Proceedings of the National Academy of Sciences, USA* **100**, 1056–61.

Thewissen, J.G.M., Cooper, L.N., Clementz, M.T., Bajpai, S. & Tiwari, B.N. 2007. Whales originated from aquatic artiodactyls in the Eocene epoch of India. *Nature* **450**, 1190–4.

Thewissen, J.G.M., Hussain, S.Y. & Arif, M. 1994. Fossil evidence for the origin of aquatic locomotion in archaeocete whales. *Science* **263**, 210–12.

Wood, A.E. 1962. The early Tertiary rodents of the family Paramyidae. *Transactions of the American Philosophical Society* **52**, 1–261.

Zhou, Z.-H., Barrett, P.M. & Hilton, J. 2003. An exceptionally preserved Lower Cretaceous ecosystem. *Nature* **421**, 807–14.

Chapter 18

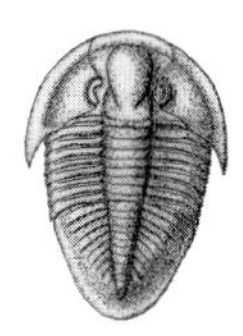

Fossil plants

Key points

- Fungi have a long fossil record, perhaps dating back to the end of the Precambrian, but they are not true plants.
- Green algae, and their relatives, are close to the origin of green plants.
- Plants moved onto land in the Ordovician and Silurian, a move enabled by the evolution of vascular tissues, waterproof cuticles and stomata, and durable spores.
- Various non-seed-bearing plants arose during the Devonian, but tree-like lycopsids, equisetopsids and groups such as ferns became established by the Carboniferous. These formed the great "coal forests".
- Palynology, the study of fossil pollen and spores, gives remarkable insights into paleoenvironments and biostratigraphy.
- The gymnosperms (seed-bearing plants) radiated in several phases: during the Carboniferous-Permian (medullosans, cordaites, cycads) and Mesozoic (conifers, ginkgos, bennettitaleans, gnetales).
- The angiosperms (flowering plants) radiated dramatically during the Cretaceous, and owed their success to fully enclosed and protected seeds, flowers and double fertilization.

What's in a name? That which we call a rose by any other name would smell as sweet.

William Shakespeare (1597) *Romeo and Juliet*

It is easy to take plants for granted, but just imagine a world without them! Not only would there be no forests and no grass, we would be unable to survive. Plants provide food for us, both directly (grains, vegetables, fruit, beans) and indirectly (through feeding farm animals that we eat), and – together with several other groups of eukaryotes and prokaryotes – they also provide the oxygen in the air we breathe by the process of **photosynthesis**. Without plants the landscape would not only be empty, there would be no soil either: soil is made of weathered rock and organic components deriving from plant remains. So, in the time before plants clothed the landscape, rates of erosion were 10 times, or higher, than they are today: a shower of rain would wash sand and debris from the landscape in a catastrophic way, as in desert wadis today. Speak to plants: let them know they are appreciated.

The study of fossil plants falls into two disciplines: **paleobotany**, which concentrates on macroscopic (visible with the naked eye) plant remains, and **palynology**, which is mainly the study of pollen and spores. Palynology is usually treated as a branch of micropaleontology (see Chapter 9) because palynologists use microscopes and much of the work is aimed specifically at biostratigraphic correlation, often for commercial purposes. We touch on palynology in this chapter, but concentrate on the history of whole plants, based on the study of leaves, roots, wood, flowers, fruits and seeds.

The fossil record of plants is rich, and a great deal of information is available about the main stages in plant evolution, from their move onto land, plants in the age of the dinosaurs, and the origin of flowering plants. Many fossil localities produce exquisitely preserved plant fossils (see p. 69), and this has allowed very detailed microscopic study of the cellular structure of ancient leaves, seeds and wood. True plants, or **metaphytes**, are considered in this chapter, together with their closest algal relatives, and the Fungi, even though modern molecular studies have shown that these groups are not particularly closely related (see pp. 190–1).

TERRESTRIALIZATION OF PLANTS

Fungi

The Fungi, represented by familiar molds and mushrooms, are not true plants. They form a separate kingdom that is more closely related to multicelled animals (Metazoa) than to multicelled plants (Metaphyta) (see pp. 190–1). However, they are included in this chapter because of botanical (and culinary!) tradition. Fungi are classified into a number of phyla on the basis of reproductive patterns. In some cases, there are specialized reproductive structures that may be identified in well-preserved fossils.

Until recently, the first good fossils of fungi were from the Devonian and Carboniferous. Now, some possible lichens have been reported by Yuan and colleagues (2005) from the Doushantuo Formation of China, a Late Neoproterozoic deposit dated at 600 Ma (see pp. 237–8). Lichens are commonly seen as scaly, grayish growths on the bark of trees, but what most people do not realize is that they are composed of two organisms, a **symbiotic** partnership between a fungus and a cyanobacterium and/or alga. Each partner contributes to the wellbeing of the lichen: the cyanobacteria or algae photosynthesize and produce glucose from carbon dioxide, while the fungus provides moisture, nutrients and protection for the consortium. The Doushantuo specimens are so well preserved, even to cellular level, that most paleobotanists are convinced by the new finds. It had long been suspected that cyanobacteria formed thin crusts on land, as they do in desert regions (e.g. Utah) today, photosynthesizing and forming thin "soils" in the Neoproterozoic. The Doushantuo lichens prove that the surface of the land, at least close to water, was already green at the end of the Precambrian, long before plants really conquered the land.

Devonian and Carboniferous fungi appear to have acted as decomposers, feeding on decaying plant material, or as parasites, infesting the tissues of living plants. In the Early Devonian Rhynie Chert, for example, fungal remains include mats of **hyphae**, branching tissue strands, some of them bearing reproductive structures (Fig. 18.1a, b), similar to those of modern oomycete Fungi. Signifi-

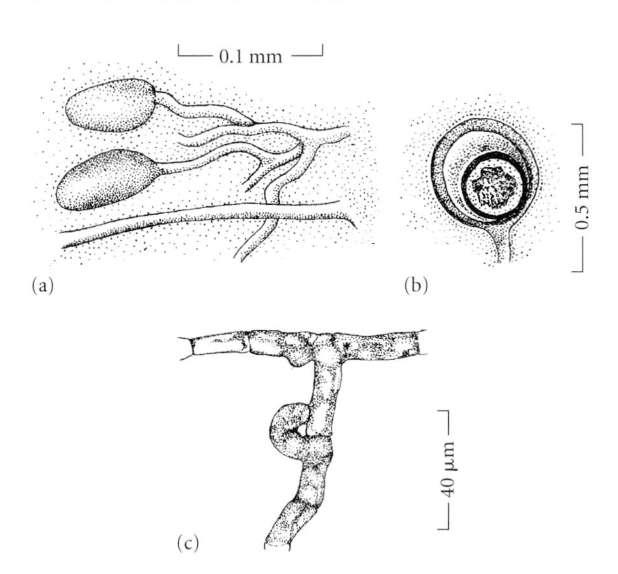

Figure 18.1 Examples of fossil fungi: (a, b) *Palaeomyces*, a possible oomycete fungus from the Early Devonian Rhynie Chert of Scotland, showing branching non-septate hyphae terminated by enlarged vesicles (a) and a resting spore (b); and (c) *Palaeancistrus*, with basidiomycete-like clamp connections, from the Pennsylvanian of North America. (a, b, courtesy of Thomas N. Taylor; c, based on Stewart & Rothwell 1993.)

cantly, the Rhynie Chert also provides the earliest evidence of symbiosis between a land plant and a fungus. Highly branched networks of thin-walled hyphae within the rhizomes of the Chert plants strongly resemble living **arbuscular mycorrhizae**, an extraordinarily widespread association of plant roots with fungal hyphae, which play a key role in the uptake of solutes in the roots of modern plants. This association between plants and fungi then goes back to the dawn of plant life on land.

Coal balls have yielded information on other fungal groups. A Carboniferous fungus, *Palaeancistrus* (Fig. 18.1c), shows extensive developments of hyphae in a mat-like structure or **mycelium**, with specialized hooked terminations on marginal hyphae called clamp connections. These are characteristic of another living fungal group, the Basidiomycotina. After the Carboniferous, there are sporadic records of fungi of different groups. Particularly abundant finds come after the radiation of flowering plants, when fungi adapted to parasitize the roots, stems and leaves of the new plant group, especially in humid tropical conditions.

The greening of the land: mosses, liverworts and hornworts

Fungi, algae and cyanobacteria may have formed crusts and thin soils from the Late Precambrian, but the land did not become green until the Ordovician at least. Algae, fungi and plants make soil by growing on the surface of rocks, and assisting their breakdown into separate grains that mix with organic debris, that in turn nourish further organic growth. The first land plants seem to have been the bryophytes. There are some 25,000 species of bryophytes today, divided into three distinctive groups. Liverworts and hornworts are flattened branching structures, some of which show differentiation into upright stems and leaves. Mosses are upright plants with slender stems and, typically, spirally arranged leaves.

Bryophytes show some special adaptations to life on land, such as a waterproof cuticle over their leaves and stems. Many hornworts and mosses have stomata, used for controlling water loss (see below) but they are absent in liverworts. A few of the larger mosses and liverworts have a very simple vascular conducting system. Some bryophytes have the unusual ability of being able to dry up completely, then rehydrating when rain falls, and continuing as normal.

The fossil record of the bryophytes is patchy. This is often attributed to low preservation potential, but fossil specimens are also difficult to distinguish from other simple land plants. The oldest recorded fossil bryophytes are Ordovician (Box 18.1) to Devonian in age, although interpretations are uncertain. For example, *Sporogonites* (Fig. 18.3) from the Lower Devonian of Belgium has been interpreted as a part of the flattened portion of a liverwort with, growing from it, the slender-stemmed spore-bearing phases of the plant. This specimen shows the unique feature of bryophytes, that their reproductive stages are the opposite of those in vascular plants. A possible Cambrian relative, *Parafunaria*, has been reported from China.

Box 18.1 The first plants on land

For years, paleontologists assumed that plants and animals moved onto land in the Silurian and Devonian: some excellent fossil examples of earliest plants and arthropods date from the Mid to Late Silurian, and these show small waterside vascular plants together with spiders, scorpions and precursors of insects and millipedes. Hints of older land plants were reported from the Ordovician. For example, Ordovician soils with root-like structures, or burrows, prove that plants were already on land: you do not get soils without plants. There are even fossil soils as old as 1.2 Ga from the Precambrian that were presumably generated by microbial or algal activity.

But something happened in the Mid Ordovician, some 470 Ma. The character of microfossil assemblages changed dramatically, with the first appearance of spores (Fig. 18.2). **Spores** are airborne microscopic cells that are characteristic of land plants. So, although these earliest land plants have not yet themselves found as fossils, they must have been there because they were producing spores already. But were these early spores produced by land plants or by their immediate green algal antecedents on their route to the land?

These Ordovician spores had been hard to understand until Charlie Wellman from the University of Sheffield, England and colleagues showed in 2003 that the spores were probably produced by small bryophytes, perhaps like liverworts. Their study of the spore walls showed some detailed similarities to those of modern liverworts, and they also found clusters of spores packaged in a type of cuticle that looked overall like a liverwort sporangium. It seems that non-vascular bryophyte-like plants invaded the land in the Mid Ordovician, and later true vascular plants evolved from within this complex.

Find web links at http://www.blackwellpublishing.com/paleobiology/.

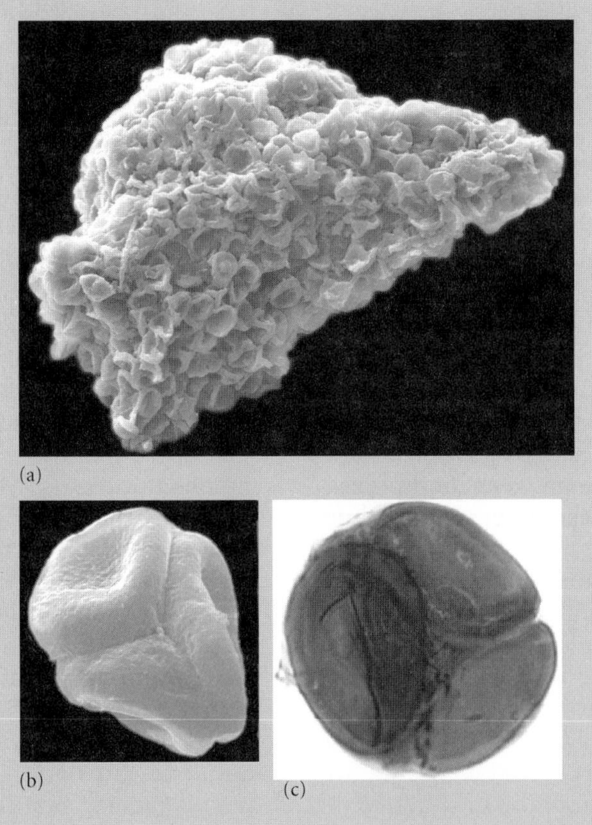

Figure 18.2 The oldest evidence of vascular plants on land? Spores from the Mid Ordovician (470 Ma) of Oman, scanning electron microscope images of a mass of spores (a) and close-up of one spore tetrad (b), and light microscope view of a spore tetrad (c). (Courtesy of Charlie Wellman.)

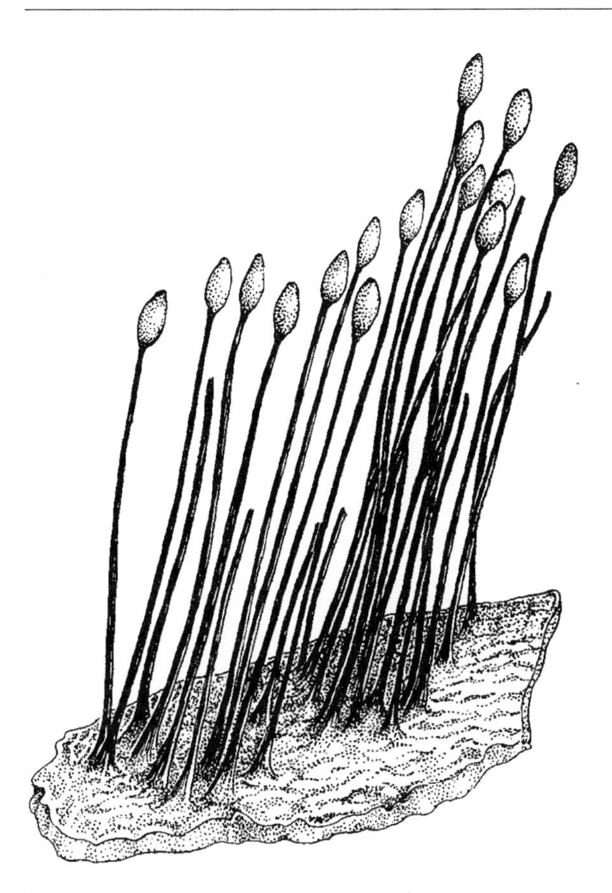

Figure 18.3 *Sporogonites*, an Early Devonian bryophyte, seemingly showing numerous slender sporophytes (20 mm tall) growing from a basal gametophyte portion. (Based on Andrews 1960.)

Relationships of green plants

Paleobotanists have long sought the origins of land plants among the Chlorophyta, the green algae, and molecular evidence confirms their close relationship (see pp. 197–8). Broader cladistic studies (Kenrick & Crane 1997) have shown that many forms traditionally classified as "algae" are close outgroups of land plants, the closest being the Charophyceae (including charophytes, see pp. 197–8), with the Chlorophyta a little more distantly related.

These algal groups, together with land plants, form a larger clade termed the Chlorobionta, or green plants (Fig. 18.4), that are all characterized by the possession of chlorophyll *b* and similarities of their flagellate cells and chloroplasts (Kenrick & Crane 1997). This clade dates back to at least 1200 Ma, the age of the red alga *Bangiomorpha* from Canada (see pp. 200–1). Chlorobiont evolution is hard to track in detail in these early stages because the fossils are, on the whole, microscopic and their diagnostic features subcellular and therefore rarely preserved. Finds improve with the diversification of land plants in the Silurian and Devonian.

The Chlorobionta are divided into various "algal" groups and the major clade Embryophyta. This latter clade is divided into two basic grades, the non-vascular plants (bryophytes) such as mosses, liverworts and hornworts that evolved in the Ordovician, and vascular plants (tracheophytes) that arose in the Mid Silurian. Bryophytes are mostly small, whereas tracheophytes have evolved into very large organisms. This might be explained by a number of reasons, including competition for light, as well as other benefits of large size such as longevity and the ability to produce many more reproductive and dispersal units per individual. Large size in vascular plants came with the evolution of the **cambium** (a lateral tissue that allows increase in girth). Hence the "race for the skies" among the early land plants that resulted in large trees by the Middle Devonian.

The tracheophytes (Box 18.2) are the vascular plants, characterized by vascular canals with secondary thickening, and include the rhyniopsids and lycophytes (lycopsids and zosterophyllopsids) as basal groups. Next up the main axis of the cladogram are the horsetails (equisetopsids) and ferns (filicopsids).

The spermatopsids (seed-bearers) are traditionally divided into gymnosperms and angiosperms (flowering plants). Flowering plants are so successful today, and they seem so different from other plants, that Charles Darwin famously referred to their origin as an "abominable mystery". Fossils suggest that gnetaleans (a group with a small number of living members) or one of several extinct Mesozoic groups, such as the Bennettitales or Caytoniales, are the sister group of angiosperms. Successively more distant outgroups are the cycads, the conifers + ginkgos, and the medullosans. Molecular studies (e.g. Bowe et al. 2000) suggest, on the other hand, that the gymnosperms form a distinct clade consisting of cycads, ginkgos, conifers and gnetaleans, none of which then is any closer to the origin of angiosperms than the others. These findings may require further study.

A phylogeny of tracheophytes (Fig. 18.5) shows the broad stratigraphic range and

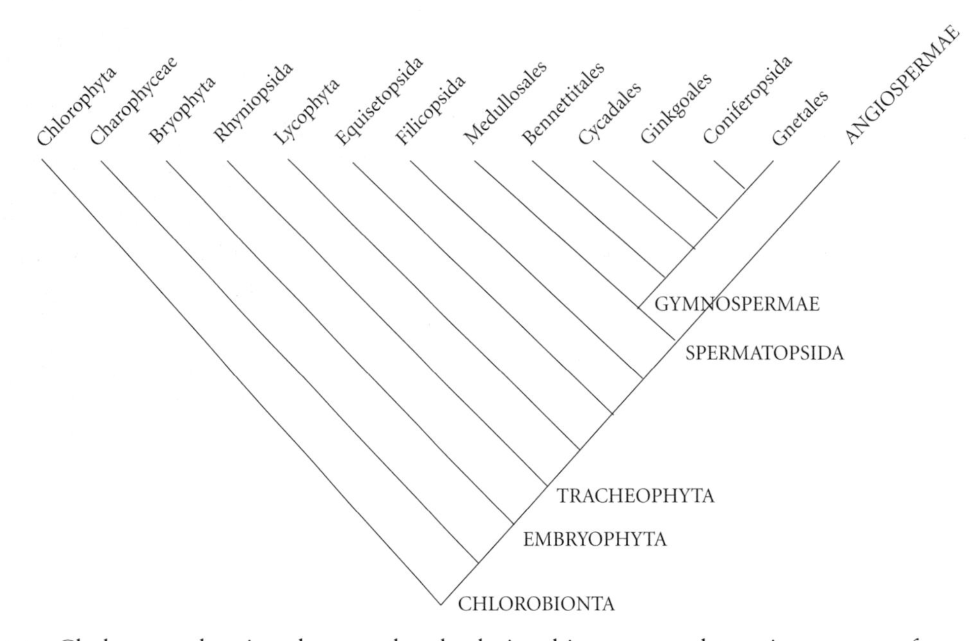

Figure 18.4 Cladogram showing the postulated relationships among the major groups of vascular land plants. Some synapomorphies that define particular nodes are: Chlorobionta (chlorophyll *b*), Charophyceae + Embryophyta (cell structure), Embryophyta (alternation of generations), Tracheophyta (vascular canals (tracheids) and secondary thickening) and Spermatopsida (seeds). Read more about the "deep green" project to establish a complete phylogeny of green plants at http://ucjeps.berkeley.edu/bryolab/GPphylo/.

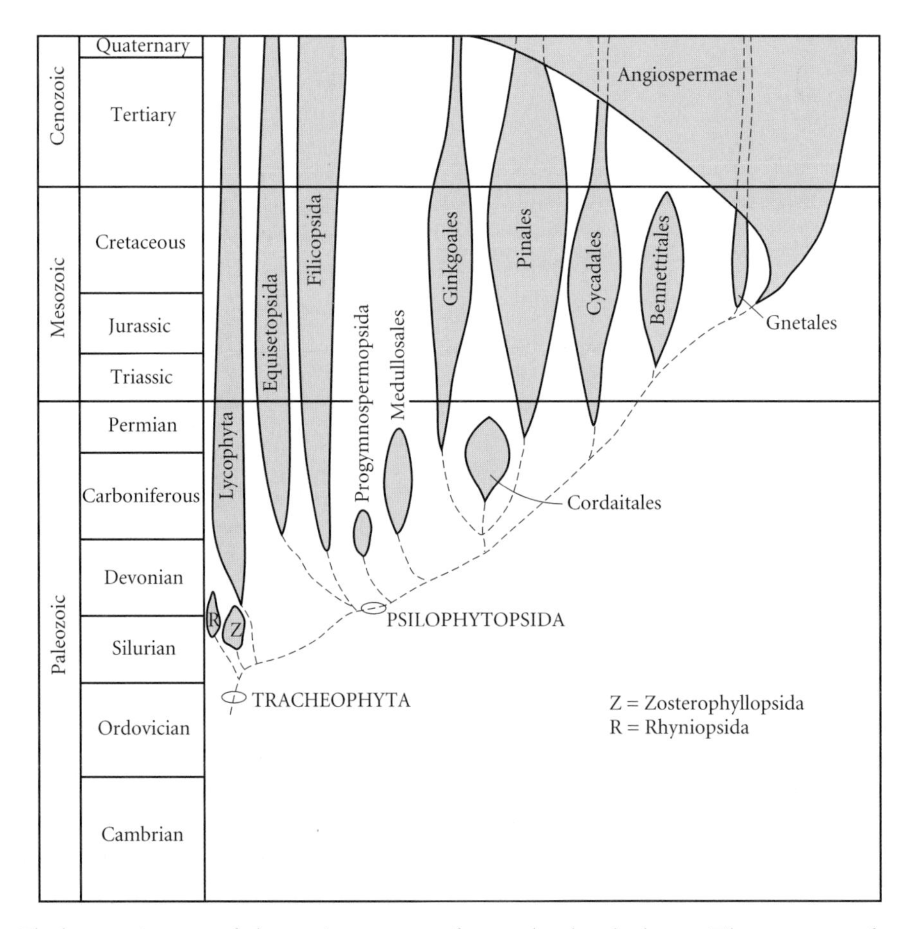

Figure 18.5 Phylogenetic tree of the main groups of vascular land plants. The pattern of postulated relationships is based on the cladogram (Fig. 18.4), and details of known stratigraphic range and species diversity are added.

Box 18.2 Classification of the tracheophytes

Tracheophytes are the vascular land plants, and include modern ferns, horsetails, conifers and flowering plants, as well as numerous extinct groups. The basal groups are distinguished in terms of branching patterns and sporangial morphology.

Division Tracheophyta

Class RHYNIOPSIDA

- Simple vascular plants with dichotomously branching stems and terminal sporangia
- Mid Silurian to Early Devonian

Class Lycophyta

- Small to large plants with lateral sporangia and (usually) small leaves
- Late Silurian to Recent

Class EQUISETOPSIDA

- Horsetails; vertical stems with jointed structure and a whorl of fused leaves at the nodes; sporangia grouped in cones
- Late Devonian to Recent

Class FILICOPSIDA

- Ferns; dichotomously-branching flat leaves which uncurl as they develop; sporangia are grouped in clusters usually on the underside of leaves
- Mid Devonian to Recent

Class PROGYMNOSPERMOPSIDA

- Plants with gymnosperm-like wood but free sporing (fern-like) reproduction. Larger members include early trees such as *Archaeopteris* from the Late Devonian

Class SPERMATOPSIDA

Subclass GYMNOSPERMAE

Order MEDULLOSALES

- Primitive seed plants with large pollen grains and unusual stem anatomy
- Mississippian to Permian

Order BENNETTITALES

- Bushy to tree-like plants with sterile scales between the seeds; frond-like leaves; flower-like cones with enclosing structures that surround ovules and pollen sacs
- Late Triassic to Late Cretaceous

Order CYCADALES

- Bushy to tree-like plants with leaf traces that girdle the stem; frond-like leaves; seeds attach to a megasporophyll stalk below a leaf-like structure
- Mississippian-Recent

Order GINKGOALES

- Trees with seed-bearing shoots and with fan-shaped or more divided leaves
- Late Triassic to Recent

Order CONIFERALES

- Conifers; trees with resin canals, and needle- or scale-like leaves
- Mississippian to Recent

Order GNETALES

- Leaves opposite each other, and vessels in the wood; male and female cones are flower-like
- Late Triassic to Recent

Subclass ANGIOSPERMAE

- Ovules are enclosed in carpels, within a flower, and fertilization is double (involving two sperm nuclei)
- Early Cretaceous to Recent

relative abundance of each group at different points in plant history. The phylogeny highlights the three major bursts of land plant evolution, the first in the Devonian (rhyniopsids, zosterophylls and other basal vascular plants), in the Carboniferous and Permian (lycopsids, ferns, horsetails, seed ferns) and in the Cretaceous (angiosperms).

Adapting to life on land

The following are the key adaptations of vascular plants for life on land:

1 Spores or seeds with durable walls to resist desiccation.
2 Surface cuticle over leaves and stems to prevent desiccation.
3 Stomata (singular, stoma), or controllable openings, to allow gas exchange through the low-permeability cuticle.
4 A vascular conducting system to pass fluids through the plant.
5 The lignification of tracheids to resist collapse. The cellulose cell walls of the conducting tubes, or **tracheids**, of vascular plants are invested with **lignin**, the tough polymer that makes up all woody tissues, providing strength and waterproofing.

These key adaptations relate to the problems a water plant must overcome when moving onto land. In water, a plant may absorb nutrients and water all over its surface, but on land all such materials must be drawn from the ground, and passed round the tissues internally. Land plants typically have specialized roots that draw moisture and nutrient ions from the soil, which are passed through water-conducting systems that connect all cells. The system is driven by **transpiration**, a process powered by the evaporation of water from leaves and stems. As water passes out of aerial parts of the plant, fluids are drawn up into the water-conducting system hydrostatically.

Water loss is a second key problem for plants on land. Whereas in water fluids may pass freely in and out of a plant, a land plant must be covered with an impermeable covering – the waxy **cuticle**. Gaseous exchange and water transport are then facilitated in many land plants by specialized openings, the **stomata** (singular, **stoma**), often located on the underside of leaves. Typically, stomata open and close depending on carbon dioxide concentration, light intensity and water stress.

The third problem of life on land is support. Water plants simply float, and the water renders them neutrally buoyant. Most land plants, even small ones, stand erect in order to maximize their uptake of sunlight for photosynthesis, and this requires some form of skeletal supporting structure. All land plants rely on a **hydrostatic skeleton**, a stiff framework supported by water in tubes, and some groups have evolved additional structural support through **lignification** of certain tissues in the wood and cortex, the process whereby lignin encrusts cellulose fibers.

Plant reproductive cycles

Plants may reproduce vegetatively and sexually. **Vegetative reproduction**, or budding, is an asexual reproductive process that involves no exchange of material from different individuals, no male and female cells. It is a property of many plants that they may multiply in this way, either naturally or by human intervention.

Algae show all kinds of reproduction, vegetative, asexual and sexual. Sexual reproduction (Fig. 18.6a) involves the combination of cellular material from two organisms of the same species. The reproductive cells, or **gametes** (sperms from the "male" and eggs from the "female"), contain a single set of n chromosomes, the **haploid** condition. When the gametes combine, forming a zygote, the chromosome number doubles to the **diploid** condition, $2n$. The diploid plant stage, the **sporophyte**, produces haploid (n) spores, each of which develops into a haploid **gametophyte** plant stage. It is the gametophyte that produces the haploid sperms and eggs.

In typical vascular plants, the green plant that we see is the sporophyte, while the gametophyte is very small (Fig. 18.6b). The opposite is the case in bryophytes, where the visible mosses and liverworts are haploid gametophytes, and the sporophyte is a small plant that depends for nourishment on the larger gametophyte (Fig. 18.6c). Hence, in *Sporogonites* (see Fig. 18.3), numerous sporophytes appear to be growing from a portion of the larger flattened gametophyte phase. Translat-

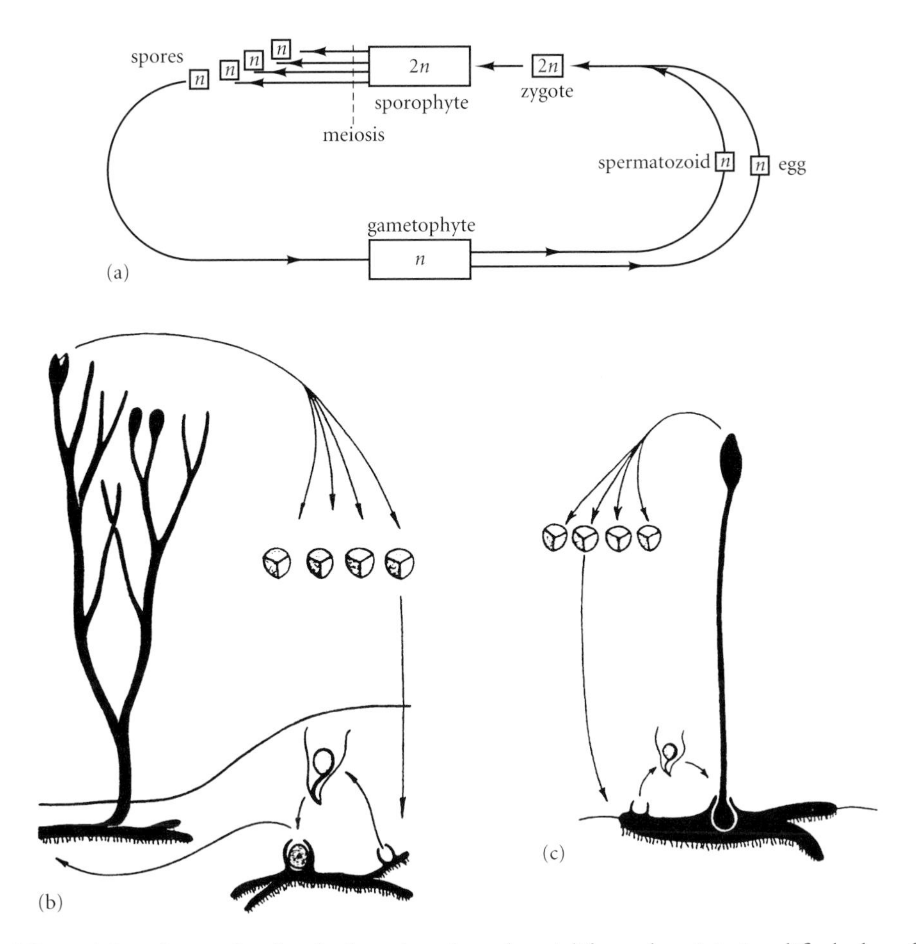

Figure 18.6 The origin of vascular land plant (tracheophyte) life cycles: (a) simplified plant life cycle showing alternation of phases; (b) life cycle of a hypothetical tracheophyte, with a dominant sporophyte phase and reduced gametophyte, in comparison with the life cycle of a hypothetical bryophyte (c), where the dominant phase is the gametophyte, and the sporophyte is a reduced dependent structure. (Based on various sources.)

ing to the human case, this would be like having the haploid sperm or egg dominant, and the diploid body (sporophyte) repressed!

Vascular plants in the Silurian and Devonian

As we have seen, non-vascular land plants had arisen at least by the Mid Ordovician, and vascular plants by the Mid to Late Silurian, some 425 Ma. Vascular plants are characterized by the possession of tracheids, true vascular conducting systems. Lignin and stomata are typical of vascular plants, but may not have been present in the earliest forms.

The oldest vascular plant is *Cooksonia* from the Mid Silurian of southern Ireland, a genus that survived until the end of the Early Devonian. *Cooksonia* (Fig. 18.7a–d) is composed of cylindrical stems that branch in two at various points and are terminated by cap-shaped **sporangia**, or spore-bearing structures, at the tip of each branch. The specimens of *Cooksonia* range from tiny Silurian examples, only a few millimeters long, to larger Devonian forms up to 65 mm long. Extraordinary anatomic detail has been revealed by studies of specimens of these tiny plants that have been freed from the rock by acid digestion, and then mounted in resin. The sporangia have been dissected to reveal that they were packed with spores, the vascular conducting tissues of Early Devonian examples have thickened walls, and there are stomata on the outer surfaces of the stems (Edwards et al. 1992).

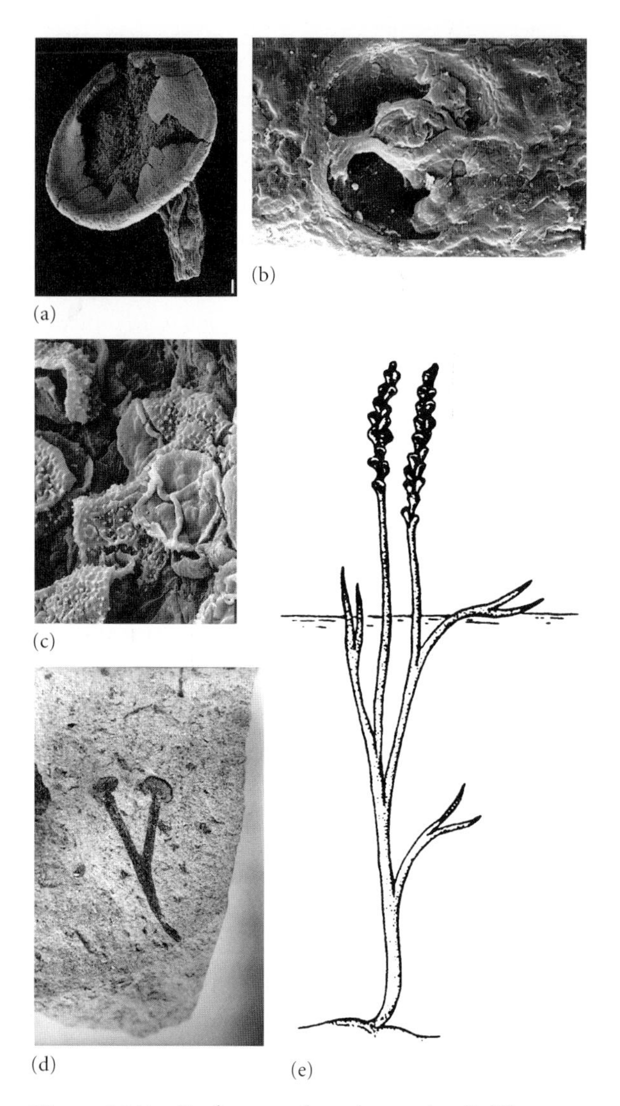

Figure 18.7 Early vascular plants. (a–d) The oldest land plant, *Cooksonia* from the Silurian to Early Devonian. Early Devonian examples from Wales, showing a complete sporangium at the end of a short stalk (a), a stoma (b) and spores (c). The sporangium is 1.6 mm wide, the stoma is 40 μm wide and the spores are 35 μm in diameter. (d) Reconstruction of *Cooksonia caledonica*, a Late Silurian form, about 60 mm tall. (e) *Zosterophyllum*, a zosterophyllopsid from the Early Devonian of Germany, 150 mm tall. (a–d, courtesy of Dianne Edwards; e, based on Thomas & Spicer 1987.)

Cooksonia is a member of the Rhyniopsida, the basal group of vascular plants, the Tracheophyta. Rhyniopsids are known most fully from the Early Devonian Rhynie Chert of northeast Scotland, a deposit that has preserved numerous plants and arthropods exquisitely in silica (Box 18.3). Some of the Rhynie rhyniopsids reached heights of 180 mm. They consisted of groups of vertical stems supported on horizontal branching structures that probably grew in the mud around small lakes.

Several other groups of vascular land plants arose in the Early Devonian. *Zosterophyllum* (Fig. 18.7e), a zosterophyllopsid, shares many features with the rhyniopsids, but has numerous lateral sporangia, instead of a single terminal one, on each vertical stem. Later in the Devonian, some basal tracheophytes became taller, as much as 3 m, the size of a shrub, and these indicate the future evolution of some vascular plants towards large size.

THE GREAT COAL FORESTS

Lycopsids, small and large

The clubmosses, Class Lycopsida, arose at the same time as the rhyniopsids and other dichotomously branching plants, but they are distinguished by having their sporangia arranged along the sides of vertical branches, instead of at the tips, and by having numerous small leaves attached closely around the stems.

Low **herbaceous** lycopsids existed throughout the Devonian and Carboniferous, and they showed considerable variation in leaf and sporangium shape, and in the nature of the spores. From the Late Devonian onwards, most lycopsids produced two kinds of spores, small and large (microspores and megaspores), that developed within terminal cones. Lycopsids are represented today by some 1100 species, all small herbaceous forms.

During the Carboniferous, several lycopsid groups achieved giant size, and these are the dominant trees seen in reconstruction scenes of the great coal swamps of that period. The best known is *Lepidodendron*, a clubmoss that reached 35 m or more in height. Fossils of *Lepidodendron* have been known for 200 years because they are commonly found in association with commercial coalfields in North America and Europe. At first, the separate parts – roots, trunk, bark, branches, leaves, cones and spores – were given different names, but over the years they have been assembled to produce a clear picture of the whole plant (Fig. 18.9).

The giant lycopsids were adapted to the wet conditions of the coal swamps, but these habitats receded at the time of a major arid phase in the latest Pennsylvanian and Early Permian. *Lepidodendron* and its like died out. Medium-sized lycopsids, about 1 m high, existed during the Mesozoic, but truly **arborescent** ("tree-like") forms never evolved again.

The horsetails

The horsetails, or equisetopsids, are familiar to gardeners as small pernicious weeds. Their upright green shoots, with a characteristic jointed structure, are linked by underground rhizome systems. The sporangia are grouped into bunches of five or 10 below an umbrella-like structure arranged along the stem to form a sort of cone, a unique feature of the group. The horsetails are a small group today, consisting of a mere 15 species, most of them small, but one reaching a height of 4 m or more. The early history of the group shows much greater diversity.

The horsetails arose during the Devonian, and Carboniferous forms flourished in disturbed streamside settings where they could

Box 18.3 The Rhynie Chert: a window on earliest land life

Rhynie is a remote village in northeast Scotland consisting of only 50 or so houses; the bus stops there once a day. In 1914, Dr William Mackie, a physician, found traces of plant fossils in some speckled black and white chert rocks. He cut thin sections and took his specimens to Glasgow, where Robert Kidston, the foremost expert in Britain on floras of the Carboniferous, confirmed that the chert contained nearly perfectly preserved plants. Kidston, together with William Lang, Professor of Botany at the University of Manchester, England, published a classic series of monographs (Kidston & Lang 1917–1921) in which they presented superb photographs of microscopic sections through the Rhynie Chert plants. These publications established the Rhynie Chert as one of the oldest land-based ecosystems on Earth.

The Rhynie fossils include the remains of seven vascular land plants, as well as algae, fungi, one species of lichen and bacteria, as well as at least six groups of terrestrial and freshwater arthropods. What is amazing is the quality of preservation: every cell and fine detail can be seen, as if frozen in an instant and preserved forever (see Fig. 3.8a).

The Rhynie ecosystem was no towering forest. If you went for a stroll in Scotland in the Early Devonian, the green rim of plants probably did not extend far from the sides of ponds and rivers, and the tallest plants would have barely brushed your knees (Fig. 18.8). To see anything, you would have to go down on your hands and knees, and peer at the stems through a magnifying glass. Most of the taller plants had smooth stems, and branched simply in two, with knob-like sporangia at the tops of their stems – just larger examples of rhyniopsid plants like the Silurian *Cooksonia* (see Fig. 18.7a-d). *Asteroxylon*, a relative of *Zosterophyllum* (see Fig. 18.7e), had small scale-like leaves growing up from the stem. Microscopic cross-sections of these plants show they had simple vascular canals, stomata and terrestrial spores. Between the plants crept spider-like trigonotarbids and insect-like arthropods, and some of these are even found within cavities in the plant stems. There were crustaceans in the warm pools.

These discoveries show how extraordinary the preservation is in the Rhynie Chert. The fossils are silicified through having been flooded by silica-rich waters from nearby hot springs. Recent work (Trewin & Rice 2004) has confirmed that Scotland in the Early Devonian was an actively volcanic zone, perhaps related to Caledonian tectonic activity associated with the closure of the Iapetus Ocean (see pp. 45–8). Rhynie in the Early Devonian was like Yellowstone National Park today, with hot geysers erupting and immersing vegetation in silica-rich waters at a temperature of 35°C – an ecosystem frozen (or boiled) in time.

Read more about the Rhynie Chert in the volume by Trewin and Rice (2004), and on web links from http://www.blackwellpublishing.com/paleobiology/.

Continued

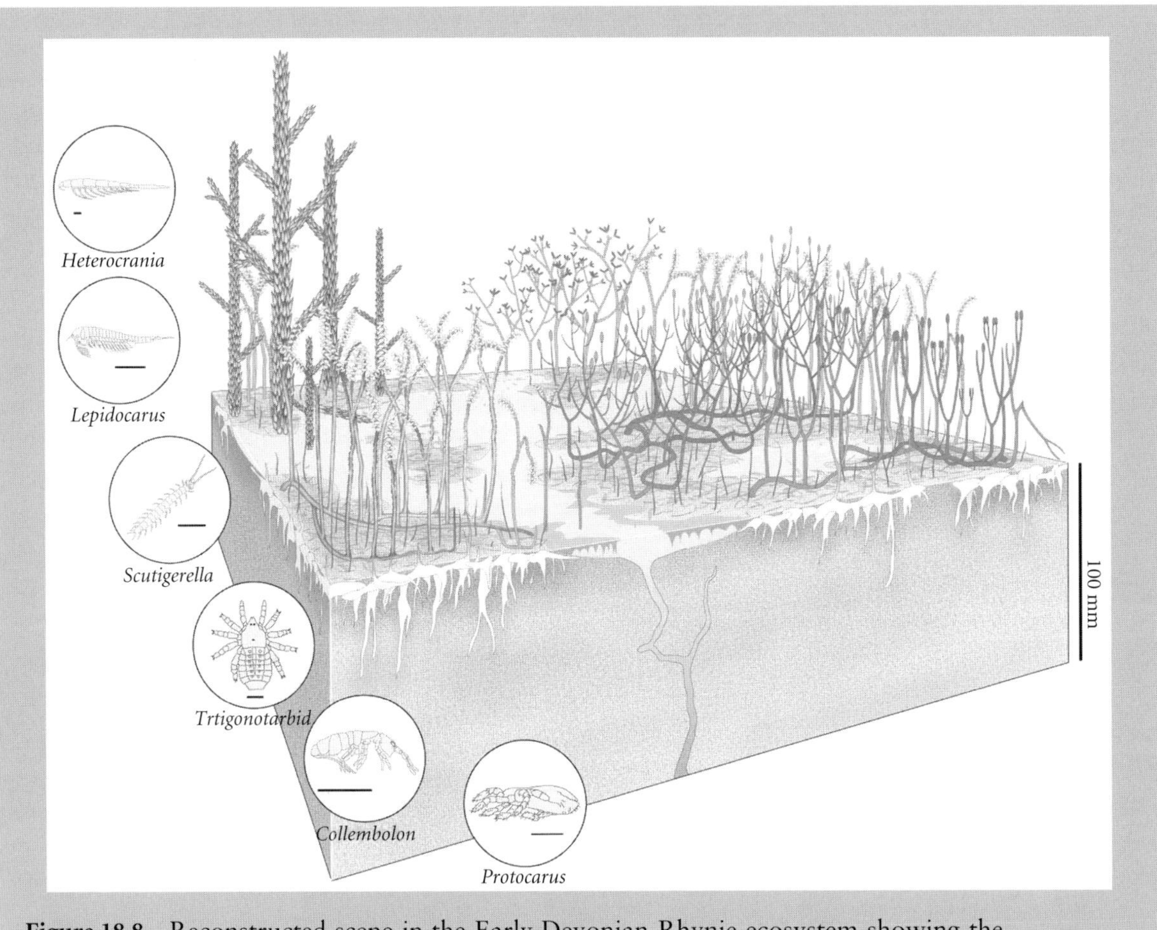

Figure 18.8 Reconstructed scene in the Early Devonian Rhynie ecosystem showing the commonest vascular plants *Rhynia* and *Asteroxylon* in the foreground, and a selection of small arthropods that lived in the water and in and on the plants (scale bars, 100 μm). (Drawing by Simon Powell, based on information from Nigel Trewin.)

resprout from underground rhizomes. They grew in incredibly dense, bamboo-like thickets with locally more than 10 trees per square meter. One form, *Calamites* (Fig. 18.10a), reached nearly 20 m in height, but shows the jointed stems and whorls of leaves at the nodes typical of modern smaller horsetails; the trunk of *Calamites* generally arose from a massive underground rhizome. Horsetail leaves formed radiating bunches at nodes along the side branch (Fig. 18.10b), and there were usually two types of cones, some bearing megaspores (Fig. 18.10c), and others bearing microspores. The giant horsetails disappeared at the end of the Pennsylvanian, as did the arborescent lycopsids. Some modest tree-like forms up to 2 m tall existed in the Permian and Triassic, but later horsetails were mainly small plants living in damp boggy areas.

Ferns – fronds or leaves?

Ferns are familiar plants today, typically with long fronds, each composed of feathery side branches that uncurl as they develop. Various fern-like plants are known in the Devonian and Carboniferous, and undisputed ferns are known in abundance from the Carboniferous onwards. As with the lycopsids and horsetails, some of the Carboniferous ferns, like *Psaronius* (Fig. 18.11) were tree-like. The fronds were borne on a vertical trunk, and they show all the features of their smaller modern tropical relatives. Other Carbonifer-

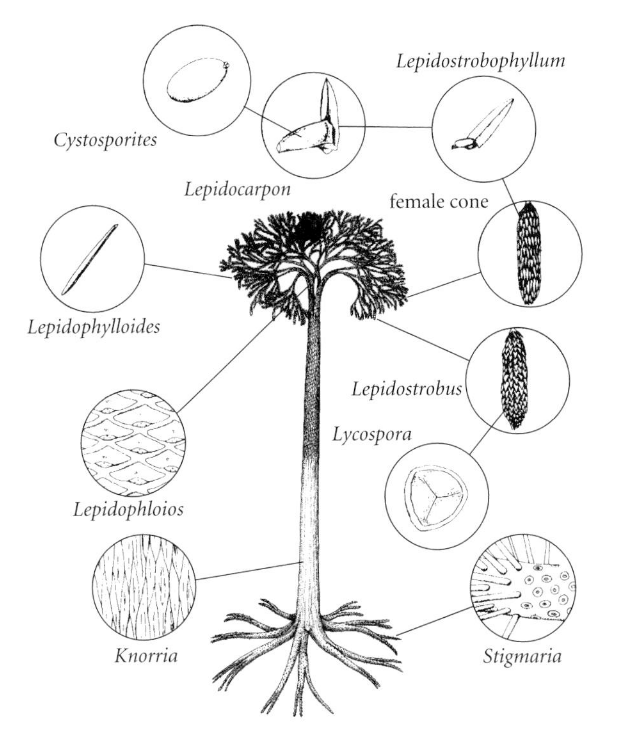

Figure 18.9 Reconstructing the arborescent lycopsid *Lepidodendron*, a 50 m-tall tree from the Carboniferous coal forests of Europe and North America. No complete specimen has ever been found, but complete root systems, *Stigmaria*, and logs from the tree trunk are relatively common. The details of the texture of the bark, branches, leaves, cones, spores and seeds are restored from isolated finds.

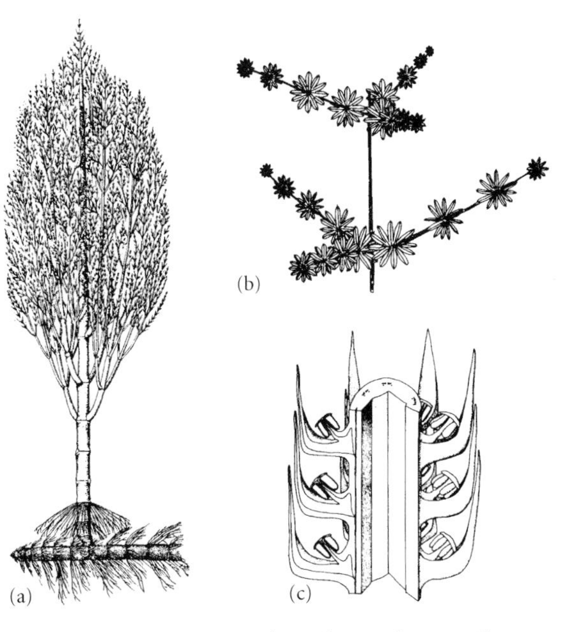

Figure 18.10 Giant Carboniferous horsetails: (a) *Calamites*, a 10 m-tall tree; (b) *Annularia*, portion of a terminal shoot bearing 10 mm-long leaves; and (c) *Palaeostachya*, diagrammatic cross-section of a cone-like structure, 15 mm in diameter, bearing small numbers of megaspores. (Based on Thomas & Spicer 1987.)

ous and Permian ferns were smaller herbaceous plants.

Ferns today are generally low-growing herbaceous plants, common in many environments. Ferns are resilient plants. After the huge eruption of Mount St Helens in 1980, the first living things to appear through the thick layers of ash were ferns. Their fronds had been burned to the ground, and yet somehow they were not killed, and they uncurled through the ash to begin the greening of the Washington State landscape within weeks of the eruption.

Is the fern frond a leaf or a branch? Technically, it is a branch, and the individual small green structures along each part of the frond are leaves. So, each frond is made from many small leaves. Leaves are common in more advanced plants, and they may have arisen by fusion of the small leaflets of fern-like plants to provide a typical leaf, an efficient broad photosynthesizing structure that can turn to face maximally toward the sun.

The ferns showed a second burst of evolutionary radiation during the Jurassic and Cretaceous, and they are the dominant plants in some Jurassic floras. Again, there were tree-like forms, as well as the more familiar low-growing ferns seen today.

Archaeopteris: the missing link?

One of the greatest developments in the evolution of plants was the seed, a key feature of the dominant modern plant groups, the gymnosperms and angiosperms. All the other plant groups considered so far – bryophytes, rhyniopsids, clubmosses, horsetails and ferns – lack seeds. Gymnosperms and angiosperms also show an advance in their woody tissues that permits the growth of very large trees; their lignified tracheids, vascular canals, can develop in a secondary system. The extraordinary early tree *Archaeopteris* from the Mid to Late Devonian, seems to represent a

Figure 18.11 The tree fern *Psaronius*, a 10 m-tall fern from the Pennsylvanian of North America. (Based on Morgan 1959.)

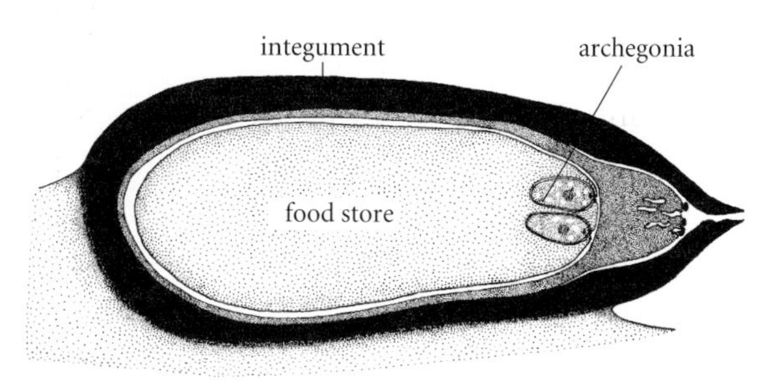

Figure 18.12 A typical gymnosperm seed, the ovule of *Pinus*, the pine, showing the archegonia (fertile female structures) surrounded by a substantial food store. Sperm enter through a narrow gap in the protective integument, and pass through pollen tubes to the archegonia.

half-way stage. It looked superficially like a tree fern, but its trunk showed the development of secondary woody tissues and growth rings, as seen in modern conifers.

SEED-BEARING PLANTS

The origin of seeds

The first plants with seeds are known from Late Devonian rocks, and seed bearers rose to prominence during the Carboniferous. After the end of the Carboniferous, and the extinction of arborescent lycopsids, ferns and horsetails, seed-bearing plants, or gymnosperms, took an increasingly dominant role in floras around the world.

Seeds in gymnosperms are naked, that is, they are not enclosed in ovaries as they are in flowering plants (angiosperms). Seeds follow from the fertilization of an ovule, the structure containing the egg. The gymnosperm ovule (Fig. 18.12) consists of the **megasporangium**, the site of the female reproductive structures, and an outer protective layer, the **integument**, with an open end through which the pollen grains enter. The pollen grains settle on the **ovule**, and may send pollen tubes into the tissues of the ovule, through which sperm head for the fertile female structures, the **archegonia**. Upon fertilization, the ovule becomes a seed, containing a viable embryo that develops within the seed coat, and feeding on the nutritive material that composed the bulk of the ovule.

Seed-bearing plants evolved from forms that lacked seeds, and produced spores that were scattered freely. Spores are commonly encountered as fossils and they form the basis of palynological studies for research and commercial purposes (Box 18.4). Free-sporing plants, such as ferns and horsetails, expel their **microspores** and **megaspores**, which develop into male and female gametophytes, respectively. The male gametes (sperm) are motile and must swim to fertilize the female egg. This is a risky business that requires at least a veneer of surface water, limiting sexual reproduction to damp conditions. Seed plants retain the egg in a watertight capsule (seed) that is fed by the vascular system. They produce sperm in tiny watertight capsules (pollen) that are blown onto the seed. After fertilization, the benefit is that the embryo has a ready-made water and food supply system (when attached to parent plant) and after dispersal is housed in a drought-resistant shell that only bursts open when soil conditions are optimum in terms of wetness and warmth. Hence seed plants can colonize drier environments.

Gymnosperms are said to have owed their success in the Carboniferous to the fact that they retained their ovules, and that the developing embryo had extra protection from the parent plant. In addition, the free-living gametophyte phase was eliminated, and water was not required for the sperm to swim through, so that gymnosperms could inhabit dry upland habitats. Gymnosperms may have had adaptive advantages in certain situations as a result of seed bearing, but it would be wrong to assume that they always prevailed. Ferns,

Box 18.4 Palynology

Palynomorphs, fossil pollen and spores, provide evidence about ancient paleoenvironments, often when other fossils are absent, and they are key tools in biostratigraphy (see pp. 26–32). Fossil pollen and spores have proved to be essential in understanding the biostratigraphy of the Late Paleozoic, Mesozoic and Cenozoic, especially in terrestrial rock sequences. Pollen analysis is also a routine part of studies of Quaternary paleoenvironments, especially those studied by archeologists. They can sometimes be used in correlating marine and non-marine rocks, because pollen and spores are easily blown from land out over lakes, rivers and shallow seas.

Palynomorphs differ from most other microfossils because of their chemical composition. They are not usually mineralized, but their polymerized, organic outer coat (**exine**) is extremely durable. This coat is resistant to nearly all acids, so hydrofluoric acid is used to dissolve surrounding sand grains and leave the pollen for microscopic examination.

Some spores show bilateral symmetry (Fig. 18.13). The proximal pole is marked by the germinal aperture, which may be a rectilinear slit (**monolete** condition) or it may have a triad of branches (**trilete** condition). The laesurae are the contact scars with neighboring spores, commonly converging at a point or commissure. When extracted in a dispersed form from sediments, an arbitrary size distinction is applied that classifies spores above 200 μm as megaspores and smaller ones as microspores.

Pollen grains are usually smaller, ranging in size from 20 to 150 μm. Inaperturate spores lack a germinal aperture (Fig. 18.13). A single aperture at the distal pole characterizes the gymnosperms, the monocotyledons and primitive dicotyledons. Acolpate or asulcate pollen grains lack an obvious germinal aperture. Many pollen grains, however, such as those of pine and spruce, are saccate, with both a body or corpus and vesicles or sacci. The terms colpus and sulcus are often used for similar depressions or furrows; strictly speaking, the sulcus refers to a furrow not crossing the equator of the pollen. Monosulcate pollen with a single distal sulcus that developed during a series of meioses is typical of gymnosperms and monocotyledon angiosperms. The tricolpate pattern, seen in dicotyledon angiosperms, has three germinal apertures or colpi arranged with triradiate symmetry.

Virtually all fossil pollen and spores are identified and classified on the basis of the morphology of the resistant outer wall or exine. As is the case with a number of other palynomorph groups, only a **parataxonomy** is possible – that is a "form system" that does not reflect evolution. In one scheme, the palynomorphs are grouped together into "turma" categories; thus spores belong to the Anteturma Sporites and pollen in the Anteturma Pollenites. However, the pollen and spores of plants are often quite distinctive, and they can be used on their own to infer the presence of families, genera, and even species.

Exploration geologists frequently describe the shapes of pollens and spores with a code that describes the exine structure, germinal aperture, outline, shape, size and ornament. Spores (S) are classified on laesurae (scar) type: c, trilete; a, monolete; b, dilete; 0, lacking laesurae. Pollen grains (P) are classified on colpation or sulcation type: a, monocolpate; c, tricolpate; 0, lacking colpation.

Continued

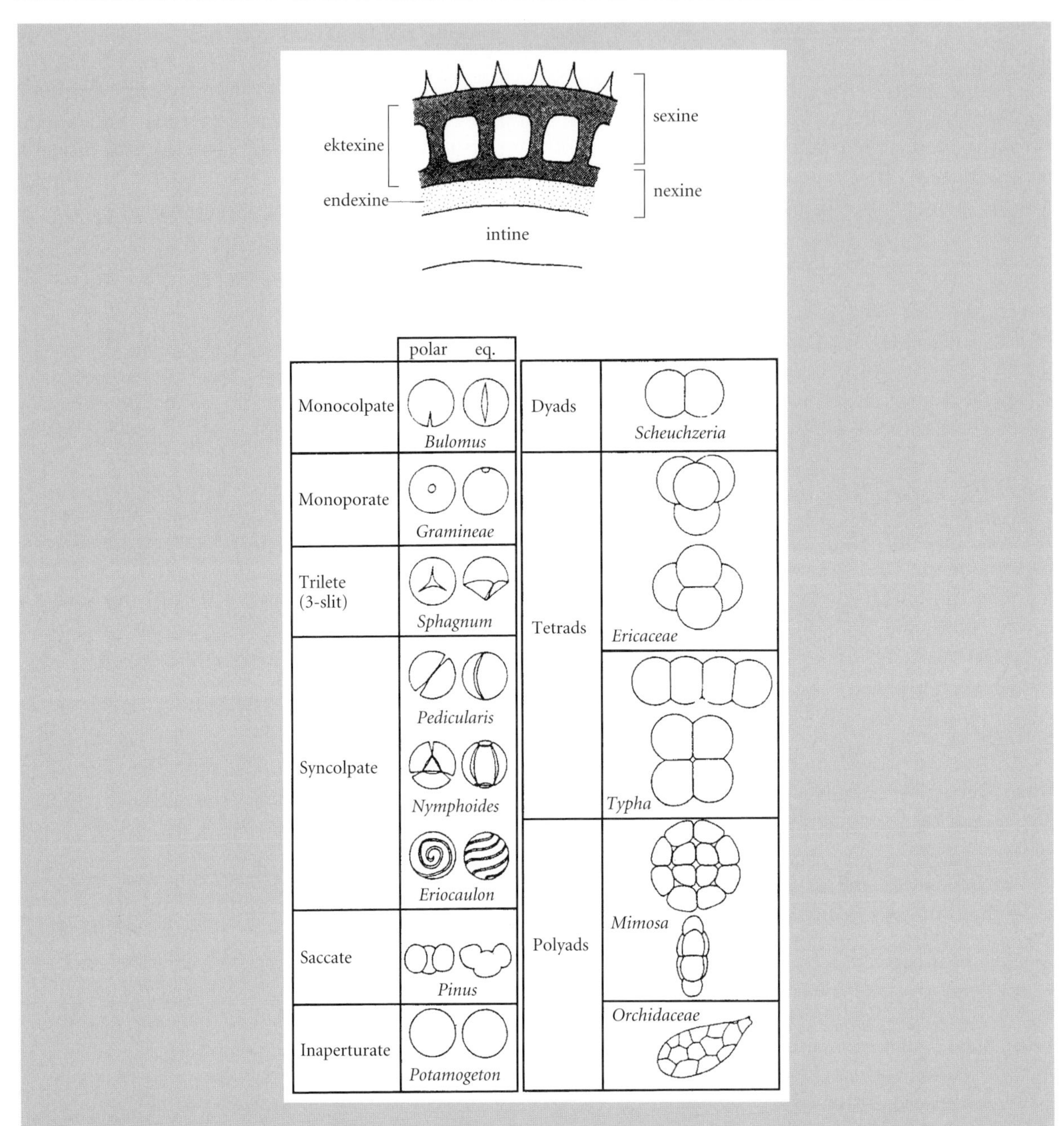

Figure 18.13 Basic morphology and terminology of spores and pollen, shown in polar and equatorial (eq.) views.

Through time, different palynomorphs came and went (Fig. 18.14). These can generally be matched with the broad outlines of plant evolution. The oldest spores are from the Ordovician (see Box 18.1). Spore diversity increased through the Silurian, when some 15 sporomorphs have been reported, including so-called cryptospores that lack monolete or trilete markings, and are commonly found in monads, dyads and tetrads, often with an outer membranaceous envelope. Smooth-walled forms dominated assemblages until the end of the Early Silurian. Some simple spore types with trilete markings may come from bryophytes, but most were probably from tracheophytes. In rare cases, spores may be found in direct association with plants, such as numerous specimens of *Ambitosporites* in the sporangia of some *Cooksonia* (see Fig. 18.7).

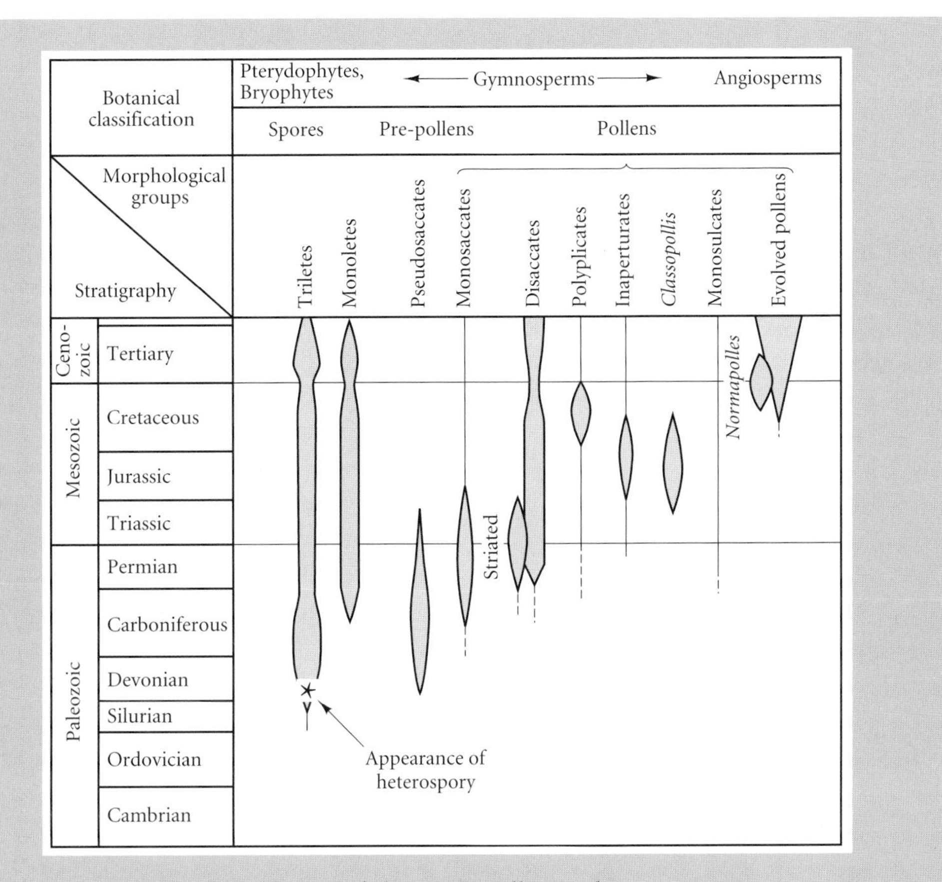

Figure 18.14 Stratigraphic distribution of the main pollen and spore types.

Devonian and Carboniferous palynofloras were much more diverse (Fig. 18.15). The plants of the Lower Devonian Rhynie Chert (see Box 18.3) were all homosporous, producing a single kind of spores, and having sculpture and spines. Monolete and trilete spores from lycopsids appeared during the Devonian. **Heterospory**, the property of having microspores and megaspores, arose independently in plants at least 11 times beginning in the Late Devonian. Lycopsid megaspores with a variety of wall sculpture appeared at this time. The seed ferns of the Carboniferous produced monolete pollen, and the conifers predominantly saccate pollen with a distal aperture. Monocolpate pollen, typical of the cycads and ginkgos, was supplemented, during the Carboniferous and Permian, by both polyplicate and saccate grains. During the Permian, spores are less common than the more dominant saccate pollens that continued through the Triassic. Gymnosperms continued to dominate the floras of the Early Jurassic (Fig. 18.16), including monosaccates from Cordaitales, disaccates from some Coniferopsida, monosulcates from Bennettitaleans, Cycadales and Ginkgoales, polyplicates from Gnetales, and inaperturates from other Coniferopsida.

Palynomorphs changed dramatically in the Cretaceous with the radiation of the angiosperms. Angiosperm pollen has a double outer wall, and the seeds also have a double protective casing. The first undoubted angiosperm pollen grains are reported from the Lower Cretaceous where morphs such as *Clavatipollenites* are oval and monosulcate. During the Cretaceous the monosulcate condition was supplemented by the tricolpate, in for example *Tricolpites*. Monocotyledon pollen is monosulcate and bilaterally symmetric, and dicotyledon pollen has both furrows and pores.

Read more about palynology from links given at http://blackwellpublishing.com/paleobiology/.

Continued

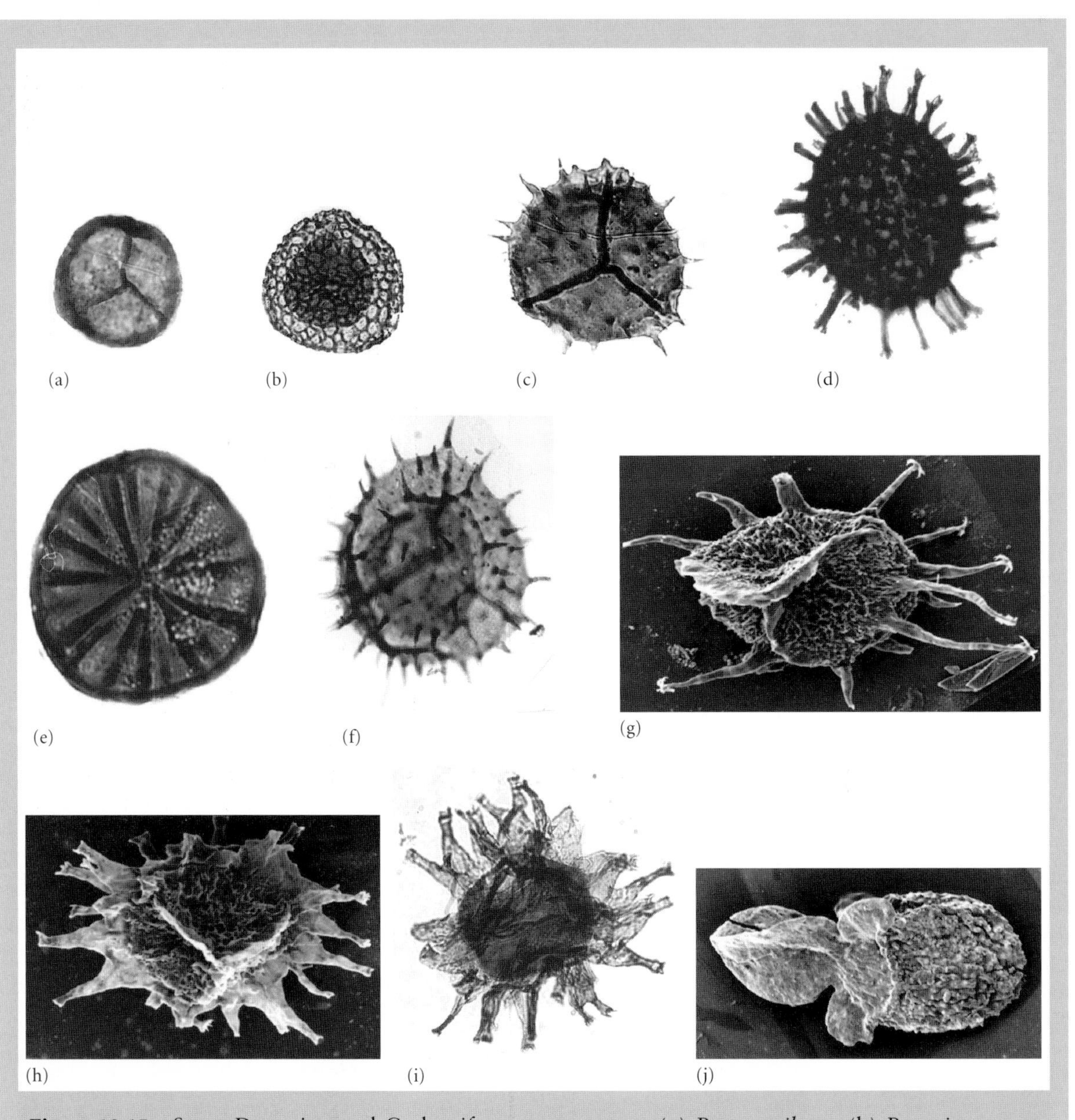

Figure 18.15 Some Devonian and Carboniferous spore taxa: (a) *Retusotriletes*, (b) *Retusispora*, (c) *Spinozonotriletes*, (d) *Raistrickia*, (e) *Emphanisporites*, (f) *Grandispora*, (g) *Hystricosporites*, (h, i) *Ancyrospora*, and (j) *Auritolagenicula*. Magnification ×400 (a–d, f, i), ×750 (e), ×90 (g), ×125 (h), ×40 (j). (Courtesy of Ken Higgs.)

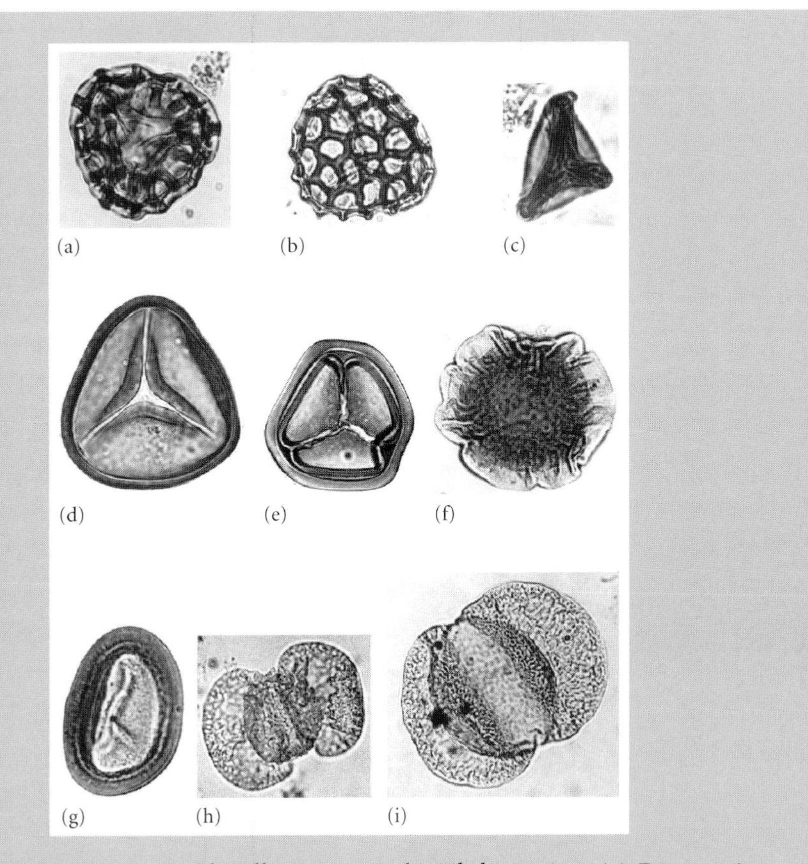

Figure 18.16 Some Jurassic spore and pollen taxa: (a, b) *Klukisporites*, (c) *Dettmanites*, (d) *Dictyophyllidites*, (e) *Retusotriletes*, (f) *Callialasporites*, (g) *Classopolis*, (h) *Podocarpidites*, and (i) *Protopinus*. Mgnification ×400 for all. (Courtesy of Ken Higgs.)

horsetails, lycopsids and more primitive plants such as mosses continued to diversify, especially in damp situations, and they continued their successful evolution without the "benefit" of seeds.

Seed ferns

The seed ferns, or "pteridosperms", have been regarded traditionally as a major gymnosperm class, but they share no unique characters, and it is clear that they are a paraphyletic or perhaps polyphyletic assemblage of gymnosperms of varied affinities. Pteridosperms were important components of Late Paleozoic and Mesozoic floras.

The Carboniferous and Permian seed ferns belong to a variety of groups, such as the Medullosales, which looked superficially like tree ferns, but bore ovules and pollen. Another group of Late Paleozoic seed ferns, the Glossopteridales, include *Glossopteris* (Fig. 18.17), a 4 m-tall tree with radiating bunches of tongue-shaped leaves. This seed fern was the key member of the famous *Glossopteris* flora that characterized Gondwana, the southern hemisphere continents, from the Pennsylvanian to Late Permian (see p. 42). The Glossopteridales existed through the Triassic, and a number of other groups of seed ferns of uncertain affinities radiated during the Triassic and Jurassic.

Plant ecology of the coal measures

Early reconstructions of Carboniferous vegetation tended to show crowds of ferns, horsetails, tree ferns and clubmosses growing in dense profusion around vegetation-filled lakes. However, detailed studies have shown that the floodplain vegetation consisted almost exclusively of clubmosses such as *Lepidodendron* and *Sigillaria*, with rare examples of horsetails such as *Calamites*. Seed ferns,

Figure 18.17 The seed fern *Glossopteris*, a 4 m-tall tree, from the Late Permian of Australia. (Based on Delevoryas 1977.)

conifers and ferns were adapted to drier conditions, and they occupied elevated locations such as levees, the banks of sand thrown up along the sides of rivers. There are hints of extensive dry upland vegetations during the Carboniferous, but the fossil record of these is barely preserved. Surprisingly, some of the best preservation came about through huge forest fires (Box 18.5).

Towards the end of the Carboniferous, the floodplain vegetation of Europe and North America changed, probably as a result of slight drying of the environment. The clubmosses were replaced to some extent by the dryland ferns and seed ferns. These boggy habitats virtually disappeared in Europe and North America by the end of the Carboniferous, but persisted to the end of the Permian in China.

Conifers

Conifers are the most successful gymnosperms, having existed since the Pennsylvanian, and being represented today by over 550 species. Living conifers include the tallest living organisms of all time, the Coastal Redwood of North America, *Sequoia sempervirens*, which can reach over 110 m tall and an estimated 1500 tonnes. Conifers have a variety of adaptations to dry conditions, including their narrow, needle-like leaves with thick cuticles and sunken stomata, all adaptations to minimize water loss. The tough needles also escape freezing in cold polar winters. The seeds are contained in tough scales grouped in spirals into cones, usually borne at the end of branches, while the pollen-producing cones are usually borne on the sides of branches.

The Cordaitales of the Carboniferous and Permian are a distinctive group of early conifers which had strap-shaped, parallel-veined leaves (Fig. 18.19). Some Cordaitales were tree-like, and bore their leaves, sometimes up to 1 m long, in tufts at the ends of lateral branches. The Voltziales, of Pennsylvanian to Jurassic age, are represented by abundant finds of leaves and cones. The cones show a variety of structures, some with a single fertile scale at the tip, showing apparent intermediate stages to the cones of modern conifers, where all or most scales are fertile.

Modern conifers radiated in the Late Triassic and Jurassic, possibly from ancestors among the Voltziales. The main families – Podocarpaceae (southern podocarps), Taxaceae (yew), Araucariaceae (monkey puzzle), Cupressaceae (cypresses, junipers), Taxodiaceae (sequoia, redwood, bald cypress), Cephalotaxaceae and Pinaceae (pines, firs, larches) – are distinguished by leaf shape and features of the cones. Podocarps and yews do not have cones.

Diverse gymnosperm groups

Compared to the conifers, the other gymnosperm groups did not radiate so widely. The ginkgos are represented today by one species, *Ginkgo biloba*, the maidenhair tree, a native

Box 18.5 Reconstructing ancient plant ecology

Some of the best evidence about fossil plants is microscopic, and indeed the scanning electron microscope (SEM) has revolutionized the levels of detail that paleobotanists can retrieve. In the tropics today there are often vast wildfires that burn up hundreds of acres of forest. Fires may be started by a carelessly thrown cigarette or a bottle that focuses the rays of the sun, but usually the causes are natural; fallen branches and leaves may just be so dry that a chance lightning strike may spark off a huge conflagration that burns for days or weeks.

Wildfires are not always destructive; indeed, many plants rely on occasional fires to clear old timber and to allow new shoots to grow. And the ash from the fire provides phosphorus and other nutrients. Wildfires were common in the past, and particularly in the tropical belt during the Carboniferous when atmospheric oxygen levels may have been higher. Howard Falcon-Lang of the University of Bristol has studied this phenomenon, and he has shown the remarkable detail that may be observed from ancient charcoal, the burned up remnants of wood. When the charcoal is examined under an SEM, it shows distinctive subcellular pits that allow identification of the precise type of tree caught in the fire (Fig. 18.18a). Fine details such tree rings are also preserved, indicating perhaps a seasonal tropical climate like present-day East Africa (Fig. 18.18b).

Close study of the distribution of charcoal and the sedimentology of typical Carboniferous beds in North America shows that fires were commonest in the higher areas, away from the banks of rivers, where plant debris could become very dry (Fig. 18.18c). The wildfires may have been set off by nearby volcanic eruptions. Careful measurements through the sediments suggest that wildfires may have been very frequent in Carboniferous times. They must have been a regular part of the growth and regrowth of forests, as well as destabilizing hill slopes thus triggering occasional landslides.

Read more about Carboniferous wildfires and plant ecosystems in Falcon-Lang (2000, 2003) and at links listed at http://www.blackwellpublishing.com/paleobiology/. Read about another astonishing discovery in the Carboniferous, a huge rain forest in Illinois, catastrophically buried by a sudden rise in sea level, in DiMichele et al. (2007).

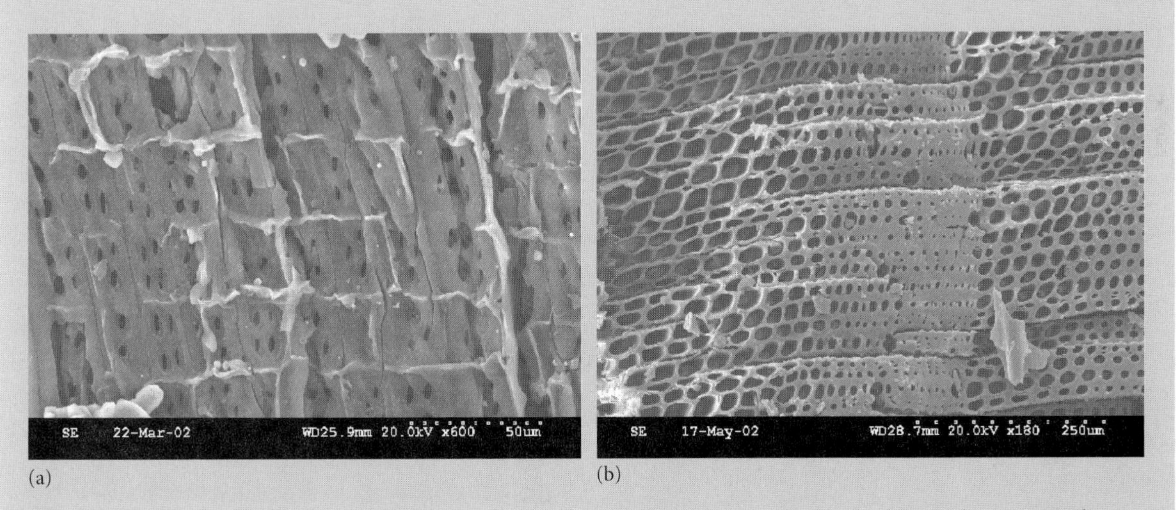

Figure 18.18 Carboniferous wildfires and the use of the SEM: (a) ancient charcoal can reveal spectacular details under the SEM, such as cross-field pitting, which provides evidence for which species of plants burned; and (b) part of a tree-ring. Note the transition from thin-walled "early wood" (left) to thick-walled "late wood" (center). The rings of growth may indicate a seasonal tropical environment like northern Australia or East Africa. Study of these plant remains and the sediments shows that wildfires happened every 3 to 35 years, and especially in drier uplands (c). PDP, poorly-drained coastal plain; WDP, well-drained coastal plain. (Courtesy of Howard Falcon-Lang.)

Continued

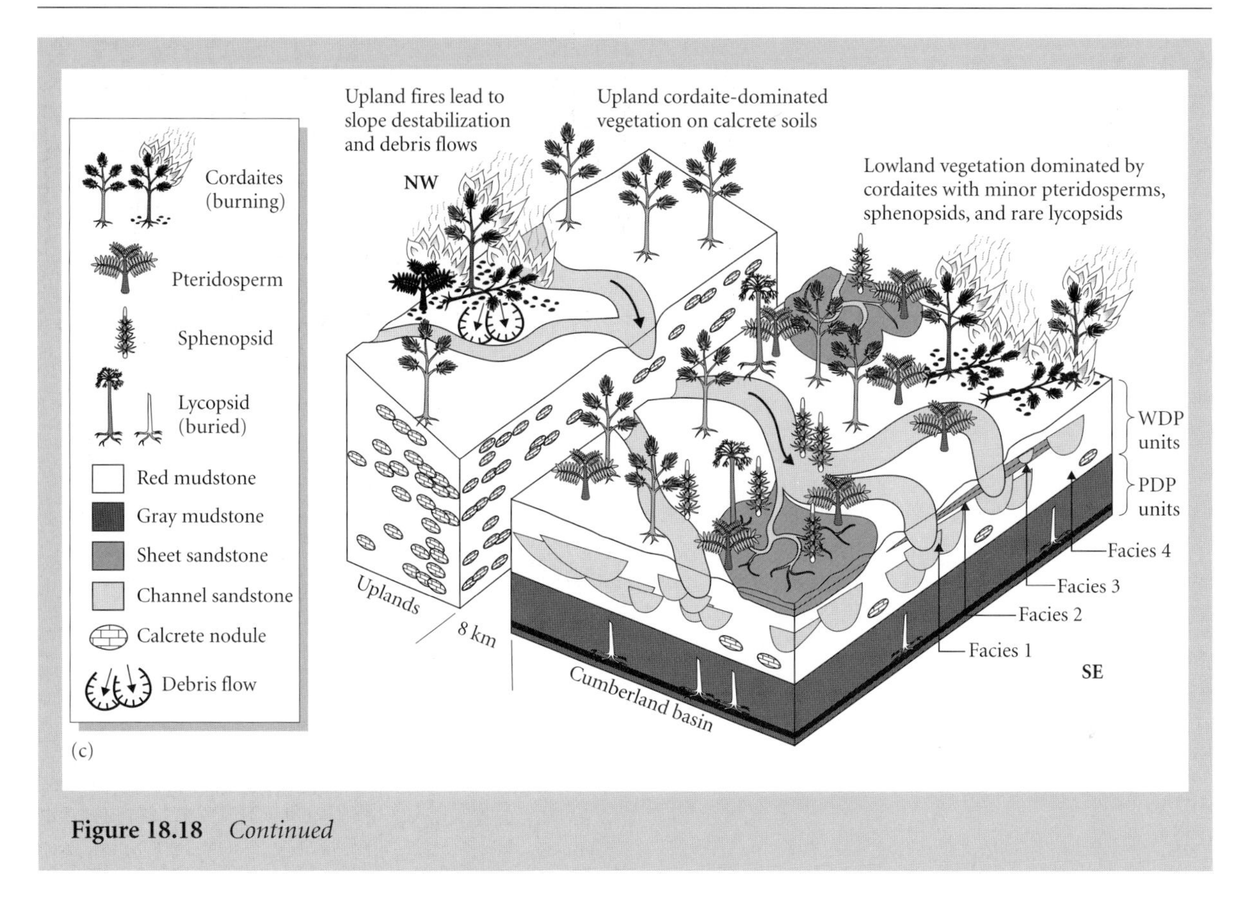

Figure 18.18 *Continued*

Figure 18.19 The early conifer *Cordaites*, about 25 m tall. (Based on Thomas & Spicer 1987.)

of China, but seen today as a typical urban tree in parts of North America and Europe. Ginkgos were more diverse in the Mesozoic. Leaf shape varies from the fan-shaped structure in the modern form, to deeply dissected leaves in some Mesozoic taxa (Fig. 18.20a, b). Catkin-like pollen organs and bulbous stalked ovules are borne in groups on separate male and female plants. The leaves in the modern *Ginkgo* are deciduous, that is they are shed in winter, and this may have been a feature of ancient ginkgos.

The cycads, represented today by 305 tropical and subtropical genera, are trees with a stem that ranges in length from a small tuber to a palm-like trunk up to 18 m tall. The leaves are provided with deep-seated leaf traces that partially girdle the stem. *Leptocycas* (Fig. 18.20c) from the Late Triassic of North Carolina has a 1.5 m-tall trunk, showing a few traces of attachment sites of leaves that had been lost as the plant grew, and a set of nine or 10 long fronds near the

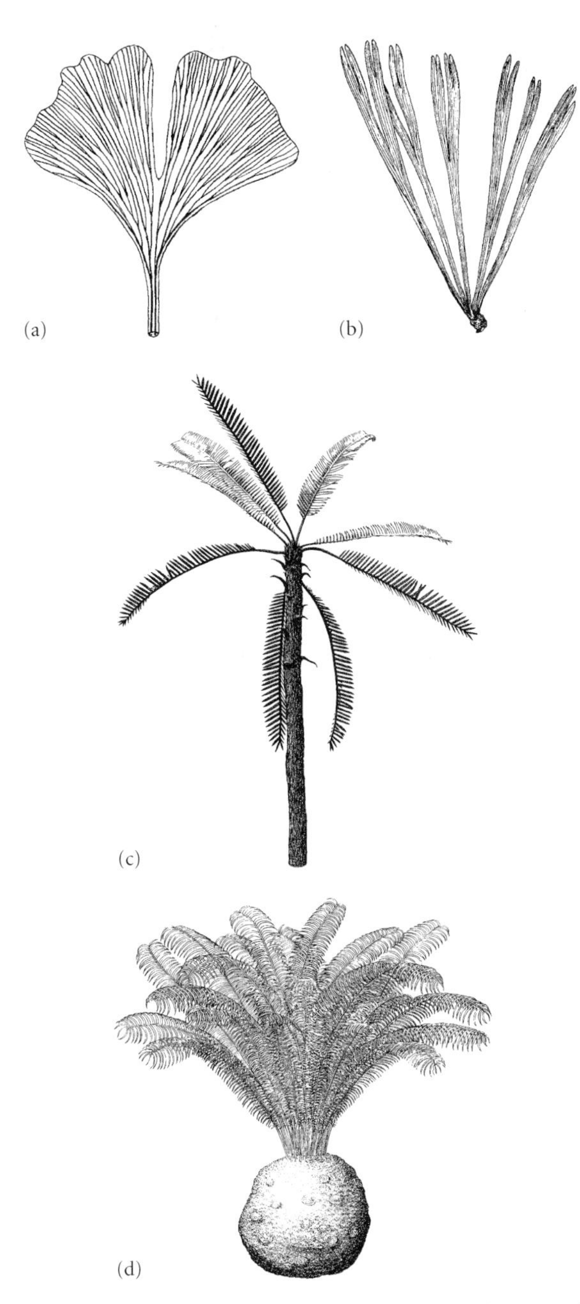

Figure 18.20 Diverse gymnosperms: (a) leaves of the modern ginkgo, *Ginkgo biloba* and (b) of the Jurassic ginkgo, *Sphenobaiera paucipartita*; (c) reconstruction of the 1.5 m-tall cycad *Leptocycas gigas* a from the Late Triassic of North America; and (d) reconstruction of the 2 m-tall bennettitalean *Cycadeoidea* from the Cretaceous of North America. (Based on Delevoryas 1977.)

top of the trunk. Many other cycads show marked leaf bases along the entire length of the trunk. Cycad fronds are typically composed of numerous parallel-sided leaflets attached to a central axis in a simple frond-like arrangement, but others had undivided leaves.

The bennettitaleans, or cycadeoids, were a Mesozoic group of bushy plants with frond-like leaves very like those of cycads. Some bennettitaleans had a trunk up to 2 m tall, with bunches of long fronds at the top of the trunk and on subsidiary branches. Other bennettitaleans like *Cycadeoidea* (Fig. 18.20d) had an irregular ball-like trunk covered in leaf bases, representing former attachment sites of fronds, and with a tight tuft of long feathery fronds on top. Some bennettitaleans had flower-like structures (Fig. 18.21a). Classic dinosaur scenes of Jurassic and Cretaceous age often picture one or other of these bennettitaleans in the background.

The gnetales have a patchy fossil record, with two Late Triassic examples, a few in the Cretaceous and Tertiary, and three living genera. Gnetales have distinctive pollen that is very abundant in Cretaceous sediments. The group was probably much more diverse at this time. Gnetales gained prominence among botanists because the group is thought by some to be the closest living gymnosperm relative of angiosperms. In particular, gnetales may have their ovules and pollen organs in cones that are rather flower-like (Fig. 18.21b).

FLOWERING PLANTS

Flowers and angiosperm success

The angiosperms are by far the most successful plants today, with over 260,000 species and occupying most habitats on land. Most of the food plants used by humans are angiosperms – wheat, barley, apples, cabbage, lentils, peas, olives, pumpkins and many more. Angiosperms arose during the Mesozoic, and radiated dramatically during the mid-Cretaceous.

The following are important characteristics of angiosperms:

1 The ovules are fully enclosed within **carpels** (Fig. 18.21c). It is believed that carpels are modified leaves that grew

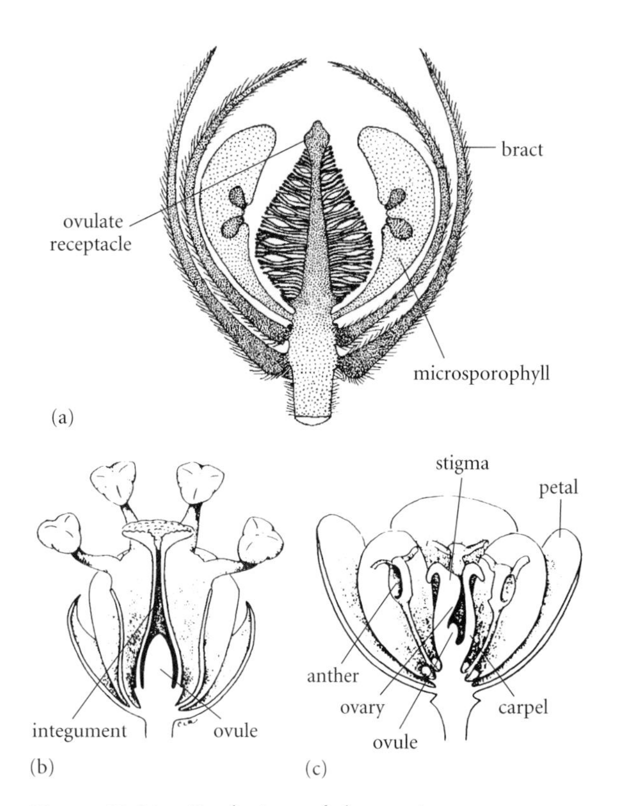

Figure 18.21 Evolution of the angiosperm flower: (a) cone of the Jurassic bennettitalean *Williamsoniella*, showing the female fertile structure, the ovule, contained in a central receptacle, and surrounded by the male fertile structures, the microsporophylls; (b) flower of the gnetale *Welwitschia*, showing the central ovule, and surrounding male elements; and (c) flower of the angiosperm *Berberis*, showing the same pattern, but with the seed enclosed in a carpel.

around the ovules, and provided a secure protective covering. In angiosperm development, the carpels grow around the ovules and fuse, although in some magnolias the carpels are not completely fused when fertilization takes place.

2 Most angiosperm ovules have two integuments, or protective casings.
3 Most angiosperms have pollen grains with a double outer wall separated by columns of tissue.
4 Angiosperms have a flower (Fig. 18.21c), a structure that is composed of whorls of **sepals** and **petals** in most. The flower includes the carpels and **stamens**, the male reproductive structures. The structure of flowers is not standard in all angiosperms.
5 Angiosperms all show double fertilization, that is, two sperm nuclei are involved in fertilization. One unites with the egg nucleus, while the other fuses with another nucleus that divides to form the food supply for the developing embryo. Double fertilization has been described also in Gnetales.
6 Most angiosperms have water-transporting vessels rather than just xylem tracheids. This feature, however, is absent in some magnolids and hamamelidids, and is present in gnetales.
7 Most angiosperms have a net-like pattern of veins in their leaves.

Most of these characters are regarded as typical of angiosperms, but many are not unique to angiosperms, nor are they present in all angiosperms. The only one that seems to be an acceptable apomorphy of the group is the possession of carpels around the ovule (character 1 above).

Flowers are certainly the most obvious feature of angiosperms, but several gymnosperm groups also had organs that bear certain resemblances to flowers. Bennettitaleans (Fig. 18.21a) and Gnetales (Fig. 18.21b) have flower-like structures with the ovule in the center, and around them structures resembling petals.

The secret of the success of the angiosperms may be the flower and the fully enclosed ovule. The carpels protect the ovule from fungal infection, desiccation and the unwelcome attentions of herbivorous insects. Double fertilization is said to offer the advantage that the parent plant does not invest energy in creating a large food store, as in gymnosperms (see Fig. 18.12), until fertilization of the ovule is assured. Pollen is produced within the **anthers**, which are typically borne on long filaments arranged around the centrally placed **ovary** or ovaries. Pollen grains are transported, by animals, often insects, or by the wind, to the **stigma**, and from it the pollen grains send pollen tubes to the ovules through which the sperm pass.

The petals, often brightly colored, with special fragrances and supplies of nectar (sugar water), are all adaptations of angiosperms to ensure fertilization by insects. Some gymnosperms show hints of this pattern: the living gnetalean *Welwitschia* has a flower with "petals" (Fig. 18.21b), and it secretes a nectar-like pollination drop as tempting food for its insect pollinator. It is clear that the evolution of angiosperm characters was paralleled by the evolution of major new groups of insects that fed from flowers and pollinated the flowers (Box 18.6).

The first angiosperms

There has been heated discussion among paleobotanists over the past century about the oldest angiosperm fossil, and about the closest relatives of angiosperms. The oldest generally accepted angiosperms are Early Cretaceous in age, but there have been repeated reports of Jurassic and Triassic angiosperms, although most of these have been highly controversial (Friis et al. 2006). Indeed, such older angiosperms might be expected if angiosperms are truly the sister group of gymnosperms, as current molecular phylogenies imply (Frohlich & Chase 2007) (see Fig. 18.4): this tree implies that angiosperms must be as old as gymnosperms, dating back to the Carboniferous.

The oldest flowers are Early Cretaceous in age, and fossils have been reported from North America, Europe and Asia. Most spectacular, and probably oldest, is *Archaefructus* from the Liaoning Formation of China (see p. 463). This fossil, named in 2002, was billed as either the first angiosperm, or the closest sister group of Angiospermae. Some of the rare and remarkable fossils of the earliest angiosperms show soft parts of flowers, such as sets of fleshy stamens with pollen grains inside, and evidence of five-fold symmetry of the flower parts, typical of modern angiosperms (Friis et al. 2006). Even more spectacular fossil specimens of flowers are known from the beginning of the Tertiary, where specimens are preserved in lithographic limestones, in chert and in amber (Fig. 18.23). Other more commonly preserved fossil evidence for the first angiosperms consists of pollen, leaves, fruits and wood.

Radiation of the angiosperms

Angiosperms radiated to a diversity of 35 families by the end of the Cretaceous (Fig. 18.24). The success of the angiosperms in the mid-Cretaceous may have been driven by environmental stresses. The early angiosperms lived in disturbed ephemeral habitats, such as riverbeds and coastal areas, and they were opportunists that could spread quickly when conditions were right. In addition, the specialized reproductive systems of angiosperms perhaps promoted rapid speciation, especially in terms of the increasing matching of flower and pollinator.

There are two competing hypotheses for angiosperm origins: the paleoherb hypothesis suggests that the basal lineages were small plants (herbs) with rapid life cycles, while the magnoliid hypothesis suggests that the basal lineages were small trees with simple flowers and slower life cycles. Older phylogenetic analyses tended to show magnolias and laurels as the basal-most angiosperms, and this seemed to suggest that flowers were not such a key innovation as had been assumed.

The paleoherb hypothesis is now confirmed by most large-scale molecular and morphological phylogenies of angiosperms (e.g. Soltis & Soltis 2004; Haston et al. 2007). These have identified the most basal living angiosperm as *Amborella*, a rare understory shrub that is found in cloud forests of New Caledonia. *Amborella* has spirally arranged floral organs and other apparently primitive features. Close to the base of the angiosperm tree are an array of paleoherbs, low plants such as the Nymphaeaceae (water lilies), monocots (gingers, grasses, palms and relatives) and Piperales (peppers and relatives). Next in the tree come the Laurales (laurels and relatives) and Magnoliales (magnolias and relatives).

The classic division of angiosperms into monocots and dicots does not now work so clearly because "dicots" are paraphyletic in the new phylogenies. The monocots form a clade, and they have one cotyledon (food-storage area of the seed), the flower parts arranged in threes and parallel leaf-venation patterns. The "dicots" – all other flowering plants – have two cotyledons, flower parts often in fours and fives, net-like venation patterns on the leaves, and specialized features of the vascular tissues in the wood. These two groups are readily distinguishable in the Cretaceous. During the Tertiary, more and more of the modern families appeared, so that at least

Box 18.6 A new career for insects: pollination

Pollinating insects existed before the Cretaceous and the radiation of the angiosperms, but their role was minor, feeding at the flowers of some of the advanced gymnosperms. During the Cretaceous, however, there is striking evidence for angiosperm–insect coevolution (Fig. 18.22). Groups of beetles and flies that pollinate various plants were already present in the Jurassic and Early Cretaceous, but the hugely successful butterflies, moths, bees and wasps are known as fossils only from the Cretaceous and Tertiary.

The composition of insect faunas changed during the Cretaceous and Tertiary. One group to evolve substantially at that time was the Hymenoptera – bees and wasps. The first hymenopterans to appear in the fossil record, the sawflies (Xyelidae), had been present since the Triassic. Some fossil specimens have masses of pollen grains in their guts, a clear indication of their preferred diet. The sphecid wasps that arose during the Early Cretaceous had specialized hairs and leg joints that show they collected pollen. Other wasps, the Vespoidea, and the true bees appear to have arisen in the Late Cretaceous.

The first angiosperms may not have had specialized relationships with particular insects, and may have been pollinated by several species. More selective plant–insect relationships probably grew up during the Late Cretaceous with the origin of vespoid wasps that today pollinate small radially symmetric flowers. These kinds of specialized relationships are shown by increasing adaptation of flowers to their pollinator in terms of flower shape, and the food rewards offered, and of the pollinator to the flower. Late Cretaceous angiosperms similar to roses had specialized features that catered for pollinators that fed on nectar as well as pollen.

Read more on web sites listed at http://www.blackwellpublishing.com/paleobiology/.

Figure 18.22 The coevolution of floral structures and of pollinating insects during the entire span of the Cretaceous and the early part of the Tertiary. Some of the major floral types are (a) small simple flowers, (b) flowers with numerous parts, (c) small unisexual flowers, (d) flowers with parts arranged in whorls of five, (e) flowers with petals, sepals and stamens inserted above the ovary, (f) flowers with fused petals, (g) bilaterally symmetric flowers, (h) brush-type flowers, and (i) deep funnel-shaped flowers. Pollinating insects include (j) beetles, (k) flies, (l) moths and butterflies, and (m–q) various groups of wasps and bees: (m) Symphyta, (n) Sphecidae, (o) Vespoidea, (p) Meliponinae, and (q) Anthophoridae. (Based on information in Friis et al. 1987.)

Figure 18.23 Fossil angiosperm remains from North America. (a) Flower of an early box-like plant, *Spanomera*, from the mid-Cretaceous of Maryland (×10). (b) Leaf of the birch, *Betula*, from the Eocene of British Columbia (×1). (Courtesy of Peter Crane.)

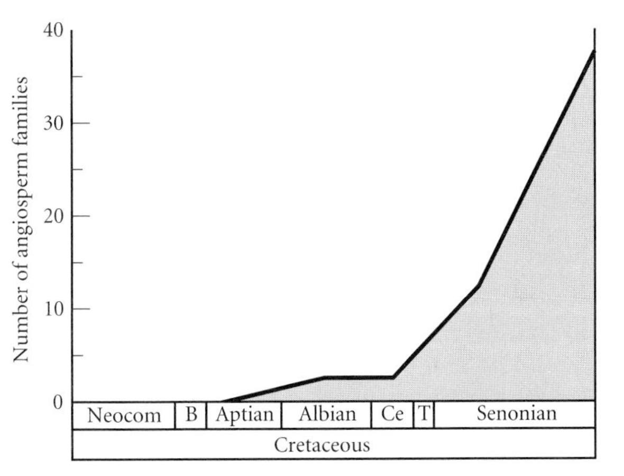

Figure 18.24 Rapid radiation of the angiosperms during the Cretaceous, shown by the rise in the number of angiosperm families, from none at the beginning of the Cretaceous to more than 35 by the end of the period. Neocom, Neocomian; B, Barremian; Ce, Cenomanian; T, Turonian. (Based on information in various sources.)

250 of the 400 or so extant families of angiosperms have a fossil record of some kind.

Angiosperms and climate

Angiosperms are highly sensitive indicators of paleoclimates on land, and they provide the best tool at present for estimating temperatures, rainfall patterns and measures of seasonality. The key to the use of angiosperms in this way is the fact that so many modern taxa may be traced well back into the Tertiary and Cretaceous, and paleobotanists assume that adaptations that are observed today had the same functions in the past.

Studies on North American Late Cretaceous angiosperm leaves have shown how precise these climatic estimates may be. Upchurch and Wolfe (1987) established ways of assessing temperatures and rainfall measures from key leaf features, such as:

1 Leaf size: largest leaves are found in tropical rain forest, and size diminishes as temperature and moisture decline.
2 Leaf margins: in tropical areas, most angiosperm leaves have entire (unbroken) margins, whereas in temperate areas there are many more tooth-margined leaves.
3 Drip tips: leaves from tropical rain forest species have elongated tips to allow water to clear the leaf during major downpours.
4 Deciduousness: the proportion of deciduous trees (those that shed all their leaves simultaneously in winter or during the dry season, that is, times of low growth rate) to evergreens is highest in temperate zones, while tropical trees are more likely to retain their leaves since they grow more continuously.
5 Lianas: certain angiosperms in tropical forests grow as long, rope-like plants that hang down from tall trees (see any Tarzan film), but such plants are uncommon in temperate forests.
6 Vessels in wood: in areas subject to freezing or drying, the vascular canals possess adaptations to prevent air filling the canals

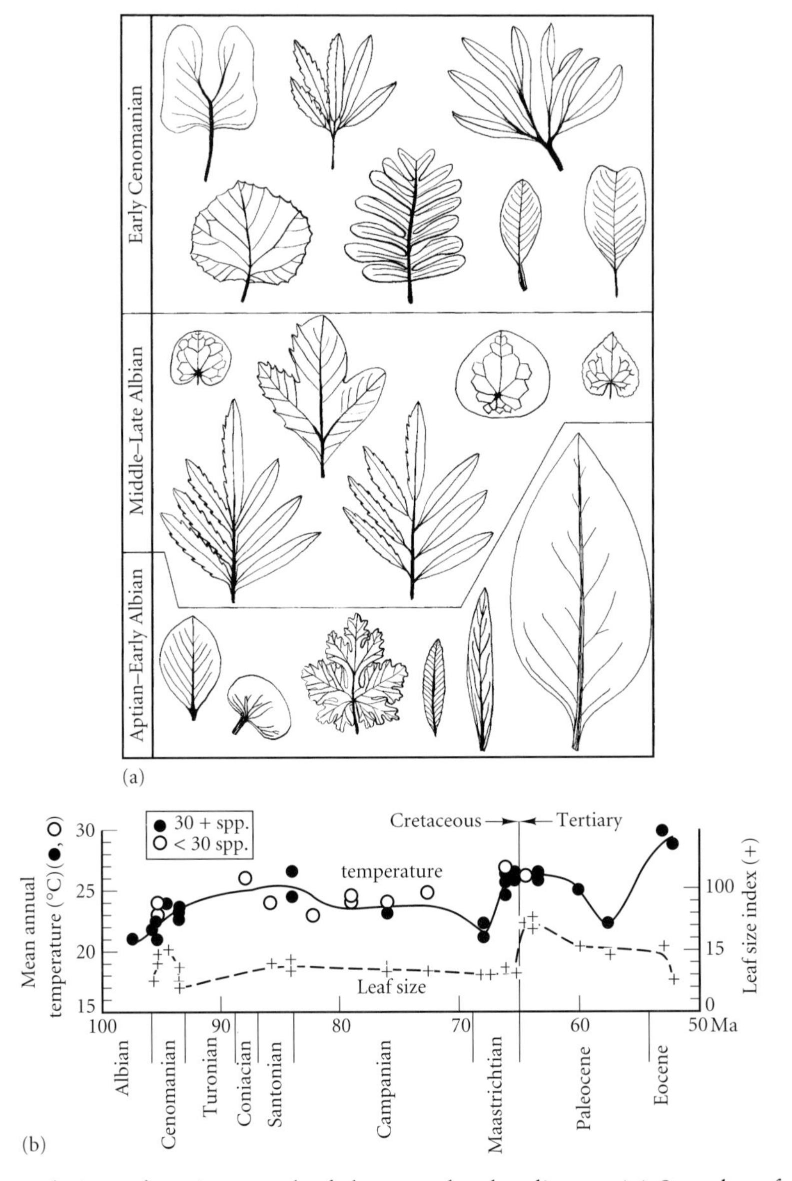

Figure 18.25 The evolution of angiosperm leaf shape and paleoclimate. (a) Samples of typical leaf shapes from North American floras spanning the mid-Cretaceous, showing variations in length, margins and shapes. The average leaf size declines, suggesting an increase in temperature. (b) The leaf size index (percentage of entire-margined species and average leaf size) for low-latitude North American floras through the Late Cretaceous shows fluctuations. These are interpreted as the result of changes in temperature. (Based on information in Upchurch & Wolfe 1987.)

when water is in short supply; the canals are narrow and densely packed.

7 Growth rings: in areas subject to highly seasonal climates, wood grows rapidly during the warm or wet season, and slows or stops growing when conditions are cold and/or dry. The variation in growth ring style indicates the degree of seasonality.

Abundant assemblages of leaves have been recovered from several hundred Late Cretaceous localities in North America, and these together show changes in leaf shapes of the sort that reflect paleoclimates (Fig. 18.25a). Measurements of the leaf-shape characters noted above, based on floras of numerous species, can give a clear plot of paleotemperature change in North America during the Late Cretaceous (Fig. 18.25b). Temperatures remained around 20–25°C, with slight variations, until the last 5 myr of the Cretaceous, when there was a dramatic rise in temperature to 27°C, then a drop, and a further rise in the Early Tertiary.

Review questions

1 What is the current best evidence for the greening of the land? Read around paleobotanical evidence for Precambrian, Cambrian and Ordovician plant fossils, and molecular evidence for Neoproterozoic dates for divergence of land plant groups. Why do the dates seem to differ so much?
2 Read about the detailed anatomy of *Cooksonia*, the first land plant to be known in any real detail. Make a detailed reconstruction, showing all the fossil evidence for the different parts of the plant.
3 When did the first land plants achieve tree size? Read about the 2007 discovery of complete trees from the Gilboa locality in New York State. How did this discovery change our views about land plant evolution?
4 What were the rain forests of the Carboniferous like? Read DiMichele et al. (2007) and related papers and web sites to find out how new information about Carboniferous ecosystems is being patched together from new studies.
5 Why have angiosperms been so successful? Trace the rise of the various angiosperm groups through the Cretaceous, and list the supposed advantageous features angiosperms have in comparison to gymnosperms. Which of these adaptations might have been most important in their major diversification?

Further reading

DiMichele, W.A., Falcon-Lang, H.J., Nelson, W.J., Elrick, S.D. & Ames, P.R. 2007. Ecological gradients within a Pennsylvanian mire forest. *Geology* **35**: 415–18.

Doyle, J.A. 1998. Phylogeny of vascular plants. *Annual Review of Ecology and Systematics* **29**, 567–99.

Falcon-Lang, H.J. 2000. Fire ecology of the Carboniferous tropical zone. *Palaeogeography, Palaeoclimatology, Palaeoecology* **164**, 339–55.

Falcon-Lang, H.J. 2003. Late Carboniferous dryland tropical vegetation in an alluvial-plain setting, Joggins, Nova Scotia, Canada. *Palaios* **18**, 197–211.

Friis, E.M., Chaloner, W.G. & Crane, P.R. 1987. *The Origins of Angiosperms and their Biological Consequences*. Cambridge University Press, Cambridge.

Kenrick, P. & Crane, P.R. 1997. The origin and early evolution of plants on land. *Nature* **389**, 33–9.

Kenrick, P. & Davis, P. 2004. *Fossil Plants*. Natural History Museum, London.

Mauseth, J.D. 2003. *Botany; An introduction to plant biology*, 3rd edn. Jones and Bartlett, Sudbury, MA.

Soltis, D.E., Soltis, P.S., Endress, P.K. & Chase, M.W. 2005. *Phylogeny and Evolution of Angiosperms*. Sinauer, Sunderland, MA.

Stewart, W.N. & Rothwell, G.W. 1993. *Paleobotany and the Evolution of Plants*, 2nd edn. Cambridge University Press, Cambridge.

Thomas, B.A. & Spicer, R.A. 1995. *Evolution and Palaeobiology of Land Plants*, 2nd edn. Chapman and Hall, London.

Trewin, N.H. & Rice, C.M. 2004. The Rhynie hot-spring system: geology, biota and mineralization. *Transactions of the Royal Society of Edinburgh: Earth Sciences* **94**, 283–521.

Willis, K.J. & McElwain, J.C. 2002. *The Evolution of Plants*. Oxford University Press, Oxford.

References

Andrews Jr., H.N. 1960. Notes on Belgian specimens of *Sporogonites*. *Palaeobotanist* **7**, 85–9.

Bowe, L.M., Coat, G. & dePamphilis, C.W. 2000. Phylogeny of seed plants based on all three genomic compartments: extant gymnosperms are monophyletic and Gnetales' closest relatives are conifers. *Proceedings of the National Academy of Sciences* **97**, 4092–7.

Delevoryas, T. 1977. *Plant Diversification*, 2nd edn. Holt, Rinehart and Winston, New York.

DiMichele, W.A., Falcon-Lang, H.J., Nelson, W.J., Elrick, S.D. & Ames, P.R. 2007. Ecological gradients within a Pennsylvanian mire forest. *Geology* **35**, 415–18.

Edwards, D., Davies, K.L. & Axe, L. 1992. A vascular conducting strand in the early land plant *Cooksonia*. *Nature* **357**, 683–5.

Friis, E.M., Chaloner, W.G. & Crane, P.R. 1987. *The Origins of Angiosperms and their Biological Consequences*. Cambridge University Press, Cambridge.

Friis, E.M., Pedersen, K.R. & Crane, P.R. 2006. Cretaceous angiosperm flowers: innovation and evolution in plant reproduction. *Palaeogeography, Palaeoclimatology, Palaeoecology* **232**, 251–93.

Frohlich, M.W. & Chase, M.W. 2007. After a dozen years of progress the origin of angiosperms is still a mystery. *Nature* **450**, 1184–9.

Haston, E., Richardson, J.E., Stevens, P.F., Chase, M.W. & Harris, D.J. 2007. A linear sequence of Angiosperm phylogeny group II families. *Taxon* **56**, 7–12.

Kenrick, P. & Crane, P.R. 1997. The origin and early evolution of plants on land. *Nature* **389**, 33–9.

Kidston, R. & Lang, W.H. 1917–1921. Old Red Sandstone plants showing structure, from the Rhynie chert bed, Aberdeenshire, Parts I–IV. *Transactions of the Royal Society of Edinburgh* **51**, 761–84; **52**, 603–27, 643–80, 831–54, 855–902.

Morgan, J. 1959. The morphology and anatomy of American species of the genus *Psaronius*. *Illinois Biological Monographs* **27**, 1–108.

Soltis, P.S. & Soltis, D.E. 2004. The origin and diversification of angiosperms. *American Journal of Botany* **91**, 1614–26.

Stewart, W.N. & Rothwell, G.W. 1993. *Paleobotany and the Evolution of Plants*, 2nd edn. Cambridge University Press, Cambridge.

Thomas, B.A. & Spicer, R.A. 1987. *The Evolution and Paleobiology of Land Plants*. Croom Helm, London.

Trewin, N.H. & Rice, C.M. 2004. The Rhynie hot-spring system: geology, biota and mineralization. *Transactions of the Royal Society of Edinburgh: Earth Sciences* **94**, 283–521.

Upchurch Jr., G.R. & Wolfe, J.A. 1987. Mid-Cretaceous to Early Tertiary vegetation and climate: evidence from fossil leaves and woods. *In* Friis, E.M., Chaloner, W.G. & Crane, P.R. (eds) *The Origins of Angiosperms and their Biological Consequences*. Cambridge University Press, Cambridge, pp. 75–105.

Wellman, C.H., Osterloff, P.L. & Mohiuddin, U. 2003. Fragments of the earliest land plants. *Nature* **425**, 282–5.

Yuan, X., Xiao, S. & Taylor, T.N. 2005. Lichen-like symbiosis 600 million years ago. *Science* **308**, 1017–20.

Chapter 19

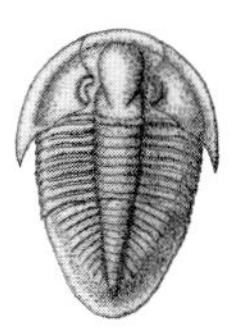

Trace fossils

Key points

- Trace fossils represent the activities of organisms.
- Trace fossils may be treated as fossilized behavior, or as biogenic sedimentary structures.
- Trace fossils include tracks and trails, burrows and borings, fecal pellets and coprolites, root penetration structures and other kinds of pellets.
- Trace fossils are named on the basis of shape and ornamentation, not on the basis of the supposed maker, environment or stratigraphy.
- One animal may produce many different kinds of trace fossils, and one trace fossil type can be produced by many different kinds of animals.
- Trace fossils may be produced within a sedimentary layer, or on the surface; trace fossils may be preserved in the round, and may be seen as molds and casts on the bottoms and tops of beds.
- Trace fossils may be classified according to the mode of behavior represented: movement, feeding, farming, dwelling, escape and resting.
- Certain trace fossil assemblages (ichnofacies) appear to repeat through time, and may give clues about the environment of deposition.
- Trace fossils often occupy particular levels (tiers) in the sediment column, and the depth of tiering has apparently increased through time.
- Trace fossils are of limited use in stratigraphy, except in some special cases.

"But one false statement was made by Barrymore at the inquest. He said that there were no traces upon the ground round the body. He did not observe any. But I did – some little distance off, but fresh and clear."
"Footprints?"
"Footprints."
"A man's or a woman's?"
Dr. Mortimer looked strangely at us for an instant, and his voice sank almost to a whisper as he answered:
"Mr. Holmes, they were the footprints of a gigantic hound!"

Arthur Conan Doyle (1901) *The Hound of the Baskervilles*

All the classic detective stories hinge on a footprint on the flowerbed, a used cigarette end, a crumpled scrap of paper. These are **traces** of what happened, and the skilled detective has an uncanny ability to read clues from them. Sherlock Holmes astounded his colleague Mr. Watson by being able to estimate the height of a felon from his footprint; but is that really so difficult?

Trace fossils are the preserved remains of the activity and behavioral patterns of organisms. Common examples are burrows of bivalves and worms that live in estuaries and shallow seas, complex feeding traces of deep-sea animals on the ocean floor, and the footprints of dinosaurs and other land animals preserved in mud and sand beside rivers and lakes. At first sight, these remains might seem rather obscure, but they can tell some remarkable stories (Box 19.1).

Every trace fossil offers us a vignette of ancient life, both the life of the organism that made the marking, as well as the environment in which it lived. Trace fossils give evidence about:

- the behavior of organisms – and so are part of the organisms' paleobiology;
- sedimentary environments – and so are like sedimentary structures.

For example, a trackway of dinosaur footprints may tell us about the shape of the soft parts of the feet of the dinosaur that made them, the pattern of scales on the skin, the running speed and the environment in which the animal lived. The dinosaur tracks can equally be used to show that the sediments were deposited on land or in shallow water, and that the climate was probably warm (appropriate for dinosaurs).

Trace fossils are common in many sedimentary rocks, and they have been observed by geologists for centuries. Indeed, many trace fossils were given zoological and botanical names from early in the 19th century, since they were thought to be fossilized seaweeds or worms. The only trace fossils that were correctly interpreted from the start were dinosaur footprints, although many of these were interpreted at first as the products of flocks of huge birds.

The modern era of trace fossil studies began in the 1950s with the work of the German paleontologist Adolf Seilacher. He established a classification of trace fossils based on behavior, and discovered that certain assemblages of trace fossils indicate particular water depths in the sea. In addition, trace fossils have been used widely by exploration geologists since the 1960s and 1970s when the study of depositional environments revolutionized understanding of the sedimentary rock record. These contributions gave a strong scientific basis to the study of trace fossils, often called **ichnology** (from the Greek *ichnos*, a trace).

UNDERSTANDING TRACE FOSSILS

Types of trace fossils

There are many kinds of trace fossils, and many of the words used to describe them (tracks, trails, burrows, borings) are in common use. There are also a variety of cryptic fossils and sedimentary structures that might be regarded as trace fossils, but perhaps should not. The main trace fossil types are given in Table 19.1.

Some ichnologists might also include other examples of biological interaction with sediments as trace fossils, such as stromatolites (see p. 191), some kinds of mud mounds, dinosaur nests, heavily **bioturbated** or reworked sediments, and the like. Not included are eggs, which are body fossils, or physical sedimentary structures such as tool marks produced by bouncing and rolling objects, including shells and pieces of wood.

Naming trace fossils: shapes not biological species

Trace fossils are given formal names, often based on Latin and Greek, just like living and fossil plants and animals (see p. 118). However, there are some fundamental differences between the nomenclature of trace fossils and that of body fossils and modern organisms. Trace fossil genera are called **ichnogenera** (singular, **ichnogenus**), and trace fossil species are called **ichnospecies**.

The key to understanding the naming of trace fossils produced by invertebrates is to realize that the names usually say nothing about the organism that made the trace. In the early days of ichnology, the common

Box 19.1 Jumping bristletails

Paleontologists need keen eyesight. The slab in Fig. 19.1 shows a clean, slightly undulating surface, with some long cracks and obscure little markings here and there. But is there anything of importance on the slab? It might not at first seem so.

The slab comes from the Lower Permian Robledo Mountains Formation of southern New Mexico (Minter & Braddy 2006), where most surfaces show tracks of one sort or another: amphibians, reptiles, scorpions, spiders, millipedes and insects. Trace fossils are most commonly preserved in red-gray siltstones to fine-grained sandstones that were deposited in a flat tidal setting; mudcracks and raindrop imprints indicate periods of exposure to the air. On looking closely at this slab, you may be able to see three arrow-shaped markings, running from bottom to top of the slab. In close up, these arrow-shaped markings show three sharp, thin lines at the top, and a fainter marking below. This trace fossil is called *Tonganoxichnus*, and it has been noted before in both the Permian and the Carboniferous. But what could have made it?

Previous authors had suggested that *Tonganoxichnus* was produced by an extinct relative of a hopping insect like a jumping bristletail. Jumping bristletails, more properly called machilids, are primitive, wingless insects that are known today from moist coastal habitats of North America and Europe. They are closely related to silverfish, commonly seen in damp carpets inside houses, and they can jump up to 100 mm at a time, 10 times their body length. Fossil machilids and their extinct relatives such as *Dasyleptus* are known from the Permian, and they fit the tracks perfectly. The insect was hopping from the bottom of the slab upwards; the sharp grooves at the top of each marking are impressions of the feeding legs at the front, and the arrow-like marking behind is an impression of the abdomen as it hit the ground and propelled the animal forward. So this seemingly obscure slab from New Mexico tells a story of how a number of small wingless insects hopped across the damp sand near a lake 270 Ma.

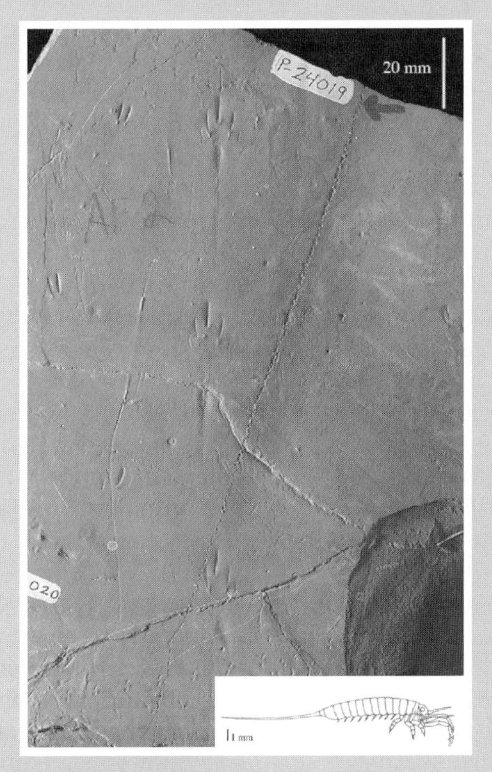

Figure 19.1 Slab of fine sandstone from the Robledo Mountains Formation (Lower Permian) of New Mexico, showing the trace fossil *Tonganoxichnus*, the hopping trace of a basal wingless insect such as *Dasyleptus* (inset). (Courtesy of Nic Minter.)

Table 19.1 The main types of trace fossils, with definitions of the key terms.

A. Traces on bedding planes
 Tracks: sets of discrete footprints, usually formed by arthropods or vertebrates
 Trails: continuous traces, usually formed by the whole body of a worm, mollusk or arthropod, either traveling or resting
B. Structures within the sediment
 Burrows: structures formed within soft sediment, either for locomotion, dwelling, protection or feeding, by moving grains out of the way
 Borings: structures formed in hard substrates, such as limestone, shells or wood, for the purpose of protection, dwelling or carbonate extraction, by cutting right through the grains. Includes bioerosion feeding traces, such as drill holes in shells produced by gastropods
C. Excrement
 Fecal pellets and fecal strings: small pellets, usually less than 10 mm in length, or strings of excrement
 Coprolites: discrete fecal masses, usually more than 10 mm in length, and usually the product of vertebrates
D. Others
 Root penetration structures: impressions of the activity of growing roots
 Non-fecal pellets: regurgitation pellets of birds and reptiles, excavation pellets of crustaceans and the like

U-shaped burrow *Arenicolites* was named after the burrow of *Arenicola*, the lugworm, and the meandering deep-sea trail *Nereites* was named after another polychaete annelid, *Nereis*. However, most *Arenicolites* burrows and most *Nereites* trails have nothing to do with the modern worms *Arenicola* and *Nereis*. Footprints made by vertebrates, on the other hand, can often be matched more readily with their producers, and track names frequently indicate the supposed affinities of the track-maker. For example, the large three-toed dinosaur track *Iguanodonichnus* was supposedly made by the ornithopod dinosaur *Iguanodon* . . . or was it?

The principle that trace fossils should not be named after the supposed maker is based on two observations:

1 One animal can make many different kinds of traces.

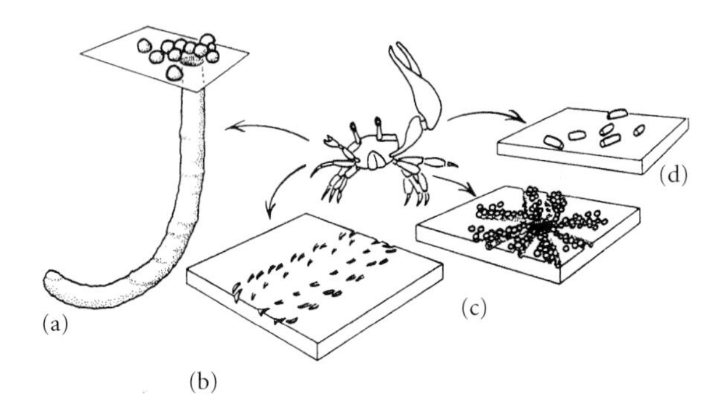

Figure 19.2 One animal may make many different kinds of trace fossils. The modern fiddler crab *Uca* makes: (a) a J-shaped living burrow (domichnion; *Psilonichnus*), (b) a walking trail (repichnion; *Diplichnites*), (c) a radiating grazing trace with balls of processed sand (pascichnion), and (d) fecal pellets (coprolites). (Based on Ekdale et al. 1984.)

2 One trace fossil may be made by many different kinds of organisms.

The fiddler crab *Uca* is observed today to make at least four quite distinct kinds of traces (Fig. 19.2): a J-shaped living burrow, a running track, a star-shaped feeding pattern and fecal pellets, each with its own name, as well as excavation pellets and feeding pellets. An example of one trace fossil made by many different animals is the ichnogenus *Rusophycus*, a bilobed resting impression marked by transverse grooves (Fig. 19.3). *Rusophycus* can be made by at least four different animals, belonging to three phyla, an annelid, a mollusk and two arthropods, but the traces are so similar that they must be given the same name.

An additional consideration is that, if trace fossils were named after their proposed makers, the name would depend on the validity of that interpretation: trace fossil names could not change at the whim of every paleobiologist who proposed a different maker for the same trace. For example, *Iguanodonichnus*, mentioned above as the supposed track of *Iguanodon*, turns out to have been made most likely by a medium-sized sauropod dinosaur. Should its name now be changed when the interpretation changes? Of course not – that would lead to endless confusion and instability. And this shows why it is best not

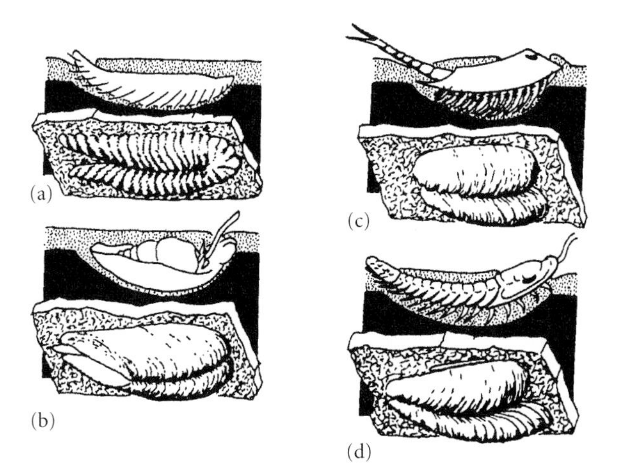

Figure 19.3 One trace fossil may be produced by many different organisms. Here, all the traces are resting impressions, cubichnia, of the ichnogenus *Rusophycus*, produced by (a) the polychaete worm *Aphrodite*, (b) a nassid snail, (c) a notostracan branchiopod shrimp, and (d) a trilobite. (Based on Ekdale et al. 1984.)

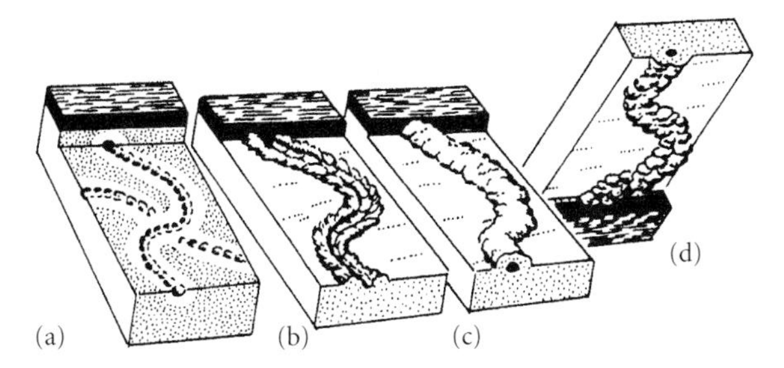

Figure 19.4 Variations in the physical nature of the sediment may create variations in the appearance of a trace fossil. Here, a subsurface, patch-feeding burrow develops different morphologies, and therefore has different names, when preserved: (a) in sand (*Scalarituba*), (b) at a sand–mud interface in firm sediment (*Nereites*), (c) at a sand–mud interface in wetter sediment (*Neonereites*), and (d) at a mud–sand interface, seen from below (*Neonereites*). (Based on Ekdale et al. 1984.)

to use names for trace fossils that imply a particular producer.

The nature of preservation of a trace fossil may affect its appearance, but the name cannot necessarily take account of this. The appearance of trails and burrows may be altered significantly by the grain size, location with respect to a fine- and coarse-grained horizon, and water content of the sediment in which they are preserved (see p. 59). This can be seen clearly with the example of the *Nereites–Scalarituba–Neonereites* complex, a series of trace fossil forms produced by a single deep-sea grazing organism (Fig. 19.4). The situation is different for many vertebrate traces. For example, it is often possible to follow a single dinosaur trackway for some distance, and the shape of individual foot and hand prints might vary substantially, depending on the sediment type and the animal's behavior. It would clearly be crazy to give each variant print in a single trackway a different name.

In conclusion, *trace fossil names should be based only on morphological features including shape and ornamentation, and not on the postulated maker or mode of preservation.*

Preservation of trace fossils

Trace fossils may be formed on bedding planes or within sedimentary horizons. The relationships of the trace fossils to the sediment, and the ways in which they are preserved must be established. Seilacher's terminology, developed in the early 1960s, is frequently used (Fig. 19.5). Burrows are three-dimensional structures, but they may be seen in different ways in the rocks: they are called **full relief** traces when they are seen in three dimensions, but **semireliefs** when just one side is seen projecting from the bedding plane. Semirelief burrows and trails may occur on the top of a bed, called **epireliefs** (*epi*, on), or on the bottom, termed **hyporeliefs** (*hypo*, under). Hyporelief preservation is very common in sedimentary sequences where sandstones and mudstones are interbedded – a feature of turbidite and storm-bed successions (Box 19.2). Here, the traces are best seen on the bottoms of sandstone beds as sole structures, because the mudstones often flake away.

It is important to realize that burrows and surface trails are not always easy to distinguish. Burrows are formed within sediment, and are thus **endogenic** (*endo*, within; *genic*, made), and they are seen both as full reliefs and as semireliefs along bedding planes. However, if subsequent erosion or weathering

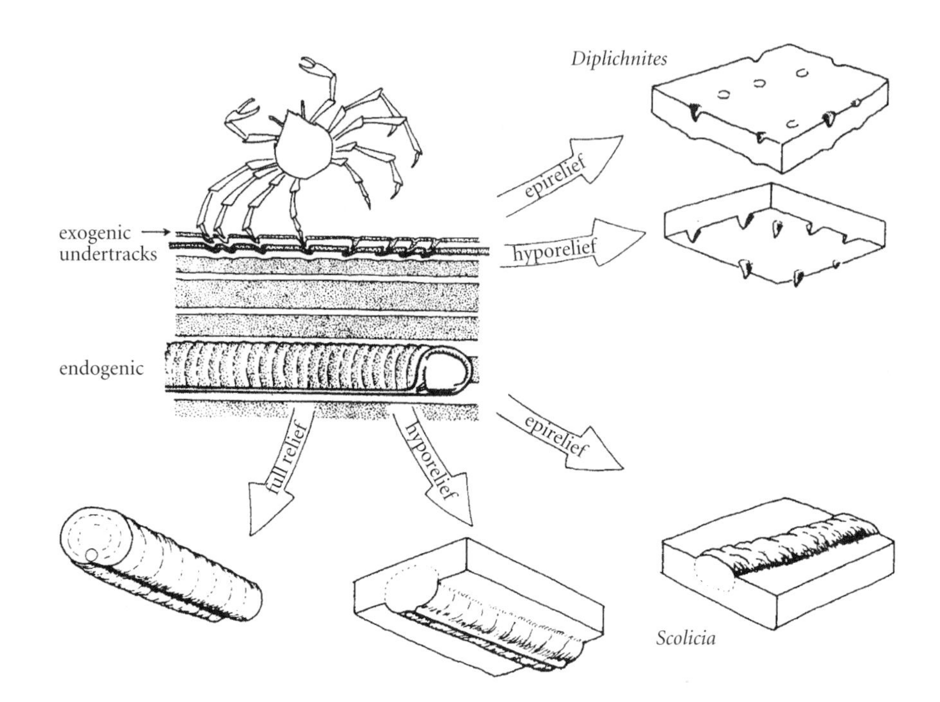

Figure 19.5 Terminology for trace fossil preservation, depending on the relationship of the trace to sediment horizons. (Based on Ekdale et al. 1984.)

Box 19.2 Turbidite timing

The mode of preservation of trace fossils can show whether they were produced before or after a major sedimentary event, such as a turbidite flow. **Turbidites** are underwater avalanches, or gravity flows, that may transport vast quantities of sediment rapidly into deeper waters. The Lower Silurian mudstones and sandstones of central Wales and the Welsh coast have long been known as a source of trace fossils that belong to different environments within the deep-ocean *Nereites* ichnofacies. Crimes and Crossley (1991) identified 25 ichnogenera from the sandstone turbidites of the Aberystwyth Grits Formation, the commonest forms being *Helminthopsis*, *Paleodictyon* and *Squamodictyon* (Fig. 19.6a, b). The finer-grained sediments yielded different ichnofaunas, consisting mainly of *Nereites*, *Dictyodora*, *Gordia* and *Helminthoida* (Fig. 19.6c, d).

One clear distinction in the Welsh Basin ichnofaunas was probably the result of minor turbidite activity at the toe of spreading fans. Pre-turbidite and post-turbidite assemblages have been identified, representing the trace fossils that are formed in normal background times, and those that were formed after a turbidity flow event. Before the flow, Orr (1995) identified an assemblage of surface trails and shallow burrows. After the passage of a low-energy turbidite flow, the top layers of the existing sediment were stripped off, casting the deeper pre-turbidite burrows as convex hyporeliefs on the sole of the turbidite sand. After the flow had waned, a post-turbidite trace fossil assemblage was developed within the turbidite sand (Fig. 19.6e).

These insights allow a clearer interpretation of the trace fossil assemblages: although *Helminthopsis* and *Paleodictyon* may be found together in the sandstones, *Helminthopsis* is an opportunistic post-turbidite form that colonized the sands soon after the turbidite flow had ceased. Only later did *Paleodictyon* and other forms move in to occupy the stable sediment.

Figure 19.6 Typical trace fossils of the Lower Silurian sediments of the Welsh Basin (*Nereites* ichnofacies): (a) *Helminthopsis*, (b) *Paleodictyon*, (c) *Nereites*, (d) *Gordia*, and (e) the pre- and post-turbidite trace fossil assemblages. (Courtesy of T. P. Crimes.)

removes the top layers of sediment above a burrow, the burrow may be seen as a semirelief. Trails are formed on the top of the sediment pile, and are thus **exogenic** (*exo*, outside) structures, typically seen as semireliefs. **Undertracks** are impressions formed on sediment layers below the surface on which the animal was moving, and it is important to distinguish these from the true track as the morphology may be different. The shapes of tracks and undertracks have been investigated in numerous experimental studies on modern animals (Box 19.3).

Interpreting ancient behavior

Trace fossils can plug gaps in knowledge when body fossils are rare. Two environments where

Box 19.3 Undertracks of the emu

Experts on dinosaur tracks have been aware for a long time about undertracks. Huge animals such as dinosaurs made very deep footprints when they walked over soft sediment. Sometimes the print shape was transmitted for a meter or more down through the layers of sediment below the layer on which the animal walked, and this means that many dinosaur tracks are actually undertracks, if they are viewed on a lower layer.

There have been many experiments to show how tracks are altered by the consistency of the sediment (grain size, water content) and the weight of the animal. Obviously, larger animals make deeper prints. Also, the larger the grain size of the sediment, the less well defined the prints are. Also, if the sediment was entirely dry when an animal moved across, the tracks might be lost. Too wet, and the sand or mud just flowed back into the footprint or trail, leaving a gloopy mess. If the sediment was just slightly wet, then an excellent impression might be preserved.

Jesper Milàn, a graduate student at the University of Copenhagen, decided to try to understand tracks and undertracks of dinosaurs by experiments with an emu (Milàn & Bromley 2006). The emu is a large flightless bird from Australia, known for its cussedness – the animal pretty much refused to run across the carefully prepared sand beds at a local emu farm, and the experimenters were soundly pecked for their efforts (Fig. 19.7a). In the end, Milàn managed to make some clean emu tracks on prepared "sediment" layers; these show very clearly how the undertrack shape changes down through the sediment (Fig. 19.7b). This is a warning to ichnologists, to be clear about identifying tracks and undertracks, and not to overinterpret the anatomy of the track-maker from a deep undertrack.

Figure 19.7 Experimental ichnology: (a) graduate student Jesper Milàn, trying to persuade an emu to walk where he wants it to walk, and (b) the tracks and undertracks of the emu – results of an experiment where an emu stepped on a package of alternating layers of concrete and sand. After the concrete hardened, the sand was flushed out and replaced with silicone rubber. The top print (left) made an impression on several layers below, shown as undertracks at depths of up to 40 mm. Notice how the impressions of the digits become wider and less well-defined along each subjacent horizon. (Courtesy of J. Milàn.)

Figure 19.8 Trace fossils of the deep ocean floor. The patch-feeding trace (pascichnia) *Helminthopsis* meanders on one horizon, and the network burrow system (agrichnia) *Paleodictyon* is seen at a different level, in this field photograph from the Lower Silurian Aberystwyth Grits, Wales. (Courtesy of Peter Crimes.)

this works very well are in the deep sea and on land. Very little is known from body fossils of the history of life in deep abyssal oceans, and indeed very little is known about life in these zones today because they are inaccessible. Trace fossils, however, are abundant in many deep oceanic settings (Fig. 19.8), and they show the diversity of trail-making and burrowing soft-bodied organisms, how many of them built complex shallow burrow systems and efficient patch-feeding trails, and how these assemblages evolved through the Phanerozoic. On land, some continental sequences preserve very few body fossils, and the only indications of animal life are abundant dinosaur and other vertebrate tracks (Box 19.4), as well as tracks and burrows made by insects and pond-living animals.

One of the major advances in trace fossil studies was Seilacher's (1967a) classification of behavioral categories. He divided trace fossils into seven behavioral types, depending on the activities represented (Fig. 19.10). Tracks and trails representing movement from A to B, such as worm trails or dinosaur trackways, are termed **repichnia** (*repere*, to creep; *ichnos*, trace). Grazing trails that involve movement and feeding at the same time are called **pascichnia** (*pascere*, to feed). These are typically coiled or tightly meandering trails found in deep oceanic sediments, where the regular pattern is an adaptation to feeding on restricted patches of food. Some unusual deep-sea horizontal burrow systems appear to have been maintained for trapping food particles, or for growing algae. These are termed **agrichnia** (*agricola*, farmer). Feeding burrows, such as those produced by earthworms, as well as many marine examples, are called **fodinichnia** (*foda*, food). Living burrows and borings are termed **domichnia** (*domus*, house). Escape structures, or **fugichnia** (*fugere*, to flee) are traces of upward movement of worms, bivalves or starfish seeking to escape from beneath a layer of sediment that has been dumped suddenly on top of them. Fugichnia are found in cases of rapid sedimentation, in beach, storm-bed and turbidite sediments. Resting traces, or **cubichnia** (*cubare*, to lie down), may be of many types, and can include impressions of the undersides of trilobites, starfish and jellyfish.

Tracks and trails can sometimes be assigned to their makers, and then it may be possible to carry out quantitative studies of their modes of locomotion. Arthropod tracks, for example, show the often complex patterns of movement of their numerous legs. Dinosaur tracks can show how fast the dinosaur was running (Box 19.5).

TRACE FOSSILS IN SEDIMENTS

Trace fossils as environmental indicators

The discovery that electrified ichnologists in the 1960s was that certain trace fossils were reliable instant guides to ancient sedimentary environments. Identify a particular trace fossil, or trace fossil assemblage, and you have pinned the water depth, tide and storm conditions, salinity and oxygen levels. And this works whatever the age of the rocks, whether Cambrian or Cretaceous. The trace fossils remained remarkably constant in appearance, even if their producers might have been quite different.

This paleoenvironmental scheme of trace fossils presented by Seilacher (1964, 1967b) has been modified and enlarged since then (Frey et al. 1990), but in principle it divides trace fossil assemblages into a number of **ichnofacies** (Fig. 19.12). The ichnofacies are named after a characteristic trace fossil, and they indicate particular sedimentary facies (Box 19.6). The ichnofacies is identified on the basis of an assemblage of trace fossils, and it may be recognized even if the name-bearing form is absent.

The classic marine ichnofacies, those named for *Nereites*, *Zoophycos*, *Cruziana* and *Skolithos*, are not simply depth-related, as Seilacher first proposed, but are associated with particular sedimentary regimes, combining aspects of water energy, bottom sediment type, temperature, chemistry and food supply. These four ichnofacies include assemblages of trace fossils typical of fair-weather, normal conditions of deposition, and those characteristic of exceptional storm and turbidite event beds. The complexity of controls on the marine ichnofacies is shown in many field-based studies where alternations between ichnofacies may be found at a single location (Box 19.7).

The *Scoyenia* ichnofacies is one of several continental trace fossil facies, and depends on the presence of shallow freshwater, while the *Psilonichnus* ichnofacies is controlled by coastal marine influence on a terrestrial setting. Not included here are some additional terrestrial ichnofacies (see McIlroy 2004). Since 1990, several ichnologists have proposed ichnofacies in ancient soils, **paleosols**, to characterize different kinds of insect burrows, nesting chambers and the like, and

Box 19.4 Dinosaur behavior

Dinosaur tracks are probably the most familiar trace fossils, and they can tell us a great deal about how the dinosaurs lived. Some dinosaur track sites cover huge areas, and may reveal hundreds or thousands of footprints, often in long trackways, sometimes representing numerous different species. It is fascinating to use these trackways to speculate about ancient behaviors – but you have to be careful! It is important to check whether all the tracks were made at the same time – do they overlap each other or not? A busy-looking track site might have been produced by just one hyperactive dinosaur trotting back and forwards around a water hole.

Three-dimensional dinosaur prints are quite rare. Normally, the dinosaur trots across firm mud or sand, and you are left with simple impressions on the top surface. In some cases, though, the dinosaurs got bogged down in soft sediment, and their feet went in a meter or more. Then, when they wanted to move on, they had to haul their feet out of the gloop, leaving odd-shaped closure traces behind.

A remarkable find from the Late Triassic of Greenland (Gatesy et al. 1999) shows this. Stephen Gatesy, from Brown University, Rhode Island, and his collaborators found strange, narrow, bird-like prints (Fig. 19.9a). Had they been made by a theropod dinosaur with feet made from wire? When they pulled apart the rock layers, they could see that the dinosaur foot had gone in, and sunk through layers of mud, so that the mud flowed back around its ankles. Then, in moving forward and pulling the foot out, the mud flowed back around the exit trace, leaving a long forward trail made by the long middle toe. Computer animations (Fig. 19.9b) demonstrated how the foot may have moved as it went into the mud, and then pulled out at the end of the stride.

Read more about dinosaur tracks on web sites linked to http://www.blackwellpublishing.com/paleobiology/.

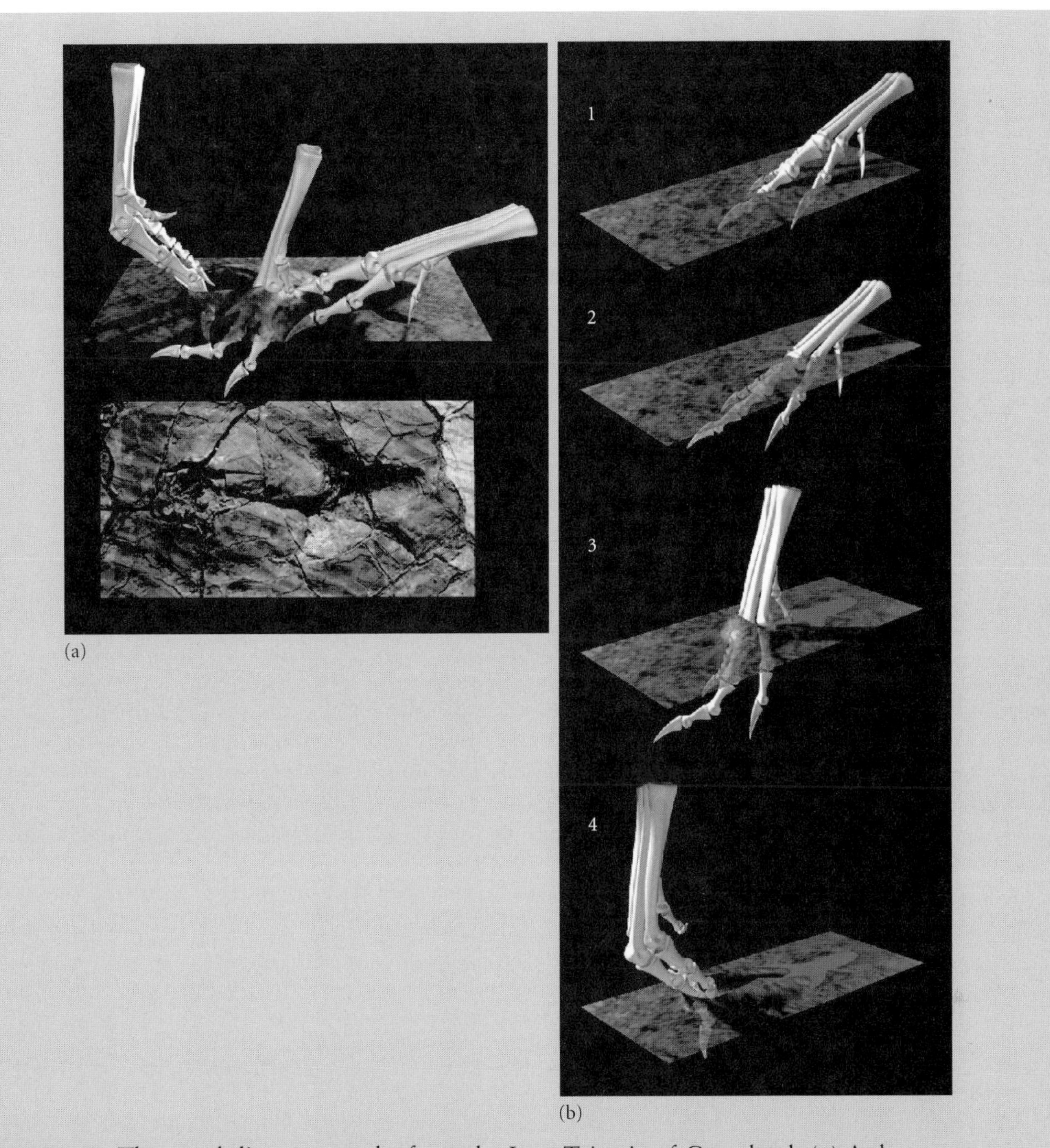

Figure 19.9 Theropod dinosaur tracks from the Late Triassic of Greenland. (a) A three-dimensional computer reconstruction (top) shows a theropod foot at three stages in the creation of a deep track, moving from right to left. A photograph of a deep Greenland footprint is shown below. (b) A three-dimensional computer image reconstructing theropod foot movements through sloppy mud. The first toe creates a rearward pointing furrow (1, 2) as it plunges down and forward. The sole of the foot leaves an impression at the back of the track (3) because it is not lifted as the foot sinks. All toes converge below the surface and emerge together from the front of the track (4). (Courtesy of Stephen Gatesy.)

in lake sediments and lake shores. Others, such as Lockley et al. (1994), have proposed ichnofacies that discriminate among different kinds of assemblages of dinosaur footprints. These proposals do not, however, cover long time spans, as do the classic Seilacher marine ichnofacies, and they are still much debated.

The *Glossifungites*, *Trypanites* and *Teredolites* ichnofacies are controlled by substrate alone, and they could theoretically occur

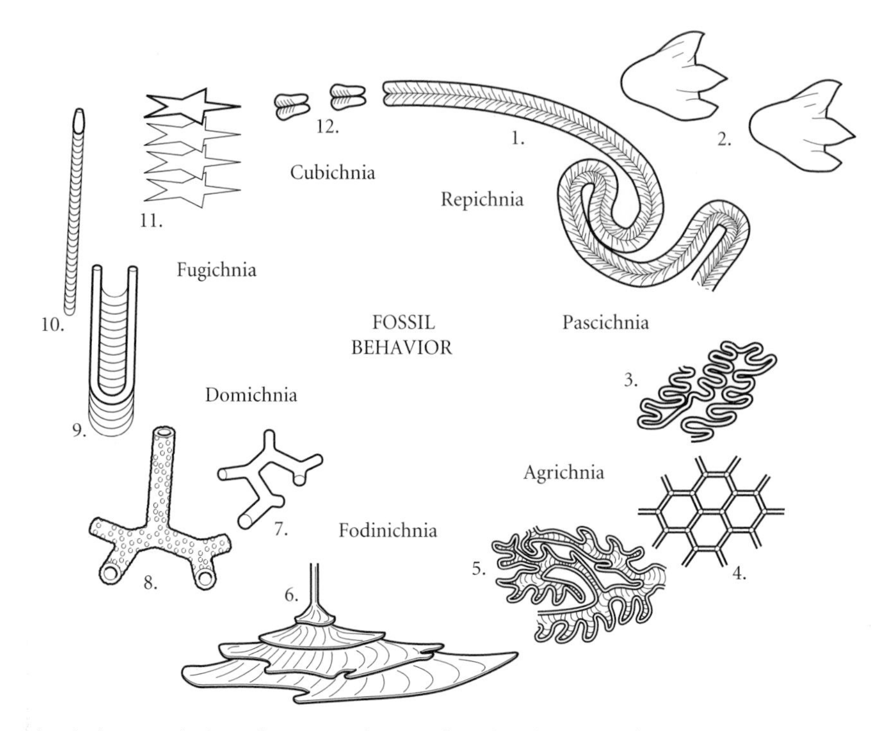

Figure 19.10 The behavioral classification of trace fossils, showing the major categories, and some typical examples of each. Illustrated ichnogenera are: 1, *Cruziana*; 2, *Anomoepus*; 3, *Cosmorhaphe*; 4, *Paleodicyton*; 5, *Phycosiphon*; 6, *Zoophycos*; 7, *Thalassinoides*; 8, *Ophiomorpha*; 9, *Diplocraterion*; 10, *Gastrochaenolites*; 11, *Asteriacites*; 12, *Rusophycus*. (Based on Ekdale et al. 1984.)

Box 19.5 Dinosaur speeds

When you walk along a beach, you leave tracks with a particular **stride length** (the distance from one foot-fall to the next by the same foot). If you begin to run, the stride length increases, and the faster you go, the longer the stride length. An English expert in biomechanics, R. McNeil Alexander, spotted something more: there was a constant relationship between stride length and speed, providing you took account of the size of the animal (measured by the height of the hip from the ground), and it did not matter whether you made the calculation for a two-legged animal like a human, or a four-legged animal like a horse.

Alexander (1976) presented his evidence and his formula, and he suggested it could also be used for estimating the speed of movement of extinct animals, such as dinosaurs:

$$u = 0.25\,g^{-0.5}d^{1.67}h^{-1.17},$$

where u is velocity, g is the acceleration of free fall (gravity), d is stride length and h is hip height (Fig. 19.11). The formula can be simplified to:

$$u = 1.4(1/h) - 0.27,$$

for rough calculations. The hip height has to be measured from a skeleton of the dinosaur that is supposed to have made the tracks. If that cannot be done, there is a fairly predictable relationship

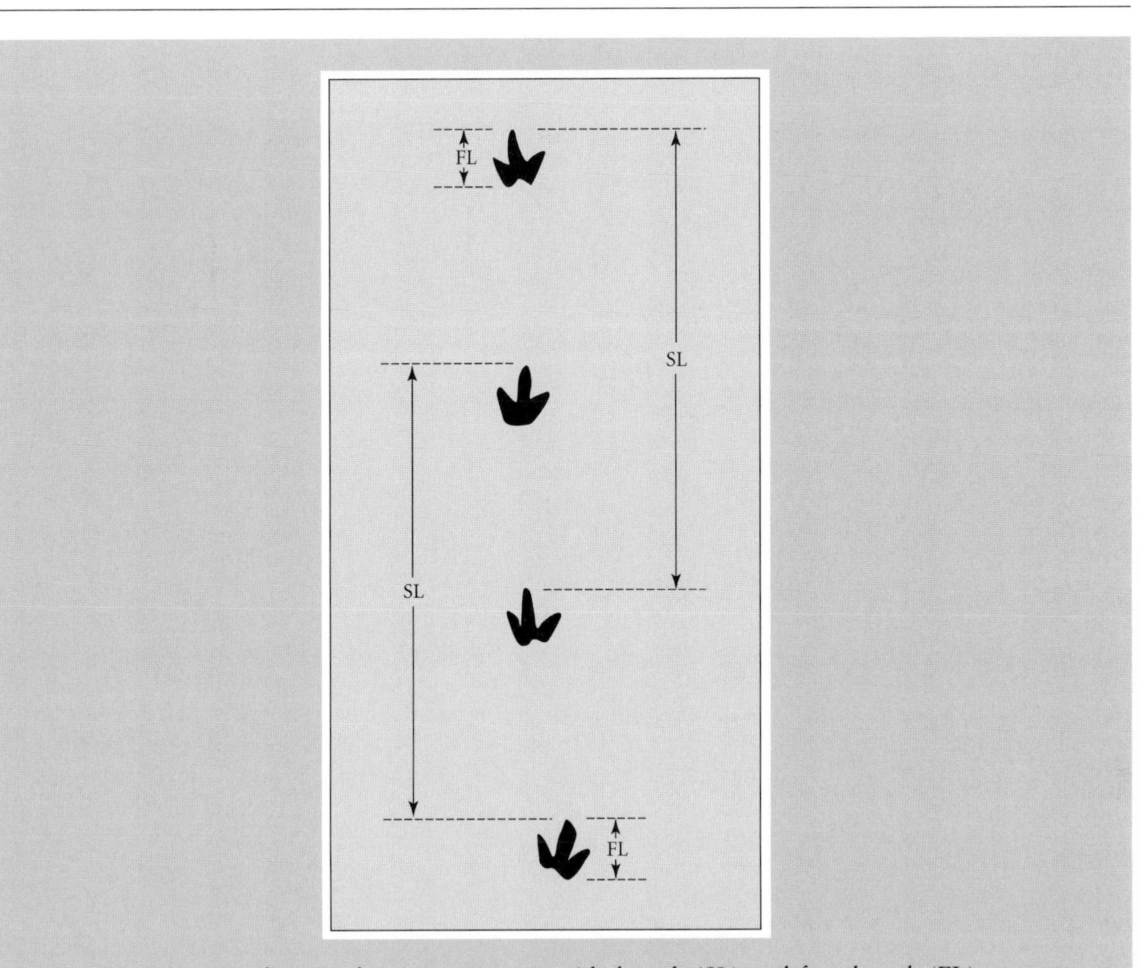

Figure 19.11 Diagram showing how to measure stride length (SL) and foot length (FL) on a dinosaur track.

between hip height and foot length for each major dinosaur group (hip height is 4 to 6 times the foot length, depending on the group), so all measurements can be made from the footprint slab if necessary.

Many paleontologists applied the Alexander formula to dinosaur tracks, and many subtle corrections have been suggested, but it seems to work pretty well. Typical calculated speeds range from 1 to 4 $m\ s^{-1}$ for walking dinosaurs (about human walking speeds), with high values of 10–15 $m\ s^{-1}$ for some smaller, flesh-eating dinosaurs that were in a hurry to catch their lunch. The maximum calculated speed of 15 $m\ s^{-1}$ is equivalent to 54 $km\ h^{-1}$, or 35 miles per hour, equivalent to a fast racehorse, or just faster than town driving speeds. Faster speeds have been claimed from some dinosaur tracks, but these are unlikely.

Calculate dinosaur speeds online via http://www.blackwellpublishing.com/paleobiology/.

across a range of the depth zones represented by Seilacher's classic bathymetric sequence of ichnofacies. In fact, they are mostly restricted to marginal marine, intertidal and shallow shelf zones, but that is related to the commonest occurrences of the required substrates.

Organisms in sediments

Trace fossils depend on sediments. The ichnofacies scheme highlights the important roles of broad sedimentary environment (marine or continental, deep oceanic, shelf or intertidal,

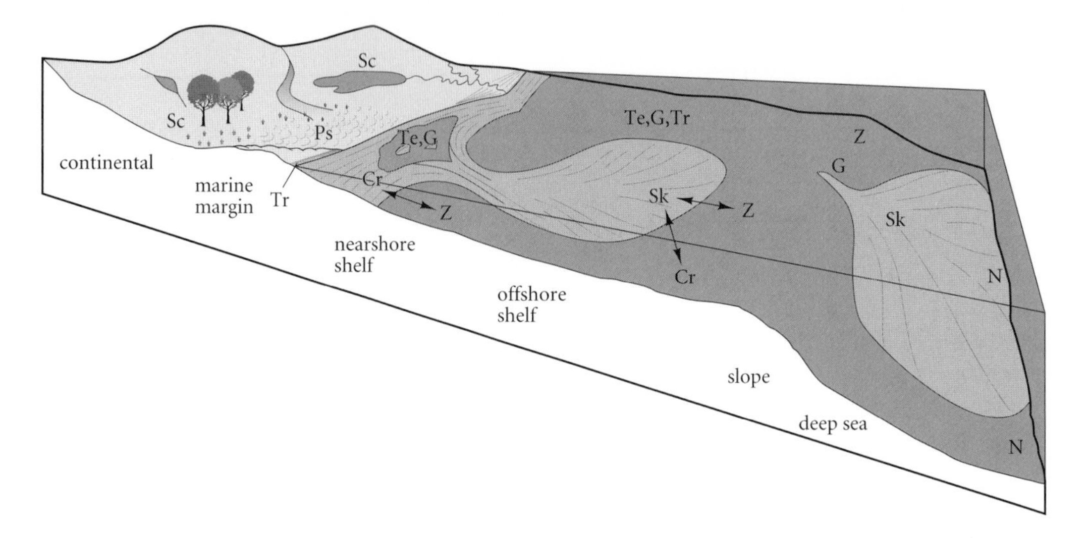

Figure 19.12 The major ichnofacies, and their typical positions in a hypothetical diagram of marine and continental environments. Typical offshore marine soft-sediment ichnofacies, from deep oceanic and basinal locations to the intertidal zone, include the *Nereites* (N), *Skolithos* (Sk), *Zoophycos* (Z) and *Cruziana* (Cr) ichnofacies, which may occur in various water depths and in different conditions of sedimentation. A storm-sand fan and a turbidite fan are indicated. The *Psilonichnus* (Ps) ichnofacies occurs in supratidal marshes and the *Scoyenia* (Sc) ichnofacies includes all lacustrine and related continental settings. The *Glossifungites* (G) ichnofacies is typical of firmgrounds, the *Trypanites* (Tr) ichnofacies consists of borings in limestone, and the *Teredolites* (Te) ichnofacies consists of borings in wood. (Modified from Frey et al. 1990, and other sources.)

lake or terrestrial), salinity and sedimentation rate. Sediments affected the ancient burrowers and crawlers (biological effects), but the sediments also affect how trace fossils look to us today (preservation effects).

The physical properties of sediments can exert controls on trace fossil distributions, and four factors are particularly important:

1 *The average grain size* affects sediment-ingesting burrowers, organisms that require particular sediment sizes to line their burrows, and filter feeders which must avoid fine suspended sediment.
2 *Sediment stability*, particularly in the *Glossifungites* and *Trypanites* ichnofacies, which depend on firm and lithified substrates, respectively. Some organisms build burrows of different morphology, depending on the stability of the sediment.
3 *Water content*, producing sediments that range from soupy in consistency to totally lithified sediments with zero porosity, whose sole trace fossils are borings (*Trypanites* ichnofacies). Firmgrounds contain relatively little water and are characterized by particular burrows of the *Glossifungites* ichnofacies.
4 *Chemical conditions* in sediments, particularly oxygen levels. Trace fossils are rare or absent in completely **anoxic** situations, but a surprising variety of animals can survive in **dysoxic** (very low oxygen) conditions. Generally, the smaller the burrow, the lower the oxygen level.

Burrowing organisms divide up the different strata of unconsolidated sediment in rather precise ways, a phenomenon called **tiering**. The top few centimeters of sediment on the seafloor, the **mixed layer**, may be a mixture of water and sediment, either loose sand that moves with the currents or soupy mud. Deeper down is the **historical layer**, the older consolidated sediments from which water has been squeezed, and between the two is the **transitional layer**. Each burrower is restricted to a particular depth of burrowing, some exploiting the near-surface oxygenated zone, and others extending ever deeper into the sediment. Interpretation of tiering can be difficult because the mixed layer is readily disturbed,

Box 19.6 The nine ichnofacies

The *Nereites* ichnofacies (Fig. 19.13a) is recognized by the presence of meandering pascichnia such as *Nereites*, *Neonereites* and *Helminthoida*, spiral pascichnia such as *Spirorhaphe*, and agrichnia such as *Paleodictyon* and *Spirodesmos*. Note that the whole concept of agrichnia is debated: the most commonly quoted example, *Paleodictyon*, has been interpreted as the mold of a deceased, soft-bodied, colonial-type organism, although others dispute this new view. Vertical burrows are almost entirely absent. This ichnofacies is indicative of deep-water environments, and includes ocean floors and deep marine basins. The trace fossils are found in muds deposited from suspension, and in the mudstones and siltstones of distal turbidites.

The *Zoophycos* ichnofacies (Fig. 19.13b) is characterized by complex fodinichnia like *Zoophycos*, and it may contain other deep traces such as *Thalassinoides* in tiered arrangements. The ichnofacies occurs in a range of water depths between the abyssal zone and the shallow continental shelf, and may be associated with low levels of oxygen. This ichnofacies may occur in normal background conditions of sedimentation, whereas the *Nereites* ichnofacies may be a matching association found at similar water depths during times of event (turbidite) deposition.

The *Cruziana* ichnofacies (Fig. 19.13c) shows rich trace fossil diversity, with horizontal repichnia (*Cruziana*, *Aulichnites*), cubichnia (*Rusophycus*, *Asteriacites*, *Lockeia*) as well as vertical burrows. This ichnofacies represents mid and distal continental shelf situations that may lie below the normal wave base, but may be much affected by storm activity.

The *Skolithos* ichnofacies (Fig. 19.13d) is recognized by the presence of a low diversity of abundant vertical burrows, domichnia-like *Skolithos*, *Diplocraterion* and *Arenicolites*, fodinichnia-like *Ophiomorpha*, and fugichnia. These all typically indicate intertidal situations where sediment is removed and deposited sporadically, and the organisms have to be able to respond rapidly in stressful conditions. The *Skolithos* ichnofacies was at first seen as occurring only in the intertidal zone, but it is also typical of other shifting-sand environments, such as the tops of storm-sand sheets and the upper reaches of submarine fan systems in deeper water.

The *Psilonichnus* ichnofacies (Fig. 19.13e) is a low-diversity assemblage, consisting of small vertical burrows with basal living chambers, *Macanopsis*, narrow sloping J-shaped and Y-shaped burrows, *Psilonichnus* (a ghost crab burrow), root traces and sometimes vertebrate footprints. It is typical of backshore, dune areas and supratidal flats on the coast.

The *Scoyenia* ichnofacies (Fig. 19.13f) is typified by a low-diversity trace fossil assemblage, mainly simple horizontal fodinichnia (*Scoyenia*, *Taenidium*), with occasional vertical domichnia (*Skolithos*) and repichnia produced by insects or freshwater shrimps (*Cruziana*) preserved in fluvial and lacustrine sediments, often in the silts and sands of red-bed sequences. Associated subaerial sediments, such as eolian sands and paleosols, representing unnamed ichnofacies, may contain domichnia and repichnia of insects, and dinosaur and other tetrapod footprints.

The *Glossifungites* ichnofacies (Fig. 19.13g) is characterized by domichnia such as *Glossifungites* and *Thalassinoides* and sometimes plant root penetration structures, but other behavioral trace fossil types are rare. The sediments are firm, but not lithified, and may occur in firm compacted muds and silts in marine intertidal and shallow subtidal zones. The firmgrounds may develop in low-energy situations such as salt marshes, mud bars or high intertidal flats, or in shallow marine environments where erosion has stripped off superficial unconsolidated layers of sediment, exposing firmer beds beneath.

The *Trypanites* ichnofacies (Fig. 19.13h) is characterized by domichnial borings of worms (*Trypanites*), bivalves (*Gastrochaenolites*), barnacles (*Rogerella*) and sponges (*Entobia*) formed in shoreline rocks or in lithified limestone hardgrounds on the seabed. In modern examples, bioerosion traces such as feeding scrapings made by gastropods and echinoids may be common, but these are rarely preserved in ancient cases.

The *Teredolites* ichnofacies (Fig. 19.13i) is identified by the presence of borings in wood (especially *Teredolites*), mostly produced by marine bivalves such as the modern shipworm, *Teredo*.

Read more about the key ichnofacies through web sites at http://www.blackwellpublishing.com/paleobiology/.

Continued

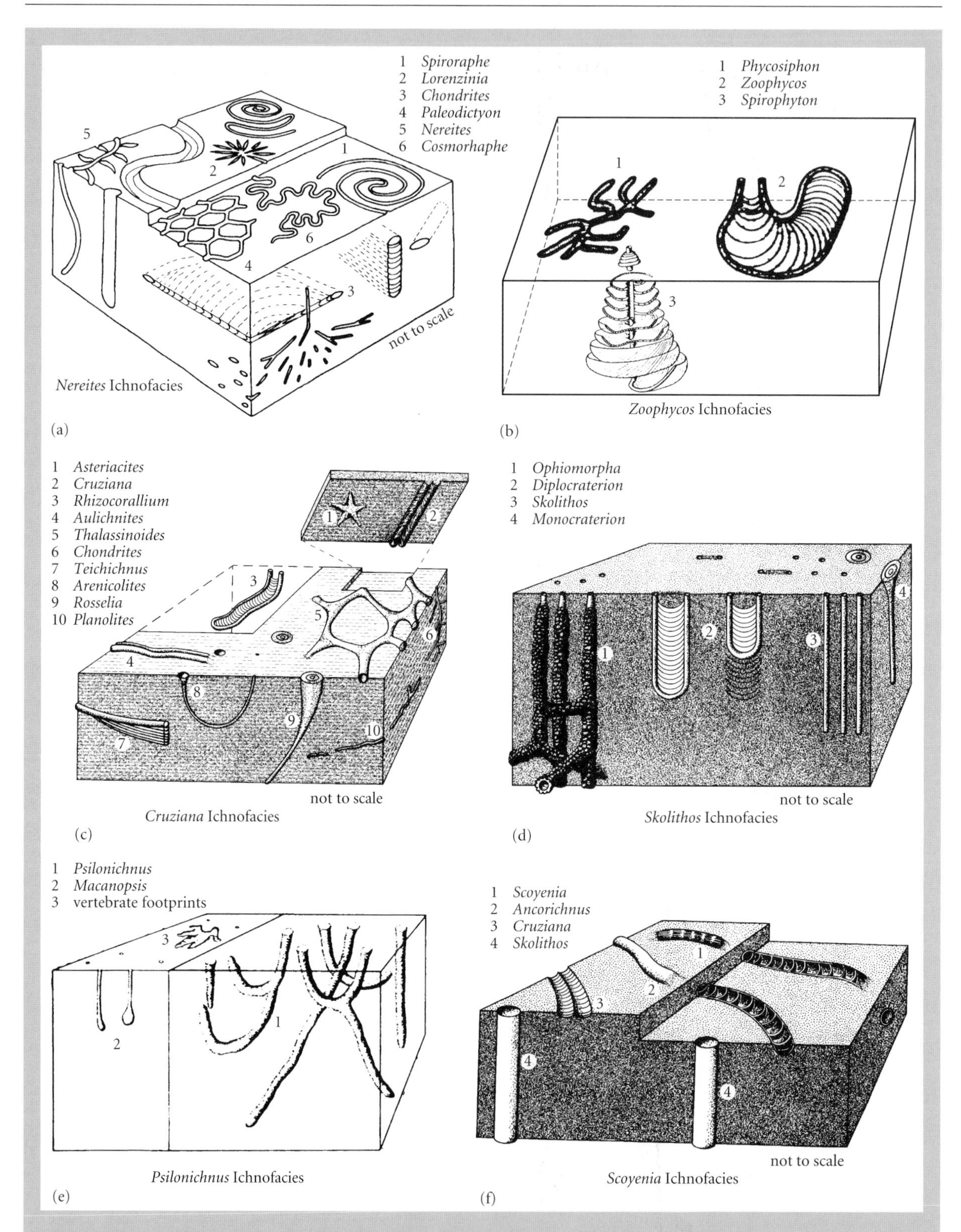

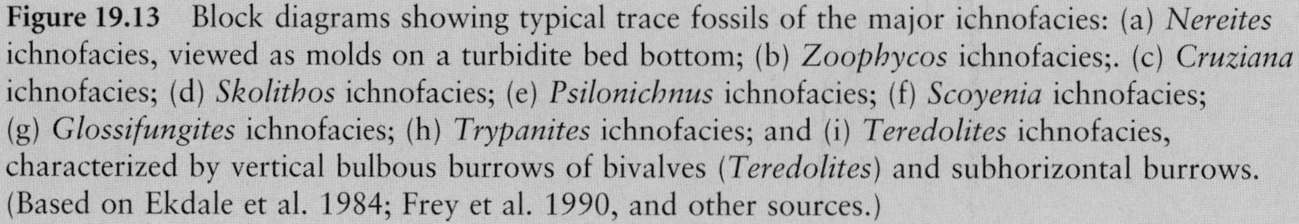
Figure 19.13 Block diagrams showing typical trace fossils of the major ichnofacies: (a) *Nereites* ichnofacies, viewed as molds on a turbidite bed bottom; (b) *Zoophycos* ichnofacies;. (c) *Cruziana* ichnofacies; (d) *Skolithos* ichnofacies; (e) *Psilonichnus* ichnofacies; (f) *Scoyenia* ichnofacies; (g) *Glossifungites* ichnofacies; (h) *Trypanites* ichnofacies; and (i) *Teredolites* ichnofacies, characterized by vertical bulbous burrows of bivalves (*Teredolites*) and subhorizontal burrows. (Based on Ekdale et al. 1984; Frey et al. 1990, and other sources.)

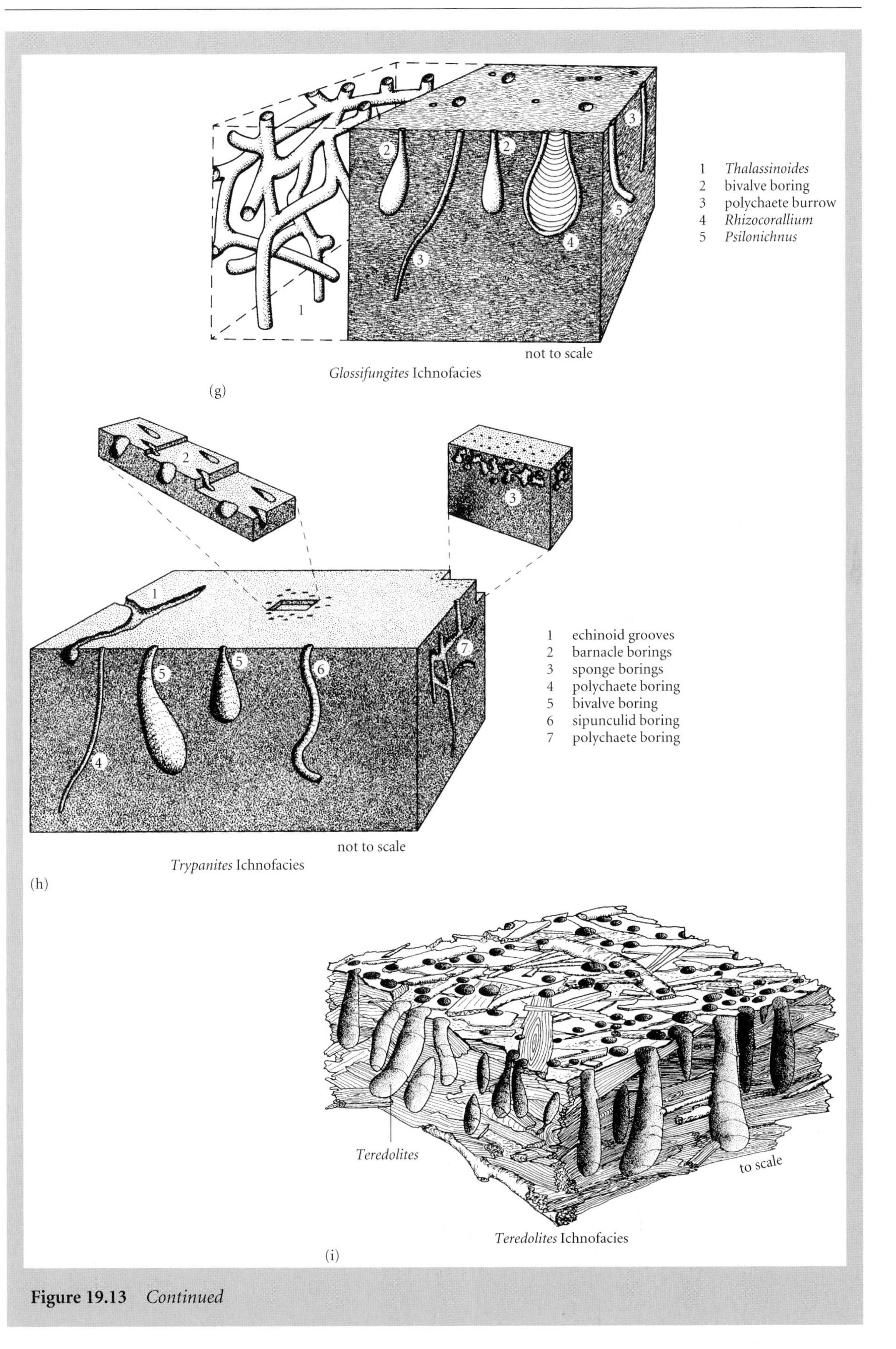

Figure 19.13 *Continued*

Box 19.7 Alternating ichnofacies

Many sedimentary sequences show a mix of ichnofacies, as would be expected, since no ichnofacies is exclusive to a single location or water depth. The Cardium Formation of Alberta, Canada has produced abundant trace fossils from a sequence of muds and sandstones (Pemberton & Frey 1984) (Fig. 19.14). The normal quiet-water sedimentation produced mud, silt and fine sand layers, and diverse trace fossils of the *Cruziana* ichnofacies, mainly representing the activities of mobile carnivores and deposit feeders exploiting relatively nutrient-rich, fine-grained sediments. These sediments were interrupted sporadically by storm beds, thick units of coarse sand washed back from the shore region into deeper water by storm-surge ebb currents. The trace fossils of these units are *Skolithos*, *Ophiomorpha*, *Diplocraterion* and various fugichnia, all typical elements of the *Skolithos* ichnofacies.

One view of this alternation between trace fossils of the *Cruziana* and *Skolithos* ichnofacies might be that there had been repeated changes in sea level from deep to shallow offshore conditions. However, the control is more probably the dramatic changes in energy of deposition. The opportunistic members of the *Skolithos* ichnofacies colonized the storm beds, probably having been washed down from the intertidal zone, and they were able to cope with the rapid fluctuations in unconsolidated sediment depth. The storm events doubtless killed off most of the members of the *Cruziana* ichnofacies, or displaced them to the margins of the affected area. After the storm-surge ebb currents waned, and slow sedimentation resumed, the surface-feeding organisms recolonized the whole area.

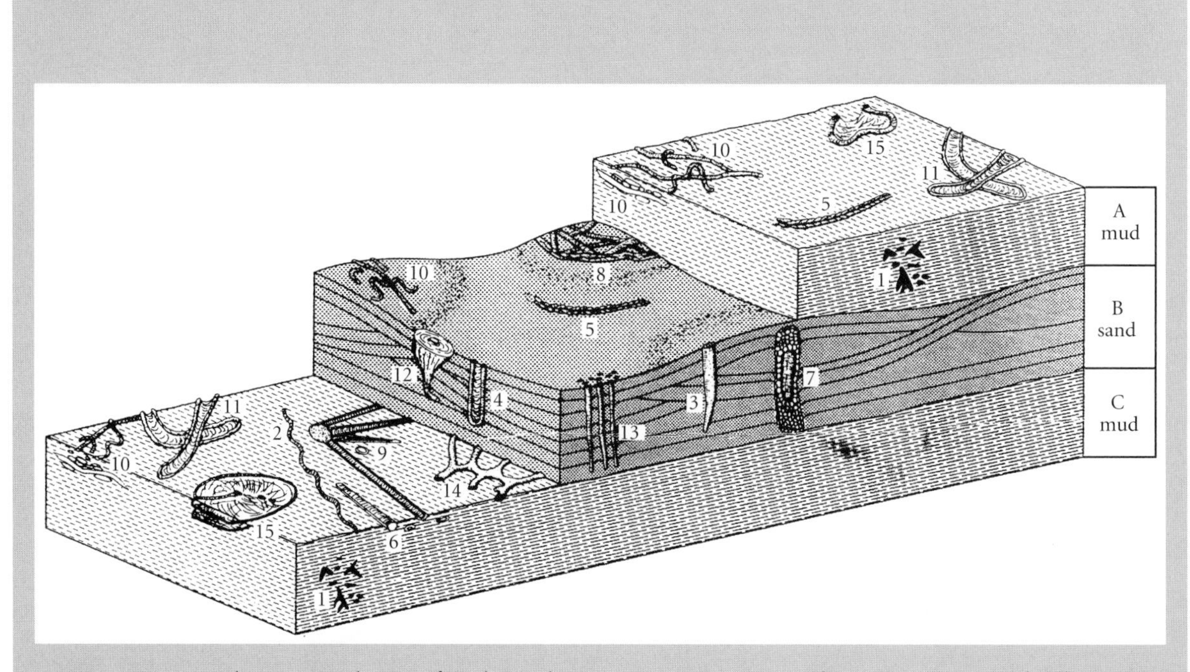

Figure 19.14 Sediments and trace fossils in the Late Cretaceous Cardium Formation of Alberta. Normal, fine-grained sediments (A, C) are associated with *Cruziana* ichnofacies trace fossils, while intermittent, coarse, sandstone, storm beds (B) show trace fossils of the *Skolithos* ichnofacies. 1, *Chondrites*; 2, *Cochlichnus*; 3, *Cylindrichnus*; 4, *Diplocraterion*; 5, *Gyrochorte*; 6, *Paleophycus*; 7, *Ophiomorpha*; 8, ?*Phoebichnus*; 9, *Taenidium*; 10, *Planolites*; 11, *Rhizocorallium*; 12, *Rosselia*; 13, *Skolithos*; 14, *Thalassinoides*; 15, *Zoophycos*. (Based on Pemberton & Frey 1984.)

and traces are lost or distorted, and the historical layer may contain traces of many different generations, perhaps representing many months or years of erosion and deposition.

Most burrowers are restricted to the mixed layer, as this minimizes the physical effort for animals that are simply moving from A to B. Organisms that feed on organic matter also favor the surface layers. Deeper burrowers are mainly those forming domichnia, where the body of the organism is large, or where it possesses long siphons, in order to keep contact with oxygenated waters above. Deeper layers are also safer from predators, whether those operating from the surface, or other burrowers. There are also feeding opportunities at depth, at the **redox layer**, where the oxygenated surface sediments meet the deeper anoxic sediments – a horizon that is characterized by unusual shelly faunas and sulfur-oxidizing bacteria.

Such tiering patterns in any particular environment increase in complexity through time. In a Middle Ordovician example (Fig. 19.15a), the subsurface layers are filled with simple horizontal burrows, *Planolites*. These are cut by branching fodinichnia, *Chondrites*, exploiting an organic-rich layer at a depth of 20–30 mm below the surface of the sediment. The deepest burrows may be *Teichichnus*, showing spreiten structure (multiple ghosts of previous burrow positions), and extending down to 100 mm at the deepest. An Early Jurassic example (Fig. 19.15b) shows a substantial increase in depth burrowed, to perhaps 0.5 m, with small *Chondrites* in the upper layers, a new large *Chondrites* extending to lower layers, and domichnia, *Thalassinoides*, at the deepest levels. Finally, in a Late Cretaceous example (Fig. 19.15c), there are at least nine tiers, three horizons of shallow burrows near the surface, *Planolites*, *Thalassinoides*, *Taenidium*, *Zoophycos*, and large and small *Chondrites* going deepest, perhaps to a maximum depth of 1 m.

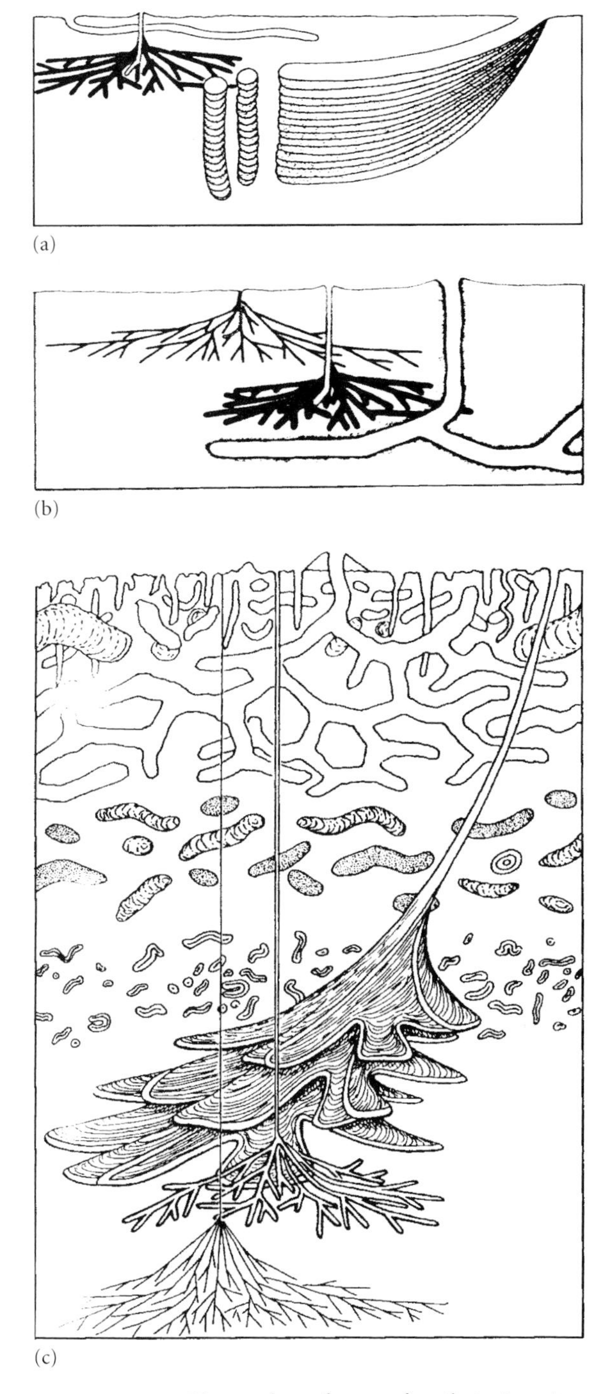

Figure 19.15 Examples of trace fossil tiering, in which burrowers choose specific depth horizons below the sediment–water interface. (a) In the Middle Ordovician limestones of Öland, Sweden, there are three tiers. (b) In the Early Jurassic Posidonienschiefer of Germany, there are also three tiers. (c) In the Late Cretaceous Chalk of Denmark, there are at least nine tiers. (Based on Ekdale & Bromley 1991, and other sources.)

Trace fossils and time

Trace fossils do not evolve in the way body fossils do, and they generally cannot be used for dating rocks. This is mainly because of their rather peculiar properties; as we have seen, trace fossils are excellent indicators of sedimentary environments just because they

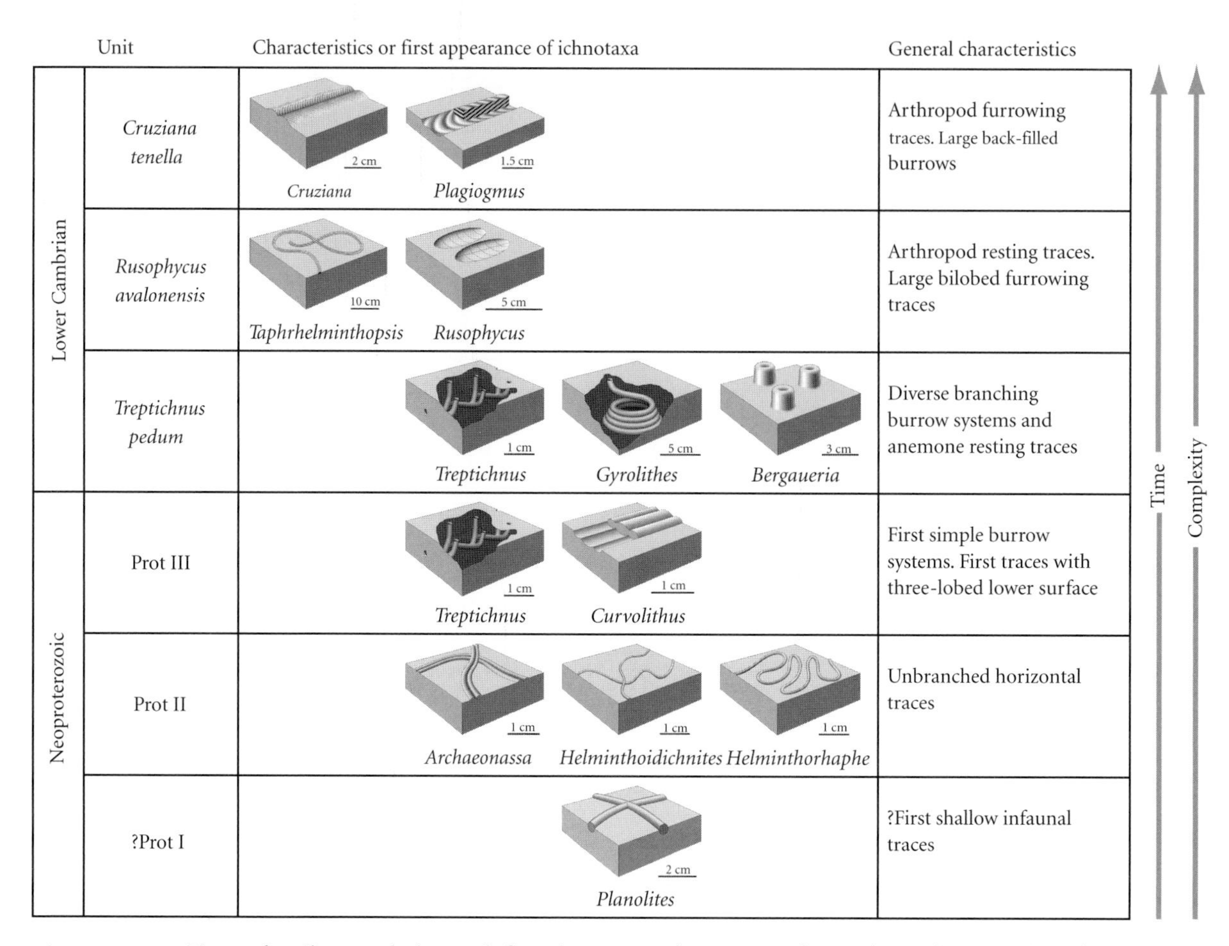

Figure 19.16 Trace fossils may help to define the Precambrian–Cambrian boundary, and to flesh out detail about the Cambrian explosion. Jensen (2003) identified seven trace fossil zones, each characterized by trace fossils of increasing complexity. Evidence for trilobites, and arthropods in general, is signaled first by trace fossils and then by body fossils. Prot, Proterozoic. (Drawing by Simon Powell.)

do *not* evolve. There are one or two exceptions, and one of these is the critical Precambrian–Cambrian boundary interval.

The timing of the Cambrian explosion has been hugely controversial, and the story from body fossils and trace fossils is different (see p. 249). The base of the Cambrian System, and therefore the Precambrian–Phanerozoic boundary, has generally been placed at the first occurrence of trilobite body fossils. But the oldest known trilobite body fossils actually occur *above* the first trilobite trace fossils, such as *Monomorphichnus*, *Rusophycus*, *Cruziana* and *Diplichnites*. Below this boundary, in the Neoproterozoic, trace fossils document how early animals were becoming more complex in the lead-up to the Cambrian explosion (Fig. 19.16). First to appear were simple shallow burrows (*Planolites*), then unbranched horizontal traces such as *Archaeonassa*, *Helminthoidichnites* and *Helminthorhaphe* (Jensen 2003). In the third Neoproterozoic, trace fossil zone are the first records of simple burrow systems (*Treptichnus*) and traces with a three-lobed lower surface ("*Curvolithus*"). The basal Cambrian is then marked by the *Treptichnus pedum* zone, characterized by *Treptichnus*, *Gyrolithes* and *Bergaueria*, examples of branching burrow systems and sea anemone resting traces. The body fossils show a sudden explosion of marine animals at the beginning of the Cambrian. The trace fossils give richer detail: a longer-term build-up of complexity in the latest Precambrian, and then the explosion of new life forms.

There is rare evidence for the evolution of certain trace fossils through time. For example, pascichnia like *Nereites* and agrichnia like *Paleodictyon* seem to have become smaller and more regular through time, perhaps evidence for improvements in feeding efficiency.

Orr et al. (2003) found another time-related aspect of animal behavior on the ocean floor. They were puzzled by the abundance of certain kinds of exceptionally preserved marine fossil assemblages in the Cambrian. The Burgess Shale is the most famous example (see p. 249), but there are many other such continental shelf and deep marine conservation Lagerstätten, where soft tissues seem to survive without damage to become biomineralized. These assemblages occur throughout the Cambrian, and are rare after that. Orr et al. (2003) suggested that this was linked with an increase in bioturbation of the ocean floor after the Cambrian, and an increase in agrichnia and pascichnia in particular. The change in trace fossils suggests that the mobile infauna in deep waters had become more diverse and voracious, and more mobile. It seems that more invertebrates moved to the deep ocean floors after the end of the Cambrian, perhaps in search of new sources of food as the shallower waters became more crowded. They then searched out any dead organisms they could find to feed on, and no Burgess Shale-type exceptional preservation could ever happen again.

Trace fossils and the oil industry

Trace fossils are powerful tools in studying large sedimentary basins. Petroleum geologists frequently want to understand the architecture of large volumes of sediment in order to determine whether they might be oil reservoirs, and yet they usually have to work from geophysical soundings across these basins and isolated boreholes. Sedimentologists who study these boreholes have to understand the significance of often subtle and rather small indicators.

Many sediments are **bioturbated**, or churned up by animal activity. Under water, the sediments may be burrowed and re-burrowed, creating sometimes dense masses of cross-cutting burrows. On land, trampling by animals may also churn up the sediments. Several measures have been devised to report the degree of bioturbation in a vertical section of sediment, such as is seen in a core, and the most widely used is the **ichnofabric index** of Droser and Bottjer (1989). This is a flash card system that can be used to assign an index to any burrowed sediment (Fig. 19.17).

Sedimentologists working for the oil industry have to identify trace fossils from chance sections in narrow borehole cores, which can

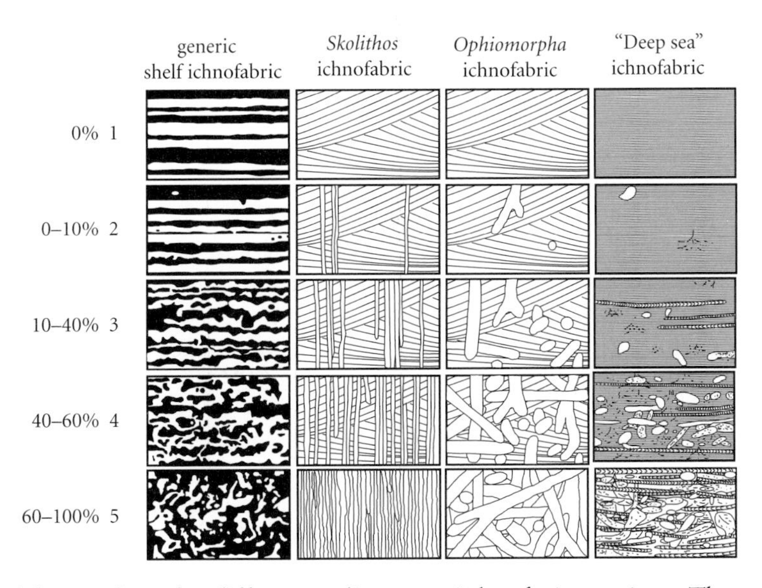

Figure 19.17 Ichnofabric indices for different sedimentary/ichnofacies settings. These diagrams show the proportions of sediment reworked by bioturbation, as seen in vertical section and numbered from top to bottom: 1 (no bioturbation) to 5 (intensely bioturbated). (Courtesy of Mary Droser and Duncan McIlroy.)

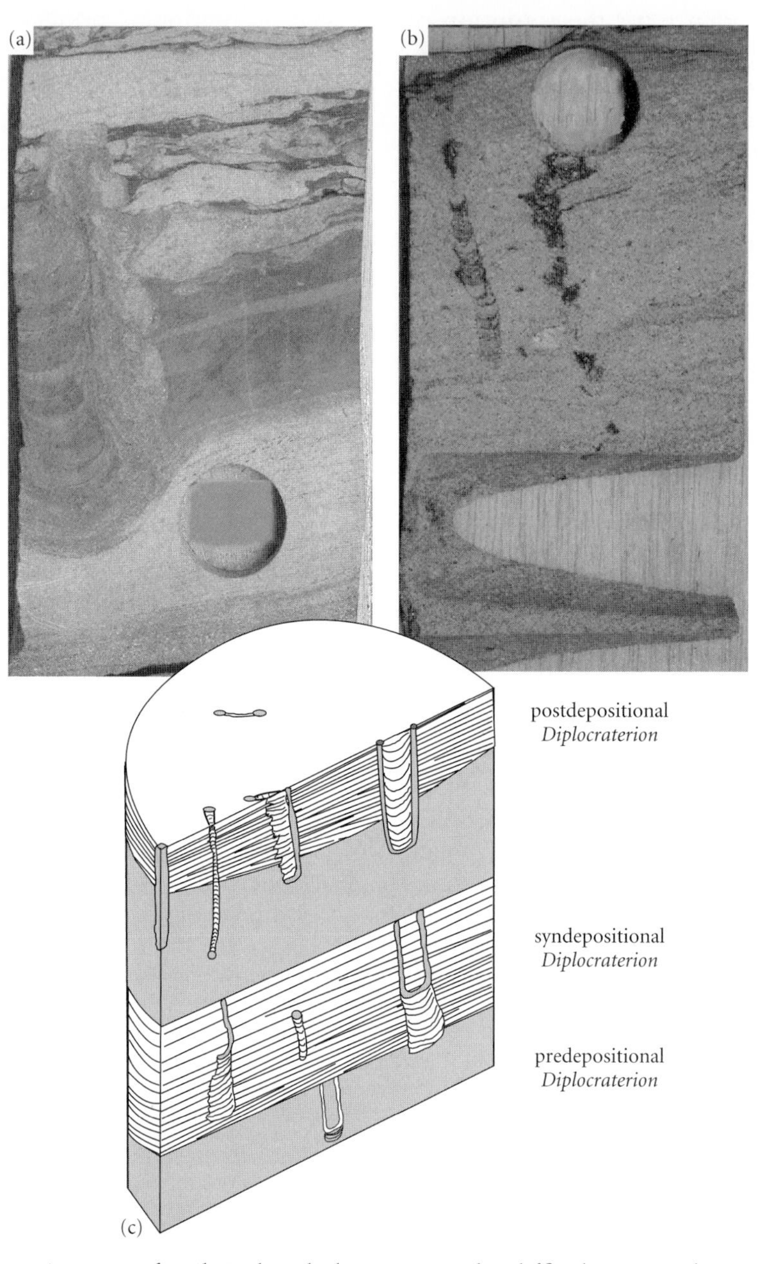

Figure 19.18 Interpreting trace fossils in borehole cores can be difficult. Vertical (a) and horizontal (b) cuts across a core may show rather obscure burrow impressions, but these make sense when interpreted in three dimensions (c). These are indications of the U-shaped burrow *Diplocraterion*, typical of the *Skolithos* ichnofacies, and so an indicator of the intertidal zone. (Courtesy of Duncan McIlroy.)

be extremely tricky (Fig. 19.18). They regularly record the following ichnological features of cores (McIlroy 2004):

- Ichnofabric index (= intensity of bioturbation).
- Diversity of trace fossils (high diversity usually means the water was well oxygenated and food was abundant).
- Relative abundance of trace fossils (does any ichnogenus dominate?).
- Trace fossil sizes (low-diameter burrows may indicate salinity stress or low oxygen levels).
- Infaunal tiering (deep tiers suggest well-oxygenated bottom-water conditions).
- Succession of burrows (did burrows pre- or postdate particular sedimentary events?).

- Colonization styles (softgrounds may be colonized from above or below, hardgrounds generally from above).

Putting this information together with sedimentological observations may allow the ichnologist to interpret the original modes of deposition throughout a core in some detail. Petroleum geologists are often able to interpret sedimentary facies from the limited information available in a narrow core.

Ichnofabric studies may provide critical evidence for the identification of key stratigraphic surfaces, the basis of sequence stratigraphy (see pp. 34–7). For example, marine flooding surfaces or exposed erosive surfaces may be indicated by distinctive ichnofabrics that can be traced laterally across a whole basin.

Trace fossils may also be important in determining oil reservoir quality. The key factors are the **porosity** (void space between grains) and **permeability** (ability to transmit fluids) of the sediments. The economic value of an oil reserve may be reduced or enhanced by trace fossil activity: porosity and permeability may be reduced by intense burrowing and mixing of sand and clay grains, and so the value is lowered; or organisms may burrow through impermeable clay layers and link porous sandy layers, so increasing the value of the oil reserve. These interactions of trace fossils and sediments are only now coming to be understood and used extensively in the oil industry. Who would have thought the price of a barrel of crude oil, such a key factor in driving the world economy, might depend on the activities of some ancient worms!

Review questions

1. What kinds of trace fossils could a human being leave behind? Think of examples of all the main categories shown in Fig. 19.10.
2. Do big dinosaurs run faster than small dinosaurs? Find 10 dinosaur track photographs on the web, check the scale and measure the stride lengths. Use the Alexander formula (see Box 19.5) to work out speeds, and plot these against estimated body sizes for the dinosaur track makers.
3. How well do the classic marine ichnofacies work (see Fig. 19.12)? Look up examples of the indicator trace fossils, such as *Nereites* and *Zoophycos*, from lakes and other settings. How many exceptions invalidate the general usefulness of a scheme like Seilacher's ichnofacies model?
4. If life has diversified through time, you might expect trace fossil diversity to increase too. Find 10 papers about deep-sea *Nereites* ichnofacies, trace fossil assemblages from the Cambrian, Ordovician, Silurian, Carboniferous, Jurassic, Cretaceous and Tertiary, and count the number of named trace fossils. Plot these against time: does deep-sea trace fossil diversity increase, decrease or stay the same through the last 500 million years? Why would your result need further work to be completely convincing?
5. Find examples of where trace fossil study has been useful in the oil industry. What ages of rocks and sedimentary facies most benefit from trace fossil studies of boreholes?

Further reading

Bromley, R.G. 1996. *Trace Fossils: Biology, taphonomy and applications*, 2nd edn. Chapman and Hall, London.

Donovan, S.K. (ed.) 1994. *The Palaeobiology of Trace Fossils*. Wiley, Chichester.

Ekdale, A.A., Bromley, R.G. & Pemberton, S.G. 1984. *Ichnology; The use of trace fossils in sedimentology and stratigraphy*. Society of Economic Paleontologists and Mineralogists, Tulsa, OK.

Lockley, M.G. 1991. *Tracking Dinosaurs*. Cambridge University Press, Cambridge.

Maples, C.G. & West, R.R. (eds) 1992. *Trace Fossils*. Short Courses in Paleontology, No. 5. Paleontological Society, Tulsa, OK.

McIlroy, D. (ed.) 2004. *The Application of Ichnology to Palaeoenvironmental and Stratigraphic Analysis*. Special Publication No. 228. Geological Society, London.

Miller III, W. 2005. *Trace Fossils; Concepts, problems, prospects*. Elsevier, Amsterdam.

Seilacher, A. 2007. *Trace Fossil Analysis*, Springer, New York.

References

Alexander, R.M. 1976. Estimates of speeds of dinosaurs. *Nature* **261**, 129–30.

Crimes, T.P. & Crossley, J.D. 1991. A diverse ichnofauna from Silurian flysch of the Aberystwyth Grits

Formation, Wales. *Geological Magazine* **26**, 27–64.

Droser, M.L. & Bottjer, D.J. 1989. Ichnofabric of sandstones deposited in high-energy nearshore environments: measurement and utilization. *Palaios* **4**, 598–604.

Ekdale, A.A. & Bromley, R.G. 1991. Analysis of composite ichnofabrics: an example in the uppermost Cretaceous Chalk of Denmark. *Palaios* **6**, 232–49.

Ekdale, A.A., Bromley, R.G. & Pemberton, S.G. 1984. *Ichnology; The use of trace fossils in sedimentology and stratigraphy*. Society of Economic Paleontologists and Mineralogists, Tulsa, OK.

Frey, R.W., Pemberton, S.G. & Saunders, T.D.A. 1990. Ichnofacies and bathymetry: a passive relationship. *Journal of Paleontology* **64**, 155–8.

Gatesy, S.M, Middleton, K.M., Jenkins Jr., F.A. & Shubin, N.H. 1999. Three-dimensional preservation of foot movements in Triassic theropod dinosaurs. *Nature* **399**, 141–4.

Jensen, S. 2003. The Proterozoic and earliest Cambrian trace fossil record; patterns, problems and perspectives. *Integrative and Comparative Biology* **43**, 219–28.

Lockley, M.G., Hunt, A.P. & Meyer, C.A. 1994. Vertebrate tracks and the ichnofacies concept: implications for paleoecology and palichnostratigraphy. *In* Donovan, S.K. (ed.) *The Palaeobiology of Trace Fossils*. Wiley, Chichester, pp. 241–68.

McIlroy, D. 2004. Some ichnological concepts, methodologies, applications and frontiers. *In* McIlroy, D. (ed.) *The Application of Ichnology to Palaeoenvironmental and Stratigraphic Analysis*. Special Publication No. 228. Geological Society, London, pp. 3–27.

Milàn, J. & Bromley, R.G. 2006. True tracks, undertracks and eroded tracks, experimental work with tetrapod tracks in laboratory and field. *Palaeogeography, Palaeoclimatology, Palaeoecology* **231**, 253–64.

Minter, N.J. & Braddy, S.J. 2006. Walking and jumping with Palaeozoic apterygote insects. *Palaeontology* **49**, 827–35.

Orr, P.J. 1995. A deep-marine ichnofaunal assemblage from Llandovery strata of the Welsh Basin, west Wales, UK. *Geological Magazine* **132**, 267–85.

Orr, P.J., Benton, M.J. & Briggs, D.E.G. 2003. Post-Cambrian closure of the deep-water slope-basin taphonomic window. *Geology* **31**, 769–72.

Pemberton, S.G. & Frey, R.W. 1984. Ichnology of storm-influenced shallow marine sequence: Cardium Formation (Upper Cretaceous) at Soebe, Alberta. *Canadian Society of Petroleum Geologists, Memoir* **9**, 281–304.

Seilacher, A. 1964. Biogenic sedimentary structures. *In* Imbrie, J. & Newell, N. (eds) *Approaches to Paleoecology*. Wiley, New York, pp. 296–316.

Seilacher, A. 1967a. Bathymetry of trace fossils. *Marine Geology* **5**, 413–28.

Seilacher, A. 1967b. Fossil behavior. *Scientific American* **217**, 72–80.

Chapter 20

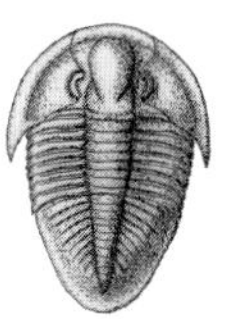

Diversification of life

Key points

- There may be 5–50 million species on Earth today, a level of diversity almost certainly higher than at any time in the past.
- There is a debate about whether life diversified according to a logistic model, reaching a global equilibrium level, or according to an exponential model in which diversity continues to expand without reaching a global carrying capacity.
- The classic logistic/equilibrium explanation for the diversification of animal life in the sea is hotly debated at present.
- Many examples of evolutionary trends, one-way changes in a feature or features, are in reality more complex.
- The idea of progress in evolution, change with improvement in competitive ability, is hard to demonstrate.
- It is important not to confuse pattern with process; too often scientists and the public assume such processes as competition, adaptation and progress without testing for alternatives.
- Major steps in evolution (e.g. evolution of wings and feathers in birds, evolution of limb loss in snakes) are well documented by fossils and evolutionary trees.
- An alternative, biological, view of the major steps focuses on fundamental subcellular systems, replicators and genetic systems.

We will move forward, we will move upward, and yes, we will move onward.

Vice-President Dan Quayle (1989)

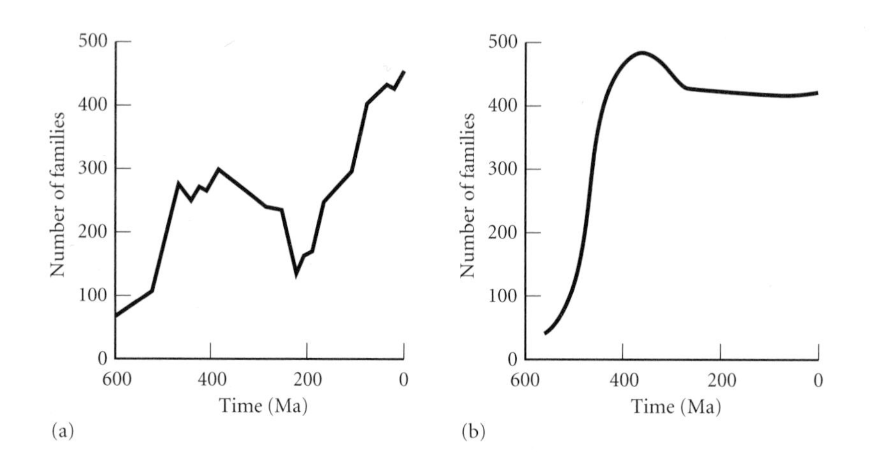

Figure 20.1 Two models for the diversification of marine invertebrate life over the past 600 myr of good-quality fossil records. (a) The empirical model, in which the data from the fossil record are plotted directly, and (b) the bias simulation model, in which corrections are made for the supposedly poor fossil record of ancient rocks. (Based on information in Valentine 1969; Raup 1972.)

Perhaps Mr Quayle was a little confused about his direction, but many have interpreted the evolution of life as a long story of progress. Others argue that evolution has probably been a little more like Mr Quayle's progress: one step forward, one sideways, one back, two forward . . . We will explore this theme further.

The record of fossils gives a rich and spectacular picture of the history of life. Paleontologists have been as successful as archeologists and historians in piecing together a detailed picture of the events of the past, even though paleontologists have a very much longer time scale to deal with and a more patchy record. It is likely that the last 200 years of paleontological research have given a broadly correct picture of the order of appearance of major groups of plants and animals through geological time, their distributions over the continents and oceans of the past, their life strategies and adaptations, and their patterns of evolution (see Chapters 2–7), despite the many gaps and inconsistencies in the fossil record (see pp. 70–7).

THE DIVERSIFICATION OF LIFE

Onward and upward

An accepted principle of evolution is that all modern and ancient life on Earth is part of a single great phylogenetic tree, and there must have been a time in the Precambrian when there was only a single species (see pp. 190–1). Today, there are between 5 and 50 million species (Box 20.1), so a plot of species numbers through time must show a pattern of phenomenal increase over the past 3500 myr. But just what sort of pattern?

It is impossible to plot an accurate diagram of species numbers through time, because so many species were never fossilized, and others are yet to be found and identified. Paleontologists have focused on those parts of the fossil record where the results might be believable. Valentine (1969) presented the first serious effort, when he plotted the numbers of families of skeletonized shallow-marine invertebrates through the Phanerozoic (Fig. 20.1a). The pattern showed a jerky increase, with several declines, and a particularly dramatic rate of increase from the Cretaceous to the present. Valentine argued that this pattern might be representative of the pattern of diversification of all of life.

However, Raup (1972) argued that the graph showed more about the sources of error in the fossil record than it did about the true pattern of the diversification of life. He suggested that the low diversity values in the Early Paleozoic reflected the fact that such ancient rocks were rare, the fossils in them were often metamorphosed or eroded away, and paleontologists devoted too little attention to them (see pp. 70–1). Raup suggested, then, that the true pattern of diversification of

Box 20.1 How many species are there today?

So far, 1.7–1.8 million species of plants and animals have been named and described formally (about 270,000 plant species and over 1 million insects). The rate of discovery of new species is highly variable within different groups of organisms. About three new bird species, about one new mammal genus and some 7250 new insect species are named each year.

It might seem easy to use such figures to produce an estimate of global diversity on the assumption that, at some time in the future, all species will have been discovered and named. The simplest approach might be to use a collector curve approach (see p. 165), and document the rate of naming of new species for different groups. This would work well for intensively studied groups, such as birds and mammals, where the analyst would decide that nearly all species have been found. But what of the insects and microbes? These hugely diverse groups are yielding new species as fast as systematists can pin them down and photograph them: their ultimate totals seem limitless. Conservatively, a collector curve of all life might predict that there are 5 million species out there waiting to be discovered.

Other scientists estimate that there are 100 million species on Earth today. They base these figures on a **sampling** approach. In other words, we can not hope to count or estimate across all of life, but we can take intense samples of certain kinds of organisms and extrapolate from those samples. The best-known example was calculated by entomologist Terry Erwin in the 1980s (e.g. Erwin 1982). Erwin sampled all the beetles from a single species of tropical tree, *Luehea seemannii* from South America. "Sampled" is a euphemism for "killed" – Erwin set a bug bomb below a tree, and the powerful insecticide knocked everything out and he collected the bodies on a sheet. To his amazement, Erwin found dozens of new species in each sample. He estimated that this one tropical tree species carried 1100 beetle species, of which about 160 were unique to each tree species. There are about 50,000 tropical tree species around the world, and if the numbers of endemic beetle species in *L. seemanni* is typical, this implies a total of 8.15 million canopy-dwelling tropical beetle species in all (50,000 × 160). Beetles typically represent about 40% of all arthropod species, and this leads to an estimate of about 20 million tropical canopy-living arthropod species. In tropical areas, there are typically twice as many arthropods in the canopy as on the ground, giving an estimate of 30 million species of tropical arthropods worldwide. This estimate came as a considerable surprise when it was published: 30 million species of tropical arthropods must imply a global diversity of all life in the region of 50 million. Some wild-eyed biologists even talked of figures of 100 million or more!

Subsequent authors have pointed out that *Luehea* was uniquely rich in its own special beetle species, and the global estimate should be nearer 15–20 million species than 50–100 million. The debate continues. It is worth noting that if is so difficult to estimate modern biodiversity, what hope do paleontologists have of providing accurate estimates for total biodiversity in the past?

Read more about modern biodiversity through http://www.blackwellpublishing.com/paleobiology/.

marine invertebrates had been a rapid rise to modern diversity levels during the Cambrian and Ordovician, and a steady equilibrium level since then (Fig. 20.1b).

The to-and-fro debate about the reality of such broad-scale diversification patterns has continued since the early 1970s, with many proposals that the pattern is broadly correct (our view), but with many current and serious challenges (Smith 2007) (see pp. 72–7). Many diversification plots show similar patterns, with slow rates of diversification at first, many set-backs, and a rapid rate of increase over the past 100 myr, with no sign of a leveling off. This is true of vertebrates, insects and plants, and the latest plots for marine animals are comparable with Valentine's original plot (Fig. 20.2). If the diversification curves show something about evolution, how should they be interpreted?

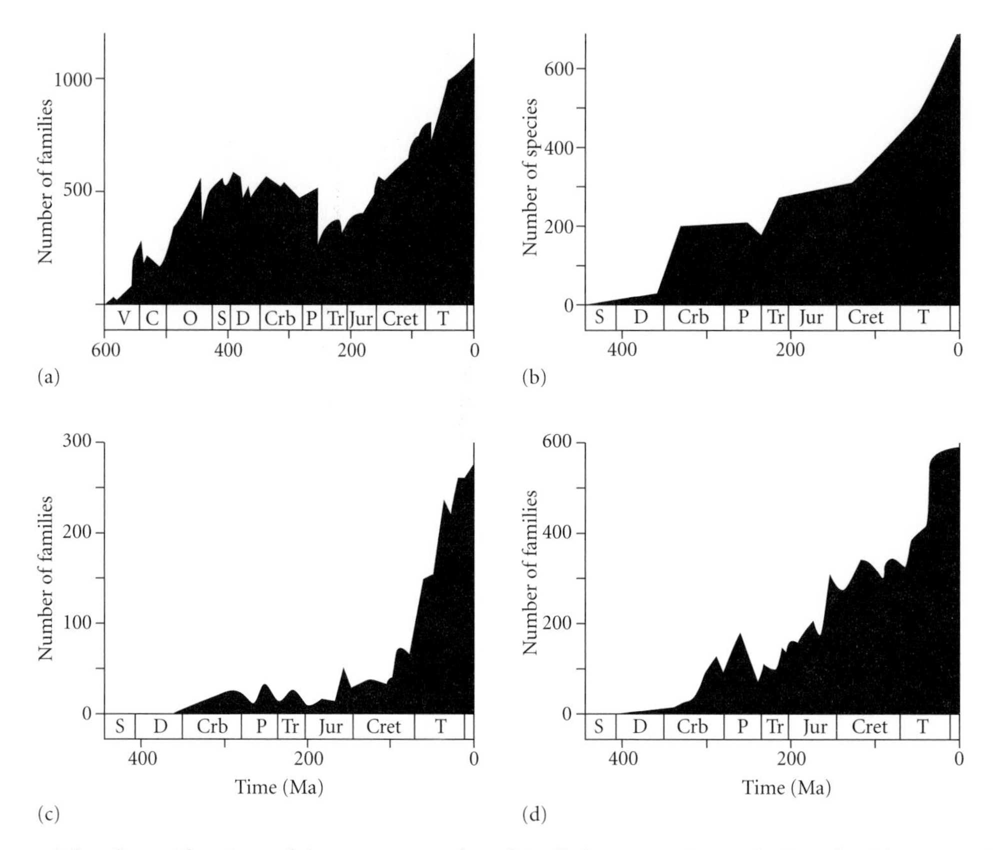

Figure 20.2 The diversification of four groups of multicellular organisms during the Phanerozoic: (a) marine animals, (b) vascular land plants, (c) non-marine tetrapods, and (d) insects. All graphs show similar shapes, with a long initial period of low diversity, and then rapid increase since the Cretaceous. Geological period abbreviations are standard, running from Vendian (V) to Tertiary (T). (Based on various sources.)

Interpreting global diversification patterns

There are several ways to go from one species to many, and these can be expressed in terms of three mathematical models, represented by a straight line, an exponential curve and a logistic curve, first as an uninterrupted increase (Fig. 20.3a), and second with some mass extinctions superimposed (Fig. 20.3b).

The **linear model** represents additive increase, the addition of a fixed number of new species in each unit of time. (The increase in this example, and the others, is a *net* increase, i.e. true increase minus extinctions.) In terms of an evolutionary branching model, additive increase would mean that, through time, speciation rates have declined, or extinction rates have increased regularly at a rate sufficient to mop up the excess speciations. The implied decline in the rate of evolution in the linear model comes about simply because the total number of species is increasing regularly, and yet the *rate* of increase across the board remains fixed. Hence, for any individual evolutionary line, the rate or probability of splitting (speciation) must decline. Such a model has generally been rejected as improbable.

The **exponential model** is more consistent with a branching mode of evolution. If speciation and extinction rates remain roughly constant, then there will be regular doubling of diversity within fixed units of time. A steady rate of evolution at the level of individual evolutionary lines scales up to an exponential rate of increase overall since total diversity is ever increasing. This model has been applied to the diversification rates of individual clades, and to the diversification of life in general (e.g. Benton 1995).

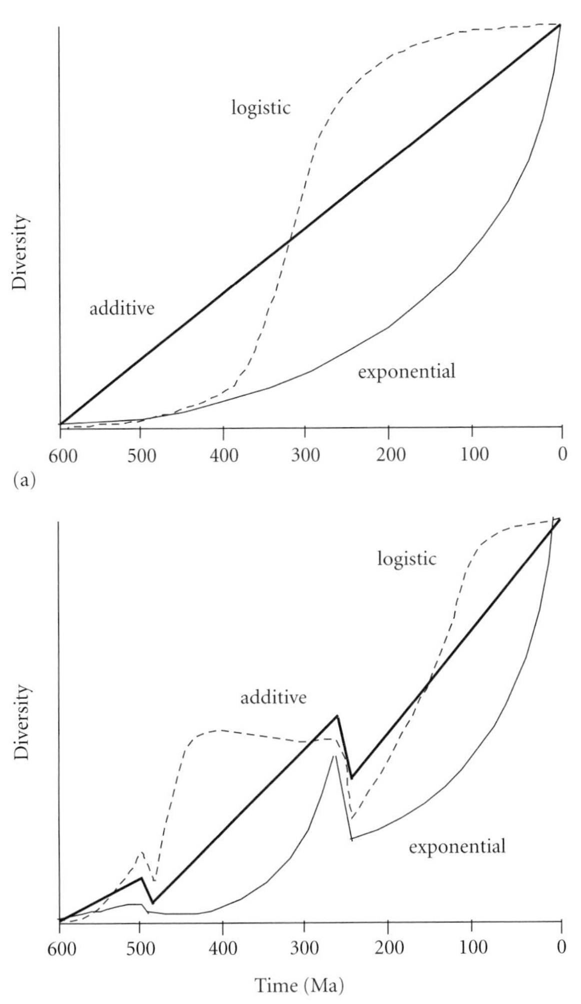

Figure 20.3 Theoretical models for the diversification of life plotted as if for the last 600 myr (a) in the absence of major perturbation and (b) with two mass extinctions superimposed.

The **logistic model** involves one or more classic S-shaped curves, each consisting of an initial period of slow diversity increase, a rapid rise, a slowing of the rate of increase as a result of diversity-dependent damping factors, and then a plateau corresponding to a limiting or equilibrium value. The logistic model has been used to explain patterns of diversification of marine organisms (Sepkoski 1984) and of plants (Niklas et al. 1983).

There is clearly no consensus on whether the exponential or logistic model best explains the diversification of major sectors of life through time, nor on whether all patterns of diversification adhere to the same model of increase. The choice of model is important because each makes profoundly different claims about evolution.

Equilibrium or expansion?

If the logistic model is correct, life has diversified in a controlled manner, reaching one or more equilibria, each of which is probably density limited. As the oceans or land filled with species, some limiting factor such as space or food came into play to stop diversification exceeding the global **carrying capacity**, the total number of species that can be accommodated. If the exponential model is correct, life has diversified in a less controlled manner, rising continually and never reaching an equilibrium level. This expansion model need not imply unfettered rates of diversification: food and space limitations can slow the rate of increase.

Equilibrium models for the diversification of life are based on an influential body of ecological theory, including classic experiments in competition where the increase of one population suppresses another that depends on the same limiting resource. In particular, Sepkoski (1984) based his logistic models (Box 20.2) on the theory of island biogeography (MacArthur & Wilson 1967), seeing the Earth's oceans as an island, arrival rates as evolutionary origination rates, and local extinction rates as global extinction rates. Like a Petri dish, or an island, the world's oceans are thereby assumed to have a fixed carrying capacity, a level that marks the limit of global species richness. Sepkoski (1984) argued for two, possibly three, equilibrium levels, each dominating the Earth's oceans for a time, and then being surpassed. In 2001, John Alroy of the National Center for Ecological Analysis and Synthesis in Santa Barbara and colleagues argued that in fact Raup (1972) had been right, and that there had really been just one equilibrium level (see Fig. 20.1a) that was achieved in the Ordovician. The apparent step-like pattern identified by Sepkoski (Box 20.2) was an artifact of the poor quality of the Paleozoic and Mesozoic fossil records.

The alternative to equilibrium is **expansion**, where there is no carrying capacity for the Earth or that carrying capacity has not yet been reached. The overall pattern of diversification of life of course incorporates the numerous constituent clades, some expanding, others diminishing and yet others remaining at constant diversity at any particular time. From an expansionist viewpoint, there is no prediction of how the individual clades affect each other, whereas an equilibrist would see clades expanding and contracting to some extent in response to each other. Global diversity may expand repeatedly by the appearance of new adaptations, habitat changes and recovery following extinction events. In the past 250 myr, the diversification of life has been dominated by the spectacular radiations of certain clades, both in the sea (decapods, gastropods, teleost fishes) and on land (insects, arachnids, angiosperms, birds, mammals). There is little evidence that these major clades have run out of steam, and nothing to indicate that they will not continue to expand into new ecospace.

Eight observations might provide tests for distinguishing equilibrium and expansion

Box 20.2 The coupled logistic model for diversification in the sea

In a series of influential papers in the 1970s and 1980s, Jack Sepkoski of the University of Chicago presented a new model for the diversification of life in the sea (Sepkoski 1984). He had labored long and hard to compile the first comprehensive database on all the families of marine animals, and this database was the source of many of his macroevolutionary studies.

Sepkoski (1984) visualized the history of life in the sea as consisting of three great evolutionary "faunas": the Cambrian, Paleozoic and Modern (Fig. 20.4). Each "fauna" consists of a set of animal groups that possessed particular arrays of adaptations, and each of which had different competitive abilities. The large-scale replacements that happened during the Ordovician and Triassic, as the Cambrian fauna gave way to the Paleozoic, and the Paleozoic to the Modern, were the result of new forms entering adaptive zones that were already occupied, taking those over, and then expanding into new modes of life. The greater adaptability of the Paleozoic fauna allowed it to reach a higher global equilibrium level, of 400 families, than the Cambrian fauna, which could not exceed 100 families. The Modern fauna has yet to reach its global equilibrium diversity level of more than 600 families.

Sepkoski (1984) modeled each of the three "faunas" as following a logistic curve: expansion rates were low at first, then rose to a steep curve, before leveling off as the global carrying capacity was achieved. The mathematical formulation was based on establishing the net pattern of origination minus extinction through time. Just as in *The Theory of Island Biogeography* (MacArthur & Wilson 1967), Sepkoski argued that origination rates would be high and extinction rates low at first (Fig. 20.5). As more and more families originated, levels of competition between the families would increase, which would have the effect of restraining origination rates and sending extinction rates up. Eventually, as the oceans become full, extinction rates exactly balance origination rates and the equilibrium level is achieved. Note that the equilibrium is dynamic: families continue to originate, but new families drive out existing families.

The full title of Sepkoski's model is the three-phase coupled logistic model, where each of the faunas has its own characteristic logistic equation, and the three are coupled, or interact, as one fauna rises and another falls. But is this model just a mathematical abstraction or does it tell us something real about ecology and evolution? Note that the model is framed at the level of families, and not at species level. Further, this kind of model only seems to make sense of the marine fossil record – the model has not been applied to the diversification of land plants, insects or vertebrates (see Fig. 20.2). Alroy (2004) reanalyzed Sepkoski's data and found that the composition of the three faunas is debatable in part and that the rise of the Modern fauna has been slower than expected. Further, Stanley (2007) has argued that the constants used by Sepkoski in his models were unrealistic, and that the global marine diversification patterns more nearly approximate a complex exponential curve.

Read about Jack Sepkoski and models of Phanerozoic biodiversity through http://www.blackwellpublishing.com/paleobiology/.

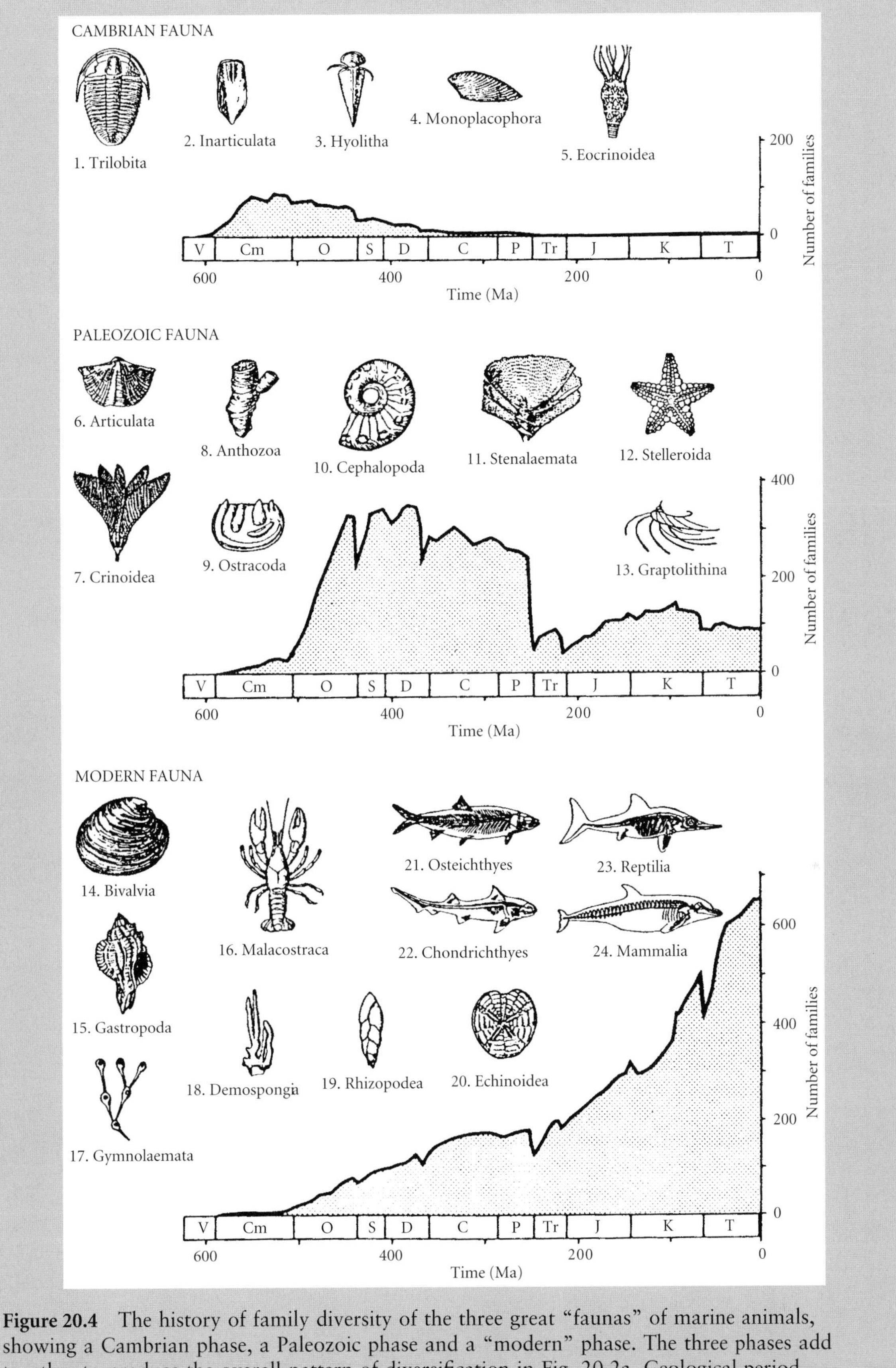

Figure 20.4 The history of family diversity of the three great "faunas" of marine animals, showing a Cambrian phase, a Paleozoic phase and a "modern" phase. The three phases add together to produce the overall pattern of diversification in Fig. 20.2a. Geological period abbreviations are standard, running from Vendian (V) to Tertiary (T). (Based on Sepkoski 1984.)

Continued

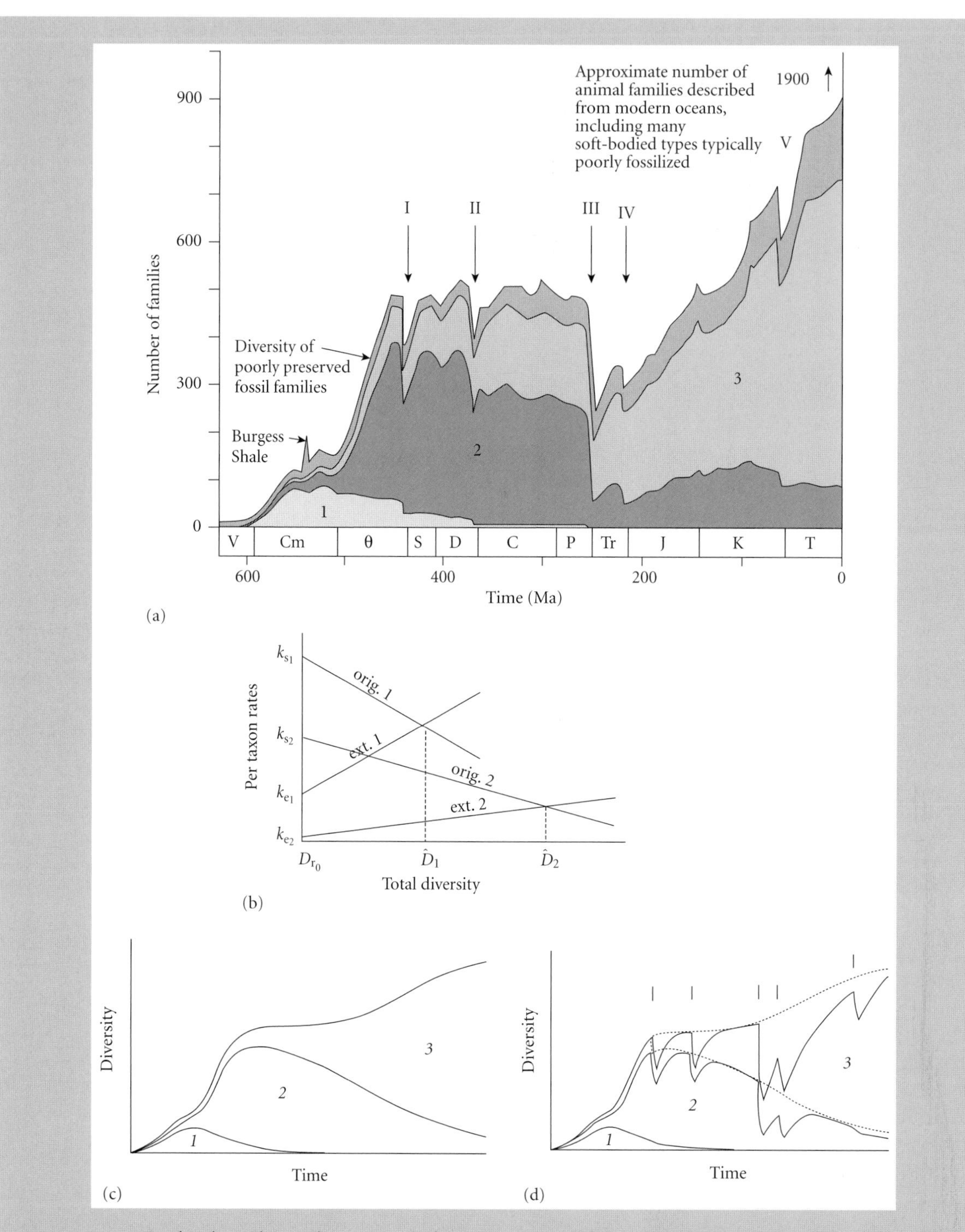

Figure 20.5 Sepkoski's three-phase coupled logistic model for diversification of animal life in the sea. (a) The family-level diversification curve for marine animals, showing the three evolutionary "faunas" from Fig. 20.4, each shaded differently. Numbers I to V are the five big mass extinctions, in sequence from left to right, Late Ordovician, Late Devonian, end-Permian, Late Triassic and Quaternary-Cretaceous. (b) The handover from the Cambrian to the Paleozoic "fauna" involved a shift in equilibrium diversity ($\hat{D}$); equilibrium diversity is achieved when origination (k_s) and extinction (k_e) rates match. (c, d) The coupled logistic model gives a simple representation of the broad outlines of the progress of the three evolutionary "faunas" 1, 2 and 3 (c), and perturbations, shown by vertical arrows, may be added to correspond to the mass extinctions (d). (Based on information in Sepkoski 1979, 1984.)

models of the diversification of life. The first four speak in favor of equilibrium, the last four more in favor of expansion.

1 There was an evolutionary explosion of marine animals during the Early Cambrian, and diversification rates slowed after this initial exponential rise. This strongly suggests a logistic/equilibrium explanation.
2 There were rapid rebounds after mass extinctions, in which local and global diversity recovered to pre-extinction levels during relatively short spans of time (see Fig. 20.5). This suggests that ecospace that had been vacated as a result of an extinction event could refill at a higher rate than entry into new ecospace. Such rapid rebounds suggest a logistic/equilibrium model of diversification.
3 Late phases of diversification cycles are associated with declining rates of origination and increasing rates of extinction, as the logistic curve approaches the equilibrium level. The marine record generally confirms such expectations – evidence for the logistic model.
4 The Paleozoic plateau in marine animal diversity is strong evidence for equilibrium. But note that the plateau appears clearly only in the family-level data compilation. At generic level, the plateau is lower and appears less regular; perhaps it disappears entirely at species level. Could the Paleozoic plateau be an artifact of the level of analysis (Benton 1997)?
5 There is debate among the supporters of equilibrium models about how many equilibria there have been. Sepkoski favored three (see Box 20.2), but Raup (1972) and Alroy et al. (2001) suggest just one. A single equilibrium level is easier to understand in terms of a global equilibrium model; otherwise each equilibrium level has to be justified as representing a complete overhaul of the evolutionary and ecological world. How many such equilibrium levels can be allowed before we accept that the pattern is really one of continual expansion (Benton 1997)?
6 There is no evidence for a global carrying capacity for species, and so a fundamental assumption behind the equilibrium model has yet to be demonstrated independently.
7 The radiation of life on land, and of certain major marine and continental clades, appears to have followed an exponential pattern, and there is no sign of slowing down in the rate of increase, nor of the occurrence of any equilibrium levels. These radiations strongly suggest patterns of expansion.
8 The Modern "fauna" radiated dramatically over the last 100 myr and shows no sign of reaching an equilibrium level; this is a weakness of any equilibrium model since the last 100 myr represent the best-sampled part of the fossil record and might be expected to show something closer to the biological pattern than earlier records.

Perhaps the best conclusion is that there is no point in seeking an overarching mathematical model to explain the diversification of life. After all, evolution happens at the level of species, and species react and interact in ever-changing ways. As environments change, families and larger groups come and go. With all the vicissitudes of history – moving continents, changing sea levels, changing atmospheres and temperatures – it could be argued that the global sum of diversification is bound to be a highly irregular pattern with no fundamental meaning or driver.

TRENDS AND RADIATIONS

Trends and progress

Does this show progress as commonly assumed? Evolution by natural selection indicates that the fittest survive, and the fittest must be better in some way than their predecessors. Also, the number of species has increased from one to many millions, and the range of life forms has increased at the same time. Plants and animals have become larger, more complex, more intelligent through the millennia. Surely evolution is progress, "onward and upward" as many have said.

However, it is important to be clear about terminology. **Progress** in evolution means change with improvement. The later forms have to be demonstrably better than their predecessors. An evolutionary **trend** is a one-way change in some feature or features in a lineage through time. So, a trend in human evolution has been for increasing brain size, and this could be classed as progress if it is assumed that a larger brain means higher intelligence and so greater evolutionary adaptability. Another trend is for increasing size and loss of toes among horses (Box 20.3). But does horse evolution show progress? That would be harder to argue: you could even turn the story around and argue that the small camouflaged, forest-dwelling *Hyracotherium* of the Eocene was "better" than the plains-dwelling, larger, modern *Equus* because it could hide and escape predation. Beware of the loose use of language: just because something lives now does not mean it is better than its ancestors.

The idea of progress in the history of life has a long and checkered history. Evolution is progressive only in that advantageous adaptations may be inherited by the offspring of successful parents. So, generation by generation, some feature may seem to show a trend of change, such as the elongation of the neck of giraffes or the increase in size of horses. The problems in transferring such ideas to the fossil record are three-fold:

1 The environment is forever changing, and it is unlikely that a selection pressure for change in a particular feature would be maintained for millions of years.
2 Paleontological evidence rarely supports the simple explanations of evolutionary trends (see Box 20.3).
3 The occurrence of changes through time does not mean progress; progress involves demonstrated improvement of adaptation.

Adaptive radiation

One of the classic observations of large-scale evolution is **adaptive radiation**, or more properly the radiation. A **radiation** is when a clade expands relatively rapidly. The adjective "adaptive" is usually tagged onto the term because there is an assumption that the radiation is happening because of some particular adaptation in the clade, a new and efficient mode of feeding or the ability to conquer a new habitat. It is wrong, however, to mix pattern and process terms.

Box 20.3 Horse evolution: the most famous example of an evolutionary trend

Fossil horses were often found by early 19th century paleontologists, such as Cuvier (see p. 12). By 1875, a convincing evolutionary story had been worked out by Marsh, based on his studies of sequences of mammalian fossils through the North American Tertiary: he could document evolution from the Eocene *Hyracotherium*, the size of a terrier, to the modern large horse, through a series of intermediates. Here was a perfect, single-line, progressive, evolutionary trend, showing how horses became ever larger and larger, and faster and faster through time.

The reality is, however, more complex. There was no single one-way pattern of change. The evolutionary tree of horses (Fig. 20.6) branches many times, and small, medium and large horse species coexisted in North America in the Miocene. The lengthening of the legs, reduction in the numbers of toes from four or five to one, and deepening of the teeth happened haphazardly across the diversity of Oligocene and Miocene horses as great prairie grasslands spread over North America. The height and single hooves were adaptations for fast running, and the deep teeth were necessary to permit horses to deal with the tough grasses. The survival of similar large species of *Equus* today – horse, donkey, zebra – is the result of chance events, such as the catastrophe in the Late Pliocene when the diverse North American native species died out. *Equus* is good at running fast on open grassy plains, and *Hyracotherium* was good at camouflaging itself in leafy woods. Is *Equus* better, or merely different?

Read more about horse evolution in MacFadden (1992), and at web sites listed on http://www.blackwellpublishing.com/paleobiology/.

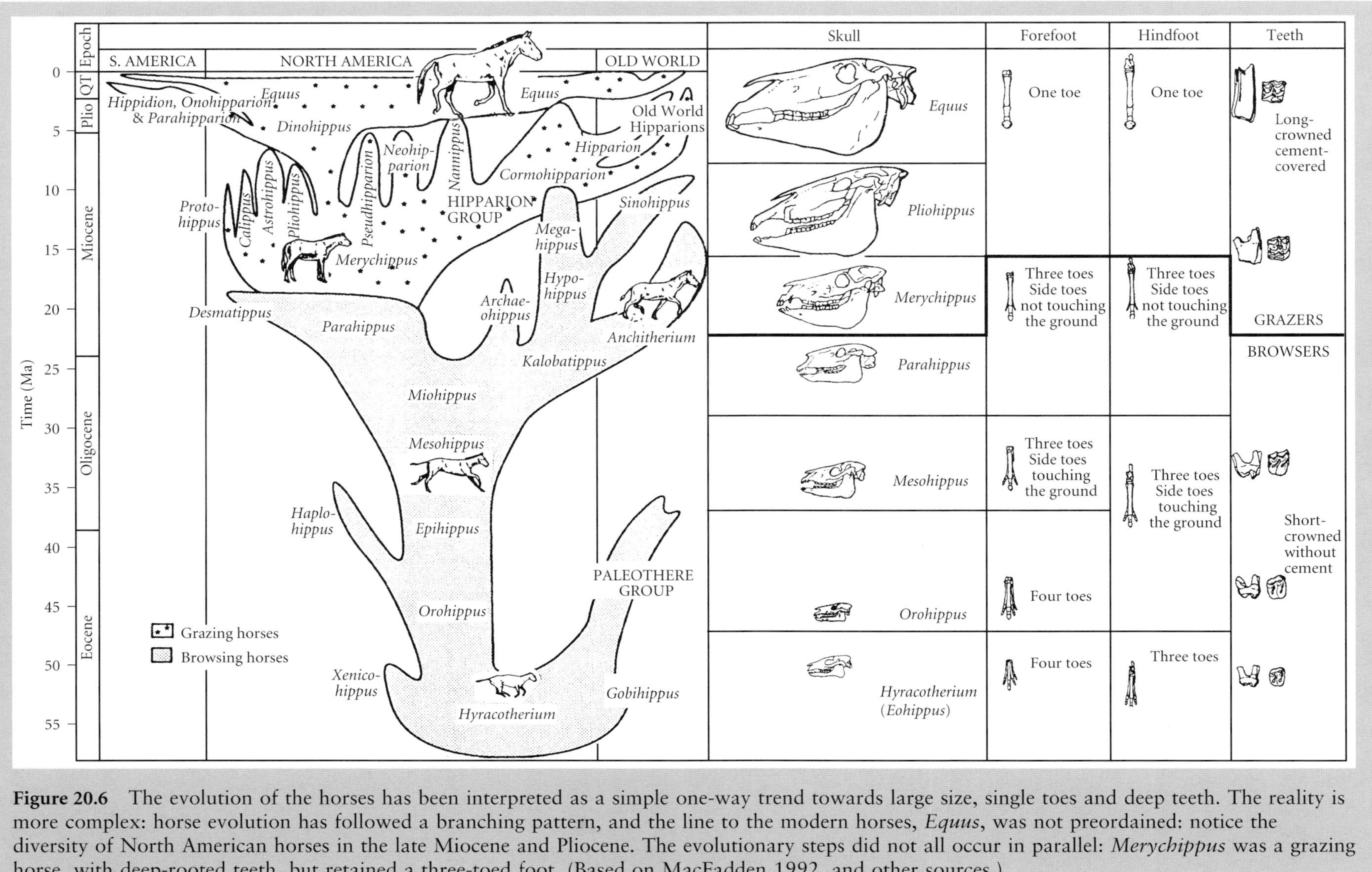

Figure 20.6 The evolution of the horses has been interpreted as a simple one-way trend towards large size, single toes and deep teeth. The reality is more complex: horse evolution has followed a branching pattern, and the line to the modern horses, *Equus*, was not preordained: notice the diversity of North American horses in the late Miocene and Pliocene. The evolutionary steps did not all occur in parallel: *Merychippus* was a grazing horse, with deep-rooted teeth, but retained a three-toed foot. (Based on MacFadden 1992, and other sources.)

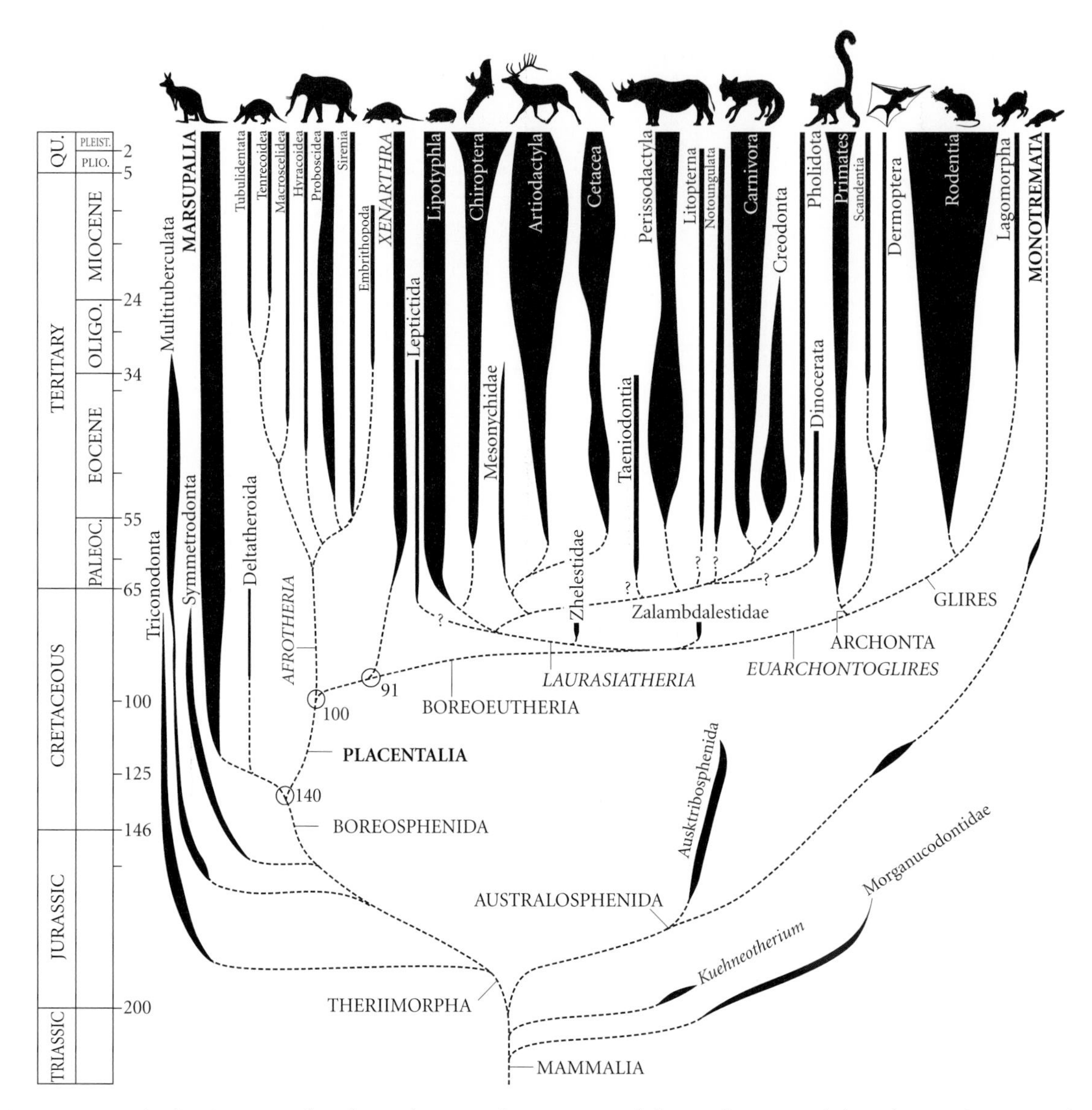

Figure 20.7 A classic example of a radiation, the pattern of diversification of the placental mammals after the Cretaceous-Tertiary mass extinction. Mammals originated in the Triassic, and diversified at a modest rate during the Jurassic and Cretaceous. Modern placental superorders originated in the Late Cretaceous, and the orders began to diversify. Only after the dinosaurs had died out did the placental mammals really diversify and become abundant worldwide. (From Benton 2005.)

Patterns are observations of the appearance and disappearance of species or occurrences in different geographic regions, for example. **Processes** are hypotheses that seek to explain the patterns. So, a radiation is a pattern; whether it was adaptive or not is a hypothesis of process. It is possible that a clade would radiate at a particular time just by chance, and with no particular adaptation. Perhaps many of the radiations after mass extinctions are a little like that: the survivors (and survival through a mass extinction is as much a result of good luck as good genes; see p. 168) are able to radiate as the world returns to normal, and they may have no particularly strong adaptations that drive their radiation. Throw the dice another way, and some other cluster of species would have survived the trauma, and the post-extinction radiation would have been started entirely differently.

Arguably the best-known radiation was the diversification of placental mammals after the Cretaceous-Tertiary (KT) event (Fig. 20.7). The placental mammals had divided into their major subdivisions and had diversified a little in the Late Cretaceous (see p. 466), but none of them were larger than a cat. Within the 10 myr of the Paleocene and early Eocene, 20 major clades evolved, and these include the ancestors of all modern orders, ranging from bats to horses, and rodents to whales. During this initial period, overall ordinal diversity was much greater than it is now: it seems that during the early parts of a radiation the founding clades may radiate rapidly, and many body forms and ecological types arise. Half of the dominant placental groups of the Paleocene became extinct soon after, during a phase of ecospace filling and competition, until a more stable community pattern became established 10 myr after the KT mass extinction.

The radiation of the placental mammals is frequently described as an adaptive radiation, and the adaptation that drove this could have been their warm-bloodedness, their differentiated teeth, their intelligence or their parental care. Any one of these, or indeed all of them, could be a reasonable explanation for the radiation and later success of placental mammals. But it is important to remember that these are hypotheses that must be tested. First, the marsupials were also around in the Late Cretaceous, and they share all of these characters with the placentals, so why did the marsupials not radiate as much as the placentals? Second, Mesozoic mammals had possessed most of these features since the Late Triassic, so why didn't they radiate then? The supposedly superior mammals originated at the same time as the dinosaurs, but somehow the dinosaurs prevailed first and kept the mammals at bay for fully 160 myr (see p. 454). It is important not to get carried away with assumptions and rhetoric, and it is especially important not to confuse pattern and process.

Biotic replacements

Biotic replacements are an obvious feature of the history of life. These are times when one group of plants or animals replaces another. The replacement of brachiopods by bivalves is a famous example. This had always been seen as a progressive process: the common view is that brachiopods are less adaptable than bivalves, and they clearly succumbed to long-term competition, perhaps lasting for tens or hundreds of millions of years. **Competition** is generally defined as any interaction between individuals of the same species or different species, in which one individual gains an advantage and the other suffers. A simple example is competition among baby birds in the nest: the big bruiser wins the attention of its parents and gets more food, and the timid smaller sibling loses out. Competition is a process, a hypothesis, and it must not be assumed in any example until it can be demonstrated.

Gould and Calloway (1980) looked closely at the brachiopods versus bivalves example, and they decided quite rightly not to assume simply that the replacement was driven by competition. Their studies suggested that the take-over was more complex. Brachiopods and bivalves had maintained fairly constant diversities through the Paleozoic, with brachiopods being more diverse (Fig. 20.8). The Permo-Triassic (PT) mass extinction 251 Ma (see pp. 170–4) drove their diversities right down. The bivalves recovered, and began to radiate rapidly during the Triassic and Jurassic, whereas the brachiopods have remained at the same low post-extinction diversity level ever since.

Other major biotic replacements have shown similar outcomes on close study. For example, the replacements of various major plant groups through time (see Chapter 18) used to be regarded as competitive, but there is limited evidence for that. The replacement of various tetrapod groups by the dinosaurs in the Late Triassic used to be seen as a process in which dinosaurs outcompeted their slower-moving and less rapacious predecessors. Restudy suggests that there was an extinction event, perhaps mediated by climate and floral change, and the initial success of the dinosaurs was more good luck than a demonstration of their overall superiority (see p. 454). The story of the "Great American interchange" (see pp. 43–4), which followed the closure of the isthmus of Panama, used to be that the North American mammals thundered into South America and slaughtered their inferior southern cousins. Close study of the data suggests the interchange was balanced, and that most invaders found new things to do and did not drive anything to extinction.

Perhaps the majority of biotic replacements were passive, mediated by extinction events that removed the old players from the field, and left the way clear for new groups to radiate. If this is the case, then it is hard to sustain a view that each new radiation of plants or animals marks a step upwards and forwards in the great relay race of evolutionary progress.

TEN MAJOR STEPS

The study of paleontology reveals a great deal about how life has achieved its present astonishing diversity. Some major events in the history of life may be identified, although these may be debated. An important study by two theoretical biologists (Box 20.4) produced a series of eight steps, based on modern understanding of molecular biology. A paleontological top 10 could be debated, but we have selected some major adaptations that enabled substantial increases in diversity to occur (Fig. 20.9).

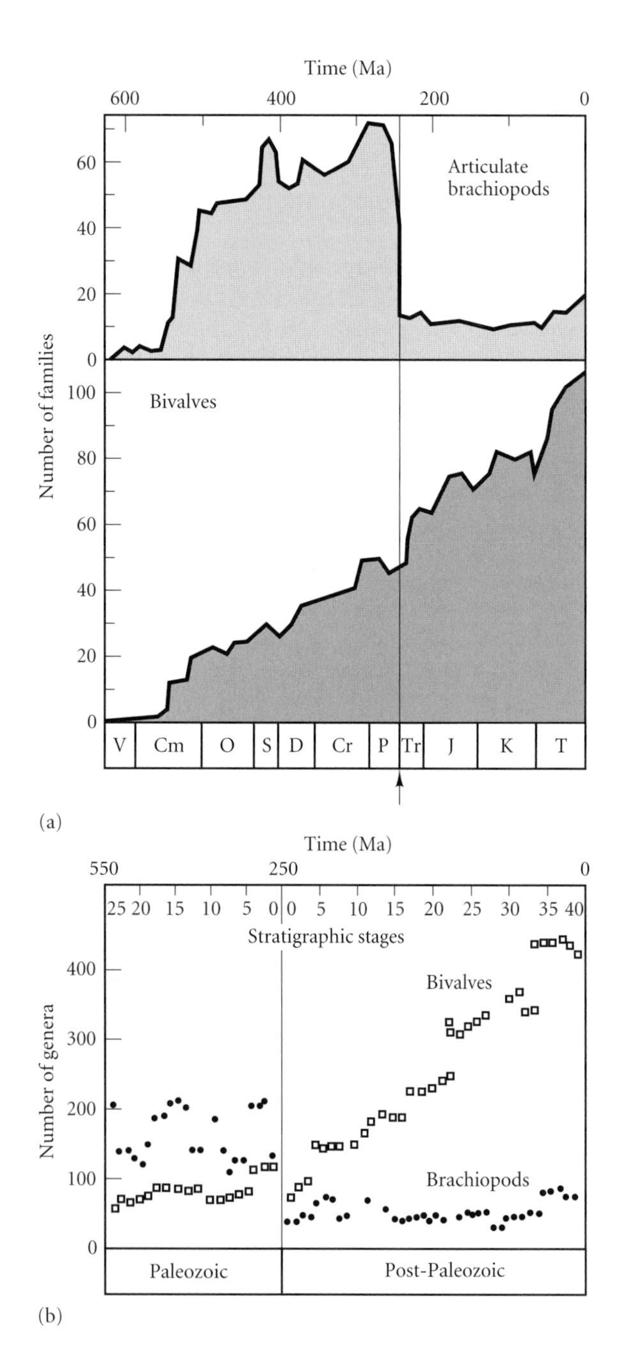

Adaptation 1: the origin of life

The most widely accepted view about the origin of life, the biochemical model (see Chapter 8), is that complex organic molecules were synthesized naturally in the Precambrian oceans. The first living organisms were probably bacteria recognized in rocks about 3500 Ma. These were prokaryotic cells, small and lacking nuclei. The initial life forms operated in the absence of oxygen, and they caused one of the most significant changes in the history of the planet – the raising of oxygen levels in the atmosphere.

Adaptation 2: eukaryotes and the origin of sex

In marked contrast to the prokaryotes, eukaryote cells are usually large, with organelles and membrane-bounded nuclei containing the chromosomes (see Chapter 8). Eukaryotes

Figure 20.8 A classic example of competitive replacement? Articulated brachiopods were the dominant seabed shelled animals in the Paleozoic, whereas bivalves take that role today. It was assumed that the bivalves competed long term with the brachiopods during the Paleozoic, even in the Permian, and eventually prevailed. (a) A plot of the long-term fates of both groups shows a steady rise in bivalve diversity, and a drop in brachiopod diversity. However, brachiopods were also diversifying during the Paleozoic, although they were hard hit by the Permo-Triassic mass extinction (arrowed). (b) The bivalves managed to recover after the mass extinction event, while the brachiopods did not. Geological period abbreviations are standard, running from Vendian (V) to Tertiary (T). (Based on information in Gould & Calloway 1980.)

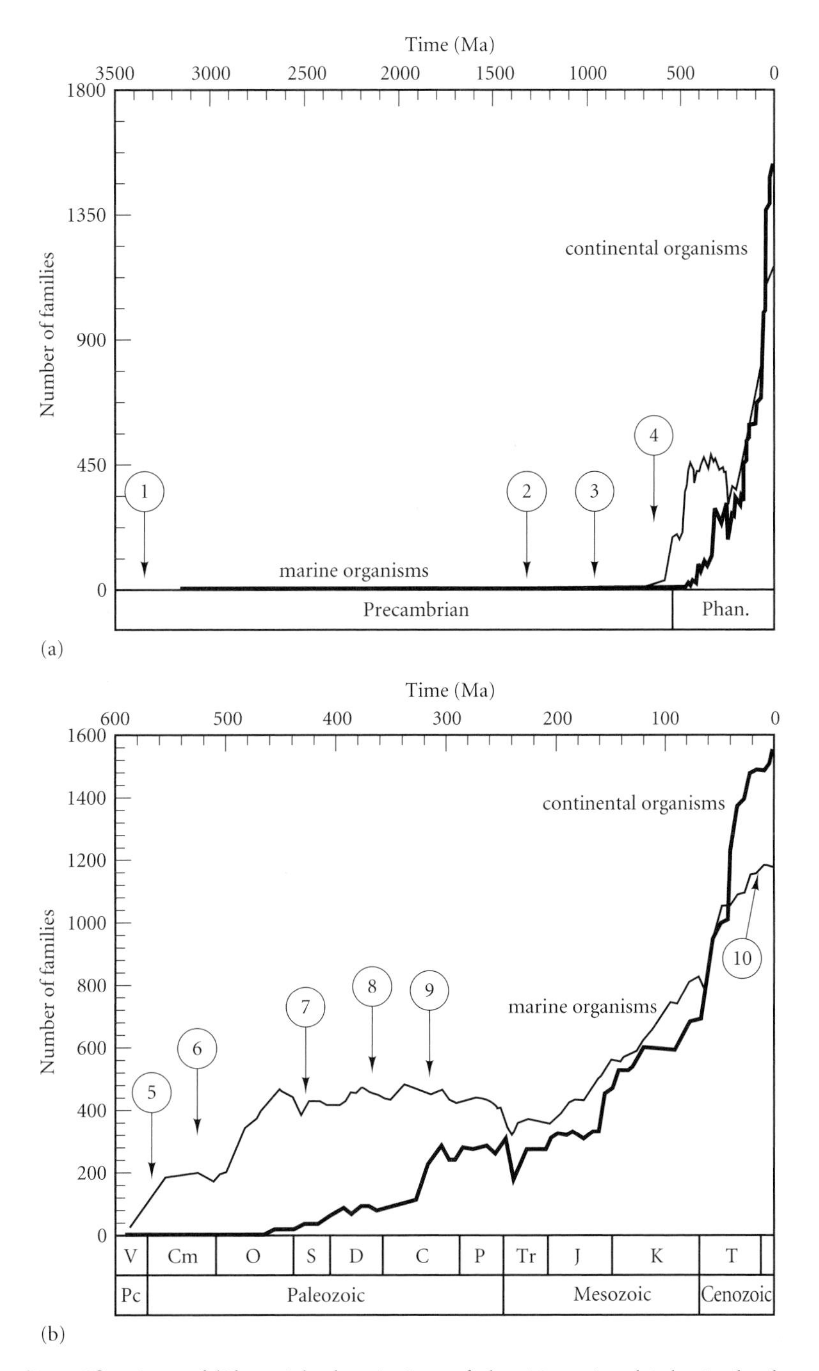

Figure 20.9 The diversification of life, with the timing of the 10 major biological advances indicated: 1, origin of life; 2, eukaryotes and the origin of sex; 3, multicellularity; 4, skeletons; 5, predation; 6, biological reefs; 7, terrestrialization; 8, trees and forests; 9, flight; 10, consciousness. The diversification of life is plotted for the whole of the past 4000 myr (a), and for the Phanerozoic (b). Geological period abbreviations are standard, running from Vendian (V) to Tertiary (T).

differ from prokaryotes in another fundamental respect: their cells reproduce sexually. The oldest fossil eukaryotes are hard to identify (see pp. 196–9), but the clade was well established by 1200 Ma. Asexual reproduction tends to propagate only clones, but the mixing of genetic material during sexual reproduction opened the door to the exchange of genetic material, mutation and recombination, and the development of variation in

Box 20.4 An alternative top eight steps

In a biologically-oriented presentation, John Maynard Smith and Eörs Szathmáry (1997) identified eight major steps from the origin of life to human societies with language:

1 *Replicating molecules*. The first objects with the properties of multiplication, variation and heredity were probably replicating molecules, similar to RNA but perhaps simpler, capable of replication, but not informational because they did not specify other structures. If evolution were to proceed further, it was necessary that different kinds of replicating molecule should cooperate, each producing effects helping the replication of others. For this to happen, populations of molecules had to be enclosed within some kind of membrane, or "compartment", corresponding to a simple cell.
2 *Independent replicators*. In existing organisms, replicating molecules, or genes, are linked together end to end to form chromosomes (a single chromosome per cell in most simple organisms). This has the effect that when one gene is replicated, all are. This coordinated replication prevents competition between genes within a compartment, and forces cooperation on them since if one fails, they all fail.
3 *RNA as gene and enzyme*. In modern organisms there is a division of labor between two classes of molecule: nucleic acids (DNA and RNA) that store and transmit information, and proteins that catalyze chemical reactions and form much of the structure of the body (for example, muscle, tendon, hair). Perhaps originally RNA molecules performed both functions. The transition from the "RNA world" (see pp. 186–8) to a world of DNA and protein required the evolution of the genetic code, whereby base sequence determines protein structure.
4 *Eukaryotes and organelles*. Prokaryotes lack a nucleus and have (usually) a single circular chromosome. They include the bacteria and cyanobacteria (blue-green algae). Eukaryotes have a nucleus containing rod-shaped chromosomes and usually other intracellular structures called organelles, including mitochondria and chloroplasts. The eukaryotes include all other cellular organisms, from the single-celled amoeba and *Chlamydomonas* up to humans.
5 *Sexual reproduction*. In prokaryotes, and in some eukaryotes, new individuals arise as asexual clones by the division of a single cell into two. In most eukaryotes, in contrast, this process of multiplication by cell division is occasionally interrupted by a process of sexual reproduction in which a new individual arises by the fusion of two sex cells, or gametes, produced by different individuals.
6 *Differentiated cells*. Among eukaryotes, protists exist either as single cells or as colonies of cells of only one or a very few kinds, whereas multicelled organisms such as animals, plants and fungi are composed of many different kinds of cells – muscle cells, nerve cells, epithelial cells and so on. Each individual, therefore, carries not one copy of the genetic information (two in a diploid) but many millions of copies. The fascinating question for biologists to understand is that although all the cells contain the same information, they are very different in shape, composition and function.
7 *Colonial living*. Most organisms are solitary, interacting with others of their species, but not dependent on them. Other animals, notably ants, bees, wasps and termites, live in colonies in which only a few individuals reproduce. Such a colony has been likened to a superorganism, analogous to a multicellular organism. The sterile workers are analogous to the body cells of an individual, and the reproducing individuals to the cells of the germline. The origin of such colonies is important; it has been estimated that one-third of the animal biomass of the Amazon rain forest consists of ants and termites, and much the same is probably true of other habitats.
8 *Primate societies* – human societies, and the origin of language and consciousness. The decisive step in the transition from ape to human society was probably the origin of language, which led in turn to consciousness in human beings, an understanding of our existence. Other animals

communicate using forms of language, but human language is more complex; information is stored and transmitted, with modification, down the generations, so human language can be compared in some ways to the genetic code. Communication holds societies together, and allows humans to control some aspects of their evolution.

Maynard Smith and Szathmáry (1997) argued that all but two of these eight transitions were unique, occurring just once in a single lineage. The two exceptions are the origins of multicellular organisms, which happened three times, and of colonial animals with sterile castes, which has happened many times. Had any of these transitions not happened, and that includes the origin of life itself (number 1), then we would not be here.

populations, the basic material for evolution.

Adaptation 3: multicellularity

Multicellular organisms are clusters of eukaryote cells organized into different tissue types and organs, where different parts of the organism are responsible for particular functions and tasks. Some of the oldest multicellular eukaryotes have been reported from rocks 1200 Ma in Canada; these red algae (see pp. 200–1) were one of nearly 20 multicellular eukaryote lineages. Molecular evidence suggests all these groups had a common origin about 1200 Ma. Some 400 myr later true metazoans had arrived.

Four lines of study have aided the search for the first metazoans. First, body fossils of possible worms, *Sinosabellidites* and *Protoarenicola* – carbonaceous tubes with growth lines – have been reported from Precambrian rocks predating the appearance of the Ediacara fauna (see pp. 242–7), dated at more than 1000 Ma. Second, trace fossils imply the development of a grade of organization and structures capable of locomotion. Ichnofossils from the Medicine Peak Quartzite of Wyoming may be 2400 Ma, hinting at the presence of a mobile metazoan around the Archean–Proterozoic boundary. This seems unlikely, and other interpretations are possible. Third, the marked decline of stromatolites about 1000 Ma suggests the presence of grazing metazoans. Finally, molecular phylogenies (see p. 242) suggest an initial divergence of metazoans 1000–800 Ma.

Adaptation 4: skeletons

During an interval of a few million years in the earliest Cambrian, a wide variety of mineralized skeletons appeared (see pp. 247–57), and these are seen spectacularly in the Mid Cambrian Burgess Shale fauna (see p. 249). Mineralized hard parts conferred distinct advantages, by providing protection, support and areas for the attachment of muscles. Predator pressure may have been the main driving force behind the acquisition of hard parts. Marked changes in ocean chemistry during the Early Cambrian marine transgression may have forced organisms to ingest and excrete large quantities of minerals, and enhanced oxygen levels in the world's oceans made precipitation of minerals easier.

Diverse mineralized tubes characterize many Early Cambrian fossil assemblages. Apart from these tubes, of uncertain taxonomic affinities, better known organisms also drew on a variety of substances to construct skeletal structures: calcite (brachiopods, bryozoans, trilobites, ostracodes, corals), aragonite (hyoliths, mollusks), apatite (conulates, lingulate brachiopods, conodonts, vertebrates), opal (hexactinellid sponges, radiolarians) and agglutinated material (foraminiferans).

Adaptation 5: predation

The Late Precambrian Ediacara fauna was soft-bodied, and it appears likely that predators and scavengers were absent. This changed during the Early Cambrian when evidence from borings in shells suggests that this form of predation was an important part of the ecosystem. The rapid diversification of armored and protective strategies was a feature of the Cambrian radiation. It is probable that predation had an important influence on evolutionary processes by the coevolution of predator–prey systems, where prey and predator organisms evolve ever-better defensive and offensive strategies, respectively. Such "arms races" became intensified several times during the Phanerozoic, most notably in the Cretaceous when new efficient predators such as crabs, hole-boring gastropods and teleost fishes caused the evolution of dramatic changes in the lifestyles of their prey (see Chapters 14 and 16).

Adaptation 6: biological reefs

Biological reefs are the marine equivalents of tropical rain forests. Modern reefs are highly diverse, colorful frameworks of brain, horn and staghorn corals, together with organ pipes, sea fans and sea whips, providing food and accommodation for thousands of species from the mantis shrimp to the carpet shark. These large carbonate structures form the basis of many types of tropical islands, from barrier reefs to fringing atolls. The origin of reefs was a major event.

Throughout the Phanerozoic, the main reef builders have changed, with major changes punctuated by mass extinctions. The first reefal frameworks appeared during the Early Cambrian, constructed first by solitary, clustered polychaete worm tubes, and then by archaeocyathans (see Chapter 10). Ordovician reefs were dominated by algae, bryozoans, stromatoporoids and rugose and tabulate corals (see Chapters 10 and 11). Tabulate corals and stromatoporoids, together with algae and bryozoans, dominated reefs from the Silurian until near the end of the Devonian. Carboniferous and Permian reefs were made from bryozoans, algae and calcareous sphinctozoan sponges. During the Mid Triassic, frameworks of algae, scleractinian corals, bryozoans and sphinctozoan sponges developed. Jurassic and Cretaceous reefs were constructed by scleractinian corals, lithothamnian algae and siliceous and sphinctozoan sponges, to which were added rudist bivalves in the Late Cretaceous. During the Tertiary, scleractinian corals expanded to their present diversity, where they now dominate biological frameworks.

Adaptation 7: terrestrialization

The colonization of the land added major new environments to those previously occupied by life. It is hard to date the first move of life on to land. Soils have been reported from Mid Precambrian sequences, and microbial life may have extended a greenish scum around the water's edge. Ordovician soils suggest that larger plants and animals had moved onto land. By the Silurian, small vascular plants such as *Cooksonia*, with a well-developed vascular system, stomata, a waxy covering and trilete spores, were well established (see Chapter 18). The new land plants relied on the soil for some of their nutrients, but they also generated modern-style soils over landscapes that had previously been bare rock. This stabilization of the land by plant growth slowed down the rate of erosion, and it was one of the most dramatic effects that life has had on the physical nature of the Earth.

Colonizing invertebrates were faced with problems of dehydration, respiration and, to a lesser extent, support. These problems were overcome by the development of waterproof skins, lungs and skeletal support. Although hydrostatic skeletons, such as those of slugs, have been successful, the toughened exoskeleton of arthropods was an ideal protective covering, providing support and attachment for the soft parts. By the Early Devonian, the low green vegetation was inhabited by myriapods, insects and possibly arachnids. During the Carboniferous, these faunas were supplemented by oligochaete worms and scorpions, together with both prosobranch and pulmonate gastropods (see Chapters 12 and 14). Vertebrates moved onto the land during the Devonian, presumably to exploit the new sources of plant and invertebrate food, and full terrestrialization occurred with the reptiles in the mid-Carboniferous, when the amniotic egg evolved (see Chapter 16).

Adaptation 8: trees and forests

The next major expansion of living space on land took place during the Carboniferous, with the development of forests. The first tree-sized plants arose in the Late Devonian, forms such as the progymnosperm *Archaeopteris*, which reached a height of 8 m and had secondary woody tissue and growth rings. Trees became abundant and diverse in the Carboniferous, with large lycopods such as *Lepidodendron* (50 m tall) and *Sigillaria* (30 m tall), equisetaleans like *Calamites* (20 m tall), ferns like *Psaronius* (3 m), progymnosperms, seed ferns like *Medullosa* (10 m), and early conifers (see Chapter 18).

The significance of the tree habit was that it created a vertically tiered range of new habitats. Trees, with their long roots, gained access to nutrients that were not available to smaller plants, and the addition of new stories of vegetation allowed more plants to pack into an available space than before. Insects and other invertebrates sheltered and fed in the bark and among the leaves and fruits, as well as in the new leaf-litter habitats in the soil. The evolution of trees and forests led to a dramatic burst in the rate of diversification of vascular land plants, as well as a radiation of insects, and their predators, such as spiders, amphibians and reptiles.

Adaptation 9: flight

The next major expansion of land life was into the air. Perhaps the evolution of trees led directly to this further leap in adaptations: having tempted various insect groups to move off the ground in search of edible leaves and fruits, gliding and true flight became inevitable. Insects arose in the Early Devonian, but the first true fliers are Carboniferous in age (see Chapter 12). There was a dramatic diversification of insect groups in the Pennsylvanian. Flight has doubtless been the clue to the vast success of insects: today there are millions of species, too many to identify and count accurately, representing perhaps 70% of all living animals.

Flight has arisen more than 30 times among vertebrates (see Chapter 17). Gliding diapsid reptiles are known from the Late Permian and Late Triassic. The first true flapping vertebrates, the pterosaurs, arose in the Late Triassic, dominating Mesozoic skies, and reaching wingspans of 11–15 m, the largest flying animals of all time. Modern vertebrate fliers include birds and bats, both successful groups, and both largely feeding on insects – the birds by day and the bats by night. The other aerial vertebrates today are gliders, animals that swoop through the air supported passively on expanded membranes of some sort: flying fishes, frogs, lizards, and snakes, as well as various gliding marsupial and placental mammals.

Plants too exploited flight and flying animals. Numerous plants use the wind for dispersal of pollen and seeds (see Chapter 18). Certain plant growth habits could also be argued to be analogous to flight, for example the lianas of tropical forests, which descend from tall trees and stretch from plant to plant, exploiting small patches of sunlight.

Adaptation 10: consciousness

Human beings probably have to feature somewhere in a list of major biotic advances, although how to view the role of humans in evolution has been a question that has dogged philosophers for centuries. What is the nature of the dramatic changes wrought by humanity? Are they to be seen in a negative light, in terms of destruction of the Earth and its inhabitants, or are they to be regarded positively, in terms of the attainment of a level of intelligence and creativity never before achieved?

Because of their brains, and the ability to adapt the environment, humans are a hugely successful species, living in vast abundance in every part of the Earth, from the poles to the equator. Humans live at all altitudes, and could theoretically create appropriate enclosed environments to allow humans to survive indefinitely underwater, at high altitudes and on other planets. *Homo sapiens* is the first species to leave the Earth and come back alive (numerous insects and microbes have doubtless been swept up into the upper reaches of the atmosphere and lost into space, but they have not benefited from the experience).

The human ability to make things is not unique because many species can make constructions, nests and the like, or even use tools (birds and apes use twigs to reach otherwise unavailable food). Language is also not unique because most animals have some form of communication, often rather complex. The extent to which humans make things and communicate is, however, unique. Perhaps unique to *H. sapiens*, but perhaps we shall never know, is consciousness, the ability to think about one's existence, about the past and about the future.

A direct result of consciousness is the ability to think ahead for more than a few minutes or hours. One person, or a team, may invent something after years of work, and that appears to be uniquely human, but the creative process may continue through many generations. No other species has invented writing, the means whereby each human brain has access to virtually infinite stores of knowledge accumulated by other people and recorded in books and in electronic form. Humans have also extended their physical capabilities in ways not achieved by any other species: it is possible to travel from New York to London in a few hours in an airplane, a useful adjunct to our rather feeble legs; it is possible to see the surface of the moon through a telescope, an extension of our poor eyesight; it is possible to speak to someone on the other side of the world without shouting; it is possible to estimate the phenetic resemblances of many specimens of brachiopods in a few minutes using a computer, a task that would take many days using the unaided brain.

Value judgments of human activities are not relevant in a purely evolutionary view of the history of the Earth. Success in evolutionary terms is measured by the abundance of a species, the scale of its role in the physical and biotic environment, and its longevity. Humans are outrageously successful on the first two counts, but the duration of the species *H. sapiens* cannot yet be judged. Consciousness seems to have permitted *H. sapiens* to achieve results that no other species has remotely approached. After all, was it possible until the geological Recent for a paleontological textbook to be written, purchased (one hopes) and read? The pursuit of knowledge of the history of life on Earth is a part of human consciousness.

Review questions

1 Read about the latest estimates of total global biodiversity. What are the main reasons why biologists find it hard to establish exactly how many species there are on Earth today?
2 Read Smith (2007) and other recent papers on the history of biodiversity and the quality of the rock record. How can paleontologists tease apart the biological (evolutionary) and geological (sampling) signals?
3 Why have evolutionists found it hard to give up the idea of progress? Or should they give it up? Read about the views of Darwin, Simpson, Dobzhansky, Mayr, Gould, Wilson and others.
4 Find 10 examples of large-scale radiations that have been described as "adaptive". What were the supposed adaptive reasons for the radiations (key adaptation; physical environmental trigger?) and what are the other possible explanations?
5 What are your top 10 major steps in evolution? How many of them match those suggested by Maynard Smith and Szathmáry (1997)?

Further reading

Briggs, D.E.G. & Crowther, P.R. 2001. *Palaeobiology, A synthesis*, 2nd edn. Blackwell, Oxford.
Foote, M. & Miller, A.I. 2007. *Principles of Paleontology*. W.H. Freeman, San Francisco.
Gaston, K.J. & Spicer, J.I. 2004. *Biodiversity*, 2nd edn. Blackwell, Oxford.
MacFadden, B.J. 1992. *Fossil Horses: Systematics, paleobiology, and evolution of the Family Equidae*. Cambridge University Press, Cambridge.

References

Alroy, J. 2004. Are Sepkoski's evolutionary faunas dynamically coherent? *Evolutionary Ecology Research* **6**, 1–32.
Alroy, J., Marshall, C.R. & Bambach, R.K. et al. 2001. Effects of sampling standardization on estimates of Phanerozoic marine diversification. *Proceedings of the National Academy of Sciences* **98**, 6261–6.
Benton, M.J. 1995. Diversification and extinction in the history of life. *Science* **268**, 52–8.
Benton, M.J. 1997. Models for the diversification of life. *Trends in Ecology and Evolution* **12**, 490–5.
Benton, M.J. 2005. *Vertebrate Palaeontology*, 3rd edn. Blackwell, Oxford.
Gould, S.J. & Calloway, C.B. 1980. Clams and brachiopods – ships that pass in the night. *Paleobiology* **6**, 383–96.
MacArthur, R.H. & Wilson, E.O. 1967. *The Theory of Island Biogeography*. Princeton University Press, Princeton, NJ.
MacFadden, B.J. 1992. *Fossil Horses: Systematics, paleobiology, and evolution of the Family Equidae*. Cambridge University Press, Cambridge.

Maynard Smith, J. & Szathmáry, E. 1997. *The Major Transitions in Evolution*. Oxford University Press, Oxford.

Niklas, K.J., Tiffney, B.H. & Knoll, A.H. 1983. Patterns in vascular land plant diversification. *Nature* **303**, 614–16.

Raup, D.M. 1972. Taxonomic diversity during the Phanerozoic. *Science* **177**, 1065–71.

Sepkoski Jr., J.J. 1979. A kinetic model of Phanerozoic taxonomic diversity. II. Early Phanerozoic families and multiple equilibria. *Paleobiology* **5**, 222–52.

Sepkoski Jr., J.J. 1984. A kinetic model of Phanerozoic taxonomic diversity. III. Post-Paleozoic families and mass extinctions. *Paleobiology* **10**, 246–67.

Smith, A.B. 2007. Marine diversity through the Phanerozoic: problems and prospects. *Journal of the Geological Society* **164**, 731–45.

Stanley, S.M. 2007. An analysis of the history of marine animal diversity. *Paleobiology* **33**, Special Paper 6, 1–55.

Valentine, J.M. 1969. Patterns of taxonomic and ecological structure of the shelf benthos during Phanerozoic time. *Palaeontology* **12**, 684–709.

Glossary

Technical terms used in describing fossils and paleontological phenomena are listed here. Use the glossary in conjunction with the index to find fuller explanations. Group names of organisms are not listed: see the index. Where necessary, the plural form (pl.), and the adjective (adj.) are given. When there are several related terms in the glossary, these are cross-referenced by the term cf. (= compare, *confere*). Many technical terms used in paleontology, as in science generally, are pure Latin or Greek: original meanings are indicated by inverted commas.

abdomen "belly"; the posterior body segment of an arthropod (cf. thorax).

aboral "opposite the mouth".

abrasion "wearing down"; removal of edges and processes by tumbling in a current.

abundance numbers of individuals of a species in a population or a sample.

abyssal "bottomless"; of the deepest oceanic zones.

accretion (in sedimentology and skeletal growth) "growth" by the addition of new material.

acelomate lacking a celom.

acid rain a rain of dilute acid, deriving from carbon dioxide and other gases that mix with water, both from volcanic emissions and from pollution.

adapical "opposite the apex" or top.

adaptation "fitting"; any feature of an organism that has a function; also, the processes behind the acquisition of an adaptation.

adductor muscle, adductor "pull towards"; a muscle that closes the valves of a brachiopod (cf. diductor muscle).

aerobic "air-containing"; oxygen-rich (cf. anaerobic, anoxic, dysoxic).

agglutinated "stuck together".

agrichnion (pl. **agrichnia**) "farming trace".

ahermatypic (of corals) living below the photic zone, generally in cold waters (cf. hermatypic).

alar "wing-like".

algal bloom a huge growth of algae, often in the surface waters of a lake.

alimentary canal the gut.

allantois the membrane inside the cleidoic egg that encloses the waste products (cf. amnion, chorion).

allochthonous "other soil"; derived from elsewhere (cf. autochthonous).

allometry "other measure"; change in proportions during growth (cf. isometry).

allopatric speciation "other homeland"; formation of new species by splitting of the geographic range occupied by the parent species.

alveolus (pl. **alveoli**) "little hollow"; a tooth socket; a deep depression, for example in the guard of a belemnite.

ambulacral area (**amb**) "walking"; one of five zones of narrower plates in the echinoid skeleton (cf. interambulacral area).

amerous "no parts"; with an undivided celom (cf. metamerous, oligomerous, pseudometamerous).

amino acid basic building block of proteins; there are 21 amino acids, which may be strung together in a huge range of different sequences to produce all the multitudes of protein types.

amnion the membrane inside the cleidoic egg that surrounds the embryo (cf. allantois, chorion).

amoeba "change"; an irregularly-shaped, single-celled organism.

anaerobic oxygen-poor (cf. aerobic, anoxic, dysoxic).

analog a feature of two or more organisms that is superficially similar in morphology or function, but which arose from different ancestries (adj. **analogous**).
anapsid "no arches"; a tetrapod skull with no temporal openings (cf. diapsid, synapsid)
anoxic oxygen-poor, or with no oxygen (cf. aerobic, anaerobic, dysoxic).
antenna (pl. **antennae**) a "feeler", or anterior long sensory appendage, in an arthropod.
anterior front (cf. posterior).
anther "flower"; the part of a flower that produces pollen.
antorbital fenestra "window in front of the eye socket"; the opening in the skull of archosaur reptiles between the nostril and eye socket.
apatite calcium phosphate ($CaPO^4$), typical mineralized constituent of skeletons of vertebrates, conodonts, some brachiopods and some worms.
aperture "opening"; specifically the opening of a gastropod shell.
apex "tip" or top (adj. **apical**).
apomorphy (pl. **apomorphies**) "from shape"; a derived character, a feature that arose once only in evolution.
apparatus in conodonts, the combination of different elements that makes a complete jaw system.
appendage a limb, or limb-like, projection from the side of the body, used mainly for locomotion and feeding.
appendiculate possessing appendages.
aptychus (pl. **aptychi**) a paired covering plate over the aperture of a cephalopod.
aragonite a form of calcium carbonate ($CaCO_3$) that occurs commonly as needles in shelly skeletons of various organisms and in lime muds.
arborescent "tree-like-growing".
archegonium (pl. **archegonia**) "founder of a race"; the fertile female structures within the ovule of a plant.
Aristotle's lantern the jaw system of echinoids, consisting of hinged plates and muscles.
articular the tiny bone at the back of a reptile lower jaw that articulates with the quadrate in the skull.
articulation "connection"; refers typically to parts of skeletons that remain in natural contact in fossils.
asexual reproduction reproduction in the absence of sex.
assemblage a collection of fossil specimens or species found together, which may partly represent a naturally-occurring community or population, but which has also been overwritten by processes of sedimentary accumulation.
astogeny growth of the compound structure of a colony.
astragalus a major ankle bone.
astrorhiza (pl. **astrorhizae**) "star root"; a star-like pattern of radiating canals on the outer surface of a stromatoporoid, the exhalent canal system.
atavism a "throw-back", an ancestral feature that reappears in an organism, often by an error in development.
atoll a reef that entirely surrounds a volcanic island which may, or may not, still be visible above the waves (cf. barrier reef, fringing reef).
authigenic preservation casting of the outer shape of a fossil.
autochthonous "self soil"; in situ, not moved from elsewhere (cf. allochthonous).
autopod the distal portion of a vertebrate limb, the hand or foot (cf. stylopod, zeugopod).
autotheca "self case"; one of the types of thecae in a dendroid graptolite; larger than a bitheca (cf. bitheca, stolotheca).
autotroph "self-feeder"; an organism that converts inorganic matter into food (cf. heterotroph).
axis the line of symmetry running through an organism; the middle stalk of a frond (adj. **axial**).

banded iron formation (BIF) Archaean rocks, which consisted of bands of iron-rich and iron-poor sediment, formed on the ocean floor in the absence of oxygen.
barrier reef a reef that lies offshore on the shelf, separated from the land by a lagoon (cf. atoll, fringing reef).
basal at the bottom; plates lying low in the calyx of a crinoid.
bathymetric "depth measure"; relating to the depth of water.
benthos "depth"; the organisms that live in and on the seabed (adj. **benthic** or **benthonic**; cf. pelagic).
bias one-way error, when observations all carry errors in the same direction, such as the generally diminishing quality of the fossil record in ever-older rocks.
biconvex "two-convex"; a shape that is convex in both directions.
bilateral "two-sided".

bilaterians triploblastic animals with a bilateral symmetry, and a digestive system with a mouth and anus.

bilobed "two-lobed".

binomen "two name"; the standard generic and specific name of an organism.

bioerosion "life removal"; the removal of skeletal materials by boring organisms using chemical and physical means.

biofilm a thin coating composed of organic material.

biogeochemical cycle the movement of carbon, and other organic chemicals, through organisms and sediments.

biogeographic province a large-scale geographic region that is inhabited by specific plants and animals.

bioherm mound-like structure built by a variety of marine organisms.

bioimmuration preservation of soft-bodied or skeletal taxa, usually by overgrowth of encrusting organisms

biological scaling principle aspects of the relative size of an organism that link to physiology, and may relate three-dimensional to two-dimensional measures.

biological species concept the definition of a species as a group of organisms that are capable of interbreeding and of producing viable offspring (cf. morphological species concept).

biomarker an organic chemical that indicates the presence of life.

biomass the mass of biological material, plant and animal, represented in a specific place at a specific time.

biomechanics the physics of how organisms move.

biostratigraphy "life stratigraphy"; the dating of rocks by means of fossils.

bioturbation "life disturbed"; disturbance of sediments by the activity of animals or plants.

bipedal "two-footed"; walking on the hindlimbs only (cf. quadrupedal).

biramous "two-branched" (cf. uniramous).

biserial "in a single row" (cf. triserial, uniserial).

bitheca one of the types of thecae in a dendroid graptolite; smaller than an autotheca (cf. autotheca, stolotheca).

blastopore "sprout opening"; a deep depression on the side of a blastula-stage embryo that eventually becomes the mouth in protostomes, or the anus in deuterostomes.

blastula "small sprout"; early embryonic phase, shaped like a hollow ball with a deep depression on one side, the blastopore.

body chamber the last-formed chamber of a cephalopod shell in which the animal lives.

body fossil the remains of an organism, usually termed simply "fossil" (cf. trace fossil).

bone the characteristic skeletal tissue of vertebrates, formed mainly from apatite and collagen.

brachial "of the arm"; plate lying at the base of the arms in the calyx of a crinoid.

brachial valve the valve in a brachiopod shell that does not contain the pedicle foramen (cf. pedicle valve).

brachiole a pore perforating the plate of a blastozoan.

byssus a bundle of fibers used for attachment to hard substrates in some bivalves.

calcified "lime-made"; invested with carbonate, either calcium carbonate or calcium phosphate.

calcite a variety of forms of calcium carbonate ($CaCO_3$) that occur commonly in the shelly skeletons of various organisms and in limestones.

calice "cup"; the upper part of the skeleton of a coral, the corallum, in which the polyp sits.

calyx "covering"; the outer covering of a flower, formed from the sepals; the cup-like skeleton containing the body of a crinoid.

cambium a layer of plant tissue that allows the stem to increase in girth.

canine teeth "dog-like"; pointed teeth in mammals, used for piercing food items (cf. cheek teeth, incisor teeth, molar teeth, premolar teeth).

capitula shell of a barnacle.

carbonate made from calcium carbonate.

cardinal major or key.

carnassial teeth "flesh-eating"; specialized cheek teeth in carnivorous mammals, used for tearing flesh.

carnivore "flesh-eater".

carpel "fruit"; the specialized structure that encloses the ovule in an angiosperm flower.

cartilage flexible supporting tissue in chordates, formed from collagen.

catastrophe a sudden event (adj. **catastrophic**; cf. gradualistic).

cateniform "chain-like".

cecum (pl. **ceca**) "blind"; a blind sac or side projection of the gut or mantle.

cellulose a carbohydrate that is the main component of the cell walls of plants.
celom "cavity"; the body cavity in animals found between the gut and the outer body surface (adj. **celomate**).
center of mass central point of the mass of an animal.
cephalic "of the head".
cephalis "head"; the upper whorl of a radiolarian test.
cephalon "head"; the anterior segment of a trilobite (cf. pygidium, thorax).
chamber a walled compartment of the cephalopod or foraminiferan shell.
cheek teeth the teeth in the back of the jaw of mammals, divided into premolars and molars, used for chewing the food (cf. canine teeth, incisor teeth, molar teeth, premolar teeth).
chelicera (pl. **chelicerae**) "crab's claw-horn"; the pincer of a chelicerate arthropod.
chitin a protein that forms most of the hard parts of arthropods.
chitinophosphate a hard tissue composed of chitin and phosphate.
chloroplast "pale green-molded"; organelle in a eukaryote plant cell that carries out photosynthesis.
choanocyte a collar cell in a sponge that moves water by beating its flagellum.
chorion the membrane inside the cleidoic egg that lines the eggshell (cf. allantois, amnion).
chromosome "color body"; a long strand of DNA, paired with a matching sequence, and typically forming an elongate X shape, composed of sequences of genes.
chronospecies "time species"; a species that is part of a lineage, whose origin and extinction are defined somewhat arbitrarily by gaps in the fossil record, or by major morphological changes.
chronostratigraphy "time stratigraphy"; the establishment of international standard divisions of geological time.
cilium (pl. **cilia**) "eyelash"; a hair-like lash borne by a cell (adj. **ciliate, ciliated**).
cirrus (pl. **cirri**) "curl"; a small flexible projection.
clade a monophyletic group.
cladistic analysis, cladistics the classification of taxa, or of biogeographic regions, in terms of shared derived characters (synapomorphies).
cladogram a dichotomously (two-way) branching diagram indicating the closeness of relationship of a number of taxa.
class a division in classification; contains one or more orders, and is contained in a phylum.
classification the process of naming organisms and arranging them in a meaningful pattern; also, the end result of such a procedure, a sequential list of organism names arranged in a way that reflects their postulated relationships.
clastic "broken"; sedimentary rocks formed from eroded and transported material.
clavate "club-shaped".
cleidoic "closed"; used in reference to the egg of amniotes.
climax community the final stable community established after a certain time of adjustment.
cluster analysis multivariate mathematical techniques for finding clusters of most-similar organisms or communities.
cnidoblast "sea anemone sprout"; a poisonous stinging cell of a cnidarian.
coaptive "fitting together".
coenosteum the skeleton of a stromatoporoid.
coevolution "evolution together" of two or more organisms that interact in some way.
collagen a flexible protein that makes up cartilage, and which forms the flexible framework of bone, upon which apatite crystals precipitate.
colonial (of corals, graptolites, bryozoans, etc.) living in fixed association with other individuals, and forming a unified "superorganism" (cf. solitary).
colpus "depression".
columella (pl. **columellae**) a pillar-like part of the skeleton of a coral that runs up the middle of the corallum.
columnal one segment of the stalk of a crinoid, an ossicle.
columnar column-like.
commensalism "feeding together"; a biological interaction where smaller species live on larger ones, and feed on debris from their eating activities, but the small organisms do not damage the larger ones (adj. **commensal**).
commissure "joining"; the line of contact of the two valves of a brachiopod or bivalve, excluding the hinge; point of contact of the laesurae of a spore.
common ancestor the individual organism, or species, that gave rise to a particular monophyletic group, or clade.
common descent the shared ancestry of all organisms, tracing back through common ancestors to the origin of life.
community a group of plants and animals that live together and interact in a specified area.

competition the interaction between two individuals or two species where both require the same limiting resource (food, space, shelter), and success by one implies a disadvantage to the other (cf. parasitism, symbiosis).
complete tree a phylogenetic tree, or cladogram, that includes all known species within its bounds.
compound eye an eye composed of many separate units, typical of arthropods (cf. ocellus).
concavoconvex concave on one side and convex on the other; hence rather C-shaped in cross-section.
concentration deposit a rich accumulation of fossils produced by physical sedimentary processes that bring the specimens together (cf. conservation deposit).
concentric repeated circular pattern, running parallel to an outer circular margin (cf. radial).
conch "shell".
concretion an irregular concentration, commonly of calcite or siderite, formed by chemical precipitation within the sediment.
congruence agreement.
congulum (pl. **congula**) sealing ring of the diatom skeleton.
conservation deposit a rich accumulation of fossils produced on the spot by processes that prevent decay and scavenging (cf. concentration deposit).
conservation trap a specific location where organisms are trapped and preserved instantly.
conterminant "near the end".
continental drift relative movements of continents and oceans through time.
coprolite "excrement stone"; fossilized dung.
corallite "coral stone"; the skeletal part of a coral.
corallum (pl. **coralla**) the skeletal part of a coral.
corpus "body".
correlation matching of rocks of equivalent age.
corrosion chemical destruction of hard tissues.
cosmopolitan "citizen of the world"; living worldwide, or nearly worldwide (cf. endemic).
costa (pl. **costae**) "rib".
costation pattern of ribbing, which gives a zigzag pattern to the commissure line in brachiopods and bivalves.
cotyledon "cup"; the food storage area of a seed.
Creationism the belief that the Earth and life were created by a divine being (cf. intelligent design).
cubichnion (pl. **cubichnia**) "resting trace".
cusp pointed projection of the top of a mammalian tooth or a conodont element.
cuticle horny protein outer covering in many plants and animals.
cyst the fertile "resting stage" of an alga.
cytology the study of cells.

"dead clade walking" the phenomenon of species that survive a mass extinction for a short time, and then die out soon after.
decay breakdown of tissues by chemical means or by microbial attack.
deciduous "fall from"; shedding leaves each winter.
declined "sloping down".
deduction drawing a conclusion from previously known information (cf. induction).
deformation distortion of rocks and fossils by physical stretching and compression.
delthyrium "delta door"; a small triangular-shaped zone in the hinge region of the pedicle valve of a brachiopod, lying between the pedicle foramen and the margin of the valve (cf. notothyrium).
deltidial plates plates that cover the delthyrium in some brachiopods.
dendrogram "tree diagram"; a branching diagram that shows relationships.
dendroid "tree-like".
dentary "tooth-bearing"; the tooth-bearing bone in the lower jaw of a vertebrate (cf. maxilla, premaxilla).
denticle "toothlet"; a small, tooth-like structure.
dentine the main constituent of teeth, a form of apatite.
dermal bone "skin"; bone formed initially within the endoderm.
deuterostome "posterior mouth"; animal in which the embryonic blastopore often develops into the anus (cf. protostome).
development growth from the egg to adult form.
developmental genes genes that direct aspects of development.
dextral "right-handed" (cf. sinistral).
diagenesis the physical and chemical processes that affect rocks and fossils after burial.
diagnosis a brief outline of the distinguishing (= diagnostic) features of an organism or a group.
diapsid "two arches"; a tetrapod skull with two temporal openings (cf. anapsid, synapsid).

dibranchiate "two-armed"; possessing one pair of gills, as in coleoids.
dichotomous "two cut"; branching in two directions.
dicyclic "double cycle" (cf. monocyclic).
diductor muscle, diductor "pull away"; a muscle that opens the valves of a brachiopod (cf. adductor muscle).
differentiated (of teeth) divided into different kinds, such as incisors, canines and cheek teeth.
digit finger or toe.
diploblastic "two-layered"; the two-layered body plan seen in cnidarians, in which the endoderm and ectoderm are separated by the mesoglea, but there is no celom (cf. triploblastic).
diploid (in cell biology) the normal double complement of chromosomes (cf. haploid).
disarticulation breaking apart and losing natural connections; typically of parts of a skeleton.
disaster taxa species that become established in the disturbed times following an extinction event.
disparity "difference"; the sum of morphological variation.
dissepiment "partition"; a horizontal, or nearly horizontal, plate of tissue supporting a tabula in an archaeocyathan or coral skeleton; a connecting structure in the rhabdosome of a dendroid graptolite.
dissepimentarium the area occupied by dissepiments in a coral.
dissoconch "apart shell"; the initial shell of a rostroconch mollusk.
dissolution chemical breakdown of a solid element or compound by dissolving in a liquid.
distal "far" from the source (cf. proximal).
diversification increase in numbers of species of a group, or of life as a whole.
diversity the number of species, genera or families in a defined geographic area, or in the world.
diverticulum "byway"; a blind-ending side branch (usually of the gut).
DNA deoxyribose nucleic acid, the organic chemical that makes up the genes and chromosomes, and which stores genetic material.
dolomite a form of limestone containing magnesium, and commonly some iron.
domichnion (pl. **domichnia**) "living trace".
dominance relative abundance of species within a community; certain species are dominant if they are much commoner than others (cf. evenness).
dorsum "back"; upper side (adj. **dorsal**; cf. venter).
doublure the thickened outer margin of the trilobite cephalon or pygidium.
durophagous feeding by crushing bones and shells.
dyad "double unit"; a double-unit spore (cf. monad, tetrad).
dysoxic "badly oxic"; containing low levels of oxygen (cf. aerobic, anaerobic, anoxic).

ecdysis molting.
ecdysozoans higher taxon (arthropod–nematode–priapulid plus others).
ecophenotypic change change in the phenotype during an organism's lifetime, induced by local environmental changes, but not coded in the genotype.
ecospace range of habitats occupied by certain organisms.
ecosystem the combination of habitats and organisms in a particular place at a particular time.
ectoderm "outside skin"; the outer skin (cf. endoderm).
ectoplasm "outside mold"; the outer layer of proteinaceous material in a cell (cf. endoplasm).
ectotherm "outside heat"; an animal that controls its temperature solely from external sources (cf. endotherm).
effect hypothesis the idea that some species-level traits might arise as a side effect of selection at the level of the individual.
Elvis taxa taxa that disappear, and are replaced some time later by close impersonators (i.e. highly convergent, but unrelated, taxa).
embryology the study of embryos.
encruster an organism that grows over the substrate, or other organisms, by creating a hard attached skeleton.
encystment turning into a cyst.
endemic "in district"; restricted to a particular geographic area (cf. cosmopolitan).
endoderm "inside skin"; the lining of the gut (cf. ectoderm).
endogastric "within the stomach" (cf. exogastric).
endogenic "formed within" the sediment, such as a burrow (cf. exogenic).
endoplasm "inside mold"; the inner portion of proteinaceous material in a cell (cf. ectoplasm).
endoskeleton "inside skeleton" (cf. exoskeleton).

endostyle "inside pen"; mucus organ in the gut of chordates.

endosymbiont "inside together life"; an organism that lives in symbiotic relationship with another, and is entirely enclosed within its structure.

endotherm "inside heat"; an animal that controls its body temperature by internal means (cf. ectotherm).

enrollment rolling up.

enteron "gut"; the gut and respiratory cavity of a cnidarian.

epifauna "top fauna"; animals that live on the seabed, not within the sediment (cf. infauna).

epirelief "top relief"; a trace fossil on the top of a bed (cf. hyporelief).

epitheca "top-case"; the upper half of a diatom theca (cf. hypotheca).

epithelium (pl. **epithelia**) cell layer forming outer tissues.

epoch a division of geological time, such as Eocene, Oligocene or Miocene; a subdivision of a period, and composed of several stages.

equilibrium "equal balance"; a fixed level.

esophagus the part of the gut between the mouth and the stomach.

eukaryote "well kernel"; single- and multicelled life form with a nucleus, including algae, fungi, plants and animals (cf. prokaryote).

eumetazoans clade including all the major metazoan groups except for the sponges.

euryotopic "wide place"; of wide ecological preferences (cf. stenotopic).

eustatic "well standing"; relating to simultaneous worldwide changes in sea level.

eutrophication "healthy nutrition"; oxygen starvation, usually in a lake, caused by decaying algae after an algal bloom.

evenness the approach to equal abundance of species within a community (cf. dominance).

evolute "rolling out"; coils of a gastropod or cephalopod shell that are all at least partially exposed (cf. involute).

evolution "unrolling"; change in organisms through time.

exceptional preservation preservation of soft parts and of soft-bodied organisms.

exine the tough outer wall of pollen and spores.

exogastric "outside the stomach" (cf. endogastric).

exogenic "formed outside"; formed on the surface of the sediment, such as a trail (cf. endogenic).

exoskeleton "external skeleton" (cf. endoskeleton).

exponential a curve that indicates geometric growth, where the *y*-value increases ever-faster in proportion to the *x*-value.

extant phylogenetic bracket (EPB) the observation that ancestors included, or bracketed, by living forms, will likely have possessed any shared characters of the bracketing forms.

extinction the disappearance of a species, genus or family.

extremophile organism adapted to life in extreme environmental conditions.

exuviae "thrown off"; cast-off molted skins.

facial suture the dividing line on the cephalon of a trilobite along which the exoskeleton splits during molting (cf. free cheek).

facies a characteristic association of sedimentary features that may indicate a particular environment of deposition.

facies fossil a fossil that is characteristic of a particular sedimentary facies.

family a division in classification; contains one or more genera, and is contained in an order.

fasciculate "bundle-like".

fauna the characteristic animals of a particular place and time.

fecal referring to excrement or dung.

femur the thigh bone (cf. fibula, tibia).

fibula one of the shin bones (cf. femur, tibia).

finite element analysis (FEA) an engineering technique that models stresses and strains in complex structures.

firmground a sea or lake floor composed of semiconsolidated calcareous sediment (cf. hardground).

flagellum (pl. **flagella**) "whip"; hair-like organelle in a eukaryote cell that is used for swimming.

flattening compression of a fossil by pressure from above.

flora characteristic plants of a particular place and time.

fluvial, fluviatile referring to rivers.

flysch an accumulation of sandstones and mudstones, generally formed in a deep basin from turbidity flows.

fodinichnion (pl. **fodinichnia**) "feeding trace".

foliated "leaf-like"; consisting of thin layers.

food chain unidirectional links between food and consumer within a community.

food pyramid the pattern of biomass distribution within a community, typically with large

biomass of primary producers, and smaller and smaller biomasses of successive consumers.
food web the complex feeding interactions among members of a community.
foramen (pl. **foramina**) "pierce"; a small opening.
foramen magnum "big opening"; the large opening at the back or base of the skull through which the spinal cord passes.
formation (in stratigraphy) a rock unit that may be identified and mapped in a regional context; subdivided into members, and combined with other formations into a group.
fossil "dug up"; the remains of a plant or animal that died in the distant past.
fragmentation breaking of a shell or skeleton into small pieces.
framework reef a reef whose basic structure is formed entirely from organic skeletons (corals, archaeocyathans, bryozoans, crinoids, etc.) (cf. reef).
free cheek the lateral portion of the cephalon of a trilobite which is divided from the central portion by a facial suture, and which separates during molting (cf. facial suture).
fringing reef a reef that lies on the margins of a landmass, with no intervening lagoon (cf. atoll, barrier reef).
frustule "a bit"; the skeleton of a diatom.
fugichnion (pl. **fugichnia**) "escape trace".
full relief (of a trace fossil) seen in three dimensions (cf. semirelief).
furca (pl. **furcae**) "fork"; a backwards-pointing flexible spine in an arthropod.
fusellar tissue the bandage-like tissues composing the periderm of graptolites.
fusiform "spindle-shaped".

gamete "marriage"; a sex cell, such as an egg or sperm.
gametophyte "marriage plant"; in plants that show alternation of generations, the stage that produces gametes and which engages in sexual reproduction (cf. sporophyte).
gas hydrates deposits of methane locked in an ice lattice, found either in deep oceans or in permafrost regions.
genal spine "cheek"; the pointed spine at the posterior lateral margin of the trilobite cephalon.
gene an identifiable sequence within a chromosome that codes for a particular feature of an organism.
gene flow the movement of genes through a population by interbreeding.
gene pool the sum total of the genotypes of all individual organisms in a defined population.
geniculation "little knee"; bent at right angles.
genotype the sum of the features of an organism, or population, contained in the genes.
genus (pl. **genera**) the category in classification above the species.
geochronometry measurement of geological time using absolute methods such as radiometric dating.
geographic range the complete area within which a species, or other taxon, lives.
germinal aperture opening for the passage of gametes in a spore or pollen grain.
gill bars bars of cartilage or bone that support the gill slits.
gill slits gill openings behind the head, found in chordates.
glabella (pl. **glabellae**) "bald"; the raised middle portion of a trilobite cephalon.
golden spike (in stratigraphy) a point in a rock section, equivalent to an instant in geological time, that marks the internationally accepted base of a stratigraphic division (e.g. member, formation, system/period or epoch).
gonad "generation"; the organ that produces sex cells; the ovary or testis.
Gondwanaland ancient supercontinent composed of South America, Africa, India, Australia and Antarctica.
gradualism *see* phyletic gradualism.
gradualistic steady change (cf. catastrophe).
greenhouse gas a gas, such as methane or carbon dioxide, that promotes heating of the atmosphere.
group (in stratigraphy) a number of formations occurring in sequence that share some broad-scale features.
guard the bullet-shaped, solid, terminal part of the belemnite shell (cf. phragomcone, pro-ostracum).

habitat the environmental setting within which a species, or a community, lives.
haploid (in cell biology) a half complement of chromosomes, as found in the sex cells (cf. diploid).
haptonema "fasten-thread"; a flagellum-like structure in coccolithophores.
hardground a sea or lake floor composed of consolidated calcareous sediment (cf. firmground).
herbaceous low-growing, bushy.
herbivore "plant-eater".

hermatypic (of corals) restricted to the photic zone, generally in tropical waters (cf. ahermatypic).

heterocercal tail "different tail"; tail fin, as in sharks, which is asymmetric and has a large upper lobe.

heterochrony "different time"; changes in the timing and rate of development that affect evolution.

heteromorph "different form" (adj. **heteromorphic**); supposed female ostracod (cf. tecnomorph).

heteropygous "different-rumped"; trilobite with a pygidium slightly smaller than the cephalon (cf. macropygous, micropygous).

heterosporous "different spores"; producing microspores and megaspores (cf. homosporous).

heterotroph "different feeder"; an organism that feeds on a variety of materials (cf. autotroph).

hierarchy a system consisting of smaller and smaller categories which, in the case of biological classifications, are inclusive, that is, the smaller units fit within larger ones.

hinge the zone of attachment about which the two valves of a brachiopod or bivalve shell open and shut.

holaspis "true shield"; the final larval stage in trilobites (cf. meraspis, nauplius larva, protaspis).

holdfast the rooting structure that fixes an archaeocyathan, a crinoid, a dendroid graptolite or a seaweed to a rock.

holochroal eye an eye with many small, closely-packed lenses, seen in trilobites and many other arthropods (cf. schizochroal eye).

homeobox genes (including ***Hox*** **genes**) "same box"; genes found in all organisms that control orientation, segmentation and limb development in embryonic development.

homeomorphic "same form".

homology a feature that arose once only; an apomorphy (adj. **homologous**).

homosporous "producing the same spores" (cf. heterosporous).

horizontal gene transfer transfer of genes, sometimes called "jumping genes", between simple organisms.

humerus the upper arm bone in vertebrates (cf. radius, ulna).

hyaline "glassy"; composed of tiny aligned calcite crystals.

hydrostatic "water-standing"; water-supported.

hyperthermophile "excessive heat lover"; a microorganism that is adapted to living in extreme heat.

hypha (pl. **hyphae**) "web"; a branching tissue strand that forms part of a fungus.

hyponome wide tube lying beneath the head of a cephalopod through which water is squirted to achieve propulsion.

hyporelief "under relief"; a trace fossil on the bottom of a bed (cf. epirelief).

hypostoma "under hole"; a plate underneath the trilobite cephalon, which may have supported the mouth region.

hypotheca "under-case"; the lower half of a diatom theca (cf. epitheca).

hypothesis a supposition or proposition that explains a number of observations.

hypothetico-deductive method the scientific method that consists of seeking to *disprove*, rather than prove, hypotheses.

ichnofacies a facies based on characteristic trace fossils.

ichnofossil "trace fossil".

ichnogenus (pl. **ichnogenera**) a genus of trace fossil.

ichnology "trace study"; the study of trace fossils.

ichnospecies a species of trace fossil.

ilium upper bone of the typical tetrapod pelvis (cf. ischium, pubis).

impendent "hanging down".

imperforate "lacking holes" (cf. perforate).

incisor teeth "cutting"; the front teeth in mammals, used for snipping food off (cf. canine teeth, cheek teeth, premolar teeth, molar teeth).

inclusive hierarchy a series in which small things fit inside larger things, such as species within genera, genera within families, and so on.

incongruence (in cladistics) a lack of matching of character sets.

induction establishing a general theory from the accumulation of many observations (cf. deduction).

infauna animals that live within the sediment (cf. epifauna).

infrabasal "below basal"; a plate lying below the basals in the calyx of a crinoid.

ingroup the organisms of interest in a cladistic study, as opposed to the outgroup, which is everything else.

integument "covering"; skin.

intelligent design (ID) the belief that the Earth and life were created by a divine being (cf. Creationism).
interambulacral area (**interamb**) one of five zones of broader plates in the echinoid skeleton (cf. ambulacral area).
interarea flattened parts of the brachiopod hinge region that are exposed externally.
intertidal "between tides"; between normal high and low water marks (cf. supratidal).
intervallum "between walls"; the space between the inner and outer walls in archaeocyathans.
involute "rolling in"; coils of a gastropod or cephalopod shell that are all concealed by the outermost coil (cf. evolute).
ischium lower posterior bone of the typical tetrapod pelvis (cf. ilium, pubis).
island dwarfing evolution to small size on an island.
isometry "same measure"; maintenance of identical proportions during growth (cf. allometry).

kingdom the highest division in classification; contains one or more phyla.

laesura (pl. **laesurae**) "scar"; contact scars between neighboring spores.
Lagerstätte (pl. **Lagerstätten**) "lying place"; a deposit containing large numbers of exceptionally preserved fossils.
lamella (pl. **lamellae**) "small thin plate"; a thin plate or layer.
lamina (pl. **laminae**) "thin plate"; a thin plate or layer.
lancet "sharp knife"; the pointed area of small plates in the calyx of a blastozoan.
lappet "small lobe"; a side flap seen in some graptolites and ammonoids.
larva (pl. **larvae**) juvenile which has a different form to the adult.
last universal common ancestor (LUCA) the organism that is ancestral to all known living groups of organisms.
lateral side.
lateral line canal a canal that runs along the side of the body in fishes, and which bears sensory cells that can detect movements in the water.
Lazarus taxa taxa that disappear and then reappear (but had clearly not died out and come back to life).
lecitotrophic larva short-lived larval phase, nonfeeding, surviving on egg yolk.
lepidotrichium (pl. **lepidotrichia**) "scale-hair"; a thin bony rod in the fin of a fish.
ligament "bind"; a bundle of fibrous tissues linking skeletal elements (in brachiopods, mollusks and vertebrates).
lignification deposition of lignin.
lignin "wood"; woody tissue.
lineage an evolving line, consisting of one or more species that have direct genetic links through time.
linear as a straight line.
lipid fatty or waxy compound of the cell.
lithostratigraphy "rock stratigraphy", the sequence and correlation of rocks.
littoral "shore"; coastal.
locus (pl. **loci**) "place".
logistic an S-shaped curve.
lophophore "crest-bearing"; a specialized feeding and respiratory organ found in brachiopods and bryozoans (adj. **lophophorate**).
lorica "leather corslet"; the outer covering of a tintinnid.
lumbar "loin"; of the lower back.
lung book the air-breathing lung of a spider, arranged in many layers like the pages of a book.

macroconch "large shell"; the larger of two morphs of a cephalopod species, probably the female (cf. microconch).
macroevolution "large evolution"; evolution at species level and above, including those evolutionary topics (speciation, lineage evolution, trends, diversification, extinction events) that may be studied by paleontologists; those parts of evolution excluded from microevolution.
macrofossil a "large fossil", one that can be seen with the naked eye, in comparison to a microfossil.
macropygous "large-rumped"; trilobite with a pygidium larger than the cephalon (cf. heteropygous, micropygous).
madreporite "mother stone"; a plate in the echinoid skeleton near the mouth that connects the water vascular system to the external environment.
magnetostratigraphy stratigraphy based on magnetic reversals.
mamelon a small "breast-like" projection on a stromatoporoid.
mantle portion of the body tissues of a mollusk involved in secretion of shell material.
mass extinction a major extinction event, typically marked by the loss of 10% or more of families, and 40% or more of species, in a short time.

massive solid.

maxilla "jawbone"; the main tooth-bearing bone in the upper jaw of a vertebrate (cf. dentary, premaxilla).

median, median "middle".

medusa (pl. **medusae**) "gorgon"; a free-swimming jellyfish-like stage in cnidarian development (cf. polyp).

megaguilds groups of organisms defined by their life mode and feeding type.

megasporangium the structure that contains megaspores or ovules.

megaspore "big spore"; the larger spore of early seed-bearing plants (cf. microspore).

meioscopic members of the meiofauna; organisms generally between 45 μm and 1 mm in size.

meiosis "diminution"; the process of cell division that involves reduction of chromosome numbers from the diploid to the haploid condition, prior to production of eggs or sperm (cf. mitosis).

member (in stratigraphy) a localized rock unit that may be mapped within a limited area; forms part of a formation.

meraspis "middle shield"; the third larval stage in trilobites (cf. holaspis, nauplius larva, protaspis).

mesentery "middle gut"; fleshy projection of the endoderm into the gut cavity of a cnidarian.

mesoderm "middle skin"; the tissue type that forms a variety of organs between the endoderm and ectoderm of many animals.

mesoglea "middle glue"; a gelatinous substance that separates the ectoderm and endoderm in diploblastic animals.

metacel "change cavity"; the body cavity of a bryozoan.

metamerous "change part"; with a segmented celom (cf. amerous, oligomerous, pseudometamerous).

metamorphism "change form"; geological processes involving high temperature and/or high pressure, usually associated with tectonic activity within the crust.

metamorphosis "change form"; change, during development, from the larval to the adult form.

metaphyte "later plant"; multicelled plant.

metasicula "later sicula"; the main part of the sicula of a graptolite (cf. prosicula).

metazoan "later animal"; multicelled animal.

methanogenesis "producing methane"; a form of respiration used by some anaerobic organisms that absorb carbon dioxide and hydrogen.

micrite "microscopic calcite"; calcite ($CaCO_3$) that occurs as small crystals (cf. sparry calcite).

microcephalic "small headed"; a person, or animal, with an unusually small head and brain.

microconch "small shell"; the smaller of two morphs of a cephalopod species, probably the male (cf. macroconch).

microevolution "small evolution"; processes of evolution below the species level, generally studied in the laboratory and in the field; those parts of evolution excluded from macroevolution.

microfossil "small fossil"; a fossil that can be seen only with a microscope.

micropygous "small-rumped"; trilobite with pygidium much smaller than the cephalon (cf. heteropygous, macropygous).

micropyle "small gate"; the opening through which the pollen tube approaches the ovule, in a flower.

microspore "small spore"; the smaller spore of early seed plants (cf. megaspore).

microsporophyll the male fertile structures of a flower.

microvertebrate a microscopic vertebrate fossil, such as a tooth or scale.

Milankovitch cycles the combined effects of the Earth's movements on its climate; named after Serbian civil engineer and mathematician Milutin Milanković.

mineralization process of formation of a mineral; in paleontology, refers typically to the formation of the hard constituent of a skeleton, or to replacement of tissues by mineral material during fossilization.

missing link popular term for an organism, usually fossil, that lies midway between two groups, such as *Archaeopteryx* which shows a mix of "reptilian" and bird-like features.

mitochondrion (pl. **mitochondria**) "thread granule"; organelle in a eukaryote cell that assists in energy transfer.

mitosis "thread"; simple cell division involved in normal growth (cf. meoisis).

Modern synthesis the current view of evolution, based on a combination of Darwin's insights into geographic variation and natural selection, paleobiology and genetics, as established in the 1930s and 1940s.

molar teeth "grinder"; one of the back teeth of a mammal, used for chewing food (cf. canine

teeth, cheek teeth, incisor teeth, premolar teeth).
molecular clock hypothesis the assumption that each protein molecule has a constant rate of amino acid substitution; the amount of difference between two homologous molecules indicates distance of common ancestry, and hence closeness of relationship.
monad "single unit"; a single-unit spore (cf. dyad, tetrad).
monocyclic "single cycle" (cf. dicyclic).
monolete of a spore with a single slit (cf. trilete).
monophyletic group "single origin"; a group that includes all the descendants of a common ancestor (cf. paraphyletic group, polyphyletic group).
morphological species concept the recognition and subdivision of species based on external appearance (cf. biological species concept).
morphology "shape study"; shape and form of an organism.
morphospace theoretical maximum range of shapes of an organism.
mosaic evolution variable rates of evolution of different parts of an organism.
motile capable of movement.
multicellular composed of more than one cell.
multilocular "many-chambered".
multimembrate "many-member" (cf. unimembrate).
mural in or of a wall.
mycelium (pl. **mycelia**) "mushroom"; the mat-like structure composed from fungal hyphae.
myophore "muscle-bearer"; an internal plate in a bivalve to which muscles attach.
myotome "muscle slice"; discrete muscle block along the trunk of a chordate.

nacreous like mother-of-pearl.
nannoplankton minute planktonic organisms measuring between 2 and 20 μm.
naris nostril.
natant "swimming".
natural selection "survival of the fittest", a process that causes evolution, first proposed by Charles Darwin in 1859; in highly variable populations, organisms with the best adaptations survive best and pass on their winning attributes to their offspring; a cumulative process, but a process that is subject to minor vicissitudes of environmental change.
nauplius larva "ship sail"; an early larval stage in many arthropods, including trilobites (cf. holaspis, meraspis, protaspis).
nekton "swimming"; organisms that swim in the open water (cf. plankton).
nema "thread"; thread-like structure at the top of the sicula of a graptolite.
nematocyst "thread bladder"; the sting within the cnidoblast of a cnidarian.
neoteny "new stretch"; pedomorphosis by retention of juvenile morphological characters in the adult.
nephridium "kidney"; a kidney-like structure for processing waste materials.
neural spine the spine on the upper surface of a vertebra.
niche lifestyle and ecological interactions of an organism.
node branching point in a cladogram.
non-parametric statistics statistics of samples that do not rely on assumptions of a normal distribution (cf. parametric statistics).
notochord "back string"; the flexible, rod-like structure that supports the body of basal chordates, and is a precursor of the backbone in vertebrates.
notothyrium "back door"; a small triangular-shaped zone in the hinge region of the brachial valve of a brachiopod, lying opposite the delthyrium (cf. delthyrium).

obrution deposit a rich accumulation of fossils produced by very rapid rates of sedimentation that bury the organisms almost instantaneously.
occipital of the back of the head.
ocellus (pl. **ocelli**) "small eye"; a single eye in an arthropod (cf. compound eye).
oligomerous "few parts"; with a celom divided longitudinally into two or three zones (cf. amerous, metamerous, pseudometamerous).
omnivore "eats all"; an animal that feeds on plant and animal food.
ontogeny development from egg to adult.
opal non-crystalline silica.
operculum (pl. **opercula**) "cover"; a cover or lid that closes an opening (adj. **opercular**).
ophiolite "snake stone"; a complex of igneous rocks associated with a subduction zone.
opisthosoma "behind body"; the abdomen of certain arthropods.
oral "of the mouth".
orbit eye socket.
order a division in classification; contains one or more families, and is contained in a class.
orogeny mountain-building (adj. **orogenic**).

orsten, stinkstone Upper Cambrian limestone nodules that have yielded an unique fauna of exceptionally-preserved animals.
osculum "little mouth"; the opening into the central cavity of a sponge.
ossicle "little bone"; one segment of the stalk of a crinoid, a columnal; or one segment of the arms of an ophiuroid, a vertebra.
ossify "turn into bone".
ostium (pl. **ostia**) "mouth"; a small perforation in the wall of a sponge.
outgroup all the organisms that lie outside the clade of interest, the ingroup.
ovary "egg"; egg-producing organ in female animals; structure that contains the ovules in plants.
ovule an undeveloped (unfertilized) seed.

paleoautecology the study of the ecology of single fossil organisms (cf. paleosynecology).
paleobotany the study of fossil plants.
paleoecology "ancient ecology"; the life and times of fossil organisms; also, the study thereof.
paleogeography "ancient geography", the layout of continents and oceans in the geological past.
paleontology "ancient life study"; the study of the life of the past.
paleosol "ancient soil".
paleosynecology the study of communities of fossil organisms (cf. paleoautecology).
pallial line "mantle"; the line that marks the outer margins of attachment of the mantle to the shell in mollusks.
pallial sinus the infolding of the pallial line in mollusks to accommodate the siphions.
palynology the study of fossil pollen and spores.
Pangea (Pangaea) "all world"; ancient supercontinent composed of all the modern continents.
paradigm shift a revolution in science, or shift from one theory to another.
paragaster "beside stomach"; the central cavity of a sponge.
parametric statistics statistics of samples that are described by a distribution, usually a normal distribution (cf. non-parametric statistics).
paraphyletic group "parallel origins"; a group that includes some, but not all, the descendants of a common ancestor (cf. monophyletic group, polyphyletic group).
parasitism "beside food"; a biological interaction where one species lives in or on another, and does it harm (cf. competition, symbiosis).
parataxonomy "parallel taxonomy"; a non-evolutionary taxonomic system.
parazoan "beside animal"; the simple body plan found in sponges in which there is no celom, and cells are not differentiated into tissue types.
parietal "of a wall".
parsimony simplicity; in cladistics, the requirement that a cladogram represents the shortest possible tree linking all taxa (adj. **parsimonious**).
pascichnion (pl. **pascichnia**) "feeding trace".
pectiniform "comb-like".
pectocaulus the linking tubes between zooid housings in pterobranchs.
pectoral "breast"; of the shoulder girdle.
pedicle "footlet"; a fleshy stalk that attaches a brachiopod to the substrate.
pedicle valve the valve in a brachiopod shell that contains the pedicle foramen (cf. brachial valve).
pedipalp "foot stroking"; a second paired appendage, or "feeler", in arthropods.
pedomorphocline a pedomorphic trend in evolution.
pedomorphosis "juvenile formation"; achievement of sexual maturity in a juvenile body (cf. peramorphosis).
pelagic of the open sea; refers to habitats and organisms that are not on the seabed (cf. benthic).
pelvic of the hip girdle.
pendent "hanging".
pentameral "five-part".
peramorphocline a peramorphic trend in evolution.
peramorphosis "overdevelopment"; the achievement of sexual maturity relatively late (cf. pedomorphosis).
perforate "possessing holes" (cf. imperforate).
periderm "surrounding skin"; the outer tissue layer of graptolites.
perignathic girdle "around the jaws"; region of the echinoid skeleton around the Aristotle's lantern.
period (in stratigraphy) the major divisions of geological time, such as Cambrian, Ordovician and Silurian, which are composed of epochs; equivalent to the system as a division of the rock column.
periostracum "around shell"; the horny outer layer of a brachiopod or mollusk shell.
periproct "around anus"; the anal opening of echinoids.

peristome "around mouth"; the mouth opening of echinoids.

permineralization near-complete replacement of the tissues of an organism by mineral material.

petrifaction "turning to rock"; fossilization by complete mineralization.

petrology "study of rocks".

phaceloid composed of numerous roughly parallel tubes.

pharynx "mouth space"; the cavity into which water is pumped in chordates; functions in feeding and in respiration.

phenotype "show type"; the sum of the externally expressed features of an organism or population.

phonetic referring to characters; analytic techniques that seek to summarize all aspects of variation in all characters of organisms or communities.

photic zone "light"; the upper parts of a water body that are penetrated by daylight; typically down to 100 m depth.

photosymbiosis "light together life"; a mutually beneficial interaction between a photosynthesizing plant or alga and some other organism.

photosynthesis "light manufacture"; the breakdown of carbon dioxide and water in the presence of sunlight to produce sugars and oxygen.

phragmocone the main part of an ammonoid shell, except the protoconch and the body chamber; the conical part of a belemnite shell between the guard and the pro-ostracum (cf. guard, pro-ostracum).

phyletic gradualism the view that evolution is continuous and gradual, and that speciation occurs as part of the gradual change within lineages (cf. punctuated equilibrium).

phylogeny "race origin"; the pattern of evolution; an evolutionary tree of all life or of some clades.

phylum (pl. **phyla**) a division in classification; contains one or more classes, and is contained in a kingdom.

phytoplankton "plant plankton"; the plant components of the plankton.

picoplankton minute planktonic organisms measuring between 0.2 and 2 μm.

pinnule "small feather"; feather-like side branches of the arms of a crinoid.

placenta "a flat cake"; the tissue structure in female mammals that transfers food and oxygen to the developing embryo.

planispiral "spiral in a plane" (cf. trochospiral).

plankton "wandering"; floating organisms that live in the top few meters of oceans and lakes (cf. nekton).

planktotrophic larva long-lived larval phase, feeding on plankton.

plate tectonics the processes within the Earth's crust and mantle that drive continental drift.

poikilotherms animals having the same body temperature as their surroundings.

pollen mobile fine-grained material produced in the anthers of flowers, and carrying the sperm.

polymerase chain reaction (PCR) the method of multiplying, or cloning, nucleic acids (DNA, RNA) from small quantities.

polymeric "many-segmented".

polymorphic "in many forms".

polyp an attached sea anemone-like stage in cnidarian development (cf. medusa).

polyphyletic group "many origins"; a group that contains members that arose from more than one ancestor (cf. monophyletic group, paraphyletic group).

porcellaneous composed of minute, randomly-oriented, calcite crystals.

posterior back (cf. anterior).

predation a biological interaction where one species feeds on another.

prehensile "seize"; flexible and grasping.

premaxilla the anterior small tooth-bearing bone in the upper jaw of a vertebrate (cf. dentary, maxilla).

premolar teeth cheek teeth of mammals, lying in front of the molars, and used for chewing food (cf. canine teeth, cheek teeth, incisor teeth, molar teeth).

proboscis "trunk"; elongate nose or snout-like projection.

process (in descriptions of morphology) projection.

progress change with improvement.

progression the sequence from simple to complex organisms through time.

prokaryote "before kernel"; basal single-celled life form with no nucleus, including bacteria and cyanobacteria (cf. eukaryote).

pro-ostracum "in front of shell"; the spatulate thin-shelled component of a belemnite shell that is attached to the phragmocone, and which supported the main part of the body (cf. guard, phragmocone).

prosicula the upper first-formed part of the sicula of a graptolite (cf. metasicula).

prosoma "before body"; the fused head and thorax found in some arthropods.

prosome "before body"; the upper part of a chitinozoan.
protaspis "first shield"; the second larval stage in trilobites (cf. holaspis, meraspis, nauplius larva).
protein a complex organic chemical composed of amino acids, the basic building block of organisms.
prothallus "before shoot"; the first stage of development of a gametophyte plant.
protoconch "first shell"; the larval portion of a shell.
protostome "first mouth"; animal in which the embryonic blastopore develops into the mouth (cf. deuterostome).
protractor muscle, protractor "pull forwards"; a muscle that pulls forwards.
provinciality the development of specific biogeographic provinces throughout the world.
proximal, proximate "near" to the source (cf. distal).
pseudocelomate possessing a "false celom", common in many embryonic animals, and found in adult nematodes.
pseudometamerous with an undivided celom, but irregularly duplicated organs (cf. amerous, metamerous, oligomerous).
pseudopodium (pl. **pseudopodia**) "false foot"; a tissue extension.
pseudopuncta (pl. **pseudopunctae**) "false hole".
pseudostome "false hole".
pubis lower anterior bone of the typical tetrapod pelvis (cf. ilium, ischium).
puncta (pl. **punctae**) "hole" (adj. **punctate**).
punctuated equilibrium the view that evolution occurs in two styles, long periods of little change (equilibrium, stasis), punctuated by short bursts of rapid change, often associated with speciation (cf. phyletic gradualism).
pygidium "small rump"; the posterior segment of a trilobite (cf. cephalon, thorax).
pygostyle "tail column"; the fused tail vertebrae of a bird.
pylome "gate"; opening in an acritarch wall.
pyriform "pear-shaped".
pyrite a form of iron sulfide (FeS), occurring as small gold-colored crystals, often associated with black mudstones and fossils deposited in anaerobic conditions.

quadrate "square"; the bone in the posterior lateral corner of a reptile skull that articulates with the articular in the lower jaw.
quadrupedal "four-footed"; walking on all fours (cf. bipedal).

radial "ray-like"; branching outwards from a central point, like the spokes in a bicycle wheel (cf. concentric); plate lying above the basals in the calyx of a crinoid.
radialian animal with a radial pattern of cells at early phases of division (cf. spiralian).
radiation (in evolution) diversification or branching of a clade.
radiometric dating dating rocks by measurement of the amount of natural radioactive decay of pairs of elements, the parent (starting element) and daughter (resultant element).
radius one of the forearm bones in vertebrates (cf. humerus, ulna).
radula "scraper"; the rasping feeding organ of mollusks.
ramified "branched".
raphe "seam"; median gash in the diatom skeleton.
rarefaction a statistical technique to standardize and compare species richness computed from samples of different sizes.
receptacle the structure that contains the ovule in a flower.
reclined "sloping back".
recumbent "lying back".
red beds red-colored sediments, generally sandstones and mudstones, formed usually in hot conditions on land.
redox "reduction–oxidation"; the junction between reducing and oxidizing conditions.
reef a wholly, or partially, organic carbonate construction (cf. framework reef).
refractory a form of carbon that does not break down readily (cf. volatile).
refugium (pl. **refugia**) a habitat that has escaped destructive environmental changes providing shelter for endangered taxa.
regression "passage back"; withdrawal of the sea from the land; may be local or global (cf. transgression).
repichnion (pl. **repichnia**) "creeping trace".
replication copying or duplication.
resupination "bent backward"; lying on the back.
reticulate "net-like".
retractor muscle, retractor "pulls back"; a muscle which pulls backwards, or pulls a structure into its protective skeleton.
retrodeformation "backwards deformation"; the process of undeforming a deformed structure, such as a metamorphosed fossil.
rhabdosome "rod body"; the whole colony of a graptolite.

rhizome "root"; an underground stem.
rostrum "beak"; the snout or anteriormost part of the head (adj. **rostral**).
ruga (pl. **rugae**) "roughness"; irregular small projections (adj. **rugose**).
ruminant a mammal that digests its food in several stages (e.g. a camel or a cow).

saccus (pl. **sacci**) "bag"; empty structure on the side of some pollen grains (adj. **saccate**).
Scala naturae the "chain of being", a sequence of organisms, from simple to complex, once interpreted as evidence for unidirectional evolution.
scandent "climbing".
scavenging feeding on organisms that are already dead.
schizochroal eye an eye with reduced numbers of large spaced lenses, seen in trilobites (cf. holochroal eye).
sclerite a "hard" skeletal plate.
scleroprotein the tough proteinaceous material that makes up the periderm of graptolites.
sclerotized "hardened".
selectivity discrimination among species, especially for survival during an extinction event, based on ecological characteristics.
selenizone "moon zone"; infilled track of the apertural slit of a gastropod.
semirelief (of a trace fossil) seen on the surface of a bed (cf. full relief).
sepal one of the outermost parts of a flower, lying outside the petals.
septum (pl. **septa**) "fence"; a dividing wall within the skeleton of various animals (adj. **septate**).
sequence stratigraphy the sedimentary sequences into major packets corresponding to times of transgression, regression and non-deposition.
sere a plant or epifaunal community that is one of a succession of unstable assemblages on the way to the establishment of a climax community.
sessile "sitting"; organisms that live on the seabed, and which do not move.
seta (pl. **setae**) "bristle"; a stiff hair.
sexual dimorphism "two forms"; differences in the morphology of males and females of a species.
sexual selection selection of traits for improving the chances of mating.
sicula the small cone that is the first part of a graptolite rhabdosome to form.
siderite a form of iron carbonate ($FeCO_3$) that occurs commonly in concretions around fossils.
Signor–Lipps effect the backwards smearing of fossil occurrences; the observation that the last fossil observed was almost certainly not the last representative of a taxon.
sinistral "left-handed" (cf. dextral).
siphon an extendable tube in a mollusk, used for sucking in water with food particles and for expelling filtered water.
siphuncle connecting strand of soft tissue that extends through the chambers of a cephalopod shell.
skeleton supporting structure in an organism, usually involving some mineralized tissues; may be internal (endoskeleton) or external (exoskeleton).
solitary an organism, usually a coral, that lives in isolation (cf. colonial).
somite "body"; a body segment in an arthropod.
sparry calcite calcite ($CaCO_3$) that occurs as large crystals (cf. micrite).
speciation the process of formation of a new species, either by splitting (branching) or by lineage evolution, from a pre-existing species.
species a group of organisms, or populations, that includes all the individuals that normally interbreed, and which can produce viable offspring; typically the smallest unit in the hierarchy of a classification of organisms.
species selection selection at the level of species.
sphincter an opening that may be closed by muscular activity.
spicule "small ear of corn"; a tiny needle-like calcareous or siliceous structure that forms part of the skeleton of a sponge.
spiracle "to breathe"; an opening near the mouth in a blastozoan; the breathing hole behind the head in a shark.
spiralian animal with an initial sequence of cell division that follows a spiral track (cf. radialian); higher taxon (mollusk–annelid–brachiopods plus most flatworm–rotifers (platyzoans)).
spongin a horny organic material that forms around the skeleton of many sponges.
spontaneous generation the idea that life could arise suddenly from non-life.
sporangium (pl. **sporangia**) spore-bearing structure in a land plant.
sporophyte "spore plant"; in plants that show alternation of generations, the stage that produces spores and that engages in asexual reproduction (cf. gametophyte).
spreite (pl. **spreiten**) "trace"; indications of former positions of a burrow.

squamosal "scale"; major bone in the side of a tetrapod skull which, in mammals, articulates directly with the dentary (lower jaw).

stage (in stratigraphy) a time unit; a subdivision of an epoch, and generally composed of several zones.

stagnation deposit a deposit of fossils preserved in anoxic conditions.

stamen "stand"; the pollen-producing structure of a flower.

stasis "standing still"; the long periods of little net evolutionary change within a lineage.

stenotopic "narrow place"; of narrow ecological preferences (cf. euryotopic).

stereom "solid"; the internal structure of echinoderm skeletal elements.

sternite "chest"; armored body covering over the underside of a segment of a eurypterid (cf. tergite).

stigma "point"; the part of the carpel in a flower that receives pollen.

stipe "post"; a branch of a graptolite rhabdosome.

stolon "sucker"; the linking tubes between the thecae in graptolites.

stolotheca "sucker case"; one of the types of thecae in a dendroid graptolite (cf. autotheca, bitheca).

stoma (pl. **stomata**) "mouth"; an opening on the underside of a leaf through which water vapor may pass.

stone canal part of the water vascular system of an echinoid.

stratification (in sedimentary geology) the layering seen within typical sediments; (in community ecology) the layering of different organisms within, typically, a forest or a reef.

stratigraphic range the time from apparent origin to apparent extinction of a fossil taxon.

stratigraphy "bedding writing"; the sequence of rocks and of events in geological time.

stratophenetics the use of stratigraphic age of specimens and their overall morphological features to draw up an evolutionary tree.

stratotype the reference section for a member or a formation, identified in a specific location.

stroma "bed"; a supporting framework of connective tissue.

stromatolite "bed/mattress rock"; a layered structure generally formed by alternating thin layers of cyanobacteria and lime mud, typically in shallow warm seawaters.

style "pen"; in a flower, the slender part above the carpels bearing the stigma.

stylopod the proximal portion of a vertebrate limb, the upper arm or thigh (cf. autopod, zeugopod).

subaerial "beneath the air"; formed on land.

subduction "pulling down"; the process whereby one tectonic plate is forced down beneath another.

substrate the underlying surface.

sulcus "furrow" (adj. **sulcate**).

superposition "positioning on top"; the observation that younger rocks lie on top of older rocks (unless they have been inverted subsequently by tectonic activity).

supratidal "above tides"; above the normal high water mark (cf. intertidal).

suspension feeder an animal that feeds on small food particles suspended in the water.

suture "stitched seam"; the firm junction between two bones; the irregular line that marks the junction between two chambers of a cephalopod shell.

symbiont a participant in a symbiotic relationship.

symbiosis "living together"; the phenomenon of species living together in close interdependence, where one or both species obtains some benefit, and neither is harmed by the relationship (adj. **symbiotic**; cf. competition, parasitism).

synapomorphy shared derived character.

synapsid "joined arch"; a tetrapod skull with one (lower) temporal opening (cf. anapsid, diapsid).

synonym "same name"; a redundant name given to an organism that has already been named (adj. **synonymous**).

synonymy equivalence of two names applied to a single species, genera or families.

synrhabdosome a group of graptolite rhabdosomes living in a linked cluster.

system (in stratigraphy) the major divisions of the rock column, such as Cambrian, Ordovician and Silurian; equivalent to the period as a division of geological time.

systematics the study of relationships of organisms and of evolutionary processes.

tabula (pl. **tabulae**) "table"; a horizontal division within the skeleton of an archaeocyathan or a coral.

tabular flattened.

taphonomy "death study"; the study of biological and geological processes that occur between the death of an organism and its final state in the rock.

taxon a group of organisms, such as a species, genus, family, order, class or phylum.
taxonomy "arrangement"; the study of the morphology and relationships of organisms.
tecnomorph supposed male ostracod (cf. heteromorph).
tectonic activity "building"; physical movements within the Earth's crust, often associated with mountain building, such as faulting and folding.
tegmen "covering"; the roof of the calyx of a crinoid.
telson the pointed tail portion of various arthropods.
temporal opening opening in the skull of a tetrapod behind the orbit.
tendon a sheet of fibrous tissue that attaches a muscle to a bone.
teratological "monstrous"; relating to abnormalities in development.
tergite "back plate"; armored body covering over the back of a segment of a eurypterid (cf. sternite).
terrane a tectonic plate that had a specific geological history.
test "pot"; the skeleton of an echinoid, foraminifer or radiolarian.
tetrad "four unit"; a four-unit spore (cf. dyad, monad).
thallus (pl. **thalli**) "young shoot"; the skeleton of a calcareous alga.
theca "case"; the skeletal wall of a coral, dinoflagellate or diatom; the calyx of a crinoid; the individual living chamber of a graptolite zooid.
theory a general explanation, or linked set of hypotheses, that explains many natural phenomena.
thermocline "temperature slope"; the level at which the water temperature in the sea, or in a lake, changes rapidly.
thermophile "heat lover"; an organism that is adapted to living in hot conditions.
thorax the middle "body" portion of an arthropod (cf. abdomen).
tibia one of the shin bones (cf. femur, fibula).
tiering a special form of stratification seen among trace fossils, where different ichnotaxa occupy different depth zones in the sediment.
time averaging the accumulation of fossils from a variety of time horizons into a single horizon.
tissue cast an impression of the tissues of an ancient organism.
tool mark impression on a sediment surface made by a transported object.
torsion "twisting".
trabecula (pl. **trabeculae**) "little beam"; rod-like structure that crosses a space.
trace fossil remains of the activity of an ancient organism, such as a burrow or track (cf. body fossil).
trachea (pl. **tracheae**) "artery"; small tube through the cuticle of an arthropod, used in respiration and water control.
tracheid "artery"; a water-conducting strand in a land plant.
transgression "passage across"; advance of the sea on to land; may be local or global (cf. regression).
transpiration "breath across"; the process whereby fluid is drawn up through a plant by the suction effect of evaporation of water from the leaves.
tree of life the phylogenetic tree that links all species.
trend (in evolution) sustained change in a feature through time.
trilete of a spore with a three-branched slit (cf. monolete).
triploblastic "three-layered"; the body arrangement found in most animals where the ectoderm and endoderm are separated by a third tissue class, the mesoderm (cf. diploblastic).
triserial "in three rows" (cf. biserial, uniserial).
trochospiral "wheel spiral"; spiral and pyramidal (cf. planispiral).
trophic of food or feeding.
tsunami a large tidal wave.
tube foot a small fleshy muscular structure that projects through the skeleton of an echinoderm and functions in cleaning, feeding and locomotion.
tubercle "small root"; a small projection from a skeleton.
turbidite a rock formed from turbidity flows, mass movements of sand and mud down a slope and into deep water.
turma (pl. **turmae**) "a troop"; the category term used in classifications of spores.
type specimen the specimen that is selected as the name-bearer, to represent all the characteristic features of a species.

ulna one of the forearm bones in vertebrates (cf. humerus, radius).
umbilicus "navel"; a cavity in the center of a gastropod shell.

umbo (pl. **umbones**) the "shoulder" region of the pedicle valve of a brachiopod; the "beak" of a bivalve.

unconformity a gap in a sequence of rocks that apparently corresponds to the passage of a considerable amount of time.

undertrack the impression of a track preserved below the surface on which the animal was moving.

uniformitarianism "the present is the key to the past"; the basic assumption in geology and paleontology that ancient phenomena may be interpreted in the light of observations of the modern world.

unimembrate "single-member" (cf. multimembrate).

uniramous "single-branched" (cf. biramous).

uniserial "in a single row" (cf. biserial, triserial).

universal tree of life (UTL) the phylogenetic tree of life.

valve one-half of a brachiopod or bivalve, each of which consists of two valves.

variation the differences between individuals that normally occur in a population, assessed either at the genotypic or phenotypic level.

venter "belly"; the underside (adj. **ventral**; cf. dorsum).

vertebra (pl. **vertebrae**) an element of the backbone of a vertebrate; or an element in the arms of an ophiuroid, an ossicle.

vesicle "bladder"; a fluid-filled sac.

vestigial structure a feature that is incomplete or has no clear function, but which appears to be homologous with something that once functioned in the ancestors.

virgella "twig"; pointed structure at the base of the sicula of a graptolite.

viscera the internal organs (adj. **visceral**).

vitrinite dark, shiny, primary component of coal derived mainly from woody tissue.

volatile "fleeing"; a form of carbon that breaks down readily (cf. refractory).

volcaniclastic deposits sedimentary deposits derived directly from volcanic eruptions.

whorl a single turn in a spiral shell; a circular array of leaves around a stem.

xenomorphic a"foreign form"; of different form in different regions.

xylem the woody tissue in vascular plants in which tracheids conduct fluids and that also acts as a support.

zeugopod the middle portion of a vertebrate limb, the forearm or calf (cf. autopod, stylopod).

zone a small unit of geological time, generally identified on the basis of one or more zone fossils, and a subdivision of a stratigraphic stage.

zone fossil a fossil species that indicates a particular unit of time.

zooarium (pl. **zooaria**) the stick-like skeleton of certain kinds of bryozoan colony.

zooecium (pl. **zooecia**) a box-like living chamber within a bryozoan colony.

zooid "small animal"; an individual animal that lives in part of a colony.

zooplankton "animal plankton"; the animal components of the plankton.

zooxanthella (pl. **zooxanthellae**) "animal yellow"; photosynthesizing alga that lives in intimate association with a coral or bivalve.

zygote "yoke"; the first stage of embryonic development, the product of the fusion of two gametes.

Appendix 1

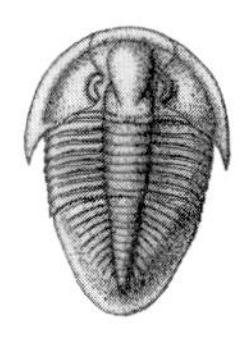

Stratigraphic chart

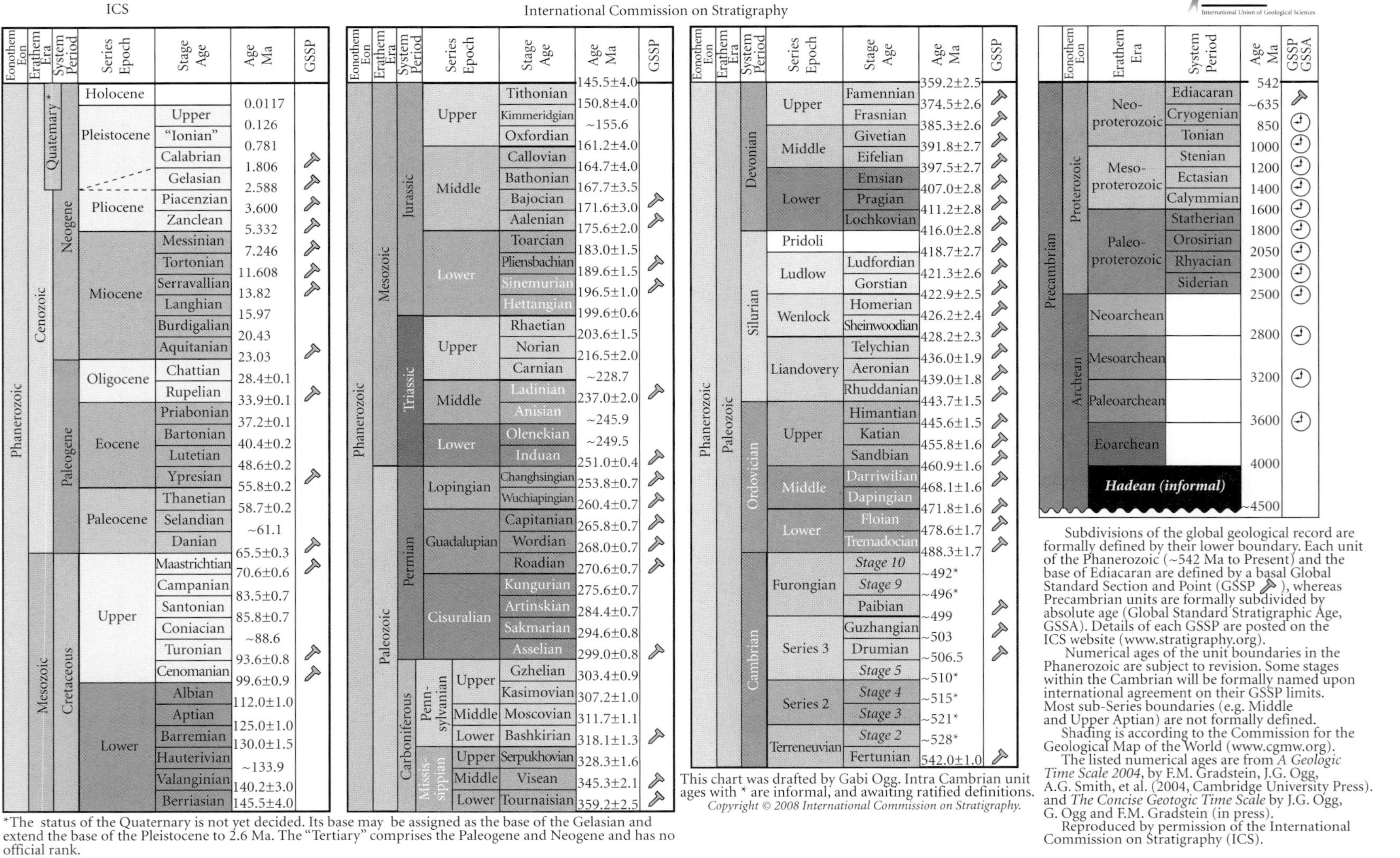

INTERNATIONAL STRATIGRAPHIC CHART

International Commission on Stratigraphy

ICS — IUGS International Union of Geological Sciences

Eonothem Eon	Erathem Era	System Period	Series Epoch	Stage Age	Age Ma	GSSP
Phanerozoic	Cenozoic	Quaternary*	Holocene		0.0117	
			Pleistocene	Upper	0.126	
				"Ionian"	0.781	
				Calabrian	1.806	GSSP
		Neogene		Gelasian	2.588	GSSP
			Pliocene	Piacenzian	3.600	GSSP
				Zanclean	5.332	GSSP
			Miocene	Messinian	7.246	GSSP
				Tortonian	11.608	GSSP
				Serravallian	13.82	GSSP
				Langhian	15.97	
				Burdigalian	20.43	
				Aquitanian	23.03	GSSP
		Paleogene	Oligocene	Chattian	28.4±0.1	
				Rupelian	33.9±0.1	GSSP
			Eocene	Priabonian	37.2±0.1	
				Bartonian	40.4±0.2	
				Lutetian	48.6±0.2	
				Ypresian	55.8±0.2	GSSP
			Paleocene	Thanetian	58.7±0.2	
				Selandian	~61.1	
				Danian	65.5±0.3	GSSP
	Mesozoic	Cretaceous	Upper	Maastrichtian	70.6±0.6	GSSP
				Campanian	83.5±0.7	
				Santonian	85.8±0.7	
				Coniacian	~88.6	
				Turonian	93.6±0.8	GSSP
				Cenomanian	99.6±0.9	GSSP
			Lower	Albian	112.0±1.0	
				Aptian	125.0±1.0	
				Barremian	130.0±1.5	
				Hauterivian	~133.9	
				Valanginian	140.2±3.0	
				Berriasian	145.5±4.0	

*The status of the Quaternary is not yet decided. Its base may be assigned as the base of the Gelasian and extend the base of the Pleistocene to 2.6 Ma. The "Tertiary" comprises the Paleogene and Neogene and has no official rank.

Eonothem Eon	Erathem Era	System Period	Series Epoch	Stage Age	Age Ma	GSSP
Phanerozoic	Mesozoic	Jurassic	Upper		145.5±4.0	
				Tithonian	150.8±4.0	
				Kimmeridgian	~155.6	
				Oxfordian	161.2±4.0	
			Middle	Callovian	164.7±4.0	
				Bathonian	167.7±3.5	
				Bajocian	171.6±3.0	GSSP
				Aalenian	175.6±2.0	GSSP
			Lower	Toarcian	183.0±1.5	
				Pliensbachian	189.6±1.5	GSSP
				Sinemurian	196.5±1.0	GSSP
				Hettangian	199.6±0.6	
		Triassic	Upper	Rhaetian	203.6±1.5	
				Norian	216.5±2.0	
				Carnian	~228.7	
			Middle	Ladinian	237.0±2.0	GSSP
				Anisian	~245.9	
			Lower	Olenekian	~249.5	
				Induan	251.0±0.4	GSSP
	Paleozoic	Permian	Lopingian	Changhsingian	253.8±0.7	GSSP
				Wuchiapingian	260.4±0.7	GSSP
			Guadalupian	Capitanian	265.8±0.7	GSSP
				Wordian	268.0±0.7	GSSP
				Roadian	270.6±0.7	GSSP
			Cisuralian	Kungurian	275.6±0.7	
				Artinskian	284.4±0.7	
				Sakmarian	294.6±0.8	
				Asselian	299.0±0.8	GSSP
		Carboniferous	Pennsylvanian Upper	Gzhelian	303.4±0.9	
				Kasimovian	307.2±1.0	
			Pennsylvanian Middle	Moscovian	311.7±1.1	
			Pennsylvanian Lower	Bashkirian	318.1±1.3	GSSP
			Mississippian Upper	Serpukhovian	328.3±1.6	
			Mississippian Middle	Visean	345.3±2.1	GSSP
			Mississippian Lower	Tournaisian	359.2±2.5	GSSP

Eonothem Eon	Erathem Era	System Period	Series Epoch	Stage Age	Age Ma	GSSP
Phanerozoic	Paleozoic	Devonian	Upper		359.2±2.5	
				Famennian	374.5±2.6	GSSP
				Frasnian	385.3±2.6	GSSP
			Middle	Givetian	391.8±2.7	GSSP
				Eifelian	397.5±2.7	GSSP
			Lower	Emsian	407.0±2.8	GSSP
				Pragian	411.2±2.8	GSSP
				Lochkovian	416.0±2.8	GSSP
		Silurian	Pridoli		418.7±2.7	GSSP
			Ludlow	Ludfordian	421.3±2.6	GSSP
				Gorstian	422.9±2.5	GSSP
			Wenlock	Homerian	426.2±2.4	GSSP
				Sheinwoodian	428.2±2.3	GSSP
			Llandovery	Telychian	436.0±1.9	GSSP
				Aeronian	439.0±1.8	GSSP
				Rhuddanian	443.7±1.5	GSSP
		Ordovician	Upper	Himantian	445.6±1.5	GSSP
				Katian	455.8±1.6	GSSP
				Sandbian	460.9±1.6	GSSP
			Middle	Darriwilian	468.1±1.6	GSSP
				Dapingian	471.8±1.6	GSSP
			Lower	Floian	478.6±1.7	GSSP
				Tremadocian	488.3±1.7	GSSP
		Cambrian	Furongian	*Stage 10*	~492*	
				Stage 9	~496*	
				Paibian	~499	GSSP
			Series 3	Guzhangian	~503	GSSP
				Drumian	~506.5	GSSP
				Stage 5	~510*	
			Series 2	*Stage 4*	~515*	
				Stage 3	~521*	
			Terreneuvian	*Stage 2*	~528*	
				Fertunian	542.0±1.0	GSSP

This chart was drafted by Gabi Ogg. Intra Cambrian unit ages with * are informal, and awaiting ratified definitions.

	Eonothem Eon	Erathem Era	System Period	Age Ma	GSSP GSSA
Precambrian	Proterozoic	Neoproterozoic		542	
			Ediacaran	~635	GSSP
			Cryogenian	850	GSSA
			Tonian	1000	GSSA
		Mesoproterozoic	Stenian	1200	GSSA
			Ectasian	1400	GSSA
			Calymmian	1600	GSSA
		Paleoproterozoic	Statherian	1800	GSSA
			Orosirian	2050	GSSA
			Rhyacian	2300	GSSA
			Siderian	2500	GSSA
	Archean	Neoarchean		2800	GSSA
		Mesoarchean		3200	GSSA
		Paleoarchean		3600	GSSA
		Eoarchean		4000	
		Hadean (informal)		~4500	

Subdivisions of the global geological record are formally defined by their lower boundary. Each unit of the Phanerozoic (~542 Ma to Present) and the base of Ediacaran are defined by a basal Global Standard Section and Point (GSSP), whereas Precambrian units are formally subdivided by absolute age (Global Standard Stratigraphic Age, GSSA). Details of each GSSP are posted on the ICS website (www.stratigraphy.org).

Numerical ages of the unit boundaries in the Phanerozoic are subject to revision. Some stages within the Cambrian will be formally named upon international agreement on their GSSP limits. Most sub-Series boundaries (e.g. Middle and Upper Aptian) are not formally defined.

Shading is according to the Commission for the Geological Map of the World (www.cgmw.org).

The listed numerical ages are from *A Geologic Time Scale 2004*, by F.M. Gradstein, J.G. Ogg, A.G. Smith, et al. (2004, Cambridge University Press). and *The Concise Geologic Time Scale* by J.G. Ogg, G. Ogg and F.M. Gradstein (in press).

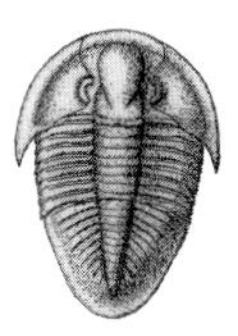

Appendix 2

Paleogeographic maps

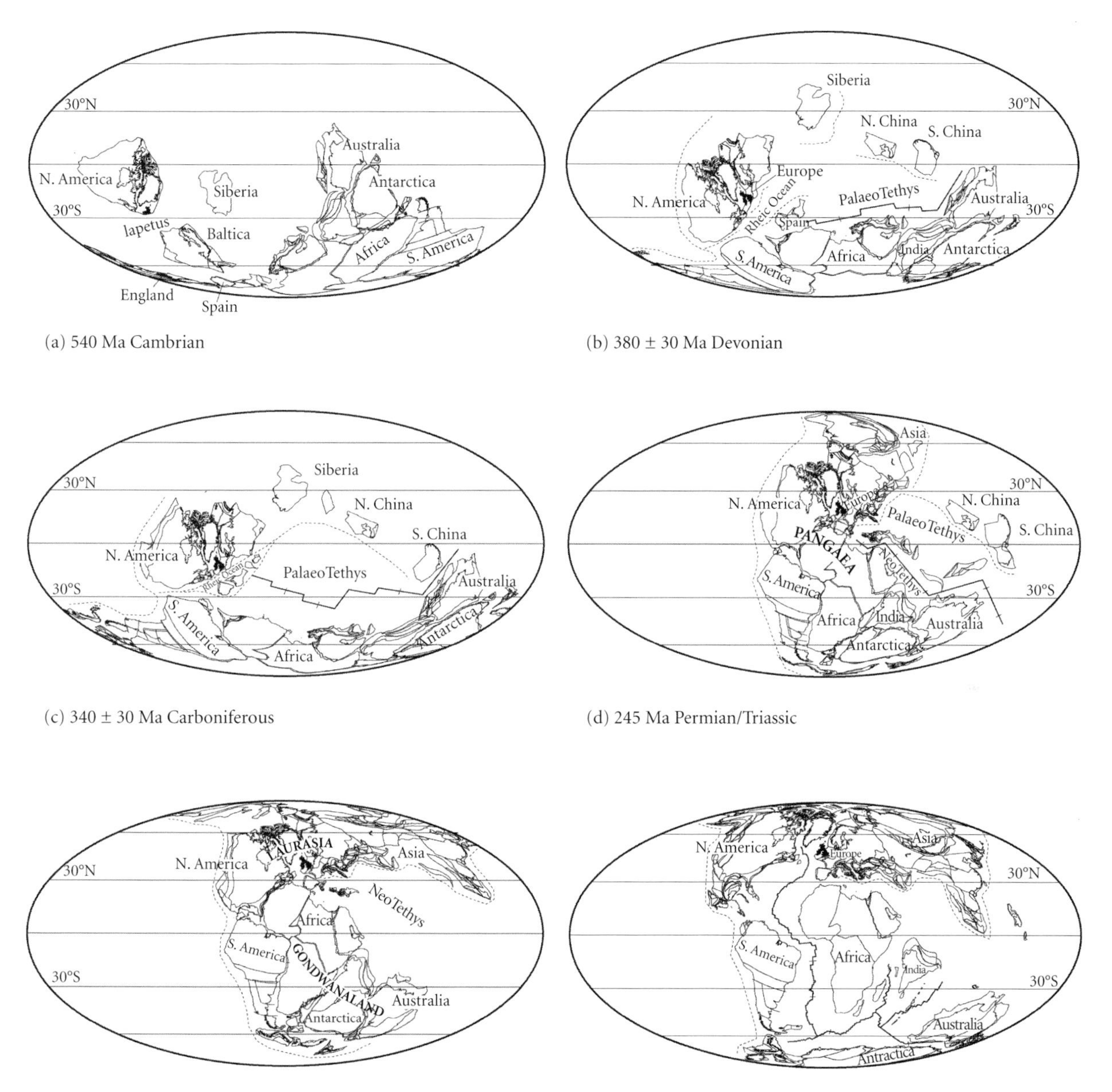

(a) 540 Ma Cambrian

(b) 380 ± 30 Ma Devonian

(c) 340 ± 30 Ma Carboniferous

(d) 245 Ma Permian/Triassic

(e) 180 Ma Jurassic

(f) 65 Ma Cretaceous/Tertiary

Collage of the main provinces through time. The Early Paleozoic was characterized by low- and high-latitude provinces separated by the Iapetus Ocean. During the Late Paleozoic, the Rheic Ocean separated the Old and New World provinces, whereas the Mesozoic was characterized by Boreal (high-latitude) and Tethyan (low-latitude) provinces. (These maps were produced by Professor Trond Torsvik, Center for Geodynamics, Geological Survey of Norway and the Center for Advanced Study, Norwegian Academy of Sciences and Letters, at the request of the authors.)

Index

Note: page numbers in **bold** refer to boxes, those in *italics* refer to figures and tables